AF357840

Fungi and Fungal Metabolites for the Improvement of Human and Animal Life, Nutrition and Health 2.0

Fungi and Fungal Metabolites for the Improvement of Human and Animal Life, Nutrition and Health 2.0

Guest Editor

Laurent Dufossé

Basel • Beijing • Wuhan • Barcelona • Belgrade • Novi Sad • Cluj • Manchester

Guest Editor
Laurent Dufossé
CHEMBIOPRO Laboratory
University of Réunion island
Sainte-Clotilde
France

Editorial Office
MDPI AG
Grosspeteranlage 5
4052 Basel, Switzerland

This is a reprint of the Special Issue, published open access by the journal *Journal of Fungi* (ISSN 2309-608X), freely accessible at: www.mdpi.com/journal/jof/special_issues/Fungi_Metabolites.

For citation purposes, cite each article independently as indicated on the article page online and using the guide below:

Lastname, A.A.; Lastname, B.B. Article Title. *Journal Name* **Year**, *Volume Number*, Page Range.

ISBN 978-3-7258-3026-8 (Hbk)
ISBN 978-3-7258-3025-1 (PDF)
https://doi.org/10.3390/books978-3-7258-3025-1

Contents

About the Editor

Laurent Dufossé

Laurent Dufossé has held the position of Professor of Food Science and Biotechnology since 2006 at the Reunion Island University, which is located on a volcanic island in the Indian Ocean, near Madagascar and Mauritius. The island is one of France's overseas territories, with almost one million inhabitants, and the university has 15,000 students. Previously, Professor Dufossé was a researcher and senior lecturer at the Université de Bretagne Occidentale, Quimper, Brittany, France.

He attended the University of Burgundy, where he received his Ph.D. in Food Science in 1993 and has been involved in the field of biotechnology of food ingredients for more than 30 years. Over the last 20 years, his research has mainly focused on microbial production of pigments, while his studies are mainly devoted to aryl carotenoids, such as isorenieratene, C50 carotenoids, azaphilones, and anthraquinones.

Journal of
Fungi

Editorial

Fungi and Fungal Metabolites for the Improvement of Human and Animal Life, Nutrition and Health

Laurent Dufossé [1,2,3]

1   Laboratoire de Chimie et Biotechnologie des Produits Naturels—CHEMBIOPRO, Université de la Réunion, 15 Avenue René Cassin, CEDEX 9, CS 92003, F-97744 Saint-Denis, Ile de la Réunion, France; laurent.dufosse@univ-reunion.fr
2   Ecole Supérieure d'Ingénieurs Réunion Océan Indien—ESIROI, 2 Rue Joseph Wetzell, F-97490 Sainte-Clotilde, Ile de la Réunion, France
3   Laboratoire ANTiOX, Université de Bretagne Occidentale, F-29000 Quimper, France

Citation: Dufossé, L. Fungi and Fungal Metabolites for the Improvement of Human and Animal Life, Nutrition and Health. *J. Fungi* **2024**, *10*, 863. https://doi.org/10.3390/jof10120863

Received: 27 November 2024
Accepted: 11 December 2024
Published: 12 December 2024

Fungi: 1, 2, 3,... 5.1 million species? Even scientists do not currently agree on how many fungi there might be on the planet Earth, but only about 140,000 have been described so far [1]. They have been grouped into a separate kingdom of organisms, as complex and diverse as plants and animals, of which only a small percentage have been named and described. Fungal biomasses and fungal metabolites have a long common history with human and animal life, nutrition, and health. Macro fungi and filamentous fungi have a large portfolio of proteins, lipids, vitamins, minerals, oligo elements, pigments, colorants, bioactive compounds, antibiotics, pharmaceuticals, etc. For example, industrially important enzymes and microbial biomass proteins have been produced from fungi for more than 50 years. Some start-ups convert by-products and side-streams rich in carbohydrates into a protein-rich fungal biomass. The biomass is then processed into a vegan meat substitute for food applications. In the last few years, there has also been a significant rise (in fact, a significant revival) in the number of publications in the international literature dealing with the production of lipids by microbial sources (the "single cell oils; SCOs" that are produced by the so-called "oleaginous" microorganisms, including "oleaginous" fungi, such as the zygomycete species, e.g., *Cunninghamella echinulata* and *Mortierella isabellina*). Fungi are potential sources of polyunsaturated fatty acids (PUFAs), as these microorganisms can accumulate large amounts of high-valued PUFAs, such as gamma-linolenic acid (GLA) and arachidonic acid (ARA).

The objective of the invitation to contribute to this MDPI Fungi Special Issue was not to give complete coverage of how fungi and fungal metabolites are able to improve human and animal life, nutrition, and health. I, as guest editor, simply wanted to encourage authors working in this field to publish their most recent work in a rapidly expanding journal in order to make a large audience discover the full potential of wonderful and beneficial fungi. Thus, this Special Issue welcomes 16 scientific contributions (10 original research papers and six reviews) on the applications of fungi and fungal metabolites, such as bioactive compounds, nephroprotectants, mycoproteins, antifungals, insecticides, antibacterials, enzymes, etc., with great potential in human and animal life, nutrition, and health.

Many original research papers of this Special Issue deal with fungal endophytes. Plants in ecosystems mostly appear to be symbiotic with fungal endophytes. This highly diverse group of fungi can have profound impacts on plant communities through increasing fitness by conferring abiotic and biotic stress tolerance, increasing biomass, and decreasing water consumption, or decreasing fitness by altering resource allocation [2]. Fungal endophytes have remarkable potential to produce a wide range of pharmacologically significant bioactive compounds that are used in disease management and human welfare. In the study by Verma et al. [3], a total of eight fungal endophytes were isolated from the leaf tissue of *Amoora rohituka*. HPTLC fingerprint analysis and antioxidant activity of ethyl acetate extract from isolated *P. oxalicum* revealed that fungal endophytes produce bioactive compounds in a host-dependent manner. More investigations are being conducted in order to have access to the chemical structures that are produced. Very large reviews are useful to newcomers in a research area. Deshmukh et al. [4] reported 451 bioactive metabolites isolated from various groups of endophytic fungi from January 2015 to April 2021, along with their antibacterial profiling, chemical structures, and mode of action. Bioactive metabolites with unique skeletons have been identified, which could be helpful in the prevention of increasing antimicrobial resistance. Antibiotic resistance is becoming a burning issue due to the frequent use of antibiotics to cure common bacterial infections, indicating that we are running out of effective antibiotics. The review by Nagarajan et al. [5] highlights the recent developments in metabolomics studies of endophytic fungi in obtaining a global picture of metabolites. Metabolomics coupled with advanced analytical tools provides a comprehensive insight into systems biology. Despite its wide scientific attention, endophytic fungi metabolomics is still relatively unexploited, and new techniques will increase the number of commercial applications.

Moving to other sources of bioactive metabolites, Meade et al. [6] describe how mushrooms have been used as traditional medicine for millennia. This literature review aims to explore recent evidence relating to the application of fungal bioactives in treating chronic mental health and chronic pain morbidities. There is now increasing interest in using fungal active compounds such as psychedelics for alleviating symptoms of mental health disorders, including major depressive disorder, anxiety, and addiction. Also related to human health, the research conducted by Sinaeve et al. [7] investigates how *Ganoderma* extracts may have nephroprotective effects. Although cisplatin is used as a first-line therapy in many cancers, its nephrotoxicity remains a real problem. Acute kidney injuries induced by cisplatin can cause proximal tubular necrosis, possibly leading to interstitial fibrosis, chronic dysfunction, and finally to the cessation of chemotherapy. The study focused on the respective in vitro effects of methanolic extracts of *Ganoderma tuberculosum* Murill. and *Ganoderma parvigibbosum* Welti & Courtec. and their association with human proximal tubular cells (HK-2) intoxicated by cisplatin. *Fusarium* fungi belong to a large genus of filamentous fungi, part of a group often referred to as hyphomycetes, widely distributed in soil and associated with plants. Rana et al. [8] emphasize that the taxonomy of the genus *Fusarium* has been in a flux because of ambiguous circumscription of species-level identification based on morphotaxonomic criteria. In this study, multigene phylogeny was conducted to resolve the evolutionary relationships of 88 Indian *Fusarium* isolates based on the internal transcribed spacer region, 28S large subunit, translation elongation factor 1-alpha, RNA polymerase second largest subunit, beta-tubulin, and calmodulin gene regions. Additionally, there was a special focus on beauvericin (BEA), and, among the 88 isolates studied, 50 were capable of producing BEA, which varied from 0.01 to 15.82 mg/g of biomass. Antifungal compounds are also very important bioactive metabolites produced by filamentous fungi. The primary purpose of the study by Kuvarina et al. [9] was to isolate and identify a hydrophobin, Sa-HFB1, from an alkaliphilic fungus, *Sodiomyces alkalinus*. The highest level of antifungal activity (MIC 1 µg/mL) was demonstrated for the clinical isolate *Cryptococcus neoformans* 297 m. Fungal metabolites may also target insects. Chemical insecticides can cause significant harm to both terrestrial and aquatic environments. New insecticides derived from microbial sources are a good option with 'no' or low environmental consequences. *Metarhizium anisopliae* (mycelia) ethyl acetate extracts were tested in [10] on larvae, pupae, and adults of *Anopheles stephensi* (Liston, 1901), *Aedes aegypti* (Meigen, 1818), and *Culex quinquefasciatus* (Say, 1823), as well as non-target species *Eudrilus eugeniae* (Kinberg, 1867) and *Artemia nauplii* (Linnaeus, 1758). Among numerous bioassay results, *Metarhizium anisopliae* extracts had remarkable toxicity on *Aedes aegypti* and *Culex quinquefasciatus*.

Erwinia mallotivora, the causal agent of papaya dieback disease, is a devastating pathogen that has caused a tremendous decrease in Malaysian papaya exports and affected papaya crops in neighboring countries. In the research conducted by Tamizi et al. [11], mycelial suspensions from five rhizospheric *Trichoderma* isolates of Malaysian origin were found to exhibit notable antagonisms against *E. mallotivora* during co-cultivation. Based on these findings, the fungal isolates are proven to be useful as potential biological control agents against *E. mallotivora*, and the genomic data opens possibilities to further explore the underlying molecular mechanisms behind their antimicrobial activity, with potential synthetic biology applications.

Antimicrobial secondary metabolites from *Fusarium oxysporum* R1 associated with the traditional Chinese medicinal plant *Rumex madaio* Makino were screened for many years by Fu et al. [12] using the one strain, many compounds (OSMAC) strategy. Two diastereomeric polyketides, neovasifuranones A and B, were obtained from their solid-rice-cultivated medium together with N-(2-phenylethyl)acetamide, 1-(3-hydroxy-2-methoxyphenyl)-ethanone, and 1,2-seco-trypacidin. Their planar structures were unambiguously determined using 1D NMR and MS spectroscopy techniques as well as comparison with the literature data. The accurate determination of the chemical structures of fungal products is of high priority.

Yeasts producing semiochemicals are increasingly used in pest management programs; however, little is known about which yeasts populate cherry fruits, and no information is available on the volatiles that modify the behavior of cherry pests, including *Rhagoletis cerasi* flies. Eighty-two compounds were extracted by Mozūraitis et al. [13] from the headspaces of eleven yeast species associated with sweet and sour cherry fruits using solid-phase microextraction. Two-choice olfactometric tests revealed that *R. cerasi* flies preferred 3-methylbutyl propionate and 3-methyl-1-butanol but avoided 3-methylbutyl acetate. Yeast-producing behaviorally active compounds indicated a potential for use in pest monitoring and control of *R. cerasi* fruit flies, an economically important pest of cherry fruits.

Another food-related paper deals with the *Issatchenkia terricola* strain WJL-G4, which has great potential in red raspberry wine fermentation [14]. In the current study, *I. terricola* WJL-G4 was applied to decrease the content of citric acid in red raspberry juice, followed by the red raspberry wine preparation by *Saccharomyces cerevisiae* fermentation, aiming to investigate the influence of *I. terricola* WJL-G4 on the physicochemical properties, organic acids, phenolic compounds, and antioxidant activities during red raspberry wine processing.

Fungal enzymes also have a huge impact on human and animal life, nutrition, and health. A comprehensive review by El-Gendi et al. [15] elaborates on the different types and structures of fungal enzymes as well as the current status of the uses of fungal enzymes in various applications. Mycoproteins are currently a hot topic. The main findings by Derbyshire [16] showed that fungal mycoproteins could contribute to an array of health benefits across the animal and human lifespan, including improved lipid profiles, glycaemic markers, dietary fiber intakes, satiety effects, and muscle/myofibrillar protein synthesis. Continued research is needed in human nutrition, which would be worthwhile at both ends of the lifespan spectrum and in specific population subgroups.

Research is also being conducted on the detrimental effects of filamentous fungi, such as those from multidrug-resistant species belonging to the *Scedosporium* genus [17] that are well recognized as saprophytic filamentous fungi found mainly in human-impacted areas and emerged as human pathogens in both immunocompetent and immunocompromised individuals. Last but not least, the final paper to be presented in this Special Issue attracts the attention of the reader to the fact that we live every day with and among filamentous fungi. Vaali et al. [18] aimed to establish an etiology-based connection between the symptoms experienced by the occupants of a workplace and the presence of toxic dampness microbiota in the building. Among the conclusions, the cytotoxicity test of the indoor air condensate is a promising tool for risk assessment in moisture-damaged buildings.

As a concluding remark, I, the Guest Editor, wish to thank all the authors and the reviewers for their significant contributions to this Special Issue and for making it a highly successful and timely collection of papers. Our acknowledgements also go to the whole MDPI team, i.e., assistant editors, editors, Editor-in-Chief, the production office, website management, etc.

Conflicts of Interest: The authors declare no conflicts of interest.

References

1. Blackwell, M. The fungi: 1, 2, 3 . . . 5.1 million species? *Am. J. Bot.* **2011**, *98*, 426–438. [CrossRef] [PubMed]
2. Rodriguez, R.J.; White, J.F., Jr.; Arnold, A.E.; Redman, A.R. Fungal endophytes: Diversity and functional roles. *New Phytol.* **2009**, *182*, 314–330. [CrossRef] [PubMed]
3. Verma, A.; Gupta, P.; Rai, N.; Tiwari, R.K.; Kumar, A.; Salvi, P.; Kamble, S.C.; Singh, S.K.; Gautam, V. Assessment of biological activities of fungal endophytes derived bioactive compounds Isolated from *Amoora rohituka*. *J. Fungi* **2022**, *8*, 285. [CrossRef] [PubMed]
4. Deshmukh, S.K.; Dufossé, L.; Chhipa, H.; Saxena, S.; Mahajan, G.B.; Gupta, M.K. Fungal endophytes: A potential source of antibacterial compounds. *J. Fungi* **2022**, *8*, 164. [CrossRef] [PubMed]
5. Nagarajan, K.; Ibrahim, B.; Ahmad Bawadikji, A.; Lim, J.W.; Tong, W.Y.; Leong, C.R.; Khaw, K.Y.; Tan, W.N. Recent developments in metabolomics studies of endophytic fungi. *J. Fungi* **2021**, *8*, 28. [CrossRef] [PubMed]
6. Meade, E.; Hehir, S.; Rowan, N.; Garvey, M. Mycotherapy: Potential of Fungal Bioactives for the Treatment of Mental Health Disorders and Morbidities of Chronic Pain. *J. Fungi* **2022**, *8*, 290. [CrossRef] [PubMed]
7. Sinaeve, S.; Husson, C.; Antoine, M.-H.; Welti, S.; Stévigny, C.; Nortier, J. Nephroprotective Effects of Two Ganoderma Species Methanolic Extracts in an In Vitro Model of Cisplatin Induced Tubulotoxicity. *J. Fungi* **2022**, *8*, 1002. [CrossRef] [PubMed]
8. Rana, S.; Singh, S.K.; Dufossé, L. Multigene Phylogeny, Beauvericin Production and Bioactive Potential of *Fusarium* Strains Isolated in India. *J. Fungi* **2022**, *8*, 662. [CrossRef] [PubMed]
9. Kuvarina, A.E.; Rogozhin, E.A.; Sykonnikov, M.A.; Timofeeva, A.V.; Serebryakova, M.V.; Fedorova, N.V.; Kokaeva, L.Y.; Efimenko, T.A.; Georgieva, M.L.; Sadykova, V.S. Isolation and Characterization of a Novel Hydrophobin, Sa-HFB1, with Antifungal Activity from an Alkaliphilic Fungus, *Sodiomyces alkalinus*. *J. Fungi* **2022**, *8*, 659. [CrossRef] [PubMed]
10. Vivekanandhan, P.; Swathy, K.; Murugan, A.C.; Krutmuang, P. Insecticidal Efficacy of *Metarhizium anisopliae* Derived Chemical Constituents against Disease-Vector Mosquitoes. *J. Fungi* **2022**, *8*, 300. [CrossRef] [PubMed]
11. Tamizi, A.-A.; Mat-Amin, N.; Weaver, J.A.; Olumakaiye, R.T.; Akbar, M.A.; Jin, S.; Bunawan, H.; Alberti, F. Genome Sequencing and Analysis of *Trichoderma* (Hypocreaceae) Isolates Exhibiting Antagonistic Activity against the Papaya Dieback Pathogen, *Erwinia mallotivora*. *J. Fungi* **2022**, *8*, 246. [CrossRef] [PubMed]
12. Fu, Z.; Liu, Y.; Xu, M.; Yao, X.; Wang, H.; Zhang, H. Absolute Configuration Determination of Two Diastereomeric Neovasifuranones A and B from *Fusarium oxysporum* R1 by a Combination of Mosher's Method and Chiroptical Approach. *J. Fungi* **2022**, *8*, 40. [CrossRef] [PubMed]
13. Mozūraitis, R.; Apšegaitė, V.; Radžiutė, S.; Aleknavičius, D.; Būdienė, J.; Stanevičienė, R.; Blažytė-Čereškienė, L.; Servienė, E.; Būda, V. Volatiles Produced by Yeasts Related to *Prunus avium* and *P. cerasus* Fruits and Their Potentials to Modulate the Behaviour of the Pest *Rhagoletis cerasi* Fruit Flies. *J. Fungi* **2022**, *8*, 95. [CrossRef] [PubMed]
14. He, H.; Yan, Y.; Dong, D.; Bao, Y.; Luo, T.; Chen, Q.; Wang, J. Effect of *Issatchenkia terricola* WJL-G4 on Deacidification Characteristics and Antioxidant Activities of Red Raspberry Wine Processing. *J. Fungi* **2022**, *8*, 17. [CrossRef] [PubMed]
15. El-Gendi, H.; Saleh, A.K.; Badierah, R.; Redwan, E.M.; El-Maradny, Y.A.; El-Fakharany, E.M. A Comprehensive Insight into Fungal Enzymes: Structure, Classification, and Their Role in Mankind's Challenges. *J. Fungi* **2022**, *8*, 23. [CrossRef] [PubMed]
16. Derbyshire, E. Fungal-Derived Mycoprotein and Health across the Lifespan: A Narrative Review. *J. Fungi* **2022**, *8*, 653. [CrossRef] [PubMed]
17. Mello, T.P.; Barcellos, I.C.; Aor, A.C.; Branquinha, M.H.; Santos, A.L.S. Extracellularly Released Molecules by the Multidrug-Resistant Fungal Pathogens Belonging to the *Scedosporium* Genus: An Overview Focused on Their Ecological Significance and Pathogenic Relevance. *J. Fungi* **2022**, *8*, 1172. [CrossRef] [PubMed]
18. Vaali, K.; Tuomela, M.; Mannerström, M.; Heinonen, T.; Tuuminen, T. Toxic Indoor Air Is a Potential Risk of Causing Immuno Suppression and Morbidity—A Pilot Study. *J. Fungi* **2022**, *8*, 104. [CrossRef] [PubMed]

Journal of
Fungi

Article

Assessment of Biological Activities of Fungal Endophytes Derived Bioactive Compounds Isolated from *Amoora rohituka*

Ashish Verma [1], Priyamvada Gupta [1,†], Nilesh Rai [1,†], Rajan Kumar Tiwari [2], Ajay Kumar [2], Prafull Salvi [3], Swapnil C. Kamble [4], Santosh Kumar Singh [1] and Vibhav Gautam [1,*]

[1]   Centre of Experimental Medicine and Surgery, Institute of Medical Sciences, Banaras Hindu University, Varanasi 221005, India; ashishambbhu@gmail.com (A.V.); priyamvada17gupta@gmail.com (P.G.); nilesh.rai17@bhu.ac.in (N.R.); singhsk71@yahoo.com (S.K.S.)
[2]   Department of Zoology, Institute of Science, Banaras Hindu University, Varanasi 221005, India; shandilya.raj31@gmail.com (R.K.T.); ajayzoo@bhu.ac.in (A.K.)
[3]   Department of Agriculture Biotechnology, National Agri-Food Biotechnology Institute, Sahibzada Ajit Singh Nagar 140306, India; prafull.salvi@nabi.res.in
[4]   Department of Technology, Savitribai Phule Pune University, Pune 411007, India; sckamble@unipune.ac.in
*   Correspondence: vibhav.gautam4@bhu.ac.in; Tel.: +91-88-6018-2113
†   These authors contributed equally.

Abstract: Fungal endophytes have remarkable potential to produce bioactive compounds with numerous pharmacological significance that are used in various disease management and human welfare. In the current study, a total of eight fungal endophytes were isolated from the leaf tissue of *Amoora rohituka*, and out of which ethyl acetate (EA) extract of *Penicillium oxalicum* was found to exhibit potential antioxidant activity against DPPH, nitric oxide, superoxide anion and hydroxyl free radicals with EC_{50} values of 178.30 ± 1.446, 75.79 ± 0.692, 169.28 ± 0.402 and 126.12 ± 0.636 µg/mL, respectively. The significant antioxidant activity of EA extract of *P. oxalicum* is validated through highest phenolic and flavonoid content, and the presence of unique bioactive components observed through high-performance thin layer chromatography (HPTLC) fingerprinting. Moreover, EA extract of *P. oxalicum* also displayed substantial anti-proliferative activity with IC_{50} values of 56.81 ± 0.617, 37.24 ± 1.26 and 260.627 ± 5.415 µg/mL against three cancer cells HuT-78, MDA-MB-231 and MCF-7, respectively. Furthermore, comparative HPTLC fingerprint analysis and antioxidant activity of *P. oxalicum* revealed that fungal endophyte *P. oxalicum* produces bioactive compounds in a host-dependent manner. Therefore, the present study signifies that fungal endophyte *P. oxalicum* associated with the leaf of *A. rohituka* could be a potential source of bioactive compounds with antioxidant and anticancer activity.

Keywords: *Amoora rohituka*; fungal endophyte; *Penicillium oxalicum*; antioxidant activity; cytotoxic activity

Citation: Verma, A.; Gupta, P.; Rai, N.; Tiwari, R.K.; Kumar, A.; Salvi, P.; Kamble, S.C.; Singh, S.K.; Gautam, V. Assessment of Biological Activities of Fungal Endophytes Derived Bioactive Compounds Isolated from *Amoora rohituka*. *J. Fungi* **2022**, *8*, 285. https://doi.org/10.3390/jof8030285

Academic Editor: Laurent Dufossé

Received: 6 February 2022
Accepted: 2 March 2022
Published: 10 March 2022

Publisher's Note: MDPI stays neutral with regard to jurisdictional claims in published maps and institutional affiliations.

1. Introduction

Medicinal plants have been a potential therapeutic and curative agent for a wealth of diseases and healthcare attributed to the phytochemicals constituted in them. The command of plants in the pharmacological evaluation is ascribed to the secondary metabolites, flavonoids, terpenoids, tannins and alkaloids [1]. It is estimated by WHO (World Health Organization) that about 80 percent of the world's population depend upon herbal plants for their natural products by virtue of their flexibility, low cost, easy availability, better compatibility with the human body, and ethical acceptability with minimal negative effect [1,2]. In contemplation of conservation, sustainable use and protection of ethnobotanical aspects of plants, researchers have been heading towards the isolation of plant associated microbial community for the extraction of bioactive compounds similar to plants. The fungal endophytes meet this demand and serve as a true alternative of plant-derived bioactive compounds. The establishment of fungal relationship with the host plant can vary from symbiotic to commensalism, and can also be parasitic [3]. The symbiotic ones during the process of co-evolution have started synthesizing the compounds similar to their host and showed therapeutic potential similar to the host's compound [4]. Endophytes have a great prospect for the production of a broad spectrum of secondary metabolites and these natural products have been reported to exhibit various biological activities [5]. The versatile synthetic tendency of fungi confers extrapolation of an extensive range of metabolites. The fungal endophyte derived metabolites are natural, affordable and less-toxic, showing a multifunction mechanism that labels them as distinctive entities. The fungal endophytes serve as a reservoir of terpenoids, phenolic acids, quinones, tannins, alkaloids, saponins, and steroids, and therefore, fungal endophyte-derived bioactive compounds display an extensive range of activities in human health and diseases, including antibiotic, antifungal, immunomodulatory, antioxidant, antidiabetic and anticancer [6,7]. A wide range of bioactive compounds have been isolated from fungal endophytes and have been studied for their chemistry using omics and mass spectrometry-based tools [8]. On account of these properties, fungal endophytes appear as an epitome of natural products with their endowment in the field of agriculture, industry and pharmaceuticals.

Amoora rohituka (Aphanamixis polystachya), also known as Pithraj tree, is a 20–30 m tall deciduous tree belonging to the Meliaceae family and is native to India. It is one of the eminent plants with pronounced medicinal properties. The seeds and stem bark have been shown to have a significant role in the treatment of tumor, splenomegaly and liver disorders [9]. The alcoholic extract of stem bark has shown significant anticancer activity against Ehrlich ascites carcinoma and Friend's leukemia in mice [10–12]. Another report has been obtained for radiation-induced chromosome damage by ethyl acetate fraction of *A. rohituka* [13]. A bioactive compound rohitukine extracted from *A. rohituka* possesses immune-modulatory, anti-inflammatory and anti-cancer properties, and also serves as a precursor for the synthesis of P-276-00 (Piramal Healthcare Limited, Mumbai, India) and flavopiridol, an anticancer drug [14,15]. The first study for isolation of rohitukine and its analog, rohitukine N-oxide has been demonstrated in endophytic fungi *Gibberella fujikuroi* MTCC 11382 obtained from *A. rohituka* [16]. In other reports, the compound rohitukine extracted from a tree *Dysoxylum binectariferum* have been shown to possess immunomodulatory, anti-inflammatory activities [17] and also reported for attenuating peptic ulcers in rat [18]. Moreover, cytotoxic activity of rohitukine has been reported in colorectal cancer and promyelocytic leukemia [19]. Previous studies have shown leaf extract of *A. rohituka* exhibit potential anticancer activity against human breast cancer cell (MCF-7) [20]. In another study, it has been reported that leaves extract of *A. rohituka* possess significant antioxidant, cytotoxic and thrombolytic activities [21]. The aforementioned studies have been performed for the isolation and characterization of natural compounds either from plant *A. rohituka* or its part including root, stem bark, leaf and seeds against various diseases including cancer. However, this approach leads to the overexploitation of medicinal plants, and therefore a better alternative is needed. The properties of fungal endophytes such as easy cultivation, scale up flexibility, high yield and unique cellular organization offer them as an excellent alternative to isolate bioactive compounds similar to host plant metabolites with numerous pharmacological properties [4]. Considering such advantages of fungal endophytes over plant tissue, the leaf of *A. rohituka*-associated fungal endophytes was assessed for the production of bioactive compounds with antioxidant and cytotoxic activity. An increasing body of evidence suggests that leaf tissue is the most prevalent part to be inhabited by fungal endophytes due to its large surface area and being more exposed to the environmental factors [5–7]. The selection of leaf tissue provides leaves of diverse stages that could be young leaf, premature leaf and mature leaf and that show an association of a wide range of fungal endophytes.

The aforementioned fact compelled us towards the isolation of fungal endophytes associated with the leaf tissue of *A. rohituka*, and further to evaluate the antioxidant and cytotoxic activity of bioactive compounds present in the EA extract of isolated fungal strains. Total phenolic and flavonoid content and HPTLC fingerprinting were performed for the qualitative screening of bioactive constituents produced by the identified fungal endophytes. The antioxidant activity of fungal endophyte derived bioactive compounds was evaluated against free radicals through DPPH free radical scavenging assay, superoxide anion scavenging assay, hydroxyl radical scavenging assay and nitric oxide scavenging assay. Moreover, based on the results of biochemical profiling and antioxidant activity of isolated fungal strains, EA extract of *P. oxalicum* was further selected to investigate the cytotoxic activity against human-derived cancer cells, T-cell lymphoma cell (HuT-78) and breast cancer cells (MDA-MB-231 and MCF-7). Host-dependent activity and production of bioactive compounds derived from fungal endophyte was also evaluated through performing the antioxidant activity and HPTLC fingerprinting profile of EA extract of fungal endophyte *P. oxalicum* isolated from the leaf of *A. rohituka* and rhizospheric soil of maize. Conclusively, our findings suggest the presence of potential antioxidant and cytotoxic compounds in EA extract *P. oxalicum* isolated from the leaf of *A. rohituka*.

2. Materials and Method

2.1. Collection of Plant Sample and Isolation of Endophytic Fungi

Fungal endophytes were isolated from the healthy leaves of plant *A. rohituka*, collected from the Department of Dravyaguna, Institute of Medical Sciences, Banaras Hindu University, Varanasi, Uttar Pradesh (India). The collected leaf samples were washed and surface disinfected according to the method discussed earlier [22]. Briefly, the leaf samples were rinsed with running tap water followed by deionized water and subsequently dipped in 70% ethanol (1–2 min) followed by sterilization in 0.1% sodium hypochlorite (2–3 min). Then, they were further dipped in 70% ethanol and finally rinsed with distilled water. Dried leaves were allowed to dry, cut aseptically into small pieces (1 cm^2) and patched onto potato dextrose agar (PDA) (Himedia, Mumbai, India) plates containing streptomycin (SRL, Mumbai, India) at a concentration of 250 μg/mL to prevent bacterial contamination. The effectiveness of surface sterilization was tested by examining the growth of epiphytes by spreading the water onto PDA plates obtained after last wash. PDA plates were placed in BOD incubator (Narang Scientific Works Pvt. Ltd., New Delhi, India) at $27 \pm 2\,^\circ$C for 7–10 days. The fungal mycelia emerging from plant parts were isolated following subculturing of the mycelia from different fungi onto fresh PDA plates until appearance of a pure individual colony.

2.2. Morphological and Molecular Identification of Isolated Fungal Strains

The morphological characterization of isolated fungal isolates was done through the preparation of semi-permanent slides of fungal mycelia. For the semi-permanent slide preparation, a fungal colony was picked from PDA plate by sterile needle and placed on a glass slide, and then Lacto-phenol cotton blue was added for staining, and observed under bright field microscope (Olympus, CX43, Tokyo, Japan) [23]. For molecular identification, the genomic DNA (gDNA) was isolated from fungal mycelia using Nucleo-pore gDNA Fungal/Bacterial mini kit followed according to the manufacturer's protocol. After extraction, conserved ITS region of fungal gDNA was amplified by general primers ITS4 (5′-TCCTCCGCTTATTGATATGC-3′) and ITS5 (5′-GGAAGTAAAAGTCGTAACAAGG-3′). PCR reaction mixture contained 2 μL of extracted gDNA, 1.5 μL of each forward primer and reverse primer (10 μM concentration), 2 μL of 10× buffer, 0.5 μL of Taq DNA polymerase enzyme (BR Biochem, New Delhi, India), 0.75 μL of 10 mM of deoxynucleotide triphosphates (BR Biochem) and volume was maintained upto 20 μL with milli-Q water for single reaction of PCR. The PCR reaction was performed under the following condition: 5 min for initial denaturation step at 94 $^\circ$C followed by 35 cycles at 94 $^\circ$C for 30 s, annealing at 60 $^\circ$C for 40 s and extension at 74 $^\circ$C for 1 min and final extension step was for 10 min at 74 $^\circ$C. After PCR cycle completion, gel electrophoresis was performed to examine the PCR products (Figure S1) by using 1% agarose gel in 1× TAE buffer. The PCR products so obtained were purified using Nucleopore Quick PCR purification kit as per manufacturer's protocol, and sequencing of purified PCR product was performed using ITS4 and ITS5 primers.

2.3. Phylogenetic Tree Construction

The identification of fungal endophytes was done based on consensus DNA sequences. Homology searches of common sequences were performed through Basic Local Alignment Search Tool (BLAST) on http://www.ncbi.nlm.gov/BLAST (accessed on 21 November 2021) in NCBI database [24]. The identification of fungal endophytes and their consensus common DNA sequences was done by comparing query sequences with the previously submitted sequences in GenBank to obtained the accession number. Both FP-ITS and RP-ITS sequences were used separately where common overlapping sequence were not available. For the comparison and alignment of query sequences and reference sequences, MUSCLE (Multiple Sequence Comparison by Log-Expectation) program was used. The study of percentage identity of aligned sequences was done using Kolmogorov–Smirnov statistical test in GeneDoc (version 2.7). Using the obtained sequences, phylogenetic analysis was performed and a phylogenetic tree was constructed through MEGA (v10.1.8) by maximum likelihood Bootstrap (MLBS) method.

2.4. Fermentation Procedure and Extraction of Crude Fungal Extract

After morphological and molecular identification of fungal isolates, the grown PDA culture of individual fungal strain was inoculated into 300 mL potato-dextrose broth in a separate Erlenmeyer flask supplemented with streptomycin (250 µg/mL) and incubated at 27 ± 2 °C for 21 days at 140 rpm. After fermentation, the mycelium was harvested through filtration using the two-layered cheese cloth and was subjected to drying overnight at 50 °C. The dried mycelium so obtained was macerated individually with liquid nitrogen and fungal metabolites were extracted using ethyl acetate (5× volume of dry weight of mycelia) as a solvent. The crude fungal extract containing the bioactive compounds was stored at 4 °C for further experimental process.

2.5. Estimation of Total Flavonoid Content and Total Phenolic Content of EA Extract of Fungal Mycelia

The colorimetric determination of total flavonoid content was done by the $AlCl_3$ method as described earlier [25]. The method involves mixing of 1 mL of EA extract isolated fungal endophytes with 3 mL methanol followed by 200 µL of $AlCl_3$ (10%) and 200 µL of potassium acetate (9.8%). Further, the reaction mixture was diluted using 5.6 mL of distilled water and incubated for 30 min. Then, the absorption was taken at 420 nm against methanol, taken as a blank. Quercetin (5–200 µg/mL) was used to plot the calibration curve. The results obtained for total flavonoid content were expressed as µg of quercetin equivalents (QE) per mg of EA extract.

The Folin–Ciocalteu method was used for the determination of total phenolic content with some modifications [26]. Briefly, the sample (1 mL) was oxidized with 100 µL of the 0.5 N FC reagent and incubated for 15 min. Further, 2.5 mL of sodium carbonate (7.5%, *w/v*) was added to neutralize the reaction mixture and was mixed thoroughly. Following incubation for 30 min, absorbance was recorded at 727 nm against methanol, taken as a blank. The calibration curve was constructed using Gallic acid (5–200 µg/mL), and the results for total phenolic content was shown as µg of gallic acid equivalents (GAE) per mg of EA extract.

2.6. HPTLC Fingerprinting Analysis-Based Metabolites Profiling of EA Extract of Fungal Mycelia

The EA extract of fungal endophytes was screened for their bioactive compounds through CAMAG HPTLC equipment (ANCHROM, Muttenz, Switzerland) composed of a Linomat-4 autosampler, CAMAG TLC scanner-4 and visualizer. The EA extract of fungal endophyte was applied as bands on the plate through the software-based applicator. The mobile phase used for the development of the silica gel 60 F_{254} TLC plate (Merck, Darmstadt,

Germany) was toluene: chloroform: ethyl alcohol (4:4:1, $v/v/v$). After application, the TLC plate was kept in Twin trough chamber CAMAG (pre-saturated with mobile phase for 30 min) for the chromatogram development up to 70 mm from lower edge of the plate. The plate was dried and images were captured using visualizer at short UV range 254 nm and long UV range 366 nm. Scanning of the TLC plate was done by TLC scanner-4 at 254 nm (D2 light source) and 366 nm (Hg light source), and analysis of area percentage and retention factor for the separated components was analyzed by winCATS Planar Chromatography Manager (version 1.4.10.0001). The separated components on the TLC plate after development were subjected to post-chromatographic derivatization using Anisaldehyde Sulphuric Acid reagent (ASR).

2.7. Antioxidant Assays

Antioxidant activity of isolated fungal extract was assessed using DPPH free radical scavenging assay, superoxide anion scavenging assay, hydroxyl radical scavenging assay and nitric oxide scavenging assay. For each extract EC_{50} value was calculated using Graph Pad Prism 8.0.2 software and data for antioxidant assays was analyzed for statistical significance using One-Way ANOVA followed by Tukey to determine statistical significance.

2.7.1. Free Radical Scavenging Assay

The free radical scavenging activity of EA extract of fungal isolates was evaluated by colorimetric assay using DPPH as the source of free radical. Briefly, a fresh 50 µg/mL of DPPH solution (prepared in methanol, SRL, India) was added to different concentrations of EA extracts (1, 5, 10, 25, 50, 100 and 200 µg/mL) of fungal endophyte, vortexed vigorously and incubated for 30 min in dark at room temperature as described earlier [27]. After incubation, the absorbance was measured at 517 nm using UV-Vis spectrophotometer. The percentage of radical scavenging potential was calculated using the formula = $[1 - ($Abs$_{(517nm)}$ of the sample/Abs$_{(517nm)}$ of the control$)] \times 100$. Ascorbic acid was used as a positive control and methanol was used as a blank.

2.7.2. Superoxide Anion Scavenging Activity

To a mixture of 1 mL of 150 µM Nitroblue Tetrazolium solution (SRL) and 1 mL NADH (468 µM in 100 mM PBS, SRL, India) 200 µL of different concentrations (1, 5, 10, 25, 50, 100 and 200 µg/mL) of EA extract of each fungal endophyte were added. Reaction was started by adding 200 µL of phenazine methosulphate (60 mM PMS in 100 mM phosphate buffer, pH 7.4) solution and incubated at 25 °C for 5 min [28]. The absorbance was taken at 560 nm was measured against methanol as a blank. The percentage of superoxide anion radical scavenging potential was calculated using the formula = $[1 - ($Abs$_{(560nm)}$ of the sample/Abs$_{(560nm)}$ of the control$)] \times 100$. Ascorbic acid was used as a positive control and methanol as a blank.

2.7.3. Hydroxyl Radical Scavenging Assay

The different concentrations of EA extract of fungal isolate were added to a mixture of 100 µM $FeCl_3$ (Merck), 100 mM EDTA, 3.75 mM 2-deoxyribose (SRL, India), and 1 mM H_2O_2 (Qualigens Fine Chemicals, Maharashtra, India). After incubation for 1 h at 37 °C, 1 mL of 2% Trichloroacetic acid (TCA) and 1% Thiobarbituric acid (TBA) were added to the reaction mixture and again incubation was done for 15 min at 90 °C boiling water-bath, and then absorbance was taken at 535 nm. The calculation for the percentage hydroxyl radical scavenging potential was done using the formula = $[1 - ($Abs$_{(535nm)}$ of the sample/Abs$_{(535nm)}$ of the control$)] \times 100$ [27]. The positive control used was ascorbic acid and methanol was used as a blank.

2.7.4. Nitric Oxide Scavenging Assay

Different concentrations (1, 5, 10, 25, 50, 100 and 200 µg/mL) of EA extract of fungal endophyte is mixed with 150 µL of (10 mM) sodium nitroprusside and following incubation of 150 min, the reaction mixture was added with 200 µL of Greiss reagent (SRL, India) and left for 30 min. The absorbance was then recorded at 546 nm. Ascorbic acid was used as a positive control and methanol was used as a blank. The percentage of nitric oxide scavenging potential was calculated using the formula = $[1 - (\text{Abs}_{(546nm)}$ of the sample/$\text{Abs}_{(546nm)}$ of the control)$] \times 100$ [29].

2.8. Culture Media Preparation and Cell Line Maintenance

Two adherent human breast cancer cell lines MDA-MB-231 and MCF-7 and one suspension cell line of human malignant T-cells (HuT-78) were used to evaluate the cytotoxic effect of EA extract of *P. oxalicum*. HuT-78 cells were grown in 10% FBS containing Roswell Park Memorial Institute (RPMI) medium supplemented with 1% penicillin-streptomycin-amphotericin B solution (CELL clone) at 37 °C in a 5% CO_2 incubator (Eppendorf, Darmstadt, Germany). MDA-MB-231 and MCF-7 cells were grown in Dulbecco's modified Eagle's medium (DMEM) supplemented with 10% fetal bovine serum (HI-FBS) (US grade, Thermo), and 1% penicillin-streptomycin-amphotericin B solution (CELL clone) and maintained at 37 °C in a 5% CO_2 incubator. The cells were observed at every 24 h for cell growth and the presence of any contaminants.

Cytotoxicity Assay

The in vitro cytotoxic activity of EA extract of *P. oxalicum* was evaluated against HuT-78, MDA-MB-231 and MCF-7 cells based on the formation of insoluble formazan salt through reduction of 3-(4,5-dimethylthiazol-2-yl)-2,5-diphenyl tetrazolium bromide (MTT) by NAD(P)H-dependent cellular oxidoreductase enzymes that corresponds to viable cells remained after treatment with EA extract. The tumor cells (5×10^4 cells/well) were seeded in 96-well culture plates for 24 h at 37 °C in a humidified CO_2 incubator. EA extract of *P. oxalicum* was dissolved in DMSO, and further stock was prepared through serial dilution with DMEM media. Tumor cells were treated with different concentrations of EA extract of *P. oxalicum* (0–200 µg/mL) for 24 h. Cells treated with 0.1% DMSO was considered as control. Thereafter, 20 µL of MTT (5 mg/mL, SRL, India) was added to each well and incubated for 4 h at 37 °C in a 5% CO_2 incubator (Panasonic, Sakata, Japan). After 4 h, the plates were centrifuged for 20 min at 3000 rpm followed by solubilization of formazan crystals using 100 µL of DMSO. Absorbance was recorded at 570 nm using Multiskan™ FC microplate photometer (ThermoFisher Scientific, Waltham, MA, USA).

2.9. Statistical Analysis

The experiments of antioxidant assays and cytotoxic assay were performed in triplicate ($n = 3$). Data are presented as mean $\pm$ S.D. in histogram. For the statistical significance of antioxidant and cytotoxic data, One-Way ANOVA (analysis of variance) followed by Tukey to determine was performed using Graph Pad Prism 8.0.2 software and mean $\pm$ S.D. of all groups were compared.

3. Results and Discussion

3.1. A Total of 8 Fungal Endophytes Were Isolated Using Culture-Dependent Approach

From the leaf segment, eight fungal isolates were obtained by culture dependent method. The isolated fungal strains were designated as AR-L1 to AR-L8, for example; AR-L1 indicates the first fungal strain associated with the leaf of *A. rohituka*. Morphological characteristics of fungal isolates were observed under a microscope, and were identified based on their key morphological characteristics like shape and pigmentation of colony, morphology of mycelium (hyphae), presence/absence of septa and shape of conidia/spores (Figure 1). The fungal strains AR-L1, AR-L2, AR-L5 and AR-L6 showed similar characteristics as of *Aspergillus* sp. The characteristics found can be defined as powdery masses of whitish cream to black spores, septate and hyaline hyphae with conidia splitting into columns imparting rough texture to colonies were seen. Globose to sub-globose conidia and uniseriate vesicles were observed, along with the thick mycelial mat underneath the colonies [30]. The morphological characteristics of AR-L3 was found to be identical with *Meyerozyma guilliermondii* with features showing glabrous, white to cream-colored colonies. The identifiable character of AR-L4 was like *Trichoderma* sp. that possess conidiophores in small cottony pustules, and it forms in the scant aerial mycelium. Conidia was smooth, and the shape of conidia ranged from ellipsoidal to oblong or tuberculate to roughened or subglobose. AR-L7 was identified as of *Penicillium* sp. based on the identical character as a compact, powdery green-colored colony of which the back side appeared to be yellowish cream in color on a PDA plate. Mycelia was colorless with an arrangement of spores similar to a broom [31]. AR-L8 was identified as of *Diaporthe* sp. due to the presence of white, reverse off-white grey olivaceous coralloid, adpressed colonies and with no aerial mycelium [32].

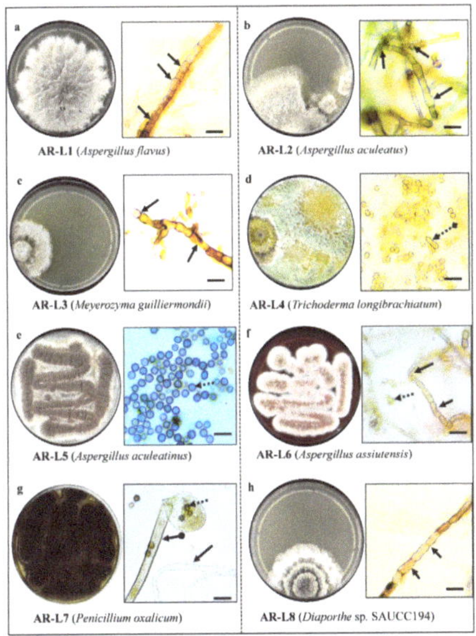

Figure 1. Pictorial representation of isolated fungal colony on PDA plate after 7 days of incubation and their microscopic view of fungal endophytes (**a**) *A. flavus*, (**b**) *A. aculeatus*, (**c**) *M. guilliermondii*, (**d**) *T. longibrachiatum*, (**e**) *A. aculeatinus*, (**f**) *A. assiutensis*, (**g**) *P. oxalicum* and (**h**) *Diaporthe* sp. SAUCC194 at 100× under bright field microscope. Mycelial septa (➝), Spores (⇢), Conidia (◆⋯▶) Conidiophore (●➤).

3.2. Eight Fungal Endophyte Species Belonging to the Division Ascomycota were Identified

After sequencing, the BLAST search was performed for obtained raw sequences for ITS region. The species level identification of isolated fungal strains along with their complete 5.8 s ribosomal RNA gene sequences was done, and strains were identified as *Aspergillus flavus*, *Aspergillus aculeatus*, *Meyerozyma guilliermondii*, *Trichoderma longibrachiatum*, *Aspergillus aculeatinus*, *Aspergillus assiutensis*, *Penicillium oxalicum* and *Diaporthe* sp. SAUCC194 (named as AR-L1 to AR-L8, respectively) (Table S1). In order to study the phylogenetic relationship between the fungal endophytes, a phylogenetic tree was constructed.

The constructed phylogenetic tree was clustered into four clades. Clade I consisted of one isolate that is *P. oxalicum* (MK332586), whereas Clade II consisted of *Diaporthe* sp. SAUCC194 (MT822596), and Clade III consisted of three species, i.e., *A. aculeatinus* (MK281555) and *T. longibrachiatum* (MT634694) along with *M. guilliermondii* (MT598067), which belong to the sub clade of Clade III. Clade IV's branch of the phylogenetic tree consisted of two closely related species, i.e., *A. flavus* (KX253948) and *A. aculeatus* (MT541887), whereas one more species belonged to the sub clade of Clade IV, i.e., *A. assiutensis* (MT640286)

(Figure S2). In the current study, being the genera of *Aspergillus*, *A. aculeatinus* does not belong to Clade IV, for such a contrasting result, the differences in the season and geographical location of the plant could be one of the possible reasons.

3.3. EA Extract of P. oxalicum Contains Highest Amount of Phenolic and Flavonoid Content

Phenolic and flavonoid compounds and their derivatives have been considered as primary free radical scavenging molecules by virtue of their aromatic rings that contribute to their antioxidant properties. The result for total flavonoid content has shown a variation of approximately 9-fold ranging from 17.70 µg QE/mg to 155.69 µg QE/mg of EA extract isolated fungal endophytes (Table 1). The highest flavonoid content has been displayed by EA extract of *P. oxalicum* with 155.69 µg QE/mg of EA extract. The lowest flavonoid content has been observed for *A. flavus* showing value of 17.70 µg QE/mg of EA extract. A linear relationship has been obtained by plotting concentration of Quercetin and its absorbance at 420 nm (R^2 = 0.9977) (Figure S3).

Table 1. Total phenolic content and flavonoid content present in EA extract of identified fungal strains associated with the leaf of *A. rohituka*.

S. No.	Species Identified	Total Phenolic Content (µg GAE/mg of EA Extract)	Total Flavonoid Content (µg QE/mg of EA Extract)
1.	*A. flavus*	45.65	17.70
2.	*A. aculeatus*	33.68	34.16
3.	*M. guilliermondii*	23.40	60.08
4.	*T. longibrachiatum*	32.42	78.88
5.	*A. aculeatinus*	69.36	32.65
6.	*A. assiutensis*	72.19	83.40
7.	*P. oxalicum*	120.99	155.69
8.	*Diaporthe* sp. SAUCC194	78.91	55.14

Total phenolic content produced by all the identified fungal endophytes revealed a variation of approximately 5-folds ranging from 23.40 µg GAE/mg to 120.99 µg GAE/mg of EA extract of isolated fungal endophytes (Table 1). The highest phenolic content value being 120.99 µg GAE/mg of EA extract for *P. oxalicum*, whereas the lowest phenolic content was observed for *M. guilliermondii* with of 23.40 µg GAE/mg of EA extract. The regression equation obtained from the calibration curve of gallic acid (R^2 = 0.9891) has been used to calculate the total content of phenolic compounds and expressed as µg GAE/mg of EA extract (Figure S4).

The phenolic compounds have been reported to prevent oxidative damages in biological systems through oxygen scavenging, free radical inhibition, metal inactivation and peroxide decomposition [33]. Among different species of fungal endophytes, the alteration in phenolic and flavonoid profiles directly affects their biological activities such as antioxidant and cytotoxic activities. The highest phenolic content of 40.30 mg GAE/g of plant dry weight in fungal extract of *Talaromyces* sp. derived from two varieties of Egyptian artichoke, namely French Hyrious and Egyptian Baladi, has evidenced the highest total antioxidant capacity (681 mg ascorbic acid equivalent per gram dry weight) [34]. Another report showed that the contents of phenolic (204 ± 6.144, 312.3 ± 2.147 and 152.7 ± 4.958 µg GAE/mg of dry extract) and flavonoid (177.9 ± 2.911, 644.1 ± 4.202 and 96.38 ± 3.851 µg RE/mg of dry extract) compounds of fungal strains *Alternaria alternata*, *Cladosporium cladosporioides* and *Alternaria brassicae* exhibited dose dependent radical scavenging activity [35]. The significant correlation of TPC and TFC with antioxidant activity has been supported through another study showing a fungal endophyte, *Nigrospora sphaerica* isolated from *Catharanthus roseus*, representing the highest TPC (0.030 ± 0.000 (mg GAE/g) and TFC (0.038 ± 0.001 (mg QE/g) with the highest reducing power activity [36]. Similarly, our findings are also in agreement with the fact that the presence of the phenolic and flavonoid content correlates with antioxidant and cytotoxic property, further suggesting that EA extract of *P. oxalicum* can serve as a promising candidate against various diseases caused by oxidative damage.

3.4. HPTLC Fingerprint Analysis Showed the Presence of Unique Bioactive Components in P. oxalicum Derived Bioactive Compounds

The HPTLC fingerprinting analysis of EA extract of all identified fungal strains was carried out for the presence of bioactive compounds. Various combinations of polar and non-polar solvents were used for HPTLC separation, but the best separation of fungal extracts was obtained with a mobile phase of toluene: chloroform: ethyl alcohol (4:4:1, *v/v/v*). Image of the plate was captured at short UV range 254 nm and long UV range 366 nm, and band for bioactive components of each fungal extracts were clearly seen in chromatogram (Figure 2). HPTLC fingerprint scanned at 254 nm (D2 light source) revealed the presence of six different components in EA extract of *A. flavus*, *M. guilliermondii*, *T. longibrachiatum* and *A. assiutensis*. Scanning of EA of *A. aculeatus* and *A. aculeatinus* revealed the presence of 7 different components in each isolate. EA extract of *Diaporthe* sp. SAUCC194 showed 5 different components according to their R_f values and area percentage. EA extract of *P. oxalicum* showed 7 different major components, out of the seven components, one novel component is found with R_f value of 0.75 and area percentage value of 6.46%.

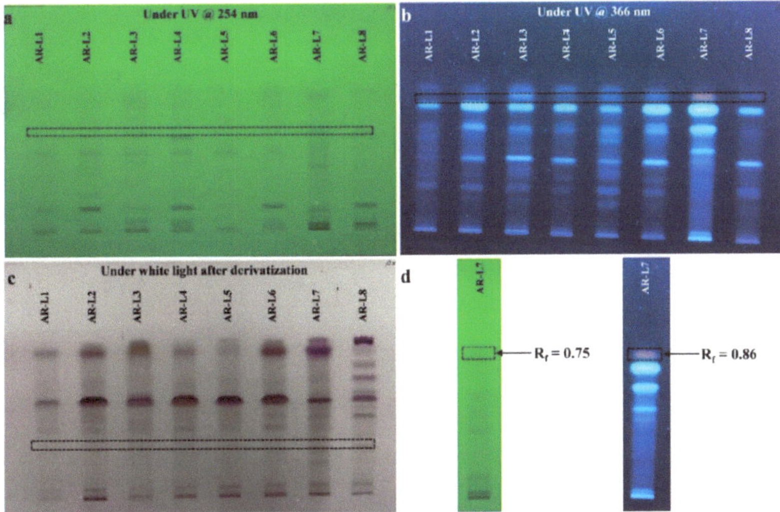

Figure 2. HPTLC fingerprint profiling of EA extract of fungal endophytes. (**a**) Image of TLC plate at 254 nm, (**b**) image of TLC plate at 366 nm, (**c**) image of TLC plate under white light after derivatization with anisaldehyde sulphuric acid reagent and (**d**) HPTLC fingerprint profiling of EA extract of *P. oxalicum* showing unique bioactive component with respective value of retention factor (R_f).

HPTLC fingerprint scanned at 366 nm (Hg lamp, fluorescence mode) for all the EA extract of fungal strains. EA extract of *A. flavus*, *A. aculeatus*, *M. guilliermondii*, *T. longibrachiatum* and *Diaporthe* sp. SAUCC194 showed four different major components. Similarly, *A. aculeatinus* possess seven different components, whereas in *A. assiutensis* and *P. oxalicum*, three different components are found. Out of three different bioactive components found in *P. oxalicum*, one unique band is observed with area percentage and R_f value of 62.50% and 0.74, respectively. Post-chromatography derivatization by anisaldehyde sulphuric acid reagent revealed the chemical nature of bioactive components such as terpenoids, steroids, sterols and saponins present in fungal extract. Out of eight EA extracts of isolated strains, the EA extract of *P. oxalicum* showed a unique band (a prominent violet color band indicates phenolic molecules [37]) that could be potential source of antioxidant and cytotoxic activity.

3.5. EA Extract of A. flavus and P. oxalicum Shows Potential Free Radical Scavenging Activity

The antioxidant activity was observed visibly through the decolorization of purple color to yellow, while the percentage inhibition was estimated through spectrophotometric analysis at 517 nm. The antioxidant activity assessed for the fungal strains displayed that some of the strains possess significant antioxidant activity and rest other strains showed less or minimum antioxidant activity. Out of eight fungal isolates *P. oxalicum* showed best antioxidant activity as compared to other strains and in a concentration dependent manner (Figure 3). The EC_{50} value for *P. oxalicum* was found to be 96.98 $\pm$ 0.270 µg/mL. Another strain with significant antioxidant activity was EA extract of *A. flavus* that displayed moderate antioxidant activity with EC_{50} value of 178.30 $\pm$ 1.446 µg/mL. The other strains were least effective against free radicals and EC_{50} value calculated was out of tested range (>200 µg/mL). The positive control used was ascorbic acid that was used to compare the antioxidant potential of fungal endophyte derived bioactive compounds. The EC_{50} value of ascorbic acid (positive control) was estimated to be 10.60 $\pm$ 0.257 µg/mL (Table 2).

Table 2. Antioxidant activity of EA extracts of identified fungal strains associated with the leaf of *A. rohituka*.

S. No.	Species	EC_{50} Value (µg/mL)			
		DPPH Free Radical Scavenging Assay	**Superoxide Anion Scavenging Activity**	**Hydroxyl Radical Scavenging Assay**	**Nitric Oxide Scavenging Assay**
1.	*A. flavus*	178.30 $\pm$ 1.446	157.52 $\pm$ 1.118	>200	>200
2.	*A. aculeatus*	>200	>200	>200	>200
3.	*M. guilliermondii*	>200	>200	>200	>200
4.	*T. longibrachiatum*	>200	170.43 $\pm$ 1.405	113.30 $\pm$ 1.206	89.14 $\pm$ 0.894
5.	*A. aculeatinus*	>200	>200	>200	>200
6.	*A. assiutensis*	>200	>200	>200	>200
7.	*P. oxalicum*	96.98 $\pm$ 0.270	169.28 $\pm$ 0.402	126.12 $\pm$ 0.636	75.79 $\pm$ 0.692
8.	*Diaporthe* sp. SAUCC194	>200	>200	>200	>200
Positive control	Ascorbic Acid	10.60 $\pm$ 0.257	33.36 $\pm$ 1.186	24.37 $\pm$ 1.116	21.75 $\pm$ 0.566

Results showed that the content of phenolic compounds was lower compared to the EA extract of *P. oxalicum*, but the antioxidant activity was found to be best among all strains. DPPH scavenging activity of EA extract of *A. flavus* was not in accordance to phenolic content and that indicated the involvement of other class of compounds responsible for displaying antioxidant activity. The method of estimation of phenolic content or interference of other compounds could be a possible reason for nonsignificant phenolic content produced by endophyte *A. flavus*. Other factors include the synergistic action of other class of compounds that showed potential antioxidant activity [38–41]. Some previous studies have also shown inconsistent correlation between phenolic content and antioxidant activity [42,43]. However, *A. flavus* have been reported with significant antioxidant activity and with a positive correlation with the phenolic content. The DPPH free radical scavenging activity of EA extract of *P. oxalicum* was in accordance with previous study, which showed that fungal endophyte *P. oxalicum* isolated from *Citrus limon* showed linear correlation between high phenolic content and DPPH scavenging activity with an EC_{50} value of 127.56 µg/mL [44]. The metabolites N-[2-(4 hydroxyphenyl) ethenyl] formamide and tuckolide isolated from *P. oxalicum* were effective against DPPH radical scavenging assay with EC_{50} values of 18.53 and 79.17 µg/mL, respectively [45].

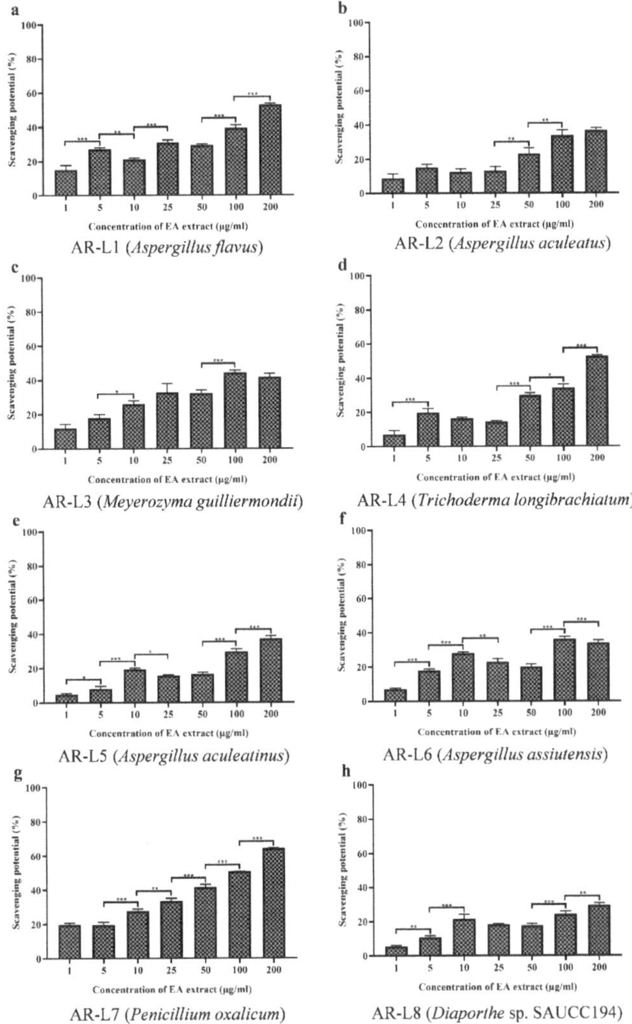

Figure 3. DPPH free radical scavenging potential of EA extract of fungal endophytes (**a**) *A. flavus*, (**b**) *A. aculeatus*, (**c**) *M. guilliermondii*, (**d**) *T. longibrachiatum*, (**e**) *A. aculeatinus*, (**f**) *A. assiutensis*, (**g**) *P. oxalicum* and (**h**) *Diaporthe* sp. SAUCC194. All the experiments were performed in triplicate. *p*-value was calculated by comparing means $\pm$ SD of the DPPH free radical scavenging potential (%), using one-way ANOVA followed by Tukey to determine statistical significance which are as follows; *** $p \leq 0.001$; ** $p \leq 0.002$; * $p \leq 0.033$.

3.6. EA Extract of A. flavus and P. oxalicum Display Potential Superoxide Anion Scavenging Activity

The superoxide radical scavenging assay performed for the eight strains showed significant inhibition of superoxide radicals. The inhibition percentage of super oxide anion radicals surged with the increasing concentration of fungal EA extract (Figure 4). Among eight fungal strains, the highest scavenging was observed for EA extract of *A. flavus* with an EC_{50} value of 157.52 $\pm$ 1.118 µg/mL, whereas the EA extract of *P. oxalicum* and *T. longibrachiatum* also exhibited a closer superoxide anion scavenging activity to EA extract of *A. flavus* with EC$_{50}$ values of 169.28 ± 0.402 µg/mL and 170.43 ± 1.405 µg/mL, respectively. EA extracts of other five fungal endophytes, such as *A. aculeatus*, *M. guilliermondii*, *A. aculeatinus*, *A. assiutensis* and *Diaporthe* sp. SAUCC194 showed no superoxide anion scavenging activity within tested range of concentration, and therefore may exhibit an EC$_{50}$ value of more than 200 µg/mL (Table 2). The standard oxidant or positive control (Ascorbic acid) exhibited EC$_{50}$ value of 33.36 ± 1.186 µg/mL. All the fungal extracts showed lower antioxidant activity than the positive control.

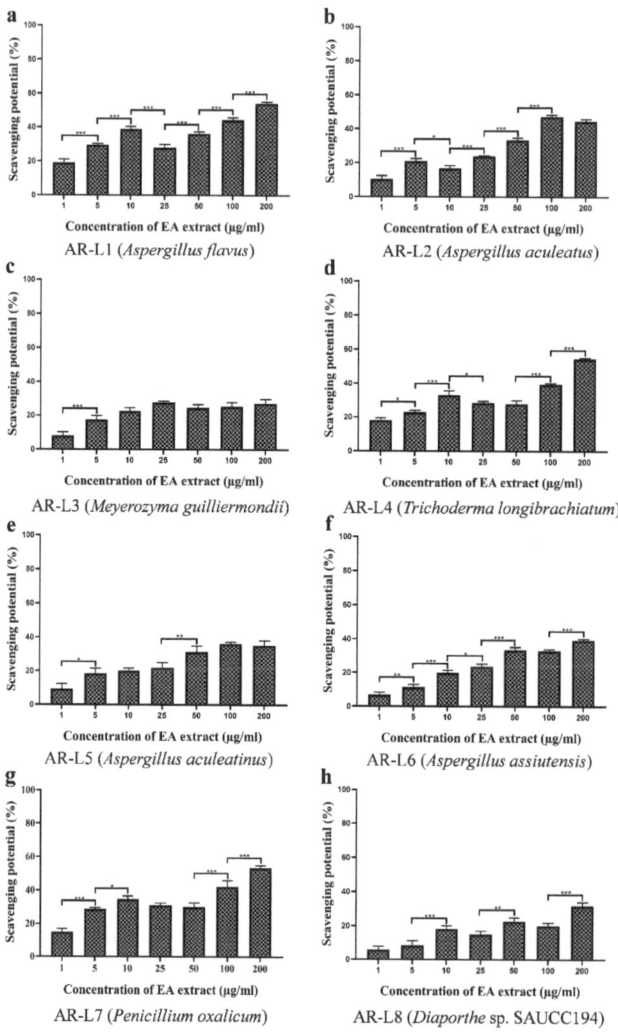

Figure 4. Superoxide anion scavenging potential of EA extract of fungal endophytes (**a**) *A. flavus*, (**b**) *A. aculeatus*, (**c**) *M. guilliermondii*, (**d**) *T. longibrachiatum*, (**e**) *A. aculeatinus*, (**f**) *A. assiutensis*, (**g**) *P. oxalicum* and (**h**) *Diaporthe* sp. SAUCC194. All the experiments were performed in triplicate. *p*-value was calculated by comparing means ± SD of the superoxide anion radical scavenging potential (%), using one-way ANOVA followed by Tukey to determine statistical significance, which are as follows; *** $p \leq 0.001$; ** $p \leq 0.002$; * $p \leq 0.033$.

The superoxide radicals are the most commonly formed radicals in the biological system through various enzymatic, non-enzymatic and autooxidation reactions and cause several disorders and pathological effects [46]. The results observed do not show linear correlation between phenolic content and antioxidant activity. Previous reports have witnessed the role of exopolysaccharide As1-1 produced by the mangrove endophytic fungus *Aspergillus* sp. Y16 in the scavenging of superoxide radicals with an EC_{50} value of 3.4 mg/mL [47]. These reports have shown that *Aspergillus* sp. exhibit superoxide anion scavenging and could be a potential natural antioxidant. The moderate superoxide anion scavenging activity of *P. oxalicum* can be related to its high phenolic content. In a report, the fungal endophyte *P. oxalicum* isolated from *Ligusticum chuanxiong* Hort (CX) was reported to exhibit superoxide anion scavenging activity with an IC_{50} value of 1.41 ± 0.22 mg/mL and contain high polyphenolic content [48]. EA extract of *T. longibrachiatum* showed moderate antioxidant activity. The previous report has shown mangrove-derived *Trichoderma* sp. to possess significant superoxide anion scavenging activity of 71.45 % at 500 μg/mL. However, the phenolic content was not reported to be significant for *T. longibrachiatum*. Besides, other strains with phenolic content did not exhibit significant antioxidant activity, and EC_{50} value was observed to be >200 μg/mL.

3.7. EA Extract of T. longibrachiatum and P. oxalicum Exhibit Significant Hydroxyl Radical Scavenging Activity

The hydroxyl radical scavenging activity was evaluated on the basis of percentage inhibition of deoxyribose degradation. The formation of pink chromogen was measured at 535 nm. The EA extract of fungal isolates showed concentration-dependent inhibition of hydroxyl radicals (Figure 5). The highest EC_{50} value was observed to be 113.30 ± 1.206 μg/mL for EA extract of *T. longibrachiatum*, which showed significant hydroxyl radicals scavenging activity, whereas EA extract of *P. oxalicum* exhibited moderate hydroxyl radical scavenging activity with an EC_{50} value of 126.12 ± 0.636 μg/mL. Beside the aforementioned strains, EA extracts of mycelia of rest other strains isolated from leaf and twig showed minimum antioxidant activity with EC_{50} value of more than 200 μg/mL (Table 2). The EC_{50} value for ascorbic acid used as positive control was reported to be 24.37 ± 1.116 μg/mL.

The strongest oxidizing power is known for hydroxyl anion, as it non-specifically targets protein, lipid and nucleic acids and leads to the development of serious diseases. The hydroxyl anion scavenging assay colorimetrically measures the formation of colored product malondialdehyde. The EA extract of *T. longibrachiatum* prevented the reaction in a concentration-dependent manner, and maximum antioxidant activity was recorded with an EC_{50} value of 113.30 ± 1.206 μg/mL, but it was not linearly correlated to its phenolic content. The linear correlation between phenolic content and hydroxyl anion scavenging has been observed for *Trichoderma* strain EMFCAS8, a mangrove endophytic fungus [49]. The considerable hydroxyl anion scavenging activity of EA extract of *P. oxalicum* has been validated through higher phenolic content. In a report, *P. oxalicum* isolated from *Citrus limon* has shown concentration dependent scavenging of hydroxyl radical that was attributed to its polyphenolic contents [43]. The fungal endophyte *P. oxalicum* isolated from *Ligusticum chuanxiong* Hort (CX) showed strong correlation between polyphenolic content and hydroxyl anion scavenging and EC_{50} value for scavenging activity was observed to be 3.38 ± 0.04 mg/mL [48]. These results suggest that the fungal endophytes *P. oxalicum* and *T. longibrachiatum* isolated from *A. rohituka* could serve as a potential candidate against oxidative damage.

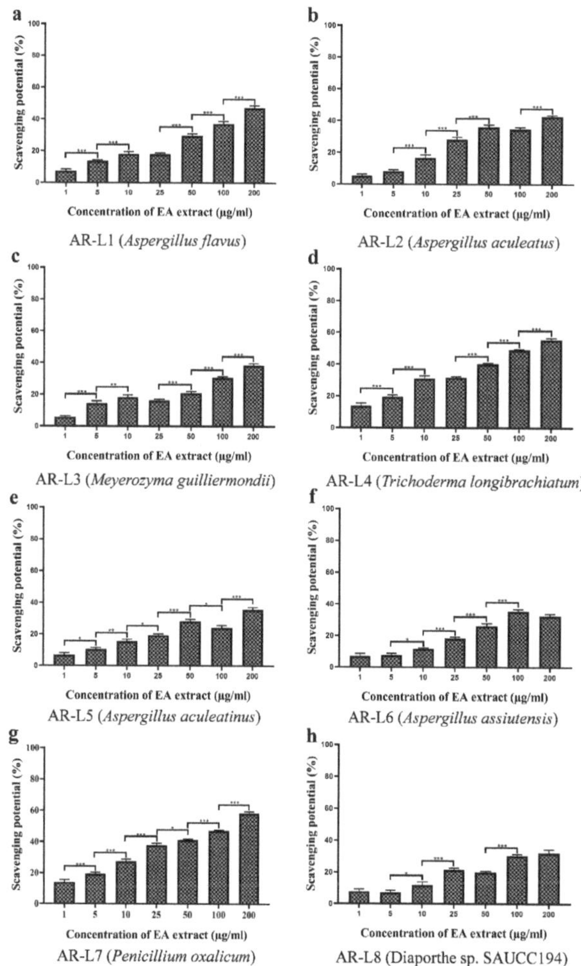

Figure 5. Hydroxyl radical scavenging potential of EA extract of fungal endophytes (**a**) *A. flavus*, (**b**) *A. aculeatus*, (**c**) *M. guilliermondii*, (**d**) *T. longibrachiatum*, (**e**) *A. aculeatinus*, (**f**) *A. assiutensis*, (**g**) *P. oxalicum* and (**h**) *Diaporthe* sp. SAUCC194. All the experiments were performed in triplicate. *p*-value was calculated by comparing means ± SD of the hydroxyl radical scavenging potential (%), using one-way ANOVA followed by Tukey to determine statistical significance, which are as follows; *** $p \leq 0.001$; ** $p \leq 0.002$; * $p \leq 0.033$.

3.8. EA Extract of T. longibrachiatum and P. oxalicum Shows Maximum Nitric Oxide Scavenging Activity

The chain of reactions leading to generation of nitrite ions was arrested through the EA extract of fungal isolates. The nitric oxide scavenging activity of EA extract was observed to be increased through the increasing concentration of extract (Figure 6). The significant antioxidant was observed for the fungal EA extracts and was comparable to the activity shown by positive control. The strongest activity was shown by the fungal isolate *P. oxalicum* with an observed EC_{50} value of 75.79 ± 0.692 µg/mL. Another strain to show significant scavenging of nitric oxide radical was *T. longibrachiatum* with an EC_{50} value of 89.14 ± 0.894 µg/mL EC_{50} value of ascorbic acid (positive control) was estimated to be 21.75 ± 0.566 µg/mL. EA extract of *A. flavus*, *A. aculeatus*, *M. guilliermondii*, *A. aculeatinus*,

A. assiutensis, and *Diaporthe* sp. SAUCC194 showed minimum antioxidative activity with EC_{50} value of more than 200 µg/mL (Table 2).

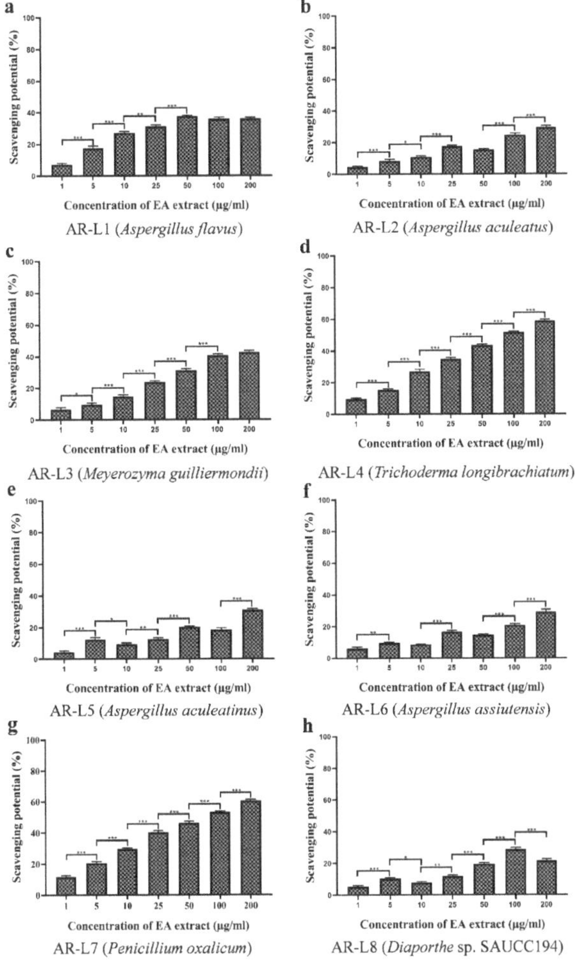

Figure 6. Nitric oxide scavenging potential of EA extract of fungal endophytes (**a**) *A. flavus*, (**b**) *A. aculeatus*, (**c**) *M. guilliermondii*, (**d**) *T. longibrachiatum*, (**e**) *A. aculeatinus*, (**f**) *A. assiutensis*, (**g**) *P. oxalicum* and (**h**) *Diaporthe* sp. SAUCC194. All the experiments were performed in triplicate. *p*-value was calculated by comparing means ± SD of the nitric oxide radical scavenging potential (%), using one-way ANOVA followed by Tukey to determine statistical significance, which are as follows; *** $p \leq 0.001$; ** $p \leq 0.002$; * $p \leq 0.033$.

The reactive oxygen species generated through various metabolic process pose deleterious effects to specific biomolecules such as lipid and proteins [50]. The presence of unpaired electron in nitric oxide causes it to act as a free radical and damage biomolecules on interaction. Natural antioxidants like endophytic fungal extract are known to be safe and bioactive for neutralizing the negative effects of free radicals. The nitric oxide scavenging assay revealed the fungal endophyte *P. oxalicum* to show highest antioxidant activity among other isolates, which is attributed to the increased phenolic and flavonoid contents. The fungal strain *P. oxalicum* derived from marine sources has been observed to exhibit strong inhibitory activities for NO production with 57.6% [51]. Another strain, *T. longibrachiatum*, showed moderate antioxidant activity with EC_{50} value of 89.14 ± 0.894 µg/mL that can be related to its moderate phenolic content. Previous report has revealed significant NO_2 radical scavenging activity of mangrove derived fungal endophyte *Trichoderma* strain EMF-CAS8 that was highly correlated with phenolic content [49]. Overall, results of antioxidant activity also show that EA extract of endophytes *A. flavus*, *A. aculeatus*, *M. guilliermondii*, *A. aculeatinus*, *A. assiutensis*, and *Diaporthe* sp. SAUCC194 exhibited minimum scavenging potential against tested synthetically generated free radicals having a certain extent of phenolic and flavonoid content. This is may be due to the antagonistic effect of phenolic and flavonoid compounds produced by fungal endophytes [52,53].

3.9. EA Extract of P. oxalicum Showed Promising Cytotoxic Activity against Cancer Cells

Among eight identified endophytes, EA extract of *P. oxalicum* exhibited potential antioxidant activity, increased contents of phenolic and flavonoid compounds and produces more and unique bioactive components therefore, further selected to evaluate the cytotoxic activity against three different human cancer cell lines (HuT-78, MDA-MB-231 and MCF-7) (Figure 7a–c). Our result showed that EA extract of *P. oxalicum* exhibited cytotoxic activity against breast cancer and T lymphoma cells in a dose dependent manner. The IC_{50} values for EA extract of *P. oxalicum* were observed to be 56.81 ± 0.671 µg/mL for HuT-78, 37.24 ± 1.26 µg/mL for MDA-MB-231 and 260.62 ± 5.415 µg/mL for MCF-7 cells (Table 3). Taken together, our findings suggest that the EA extract of *P. oxalicum* has potential anticancer activity and thus might be considered as a therapeutic option for cancer treatment. Cancer is caused due to the development of abnormal cells that grow beyond their limits and through metastasis invade other organs of the body and pose damages. The proliferation of cells leading to cancer is significantly inhibited by fungal endophytes derived bioactive compounds. Previous studies have shown anticancer potential of fungal endophyte isolated from *Datura stramonium* against UMG87 glioblastoma and A549 lung carcinoma cell lines using MTT assay [54]. In another report, a bioactive compound flavipin isolated from an endophytic fungus associated with the *Chaetomium globosum* inhibited the growth of A549, MCF-7, and HT-29 cancer cell line with an IC_{50} value of 9.549 µg/mL, 54 µg/mL and 18 µg/mL, respectively [55]. In a study, it was found that a compound ergoflavin isolated from an endophytic fungus *P. oxalicum* caused cytotoxicity in HCT116, Panc1, ACHN and H460 cancer cell line with IC_{50} values of 8.0 ± 0.45 µM, 2.4 ± 0.02 µM, 1.2 ± 0.20 µM and 4.0 ± 0.08 µM, respectively [56]. In one of the recent reports, it was found that a secondary metabolite obtained from marine endophytic fungi *P. chrysogenum* isolated from green algae showed anticancer activity against the MCF-7 cancer cell line with an IC_{50} value of 101.1 µg/mL [57].

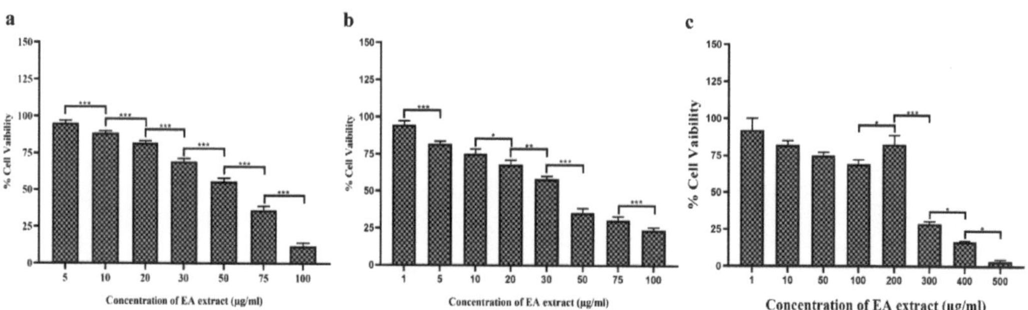

Figure 7. (**a**) Cytotoxic effect of EA extract of *P. oxalicum* against HuT-78 cell line (Human T cell lymphoma cell line), (**b**) MDA-MB-231 (breast cancer cell line) and (**c**) MCF-7 cell line (Human breast cancer cell line). All experiments were performed in triplicate. *p*-value was calculated by comparing means ± SD of percentage of cell viability of cancer cell, using one-way ANOVA followed by Tukey to determine statistical significance, which are as follows; *** $p \leq 0.001$; ** $p \leq 0.002$; * $p \leq 0.033$.

Table 3. Cytotoxic activity of EA extract of *P. oxalicum* associated with the leaf of *A. rohituka*.

Isolated Strains	Species	IC$_{50}$ Value (μg/mL)		
		T-Cell Lymphoma Cancer Cell (HuT-78)	Human Breast Cancer Cell (MDA-MB-231)	Human Breast Cancer Cell (MCF-7)
AR-L7	*P. oxalicum*	56.81 ± 0.617	37.24 ± 1.26	260.627 ± 5.415

3.10. Comparative Analysis between EA Extract of AR-L7 and P. oxalicum (PO-01) Isolated from Rhizospheric Soil of Maize

Results of phenolic and flavonoid contents, antioxidant and cytotoxic activity of EA extract of fungal endophyte *P. oxalicum* have compelled us to perform a host-dependent activity and biochemical profiling of *P. oxalicum* isolated from two different hosts. In the present study, we have compared the antioxidant activity of EA extract of fungal endophyte *P. oxalicum* associated with leaf of *A. rohituka* (AR-L7) and rhizospheric soil of maize (PO-01). The results showed the fungal strain AR-L7 exhibit significant scavenging potential against all the synthetically generated oxidative radicals such as DPPH free radicals, super oxide anion radicals, hydroxyl radicals and nitric oxide radicals. The EC$_{50}$ value of EA extract of AR-L7 was found to be 83.76 ± 1.14 μg/mL, 162.12 ± 1.185 μg/mL, 123.62 ± 2.01 μg/mL and 75.60 ± 1.31 μg/mL, for DPPH radical scavenging (Figure 8a,b), superoxide anion scavenging (Figure 8c,d), hydroxyl radical scavenging (Figure 8e,f) and nitric oxide scavenging assay (Figure 8g,h), respectively. The fungal strain PO-01 isolated from rhizospheric soil of maize failed to scavenge 50% of oxidative radicals within tested range of concentration and therefore, possessed EC$_{50}$ value more than 200 μg/mL (Table 4). The screening of bioactive components has been done through HPTLC fingerprinting that showed *P. oxalicum* derived from *A. rohituka* (AR-L7) to possess more bioactive components as compared to the similar strain isolated from rhizospheric soil of maize (PO-01). HPTLC fingerprint scanned at 254 nm (D2 light source) revealed the presence of six different major components in EA extract of AR-L7 whereas, in PO-01, there were five major components according to their R$_f$ values and area percentage. HPTLC fingerprint scanned at 366 nm (Hg lamp, fluorescence mode) for the EA extract of AR-L7 showed three different major components (Figure 8i,j) whereas, PO-01 showed that four major components were present according to their R$_f$ values and area percentage.

The comparative study revealed that fungal endophyte *P. oxalicum* isolated from *A. rohituka* produces unique bioactive components and shows potential antioxidant activity as compared to a similar strain isolated from rhizospheric soil of maize. Our result reveals the host-dependent activity of fungal endophyte *P. oxalicum*. The fungal endophyte and its host establish a flexible relationship whose directionality is driven by the host genotype. Host genotypes shape the colonization of fungal endophytes and also influence their biological activity. In addition, host recognition and response to the colonized fungal endophytes and fungal gene expressions determine the type of interaction that may be either positive, negative and neutral [58–60]. The activity of fungal endophytes in different hosts vary in response to ecological, physiological and genetic factors [61]. The biochemical characteristics of fungal endophytes is directed through their association with the host. The bioactive components through HPTLC fingerprinting revealed that in the EA extract of AR-L7 (*P. oxalicum* isolated from *A. rohituka*), six major bioactive components were present, whereas it revealed the presence of five major components in EA extract of PO-01 (*P. oxalicum* isolated from rhizospheric soil of maize). The association of fungal endophytes with the host drives the qualitative and quantitative synthesis of bioactive compounds, and therefore affects their biological activity. The significant antioxidant activity of AR-L7 may be attributed due to the presence of unique bioactive component observed through HPTLC fingerprint analysis. Previous studies have shown a toxic metabolite, Secalonic acid D derived from *P. oxalicum* cause acute toxicoses in rats, mice and ducklings and also showed toxicity to male Swiss albino rats [62]. In other reports, fungal endophyte *P. oxalicum* have been reported for the presence of cytotoxic compounds dihydrothiophene-condensed chromones [63], production of potential antibiotics against pathogenic fungal strains [64] and antibacterial and antitumor activities [65]. Therefore, it can be concluded that the diversity, distribution and biological activity of fungal endophytes is determined through their response against interaction with host and that is attributed to their genetic background, ecological and physiological factors.

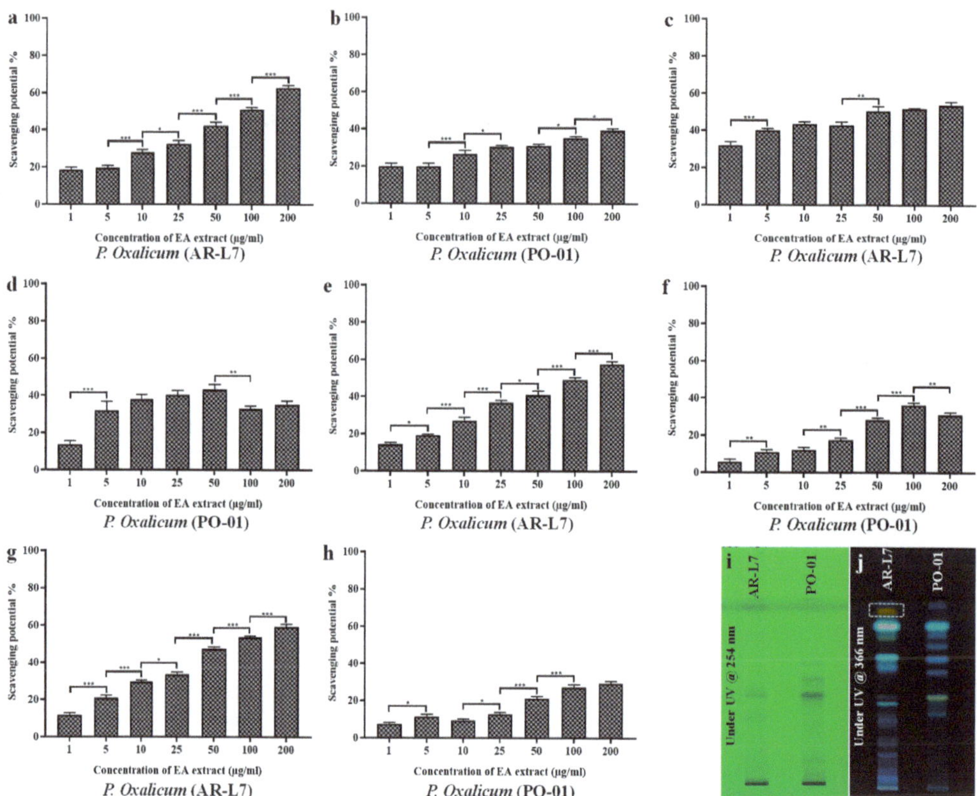

Figure 8. Comparative evaluation of antioxidant potential (**a,b**) DPPH free radical scavenging, (**c,d**) superoxide anion scavenging, (**e,f**) hydroxyl radical scavenging, and (**g,h**) nitric oxide scavenging for the EA extract of fungal strains *P. Oxalicum* (AR-L7) isolated from *A. rohituka* and *P. Oxalicum* (PO-01) isolated from rhizospheric soil of maize. All the experiments were performed in triplicate. *p*-value was calculated by comparing means ± SD of the free radical scavenging potential (%), using one-way ANOVA followed by Tukey to determine statistical significance. Statistical significances are as follows; *** $p \leq 0.001$; ** $p \leq 0.002$; * $p \leq 0.033$. HPTLC fingerprint profiling of EA extract of fungal endophytes *P. Oxalicum* (AR-L7) and *P. Oxalicum* (PO-01) (**i**) image of TLC plate at 254 nm, (**j**) image of TLC plate at 366 nm.

Table 4. Host-dependent antioxidant activity of EA extract of fungal endophyte *P. oxalicum* (AR-L7) and PO-01.

Fungal Strains	Host Plant	EC$_{50}$ Value (µg/mL)			
		DPPH Free Radical Scavenging Assay	Superoxide Anion Scavenging Assay	Hydroxyl Radical Scavenging Assay	Nitric Oxide Scavenging Activity
AR-L7 (*P. oxalicum*)	Leaf of *A. rohituka*	>200	>200	>200	>200
PO-01 (*P. oxalicum*)	Rhizospheric soil of maize	83.760 ± 1.14 µg/mL	162.12 ± 1.185 µg/mL	123.62 ± 2.01 µg/mL	75.60 ± 1.31 µg/mL

4. Conclusions

The present study revealed eight fungal endophytes isolated from the medicinal plant *A. rohituka* and their bioactive compounds were assessed for various antioxidant and cytotoxic activity. The fungal endophyte, *P. oxalicum* has been identified to exhibit potential biological activity in terms of antioxidant activity by virtue of its significant flavonoid and phenolic content. The unique band obtained for *P. oxalicum* through HPTLC fingerprinting also validates the significant activity of this fungal endophyte. The cytotoxic assay performed for EA extract of *P. oxalicum* has showed effective and significant cytotoxicity against human breast cancer cells (MDA-MB-231). Host-dependent activity revealed that EA extract of fungal endophyte *P. oxalicum* isolated from the leaf of *A. rohituka* produces more bioactive components and, therefore, possessed significant antioxidant activity over *P. oxalicum* isolated from rhizospheric soil of maize. Conclusively, the present study further directs the characterization and identification of bioactive compounds present in the EA extract of *P. oxalicum* with antioxidant and anticancer activity, and further appeals for its future examination in the improvement of human health.

Supplementary Materials: The following supporting information can be downloaded at: https://www.mdpi.com/article/10.3390/jof8030285/s1. Table S1. List of leaf associated fungal strains identified through DNA sequencing of ITS region and their accession number. Figure S1. Gel image of PCR products of conserved ITS region of fungal strains associated with the leaf of *Amoora rohituka*. Figure S2. Phylogenetic tree of fungal endophytes constructed by maximum likelihood bootstrap (MLBS) method. Figure S3. Calibration curve for Quercetin (Standard). Figure S4. Calibration curve for Gallic acid (Standard). Refs. [66–70] are cited in Table S1.

Author Contributions: V.G.: Conceptualizing the work, funding acquisition and supervision of the study. A.V.: Majorly performed the experiments, data analysis and wrote the manuscript under the supervision of V.G. P.G. and N.R.: Contributed in the experiments, revision and editing of the manuscript. R.K.T.: Performed the MTT assay on HuT-78 under the supervision of A.K. A.K.: Proofreading of the manuscript. S.C.K.: Experiment and writing. P.S. and S.K.S.: Revision and editing of the manuscript. All authors have read and agreed to the published version of the manuscript.

Funding: This work is supported by Institution of Eminence (IoE)-Seed Grant, Banaras Hindu University, Varanasi, India to V.G.

Institutional Review Board Statement: Not applicable.

Informed Consent Statement: Not applicable.

Data Availability Statement: All the data are mentioned in the manuscript and as supplementary materials.

Acknowledgments: A.V. would like to thank the Council of Scientific and Industrial Research, New Delhi, India for the Junior Research Fellowship. P.G. would like to thank Science and Engineering Research Board (SERB), India for Junior Research Fellowship under Empowerment and Equity Opportunities for Excellence in Science (EMEQ scheme; EEQ/2019/000025). N.R. would like to thank University Grants Commission, New Delhi, India, for the Junior Research Fellowship. V.G. also acknowledges funding support to his lab from Science and Engineering Research Board (SERB), India under Empowerment and Equity Opportunities for Excellence in Science (EMEQ scheme; EEQ/2019/000025) and University Grants Commission (for Startup Grant), New Delhi, India. S.K.S. acknowledges Institute of Medical Sciences, Banaras Hindu University, Varanasi, India for providing funding support to his lab.

Conflicts of Interest: The corresponding author on behalf of all the authors declares no potential conflict of interest.

References

1. Oladeji, O. The characteristics and roles of medicinal plants: Some important medicinal plants in Nigeria. *Nat. Prod. Ind. J.* **2016**, *12*, 102.
2. Hamilton, A.C. Medicinal plants, conservation and livelihoods. *Biodivers. Conserv.* **2004**, *13*, 1477–1517. [CrossRef]
3. Kaul, S.; Gupta, S.; Ahmed, M.; Dhar, M.K. Endophytic fungi from medicinal plants: A treasure hunt for bioactive metabolites. *Phytochem. Rev.* **2012**, *11*, 487–505. [CrossRef]
4. Keshri, P.K.; Rai, N.; Verma, A.; Kamble, S.C.; Barik, S.; Mishra, P.; Singh, S.K.; Salvi, P.; Gautam, V. Biological potential of bioactive metabolites derived from fungal endophytes associated with medicinal plants. *Mycol. Prog.* **2021**, *20*, 577–594. [CrossRef]
5. Rai, N.; Kumari Keshri, P.; Verma, A.; Kamble, S.C.; Mishra, P.; Barik, S.; Kumar Singh, S.; Gautam, V. Plant associated fungal endophytes as a source of natural bioactive compounds. *Mycology* **2021**, *12*, 139–159. [CrossRef]
6. Fadiji, A.E.; Babalola, O.O. Elucidating mechanisms of endophytes used in plant protection and other bioactivities with multi-functional prospects. *Front. Bioeng. Biotechnol.* **2020**, *8*, 467. [CrossRef]
7. Ranjan, A.; Singh, R.K.; Khare, S.; Tripathi, R.; Pandey, R.K.; Singh, A.K.; Gautam, V.; Tripathi, J.S.; Singh, S.K. Characterization and evaluation of mycosterol secreted from endophytic strain of *Gymnema sylvestre* for inhibition of α-glucosidase activity. *Sci. Rep.* **2019**, *9*, 17302. [CrossRef]
8. Gupta, P.; Verma, A.; Rai, N.; Singh, A.K.; Singh, S.K.; Kumar, B.; Kumar, R.; Gautam, V. Mass Spectrometry-Based Technology and Workflows for Studying the Chemistry of Fungal Endophyte Derived Bioactive Compounds. *ACS Chem. Biol.* **2021**, *16*, 2068–2086. [CrossRef]
9. Chopra, R.N. *Glossary of Indian Medicinal Plants*; Council of Scientific & Industrial Research: Delhi, India, 1956.
10. Jagetia, G.C.; Venkatesha, V. Preclinical determination of the anticancer activity of rohituka (*Aphanamixis polystachya*) in Ehrlich ascites tumor-bearing mice. *Med. Aromat. Plant Sci. Biotechnol.* **2012**, *6*, 42–51.
11. Dhar, M.; Dhar, M.; Dhawan, B.; Mehrotra, B.; Ray, C. Screening of Indian plants for biological activity: Part I. *Indian J. Exp. Biol.* **1968**, *6*, 232–472.
12. Jagetia, G.C.; Venkatesha, V.A.K. Enhancement of radiation effect by *Aphanamixis polystachya* in mice transplanted with Ehrlich ascites carcinoma. *Biol. Pharm. Bull.* **2005**, *28*, 69–77. [CrossRef] [PubMed]
13. Jagetia, G.C.; Venkatesha, V. Treatment of mice with stem bark extract of *Aphanamixis polystachya* reduces radiation-induced chromosome damage. *Int. J. Radiat. Biol.* **2006**, *82*, 197–209. [CrossRef] [PubMed]
14. Kumar, V.; Hart, A.; Wimalasena, T.; Tucker, G.; Greetham, D. Expression of RCK2 MAPKAP (MAPK-activated protein kinase) rescues yeast cells sensitivity to osmotic stress. *Microb. Cell Factories* **2015**, *14*, 85. [CrossRef] [PubMed]
15. Boericke, W. *New Manual of Homoeopathic Materia Medica and Repertory*; B. Jain Publishers: Uttar Pradesh, India, 2001.
16. Harmon, A.D.; Weiss, U.; Silverton, J. The structure of rohitukine, the main alkaloid of *Amoora rohituka* (syn. Aphanamixis polystachya) (Meliaceae). *Tetrahedron Lett.* **1979**, *20*, 721–724. [CrossRef]
17. Naik, R.G.; Kattige, S.; Bhat, S.; Alreja, B.; De Souza, N.; Rupp, R. An antiinflammatory cum immunomodulatory piperidinylben-zopyranone from *Dysoxylum binectariferum*: Isolation, structure and total synthesis. *Tetrahedron* **1988**, *44*, 2081–2086. [CrossRef]
18. Singh, N.; Singh, P.; Shrivastva, S.; Mishra, S.K.; Lakshmi, V.; Sharma, R.; Palit, G. Gastroprotective effect of anti-cancer compound rohitukine: Possible role of gastrin antagonism and H+ K+-ATPase inhibition. *Naunyn-Schmiedeberg's Arch. Pharmacol.* **2012**, *385*, 277–286. [CrossRef]
19. Ismail, I.S.; Nagakura, Y.; Hirasawa, Y.; Hosoya, T.; Lazim, M.I.M.; Lajis, N.H.; Shiro, M.; Morita, H. Chrotacumines A−D, Chromone Alkaloids from *Dysoxylum acutangulum*. *J. Nat. Prod.* **2009**, *72*, 1879–1883. [CrossRef]
20. Singh, R.K.; Ranjan, A.; Srivastava, A.K.; Singh, M.; Shukla, A.K.; Atri, N.; Mishra, A.; Singh, A.K.; Singh, S.K. Cytotoxic and apoptotic inducing activity of *Amoora rohituka* leaf extracts in human breast cancer cells. *J. Ayurveda Integr. Med.* **2020**, *11*, 383–390. [CrossRef]
21. Apu, A.S.; Pathan, A.H.; Jamaluddin, A.T.M.; Ara, F.; Bhuyan, S.H.; Islam, M.R. Phytochemical analysis and bioactivities of *Aphanamixis polystachya* (Wall.) R. Parker leaves from Bangladesh. *J. Biol. Sci.* **2013**, *13*, 393–399. [CrossRef]
22. Petrini, O.; Sieber, T.N.; Toti, L.; Viret, O. Ecology, metabolite production, and substrate utilization in endophytic fungi. *Nat. Toxins* **1993**, *1*, 185–196. [CrossRef]
23. Nisa, H.; Kamili, A.N.; Nawchoo, I.A.; Bhat, M.S.; Nazir, R. Isolation and identification of endophytic fungi from Artemisia scoparia (Asteraceae). *Int. J. Theor. Appl. Sci.* **2018**, *10*, 83–88.
24. Altschul, S.F.; Gish, W.; Miller, W.; Myers, E.W.; Lipman, D.J. Basic local alignment search tool. *J. Mol. Biol.* **1990**, *215*, 403–410. [CrossRef]
25. Cai, Y.; Luo, Q.; Sun, M.; Corke, H. Antioxidant activity and phenolic compounds of 112 traditional Chinese medicinal plants associated with anticancer. *Life Sci.* **2004**, *74*, 2157–2184. [CrossRef]
26. Kumar, S.; Pandey, A.K. Chemistry and biological activities of flavonoids: An overview. *Sci. World J.* **2013**, *2013*, 162750. [CrossRef]
27. Singh, N.; Rajini, P. Free radical scavenging activity of an aqueous extract of potato peel. *Food Chem.* **2004**, *85*, 611–616. [CrossRef]
28. Nishikimi, M.; Rao, N.A.; Yagi, K. The occurrence of superoxide anion in the reaction of reduced phenazine methosulfate and molecular oxygen. *Biochem. Biophys. Res. Commun.* **1972**, *46*, 849–854. [CrossRef]

29. Gupta, P.; Jain, V.; Pareek, A.; Kumari, P.; Singh, R.; Agarwal, P.; Sharma, V. Evaluation of effect of alcoholic extract of heartwood of Pterocarpus marsupium on in vitro antioxidant, anti-glycation, sorbitol accumulation and inhibition of aldose reductase activity. *J. Tradit. Complementary Med.* **2017**, *7*, 307–314. [CrossRef]

30. Okayo, R.O.; Andika, D.O.; Dida, M.M.; K'Otuto, G.O.; Gichimu, B.M. Morphological and Molecular Characterization of Toxigenic Aspergillus flavus from Groundnut Kernels in Kenya. *Int. J. Microbiol.* **2020**, *2020*, 8854718. [CrossRef]

31. Zhao, R.-B.; Bao, H.-Y.; Liu, Y.-X. Isolation and characterization of *Penicillium oxalicum* ZHJ6 for biodegradation of methamidophos. *Agric. Sci. China* **2010**, *9*, 695–703. [CrossRef]

32. Gomes, R.; Glienke, C.; Videira, S.; Lombard, L.; Groenewald, J.; Crous, P. Diaporthe: A genus of endophytic, saprobic and plant pathogenic fungi. *Pers. Mol. Phylogeny Evol. Fungi* **2013**, *31*, 1–41. [CrossRef]

33. Oberoi, H.; Sandhu, S. Therapeutic and Nutraceutical Potential of Bioactive Compounds Extracted from Fruit Residues AU—Babbar, Neha. *Crit. Rev. Food Sci. Nutr* **2015**, *55*, 319–337.

34. Nermien, H.; Thabet, A.; Markeb, A.; Sayed, D.; El-Maali, N. In vitro-exploration of fungal endophytes of Egyptian *Cynara scolymus* L. (artichoke) and investigation of some their bioactive potentials. *Glob. NEST J.* **2019**, *21*, 296–308.

35. Gauchan, D.P.; Kandel, P.; Tuladhar, A.; Acharya, A.; Kadel, U.; Baral, A.; Shahi, A.B.; García-Gil, M.R. Evaluation of antimicrobial, antioxidant and cytotoxic properties of bioactive compounds produced from endophytic fungi of Himalayan yew (*Taxus wallichiana*) in Nepal. *FResearch* **2020**, *9*, 379.

36. Ayob, F.W.; Mohamad, J.; Simarani, K. Antioxidants and Phytochemical Analysis of Endophytic Fungi Isolated from a Medicinal Plant Catharanthus roseus. *Borneo J. Sci. Technol.* **2019**, *1*, 62–68.

37. Agatonovic-Kustrin, S.; Kustrin, E.; Gegechkori, V.; Morton, D.W. High-performance thin-layer chromatography hyphenated with microchemical and biochemical derivatizations in bioactivity profiling of marine species. *Mar. Drugs* **2019**, *17*, 148. [CrossRef]

38. Randhir, R.; Shetty, K. Mung beans processed by solid-state bioconversion improves phenolic content and functionality relevant for diabetes and ulcer management. *Innov. Food Sci. Emerg. Technol.* **2007**, *8*, 197–204. [CrossRef]

39. Zheng, W.; Wang, S.Y. Oxygen radical absorbing capacity of phenolics in blueberries, cranberries, chokeberries, and lingonberries. *J. Agric. Food Chem.* **2003**, *51*, 502–509. [CrossRef]

40. De Pinedo, A.T.; Peñalver, P.; Morales, J.C. Synthesis and evaluation of new phenolic-based antioxidants: Structure–activity relationship. *Food Chem.* **2007**, *103*, 55–61. [CrossRef]

41. Sousa, B.; Correia, R. Phenolic content, antioxidant activity and antiamylolytic activity of extracts obtained from bioprocessed pineapple and guava wastes. *Braz. J. Chem. Eng.* **2012**, *29*, 25–30. [CrossRef]

42. Lim, S.M.; Agatonovic-Kustrin, S.; Lim, F.T.; Ramasamy, K. High-performance thin layer chromatography-based phytochemical and bioactivity characterisation of anticancer endophytic fungal extracts derived from marine plants. *J. Pharm. Biomed. Anal.* **2021**, *193*, 113702. [CrossRef]

43. Piluzza, G.; Bullitta, S. Correlations between phenolic content and antioxidant properties in twenty-four plant species of traditional ethnoveterinary use in the Mediterranean area. *Pharm. Biol.* **2011**, *49*, 240–247. [CrossRef]

44. Kaur, R.; Kaur, J.; Kaur, M.; Kalotra, V.; Chadha, P.; Kaur, A.; Kaur, A. An endophytic *Penicillium oxalicum* isolated from Citrus limon possesses antioxidant and genoprotective potential. *J. Appl. Microbiol.* **2020**, *128*, 1400–1413. [CrossRef]

45. Abrol, V.; Kushwaha, M.; Arora, D.; Mallubhotla, S.; Jaglan, S. Mutation, Chemoprofiling, Dereplication, and Isolation of Natural Products from *Penicillium oxalicum*. *ACS Omega* **2021**, *6*, 16266–16272. [CrossRef]

46. Maddu, N. Diseases related to types of free radicals. In *Antioxidants*; IntechOpen: London, UK, 2019.

47. Chen, Y.; Mao, W.; Tao, H.; Zhu, W.; Qi, X.; Chen, Y.; Li, H.; Zhao, C.; Yang, Y.; Hou, Y. Structural characterization and antioxidant properties of an exopolysaccharide produced by the mangrove endophytic fungus *Aspergillus* sp. Y16. *Bioresour. Technol.* **2011**, *102*, 8179–8184. [CrossRef]

48. Tang, Z.; Qin, Y.; Chen, W.; Zhao, Z.; Lin, W.; Xiao, Y.; Chen, H.; Liu, Y.; Bu, T.; Li, Q. Diversity, Chemical Constituents, and Biological Activities of Endophytic Fungi Isolated from *Ligusticum chuanxiong* Hort. *Front. Microbiol.* **2021**, *12*, 771000–771000. [CrossRef]

49. Kandasamy, S.; Kandasamy, K. Antioxidant activity of the mangrove endophytic fungus (*Trichoderma* sp.). *J. Coast. Life Med.* **2014**, *2*, 566–570.

50. Bonekamp, N.A.; Völkl, A.; Fahimi, H.D.; Schrader, M.J.B. Reactive oxygen species and peroxisomes: Struggling for balance. *Biofactors* **2009**, *35*, 346–355. [CrossRef]

51. Yang, X.; Kang, M.-C.; Li, Y.; Kim, E.-A.; Kang, S.-M.; Jeon, Y.-J. Asperflavin, an anti-inflammatory compound produced by a marine-derived fungus, *Eurotium amstelodami*. *Molecules* **2017**, *22*, 1823. [CrossRef]

52. Freeman, B.L.; Eggett, D.L.; Parker, T.L. Synergistic and antagonistic interactions of phenolic compounds found in navel oranges. *J. Food Sci.* **2010**, *75*, C570–C576. [CrossRef]

53. Hidalgo, M.; Sánchez-Moreno, C.; de Pascual-Teresa, S. Flavonoid–flavonoid interaction and its effect on their antioxidant activity. *Food Chem.* **2010**, *121*, 691–696. [CrossRef]

54. Tapfuma, K.I.; Uche-Okereafor, N.; Sebola, T.E.; Hussan, R.; Mekuto, L.; Makatini, M.M.; Green, E.; Mavumengwana, V. Cytotoxic activity of crude extracts from Datura stramonium's fungal endophytes against A549 lung carcinoma and UMG87 glioblastoma cell lines and LC-QTOF-MS/MS based metabolite profiling. *BMC Complementary Altern. Med.* **2019**, *19*, 330. [CrossRef]

55. Kumar, V.S.; Kumaresan, S.; Tamizh, M.M.; Islam, M.I.H.; Thirugnanasambantham, K. Anticancer potential of NF-κB targeting apoptotic molecule "flavipin" isolated from endophytic *Chaetomium globosum*. *Phytomedicine* **2019**, *61*, 152830. [CrossRef]

56. Deshmukh, S.K.; Mishra, P.D.; Kulkarni-Almeida, A.; Verekar, S.; Sahoo, M.R.; Periyasamy, G.; Goswami, H.; Khanna, A.; Balakrishnan, A.; Vishwakarma, R. Anti-inflammatory and anticancer activity of ergoflavin isolated from an endophytic fungus. *Chem. Biodivers.* **2009**, *6*, 784–789. [CrossRef]

57. Parthasarathy, R.; Chandrika, M.; Rao, H.Y.; Kamalraj, S.; Jayabaskaran, C.; Pugazhendhi, A. Molecular profiling of marine endophytic fungi from green algae: Assessment of antibacterial and anticancer activities. *Process Biochem.* **2020**, *96*, 11–20. [CrossRef]

58. Unterseher, M.; Schnittler, M. Species richness analysis and ITS rDNA phylogeny revealed the majority of cultivable foliar endophytes from beech (*Fagus sylvatica*). *Fungal Ecol.* **2010**, *3*, 366–378. [CrossRef]

59. Moricca, S.; Ragazzi, A. Fungal endophytes in Mediterranean oak forests: A lesson from *Discula quercina*. *Phytopathology* **2008**, *98*, 380–386. [CrossRef]

60. Redman, R.S.; Dunigan, D.D.; Rodriguez, R.J. Fungal symbiosis from mutualism to parasitism: Who controls the outcome, host or invader? *New Phytol.* **2001**, *151*, 705–716. [CrossRef]

61. Kogel, K.-H.; Franken, P.; Hückelhoven, R. Endophyte or parasite–what decides? *Curr. Opin. Plant Biol.* **2006**, *9*, 358–363. [CrossRef]

62. Steyn, P. The isolation, structure and absolute configuration of secalonic acid D, the toxic metabolite of *Penicillium oxalicum*. *Tetrahedron* **1970**, *26*, 51–57. [CrossRef]

63. Sun, Y.-L.; Bao, J.; Liu, K.-S.; Zhang, X.-Y.; He, F.; Wang, Y.-F.; Nong, X.-H.; Qi, S.-H. Cytotoxic dihydrothiophene-condensed chromones from the marine-derived fungus *Penicillium oxalicum*. *Planta Med.* **2013**, *79*, 1474–1479. [CrossRef]

64. Yang, L.; Xie, J.; Jiang, D.; Fu, Y.; Li, G.; Lin, F. Antifungal substances produced by *Penicillium oxalicum* strain PY-1—Potential antibiotics against plant pathogenic fungi. *World J. Microbiol. Biotechnol.* **2008**, *24*, 909–915. [CrossRef]

65. Shen, S.; Li, W.; Wang, J. Antimicrobial and antitumor activities of crude secondary metabolites from a marine fungus *Penicillium oxalicum* 0312F1. *Afr. J. Microbiol. Res.* **2014**, *8*, 1480–1485.

66. Nyongesa, B.W.; Okoth, S.; Ayugi, V. Identification key for *Aspergillus* species isolated from maize and soil of Nandi County, Kenya. *Adv. Microbiol.* **2015**, *5*, 205. [CrossRef]

67. Wickerham, L.J. Validation of the species *Pichia guilliermondii*. *J. Bacteriol.* **1966**, *92*, 1269. [CrossRef]

68. Prameeladevi, T.; Prabhakaran, N.; Kamil, D.; Toppo, R.S.; Tyagi, A. *Trichoderma pseudokoningii* identified based on morphology was re-identified as *T. longibrachiatum* through molecular characterization. *Indian Phytopathol.* **2018**, *71*, 579–587. [CrossRef]

69. Noonim, P.; Mahakarnchanakul, W.; Varga, J.; Frisvad, J.C.; Samson, R.A. Two novel species of *Aspergillus* section Nigri from Thai coffee beans. *Int. J. Syst. Evol. Microbiol.* **2008**, *58*, 1727–1734. [CrossRef]

70. Moubasher, A.; Soliman, Z. *Aspergillus assiutensis*, a new species from the air of grapevine plantation, Egypt. *J. Basic Appl. Mycol. Egypt* **2011**, *2*, 83–90.

Journal of
Fungi

Fungal Endophytes: A Potential Source of Antibacterial Compounds

Sunil K. Deshmukh [1,2,*], Laurent Dufossé [3,*], Hemraj Chhipa [4], Sanjai Saxena [2,5], Girish B. Mahajan [6] and Manish Kumar Gupta [7]

[1] TERI-Deakin Nano Biotechnology Centre, The Energy and Resources Institute, Darbari Seth Block, IHC Complex, Lodhi Road, New Delhi 110003, Delhi, India

[2] Agpharm Bioinnovations LLP, Incubatee: Science and Technology Entrepreneurs Park (STEP), Thapar Institute of Engineering and Technology, Patiala 147004, Punjab, India; ssaxena@thapar.edu

[3] Chimie et Biotechnologie des Produits Naturels (CHEMBIOPRO Lab) & ESIROI Agroalimentaire, Université de la Réunion, 15 Avenue René Cassin, 97744 Saint-Denis, France

[4] College of Horticulture and Forestry, Agriculture University Kota, Jhalawar 322360, Rajasthan, India; hrchhipa8@gmail.com

[5] Department of Biotechnology, Thapar Institute of Engineering and Technology, Patiala 147004, Punjab, India

[6] HiMedia Laboratories Pvt. Ltd., Mumbai 400086, Maharashtra, India; girishbm2000@gmail.com

[7] SGT College of Pharmacy, SGT University, Gurugram 122505, Haryana, India; mkgupta5@gmail.com

[*] Correspondence: sunil.deshmukh1958@gmail.com (S.K.D.); laurent.dufosse@univ-reunion.fr (L.D.); Tel.: +91-9820510292 (S.K.D.); +33-668731906 (L.D.)

Abstract: Antibiotic resistance is becoming a burning issue due to the frequent use of antibiotics for curing common bacterial infections, indicating that we are running out of effective antibiotics. This has been more obvious during recent corona pandemics. Similarly, enhancement of antimicrobial resistance (AMR) is strengthening the pathogenicity and virulence of infectious microbes. Endophytes have shown expression of various new many bioactive compounds with significant biological activities. Specifically, in endophytic fungi, bioactive metabolites with unique skeletons have been identified which could be helpful in the prevention of increasing antimicrobial resistance. The major classes of metabolites reported include anthraquinone, sesquiterpenoid, chromone, xanthone, phenols, quinones, quinolone, piperazine, coumarins and cyclic peptides. In the present review, we reported 451 bioactive metabolites isolated from various groups of endophytic fungi from January 2015 to April 2021 along with their antibacterial profiling, chemical structures and mode of action. In addition, we also discussed various methods including epigenetic modifications, co-culture, and OSMAC to induce silent gene clusters for the production of noble bioactive compounds in endophytic fungi.

Keywords: endophytic fungi; antibacterial compound; natural product; drug resistance; medicinal plant; AMR

Citation: Deshmukh, S.K.; Dufossé, L.; Chhipa, H.; Saxena, S.; Mahajan, G.B.; Gupta, M.K. Fungal Endophytes: A Potential Source of Antibacterial Compounds. *J. Fungi* **2022**, *8*, 164. https://doi.org/ 10.3390/jof8020164

Academic Editor: Ulrich Kück

Received: 10 December 2021
Accepted: 5 February 2022
Published: 8 February 2022

Publisher's Note: MDPI stays neutral with regard to jurisdictional claims in published maps and institutional affiliations.

1. Introduction

Over the decades since the discovery of the first antibiotics, resistance to those has been a curse that is being dragged along with every discovery of new antibiotics. This has kept all scientists, professionals, and clinical specialists working on antibiotics on their toes. The quest for new antibiotics scaffolds and repurposing of existing molecules has been persistent for the past nine decades. Getting a new and right scaffold is a herculean task, especially with the least ability to induce mutations in the target bacteria. As examined in some of the earlier reviews [1,2] there are several ways of getting new scaffolds and classes of antimicrobial bioactive compounds. In the domain of natural products, one of the most demonstrated ways is studying less explored species and genera of microbes [3–5]. Investigating unexplored ecological units on the globe synergizes with the concept of investigating the least or not explored species of microbes.

In the current review, we present the latest ways of exploring the credentials of such microbial sources, especially endophytic fungi, as a main stream of novel antimicrobial scaffolds. Bioactive compounds are mainly responsible for the activity profiles displayed by endophytic fungi. These metabolites belong to a wide range of scaffolds such as alkaloids, benzopyranones, chinones, peptides, phenols, quinones, flavonoids, steroids, terpenoids, tetralones, xanthones, and others. Moreover, they, in the pure form, have demonstrated abundant biological activities, including antibacterial, antifungal, anticancer, antiviral, antioxidant, immunosuppressant, anti-inflammatory, and antiparasitic properties [6–15]. Even though there are a few specialized reviews on the bioactive compounds from fungi, actinomycetes and other microbes [16,17], the amount of work done in the area is quite versatile, tenacious and significant. There is a need to comprehend these topics periodically to have its effective output for future research keeping in mind the probability of success of any newly discovered bioactive compound in clinical studies has been 0.01 to 1 % based on therapeutic area and type of scaffold. This demands that the base of such scaffolds in the ladder of clinical development should be wider. This width can be increased by exploring such less-tapped resources, the endophytic fungi.

In our previous review, we have covered antibacterials reported from endophytic fungi up to 2014 [1]. This review describes some bioactive molecules isolated from 2015 onwards to early 2021 from various endophytic fungi from terrestrial plants and designated as antibacterials. The antibacterial activity against various pathogenic organisms is listed in Table 1.

2. Antibacterials from Various Class of Endophytic Fungi

2.1. Ascomycetes

Ascomycetes are the fungi characterized by the formation of ascospores and some of the genera belonging to this class are known to produce chemically diverse metabolites. The important genera include *Diaporthe*, *Xylaria*, *Chaetomium*, *Talaromyces*, and *Paraphaeosphaeria* and are known to produce terpenoids, cytochalasins, mellein, alkaloids, polyketides, and aromatic compounds. Here we report the antibacterial from ascomycetes.

2.1.1. *Diaporthe* (Asexual State: *Phomopsis*)

The genus *Diaporthe* (asexual state: *Phomopsis*) has been thoroughly investigated for secondary metabolites that have various pathogenic, endophytic and saprobic species of temperate and tropical habitats. Two natural bisanthraquinone, (+)-1,1′-bislunatin (bis) (**1**) and (+)-2,2′-epicytoskyrin A (epi) (**2**, Figure 1), were extracted from endophytic fungi, *Diaporthe* sp. GNBP-10 is associated with plant *Uncaria gambir*. Compounds (bis)-(**1**) and (epi)-(**2**) showed promising anti-tubercular activity, against *Mycobacterium tuberculosis* strains H37Rv (Mtb H37Rv) with MIC values of 0.422 and 0.844 μM, respectively. Both compounds have the ability to combat nutrient-starvation and biofilms of the Mtb model with relatively moderate activity in bacterial reduction with between 1–2 fold log reduction. Both compounds could reduce the number of Mtb infected into macrophages with 2-fold log reduction. The in-silico results via a docking study show that both compounds have a good affinity with pantothenate kinase (PanK) enzyme with a Glide score of −8.427 kcal/mol and −7.481 kcal/mol for the epi and bis compounds, respectively [18].

An endophytic fungus, *Diaporthe* sp. GDG-118, associated with *Sophora tonkinensis* collected from Hechi City (China) yielded a new compound 21-acetoxycytochalasin J3 (**3**, Figure 1) and inhibited the pathogens *Bacillus anthraci* and *E. coli* at 12.5 μg/mL concentration (6 mm sterile filter paper discs were impregnated with 20 μL (50 μg) of each compound) [19].

Two novel naphthalene derivatives, 1-(3-hydroxy-1-(hydroxymethyl)-2-methoxy-6-methylnaphthalen-7-yl) propan-2-one (**4**) and 1-(3-hydroxy-1-(hydroxymethyl)-6-methyl-naphthalen-7-yl)propan-2-one (**5**, Figure 1), were obtained from the *Phomopsis fukushii*. Compounds **4** and **5** displayed poor anti-methicillin-resistant *Staphylococcus aureus* (anti-MRSA) activity, with zones of inhibition of 10.2 and 11.3 mm, respectively (6 mm sterile filter paper discs were impregnated with 20 μL (50 μg) of each compound) [20].

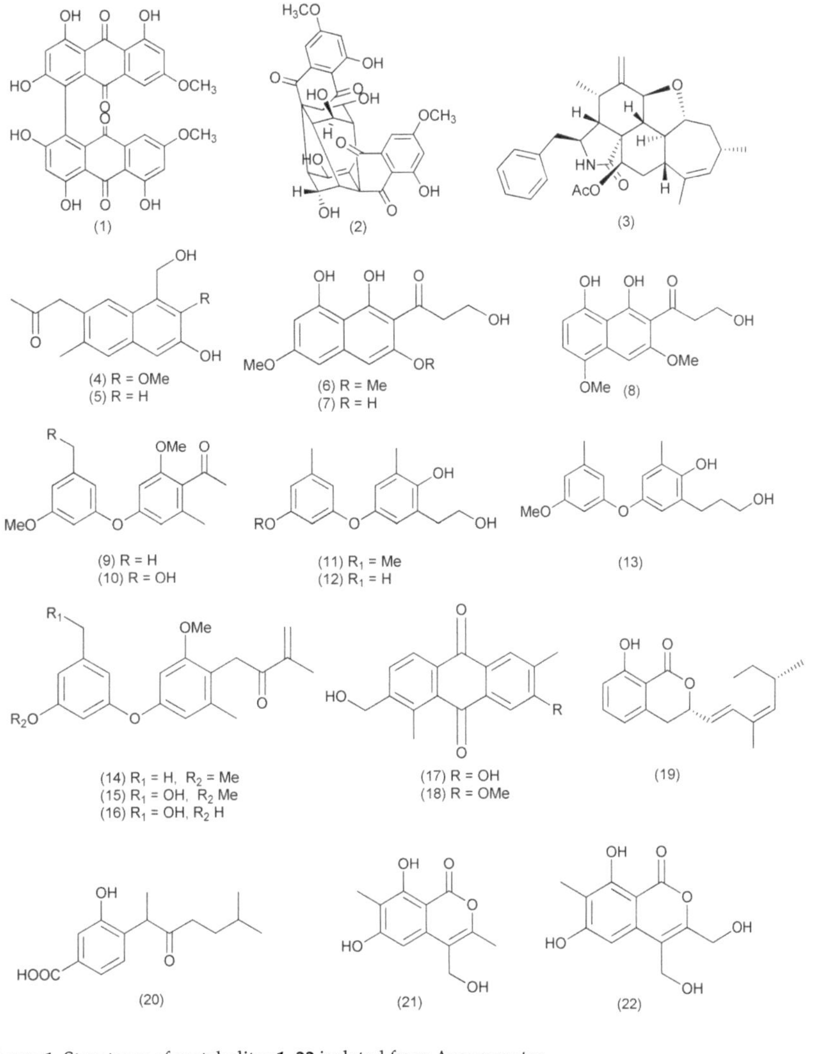

Figure 1. Structures of metabolites **1–22** isolated from Ascomycetes.

Earlier *Phomopsis fukushii (Diaporthe fukushii)* isolated from the rhizome of *Paris polyphylla var. yunnanensis* was the source of three new compounds namely 3-hydroxy-1-(1,8-dihydroxy-3,6-dimethoxynaphthalen-2-yl)propan-1-one (**6**), 3-hydroxy-1-(1,3,8-trihydroxy-6-methoxynaphthalen-2-yl)propan-1-one (**7**) and 3-hydroxy-1-(1,8-dihydroxy3,5-dimethoxy naphthalen-2-yl) propan-1-one (**8**, Figure 1). Compounds **6–8** exhibited anti-MRSA-ZR11 activity, with MIC values of 8, 4, and 4 µg/mL, respectively [21]. Later two new di-Ph ethers, 1-[2-methoxy-4-(3-methoxy-5-methylphenoxy)-6-methylphenyl]-ethanone (**9**) and 1-[4-(3-(hydroxymethyl)-5-methoxyphenoxy)-2-methoxy-6-methylphenyl]-ethanone (**10**, Figure 1), were also purified from the same fungus. Compounds **9–10** exhibited anti-MRSA activity with good inhibition (zones of 13.8 and 14.6 mm, respectively) [22].

Three new di-Ph ethers, 4-(3-methoxy-5-methylphenoxy)-2-(2-hydroxyethyl)-6-methyl phenol (**11**), 4-(3-hydroxy-5-methylphenoxy)-2-(2-hydroxyethyl)-6-methylphenol (**12**) and 4-(3-methoxy-5-methylphenoxy)-2-(3-hydroxypropyl)-6-methylphenol (**13**, Figure 1) were purified from *Phomopsis fukushii* associated with the rhizome of *Paris polyphylla* var. *yunna-*

nensis. Compounds **11–13**, exhibited potent anti-MRSA activity, with 20.2, 17.9 and 15.2 mm inhibition zones, respectively, when tested at 50 μg concentration in 6 mm discs [23].

Phomopsis fukushii isolated from the rhizome of *Paris polyphylla* var. *yunnanensis* yielded three new isopentylated diphenyl ethers, 1-(4-(3-methoxy-5-methylphenoxy)-2-methoxy-6-methylphenyl)-3-methylbut-3-en-2-one (**14**), 1-(4-(3-(hydroxymethyl)-5-methoxyphenoxy)-2-methoxy-6-methylphenyl)-3-methylbut-3-en-2-one (**15**) and 1-(4-(3-hydroxy-5-(hydroxym ethyl) phenoxy)-2-methoxy-6-methylphenyl)-3-methylbut-3-en-2-one (**16**, Figure 1). Compounds **14–16** displayed anti-MRSA activity with 21.8, 16.8 and 15.6 mm inhibition zones, respectively (50 μg/6 mm disc) [24].

Two new anthraquinones, 3-hydroxy-6-hydroxymethyl-2,5-dimethylanthraquinone (**17**) and 6-hydroxymethyl-3-methoxy-2,5-dimethylanthraquinone (**18**, Figure 1), were purified from the endophytic fungus *Phomopsis* sp. and displayed good anti-MRSA activity with inhibition zone diameters (IZDs) of 14.2 and 14.8 mm, respectively [25].

A new dihydroisocoumarin derivative diaporone A (**19**, Figure 1), was purified from *Diaporthe sp.* an endophyte of *Pteroceltis tatarinowii*. Compound **19** showed MIC at 66.7 μM against *Bacillus subtilis* [26].

A pair of new phenolic bisabolane-type sesquiterpenoid enantiomers (±)-phomoterp enes A and B [(±)-1] (**20**) along with two new isocoumarins, phomoisocoumarins C-D (**21–22**, Figure 1) were purified from an endophytic fungus *Phomopsis prunorum* (F4-3). Compounds (+)-1 (**20 and 22**) exhibited average antimicrobial activity against *Pseudomonas syringae pv. lachrymans* with MIC values of 15.6 μg/mL, and compounds (−)-1 (**20 and 21**) displayed poor activity with MICs of 31.2 μg/mL each. Compounds (−)-1, (+)-1, (**20, 21, 22**) showed antibacterial activity against *Xanthomonas citri pv. phaseoli* var. *fuscans* with MIC values of 31.2, 62.4, 31.2, and 31.2 μg/mL, respectively [27].

The fungus *Diporthe vochysiae* LGMF1583 isolated from *Vochysia divergens* yielded two new carboxamides, vochysiamides A (**23**), and B (**24**, Figure 2). Compound **24** inhibited *Klebsiella pneumoniae* carbapenemase-producing (KPC), MSSA, and MRSA with MIC of 0.08, 1.0, and 1.0 μg/mL, respectively, and compound **23** was active against KPC with a MIC of 1.0 μg/mL. KPC is of public health concern due to the presence of antimicrobial resistance carbapenemases [28].

An endophyte *Phomopsis asparagi* obtained from the rhizome of *Paris polyphylla* var. *yunnanensis* was the source of two new di-Ph ethers, 4-(3-methoxy-5-methylphenoxy)-2-(2-hydroxyethyl)- 6-(hydroxymethyl)phenol (**25**), and 4-(3-hydroxy-5-methylphenoxy)-2-(2-hydroxyethyl)-6-(hydroxymethyl)phenol (**26**, Figure 2). Compounds **25** and **26** exhibited potent anti-MRSA activity with 10.8 and 11.4 mm inhibition zones, respectively [29].

Two new naphthalene derivatives, 5-methoxy-2-methyl-7-(3-methyl-2-oxobut-3-enyl)-1-naphthaldehyde (**27**) and 2-(hydroxymethyl)-5-methoxy-7-(3-methyl-2-oxobut-3-enyl)-1-naphthaldehyde (**28**, Figure 2), were characterized from *Phomopsis* sp., an endophyte of *Paris polyphylla* var. *yunnanensis*. Compounds **27** and **28** displayed potent antibacterial activity with 14.5 and 15.2 mm zones of inhibition, respectively, against MRSA [30].

The endophytic fungus *Diaporthe terebinthifolii* LGMF907 associated with the plant *Schinus terebinthifolius* yielded diaporthin (**29**) and orthosporin (**30**, Figure 2). Compound **29** displayed antimicrobial activity against various pathogens like *E. coli*, *Micrococcus luteus*, MRSA, and *S. aureus* with 1.73, 2.47, 9.50, and 9.0 mm zones of inhibition, respectively at 100 μg/disk concentration. Compound **30** inhibited *E. coli*, *M. luteus*, MRSA, and *S. aureus* with 1.03, 1.53, 9.0 and 9.33 mm zones of inhibition, respectively, when tested at 100 μg/disk [31].

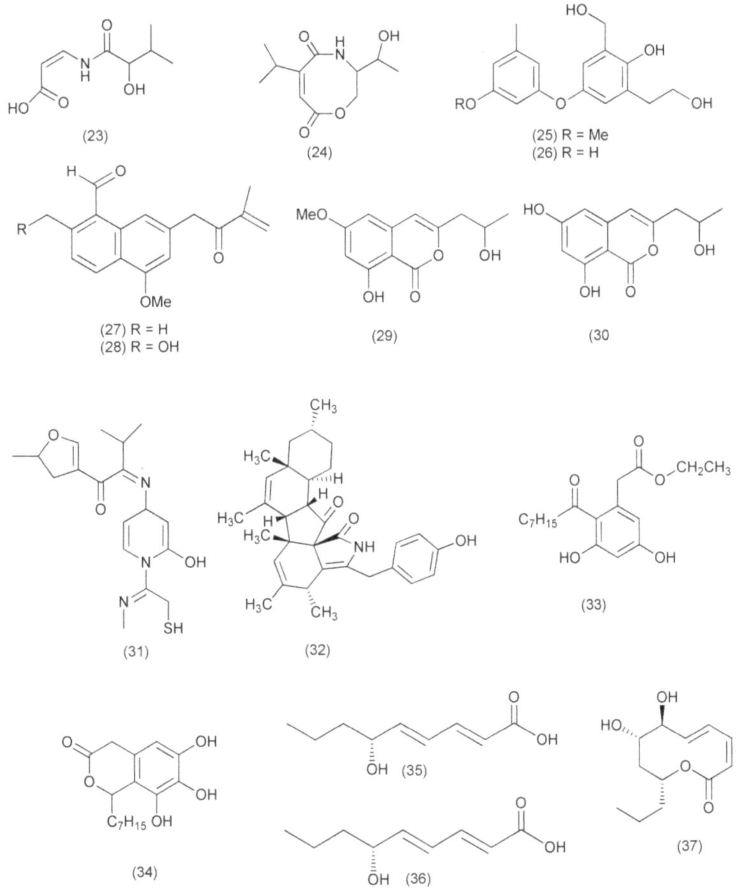

Figure 2 shows chemical structures labeled (23) through (37).

Figure 2. Structures of metabolites **23–37** isolated from Ascomycetes.

A pyrimidine iminomethylfuran derivative, (2Z)-2-(1,4-dihydro-2-hydroxy-1-((*E*)-2-mercapto-1-(methylimino)ethyl)pyrimidine-4-ylimino)-1-(4,5-dihydro-5-methylfuran-3-yl)-3-methylbutane-1-one (**31**, Figure 2) was extracted from *Phomopsis/Diaporthe* sp. GJJM 16 is associated with *Vitex negundo* and inhibited *S. aureus*, and *P. aeroginosa* with MICs of 1.25 µg/mL each [32].

Phomopsis sp. PSU-H188 associated with *Hevea brasiliensis*, yielded the known compounds diaporthalasin (**32**), cytosporones B (**33**) and cytosporones D (**34**, Figure 2). Compound **32**, displayed antibacterial activity against *S. aureus* and MRSA with equal MIC values of 4 µg/mL, but compound **33** inhibited *S. aureus* and MRSA with MIC values of 32 and 16 µg/mL, respectively. Compound **34** also inhibited *S. aureus* and MRSA with MIC values at higher concentrations of 64 and 32 µg/mL, respectively [33].

An endophyte, *Diaporthe terebinthifolii* GG3F6, associated with *Glycyrrhiza glabra* yielded two new hydroxylated unsaturated fatty acids namely diapolic acid A–B (**35–36**) and the known molecules xylarolide (**37**, Figure 2) and phomolide G (**38**, Figure 3). Compounds **35–38** inhibited *Yersinia enterocolitica* with an IC$_{50}$ values of 78.4, 73.4, 72.1 and 69.2 µM, respectively [34].

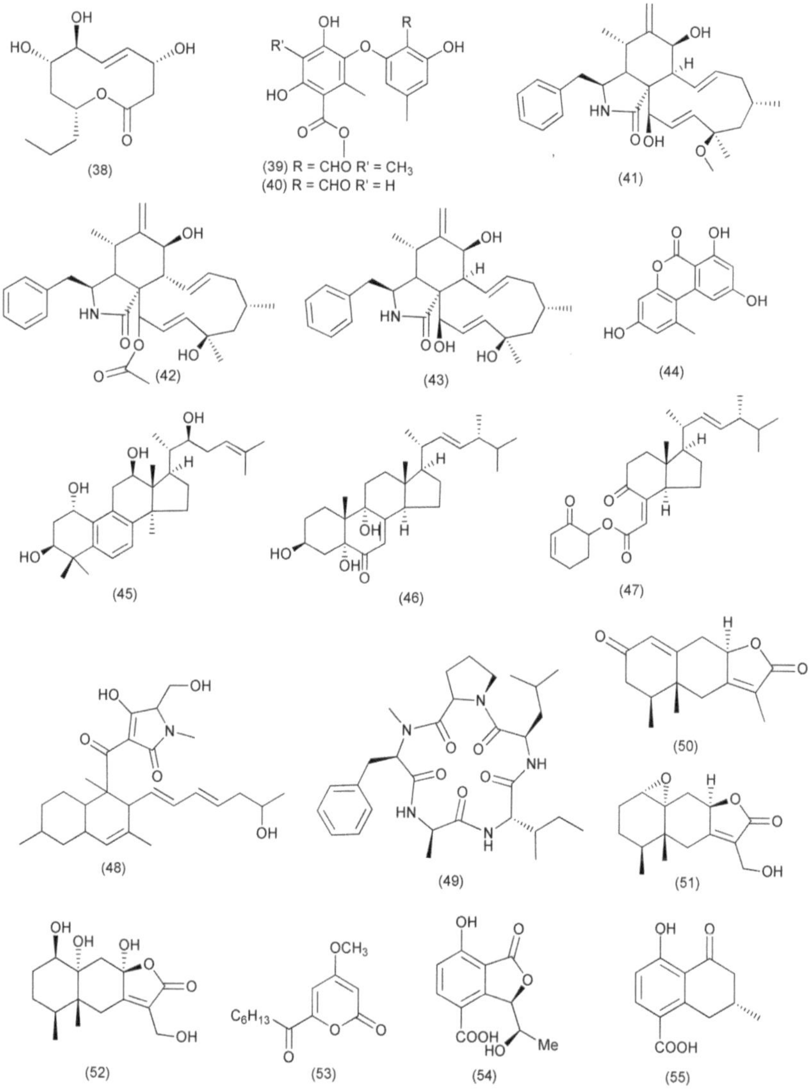

Figure 3. Structures of metabolites **38–55** isolated from Ascomycetes.

The compounds phomosine A (**39**), and phomosine C (**40**, Figure 3), were obtained from *Diaporthe* sp. F2934 from *Siparuna gesnerioides*. Compound **39** was found to be active against *Bordetella bronchiseptica*, *Enterococcus faecalis*, *Enterococcus cloacae*, *S. aureus*, and *Streptococcus oralis* with 10, 10, 10, 12 and 9 mm inhibition zones at 4 µg/mL concentration, respectively. Compound **40** inhibited *S. aureus*, *M. luteus*, *S. oralis*, *E. faecalis*, *E. cloacae*, and *B. bronchiseptica*, with 9, 6, 8, 8, 8 and 9 mm inhibition zones at 4 µg/mL concentration, respectively [35].

Known cytochalasins 18-methoxycytochalasin J (**41**), cytochalasins H (**42**), J (**43**) and alternariol (**44**, Figure 3) were extracted from *Phomopsis* sp., residing inside *Garcinia kola* nuts. Compounds **41–44** were found to be active against *Shigella flexneri* (MIC, 128 µg/mL each). Compounds **41** and **42** showed activity against *S. aureus* with MIC values of 128 and 256 µg/mL, respectively [36].

The fungal culture *Diaporthe* sp. LG23, an endophyte of *Mahonia fortune*, yielded some new lanostanoids, 19-nor-lanosta-5(10),6,8,24-tetraene-1α,3β,12β,22S-tetraol (**45**), 3β,5α,9α-trihydroxy-(22E,24R)-ergosta-7,22-dien-6-one (**46**), and chaxine C (**47**, Figure 3). Compound **45** was found to be active against *S. aureus*, *E. coli*, *B. subtilis*, *Pseudomonas aeruginosa*, and *Streptococcus pyogenes*, with MIC values of 5.0, 5.0, 2.0, 2.0 and 0.1 µg/mL, respectively. Compounds **46** and **47** were active against *B. subtilis* with MIC values of 5.0 µg/mL each [37].

The known compound, pyrrolocin A (**48**, Figure 3), was purified from *Diaporthales* sp. E6927E isolated from *Ficus sphenophyllum*. Pyrrolocin A (**48**) displayed inhibition against *S. aureus* and *E. faecalis* with MICs of 4 and 5 µg/mL, respectively [38].

2.1.2. Xylaria

The genus *Xylaria* comprises various endophytic species associated with both vascular and nonvascular plants. For example, ellisiiamide A (**49**, Figure 3) was isolated from *Xylaria ellisii* from *Vaccinium angustifolium* and was chemically characterized using 1D and 2D NMR, HRMS/MS data. It showed modest inhibitory activity against *E. coli* (MIC, 100 µg/mL) [39].

Xylareremophil (**50**), a new eremophilane sesquiterpene, along with the already reported eremophilanes mairetolides B (**51**) and G (**52**, Figure 3) were extracted from *Xylaria* sp. GDG-102 residing inside *S. tonkinensis*. Compound **50** displayed moderate activity against *Proteus vulgaris* and *Micrococcus luteus* (MIC, of 25 µg/mL each). Compound **51** was found to be active against *M. luteus*, with a MIC value of 50 µg/mL. Compound **52** inhibited *P. vulgaris* with a MIC value of 25 µg/mL and *M. luteus* with a MIC value of 50 µg/mL. Compounds **50–52** also displayed inhibition of *B. subtilis* and *Micrococcus lysodeikticus* with MIC values of 100 µg/mL, respectively [40].

A new compound, 6-heptanoyl-4-methoxy-2H-pyran-2-one (**53**, Figure 3), was purified from *Xylaria* sp. (GDG-102) an endophyte of *S. tonkinensis* and displayed antibacterial activity against *E. coli* as well as *S. aureus* (MIC, 50 µg/mL) [41].

The phthalide derivative xylarphthalide A (**54**) and known compounds (−)-5-carboxyl mellein (**55**, Figure 3) and (−)-5-methylmellein (**56**, Figure 4) were extracted from *Xylaria* sp. (GDG-102) associated with *S. tonkinensis*. Compound **54** inhibited *Bacillus anthracis*, *B. megaterium*, *B. subtilis*, *S. aureus*, *E. coli*, *Shigella dysenteriae* and *Salmonella paratyphi*, with the MICs of 50, 25, 12.5, 25, 12.5, 25 and 25 µg/mL, respectively. Compound **55** showed antibacterial activity with MIC of values of 25, 25, 12.5, 25, 25, 25 and 25 µg/mL against *B. anthracis*, *B. megaterium*, *B. subtilis*, *S. aureus*, *E. coli*, *S. dysenteriae* and *S. paratyphi*, respectively. Compound **56** displayed antibacterial activity with MIC values of 25, 12.5, 12.5, 25, 25, and 50 µg/mL against *B. megaterium*, *B. subtilis*, *S. aureus*, *E. coli*, *S. dysenteriae* and *S. paratyphi*, respectively [42].

A novel compound 3,7-dimethyl-9-(-2,2,5,5-tetramethyl-1,3-dioxolan-4-yl)nona-1,6-dien-3-ol (**57**), and previously reported compound nalgiovensin (**58**, Figure 4) were purified from *Xylaria* sp., associated with *Taxus mairei*. Compound **57** exhibited strong inhibition against *B. subtilis* (48.1%), *B. pumilus* (31.6%) and *S. aureus* (47.1%). Compound **58** exhibited broad inhibition against *S. aureus* (42.1%), *B. subtilis* (36.8%), *B. pumilus* (47.1%) and *E. coli* (41.2%) [43].

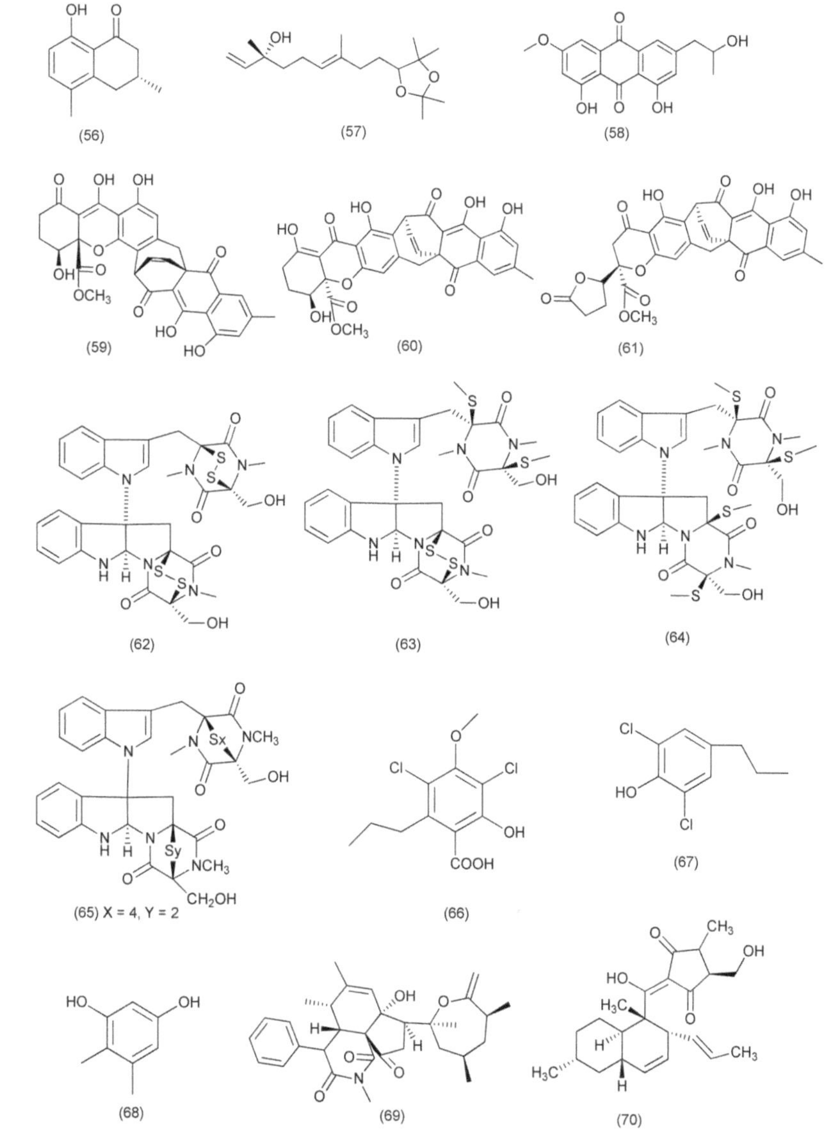

Figure 4. Structures of metabolites **56–70** isolated from Ascomycetes.

2.1.3. *Chaetomium*

The genus *Chaetomium* has been included among the genera producing various bioactive compounds and more than 200 secondary metabolites belonging to diverse structural types such as anthraquinones, azaphilones, chaetoglobosins, chromones, depsidones, epipolythiodioxopiperazines, terpenoids, and steroids and xanthones have beenrecorded, making it a rich source of novel bioactive metabolites. Most of these fungal metabolites exhibited antitumor, cytotoxic, antimalarial, enzyme inhibitory, antibiotic, and other activities [44]. Here we report the antibacterial compounds isolated from the genus *Chaetomium*.

A new xanthoquinodin B9 (**59**), along with previously reported two xanthoquinodins, xanthoquinodin A1 (**60**) and xanthoquinodin A3 (**61**), and three epipolythio- dioxopiperazines, chetomin (**62**), chaetocochin C (**63**) and dethiotetra(methylthio)chetomin (**64**, Figure 4), were obtained from *C. globosum* 7s-1, associated with *Rhapis cochinchinensis*. Xanthoquinodins **59–61** displayed potent antibacterial activity, with MIC values of 0.87, 0.44 and 0.22 μM against *B. cereus*, respectively. Compounds **59–61** were also found active against *S. aureus* and MRSA (MICs in the range of 0.87 to 1.75 μM). Epipolythiodioxopiperazines **62–64** exhibited potent activity against *B. cereus*, *S. aureus*, and MRSA (MICs in the range of 0.02 pM to 10.81 mM). Compound **62** showed the highest activity towards *B. cereus*, *S. aureus* and MRSA (MICs of 0.35 μM, 10.74 and 0.02 pM). Compounds **59–64** showed poor activity against *E. coli*, *P. aeruginosa*, and *Salmonella typhimurium* (MICs of 45.06 to >223.72 μM). Epipolythiodioxopiperazines **62–64** showed activity against *Mycobacterium tuberculosis* with MICs of 0.55, 4.06 and 8.11 μM, respectively [45].

Known compounds chaetocochin C (**63**), chetomin A (**65**) and chetomin (**62**, Figure 4) were extracted from *Chaetomium* sp. SYP-F7950 residing inside *Panax notoginseng*. Compounds **62**, **63** and **65** displayed potent activity against *B. subtilis*, *S. aureus*, and *Enterococcus faecium*, with MIC values ranging from 0.12 to 19.3 μg/mL. The length of *B. subtilis* was increased up to 1.8-fold after treatment with compounds **62**, **63** and **65**. These compounds also showed good interactions with the filamentous temperature-sensitive protein Z (FtsZ) of *B. subtilis* in an in silico molecular docking study. These results revealed that inhibition of pathogenic *B. subtilis* could be achieved by combination with FtsZ and inhibition of cell division [46].

Compounds differanisole A (**66**), 2,6-dichloro-4-propylphenol (**67**) and 4,5-dimethylresorcinol (**68**, Figure 4), were purified from *Chaetomium* sp. HQ-1, isolated from *Astragalus chinensis*. Compounds **66–68** displayed average activity against *Listeria monocytogenes*, *S. aureus*, and MRSA (MICs ranging from 16 to 128 μg/mL). Compound **66** showed a MIC of 16 μg/mL for *L. monocytogenes* and a MIC of 128 μg/mL for *S. aureus* and MRSA. Compounds **67** and **68** could suppress the growth of *L. monocytogenes* with MICs of 64 and 32 μg/mL, respectively [47].

A novel cytochalasan, chamiside A (**69**, Figure 4), was obtained from *Chaetomium nigricolor* F5, an endophytic fungus associated with *Mahonia fortune* collected from Qingdao (China) and showed inhibition of *S. aureus* with a MIC of 25 μg/mL [48].

A known compound, equisetin (**70**, Figure 4), was purified from *C. globosum* of *Salvia miltiorrhiza*. Compound **70** displayed activity against multidrug-resistant *E. faecalis*, *E. faecium*, *S. aureus*, and *S. epidermidis* with MIC values of 3.13, 6.25, 3.13, and 6.25 μg/mL, respectively [49].

Chaetomium sp. Eef-10, from *Eucalyptus exserta* yielded a new depsidone mollicellin O (**71**), along with the known compounds mollicellin H (**72**) and mollicellin I (**73**, Figure 5). Mollicellin H (**72**) displayed potent activity against *S. aureus* and *S. aureus* N50, with IC_{50} values of 5.14 and 6.21 μg/mL, respectively. Mollicellin O (**71**) exhibited antibacterial activities against *S. aureus* and *S. aureus* N50, with IC_{50} values of 79.44 and 76.35 μg/mL, respectively, while mollicellin I (**73**) exhibited activity against *S. aureus* and *S. aureus* N50 with IC_{50} values of 70.14 and 63.15 μg/mL, respectively [50].

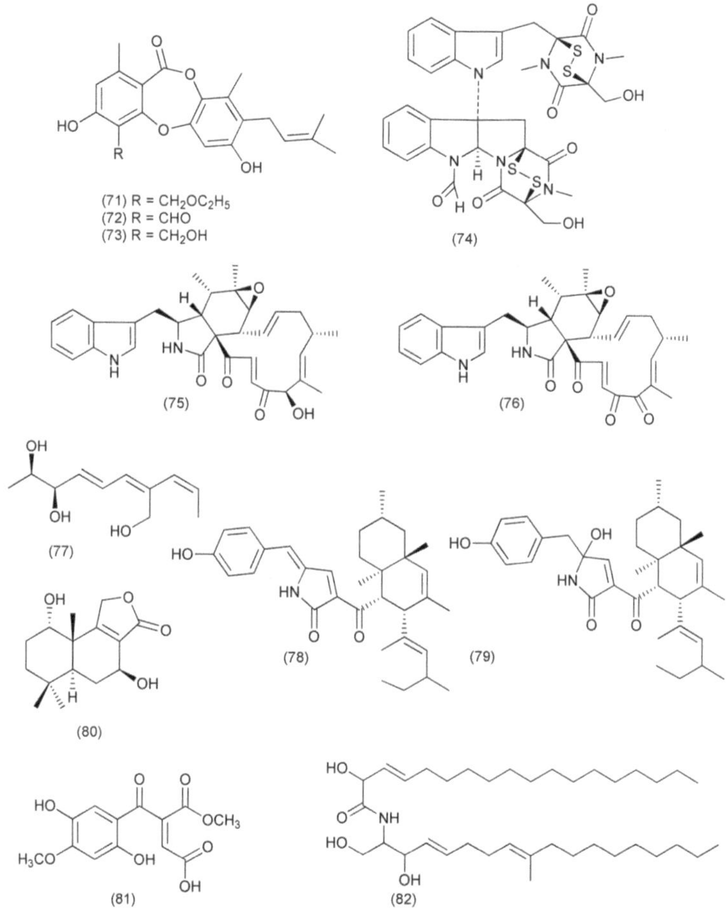

Figure 5. Structures of metabolites **71–82** isolated from Ascomycetes.

A new compound, 6-formamidochetomin (**74**, Figure 5) was isolated from *Chaetomium* sp. M336 an endophyte of *Huperzia serrata*. Compound **74** inhibited *E. coli*, *S. aureus*, *S. typhimurium* and *E. faecalis* with MIC values of 0.78 μg/mL [51].

Two known cytochalasans, chaetoglobosin A (**75**) and C (**76**, Figure 5), were purified from *Chaetomium globosum*, an endophyte of *Nymphaea nouchali*. Compound **75** inhibited *B. subtilis*, *S. aureus*, and MRSA with MIC values of 16, 32 and 32 μg/mL, respectively, and the MIC values for compound **76** were >64 μg/mL for all the microorganisms tested [52].

2.1.4. *Talaromyces*

An endophytic fungus *Talaromyces pinophilus* XL-1193 residing inside the plant *Salvia miltiorrhiza* yielded a new polyene, pinophol A (**77**, Figure 5). Pinophol A (**77**) exhibited low activity against *Bacterium paratyphosum* B with a MIC value of 50 μg/mL [53].

The compounds talaroconvolutin A (**78**) and talaroconvolutin B (**79**, Figure 5), were discovered in *Talaromyces purpureogenus* XL-25, an endophyte associated with *Panax notoginseng*. Compound **78** showed pronounced activity against *B. subtilis* (MIC, 1.56 μM). Compound **79** had a certain inhibitory activity against *Micrococcus lysodeikticus* (MIC = 0.73 μM) and *Vibrio parahaemolyticus* (MIC = 0.18 μM) [54].

A drimane sesquiterpenoid (1*S*,5*S*,7*S*,10*S*)-dihydroxyconfertifolin (**80**, Figure 5) was purified from *Talaromyces purpureogenus* residing inside the plant *Panax notoginseng*. Compound **80** inhibited *E. coli* with a MIC value of 25 μM/L [55].

A novel polyketide, talafun (**81**), and a new compound, N-(2′-hydroxy-3′-octadecenoyl)-9-methyl-4,8-sphingadienin (**82**, Figure 5), were purified from *Talaromyces funiculosus* - Salicorn 58 together with some previously reported compounds, chrodrimanin A (**83**), and chrodrimanin B (**84**, Figure 6). Compound **81** exhibited potent activity against *E. coli* (MIC, 18 μM) but poor activity toward *S. aureus* (MIC, 93 μM). Compound **82** was found to be active against *Mycobacterium smegmatis*, *S. aureus*, *Micrococcus tetragenus*, and *E. coli*, with MIC values of 85, 90, 24, and 68, 93 μM, respectively. Compound **83** inhibited *S. aureus*, *M. tetragenus*, *Mycobacterium phlei*, and *E. coli* (MICs of 67, 28, 47, and 26 μM). However, compound **84** showed only moderate activity against *E. coli* with a MIC of 43 μM [56].

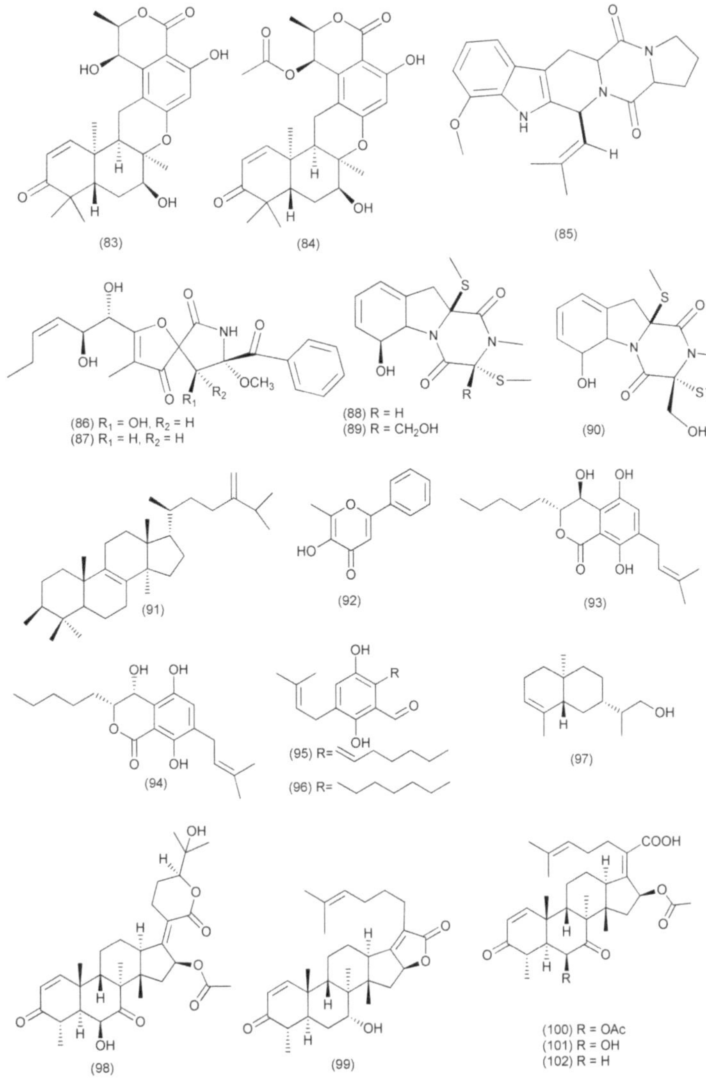

Figure 6. Structures of metabolites **83–102** isolated from Ascomycetes.

Alkaloids **85–90** (Figure 6), were extracted from *Talaromyces* sp. LGT-2, from *Tripterygium wilfordii*. Compounds **85–90** inhibited *E. coli*, *P. aeruginosa*, *S. aureus*, *Bacillus licheniformis*, and *Streptococcus pneumoniae*, with MIC values in the range of 0.125 to 1.0 50 μg/mL [57].

2.1.5. Minor Taxa of the *Ascomycetes*

The known compound euphorbol (**91**, Figure 6) was isolated from *Rhytidhysteron* sp. BZM-9, an endophyte isolated from the leaves of *Leptospermum brachyandrum*. Compound **91** displayed weak antibacterial activity against MRSA, with a MIC value of 62.5 μg/mL (positive control vancomycin MIC 1.25 μg/mL) [58].

A new natural product, stagonosporopsin C (**92**, Figure 6) was purified from an endophytic fungus, *Stagonosporopsis oculihominis*, isolated from *Dendrobium huoshanense*. Stagonosporopsin C (**92**) exhibited moderate inhibitory activity against *S. aureus* sub sp. *aureus* ATCC29213 with a MIC_{50} value of 41.3 μM (positive control penicillin G, MIC_{50} value 1.963 μM) [59].

Two new compounds eutyscoparols H-I (**93**, **94**) together with the related known ones tetrahydroauroglaucin (**95**) and flavoglaucin (**96**, Figure 6), were isolated from the endophytic fungus *Eutypella scoparia* SCBG-8. Compounds **93–96** displayed growth inhibition against *S. aureus* and MRSA, with MIC values ranging from 1.25 to 6.25 μg/mL [60].

A new sesquiterpene eutyscoparin G (**97**, Figure 6) was purified from an endophytic fungus *Eutypella scoparia* SCBG-8 isolated from leaves of *Leptospermum brachyandrum* from the South China Botanical Garden (SCBG, Chinese Academy of Sciences, Guangzhou, China). Compound **97** exhibited antibacterial activity against *S. aureus* and MRSA with MIC values of 6.3 μg/mL [61].

Two new helvolic acid derivatives named sarocladilactone A (**98**), sarocladilactone B (**99**), along with the previously reported compounds helvolic acid (**100**), helvolinic acid (**101**), 6-desacetoxyhelvolic acid (**102**, Figure 6), and 1,2-dihydrohelvolic acid (**103**, Figure 7), were isolated from *Sarocladium oryzae* DX-THL3, associated with leaves of *Oryza rufipogon* Griff. Compounds **98–103** showed antibacterial activity against *S. aureus* with MIC values of 64, 4, 8, 1, 4 and 16 μg/mL, respectively (positive control tobramycin MIC 1 μg/mL), while compound **101** also showed antibacterial activity against *B. subtilis* with a MIC value of 64 μg/mL (positive control tobramycin, MIC 64 μg/mL). Compounds **98**, **101**, **103**, showed some potent antibacterial activity against *E. coli* with MIC 64 μg/mL [62].

The diketopiperazine cyclo(L-Pro-L-Phe) (**104**, Figure 7), was purified from *Paraphaeosphaeria sporulosa*, associated with *Fragaria x ananassa*. Compound **104** displayed activity against *Salmonella* strains, S1 and S2, with IC_{50} values of 7.2 and 7.9 μg/mL and MICs of 71.3 and 78.6 μg/mL, respectively [63].

A fungal culture of *Aplosporella javeedii* isolated from *Orychophragmus violaceus* was the source of terpestacin (**105**) fusaproliferin (**106**), 6,7,9,10-tetrahydromutolide (**107**) and mutolide (**108**, Figure 7). Compounds **105**, **106**, **108** showed poor activities against *M. tuberculosis* H37Rv and compound **107** against *S. aureus*, respectively, with MICs of 100 μM [64].

A new chlamydosporol derivative pleospyrone E (**109**, Figure 7), was extracted from *Pleosporales* sp. Sigrf05, residing inside the tuberous roots of *Siraitia grosvenorii*. Compound **109** exhibited weak inhibition against *Agrobacterium tumefaciens*, *B. subtilis*, *R. solanacearum*, and *X. vesicatoria* with the same MIC value of 100.0 μM [65].

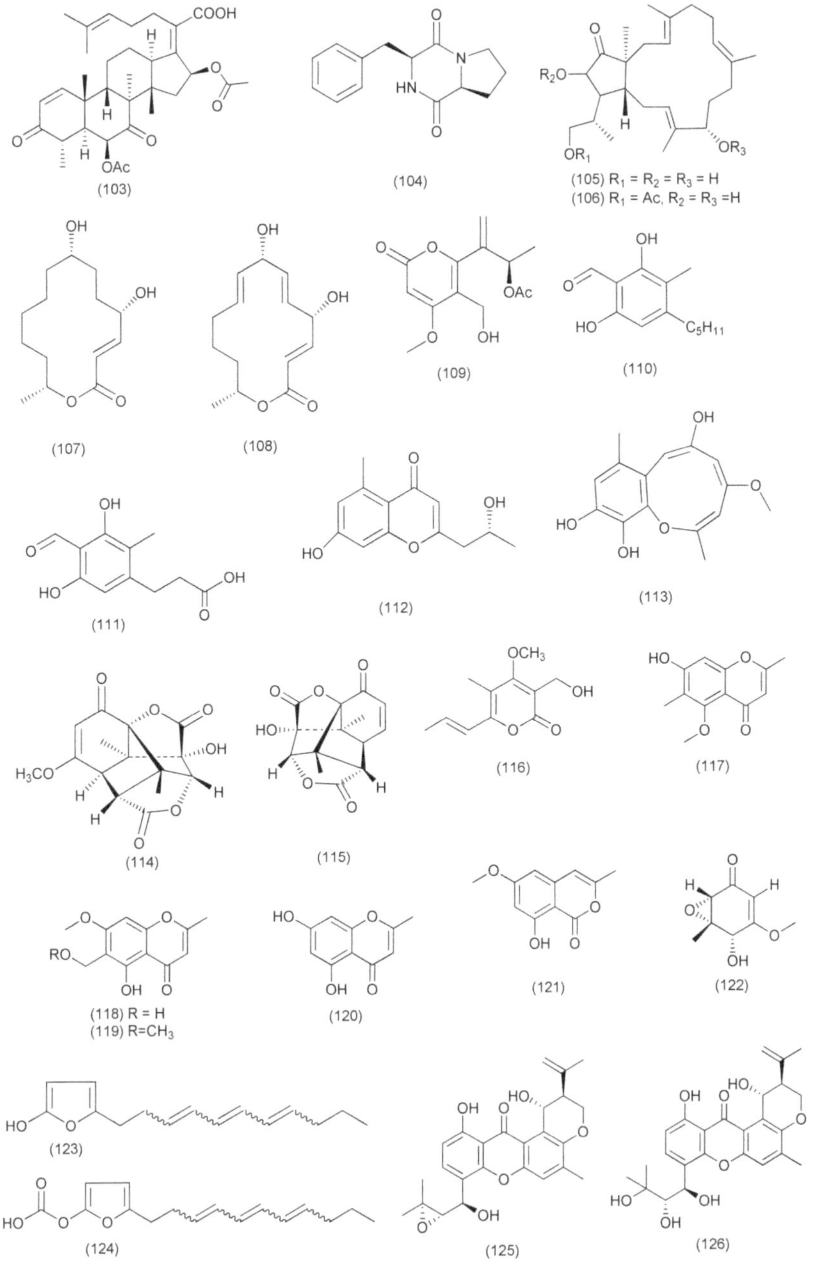

Figure 7. Structures of metabolites **103–126** isolated from Ascomycetes.

New polyketides aplojaveediins A and F (**110, 111**, Figure 7) were purified from the *Aplosporella javeedii* associated with the *Orychophragmus violaceus*. Compound **110** exhibited average activity against the sensitive *Staphylococcus aureus* strain ATCC 29213, the methicillin-resistant and vancomycin-intermediate sensitive (MRSA/VISA) *S. aureus* strain ATCC 700699 and *B. subtilis* (ATCC 169) with MICs of 50, 50 and 25 µM, respectively. Compound **111** also exhibited moderate inhibition against *S. aureus* ATCC 29213 and ATCC 700699 with MICs of 25 and 50 µM, respectively [66].

A new chromone, lawsozaheer (**112**, Figure 7), was isolated from *Paecilomyces variotii* from *Lawsonia alba*. Compound **112** showed activity against *S. aureus* (NCTC 6571) with 84.26% inhibition at 150 µg/mL [67].

A known polyketide, setosol (**113**, Figure 7), was extracted from an endophytic fungus *Preussia isomera* in *Panax notoginseng* from Wenshan, by using an OSMAC strategy. Compound **113** displayed potent activity against multidrug-resistant *E. faecium*, methicinllin-resistant *S. aureus* and multidrug-resistant *E. faecalis* with MIC values of 25 µg/mL [68].

A pair of enantiomeric norsesquiterpenoids, (+)- (**114**) and (−)-preuisolactone A (**115**, Figure 7) featuring an unprecedented tricyclo[4.4.01,6.02,8]decane carbon scaffold were isolated from *Preussia isomera*. XL-1326, obtained from the stems of *Panax notoginseng*. Compounds (+)-I and (−)-II are 2 rare naturally occurring sesquiterpenoidal enantiomers. Compounds **114** and **115** exhibited potent antibacterial activity against *Micrococcus luteus* and *B. megaterium* with MIC values of 10.2 and 163.4 µM, respectively [69].

A new α-pyrone derivative, udagawanone A (**116**, Figure 7) was isolated from *Neurospora udagawae* associated with *Quercus macranthera*, and displayed moderate inhibition against *S. aureus* (MIC = 66 µg/mL) [70].

Five chromone derivatives, including 2,6-dimethyl-5-methoxy-7-hydroxychromone (**117**), 6-hydroxymethyleugenin (**118**), 6-methoxymethyleugenin (**119**), and isoeugenitol (**120**), and isocoumarin congeners, 8-hydroxy-6-methoxy-3-methylisocoumarin (**121**, Figure 7) and diaporthin (**29**), were purified from *Xylomelasma* sp. Samif07, an endophyte of *Salvia miltiorrhiza*. Compound **120** showed good activity against *M. tuberculosis* (MIC 10.31 µg/mL). Compounds **29**, **117–121** displayed inhibitory activities against *B. subtilis*, *Staphylococcus haemolyticus*, *A. tumefaciens*, *Erwinia carotovora*, and *X. vesicatoria* (with MICs ranging from 25 ~ 100 µg/mL). Compounds **117** and **29** showed inhibition against only *E. carotovora* (MIC, 100 µg/mL), and *B. subtilis* (MIC, 50 µg/mL), respectively. Compounds **118**, **119**, **29** were found active against *S. haemolyticus* and *E. carotovora* (MIC of 75 µg/mL), whereas compound **121** exhibited stronger inhibition against *B. subtilis*, *A. tumefaciens*, and *X. vesicatoria*, with MICs of 25, 75, and 25 µg/mL, respectively [71].

The compound (4*S*,5*S*,6*S*)-5,6-epoxy-4-hydroxy-3-methoxy-5-methylcyclohex-2-en-1-one (**122**, Figure 7) was purified from *Amphirosellinia nigrospora* JS-1675, an endophytic fungus isolated from the stem tissue of *Pteris cretica*. Compound **122** showed high to moderate in vitro antibacterial activity, with MIC values ranging between 31.2 and 500 µg mL^{-1} against *Pectobacterium carotovorum* subsp. *Carotovorum*, *Agrobacterium konjaci*, *Burkholderia glumae*, *Clavibacter michiganensis* subsp. *michiganensis*, *A. tumefaciens*, *Pectobacterium chrysanthemi*, *R. solanacearum*, *Acidovorax avenae* subsp. *cattlyae*, *Xanthomonas arboricola* pv. *pruni*, *X. euvesicatoria*, *X. axonopodis* pv. *Citri*, *X. oryzae* pv. *oryzae* [72].

Two new alkylated furan derivatives, 5-(undeca-3′,5′,7′-trien-1′-yl)furan-2-ol (**123**) and 5-(undeca-3′,5′,7′-trien-1′-yl)furan-2-carbonate (**124**, Figure 7), were isolated from *Emericella* sp. XL029, an endophyte of *Panax notoginseng*. Compounds **123**, **124** inhibited *B. subtilis*, *B. cereus*, *S. aureus*, *B. paratyphosum* B, *S. typhi*, *P. aeruginosa*, *E. coli*, and *E. aerogenes* with MIC values ranging from 6.3 to 50 µg/mL [73].

Four new compounds, 14-hydroxytajixanthone (**125**), 14-hydroxyltajixanthone hydrate (**126**, Figure 7), 14-hydroxy-15-chlorotajixanthone hydrate (**127**) and epitajixanthone hydrate (**128**), along with known compounds tajixanthone hydrate (**129**), 14-methoxyltajixanthone-25-acetate (**130**), and 15-chlorotajixanthone hydrate (**131**), questin (**132**) and carnemycin B (**133**, Figure 8), were purified from *Emericella* sp. XL029 residing inside the leaves of *Panax notoginseng*. Compounds **125–127**, **130**, **132**, **133** exhibited potent activity against *M. luteus*, *S. aureus*, *B. megaterium*, *B. anthracis*, and *B. paratyphosum* B (MIC values ranging from 12.5 and 25 µg/mL). Compound **128** exhibited potent activity against *M. luteus*, *S. aureus*, *B. megaterium*, and *B. paratyphosum* B (MIC 25 µg/mL each), while compounds **129**, **131** inhibited *S. aureus*, *B. megaterium*, and *B. paratyphosum* B (MIC 25 and 12.5 µg/mL).

Compounds **125**, **128**, **133** displayed average activity against drug-resistant *S. aureus* (MICs 50 µg/mL each). All isolated compounds **125–133** displayed moderate activity against *P. aeruginosa*, *E. coli*, and *E. aerogenes* (MIC 50 µg/mL) [74].

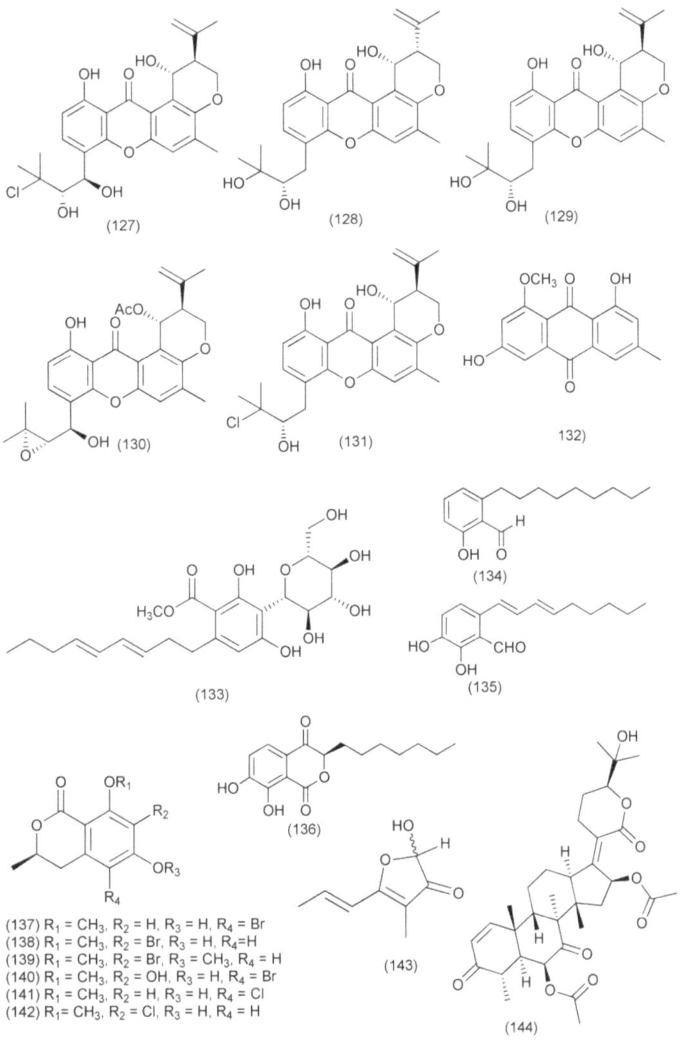

Figure 8. Structures of metabolites **127–144** isolated from Ascomycetes.

An endophytic fungus *Byssochlamys spectabilis* from the plant *Edgeworthia chrysantha* yielded bysspectin C (**134**, Figure 8) which was active against *E. coli* and *S. aureus* with MIC values of 32 and 64 µg/mL, respectively [75].

Two new compounds, sydowianumols A (**135**), and B (**136**, Figure 8), were isolated from *Poculum pseudosydowianum* (TNS-F-57853), an endophytic fungus associated with the petiole of *Quercus crispula* var. *crispula* in Yoshiwa. Compounds **135** and **136** exhibited anti-MRSA activity, with MIC$_{90}$ values of 12.5 µg/mL [76].

Six previously undescribed halogenated dihydroisocoumarins, palmaerones A–C, (**137–139**) and E–G (**140–142**, Figure 8) were purified from *Lachnum palmae*, an endophytic fungus from *Przewalskia tangutica* by exposure to a histone deacetylase inhibitor SAHA.

Compounds **137**, **138**, **140–142** were active against *B. subtilis*, with MIC values of 35, 30, 10, 50, and 55 µg/mL, respectively, while compounds **137–140**, were found active against *S. aureus* with MIC values of 65, 55, 60, and 55 µg/mL, respectively [77].

The polyketide nemanifuranone A (**143**), a nordammarane triterpenoid, was isolated from *Nemania serpens*, an endophyte of *Vitis vinifera*. Additionally, a known metabolite **144**, also a nordammarane triterpenoid (Figure 8) was isolated from the mycelium. Nemanifuranone A (**143**) showed modest activity against *E. coli*, with a MIC of 200 µg/mL, and significant inhibition (>75% inhibition) against *S. aureus*, *B. subtilis* and *M. luteus* at a concentration of 100–200 µg/mL. However, **144** showed significant inhibition (>75% inhibition) of *M. luteus* at a concentration of 100 µg/mL [78].

A sesquiterpene, variabilone (**145**, Figure 9), with a new skeleton, was isolated from the endophytic fungus *Paraconiothyrium variabile* isolated from *Cephalotaxus harringtonia*. Compound **145** behaved as a potent growth inhibitor of *B. subtilis* at an IC_{50} of 2.13 µg/mL after 24 h [79].

A new 4-hydroxycinnamic acid derivative compound, methyl 2-{(*E*)-2-[4-(formyloxy) phenyl]ethenyl}-4-methyl-3-oxopentanoate (**146**), along with the known compounds (3*R*,6*R*)-4-methyl-6-(1-methylethyl)-3-phenylmethylperhydro-1,4-oxazine-2,5-dione (**147**), (3*R*,6*R*)-N-methyl-N-(1-hydroxy-2-methylpropyl)-phenylalanine (**148**), siccanol (**149**), sambutoxin (**150**, Figure 9) and fusaproliferin (**106**), were extracted from *Pyronema* sp. an endophyte of the *Taxus mairei*. Compounds **106**, **146–150** also exhibited potential inhibitory activity, with IC_{50}s of 64, 59, 57, 84, 43 and 32 µM against *Mycobacterium marinum*, respectively [80].

Three new natural furanones, pulvinulin A (**151**), graminin C (**152**), and *cis*-gregatin B (**153**), together with the known fungal metabolite, graminin B (**154**, Figure 9), were isolated from *Pulvinula* sp. 11120, an endophyte of the leaves of *Cupressus arizonica*. Compounds **151–154** displayed antibacterial against *E. coli* with 12, 18, 16, and 14 mm zones of inhibition [81].

Stelliosphaerols A (**155**) and B (**156**, Figure 9), new sesquiterpene−polyol conjugates were purified from a *Stelliosphaera formicum* endophytic fungus associated with the plant *Duroia hirsuta*. Compounds **155** and **156** inhibited *S. aureus* with MIC values of 250 µg/mL [82].

Two novel polyketides, *cis*-4-acetoxyoxymellein (**157**) and 8-deoxy-6-hydroxy-*cis*-4-acetoxyoxymellein (**158**, Figure 9) were extracted from an unidentified ascomycete, associated with *Melilotus dentatus*. Compound **157** was found to be active against *E. coli* and *B. megaterium* with 10 and 10 (partial inhibition) zones of inhibition at 0.05 mg concentration. Compound **158** displayed antibacterial activity against *E. coli* and *B. megaterium* with 9 and 9 (partial inhibition) zones of inhibition at a concentration of 0.05 mg [83].

2.2. Anamorphic Ascomycetes

Anamorphic Ascomycetes are the fungi that are the asexual form of ascomycetes. The first antibiotic penicillin-producing fungi belonged to this group. Fungi belonging to this group are prolific producers of bioactives metabolites. After the discovery of penicillin, this group is extensively screened for bioactives. Some important genera in this group are *Penicillium*, *Aspergillus*, *Fusarium*, *Pestalotiopsis*, *Phoma* and *Colletotrichum*. Here we report the antibacterials compounds from this group of fungi.

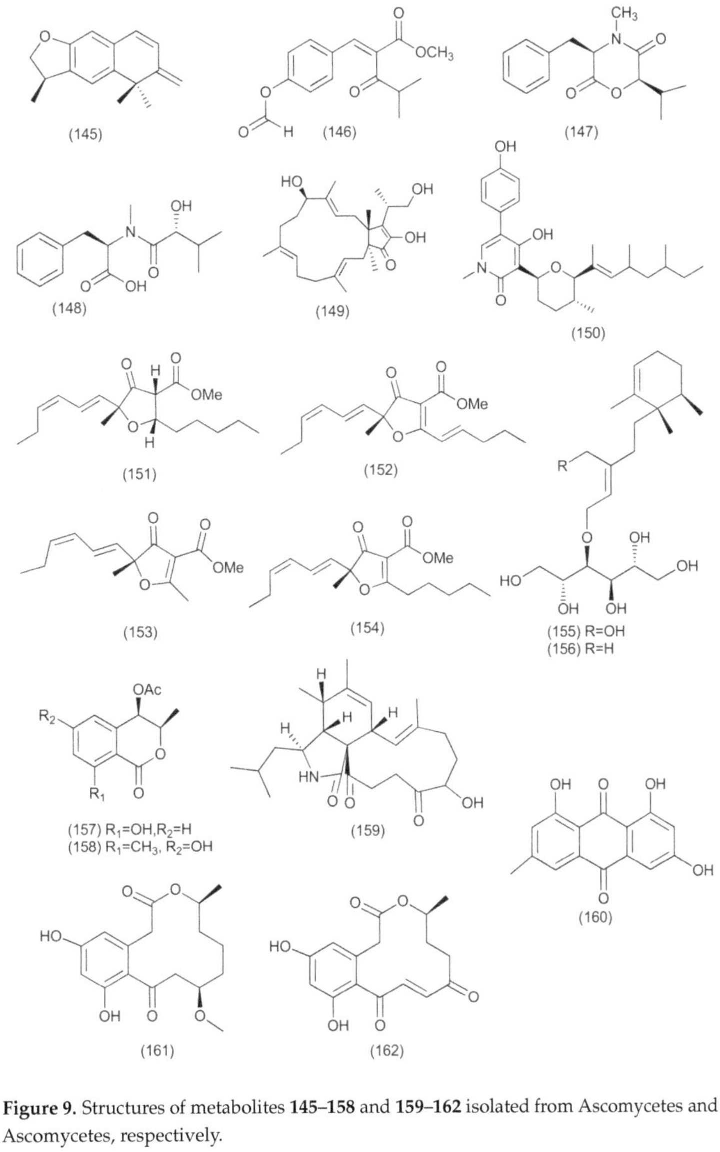

Figure 9. Structures of metabolites **145–158** and **159–162** isolated from Ascomycetes and Anamorphic Ascomycetes, respectively.

2.2.1. *Aspergillus*

Aspergillus is one of the important fungal genera and some of the antibacterials from this genus such as aspochalasin P (**159**), alatinone (**160**), β-11-methoxycurvularine (**161**), and 12-keto-10,11-dehydrocurvularine (**162**, Figure 9) were purified from *Aspergillus* sp. FT1307 associated with plant *Heliotropium* sp. Compounds **159–162** showed weak activity against *Staphylococcus aureus* ATCC12600, *Bacillus subtilis* ATCC6633 and MRSA ATCC43300 with MICs in the range of 40 to 80 μg/mL [84].

A new polyketide, aspergillone A (**163**, Figure 10), was isolated from *Aspergillus crista-tus* associated with *Pinellia ternata*. Aspergilline A (**163**) is the first example of a bicyclo[2.2.2] diazaoctane indole alkaloid where the diketopiperazine structure is constructed from tryptophan and alanine. Aspergillone A (**163**) exhibited average antibacterial activities against *B. subtilis* and *S. aureus*, with MIC$_{50}$ values of 8.5 and 32.2 μg/mL, respectively [85].

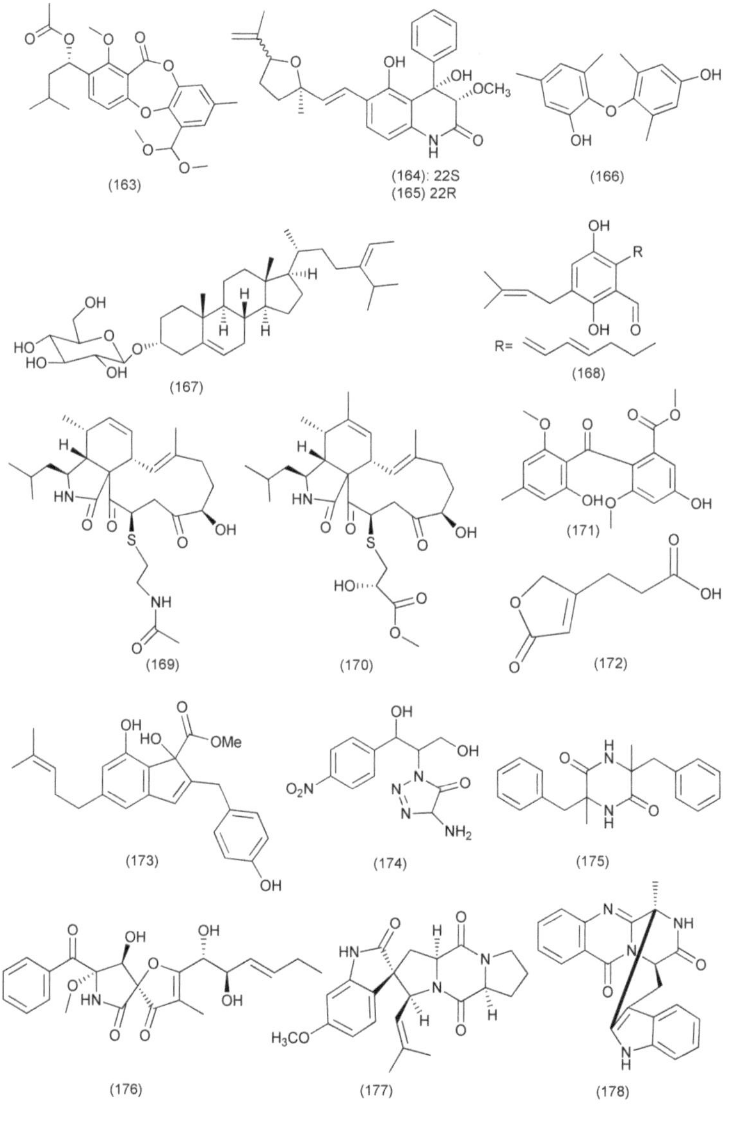

Figure 10. Structures of metabolites **163–178** isolated from Anamorphic Ascomycetes.

A new quinolone derivative, (22*S*)-aniduquinolone A (**164**) and its known isomer (22*R*)-aniduquinolone A (**165**, Figure 10) were purified from the endophytic fungus *Aspergillus versicolor* strain Eich.5.2.2 from the petals of flowers of *Eichhornia crassipes*. The epimers **164/165** together exhibited significant antibacterial activity against *S. aureus*, with a MIC of 0.4 μg/mL [86].

A new diaryl ether derivative aspergillether B (**166**, Figure 10) was separated from *Aspergillus versicolor* residing inside the roots of *Pulicaria crispa*. Compound **166** exhibited significant antibacterial capacity towards *S. aureus, Bacillus cereus*, and *E. coli* with MICs values of 4.3, 3.7, and 3.9 µg/mL, respectively [87].

The known compound 3-O-β-D-glucopyranosyl stigmasta-5(6),24(28)-diene (**167**, Figure 10) was extracted from an endophytic fungus *Aspergillus ochraceus* SX-C7 eus SX-C7 from *Setaginella stauntoniana* and displayed inhibitory activity against *B. subtilis* with a MIC value of 2 µg/mL [88].

A prenylated benzaldehyde derivative, dihydroauroglaucin (**168**, Figure 10), was isolated from *Aspergillus amstelodami* (MK215708) an endophytic fungi of *Ammi majus*, a plant indigenous to Egypt. Compound **168** showed activity against *E. coli, Streptococcus mutans* and *S. aureus*, with MICs of 1.95, 1.95 and 3.9 µg/mL, respectively. The highest antibiofilm activity at concentrataion 7.81 µg/mL against *S. aureus* and *E. coli* biofilms, at 15.63 µg/mL concentration against *S. mutans* and moderate activity (MBIC = 31.25 µg/mL) against *P. aeruginosa* biofilm was measured [89].

Two cysteine residue-containing merocytochalasans, cyschalasins A (**169**) and B (**170**, Figure 10) were isolated from *Aspergillus micronesiensis* associated with the root of *Phyllanthus glaucus*. Compounds **169** and **170** displayed anti-MRSA activity with MIC$_{50}$ values of 17.5 and 10.6 µg/mL and MIC$_{90}$ values of 28.4 and 14.7 µg/mL, respectively [90].

Methylsulochrin (**171**, Figure 10) is a diphenyl ether derivative isolated from *A. niger* associated with the stems of *Acanthus montanus*. It inhibits *Enterobacter cloacae, Enterobacter aerogenes* and *S. aureus* with MIC values of 7.8, 7.8 and 15.6 µg/mL, respectively [91].

A new furan derivative named 3-(5-oxo-2,5-dihydrofuran-3-yl) propanoic acid (**172**, Figure 10) was purified from *Aspergillus tubingensis*, an endophyte from the stems of *Decaisnea insignis*. Compound **172** inhibited *Streptococcus lactis* with MIC value of 32 µg/mL [92].

A new compound, methyl 2-(4-hydroxybenzyl)-1,7-dihydroxy-6-(3-methylbut-2-enyl)-1*H*-indene-1-carboxylate (**173**, Figure 10) was extracted from *Aspergillus flavipes* Y-62, associated with the plant *Suaeda glauca*. Compound **173** showed poor activity against MRSA, with an MIC value of 128 µg/mL, and against *K. pneumoniae* and *P. aeruginosa* with equal MIC values of 32 µg/mL [93].

The alkaloids 4-amino-1-(1,3-dihydroxy-1-(4-nitrophenyl)propan-2-yl)-1*H*-1,2,3-triazole-5(4H)one (**174**) and 3,6-dibenzyl-3,6-dimethylpiperazine-2,5-dione (**175**, Figure 10) were obtained from *Aspergillus* sp. isolate of *Zingiber cassumunar* rhizome. Compounds **174** and **175** exhibited inhibitory activity against *X. oryzae* and *E. coli*, with a 16–30 mm zone of inhibition [5].

Aspergillus fumigatus, an endophyte associated with *Edgeworthia chrysantha*, was the source of pseurotin A (**176**) and spirotryprostatin A (**177**, Figure 10). Compounds **176**, **177** displayed good antibacterial activity against *S. aureus* (MIC 0.39 µg/mL each). Compound **177** also showed potent antibacterial activity against *E. coli* (MIC of 0.39 µg/mL) [94].

Six compounds, fumiquinazoline J (**178**, Figure 10), fumiquinazoline I (**179**), fumiquinazoline C (**180**), fumiquinazoline H (**181**), fumiquinazoline D (**182**), and fumiquinazoline B (**183**, Figure 11) were extracted from *Aspergillus* sp., residing inside the plant *Astragalus membranaceus*. Compounds **178**, **180**–**182** displayed potent activity against *B. subtilis, E. coli, P. aeruginosa* and *S. aureus* (MICs in the range of 0.5–8 µg/mL). Compounds **179**, **183** displayed moderate activity against *B. subtilis, E. coli, P. aeruginosa* and *S. aureus* with MICs of 4–16 µg/mL [95].

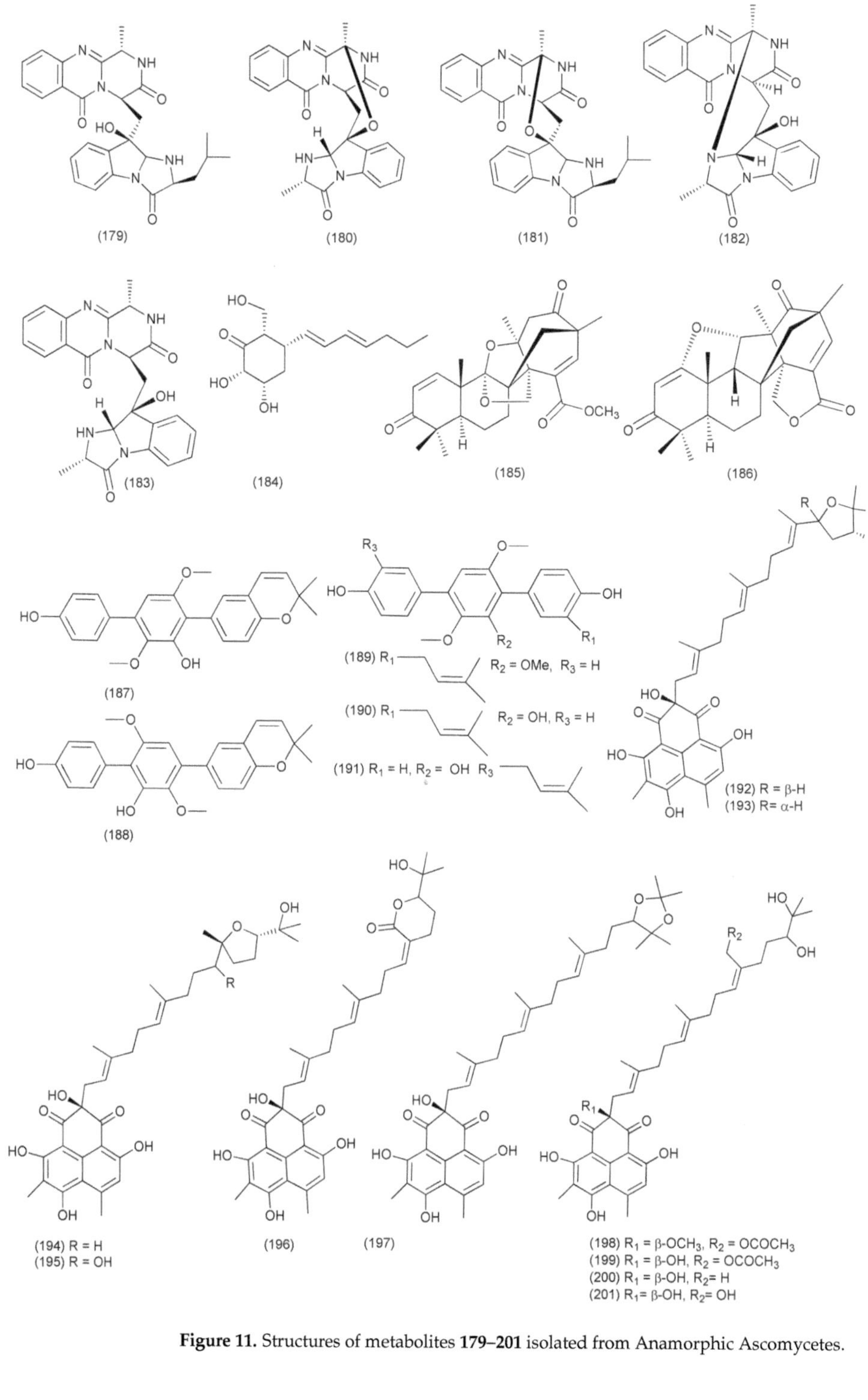

Figure 11. Structures of metabolites 179–201 isolated from Anamorphic Ascomycetes.

An antibacterial polyketide named (-) palitantin (**184**, Figure 11) was isolated from *Aspergillus fumigatiaffnis*, an endophyte of the medicinal plant *Tribulus terrestris*, which displayed antibacterial activity against *E. faecalis* UW 2689 and *S. pneumoniae* with MIC values of 64 µg/mL each [96].

A novel terpene-polyketide hybrid, i.e., a meroterpenoid, aspermerodione (**185**), and a new heptacyclic analog and iconin C (**186**, Figure 11) were purified from *Aspergillus* sp. TJ23 residing inside the plant *Hypericum perforatum*. Compound **185** showed antibacterial activity against MRSA (MIC of 32 µg/mL), whereas compound **186** showed poor anti- MRSA activity (>100 µg/mL). Aspemerodione (**186**) worked synergistically with the antibiotics oxacillin and piperacillin against MRSA and was found to be a potential inhibitor of PBP2a [97].

Aspergillus sp. YXf3, an endophyte residing inside the leaves of *Ginkgo biloba*, yielded some novel *p*-terphenyls named prenylterphenyllin D (**187**), prenylterphenyllin E (**188**), and 2′-O-methylprenylterphenyllin (**189**), along with the known compounds prenylterphenyllin (**190**) and prenylterphenyllin B (**191**, Figure 11). Compounds **187–191** displayed antibacterial activity against *X. oryzae* pv. *oryzicola* and *E. amylovora* with the same MIC values of 20 µg/mL, while compound **191** exhibited activity against *E. amylovora* with a MIC value of 10 µg/mL [98].

Nine new phenalenone derivatives, aspergillussanone D (**192**), aspergillussanone E (**193**), F (**194**) G (**195**) H (**196**), I (**197**), J (**198**), K (**199**), along with two known analogues, the aspergillussanones L (**200** and **201**, Figure 11) were extracted from *Aspergillus* sp. residing inside the plant *Pinellia ternate*. Compound **200** exhibited good antimicrobial activity against *P. aeruginosa*, *S. aureus*, and *B. subtilis* (MIC$_{50}$ values of 1.87, 2.77, and 4.80 µg/mL). Compound **192** exhibited the antibacterial activity against *P. aeruginosa*, and *S. aureus*, (MIC$_{50}$ of 38.47 and 29.91 µg/mL). Compound **193** was found to be selectively active against *E. coli* (MIC$_{50}$ of 7.83 µg/mL). Compound **194** exhibited antimicrobial activity against *P. aeruginosa*, and *S. aureus*, (MIC$_{50}$ values of 26.56, 3.93 and 16.48 µg/mL). Compound **195** inhibited *P. aeruginosa*, and *S. aureus*, (MIC$_{50}$ values of 24.46 and 34.66 µg/mL). Compound **196** inhibited *P. aeruginosa*, and *E. coli*, (MIC$_{50}$ values of 8.59 and 5.87 µg/mL). Compound **197** selectively inhibited *P. aeruginosa*, (MIC$_{50}$ of 12.0 µg/mL). Compound **198** exhibited activity against *P. aeruginosa*, *E. coli* and *S. aureus* with MIC$_{50}$ values of 28.50, 5.34 and 29.87 µg/mL, respectively. Compound **199** exhibited antibacterial activity against *P. aeruginosa*, and *S. aureus*, (MIC$_{50}$ values of 6.55 and 21.02 µg/mL). Compound **201** inhibited *P. aeruginosa*, and *E. coli*, with MIC$_{50}$ values of 19.07 and 1.88 µg/mL, respectively [99].

The compound terrein (**202**, Figure 12), a polyketide, was extracted from *Aspergillus terreus* JAS-2 associated with *Achyranthus aspera*. Terrein (**202**) exhibited antibacterial activity with an IC$_{50}$ value of 20 µg/mL against *E. faecalis*, and more than 20 µg/mL against *Aeromonas hydrophila* and *S. aureus*, as the compound showed only 48% and 38.3% inhibition [100].

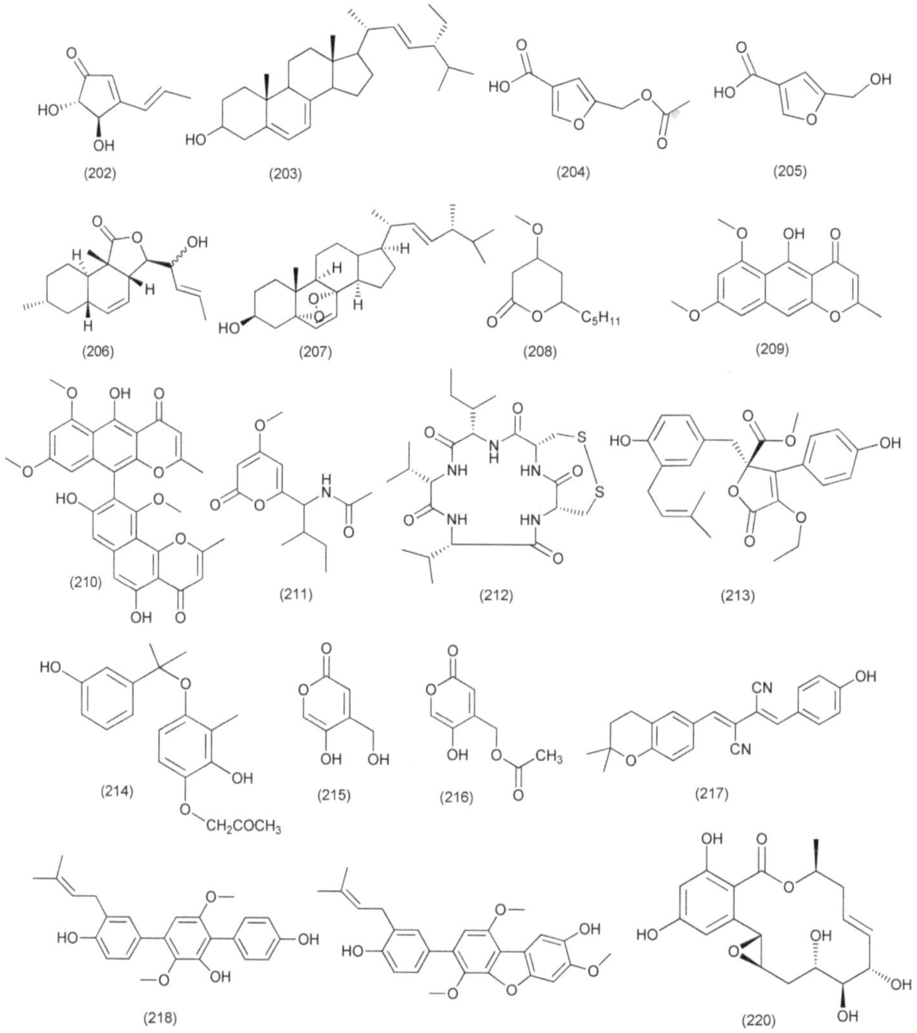

Figure 12. Structures of metabolites **202–220** isolated from Anamorphic Ascomycetes.

A known compound (22*E*,24*R*)-stigmasta-5,7,22-trien-3-β-ol (**203**, Figure 12), was purified from the *Aspergillus terreus* isolate of *Carthamus lanatus*. Compound **203** displayed potent anti-MRSA activity, with IC_{50} values of 2.29 μM compared to ciprofloxacin (IC_{50} 0.21 μM) [101].

A new furan derivative named 5-acetoxymethylfuran-3-carboxylic acid (**204**), along with the furan compound 5-hydroxymethylfuran-3-carboxylic acid (**205**, Figure 12), were obtained from *Aspergillus flavus*, isolated from *Cephalotaxus fortunei*. The compounds **204–205** inhibited *S. aureus* with MIC values of 15.6 and 31.3 μg/mL, respectively [102].

A new compound, allahabadolactone B (**206**), and the known compound ergosterol peroxide (**207**, Figure 12) were purified from *Aspergillus allahabadii* BCC45335 residing inside the roots of *Cinnamomum subavenium*. Compounds **206–207** displayed antimicrobial activity against *B. cereus* with IC_{50} values of 12.50 and 3.13 μg/mL, respectively [103].

A new pyrone named 6-isovaleryl-4-methoxy-pyran-2-one (**208**), along with three known pyrone compounds, rubrofusarin B (**209**), asperpyrone A (**210**) and campyrone A (**211**, Figure 12), was purified from *Aspergillus tubingensis* isolated from the roots of *Lycium ruthenicum*. Compound **209** possessed potent activity against *E. coli* with a MIC of 1.95 µg/mL while the compounds **208**, **210**, **211** showed poor activity against *E. coli*, *P. aeruginosa*, *S. aureus* and *Streptococcus lactis* [104].

A new cyclic pentapeptide, malformin E (**212**, Figure 12), was extracted from *Aspergillus tamarii* FR02 associated with *Ficus carica*. Compound **212** displayed potent activity against *B. subtilis*, *S. aureus*, *P. aeruginosa*, and *E. coli* with MIC values of 0.91, 0.45, 1.82, and 0.91 µM, respectively [105].

A new butyrolactone, aspernolide F (**213**), together with a known stigmasterol deriva-tive, (22*E*,24*R*)-stigmasta-5,7,22-trien-3-β-ol (**203**, Figure 12), were purified from *Aspergillus terreus*, an endophyte of *Carthamus lanatus*. Compound **203** displayed a potent anti-MRSA activity, with an IC_{50} value of 0.96µg/mL while compound **213** displayed poor anti-MRSA activity (IC_{50} 6.39µg/mL) [106].

The metabolites 1-(3,8-dihydroxy-4,6,6-trimethyl-6*H*-benzochromen-2-yloxy)propane-2-one (**214**), 5-hydroxy-4-(hydroxymethyl)-2*H*-pyran-2-one (**215**) and 5-hydroxy-2-oxo-2*H*-pyran-4-yl)methyl acetate (**216**, Figure 12) were purified from *Aspergillus* sp. (SbD5) associated with the plant *Andrographis paniculata*. Compounds **214–216** displayed poor to average activity against *S. aureus*, *E. coli*, *S. dysenteriae* and *Salmonella typhi* with an inhibition zone diameter ranging from 8.1 to 12.1 mm at a concentration 500 µg/mL [107].

The compounds xanthoascin (**217**), prenylterphenyllin B (**218**) and prenylcandidusin (**219**, Figure 12), were extracted from *Aspergillus* sp. IFB-YXS, associated with the leaves of *Ginkgo biloba*. Compound **217** displayed antibacterial activity against *X. oryzae* pv. *oryzicola*, *E. amylovora*, *P. syringae* pv. *lachrymans* and *C. michiganense* subsp. *sepedonicus* with MICs of 20, 10, 5.0 and 0.31 µg/mL, respectively. Compound **218** exhibited antibiotic activities with MICs of 20 µg/mL each towards *X. oryzae* pv. *oryzicola*, *E. amylovora*, *P. syringae* pv. *lachrymans*, respectively. Compound **219** was found to be effective against *X. oryzae* pv. *oryzae* and *X. oryzae* pv. *oryzicola* (MIC of 10 and 20 µg/mL). It was observed that compound **217** can change the permeability and cause nucleic acid leakage of the cytomembrane of the phytopathogen [108].

2.2.2. *Penicillium*

New β-resorcylic acid lactones, including 4-O-desmethyl-aigialomycin B (**220**, Figure 12), and penochrochlactones C (**221**), and D (**222**, Figure 13), were purified from *Penicillium ochrochloron* SWUKD4.1850 from the medicinal plant *Kadsura angustifolia*. Compounds **220–222** exhibited moderate activities against *S. aureus*, *B. subtilis*, *E. coli*, and *P. aeruginosa* with MIC values between 9.7 and 32.0 µg/mL [109].

The compound *p*-hydroxybenzaldehyde (**223**, Figure 13), was isolated from *Penicillium brefeldianum*, an endophyte residing inside the root bark of *Syzygium zeylanicum*. Compound **223** was found to be active against *S. typhi*, *E. coli*, and *B. subtilis* with MIC values of 64 g/mL. *p*-Hydroxybenzaldehyde was also reported from *Syzygium zeylanicum* [110].

An endophytic fungus, *Penicillium vulpinum* GDGJ-91, from the roots of *Sophorae tonkinensis*, yielded the new compound 10-demethylated andrastone A (**224**), and four known analogs, 15-deacetylcitreohybridone E (**225**), citreohybridonol (**226**) and andrastins A (**227**) and B (**228**, Figure 13). Compounds **224** and **227** displayed good activity against *Bacillus megaterium* (MIC value of 6.25 µg/mL), and compounds **225**, **226**, **228** showed average activity against *Bacillus megaterium* (MIC of 25, 12.5 and 25 µg/mL). Compound **226** showed potent antibacterial activity against *B. paratyphosus* B at 6.25 µg/mL, while the other compounds showed average activities against *B. paratyphosus* B at 12.5 or 25 µg/mL and compound **226** also exhibited moderate activities against *E. coli* and *S. aureus* with MIC values of 25 µg/mL [111].

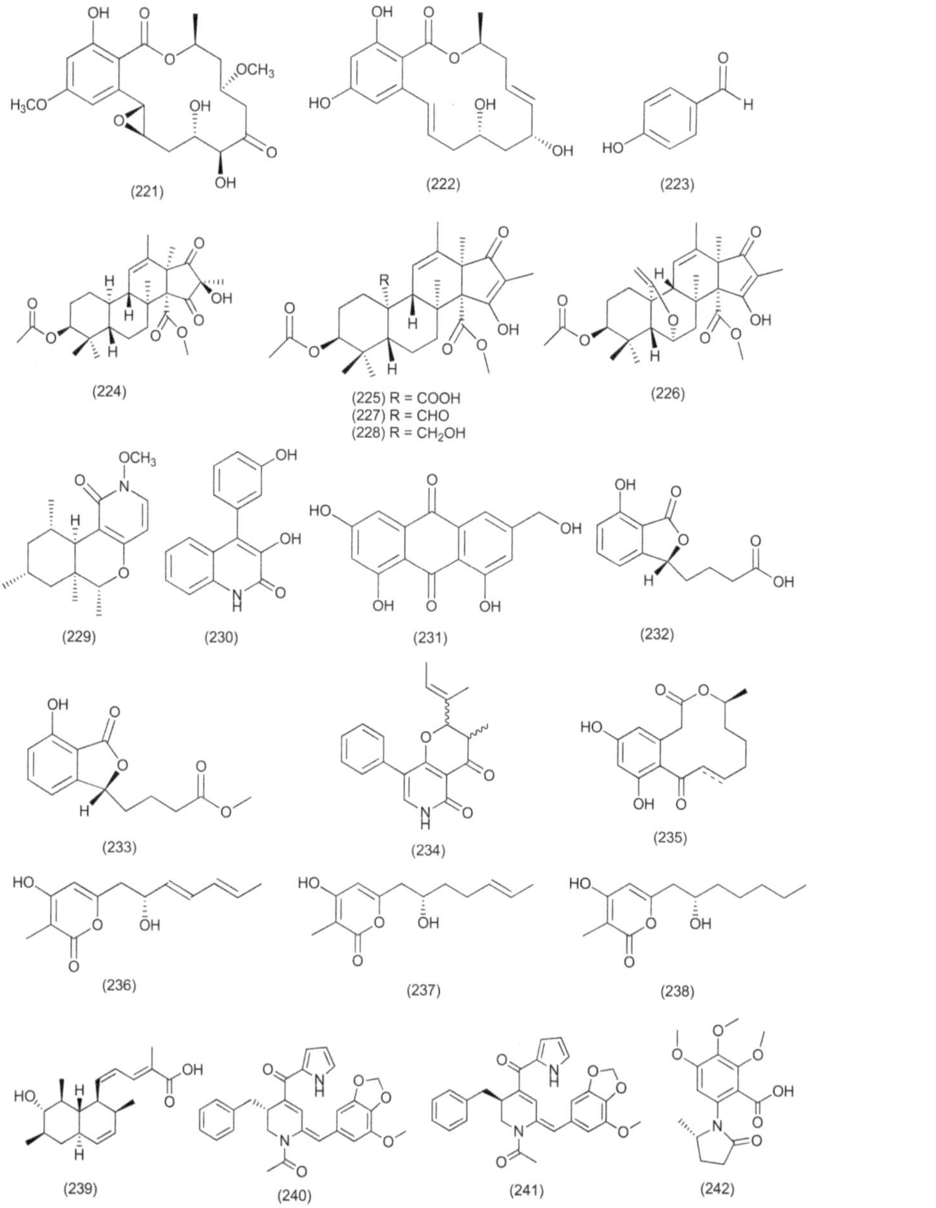

Figure 13. Structures of metabolites **221–242** isolated from Anamorphic Ascomycetes.

A novel N-methoxy-1-pyridone alkaloid, chromenopyridin A (**229**), and the already reported compound viridicatol (**230**, Figure 13) were purified from *Penicillium nothofagi* P-6, residing inside the bark of *Abies beshanzuensis*. Compounds **229** and **230** exhibited antibacterial activity against *S. aureus*, with MIC values of 62.5 and 15.6 μg/mL, respectively [112].

ω-Hydroxyemodin (**231**, Figure 13) a polyhydroxy anthraquinone, was extracted from *Penicillium restrictum* (strain G85) from *Silybum marianum*. Compound **231** showed inhibition against MRSA as a quorum sensing inhibitor in both in vitro and in vivo systems [113].

Two new phthalide derivatives, (−)-3-carboxypropyl-7-hydroxyphthalide (**232**) and (−)-3-carboxypropyl-7-hydroxyphthalide methyl ester (**233**, Figure 13), were isolated from *Penicillium vulpinum* residing inside the plant *S. tonkinensis*. Compound **232** exhibited a medium inhibition against *Shigella dysenteriae*, *Enterobacter areogenes*, *B. subtilis*, *B. megaterium*, and *Micrococcus lysodeikticus* with MIC value between 12.5–50 μg/mL. Compound **233** showed average activity against *E. areogenes* with MIC value of 12.5 μg/mL, and showed poor activity against *B. subtilis*, *B. megaterium* and *M. lysodeikticus* with MIC values of 100 μg/mL [114].

Citridone E (**234**), a new phenylpyridone derivative, and the previously reported compound (–)-dehydrocurvularin (**235**, Figure 13) were purified from *Penicillium sumatrense* GZWMJZ-313 associated with the plant *Garcinia multiflora*. Compounds **234** and **235** showed antibacterial activity against *S. aureus*, *P. aeruginosa*, *Clostridium perfringens*, and *E. coli* (with MICs ranging from 32 to 64 μg/mL) [115].

Three new 3,4,6-trisubstituted α-pyrone derivatives, namely 6-(2′R-hydroxy-3′E,5′E-diene-1′-heptyl)-4-hydroxy-3-methyl-2H-pyran-2-one (**236**), 6-(2′S-hydroxy-5′E-ene-1′-heptyl)-4-hydroxy-3-methyl-2H-pyran2-one (**237**), and 6-(2′S-hydroxy-1′-heptyl)-4-hydroxy-3-methyl-2H-pyran-2-one (**238**), along with the previously reported compound trichodermic acid (**239**, Figure 13), were purified from *Penicillium ochrochloron* associated with *Taxus media*. Compounds **236–239** displayed antimicrobial activity with MIC values ranging from 25 to 50 μg/mL against *B. subtilis*, *B. megaterium*, *E. coli*, *Enterobacter aerogenes*, *Micrococcus luteus*, *Proteusbacillm vulgaris*, *P. aeruginosa*, *S. aureus*, *Salmonella enterica*, and *Salmonella typhi* [116].

Three new compounds, brasiliamide J-a (**240**), brasiliamide J-b (**241**) and peniciolidone (**242**, Figure 13), as well as the known compound austin (**243**, Figure 14), were isolated from *Penicillium janthinellum* SYPF 7899 associated with the plant *Panax notoginseng*. Compound **240** exhibited potent activity against *B. subtilis* and *S. aureus* (MICs of 15 and 18 μg/mL). Compounds **241** and **243** showed average inhibitory activities against *B. subtilis* (MIC 35 μg/mL and 50 μg/mL, respectively) and *S. aureus* (MIC 39 μg/mL and 60 μg/mL, respectively). In addition, compound **240** also affected the length of *B. subtillius*. Similarly, coccoid cells of *S. aureus* also swelled 2-fold after treatment with compound **240**. Compounds **240, 241, 242** showed high binding energies, strong H-bond interactions and hydrophobic interactions with filamentous temperature-sensitive protein Z (FtsZ) [117].

The new compounds penicimenolidyu A (**244**), and penicimenolidyu B (**245**) and the known compound rasfonin (**246**, Figure 14) were purified from *Penicillium cataractarum* SYPF 7131 obtained from the plant *Ginkgo biloba*. Compound **246** exhibited good antibacterial activity against *S. aureus*, with a MIC value of 10 μg/mL. Compounds **245** and **246** showed moderate inhibitory activity against *S. aureus* (MIC 65 μg/mL and 59 μg/mL). The docking results revealed that compounds **244–246** possess high binding energies, strong H-bond interactions and hydrophobic interactions with FtsZ from *S. aureus*, validating the observed antimicrobial activity [118].

A rare dichloroaromatic polyketide, 3′-methoxycitreovirone (**247**) along with known metabolites *cis*-bis-(methylthio)-silvatin (**248**), citreovirone (**249**), trypacidin A (**250**, Figure 14) and helvolic acid (**100**), were obtained from endophytic *Penicillium* sp. of *Pinellia ternate*. Compound **100** displayed potent antibacterial activity against *S. aureus* and *P. aeruginosa* (MIC = 5.8 and 4.6 μg/mL) as well as mild activity against *B. subtilis* and *E. coli* (MIC = 42.2 and 75.0 μg/mL). Compounds **247** and **249** were found to have moderate antibacterial activity against *E. coli* and *S. aureus* (MIC = 62.6 and 76.6 μg/mL). Compounds **248** and **250** exhibited poor antibacterial activity against *S. aureus* with MIC values of 43.4 and 76.0 μg/mL and **250** also displayed effect against *B. subtilis* (MIC = 54.1 μg/mL) [119].

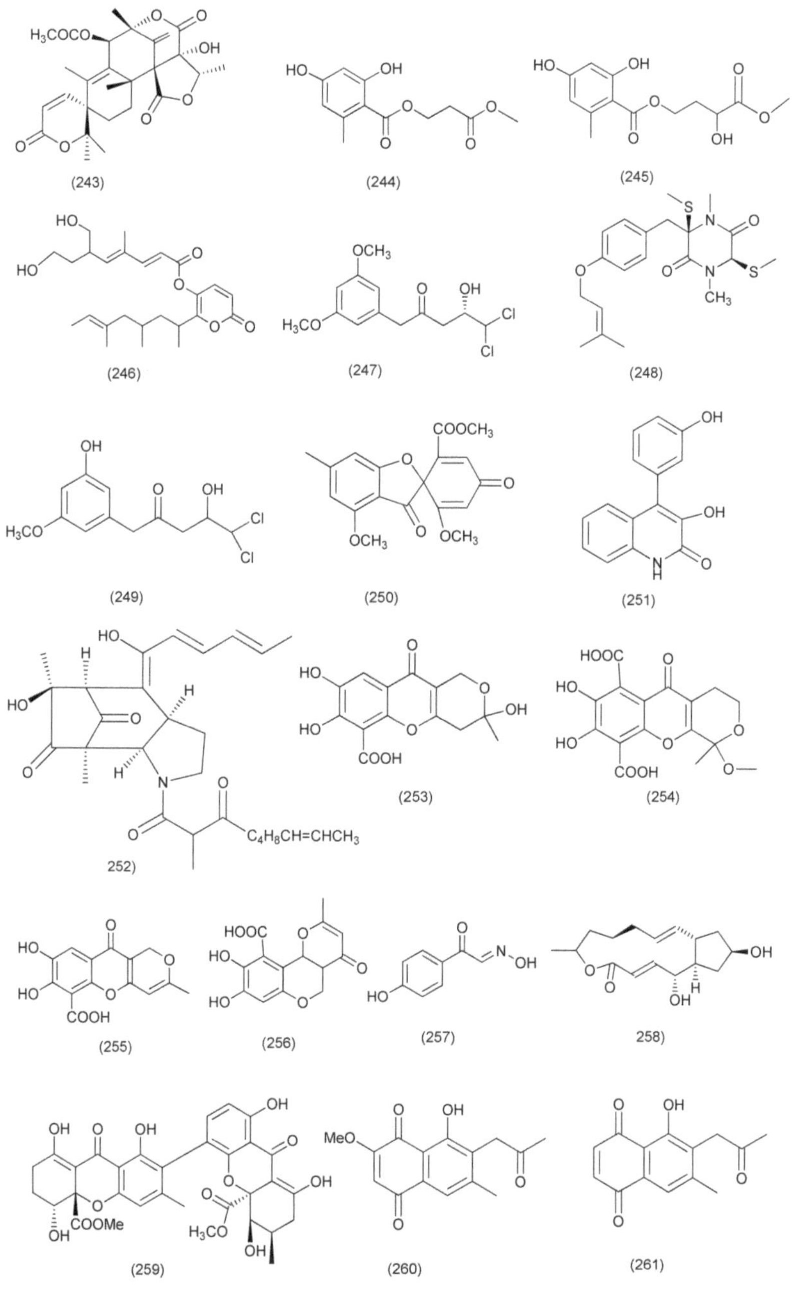

Figure 14. Structures of metabolites **243–261** isolated from Anamorphic Ascomycetes.

A known quinolinone alkaloids viridicatol (**251**, Figure 14) was obtained from *Penicillium* sp. R22 was associated with *Nerium indicum* and displayed potent antibacterial activity against *S. aureus* with MIC value of 15.6 µg/mL [120]. The novel compound penicitroamide (**252**, Figure 14), was purified from *Penicillium* sp. (NO. 24) isolated from the leaves of *Tapis-*

cia sinensis. Compound **252** displayed potent antibacterial activity against plant pathogens, *Erwinia carotovora* sub sp. *carotovora* (Jones) Bersey, et al. with MIC_{50} at 45 μg/mL [121].

Penialidins A-C (**253–255**), citromycetin (**256**), *p*-hydroxyphenylglyoxalaldoxime (**257**) and brefelfin A (**258**, Figure 14) were purified from the *Penicillium* sp. CAM64 a fungus associated with the plant *Garcinia nobilis*. Compounds **253–258**, exhibited antibacterial activity against *Vibrio cholerae* SG24 (1), *V. cholerae* CO6, *V. cholerae* NB2, *V. cholerae* PC2, *S. flexneri* SDINT (MIC = 0.50–128 μg/mL). Compound **255** exhibited potent activity against *V. cholerae* SG24 (1), *V. cholerae* CO6, *V. cholerae* NB2, *V. cholerae* PC2, *S. flexneri* SDINT, with MIC values of 0.50, 16, 8, 0.50 and 8 μg/mL, respectively following in decreasing order of activity by compound **254** (MIC = 4–32 μg/mL), compound **257** (MIC = 8–32 μg/mL), compound **257** (MIC = 32–64 μg/mL) and compounds **256** and **258** (MIC = 64–128 μg/mL) [122].

Purpureone (**259**, Figure 14) was extracted from *Purpureocillium lilacinum*, residing inside the roots of *Rauvolfia macrophylla*. Compound **259** displayed antibacterial activity with the zone of inhibition of 10.6, 12.3, 13.0, 8.7, 12.3, and 10.0, mm against *B. cereus*, *L. monocytogenes*, *E. coli*, *K. pneumoniae*, *P. stuartii*, and *P. aeruginosa* (6 mm filter paper disks impregnated with 10 μL of compound) [123].

2.2.3. *Fusarium*

Secondary metabolites identified as 2-methoxy-6-methyl-7-acetonyl-8-hydroxy-1,4-maphthalenedione (**260**) 5,8-dihydroxy-7-acetonyl-1,4-naphthalenedione (**261**, Figure 14), anhydrojavanicin (**262**), and fusarnaphthoquinone B (**263**, Figure 15), were purified from *Neocosmospora* sp. MFLUCC 17-0253 associated with *Rhizophora apiculata*. All three compounds showed potent antibacterial against *Acidovorax citrulli* (responsible for bacterial fruit blotch (BFB) a bacterial disease of *Cucurbitaceae* crops) with MIC values of 0.0075 mg/mL (mixture of **260**, **261**), 0.004 mg/mL (**262**), 0.025 mg/mL (**263**). Compounds **260–263** significantly inhibited biofilm development of *Acidovorax citrulli*, thus demonstrating that these metabolites can be used for biological control of bacterial fruit blotch of watermelon and melon [124].

A new aminobenzamide derivative, namely fusaribenzamide A (**264**, Figure 15), was purified from *Fusarium* sp. of *Mentha longifolia*. Compound **264** displayed antibacterial activity against *S. aureus* and *E. coli* with MIC values of 62.8 and 56.4 μg/disc, respectively [125].

Two alkaloids, indol-3-acetic acid (**265**), bassiatin (**266**), a depsipeptide, beauvericin (**267**), two sesquiterpenoids, cyclonerodiol (**268**), epicyclonerodiol oxide (**269**), four 1,4-naphthoquinones, 5-O-methylsolaniol (**270**), 5-O-methyljavanicin (**271**), fusarubin methyl ether (**272**), and anhydrojavanicin (**273**, Figure 15) and a sesterterpene, fusaproliferin (**106**), were separated from the green Chinese onion-derived fungus *F. proliferatum* AF-04. Compounds **270–273** displayed good antibacterial activity against *B. megaterium* with MICs of 25 μg/mL each; compounds **265**, **267**, **269** displayed moderate activity with MICs of 50 μg/mL each and compound **268**, displayed activity with an MIC of 12.50 μg/mL. Compounds **266**, **270–272** displayed good antibacterial activity against *B. subtilis*, with MICs of 50 μg/mL each. Compounds **269** and **272** were found to be active against *E. coli* with MIC values of 50 μg/mL each and compounds **270**, **271**, **273** with MIC values of 25 μg/mL, respectively. Compounds **269–272** displayed antibacterial activity against *Clostridium perfringens* with MIC values of 50, 50, 12.5 and 50 μg/mL, respectively. Compounds **267**, **106**, **270–273** displayed anti-MRSA activity with MIC values of 50, 50, 12.5, 12.5, 12.5, and 25μg/mL, respectively. Compounds **270–273** displayed antibacterial activity against RN4220 (MICs of 50 μg/mL each). Compounds **272**, **273** showed inhibition against NewmanWT (MICs of 50 μg/mL each). Compound **266** displayed antibacterial activity against NewmanWT with a MIC value of 50 μg/mL each. [126].

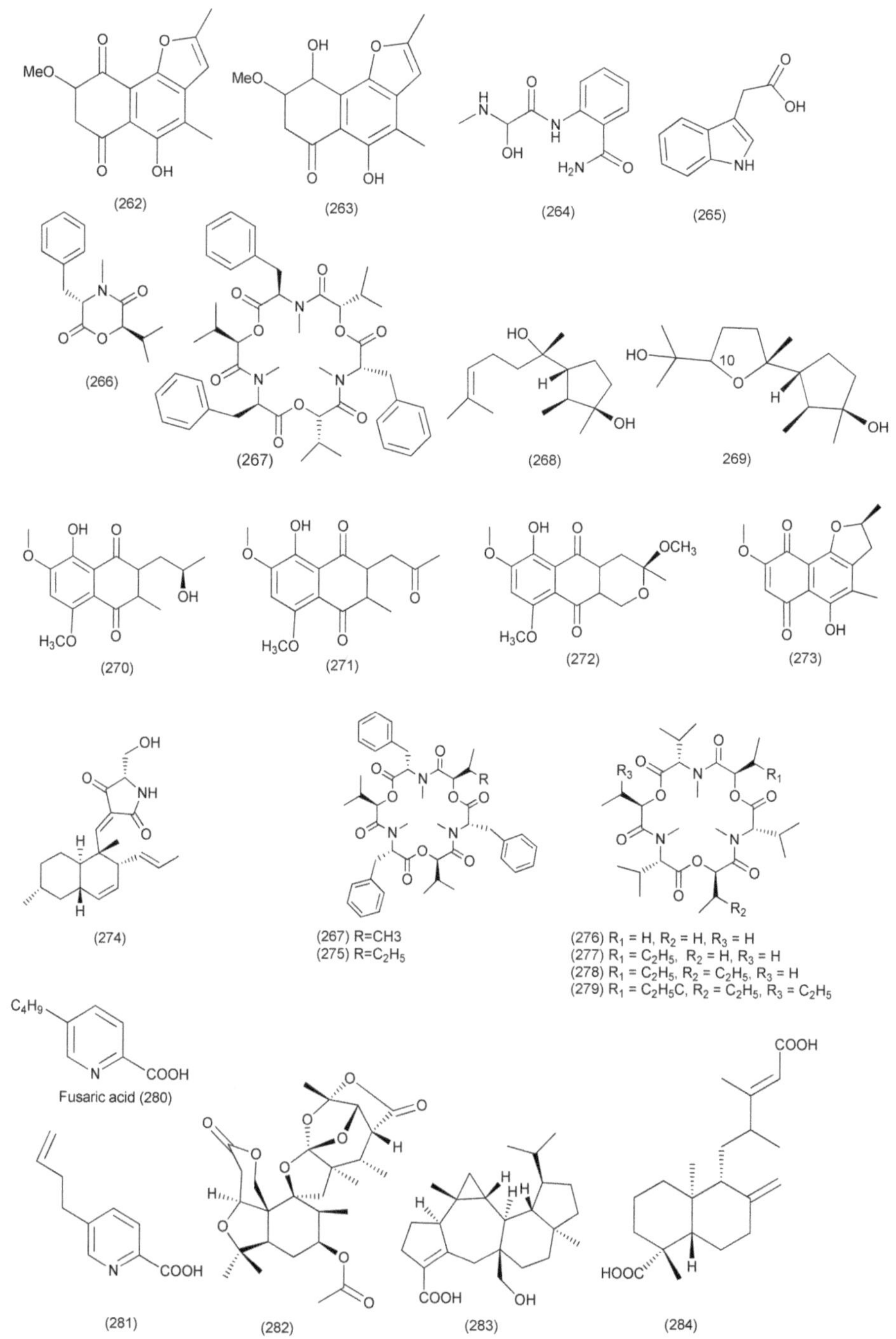

Figure 15. Structures of metabolites **262–284** isolated from Anamorphic Ascomycetes.

Fusarium sp. TP-G1 an endophyte of *Dendrobium officinable*, was the source of the compounds trichosetin (**274**), beauvericin A (**275**), enniatin B (**276**), enniatin H (**277**), enniatin I (**278**), enniatin MK1688 (**279**), fusaric acid (**280**) and dehydrofusaric acid (**281**, Figure 15) and beauvericin (**267**). Compounds **267**, **274**, **275**, **277–279** displayed antibacterial activity against *S. aureus* and MRSA with IC_{50} values in the range of 2–32 µg/mL. Compounds **280**, **281** displayed antimicrobial activity against *Acinetobacter baumannii* with a MIC value of 64 µg/mL and 128 µg/mL, respectively. Compound **276** inhibited *S. aureus* and MRSA with IC_{50} value of 128 µg/mL each [127].

A new spiromeroterpenoid, namely fusariumin A (**282**), together with the previously reported terpenoids asperterpenoid A (**283**) and agathic acid (**284**, Figure 15), were purified from *Fusarium* sp. YD-2 associated with the plant *Santalum album*. Compound **282** showed antibacterial activity against pathogenic *S. aureus* and *P. aeruginosa* (MIC of 6.3 µg/mL), and compound **283** showed average activity against pathogenic *Salmonella enteritidis* and *Micrococcus luteus* (MICs of 25.2 and 6.3 µg/mL). Compound **284** showed moderate activities against *B. cereus* and *M. luteus*, with MIC values of and 12.5 and 25.4 µg/mL, respectively [128].

A new aminobenzamide derivative, namly fusarithioamide B (**285**, Figure 16), was separated from *Fusarium chlamydosporium* an endophyte of *Anvillea garcinii* and exhibited antibacterial activity against *E. coli*, *B. cereus*, and *S. aureus* (MIC values of 3.7, 2.5 and 3.1 µg/mL) [129].

The compounds 3,6,9-trihydroxy-7-methoxy4,4-dimethyl-3,4-dihydro-1*H*-benzo[g] isochromene-5,10-dione (**286**), fusarubin (**287**), 3-O-methylfusarubin (**288**) and javanicin (**289**, Figure 16) were extracted from *Fusarium solani* A2 residing inside the plant *Glycyrrhiza glabra*. Compounds **286–289** showed inhibition of *B. subtilis*, *B. cereus*, *E. coli*, *S. aureus*, *K. pneumonia*, *S. pyogenes*, and *Micrococcus luteus* (MICs in the range of < 1 to 256 µg/mL). Fusarubin (**287**) showed good activity against *M. tuberculosis* strain H37Rv with a MIC value of 8 µg/mL, whereas compounds **286**, **288**, **289** exhibited moderate activity with MIC values of 256, 64, 32 µg/mL, respectively [130].

A new benzamide derivative, fusarithioamide A (**290**, Figure 16) was characterized from *Fusarium chlamydosporium*, an endophyte of *Anvillea garcinii*. Compound **290** had antibacterial potential towards *B. cereus*, *S. aureus*, and *E. coli* with MIC values of 3.1, 4.4, and 6.9 µg/mL, respectively [131].

The polyketide javanicin (**289**, Figure 16) was purified from *Fusarium* sp. associated with *Rhoeo spathacea*, and displayed activity against *M. tuberculosis* with a MIC value of 25 µg/mL and *M. phlei* with a MIC value of 50 µg/mL [132].

Helvolic acid methyl ester (**291**, Figure 16), a new helvolic acid derivative, together with previously reported hydrohelvolic acid (**292**, Figure 16), and helvolic acid (**100**) were isolated from a *Fusarium* sp. residing inside the plant *Ficus carica*. Compound **291** was found to be active against *B. subtilis*, *S. aureus*, *E. coli* and *P. aeruginosa* (MIC between 3.13 to 12.5, µg/mL). Compound **100** displayed activity against *B. subtilis*, *S. aureus*, *E. coli* and *P. aeruginosa* (MICs between 3.13 to 6.25 µg/mL). Compound **292** displayed activity against *B. subtilis*, *S. aureus*, *E. coli* and *P. aeruginosa* with MIC values between 3.13 to 12.5 µg/mL [133].

The compounds colletorin B (**293**) and 4,5-dihydroascochlorin (**294**, Figure 16) were purified from an endophytic *Fusarium* sp. fungus. Compounds **293** and **294** exhibited potent antibacterial activity towards *B. megaterium*, with 5 and 10 mm zones of inhibition at a concentration of 10 µg/mL [134].

The tetramic acid derivative equisetin (**295**, Figure 16) was isolated from a *Fusarium* sp. associated with *Opuntia dillenii*, and displayed antibacterial activity against *B. subtilis* with a MIC value of 8 and MICs of 16 µg/mL against *S. aureus* and MRSA [135].

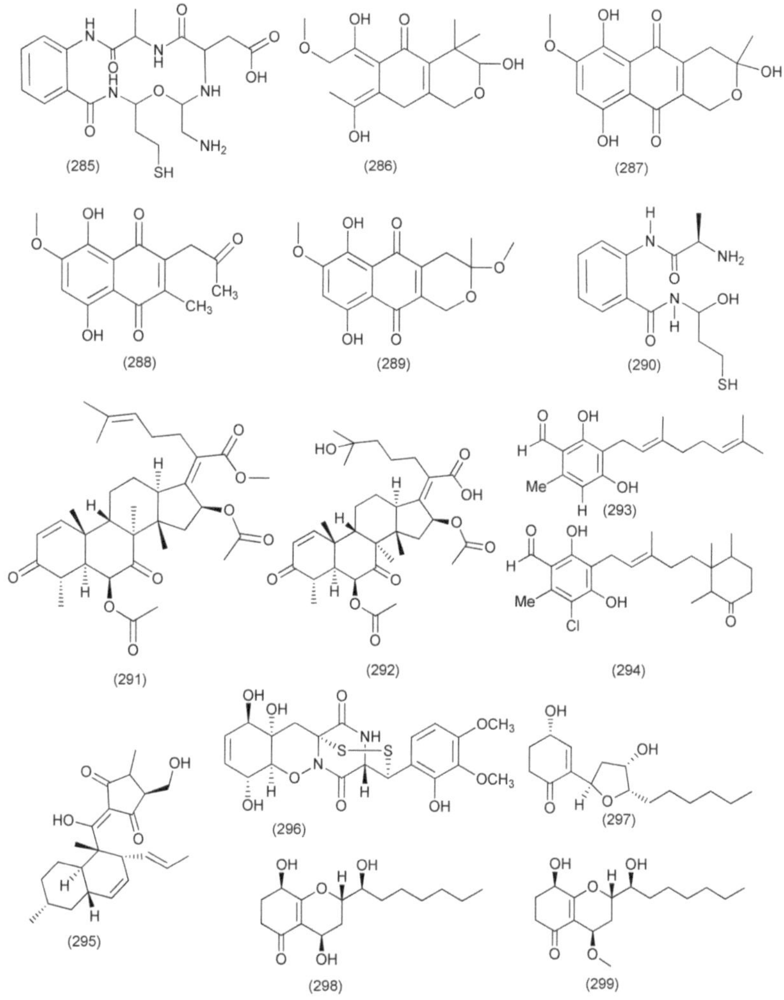

Figure 16. Structures of metabolites **285–299** isolated from Anamorphic Ascomycetes.

2.2.4. *Trichoderma*

Pretrichodermamide A (**296**, Figure 16), a known compound, was isolated from *Trichoderma harzianum*, an endophyte of *Zingiber officinale* and displayed antimycobacterial activity towards *M. tuberculosis* with a MIC value of 25 μg/mL (50 μM) [136].

A new compound named koninginin W (**297**) and four known polyketides, namely koninginin D (**298**), 7-O-methylkoninginin D (**299**, Figure 16), koninginin T (**300**) and koninginin A (**301**, Figure 17) were isolated from the endophytic fungus *Trichoderma koningiopsis* YIM PH30002 of *Panax notoginseng*. Compounds **297**, **298**, **301**, showed the weak activity against *B. subtilis* with MICs of 128 μg/mL. Compounds **297** and **299**, showed weak activity against *S. typhimurium*, with MIC values of 64 and 128 μg/mL; Compounds **297** and **300**, showed the weak activity against *E. coli* with MICs of 128 μg/mL. [137].

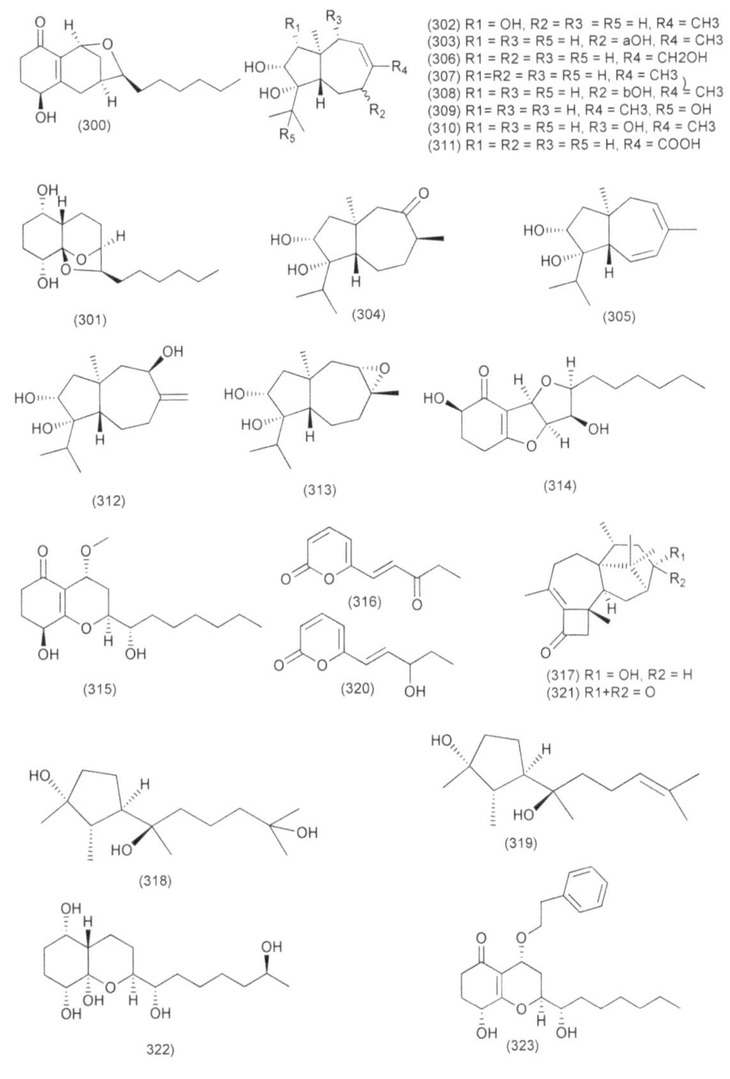

Figure 17. Structures of metabolites **300–323** isolated from Anamorphic Ascomycetes.

Five new carotane sesquiterpenes, trichocarotins I–M (**302–306**), which have diverse substitution patterns, and seven known related analogues including CAF-603 (**307**), 7β-hydroxy CAF-603 (**308**), trichocarotins E–H (**309–312**), and trichocarane A (**313**, Figure 17) were purified from *Trichoderma virens* QA-8, an endophytic fungus associated with the inner root tissue of *Artemisia argyi*. Compounds **302–313** displayed antibacterial activity against *E. coli* EMBLC-1, with MIC values ranging from 0.5 to 32 μg/mL, while 7β-hydroxy CAF-603 (**308**) displayed potent activity against *Micrococcus luteus* QDIO-3 (MIC = 0.5 μg/mL) [138].

Three new polyketides, trichodermaketone E (**314**), 4-*epi*-7-O-methylkoninginin D (**315**), and trichopyranone A (**316**), two new terpenoids, 3-hydroxyharziandione (**317**) and 10,11-dihydro-11-hydroxycyclonerodiol (**318**), together with three related known congeners, cyclonerodiol (**319**), 6-(3-hydroxypent-1-en-1-yl)-2*H*-pyran-2-one (**320**), and harziandione (**321**, Figure 17) were isolated from the endophytic fungus *Trichoderma koningiopsis* QA-3 associated with the plant *Artemisia argyi*. Compounds **314**, **316–318**, **321** displayed potent activities against *E. coli*, with MIC values ranging from 0.5 to 64 µg/mL, while compounds **316–321** showed inhibitory activities against *M. luteus* with MIC values ranging from 1 to 16 µg/mL, compounds **314**, **315**, **317–321**, showed inhibitory activities against *P. aeruginosa* with MIC values ranging from 4 to 16 µg/mL, and compounds **314**, **318–321** showed activities against *V. parahaemolyticus* with MIC values ranging from 4 to 16 µg/mL. Among the compounds tested, compound **317** showed the strongest activity against *E. coli*, with a MIC value of 0.5 µg/mL and compound **320** showed the strongest activity against *M. luteus*, with a MIC value of 1 µg/mL, comparable to that of the positive control chloramphenicol [139].

New highly oxygenated polyketides, 15-hydroxy-1,4,5,6-tetra-*epi*-koninginin G (**322**), koninginin U (**323**, Figure 17) and 14-ketokoninginin B (**324**, Figure 18), were isolated from *Trichoderma koningiopsis* QA-3, isolated from *Artemisia argyi*. Compound **322** displayed good activity against the aquatic pathogen *Vibrio alginolyticus*, with a MIC value of 1 µg/mL. Compounds **323**, **324** exhibited activity against aquatic bacteria *Vibrio harveyi* and *Edwardsiella tarda* with MICs of 4 and 2 µg/mL, respectively [140].

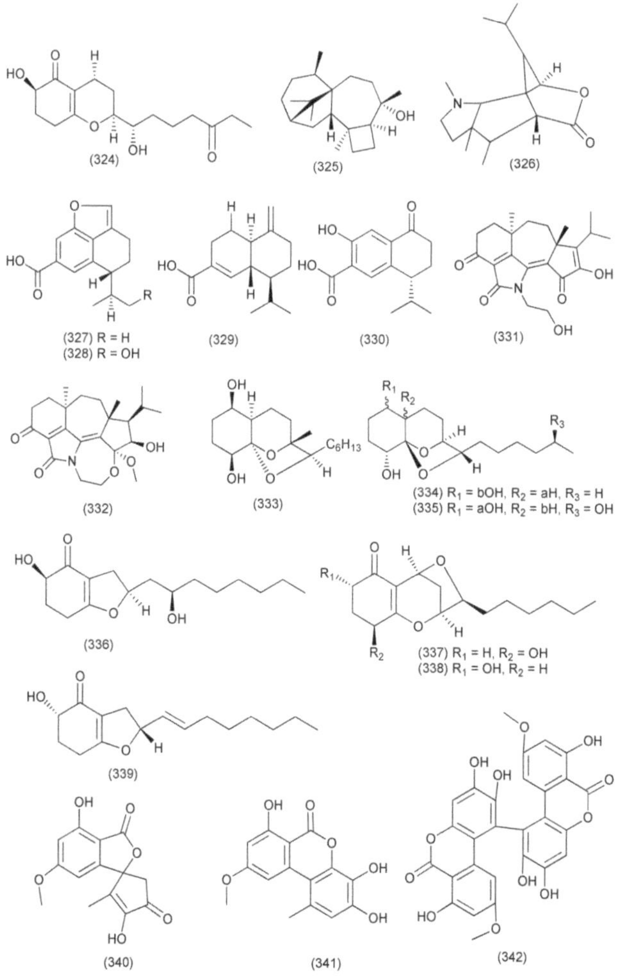

Figure 18. Structures of metabolites **324–342** isolated from Anamorphic Ascomycetes.

A new harziane diterpenoid with a 4/7/5/6 tetracyclic scaffold, harzianol I (**325**, Figure 18) was isolated from *Trichoderma atroviride* B7, an endophyte associated with the plant *Colquhounia coccinea* var. *mollis*. Compound **325** exhibited potent inhibitory activity against *S. aureus*, *B. subtilis*, and *M. luteus*, with EC_{50} values of 7.7, 7.7, and 9.9 µg/mL, respectively [141].

The compound dendrobine (**326**, Figure 18) was purified from *Trichoderma longibrachiatum* MD33, an endophyte of *Dendrobium nobile*. Compound **326** inhibited *Bacillus mycoides*, *B. subtilis*, and *Staphylococcus* spp., with zones of inhibition of 9, 12 and 8 mm, respectively [142].

Trichocadinins B-D and G (**327–330**, Figure 18), new cadinane-type sesquiterpene derivatives, were isolated from *Trichoderma virens* QA-8 residing inside the plant *Artemisia argyi*. Compounds **327–330** displayed antibacterial activity against *E. coli*, *Aeromonas hydrophilia* QDIO-1, *Edwardsiella tarda*, *E. ictarda*, *Micrococcus luteus*, *P. aeruginosa*, *Vibrio alginolyticus*, *V. anguillarum*, *V. harveyi*, *V. parahemolyticus*, and *V. vulnificus* (MICs in the range of 8–64 µg/mL). Compound **330** inhibited *Ed. tarda* and *V. anguillarum* with MIC values of 1 and 2 µg/mL, respectively [143].

New diterpenes koninginols A (**331**) and B (**332**, Figure 18) were isolated from *Trichoderma koningiopsis* A729, an endophyte of *Morinda officinalis*. Compounds **331–332** exhibited potent inhibition against *B. subtilis*, with MIC values of 10 and 2 µg/mL, respectively [144].

Trichoderma koningiopsis QA-3, isolated from the plant *Artemisia argyi*, produced five new polyketides: ent-koninginin A (**333**), 1,6-di-*epi*-koninginin A (**334**), 15-hydroxykoninginin A (**335**), 10-deacetylkoningiopisin D (**336**) and koninginin T (**337**) and two known analogs, koninginin L (**338**), trichoketide A (**339**, Figure 18). Compounds **333** and **339** inhibited the aquatic bacteria *E. tarda*, *V. anguillarum*, and *V. parahemolyticus*, and the human pathogen *E. coli* (MICs ranging from 8 to 64 µg/mL). Compound **333** also showed activity against the aquatic bacteria *M. luteus* and *P. aeruginosa* and agropathogens. Compounds **333–339** were found to be active against *E. coli* (each with MIC values of 64 µg/mL) and *E. tarda*, *V. alginolyticus*, and *V. anguillarum* (MICs ranging from 8 to 64 µg/mL) while compounds **333** and **339** also showed antimicrobial activity against *M luteus*, *V. parahemolyticus*, and *V. vulnificus* (MIC values ranging from 4 to 64 µg/mL). Compound **333** was also found active against *V. vulnificus* with a MIC of 4 µg/mL [145].

2.2.5. Alternaria

A novel polyketide derivative, isotalaroflavone (**340**), along with the known compounds 4-hydroxyalternariol-9-methyl ether (**341**) and verrulactone A (**342**, Figure 18) were obtained from *Alternaria alternata* ZHJG5 that was isolated from the leaves of *Cercis chinensis* collected from Nanjing Botanical Garden (Nanjing, China). Compounds **340–342** were found to be active against *Xanthomonas oryzae* pv. *oryzae* (Xoo), *Xanthomonas oryzae* pv. *oryzicola* (Xoc) and *Ralstonia solanacearum* (Rs) with MICs ranging from 0.5 to 64 µg/mL. In addition, compound **340** showed a potent protective effect against rice bacterial leaf blight caused by Xoo with a protective efficacy of 75.1% at a concentration of 200 µg/mL [146].

A new biphenyl compound altertoxin VII (**343**), and the related compounds altenuisol (**344**, Figure 19), alternariol (**44**), were purified from *Alternaria* sp. PfuH1 is associated with *Pogostemon cablin*. Compounds **44**, **343**, **344** showed activity against *S. agalactiae* with MIC values of 9.3, 17.3, and 85.3, µg/mL, respectively, and compound **343** also showed poor activity against *E. coli* with MIC value of 128 µg/mL [147].

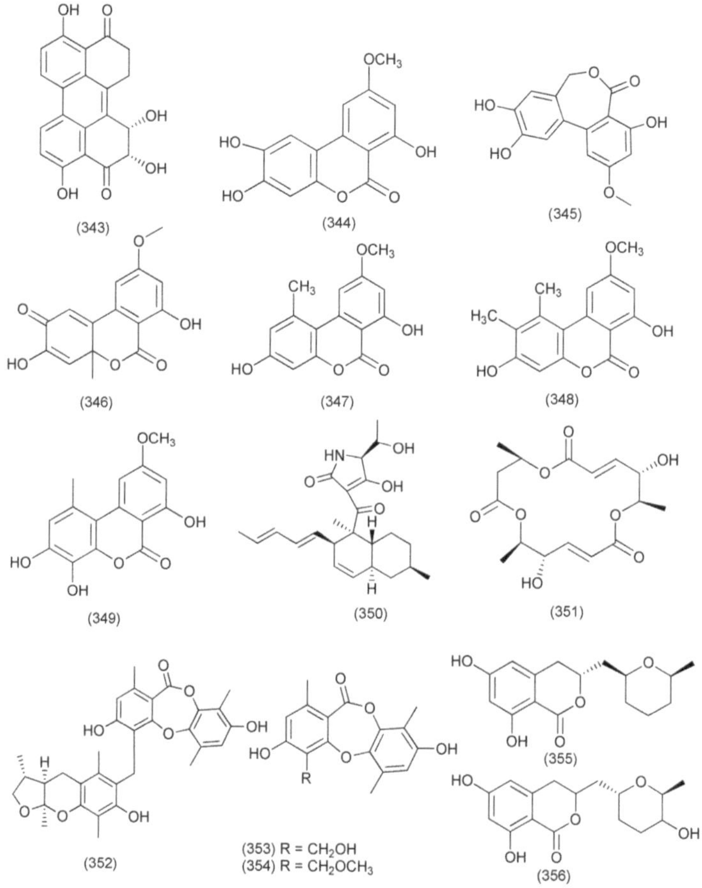

Figure 19. Structures of metabolites **343–356** isolated from Anamorphic Ascomycetes.

Known metabolites altenuisol (**344**), alterlactone (**345**), and dehydroaltenusin (**346**, Figure 19) and alternariol (**44**), were isolated from *Alternaria alternata* ZHJG5 residing inside the leaves of *Cercis chinensis*. The compounds **44**, **344**, **345**, **346**, showed inhibitory activities on FabH of *X. oryzae* pv. *oryzae* (Xoo) with IC$_{50}$ values ranging from 29.5 to 74.1 µM and also displayed a varying degree of antibacterial activities against *X. oryzae* pv. oryzae (Xoo) with MIC values ranging from 4 to 64 µg/mL. Molecular modeling was then used to picture how these compounds interact with XooFabH. Compounds **44**, and **343**, displayed significant bactericidal activity against rice bacterial leaf blight with a protective efficiency of 66.2 and 82.5% at concentration of 200 µg/mL, respectively [148].

The compound alternariol 9-Me ether (**347**, Figure 19) was purified from *Alternaria alternata* MGTMMP031 associated with *Vitex negundo*. Compound **347** exhibited potential activity against *B. cereus*, *Klebsiella pneumoniae* with a MIC at 30 µM/L. The compound inhibited the growth of *E. coli*, *Salmonella typhi*, *Proteus mirabilis*, *S. aureus* and *S. epidermidis* at a MIC of 35 µM/L [149].

An endophytic fungus, *Alternaria alternata*, associated with *Grewia asiatica* yielded a new structural isomer of alternariol, i.e., 3,7-dihydroxy-9-methoxy-2-methyl-6*H*-benzo[c]-chromen-6-one (**348**, Figure 19), along with alternariol (**44**). Compound **44** inhibited *S. aureus*, VRE, and MRSA with MIC values of 32, 32 and 8 µg/mL, respectively. Compound

348 also inhibited *S. aureus*, VRE, and MRSA with MIC values of 128, 128, and 64 µg/mL, respectively [150].

The compounds 4-hydroxyalternariol-9-methyl ether (**349**, Figure 19) altenuisol (**344**), and alternariol (**44**) were purified from *Alternaria* sp. Samif01, an endophytic fungus of *Salvia miltiorrhiza*. Compounds **44**, **344**, and **349** showed inhibition against *A. tumefaciens*, *B. subtilis*, *P. lachrymans*, *R. solanacearum*, *Staphylococcus hemolyticus* and *Xanthomonas vesicatorya* with MIC values in the range of 86.7–364.7 µM [151]. Previously alternariol 9-Me ether (**347**, Figure 19) was isolated the same fungus and was found active against *B. subtilis*, *S. haemolyticus*, *A. tumefaciens*, *P. lachrymans*, *R. solanacearum*, and *X. vesicatoria* with IC$_{50}$ values ranging from 16.00 to 38.27 g/mL [152].

An endophytic fungus *Alternaria* sp. and *Pyrenochaeta* sp., purified from *Hydrastis canadensis* yielded altersetin (**350**) and macrosphelide A (**351**, Figure 19). Compounds **350** and **351** displayed antibacterial activity against *S. aureus* with MIC values of 0.23 and 75 µg/mL, respectively [153].

2.2.6. Simplicillium

The fungal strain *Simplicillium lanosoniveum* associated with *Hevea brasiliensis*, yielded a new depsidone, simplicildone K (**352**), together with the known compounds botryorhodine C (**353**), and simplicildone A (**354**, Figure 19). Compounds **353** and **354** displayed activity against *S. aureus*, MRSA with equal MIC values of 32 µg/mL, whereas **352** exhibited 4-fold less activity against both strains (MIC values of 128 µg/mL) [154].

The compounds botryorhodine C (**353**), and simplicildone A (**354**, Figure 19), were purified from *Simplicillium* sp. PSU-H41 which is associated with the leaves of *Hevea brasiliensis*. Compounds **353** and **354** exhibited poor activity against *S. aureus* (MIC of 32 µg/mL each). Compound **353** was found to be active against MRSA with the same MIC value [155].

2.2.7. Cladosporium

An endophytic fungus, *Cladosporium cladosporioides*, residing inside the leaves of *Zygophyllum mandavillei* yielded isocladosporin (**355**), 5′-hydroxyasperentin (**356**, Figure 19), 1-acetyl-17-methoxyaspidospermidin-20-ol (**357**), and 3-phenylpropionic acid (**358**, Figure 20). Compounds **355–358** displayed antibacterial activity against *X. oryzae* and *Pseudomonas syringae* with MIC values in the range of 7.81 to 125 µg/mL [156].

A new hybrid polyketide, named cladosin L (**359**, Figure 20) was discovered in the endophytic fungus *Cladosporium sphaerospermum* WBS017 associated with the bulbs of *Fritillaria unibracteata* var. *wabuensis*. Compound **359** inhibited *S. aureus* ATCC 29213 and *S. aureus* ATCC 700699 with MICs of 50 and 25 mM, respectively [157].

A naphthoquinone Me ether of fusarubin (**360**, Figure 20), was purified from a *Cladosporium* sp. associated with the *Rauwolfia serpentina*. Compound **360** (40 µg/disk) displayed potent activity against *S. aureus*, *E. coli*, *P. aeruginosa* and *B. megaterium* with 27, 25, 24 and 22 mm zones of inhibition, respectively and the activities were compared with kanamycin (30 µg/disk) [158].

2.2.8. Pestalotiopsis

The genus *Pestalotiopsis* is reported as an endophyte from rain forests in almost all parts of the world and is a prolific producer of chemically diverse bioactive compounds. One such compound is the new drimane sesquiterpenoid 11-dehydro-3a-hydroxyisodrimeninol (**361**, Figure 20), produced by *Pestalotiopsis* sp. M-23, an endophytic fungus of *Leucosceptrum canum*. Compound **361** displayed poor inhibitory effect against *B. subtilis* with IC$_{50}$ value of 280.27 µM [159].

The compounds (1*S*,3*R*)-austrocortirubin (**362**), (1*S*,3*S*)-austrocortirubin (**363**), and 1-deoxyaustrocortirubin (**364**, Figure 20), were obtained from *Pestalotiopsis* sp., an endophyte of *Melaleuca quinquenervia*. Compounds **362–364** displayed with poor antibacterial activity (100 µM) against Gram-positive isolates [160].

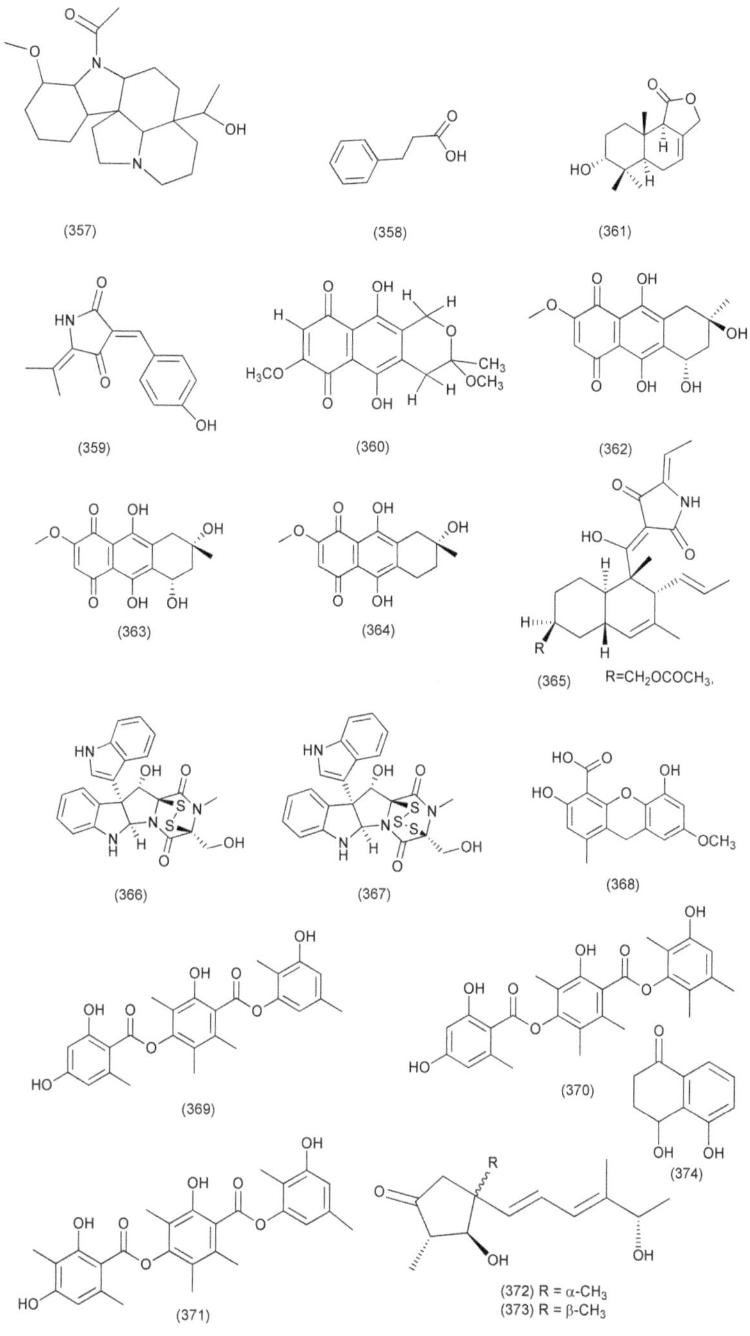

Figure 20. Structures of metabolites **357–374** isolated from Anamorphic Ascomycetes.

A new tetramic acid analog, neopestalotin B (**365**, Figure 20), was extracted from *Neopestalotiopsis* sp. and inhibited *B. subtilis*, *S. aureus*, *S. pneumoniae*, with MIC values of 10, 20, and 20 µg/mL, respectively [161].

2.2.9. Phoma

Two known thiodiketopiperazine derivatives **366** and **367** (Figure 20) were purified from *Phoma cucurbitacearum* (now known as *Stagonosporopsis cucurbitacearum*), an endophyte of *Glycyrrhiza glabra*. Compounds **366** and **367** were found to inhibit the battery of bacterial pathogens, including *S. aureus* and *Streptococcus pyogenes* with IC_{50} values of <10 µM. Both compounds potentially inhibited biofilm formation in *S. aureus* and *S. pyogenes* and acted synergistically with streptomycin and inhibited transcription/translation. It was also observed that the sea gene was overexpressed by several fold on treatment with compound **366** while its expression was not affected significantly with compound **367**. The expression of agrA gene was also not affected significantly in *S. aureus* with the treatment of either of the compounds [162].

Barceloneic acid C (**368**, Figure 20), purified from a *Phoma* sp. JS752 residing inside *Phragmites communis*. Compound (**368**) exhibited average antibacterial activities against *Listeria monocytogenes* and *Staphylococcus pseudintermedius*, (MIC of 1.02 µg/mL each) [163].

The polyketides thielavins T (**369**), U (**370**), and V (**371**, Figure 20) were purified from *Setophoma* sp., an endophytic fungus of *Psidium guajava*. Compounds **369–371** displayed antibacterial activity against pathogenic *S. aureus* with MIC values of 6.25, 50, and 25 µg/mL, respectively [164].

2.2.10. Colletotrichum

Two new γ-butyrolactone derives., colletolides A and B (**372**, **373**), together with the already reported compounds sclerone (**374**, Figure 20), and 3-methyleneisoindolinon (**375**, Figure 21) were purified from *Colletotrichum gloeosporioides* B12, an endophyte of plant *Illigera rhodantha*. Compounds **372**, **373**, **375** were found to be active against *Xanthomonas oryzae* pv. *oryzae*, with the same MIC values of 128 µg/mL, while compound **374** was found active against *X. oryzae* pv. *oryzae* with MIC values of 64 µg/mL [165].

The new compounds colletotrichones A (**376**), B (**377**), and C (**378**, Figure 21) were purified from *Colletotrichum* sp. BS4 residing inside the leaves of *Buxus sinica*. Compound **376** inhibits *E. coli* and *B. subtilis* with MIC values 1.0 and 0.1 µg/mL, respectively. Compound **377** inhibited *S. aureus* with a MIC value of 5.0 µg/mL. Compound **378** has shown antibacterial activity against *E. coli* with a MIC value of 5.0 µg/mL [166].

2.2.11. Minor Taxa of Anamorphic Ascomycetes

New dibenzo-α-pyrones, rhizopycnolide A (**379**), rhizopcnin C (**380**) and rhizopycnin D (**381**), together with known congeners TMC-264 (**382**), palmariol B (**383**) penicilliumolide D (**384**, Figure 21) alternariol 9-methyl ether (**347**) and alternariol (**44**) and were purified from *Rhizopycnis vagum* (now known as *Acrocalymma vagum*) isolated from *Nicotiana tabacum*. Compounds **380**, **384**, **44** inhibited *A. tumefaciens*, *B. subtilis*, *Pseudomonas lachrymans*, *R. solanacearum*, *Staphylococcus hemolyticus*, and *Xanthomonas vesicatoria*, with MICs in the 25−100 µg/mL range. Rhizopycnolide A (**379**) was active against *A. tumefaciens*, *B. subtilis*, and *P. lachrymans*, with MIC values of 100, 75, and 100 µg/mL, respectively. Rhizopycnin D (**381**) was found to be active against *A. tumefaciens*, *B. subtilis*, and *R. solanacearum*, with an equal MIC value of 50 µg/mL, and against *X. vesicatoria*, with a MIC value of 75 µg/mL. TMC-264 (**382**) was selectively active against *B. subtilis* (MIC value of 50 µg/mL). Compounds **383** and **347** inhibited *A. tumefaciens*, *B. subtilis*, *P. lachrymans*, *R. solanacearum*, and *X. vesicatoria*, with IC_{50} values in the range 16.7−34.3 µg/mL [167].

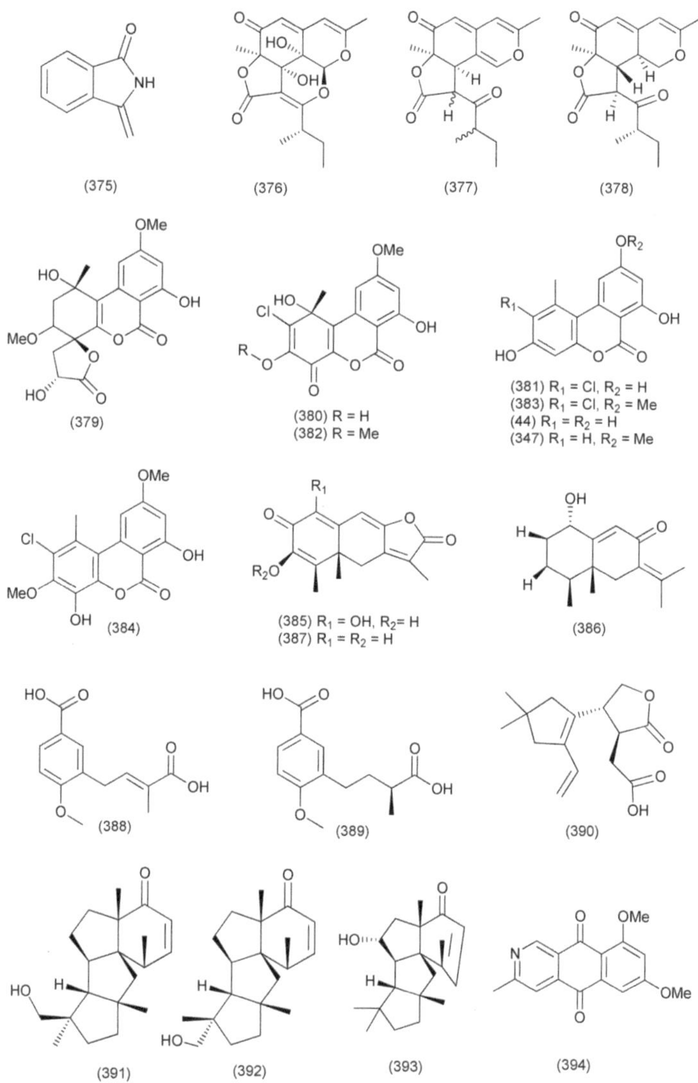

Figure 21. Structures of metabolites **375–378** isolated from Anamorphic Ascomycetes and **379–394** from Minor Anamorphic Ascomycetes.

Rhizoperemophilane K (**385**), 1α-hydroxyhydroisofukinon (**386**) and 2-oxo-3-hydroxy-eremophila-1(10),3,7(11),8-tetraen-8,12-olide (**387**, Figure 21) were purified from *Rhizopycnis vagum* (now known as *Acrocalymma vagum*), an endophyte of *Nicotiana tabacum*. Compounds **385**, **386** and **387** displayed inhibition against *A. tumefaciens*, *B. subtilis*, *P. lachrymans*, *Ralstonia solanacearum*, *S. haemolyticus*, and *X. vesicatoria*, with MIC values in the range of 32~128 μg/mL [168].

Rhizopycnis acids A (**388**) and B (**389**, Figure 21), were purified from *Rhizopycnis vagum* (now known as *Acrocalymma vagum*) an endophyte of *Nicotiana tabacum* from China Agricultural University (Beijing, China). Compound **388** inhibited *A. tumefaciens*, *B. subtilis*, *P. lachrymans*, *R. solanacearum*, *S. hemolyticus* and *X. vesicatoria* with MIC values of 20.82,

16.11, 23.48, 29.46, 21.11, and 24.31 µg/mL, respectively. Compound **389** also inhibited *A. tumefaciens, B. subtilis, P. lachrymans, R. solanacearum, S. haemolyticus,* and *X. vesicatoria* with MIC values of 70.89, 81.28, 21.23, 43.40, 67.61, and 34.86 µg/mL, respectively [169].

Leptosphaeria sp. XL026 associated with *Panax notoginseng* yielded a new sesquiterpenoids, leptosphin B (**390**), along with three known diterpenes, conidiogenone C (**391**), conidiogenone D (**392**) and conidiogenone G (**393**, Figure 21). The site of the collection was Shijiazhuang (Hebei Province, China). Compounds **390–393** showed average antibacterial activity against *B. cereus*, with MIC values of 12.5–6.25 µg/mL and compound **392** also showed antibacterial activity against *P. aeruginosa* with a MIC value of 12.5 µg/mL [170].

Two 2-azaanthraquinones, scorpinone (**394**, Figure 21) and 5-deoxybostrycoidin (**395**, Figure 22), were purified from *Lophiostoma* sp. Eef-7 is associated with *Eucalyptus exserta*. Compounds **394** and **395** displayed poor antibacterial activity against *Ralstonia solanacearum* with 9.86 and 9.58 mm zones of inhibition when 64 µg was added (positive control was streptomycin sulfate with a 13.03 mm zone of inhibition at an added amount of 6.25 µg) [171].

Two new cytochalasan alkaloids, cytochrysins A and C (**396** and **397**, Figure 22), were isolated from *Cytospora chrysosperma*, an endophytic fungus isolated from *Hippophae rhamnoides*. Compound **396** showed significant antibacterial activity against multi-drug resistant *Enterococcus faecium* with MIC value of 25 µg/mL, and compound **397** was active against MRSA with a MIC value of 25 µg/mL [172].

Two known α-pyridones, (8*R*,9*S*)-dihydroisoflavipucine (**398**) and (8*S*,9*S*)-dihydroi soflavipucine (**399**, Figure 22) were isolated from *Lophiostoma* sp. Sigrf10 is associated with *Siraitia grosvenorii*. Compounds **398** and **399** were active against *B. subtilis, A. tumefaciens, R. solanacearum,* and *X. vesicatoria*, with IC$_{50}$ values in the range of 35.68–44.85 µM [173].

Microsphaerol (**400**), a novel polychlorinated triphenyl diether was extracted from *Microsphaeropsis* sp and seimatorone (**401**, Figure 22), a new naphthalene derivative, was purified from the endophyte *Seimatosporium* sp. Compound **400** displayed potent antibacterial activity against *B. megaterium* and *E. coli*, with 8 and 9 mm zones of inhibition at 0.05 mg concentration (50 mL of 1 mg/mL). Compound **401** exhibited moderate antibacterial activity against *B. megaterium* and *E. coli*, with 3 and 7 (partial inhibition) mm zones of inhibition at a 0.05 mg concentration (50 mL of 1 mg/mL) [174].

Known compounds epicocconigrone A (**402**), epipyrone A (**403**), and epicoccolide B (**404**, Figure 22) were purified from *Epicoccum nigrum* MK214079 associated with *Salix* sp. Compounds **402–404** exhibited moderate activity against *S. aureus*, with MICs ranging from 25 to 50 µM [175].

The known compounds *p*-hydroxybenzaldehyde (**223**), indole-3-carboxylic acid (**405**) and quinizarin (**406**, Figure 22) and beauvericin (**267**), were isolated from *Epicoccum nigrum* associated with the *Entada abyssinica*. Compound **267** displayed activity against *S. aureus, B. cereus,* and *Salmonella typhimurium*, with MIC values of 3.12, 12.5, and 12.5 µg/mL. Compounnd (**223**) displayed activity against *S. aureus, B. cereus, P. aeruginosa,* and *E. coli* with MIC values of 50, 25, 50, and 25 µg/ml. Compound **405** was found to be active against *S. aureus* and *E. faecalis* (MICs of 6.25 and 50 µg/mL) while compound **406** displayed activity against *S. aureus, B. cereus* St (MICs of 50 µg/mL each) [176].

The endophytic fungus *Stemphylium lycopersici* from *S. tonkinensis* yielded xylapeptide B (**407**), cytochalasin E (**408**), 6-heptanoyl-4-methoxy-2*H*-pyran2-one (**409**) and (–)-5-carboxymellein (**410**, Figure 22). Compound **407** showed average inhibition against *B. subtilis* with a MIC value of 12.5 µg/mL, and against *S. aureus and E. coli* with MIC values of 25 µg/mL. Compound **408** inhibited *B. subtilis, S. aureus, B. anthracis, S. dysenteriae,* and *E. coli* with MIC values ranging from 12.5 to 25 µg/mL. Compound **409** inhibited *S. paratyphi* B with MIC value of 12.5 µg/mL. Compound **410** inhibited *B. subtilis, S. aureus, B. anthracis, S. dysenteriae, S. paratyphi, E. coli* and *S. paratyphi* B with MIC values ranging from 12.5 to 25 µg/mL [177].

A new tetrahydroanthraquinone derivative, dihydroaltersolanol C (**411**, Figure 22) was purified from *Stemphylium globuliferum* residing inside the plant *Juncus acutus*. Compound **411** exhibited moderate growth inhibition effects against *S. aureus* with a MIC of 49.7 µM [178].

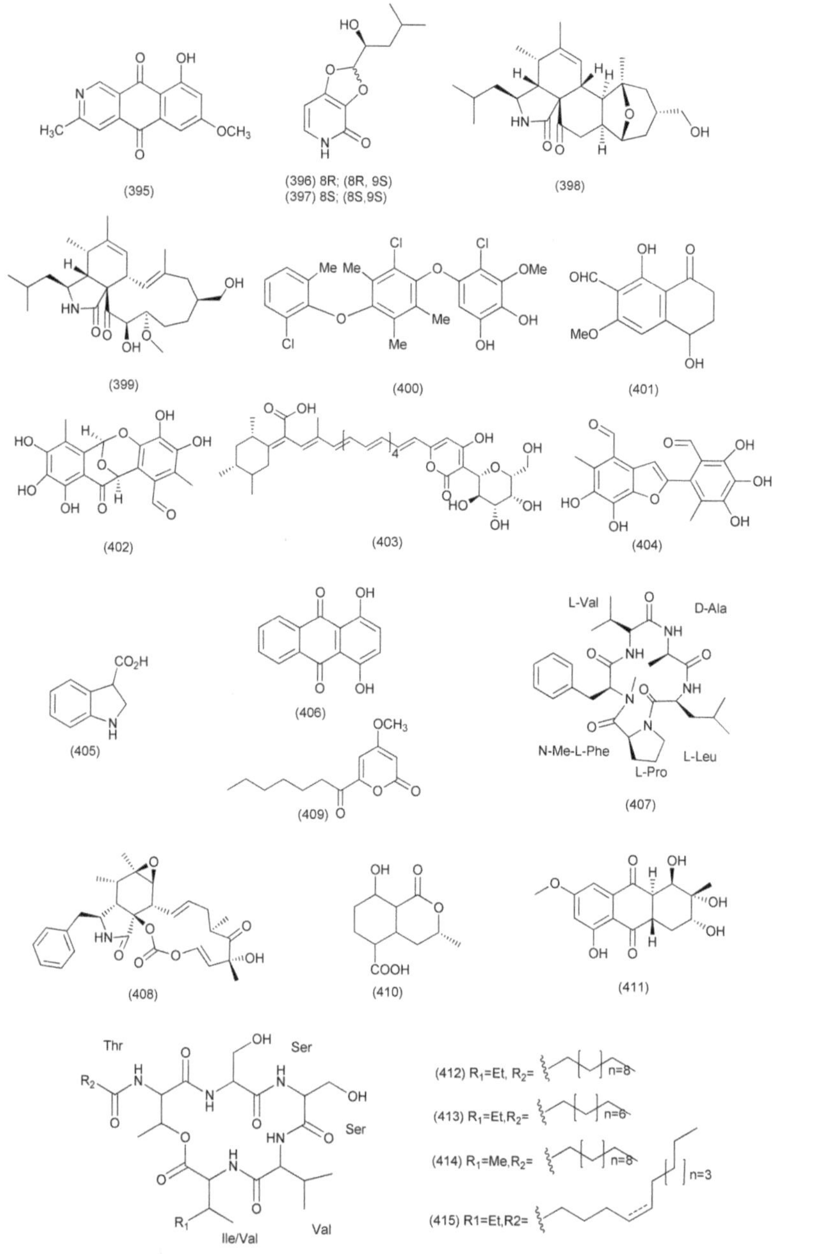

Figure 22. Structures of metabolites **395–415** isolated from Minor Anamorphic Ascomycetes.

An endophytic fungus *Lecanicillium* sp. (BSNB-SG3.7 Strain) associated with *Sandwithia guyanensis* yielded stephensiolides I (**412**), D (**413**), G (**414**), and stephensiolide F (**415**, Figure 22). Compounds **412–415** displayed anti-MRSA activity with MIC values of 4, 32, 16 and 32 µg/mL, respectively [179].

The compound phomalactone (**416**, Figure 23) was isolated from the endophyte *Nigrospora sphaerica* associated with *Adiantum philippense*. Compound **416** displayed good antibacterial activity against *E. coli* and *X. campestris* with MIC values of 3.12 µg/mL and moderate activity against *S. typhi*, *B. subtilis*, *B. cereus*, and *K. pneumonia* with a MIC value of 6.25 µg/mL. A MIC of 12.5 µg/mL was found against *S. aureus*, and *S. epidermidis* [180].

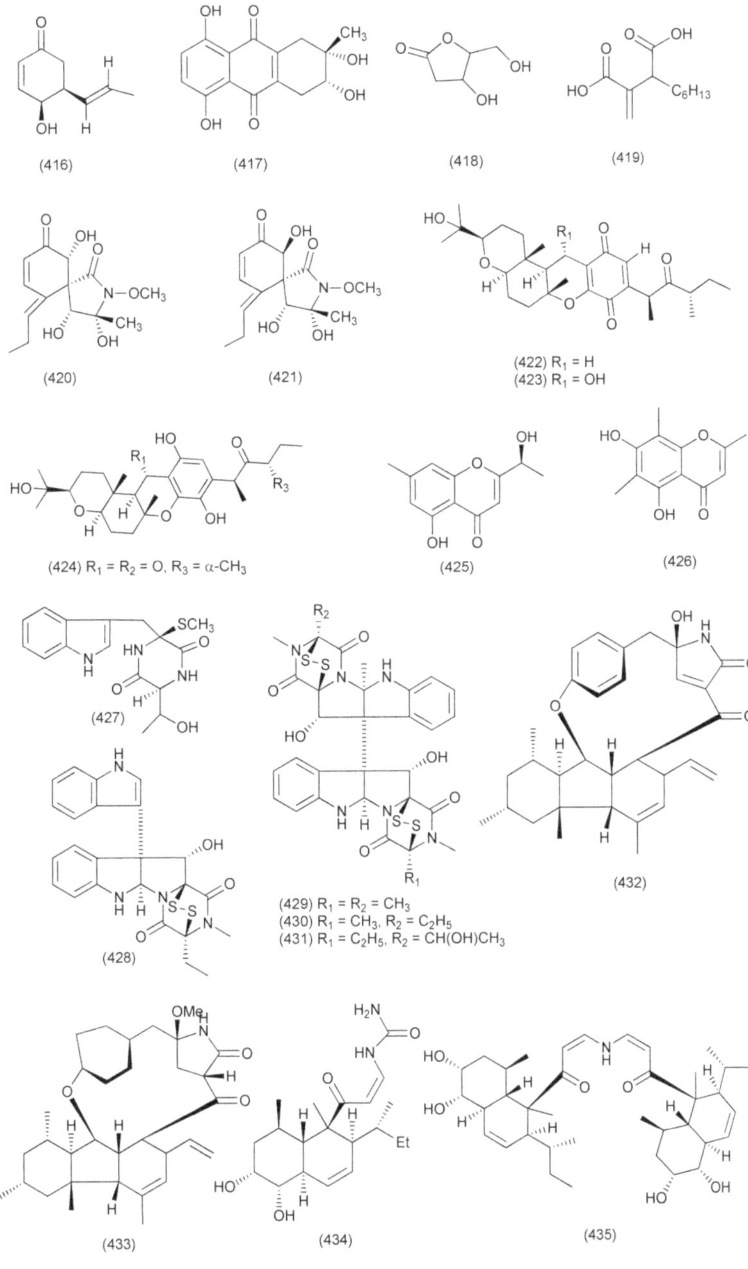

Figure 23. Structures of metabolites **416–435** isolated from Minor Anamorphic Ascomycetes.

A new naturally occurring compound, nigrosporone B (**417**, Figure 23), was purified from *Nigrospora* sp. BCC 47789 associated with the leaves of *Choerospondias axillaris*. Compound **417** exhibited antibacterial activity against *M. tuberculosis*, *B. cereus* and *E. faecium* with MIC values of 172.25, 21.53 and 10.78 μM, respectively [181].

Two bioactive compounds, 2''-deoxyribolactone (**418**) and hexylitaconic acid (**419**, Figure 23) were purified from *Curvularia sorghina* BRIP 15900 associated with the stem bark of *Rauwolfia macrophylla*. Compounds **418** and **419** inhibited *Staphylococcus warneri E. coli*, *Pseudomonas agarici* and *Micrococcus luteus*, with MICs ranging between 0.17 μg/mL and 0.58 μg/mL [182].

Known compounds, namely the triticones E (**420**) and F (**421**, Figure 23), were purified from *Curvularia lunata*, isolated from healthy capitula of *Paepalanthus chiquitensis*. Compounds **420** and **421** showed good antibacterial activity for *E. coli*, with MIC values of 62.5 μg/mL [183].

The known compounds cochlioquinones B (**422**), C (**423**), and isocochlioquinone C (**424**, Figure 23) were purified from *Bipolaris* sp. L1-2 which is associated with the leaves of *Lycium barbarum*. Compounds **422–424** showed antimicrobial activity against *B. subtilis*, *C. perfringens*, and *P. viridiflava*, with MICs of 26 μM [184].

A new previously undescribed chromone, (*S*)-5-hydroxyl-2-(1-hydroxyethyl)-7-methylchromone (**425**) and the known sativene-type sesquiterpenoid 5,7-dihydroxy-2,6,8-trimethylchromone (**426**, Figure 23), were purified from *Bipolaris eleusines* associated with potatoes from Yunnan Agricultural University (Kunming, Yunnan, China). Compounds **425** and **426** displayed poor inhibitory activities against *S. aureus* sub sp. *aureus* with the inhibition rates of 56.3 and 32 %, respectively, at the concentration of 128 μg/mL (penicillin G: 99.9% at 5 μg/mL) [185].

Two new diketopiperazines, bionectin D (**427**) and bionectin E (**428**) and the known compounds verticillin A (**429**) sch 52901 (**430**) and gliocladicillin C (**431**, Figure 23) were purified from *Bionectria* sp. Y1085, isolated from *Huperzia serrata*. Bionectin D (**427**) is a rare diketopiperazine with a single methylthio substitution at the α-carbon of a cyclized amino acid residue. Compounds **427–331** exhibited antibacterial activity against *E. coli*, *S. aureus*, and *S. typhimurium*, with MIC values ranging from 6.25–25 μg/mL [186].

Known compounds pyrrocidine A (**432**) and 19-O-methylpyrrocidine B (**433**, Figure 23) were extracted from the endophytic fungus, *Cylindrocarpon* sp., isolated from *Sapium ellipticum*. Compound **433** exhibited moderate antibacterial activity against *S. aureus* ATCC 25923 and ATCC 700699 with MIC values of 50 and 25 μM, respectively. Compound **432** showed strong to moderate inhibitory effects against *S. aureus* strain ATCC 25923 and ATCC 700699, *E. faecalis* strain ATCC 29212 and ATCC 51299, *E. faecium* strain ATCC 35667 and ATCC 700221 with MIC values ranging from 0.78 to 25 μM [187].

Two new decalin-containing compounds, eupenicinicols C (**434**), and D (**435**, Figure 23), along with two biosynthetically-related known metabolites, eujavanicol A (**436**), and eupenicinicol A (**437**, Figure 24) were obtained from *Eupenicillium* sp. LG41.9 (now considered as *Penicillium*) residing inside the roots of *Xanthium sibiricum* when treated with the HDAC inhibitor nicotinamide (15 mg/100 mL). Compound **435** exhibited pronounced efficacy against *S. aureus* with a MIC of 0.1 μg/mL, and compound **436**, was active against *E. coli* with a MIC of 5.0 μg/mL [188].

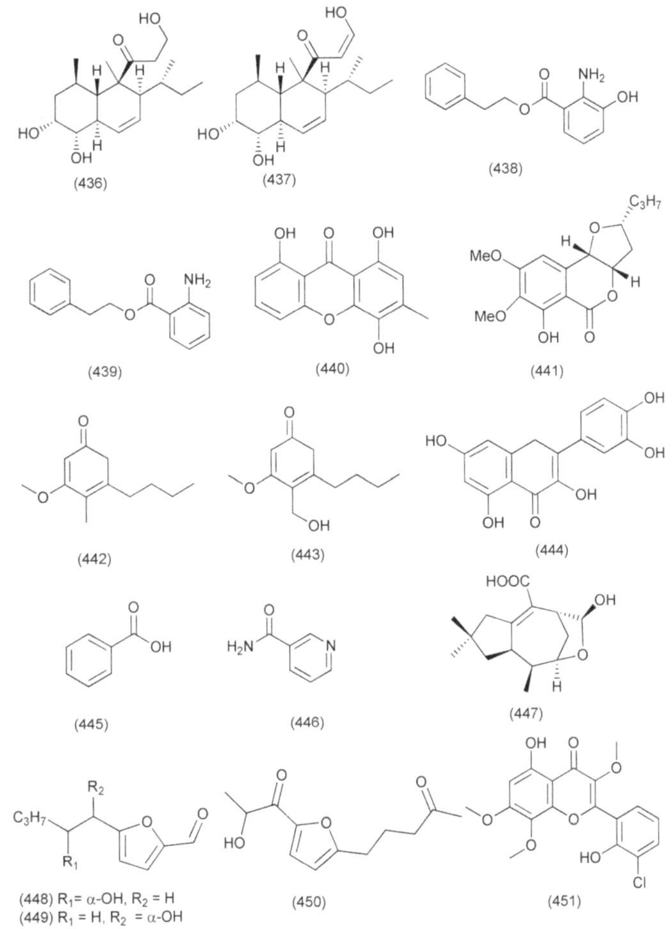

Figure 24. Structures of metabolites **436–443** isolated from Minor Anamorphic Ascomycetes, **444–450** from Basidiomycetes and **451** from Zygomycetes.

A new anthranilic acid derivative, 2-phenylethyl 3-hydroxyanthranilate (**438**) and 2-phenylethyl anthranilate (**439**, Figure 24) were extracted from *Dendrothyrium variisporum* extracted from the roots of *Globularia alypum*. Metabolite **438** was found to be active against *B. subtilis* and *M. luteus* (MICs of 8.33 and 16.66 µg/mL). Compound **439** showed potent activity against *B. subtilis* and *S. aureus* with MIC values of 66.67 µg/mL each [189].

Ravenelin (**440**, Figure 24) was extracted from *Exserohilum rostratum*, an endophyte of *Phanera splendens*, an endemic medicinal plant of the Amazon region. Ravenelin (**440**) displayed antibacterial activity against *B. subtilis* and *S. aureus* with MIC values of 7.5 and 484 µM, respectively (amoxicillin MIC against *B. subtilis* and *S. aureus* 1.3 and 21.4 µM; another positive control terramycin MIC against *B. subtilis* and *S. aureus* 16.3 and 16.3 µM, respectively) [190].

The compounds monocerin (**441**), annularin I (**442**), and annularin J (**443**, Figure 24) were purified from *Exserohilum rostratum* isolated from *Bauhinia guianensis*. Compound **441** displayed antibacterial activity with MIC values of 62.5 µg/mL against *P. aeruginosa*. Compound **442** exhibited antibacterial activity with MIC values of 62.50 and 31.25 µg/mL against *E. coli* and *B. subtilis*, respectively. Compound **443** displayed weak activity against *E. coli* and *B. subtilis* with MIC values of 62.50 µg/mL each [191].

2.3. Basidiomycetes

The compounds quercetin (**444**), carboxybenzene (**445**), and nicotinamide (**446**, Figure 24) were purified from *Psathyrella candolleana* residing inside the seeds of *Ginkgo biloba*. Compounds **444–446** have antibacterial activity against *S. aureus* (MIC 0.3906, 0.7812 and 6.25 µg/mL) [192].

A new tremulane sesquiterpene, irpexlacte A (**447**), and three new furan derivatives, irpexlactes B-D (**448–450**, Figure 24), were isolated from the endophytic fungus *Irpex lacteus* DR10-1 of the waterlogging-tolerant plant *Distylium chinense*. Compounds **447–450** showed moderate antibacterial activity against *P. aeruginosa* with MIC values ranging from 23.8 to 35.4 µM [193].

2.4. Zygomycetes

A flavonoid compound, chlorflavonin (**451**, Figure 24) was purified from the endophytic fungus *Mucor irregularis*, isolated from *Moringa stenopetala*. It has shown antibacterial activity (MIC_{90}) against *M. tuberculosis* at a 1.56 µM concentration. Chlorflavonin also had shown synergistic effects with isoniazid and delamanid in combination treatment experiments. Various molecular and docking techniques have shown that chlorflavonin interacts with the acetohydroxyacid synthase catalytic subunit IlvB1 and inhibits their activity. Recently, Rehberg et al. [194] found the antimicrobial activity of chlorflavonin (**451**) to be higher in comparison to streptomycin treatment against macrophages infected with *M. tuberculosis*.

3. Volatile Organic Compounds (VOCs)

Volatile organic compounds (VOCs) are chemical entities which have low molecular weights and typically evaporate or get into the vapor phase at normal temperature and pressure. They generally possess a characteristic odor [195]. Several reviews have emphasized the production of biogenic VOCs as possible signal molecules in the course of interaction with a host or that play a role in the process of host integration. At times they are also identified as indicators of fungal growth [196–198]. Fungal VOCs largely comprise aliphatic as well as aromatic hydrocarbons, aldehydes, mono-, di- and sesquiterpenes, esters and ketones. Some of the interesting aspects of fungal volatiles is their possible role during interactions among the microbes i.e., with bacteria as well as fungi. However, the application of fungal VOCs as an arsenal to kill bacteria and fungi has not been extensively explored.

The discovery of the endophytic fungus *Muscodor albus* Cz 620 which exhibited potent antibiotic type activity, wiping out all the microbes in its vicinity was serendipitous. This was attributed due to the volatile cocktail produced by *Muscodor albus* Cz 620. This marked the beginning of the exploration of fungal endophytes with the potential to produce volatile antibiotics. The genus *Muscodor* has expanded in the last two decades owing to the addition of novel members that were largely based on the chemical signatures and genetic profiles. Presently there are ~22 known type species that have been documented [199]. Uniquely, all the species of *Muscodor* reported to date are sterile in nature and exhibit a characteristic spectrum of antibacterial as well as anti-fungal activities largely driven by the chemical composition of their volatile gas mixtures. It has also been shown that a single component of the volatile gas is unable to mimic the anti-microbial action suggesting it to be a synergistic action of the finely tuned composition of different VOCs [200]. The pharmaceutical importance of the VOCs produced by *Muscodor* species was exemplified by the anti-bacterial and anti-fungal potential of the VOCs emitted by the fungus. VOCs of *Muscodor albus* Cz620 inhibited *E. coli* and *Bacillus subtilis* while only *E. coli* was inhibited in the presence of volatiles of other isolates of *Muscodor albus* viz. KN-26, KN-27, GP-100, GP-115, TP-21, which inhibited only *E. coli* [201]. The volatiles of *M. albus* I-41.3s on the other hand inhibited *Bacillus subtilis*, *E. coli*, and *Salmonella typhi*. All the VOC emissions were predominantly bacteriostatic and not bactericidal [202].

Muscodor crispans (B-23) has a characteristic VOC spectrum which exhibited anti-mycobacterial activity i.e., against *Mycobacterium marianum* apart from *S. aureus* ATCC6538, *Salmonella cholereasus*, and *Yersinia pestis* [203]. *Muscodor fengyangensis* exclusively inhibited *E. coli* [204]. The volatiles produced by *Muscodor kashayum* has a potent bactericidal activity towards *E. coli, Pseudomonas aeruginosa, Salmonella typhi* and *S. aureus* [205]. Four isolates of *Muscodor* reported from Southeast Asia, viz. *M. oryzae, M. musae, M. suthepensis* and *M. equisetii*, exerted bactericidal activity against *Enterococcus faecalis, E. coli, Proteus mirabilis, S. aureus* and *Pseudomonas pneumoniae* [206]. The VOCs of *Muscodor* have also inspired development of a veterinary medicine formulation which is used as an anti-diarrhoeal product. The formulation is called Sx calf, that is currently being produced and marketed by Ecoplanet Environment LLC (Belgrade, MT, USA) [207]. Similarly, the volatiles of *Muscodor cinnamomi* was found to be effective against *Staphylococcal* spp., *Salmonella* sp., *E. coli, Klebsiella* spp., *Streptococcus* spp. and *Enterococcus* species which contaminate eggs thereby not only affecting their shelf life but also making them unfit for human consumption [208]. The volatile cocktail of *Muscodor crispans* (B-23) was found to kill the bacterial pathogen of citrus *Xanthomonas axonopodis* pv. *citri* [203].

The introspection of the spectrum of the volatile organic mixture from different *Muscodor* species has revealed the antibacterial spectrum of some commonly occurring entities such as isobutyric acid [209–211], β-bisabolol and azulene and its derivatives [212]. Thus, creating artificial mixtures and evaluating them for their anti-bacterial activities may prove to be very useful for preventing drug-resistant film-forming bacteria from causing infections in clinical as well as non-clinical settings. Hence the present study, opens avenues to explore higher numbers of fungal endophytes for their unique volatile signatures and assess them for anti-bacterial activities for developing interventions that could check the spread and infections caused by the drug-resistant bacteria by using them in volatile form or as gaseous sprays.

Table 1. Anti-bacterial metabolites reported from endophytic fungi.

Sr. No.	Fungus	Source	Locality	Compounds Isolated	Biological Target	Biological Activity (MIC/IC$_{50}$/ID$_{50}$)	Reference
Ascomycetes							
Diaporthe							
1	*Diaporthe* sp.	*Uncaria gambier*		(+)-1,1′-Bislunatin (**1**) and (+)-2,2′-epicytoskyrin A (**2**)	*Mycobacterium tuberculosis* strains H37Rv	MICs 0.422 and 0.844 µM	[18]
2	*Diaporthe* sp. GDG-118	*Sophora tonkinensis*	Hechi City, China	21-Acetoxycytochalasin J$_3$ (**3**)	*Bacillus anthraci* and *E. coli*	inhibited at 12.5 µg/mL concentration	[19]
3	*Phomopsis fukushii.*			1-(3-Hydroxy-1-(hydroxymethyl)-2-methoxy-6-methylnaphthalen-7-yl) propan-2-one (**4**) and 1-(3-hydroxy-1- (hydroxymethyl)-6-methylnaphthalen-7-yl)propan-2-one (**5**)	MRSA	Zone of inhibition of 10.2 and 11.3 mm (6 mm strile filterpaper disc were impregnated with 20µL (50 µg) of each compound)	[20]
4	*Phomopsis fukushii*	*Paris polyphylla* var. *yunnanensis*	Kunming, Yunnan, China	3-Hydroxy-1-(1,8- dihydroxy-3,6-dimethoxynaphthalen-2-yl)propan-1-one (**6**), 3-hydroxy-1-(1,3,8-trihydroxy-6-methoxynaphthalen-2-yl)propan-1-one (**7**) and 3-hydroxy-1-(1,8-dihydroxy3,5-dimethoxynaphthalen-2-yl) propan-1-one (**8**)	MRSA- ZR11	MIC, 8, 4, and 4 µg/mL,	[21]
5	*Phomopsis fukushii*	*Paris polyphylla* var. *yunnanensis*	Kunming, Yunnan, China	1-[2-Methoxy-4-(3-methoxy-5-methylphenoxy)-6-methylphenyl]-ethanone (**9**) and 1-[4-(3-hydroxymethyl)-5-methoxyphenoxy)-2-methoxy-6-methylphenyl]-ethanone (**10**)	MRSA	Zone of inhibition 13.8 and 14.6 mm	[22]

Table 1. *Cont.*

Sr. No.	Fungus	Source	Locality	Compounds Isolated	Biological Target	Biological Activity (MIC/IC$_{50}$/ID$_{50}$)	Reference
6	*Phomopsis fukushii*	*Paris polyphylla* var. *yunnanensis*	Kunming, Yunnan, P. R. China	4-(3-Methoxy-5-methylphenoxy)-2-(2-hydroxyethyl)-6-methylphenol (**11**), 4-(3-Hydroxy-5-methylphenoxy)-2-(2-hydroxyethyl)-6-methylphenol (**12**) and 4-(3-methoxy-5-methylphenoxy)-2-(3-hydroxypropyl)-6-methylphenol (**13**)	MRSA	Zone of inhibition of 20.2, 17.9 and 15.2 mm (tested at 50µg/6 mm disc)	[23]
7	*Phomopsis fukushii*	*Paris polyphylla* var. *yunnanensis*	Kunming, Yunnan, China.	1-(4-(3-Methoxy-5-methylphenoxy)-2-methoxy-6-methylphenyl)-3-methylbut-3-en-2-one (**14**), 1-(4-(3-(hydroxymethyl)-5-methoxyphenoxy)-2-methoxy-6-methylphenyl)-3-methylbut-3-en-2-one (**15**), 1-(4-(3-hydroxy-5-(hydroxymethyl)phenoxy)-2-methoxy-6-methylphenyl)-3-methylbut-3-en-2-one (**16**)	MRSA	Zone of inhibition of 21.8, 16.8 and 15.6 mm, (50 µg/6 mm disc)	[24]
8	*Phomopsis* sp.	-	-	3-Hydroxy-6-hydroxymethyl-2,5-dimethylanthraquinone (**17**), 6-hydroxymethyl-3-methoxy-2,5-dimethylanthraquinone (**18**)	MRSA	IZD 14.2 and 14.8 mm	[25]
9	*Diaporthe* sp.	*Pteroceltis tatarinowii*	Mufu Mountain of Nanjing, China.	Diaporone A (**19**)	*B. subtilis*	MIC, 66.7 µM,	[26]
10	*Phomopsis prunorum* (F4-3).	-	-	(−)-1 and (+)- Phomoterpenes A and B (**20**) phomoisocoumarins C (**21**), D (**22**)	*X. citri* pv. *phaseoli* var. *fuscans*	MIC, 31.2, 62.4, 31.2, and 31.2 µg/mL,	[27]
					Pseudomonas syringae pv. *Lachrymans*	MIC, 31.2, 15.6, 31.2 and 15.6 µg/mL	

Table 1. *Cont.*

Sr. No.	Fungus	Source	Locality	Compounds Isolated	Biological Target	Biological Activity (MIC/IC$_{50}$/ID$_{50}$)	Reference
11	*Diporthe vochysiae* LGMF1583	*Vochysia divergens*	-	Vochysiamides A (**23**)	KPC (*Klebsiella pneumoniae* carbapenemase producing).	MIC, 1.0 µg/mL	[28]
				Vochysiamides B (**24**)	KPC, MSSA, MRSA	MIC, 0.08, 1.0, and 1.0 µg/mL	
12	*Phomopsis asparagi*	*Paris polyphylla* var. *yunnanensis*	Kunming, Yunnan, China	4-(3-Methoxy-5-methylphenoxy)-2-(2-hydroxyethyl)-6-(hydroxymethyl)phenol (**25**), 4-(3-Hydroxy-5-methylphenoxy)-2-(2-hydroxyethyl)-6-(hydroxymethyl)phenol(**26**)	MRSA	Zone of inhibition of 10.8 and 11.4 mm	[29]
13	*Phomopsis* sp.	*Paris polyphylla* var. *yunnanensis*	ShiZhong, Yunnan, China	5-Methoxy-2-methyl-7-(3-methyl-2-oxobut-3-enyl)-1-naphthaldehyde (**27**), 2-(hydroxymethyl)-5-methoxy-7-(3-methyl-2-oxobut-3-enyl)-1-naphthaldehyde (**28**)	MRSA	Zone of inhibition of 14.5 and 15.2 mm	[30]
14	*Diaporthe terebinthifolii* LGMF907	*Schinus terebinthifolius*	Curitiba, Paraná, Brazil	Diaporthin (**29**)	*E. coli*, *Micrococcus luteus*, MRSA, and *S. aureus*	Zone of inhibition 1.73, 2.47, 9.50, and 9.0 mm tested at 100 µg/disk.	[31]
				Orthosporin (**30**)		Zone of inhibition of 1.03, 1.53C, 9.0, and 9.33 mm	
15	*Phomopsis/Diaporthe* sp. GJJM 16	*Vitex negundo*	Azhiyar, Pollachi, Tamilnadu, India	(2Z)-2-(1,4-dihydro-2-hydroxy-1-((E)-2-mercapto-1 (methylimino)ethyl) pyrimidine-4-ylimino)-1-(4,5-dihydro-5-methylfuran-3-yl)-3-methylbutane-1-one (**31**)	*S. aureus*, and *P. aeroginosa*	MIC of 1.25 µg/mL against each organism	[32]

Table 1. *Cont.*

Sr. No.	Fungus	Source	Locality	Compounds Isolated	Biological Target	Biological Activity (MIC/IC$_{50}$/ID$_{50}$)	Reference
16	*Phomopsis* sp. PSU-H188	*Hevea brasiliensis*	Trang Province, Thailand.	Diaporthalasin (**32**)	*S. aureus* ATCC25923, MRSA	MIC, 4 μg/mL each	[33]
				Cytosporone B (**33**)		MIC, 32 and 16 μg/mL	
				Cytosporone D (**34**)		MIC, 64 and 32 μg/mL	
17	*Diaporthe terebinthifolii* GG3F6	*Glycyrrhiza glabra*	Jammu, J & K, India	Diapolic acid A (**35**), B (**36**) xylarolide (**37**) phomolide G (**38**)	*Yersinia enterocolitica*	IC$_{50}$, 78.4, 73.4, 72.1 and 69.2 μM	[34]
18	*Diaporthe* sp. F2934	leaves of *Siparuna gesnerioides*	Chagres National Park, a protected area of Panama	Phomosine A (**39**)	*S. aureus* (ATCC 25923), *Streptococcus oralis* (ATTC 35037), *Enterococcus faecalis* (ATCC 19433), *Enterococcus cloacae* (ATCC 13047), *Bordetella bronchiseptica* (CECT 440),	Zone of Inhibition 12, 9, 10, 11, 10 and 10 mm at 4 μg/mL concentration	[35]
				Phomosine C (**40**)		Zone of Inhibition 9, 6, 8, 8, 8 and 9 mm at 4 μg/mL concentration	
19	*Phomopsis* sp.,	*Garcinia kola* nuts	bought at Mokolo local market in Yaounde (Cameroon)	18-Methoxycytochalasin J (**41**), cytochalasins H (**42**) and J (**43**), alternariol (**44**)	*Shigella flexneri*	MIC, 128 μg/mL each	[36]
				18-Methoxycytochalasin J (**41**), cytochalasins H (**42**)	*S. aureus* ATCC 25923	MIC, 128 and 256 μg/mL	
20	*Diaporthe* sp. LG23	*Mahonia fortunei*	Shanghai, China	19-nor-Lanosta-5(10),6,8,24-tetraene-1α,3β,12β,22S-tetraol (**45**)	*S. aureus, E. coli, Bacillus subtilis, P. aeruginosa, Streptococcus pyogenes*	MIC, 5.0, 5.0, 2.0, 2.0 and 0.1 μg/mL	[37]
				3β,5α,9α-Trihydroxy-(22E,24R)-ergosta-7,22-dien-6-one (**46**), and chaxine C (**47**)	*B. subtilis*	MIC, 5.0 μg/mL each	
21	*Diaporthales* sp. E6927E	*Ficus sphenophyllum*	Ecuadorean dry forest near the Napo River, USA	Pyrrolocin A (**48**)	*S. aureus* and *E. faecalis*	MICs 4 and 5 μg/mL	[38]

Table 1. *Cont.*

Sr. No.	Fungus	Source	Locality	Compounds Isolated	Biological Target	Biological Activity (MIC/IC$_{50}$/ID$_{50}$)	Reference
	Xylaria						
22	*Xylaria ellisii*	Blueberry (*Vaccinium angustifolium*)		Ellisiiamide (**49**)	*Escherichia coli*	MIC, 100 μg/mL	[39]
23	*Xylaria* sp. GDG-102	*S. tonkinensis*	Hechi, Guangxi province, China	Xylareremophil (**50**)	*Micrococcus luteus* and *Proteus vulgaris*	MIC 25 μg/mL each	[40]
				Mairetolides B (**51**)	*M. luteus*	MIC, 50 μg/mL	
				Mairetolide G (**52**)	*P. vulgaris M. luteus*	MIC 25 and 50 μg/mL	
				Xylareremophil (**50**), mairetolides B (**51**), and G (**52**)	*Micrococcus lysodeikticus* and *Bacillus subtilis*	MIC 100 μg/mL	
24	*Xylaria* sp. (GDG-102)	Leaves of *S. tonkinensis*		6-Heptanoyl-4-methoxy-2H-pyran-2-one (**53**)	*E. coli* as well as *S. aureus*	MIC, 50 μg/mL	[41]
25	*Xylaria* sp. GDG-102	*S. tonkinensis*	Hechi, Guangxi province, China	Xylarphthalide A (**54**)	*B. subtilis* and *E. coli*,	MIC, 12.5 μg/mL each	[42]
					B. megaterium, S. aureus, S. dysenteriae and *S. paratyphi*	MIC, 25 μg/mL each	
				(−)-5-Carboxymellein (**55**)	*B. Subtilis*	MIC, 12.5 μg/mL	
					B. anthracis, B. megaterium, S. aureus, E. coli, S. dysenteriae and *S. paratyphi* B	MIC, 25 μg/mL	
				(−)-5-Methylmellein (**56**)	*B. subtilis* and *S. aureus*	MIC, 12.5 μg/mL	
					B. megaterium, E. coli and *S. dysenteriae*	25 μg/mL	

Table 1. *Cont.*

Sr. No.	Fungus	Source	Locality	Compounds Isolated	Biological Target	Biological Activity (MIC/IC$_{50}$/ID$_{50}$)	Reference
26	*Xylaria* sp.,	*Taxus mairei.*		3,7-Dimethyl-9-(-2,2,5,5-tetramethyl-1,3-dioxolan-4-yl) nona-1,6-dien-3-ol (**57**)	*B. subtilis* ATCC 9372, *B. pumilus* 7061 and *S. aureus* ATCC 25923	48.1, 31.6 and 47.1% inhibition.	[43]
				Nalgiovensin (**58**)	*S. aureus* ATCC 25923, *B. subtilis* ATCC 9372, *B. pumilus* ATCC 7061 and *E. coli* ATCC 25922	42.1, 36.8, 47.1 and 41.2% inhibition.	
	Chaetomium						
27	*C. globosum* 7s-1,	*Rhapis cochinchinensis*		Xanthoquinodin B9 (**59**), xanthoquinodin A1 (**60**), xanthoquinodin A3 (**61**)	*B. cereus*	MICs of 0.87, 0.44 and 0.22 μM,	[45]
				Xanthoquinodin B9 (**59**), xanthoquinodin A1 (**60**), xanthoquinodin A3 (**61**)	*S. aureus* and MRSA	MIC values ranging from 0.87 to 1.75 μM	
				3-Epipolythiodioxopiperazines, chetomin (**62**), chaetocochin C (**63**) and dethio-tetra(methylthio) chetomin (**64**)	*B. cereus* ATCC 11778, *S. aureus* ATCC 6538, and MRSA	MIC values ranging from 0.02 pM to 10.81 μM.	
				Chetomin (**62**)	*B. cereus, S. aureus* and MRSA	MICs, 0.35 μM, 10.74 and 0.02 pM	
				Compounds **59–64**	*E. coli* ATCC 25922, *P. aeruginosa* ATCC 27853, and *Salmonella typhimurium* ATCC 13311	MICs of 45.06 to >223.72 μM	
				Epipolythiodioxopiperazines (**62–64**)	*Mycobacterium tuberculosis*	MICs, 0.55, 4.06 and 8.11 μM,	
28	*Chaetomium* sp. SYP-F7950	*Panax notoginseng*	Wenshan, Yunnan, China	Chaetocochin C (**63**), chetomin A (**65**), and chetomin (**62**)	*S. aureus, B. subtilis, Enterococcus faecium*	MIC values ranging from 0.12 to 19.3 μg/mL	[46]

Table 1. *Cont.*

Sr. No.	Fungus	Source	Locality	Compounds Isolated	Biological Target	Biological Activity (MIC/IC$_{50}$/ID$_{50}$)	Reference
29	*Chaetomium* sp. HQ-1,	*Astragalus chinensis*	Tai'an, Shandong Province, China	Differanisole A (**66**)	*L. monocytogenes S. aureus* and MRSA,	MIC, 16, 128, 128 µg/mL	[47]
				2,6-Dichloro-4-propylphenol (**67**), 4,5-dimethylresorcinol (**68**)	*L. monocytogenes*	MICs of 64 and 32 µg/mL,	
30	*Chaetomium nigricolor* F5,	*Mahonia fortune*	Qingdao, People's Republic of China	Chamiside A (**69**)	*S. aureus*	MIC of 25 µg/mL	[48]
31	*C. globosum*	*Salvia miltiorrhiza*	Shenyang, Liaoning province, China	Equisetin (**70**)	Multidrug-resistant *E. faecalis, E. faecium, S. aureus,* and *S. epidermidis*	MIC values of 3.13, 6.25, 3.13, and 6.25 µg/mL	[49]
32	*Chaetomium* sp. Eef-10,	*Eucalyptus exserta*	Guangdong Province, China	Mollicellins H (**71**)	*S. aureus* ATCC29213, *S. aureus* N50, MRSA,	IC$_{50}$, 5.14, and 6.21 µg/mL	[50]
				Mollicellin O (**72**)	*S. aureus* ATCC29213 and *S. aureus* N50	IC$_{50}$, 79.44 and 76.35 µg/mL	
				Mollicellin I (**73**)		IC$_{50}$, 70.14 and 63.15 µg/mL	
33	*Chaetomium* sp. M336	*Huperzia serrata*	Xichou County, Yunnan Province, China	6-Formamidochetomin (**74**)	*E. coli, S. aureus, S. typhimurium* ATCC 6539 and *E. faecalis*	MIC, 0.78 µg/mL	[51]
34	*Chaetomium globosum*	*Nymphaea nouchali*	Udugampola in the Gampaha District, Sri Lanka	Chaetoglobosin A (**75**)	*B. subtilis, S. aureus,* and MRSA	MIC, 16, 32 and 32 µg/mL	[52]
				Chaetoglobosin B (**76**)		>64 µg/mL	
	Talaromyces						
35	*Talaromyces pinophilus* XL-1193	*Salvia miltiorrhiza*	Shenyang, Liaoning province, China	Pinophol A (**77**)	*Bacterium paratyphosum* B	MIC, 50µg/mL	[53]
36	*Talaromyces purpureogenus* XL-25	*Panax notoginseng*	Shijiazhuang, Hebei Province, China	Talaroconvolutin A (**78**)	*B. subtilis Micrococcus lysodeikticus, Vibrio parahaemolyticus*	MIC value of 1.56 µM	[54]
				Talaroconvolutin B (**79**)		MIC = 0.73 and 0.18 µM	

Table 1. Cont.

Sr. No.	Fungus	Source	Locality	Compounds Isolated	Biological Target	Biological Activity (MIC/IC$_{50}$/ID$_{50}$)	Reference
37	*Talaromyces purpureogenus*	*Panax notoginseng*		(1*S*,5*S*,7*S*,10*S*)-dihydroxyconfertifolin (**80**)	*E. coli*	MIC, 25 µM	[55]
38	*Talaromyces funiculosus* -Salicorn 58.			Talafun (**81**)	*E. coli, S. aureus*	MIC, 18 and 93 µM	[56]
				N-(2′-hydroxy-3′-octadecenoyl)-9-methyl-4,8-sphingadienin (**82**)	*Mycobacterium smegmatis, S. aureus, Micrococcus tetragenus,* and *E. coli*	MIC, 85, 90, 24, and 68, 93 µM	
				Chrodrimanin A (**83**)	*S. aureus, M. tetragenus, Mycobacterium phlei,* and *E. coli*	MIC, 67, 28, 47, and 26 µM	
				Chrodrimanin B (**84**)	*E.coli*	MIC, 43 µM.	
39	*Talaromyces* sp. LGT-2	*Tripterygium wilfordii.*		Alkaloids **85–90**	*E. coli, P. aeruginosa, S. aureus, Bnfillus licheniformis,* and *Streptococcus pneumoniae*	MICs in the range of 0.125 to 1.0 50 µg/mL	[57]
40	*Rhytidhysteron* sp. BZM-9	*Leptospermum brachyandrum*		Euphorbol (**91**)	MRSA	MIC, 62.5 ug/mL	[58]
41	*Stagonosporopsis oculihominis*	*Dendrobium huoshanense.*		Stagonosporopsin C (**92**)	*Staphylococcus aureus* subsp. *aureus* ATCC29213	MIC$_{50}$, 41.3 µM	[59]
42	*Eutypella scoparia* SCBG-8.	*Leptospermum brachyandrum*	SCBG, Chinese Academy of Sciences, China	Eutyscoparols H (**93**), I (**94**), tetrahydroauroglaucin (**95**), flavoglaucin (**96**)	*Staphylococcus aureus* and MRSA	MICs in the range of 1.25 to 6.25 µg/mL	[60]
43	*Eutypella scoparia* SCBG-8	*Leptospermum brachyandrum*	SCBG, Chinese Academy of Sciences, Guangzhou 510650, China	Eutyscoparin G (**97**)	*S. aureus* and MRSA	MIC values of 6.3 µg/mL	[61]

Table 1. *Cont.*

Sr. No.	Fungus	Source	Locality	Compounds Isolated	Biological Target	Biological Activity (MIC/IC$_{50}$/ID$_{50}$)	Reference
44	*Sarocladium oryzae* DX-THL3,	*Oryza rufipogon* Griff.		Sarocladilactone A (**98**), sarocladilactone B (**99**), helvolic acid (**100**), helvolinic acid (**101**), 6-desacetoxy-helvolic acid (**102**), 1,2-dihydrohelvolic acid (**103**)	*S. aureus*	MIC values of 64, 4, 8, 1, 4 and 16 µg/mL	[62]
				Compound **101**	*B. subtilis*	MIC, 64 µg/mL	
				Compounds **99, 101, 103**	*E. coli*	MIC 64 µg/mL each	
45	*Paraphaeosphaeria sporulosa*	*Fragaria x ananassa*	Caserta province, Southern Italy	Cyclo(L-Pro-L-Phe) (**104**)	*Salmonella* strains, S1 and S2	MIC 71.3 and 78.6 µg/mL	[63]
46	*Aplosporella javeedii*	*Orychophragmus violaceus*	Beijing, China	Terpestacin (**105**), fusaproliferin (**106**), mutolide (**108**)	*M. tuberculosis* H37Rv	MICs of 100 µM	[64]
				6,7,9,10-Tetrahydromutolide (**107**)	*S. aureus*,	MICs of 100 µM	
47	*Pleosporales* sp. Sigrf05	roots of *Siraitia grosvenorii*	Guangxi Province of China	Pleospyrone E (**109**)	*B. subtilis*, *Agrobacterium tumefaciens*, *Ralstonia solanacearum*, and *Xanthomonas vesicatoria*	MIC 100.0µM each	[65]
48	*Aplosporella javeedii*	*Orychophragmus violaceus*	Beijing, China	Aplojaveediin A (**110**)	*Staphylococcus aureus* strain ATCC 29213, *S. aureus* strain ATCC 700699 and *Bacillus subtilis* (ATCC 169)	MICs 50, 50 and 25 µM,	[66]
				Aplojaveediin F (**111**)	*S. aureus* ATCC 29213 and ATCC 700699	MICs of 25 and 50 µM	
49	*Paecilomyces variotii*	*Lawsonia Alba*	University of Karachi, Pakistan	Lawsozaheer (**112**)	*S. aureus* (NCTC 6571)	84.26% inhibition at 150 µg/mL	[67]

Table 1. *Cont.*

Sr. No.	Fungus	Source	Locality	Compounds Isolated	Biological Target	Biological Activity (MIC/IC$_{50}$/ID$_{50}$)	Reference
50	*Preussia isomera* OSMAC strategy	*Panax notoginseng*	Wenshan, Yunnan Province, China	Setosol (**113**)	Multidrug-resistant *E. faecium*, methicinllin-resistant *S. aureus* and multidrug-resistant *E. faecalis*	MIC 25 µg/mL	[68]
	Preussia isomera. XL-1326,	*Panax notoginseng*		(+)- and (−)-Preuisolactone A (**114, 115**)	*Micrococcus luteus* and *B. megaterium*	MIC, 10.2 and 163.4 µM	[69]
51	*Neurospora udagawae*	*Quercus macranthera*	Kaleybar region in northwestern Iran	Udagawanones A (**116**)	*S. aureus*	MIC, 66 µg/mL	[70]
52	*Xylomelasma* sp. Samif07	Salvia miltiorrhiza Bunge		2,6-Dimethyl-5-methoxy-7-hydroxychromone (**117**), 6-hydroxymethyleugenin (**118**), 6-methoxymethyleugenin (**119**), isoeugenitol (**120**), diaporthin (**29**), 8-hydroxy-6-methoxy-3-methylisocoumarin (**121**)	*Bacillus subtilis, Staphylococcus haemolyticus, A. tumefaciens, Erwinia carotovora,* and *Xanthomonas vesicatoria*	MIC values at the range of 25 ~ 100 µg/mL	[71]
				2,6-Dimethyl-5-methoxy-7-hydroxychromone (**117**), diaporthin (**29**)	*B. subtilis, E. carotovora*	MIC, 50 and 100 µg/mL	
				6-Hydroxymethyleugenin (**118**), 6-methoxymethyleugenin (**119**), isoeugenitol (**120**), diaporthin (**29**)	*S. haemolyticus* and *E. carotovora*	MIC, 75 µg/mL each	
				8-Hydroxy-6-methoxy-3-methylisocoumarin (**121**)	*B. subtilis, A. tumefaciens,* and *X. vesicatoria,*	MICs 25, 75, and 25 µg/mL,	

Table 1. *Cont.*

Sr. No.	Fungus	Source	Locality	Compounds Isolated	Biological Target	Biological Activity (MIC/IC$_{50}$/ID$_{50}$)	Reference
53	*Amphirosellinia nigrospora*JS-1675	*Pteris cretica*		(4S,5S,6S)-5,6-epoxy-4-hydroxy-3-methoxy-5-methylcyclohex-2-en-1-one (**122**)	*Acidovorax avenae* subsp. *cattlyae*, *Agrobacterium konjaci*, *A. tumefaciens*, *Burkholderia glumae*, *Clavibacter michiganensis* subsp. *michiganensis*, *Pectobacterium carotovorum* subsp. *carotovorum*, *Pectobacterium chrysanthemi*, *Ralstonia solanacearum*, *Xanthomonas arboricola* pv. *pruni*, *Xanthomonas axonopodis* pv. Citri, *Xanthomonas euvesicatoria*, *Xanthomonas oryzae* pv. *oryzae*	MICs ranging between 31.2 and 500 µg/ml	[72]
54	*Emericella* sp. XL029	*Panax notoginseng*		5-(Undeca-3′,5′,7′-trien-1′-yl)furan-2-ol (**123**) and 5-(undeca-3′,5′,7′-trien-1′-yl)furan-2-carbonate (**124**)	*B. subtilis*, *B. cereus*, *S. aureus*, *B. paratyphosum* B, *S. typhi*, *P. aeruginosa*, *E. coli*, and *E. aerogenes*	MIC values ranging from 6.3 to 50 µg/mL	[73]
56	*Emericella* sp. XL029	*Panax notoginseng*	Shijiazhuang, Hebei Province, China	14-Hydroxytajixanthone (**125**), 14-hydroxytajixanthonehydrate (**126**), 14-hydroxy-15-chlorota jixanthone hydrate (**127**), 14-methoxytajixanthone-25-acetate (**130**), questin (**132**), and carnemycin B (**133**)	*M. luteus*, *S. aureus*, *B. megaterium*, *B. anthracis*, and *B. paratyphosum* B	MIC, in the range of of 12.5 and 25µg/mL	[74]
				Epitajixanthone hydrate (**128**)	*M. luteus*, *S. aureus*, *B. megaterium*, and *B. paratyphosum* B	MIC 25 µg/mL	
				Tajixanthone hydrate (**129**), 15-chlorotajixanthone hydrate (**131**)	*S. aureus*, *B. megaterium*, and *B. paratyphosum* B	MICs 25 and 12.5 µg/mL,	
				14-Hydroxytajixanthone (**125**) Epitajixanthone hydrate (**128**), carnemycin B (**133**)	drug resistant *S. aureus*	MIC 50 µg/mL	
				Compounds **125–133**	*P. aeruginosa*, *E. coli*, and *E. aerogenes*	MIC 50 µg/mL	

Table 1. *Cont.*

Sr. No.	Fungus	Source	Locality	Compounds Isolated	Biological Target	Biological Activity (MIC/IC$_{50}$/ID$_{50}$)	Reference
57	*Byssochlamys spectabilis*	*Edgeworthia chrysantha*	Hangzhou Bay, Hangzhou, Zhejiang Province, China	Bysspectin C (**134**)	*E. coli* ATCC 25922 and *S. aureus* ATCC 25923	MIC, 32 and 64 µg/mL	[75]
58	*Poculum pseudosydowianum* (TNS-F-57853),	*Quercus crispula* var. *crispula*	Yoshiwa, Hatsukaichi, Hiroshima prefecture, Japan	Sydowianumols A (**135**), and B (**136**)	MRSA	MIC90 values of 12.5 µg/mL	[76]
59	*Lachnum palmae* exposure to a HDAC inhibitor SAHA	*Przewalskia tangutica*	Linzhou Country of the Tibet Autonomous Region, China	Palmaerones A-B, E-G (**137**, **138**, **140**, **141**, **142**)	*B. subtilis*	MICs, 35, 30, 10, 50, and 55 µg/mL	[77]
				Palmaerones A-C, E (**137**, **138**, **139**, **140**)	*S. aureus*	MICs 65, 55, 60, and 55, µg/mL	
60	*Nemania serpens*	*Vitis vinifera*	Canada's Niagara region	Nemanifuranone A (**143**)	*E. coli*	MIC 200 µg/mL	[78]
					S. aureus, B. subtilis and *M. luteus*	>75% inhibition at a concentration of 100–200 µg/mL	
				Triterpenoid **144**	*S. cerevisiae*	(>25% inhibition) against at 200 µg/mL	
					M. luteus	(>75% inhibition) of at a concentration of 100 µg/mL	
61	*Paraconiothyrium variabile*	*Cephalotaxus harringtonia*		Variabilone (**145**)	*B. subtilis*	IC$_{50}$ of 2.13 µg/mL after 24 h (0.36 µg/mL for kanamycin)	[79]

Table 1. *Cont.*

Sr. No.	Fungus	Source	Locality	Compounds Isolated	Biological Target	Biological Activity (MIC/IC$_{50}$/ID$_{50}$)	Reference
62	*Pyronema* sp. (A2-1 & D1-2)	*Taxus mairei*	Shennongjia National Nature Reserve, Hubei province, China.	Methyl 2-{(*E*)-2-[4-(formyloxy)phenyl]ethenyl}-4-methyl-3-oxopentanoate (**146**), (3*R*,6*R*)-4-methyl-6-(1-methylethyl)-3-phenylmethyl-perhydro-1,4-oxazine-2,5-dione (**147**), (3*R*,6*R*)-N-methyl-N-(1-hydroxy-2-methylpropyl)-phenylalanine (**148**), siccanol (**149**), fusaproliferin (**106**), and sambutoxin (**150**)	*Mycobacterium marinum* ATCCBAA-535,	IC$_{50}$ of 64, 59, 57, 84, 43 and 32 μM, (positive control rifampin IC$_{50}$ of 2.1 μM)	[80]
63	*Pulvinula* sp. 11120	*Cupressus arizonica*	Tucson, AZ, USA	Pulvinulin A (**151**), graminin C (**152**), cis-gregatin B (**153**), and graminin B (**154**)	*E. coli*	12, 18, 16 and14 mm zone of inhibition at 100 μg/mL	[81]
64	*Stelliosphaera formicum*	*Duroia hirsuta*	Yasuni' National Park off the Napo River in Ecuador	Stelliosphaerols A (**155**) and B (**156**)	*S. aureus*	MIC values of 250 μg/mL	[82]
65	Unidentified Ascomycete	*Melilotus dentatus*		cis-4-Acetoxyoxymellein (**157**)	*E. coli* and *B. megaterium*	Zone of inhibition of 10 and 10 mm (Partial inhibition) at a concentration of 0.05 mg	[83]
				8-Deoxy-6-hydroxy-*cis*-4-acetoxyoxymellein (**158**)	*E. coli* and *B. megaterium*	Zone of inhibition of 9 and 9 mm (Partial inhibition) at a concentration of 0.05 mg	
	Anamorphic Ascomycetes						
	Aspergillus						
66	*Aspergillus* sp. FT1307	*Heliotropium* sp.		Aspochalasin P (**159**), alatinone (**160**), β-11-methoxy curvularine (**161**), 12-keto-10,11-dehydrocurvularine (**162**)	*S. aureus* ATCC12600, *B. subtilis* ATCC6633 and MRSA ATCC43300	MIC in the range of 40 to 80 μg/mL	[84]

Table 1. *Cont.*

Sr. No.	Fungus	Source	Locality	Compounds Isolated	Biological Target	Biological Activity (MIC/IC$_{50}$/ID$_{50}$)	Reference
67	*Aspergillus cristatus*	*Pinellia ternata*		Aspergillone A (**163**)	*B. subtilis* and *S. aureus*	MIC$_{50}$, 8.5 and 32.2 µg/mL	[85]
68	*Aspergillus versicolor* strain Eich.5.2.2	*Eichhornia crassipes*	El-Kanater El-Khayriah in Egypt	22S-Aniduquinolone A (**164**), 22R-aniduquinolone A (**165**)	*S. aureus* (ATCC700699)	MIC, 0.4 µg/mL	[86]
69	*Aspergillus versicolor*	roots of *Pulicaria crispa*	Saudi Arabia	Aspergillether B (**166**)	*S. aureus, B. cereus,* and *E. coli*	MICs, 4.3, 3.7, and 3.9 µg/mL	[87]
70	*Aspergillus ochraceus* SX-C7 *eus* SX-C7	*Setaginella stauntoniana*		3-O-β-D-Glucopyranosyl stigmasta-5(6),24(28)-diene (**167**)	*Bacillus subtilis*	MIC, 2 µg/mL	[88]
71	*Aspergillus amstelodami* (MK215708)	Ammi majus	Egypt	Dihydroauroglaucin (**168**)	*E. coli, Streptococcus mutans, S. aureus*	MIC, 1.95, 1.95 and 3.9 µg/mL	[89]
					S. aureus, E. coli, Streptococcus mutans, P. aeruginosa	Minimum biofilm inhibitory concentration (MBIC) = 7.81, 7.81, 15.63 and 31.25 µg/mL	
72	*Aspergillus micronesiensis*	*Phyllanthus glaucus*	LuShan Mountain, Jiangxi Province, China	Cyschalasins A (**169**) and B (**170**)	MRSA	MIC$_{50}$, 17.5 and 10.6 µg/mL: MIC90, 28.4 and 14.7 µg/mL	[90]
73	A. niger	*Acanthus montanus*	Kala Mountain neighborhood of Yaoundé, Africa	Methylsulochrin (**171**)	*S. aureus, Enterobacter cloacae* and *Enterobacter aerogenes*	MIC, 15.6, 7.8 and 7.8 µg/mL	[91]
74	*Aspergillus tubingensis*	stem of *Decaisnea insignis*	Qinling Mountain, Shaanxi Province, China	3-(5-Oxo-2,5-dihydrofuran-3-yl) propanoic acid (**172**)	*Streptococcus lactis*	MIC value of 32 µg/mL	[92]
75	*Aspergillus flavipes* Y-62	*Suaeda glauca*	Zhoushan coast, Zhejiang province, East China	Methyl 2-(4-hydroxybenzyl)-1,7-dihydroxy-6-(3-methylbut-2-enyl)-1H-indene-1-carboxylate (**173**)	MRSA	MIC, 128 µg/mL	[93]
					K. pneumoniae and *P. aeruginosa*	MIC, of 32 µg/mL each	

Table 1. *Cont.*

Sr. No.	Fungus	Source	Locality	Compounds Isolated	Biological Target	Biological Activity (MIC/IC$_{50}$/ID$_{50}$)	Reference
76	*Aspergillus* sp.	Rhizome of *Zingiber cassumunar*		4-Amino-1-(1,3-dihydroxy-1-(4-nitrophenyl)propan-2-yl)-1*H*-1,2,3-triazole-5(4H)one (**174**)	*Xanthomonas oryzae*, *Bacillus subtilis* and *E. coli*	Zone of inhibition 37, 30 and 27 mm	[5]
				3,6-Dibenzyl-3,6-dimethy lpiperazine-2,5-dione (**175**)	*E. coli* and *X. oryzae*	Zone of inhibition 21 and 16 mm.	
77	*Aspergillus fumigatus*	*Edgeworthia chrysantha*	Hangzhou Bay (Hangzhou, China)	Pseurotin A (**176**), spirotryprostatin A (**177**)	*S. aureus*	MIC of 0.39 µg/mL each	[94]
				Spirotryprostatin A (**177**)	*E. coli*	MIC, 0.39 µg/mL	
78	*Aspergillus* sp.,	*Astragalus membranaceus*		Fumiquinazoline J (**178**), fumiquinazoline C (**180**), fumiquinazoline H (**181**), fumiquinazoline D (**182**)	*B. subtilis*, *S. aureus*, *E. coli* and *P. aeruginosa*	MICs in the range of 0.5–8 µg/mL	[95]
				Fumiquinazoline I (**179**), fumiquinazoline B (**183**)		MICs in the range of 4–16 µg/mL	
79	*Aspergillus fumigatiaffnis*	*Tribulus terestris*		(−)-Palitantin (**184**)	*E. faecalis* UW 2689 and *Streptococcus pneumoniae*	MIC, 64µg/mL	[96]
80	*Aspergillus* sp. TJ23	*Hypericum perforatum* (St John' Wort)	Shennongjia areas of Hubei Province, China	Aspermerodione (**185**)	MRSA	MIC, 32 µg/mL/potential inhibitor of PBP2a	[97]
				Andiconin C (**186**)		marginal antimicrobial activity (>100µg/mL)	
81	*Aspergillus* sp. YXf3	*Ginkgo biloba*		Prenylterphenyllin D (**187**), prenylterphenyllin E (**188**), 2′-O-Methylprenylterphenyllin (**189**), prenylterphenyllin (**190**)	*X. oryzae* pv. *oryzicola* Swings and *E. amylovora*	MIC, 20 µg/mL each	[98]
				Prenylterphenyllin B (**191**)	*E. amylovora*	MIC, 10 µg/mL	

Table 1. *Cont.*

Sr. No.	Fungus	Source	Locality	Compounds Isolated	Biological Target	Biological Activity (MIC/IC$_{50}$/ID$_{50}$)	Reference
82	*Aspergillus* sp.	*Pinellia ternata*	Nanjing, Jiangsu Province, China	Aspergillussanone D (**192**)	*P. aeruginosa*, and *S. aureus*	MIC$_{50}$, 38.47 and 29.91 μg/mL	[99]
				Aspergillussanone E (**193**)	*E. coli*	MIC$_{50}$, 7.83 μg/mL	
				Aspergillussanone F (**194**)	*P. aeruginosa*, and *S. aureus*	MIC$_{50}$, 26.56, 3.93 and 16.48 μg/mL	
				Aspergillussanone G (**195**)	*P. aeruginosa*, and *S. aureus*,	MIC$_{50}$, 24.46 and 34.66 μg/mL	
				Aspergillussanone H (**196**)	*P. aeruginosa*, and *E. coli*,	MIC$_{50}$, 8.59 and 5.87 μg/mL	
				Aspergillussanone I (**197**)	*P. aeruginosa*,	MIC$_{50}$, 12.0 μg/mL	
				Aspergillussanone J (**198**)	*P. aeruginosa*, *E. coli* and *S. aureus*	MIC$_{50}$, 28.50, 5.34 and 29.87 μg/mL	
				Aspergillussanone K (**199**)	*P. aeruginosa*, and *S. aureus*,	MIC$_{50}$, 6.55 and 21.02 μg/mL	
				Aspergillussanone L (**200**)	*P. aeruginosa*, *S. aureus*, and *B. subtilis*	MIC$_{50}$, 1.87, 2.77, and 4.80 μg/mL,	
				Compound **201**	*P. aeruginosa*, and *E. coli*,	MIC$_{50}$, 19.07 and 1.88 μg/mL	
83	*Aspergillus terreus* JAS-2	*Achyranthus aspera*	Varanasi, India	Terrein (**202**)	*E. faecalis*	IC$_{50}$, 20 μg/mL	[100]
					S. aureus and *Aeromonas hydrophila*	20 μg/mL	
84	*Aspergillus terreus*	roots of *Carthamus lanatus*	Al-Azhar University campus in Cairo, Egypt	(22E,24R)-Stigmasta-5,7,22-trien-3-β-ol (**203**)	MRSA	IC$_{50}$, 2.29 μM	[101]

Table 1. *Cont.*

Sr. No.	Fungus	Source	Locality	Compounds Isolated	Biological Target	Biological Activity (MIC/IC$_{50}$/ID$_{50}$)	Reference
85	*Aspergillus flavus*	*Cephalotaxus fortunei*	Taibai Mountains, Shaanxi Province, China	5-Hydroxymethylfuran-3-carboxylic acid (**204**), 5-acetoxymethylfuran-3-carboxylic acid (**205**)	*S. aureus*	MIC, 31.3 and 15.6 µg/mL	[102]
86	*Aspergillus allahabadii* BCC45335	root of *Cinnamomum subavenium*	Khao Yai National Park, Nakhon Ratchasima Province, Thailand	Allahabadolactone B (**206**), (22*E*)-5α,8α-epidioxyergosta-6,22-dien-3β-ol (**207**)	*B. cereus*	IC$_{50}$, 12.50 and 3.13 µg/mL.	[103]
87	*Aspergillus tubingensis*	*Lycium ruthenicum*		6-Isovaleryl-4-methoxypyran-2-one (**208**), asperpyrone A (**210**), campyrone A (**211**)	*E. coli, Pseudomonas aeruginosa, Streptococcus lactis* and *S. aureus*	MIC values ranging from 62.5 to 500 µg/mL	[104]
				Rubrofusarin B (**209**)	*E. coli*	MIC, 1.95 µg/mL	
88	*Aspergillus tamarii* FR02	roots of *Ficus carica*	Qinling Mountain in China's Shaanxi province	Malformin E (**212**)	*B. subtilis, S. aureus, P. aeruginosa,* and *E. coli*	MIC, 0.91, 0.45, 1.82, and 0.91 µM	[105]
89	*Aspergillus terreus*	Roots of *Carthamus lanatus*	Al-Azhar University campus, Egypt	(22*E*,24*R*)-Stigmasta-5,7,22-trien-3-β-ol (**203**)	MRSA	IC$_{50}$, 0.96µg/mL	[106]
				Aspernolide F (**213**)		IC$_{50}$ 6.39µg/mL	
90	*Aspergillus* sp. (SbD5)	Leaves of *Andrographis paniculata*	Indralaya, Ogan Ilir, South Sumatra.	1-(3,8-Dihydroxy-4,6,6-trimethyl-6*H*-benzochromen-2-yloxy)propane-2-one (**214**), 5-hydroxy-4-(hydroxymethyl)-2*H*-pyran-2-one (**215**), (5-hydroxy-2-oxo-2*H*-pyran-4-yl)methyl acetate (**216**)	*S. aureus, E. coli, S. dysenteriae* and *Salmonella typhi*	Zone of inhibition diameters ranging from 8.1 to 12.1 mm at a concentration 500 µg/mL.	[107]

Table 1. *Cont.*

Sr. No.	Fungus	Source	Locality	Compounds Isolated	Biological Target	Biological Activity (MIC/IC$_{50}$/ID$_{50}$)	Reference
91	*Aspergillus* sp. IFB-YXS	*Ginkgo biloba*		Xanthoascin (**217**)	*X. oryzae* pv. *oryzicola*, Swings, *E.amylovora*, *P. syringae* pv. *Lachrymans* and *C. michiganense* subsp. *sepedonicus*	MICs, 20, 10, 5.0 and 0.31 µg/mL	[108]
				Prenylterphenyllin B (**218**)	*X. oryzae* pv.*oryzicola* Swings, *E.amylovora*, *P. syringae* pv. *Lachrymans*,	MICs of 20 µg/mL each	
				Prenylcandidusin (**219**)	*X. oryzae* pv.*oryzae* Swings *X. oryzae* pv. *oryzicola* Swings	MIC values of 10 and 20 µg/mL	
	Penicillium						
92	*Penicillium ochrochloron* SWUKD4.1850	*Kadsura angustifolia*		4-O-Desmethylaigialomycin B (**220**), penochrochlactones C (**221**) and D (**222**)	*Staphylococcus aureus*, *Bacillus subtilis*, *Escherichia coli*, and *Pseudomonas aeruginosa*	MIC values between 9.7 and 32.0 µg/mL	[109]
93	*Penicillium brefeldianum*	*Syzygium zeylanicum*		*p*-Hydroxybenzaldehyde (**223**),	*S. typhi*, *E. coli*, and *B. subtilis*	MIC values of 64 g/mL	[110]
94	*Penicillium vulpinum* GDGJ-91	*Sophorae tonkinensis*	Baise, Guangxi Province, China	10-Demethylated andrastone A (**224**), andrastin A (**227**)	*Bacillus megaterium*	MIC value of 6.25 µg/mL	[111]
				Citreohybridone E (**225**), citreohybridonol (**226**), citreohybridone B (**228**)	*B. megaterium*	MIC values of 25, 12.5 and 25 µg/mL	
				Citreohybridonol (**226**)	*B. paratyphosus* B, *E. coli* and *S. aureus*	MIC, 6.25, 25 and 25 µg/mL	
				10-Demethylated andrastone A (**224**), citreohybridone E (**225**), andrastin A (**227**), andrastin B (**228**)	*B. paratyphosus* B	MIC, 12.5 or 25 µg/mL.	

Table 1. *Cont.*

Sr. No.	Fungus	Source	Locality	Compounds Isolated	Biological Target	Biological Activity (MIC/IC$_{50}$/ID$_{50}$)	Reference
95	*Penicillium nothofagi* P-6,	*Abies beshanzuensis*	Baishanzu Mountain in Lishui, Zhejiang Province of China	Chromenopyridin A (**229**), viridicatol (**230**)	*S. aureus* ATCC29213	MIC, 62.5 and 15.6 µg/mL	[112]
96	*Penicillium restrictum* (strain G85)	*Silybum marianum*	Horizon Herbs, LLC (Williams, OR, USA).	ω-Hydroxyemodin (**231**)	Clinical isolates of MRSA	Quorum-sensing inhibition in both in vitro and in vivo models	[113]
97	*Penicillium vulpinum*	*S. tonkinensis*	Baise, Guangxi Province, China	(−)-3-Carboxypropyl-7-hydroxyphthalimide (**232**)	*Shigella dysenteriae* and *Enterobacter areogenes*	MIC, 12.5 µg/mL each	[114]
					B. subtilis	MIC, 25 µg/mL	
					B. megaterium and *Micrococcus lysodeikticus*	MIC, 50 µg/mL	
				(−)-3-Carboxypropyl-7-hydroxyphthalide methyl ester (**233**)	E. areogenes	MIC, 12.5 µg/mL	
					B. subtilis, B. megaterium and *M. lysodeikticus*	MIC, 100 µg/mL.	
98	*Penicillium sumatrense* GZWMJZ-313	Leaf of *Garcinia multiflora*	Libo, Guizhou Province of China	Citridone E (**234**), (–)-dehydrocurvularin (**235**)	*S. aureus, P. aeruginosa, Clostridium perfringens,* and *E. coli*	MIC values ranging from 32 to 64 µg/mL	[115]
99	*Penicillium ochrochloronthe*	Roots of Taxus media	Qingfeng Mountain, Chongqing, China	3,4,6-Trisubstituted α-pyrone derivatives, namely 6-(2′R-hydroxy-3′E,5′E-diene-1′-heptyl)-4-hydroxy-3-methyl-2H-pyran-2-one (**236**), 6-(2′S-hydroxy-5′E-ene-1′-heptyl)-4-hydroxy-3-methyl-2H-pyran2-one (**237**), 6-(2′S-hydroxy-1′-heptyl)-4-hydroxy-3-methyl-2H-pyran-2-one (**238**), trichodermic acid (**239**)	*B. subtilis, Micrococcus luteus, S. aureus, B. megaterium, Salmonella enterica, Proteusbacillm vulgaris, Salmonella typhi, P. aeruginosa, E. coli* and *Enterobacter aerogenes*	MIC values ranging from 25 to 50 µg/mL	[116]

Table 1. *Cont.*

Sr. No.	Fungus	Source	Locality	Compounds Isolated	Biological Target	Biological Activity (MIC/IC$_{50}$/ID$_{50}$)	Reference
100	*Penicillium janthinellum* SYPF 7899	*Panax notoginseng*	Wenshan region, Yunnan province, China	Brasiliamide J-a (**240**), brasiliamide J-b (**241**)	*B. subtilis* and *S. aureus*	MIC, 15 and 18 µg/mL,	[117]
				Peniciolidone (**242**), austin (**243**)	*B. subtilis*	MIC, 35 and 50 µg/mL	
					S. aureus	MIC 39, and 60 µg/mL	
101	*Penicillium cataractum* SYPF 7131	*Ginkgo biloba*		Penicimenolidyu A (**244**), penicimenolidyu B (**245**) and rasfonin (**246**)	*S. aureus*	MIC 65, 59 and 10 µg/mL	[118]
102	*Penicillium* sp.,	Tubers of *Pinellia ternata*	suburb of Nanjing, Jiangsu, China.	3′-Methoxycitreovirone (**247**), citreovirone (**249**)	*E. coli* and *S. aureus*	MIC = 62.6 and 76.6 µg/mL	[119]
				Helvolic acid (**100**)	*S. aureus, P. aeruginosa, B. subtilis* and *E. coli*	MIC = 5.8, 4.6, 42.2 and 75.0 µg/mL	
				cis-bis-(Methylthio)-silvatin (**248**), trypacidin A (**250**)	*S. aureus*	MIC values of 43.4 and 76.0 µg/mL	
				Trypacidin A (**250**)	*B. subtilis*	MIC = 54.1 µg/mL	
103	*Penicillium* sp. R22	*Nerium indicum*	Qinling Mountain, Shaanxi Province, China	Viridicatol (**251**)	*S. aureus*	MIC value of 15.6 µg/mL	[120]
104	*Penicillium* sp. (NO. 24)	*Tapiscia sinensis*	Shennongjia National Forest Park China	Penicitroamide (**252**)	*Erwinia carotovora* subsp. Carotovora	MIC$_{50}$ at 45 µg/mL	[121]
105	*Penicillium* sp. CAM64	Leaves of *Garcinia nobilis*	Mount Etinde, Southwest region Cameroon	Penialidin A (**253**)	*Vibrio cholerae* SG24 (1), *V. cholerae* CO6, *V. cholerae* NB2, *V. cholerae* PC2, *S. flexneri* SDINT,	MIC, 8–32 µg/mL	[122]
				Penialidin B (**254**)		MIC, 4–32 µg/mL	
				Penialidin C (**255**)		MIC, 0.50, 16, 8, 0.50 and 8 µg/mL	
				Citromycetin (**256**), brefelfin A (**258**)		MIC, 64–128 µg/mL	
				p-Hydroxyphe nylglyoxalaldoxime (**257**)		MIC, 32–64 µg/mL	

Table 1. *Cont.*

Sr. No.	Fungus	Source	Locality	Compounds Isolated	Biological Target	Biological Activity (MIC/IC$_{50}$/ID$_{50}$)	Reference
106	*Purpureocillium lilacinum*	roots of *Rauvolfia macrophylla*	Mount Kalla in the Center Region of Cameroon	Purpureone (**259**)	*B. cereus, L. monocytogenes, E. coli* ATCC 8739, *K. pneumoniae* ATCC 1296, *P. stuartii* ATCC 29916, *P. aeruginosa* ATCC PA01	Zone of inhibition of 10.6, 12.3, 13.0, 8.7, 12.3, and 10.0 mm against (10 µL/6 mm Filter paper disks).	[123]
	Fusarium						
	Neocosmospora sp. MFLUCC 17-0253	*Rhizophora apiculate.*		Mixture of 2-methoxy-6-methyl-7-acetonyl-8-hydroxy-1,4-naphthalenedione (**260**), and 5,8-dihydroxy-7-acetonyl-1,4-naphthalenedione (**261**)	*Acidovorax citrulli*	MIC value of 0.0075 mg/mL	[124]
				Anhydrojavanicin (**262**)		0.004 mg/mL	
				Fusarnaphthoquinone (**263**)		0.025 mg/mL	
107	*Fusarium* sp.	*Mentha longifolia*	Al Madinah Al Munawwarah, Saudi Arabia.	Fusaribenzamide A (**264**)	*S. aureus* and *E. coli*	MICs, 62.8 and 56.4 µg/disc	[125]

Table 1. *Cont.*

Sr. No.	Fungus	Source	Locality	Compounds Isolated	Biological Target	Biological Activity (MIC/IC$_{50}$/ID$_{50}$)	Reference
108	*F. proliferatum* AF-04	Green Chinese onion		5-O-Methylsolaniol (**270**), 5-O-methyljavanicin (**271**), methyl ether fusarubin (**272**), anhydrojavanicin (**273**)	*B. megaterium*	MICs 25 µg/mL each.	[126]
				5-O-Methylsolaniol (**270**), 5-O-methyljavanicin (**271**), methyl ether fusarubin (**272**)	*B. subtilis*	MICs, 50 µg/mL each.	
				Indol-3-acetic acid (**265**), beauvericin (**267**), epicyclonerodiol oxide (**269**)	*B. megaterium*	MICs 50 µg/mL each	
				Cyclonerodiol (**268**)	*B. megaterium*	MIC 12.50 µg/mL.	
				epi-Cyclonerodiol oxide (**269**), methyl ether fusarubin (**272**)	*E. coli*	MIC 50 µg/mL	
				5-O-Methylsolaniol (**270**), 5-O-methyljavanicin (**271**), anhydrojavanicin (**273**)	*E. coli*	MIC 25 µg/mL	
				epi-Cyclonerodiol oxide (**269**), 1,4-naphthoquinones, 5-O-methylsolaniol (**270**), 5-O-methyljavanicin (**271**), methyl ether fusarubin (**272**)	*Clostridium perfringens*	MICs 50, 50, 12.5 and 50 µg/mL	
				Beauvericin (**267**), fusaproliferin (**106**), 5-O-methylsolaniol (**270**), 5-O-methyljavanicin (**271**), methyl ether fusarubin (**272**), anhydrojavanicin (**273**)	MRSA	MIC value of 50, 50, 12.5, 12.5, 12.5, and 25 µg/mL respectively.	
				5-O-Methyljavanicin (**271**), methyl ether fusarubin (**272**), anhydrojavanicin (**273**)	RN4220	MIC value of 50 µg/mL each.	
				Methyl ether fusarubin (**272**), anhydrojavanicin (**273**)	NewmanWT	MIC value of 50 µg/mL each.	
				Bassiatin (**266**)	NewmanWT	MIC, 50 µg/mL	

Table 1. *Cont.*

Sr. No.	Fungus	Source	Locality	Compounds Isolated	Biological Target	Biological Activity (MIC/IC$_{50}$/ID$_{50}$)	Reference
109	*Fusarium* sp. TP-G1	*Dendrobium officinable*	Chongqing Academy of Chinese Materia Medica in China	Trichosetin (**274**), beauvericin (**267**), beauvericin A (**275**), enniatin H (**277**), enniatin I (**278**), enniatin MK1688 (**279**)	*S. aureus* and MRSA	IC$_{50}$ values in the range of 2–32 µg/mL	[127]
				Enniatin B (**276**)	*S. aureus* and MRSA	IC$_{50}$, 128 µg/mL each	
				Fusaric acid (**280**), dehydrofusaric acid (**281**)	*Acinetobacter baumannii*	MIC, 64 and 128 µg/mL	
	Fusarium sp. YD-2	*Santalum album*	Dongguan, Guangdong Province, China	Fusariumin A (**282**)	*S. aureus* and *P. aeruginosa*	MIC, 6.3 µg/mL	[128]
				Asperterpenoid A (**283**)	*Salmonella enteritidis* and *Micrococcus luteus*	MIC, 25.2 and 6.3 µg/mL	
				Agathic acid (**284**)	*B. cereus* and *M. luteus*	MIC, 12.5 and 25.4 µg/mL	
110	*Fusarium chlamydosporium*	Leaves of *Anvillea garcinii*	Al-Azhar University campus, Egypt	Fusarithioamide B (**285**)	*E. coli*, *B. cereus*, and *S. aureus*	MIC value of 3.7, 2.5 and 3.1 µg/mL	[129]
111	*Fusarium solani* A2	*Glycyrrhiza glabra*	Kashmir Himalayas of Jammu and Kashmir State, India	3,6,9-Trihydroxy-7-methoxy-4,4-dimethyl-3,4-dihydro-1*H*-benzo[g]-isochromene-5,10-dione (**286**), fusarubin (**287**), 3-*O*-methylfusarubin (**288**), javanicin (**289**)	*S. aureus* (MTCC 96), *K. pneumonia* (MTCC 109), *S. pyogenes* (MTCC 442), *B. subtilis* (MTCC 121), *B. cereus* (IIIM 25), *Micrococcus luteus* (MTCC 2470) and *E. coli* (MTCC 730)	MIC values in the range of <1 to 256 µg/mL.	[130]
				Fusarubin (**287**)	*Mycobacterium tuberculosis* strain H37Rv	MIC, 8 µg/mL,	
				3,6,9-Trihydroxy-7-methoxy-4,4-dimethyl-3,4-dihydro-1*H*-benzo[g]-isochromene-5,10-dione (**286**), 3-*O*-methylfusarubin (**288**), javanicin (**289**)		MIC values of 256, 64, 32 µg/mL	

Table 1. *Cont.*

Sr. No.	Fungus	Source	Locality	Compounds Isolated	Biological Target	Biological Activity (MIC/IC$_{50}$/ID$_{50}$)	Reference
112	*Fusarium chlamydosporium*	*Anvillea garcinii*	Al-Azhar University, Saudi Arabia	Fusarithioamide A (**290**)	*B. cereus, S. aureus,* and *E. coli*	MICs values of 3.1, 4.4, and 6.9 µg/mL	[131]
113	*Fusarium* sp.	*Rhoeo spathacea*	Pondok Cabe, Banten, Indonesia.	Javanicin (**289**)	*M. tuberculosis* and *M. phlei*	MIC 25 and 50 µg/mL	[132]
114	*Fusarium* sp.	*Ficus carica*	Qinling Mountain, Shaanxi Province, China	Helvolic acid Me ester (**291**)	*B. subtilis, S. aureus, E. coli* and *P. aeruginosa*	MIC, 6.25, 12.5, 6.25, and 3.13 µg/mL	[133]
				Helvolic acid (**100**)		MICs 6.25, 6.25, 6.25, and 3.13 µg/mL	
				hydrohelvolic acid (**292**)		MICs 6.25, 12.5, 6.25, and 3.13 µg/mL	
115	*Fusarium* sp.	-	-	Colletorin B (**293**), 4,5-dihydroascochlorin (**294**)	*B. megaterium*	5 and 10 mm zone of inhibition at 10 µg/mL concentration of	[134]
116	*Fusarium* sp.	*Opuntia dillenii*	South-Eastern arid zone of Sri Lanka	Equisetin (**295**)	*B. subtilis*	MIC, 8 µg/mL	[135]
					S. aureus and MRSA.	MIC, 16 µg/mL	
117	*Trichoderma harzianum*	*Zingiber officinale*	Banyumas, Central Java, Indonesia	Pretrichodermamide A (**296**)	*M. tuberculosis*	MIC, 25 µg/mL (50 µM)	[136]
118	*Trichoderma koningiopsis* YIM PH30002	*Panax notoginseng*		Koninginin W (**297**), koninginin D (**298**), 7-O- and koninginin A (**301**)	*B. subtilis*	MIC of 128 µg/mL.	[137]
				Koninginin W (**297**), 7-O-methylkoninginin D (**299**)	*S. typhimurium*	MIC, 64 and 128 µg/mL;	
				Koninginin W (**297**), koninginin (**300**)	*E. coli*	MIC of 128 µg/mL.	
119	*Trichoderma virens* QA-8	*Artemisia argyi*		Trichocarotins I–M (**302–306**), CAF-603 (**307**), 7β-hydroxy CAF-603 (**308**), trichocarotins E–H (**309–312**), and trichocarane A (**313**)	*E. coli* EMBLC-1,	MIC values ranging from 0.5 to 32 µg/mL MIC = 0.5 µg/mL	[138]
				7β-Hydroxy CAF-603 (**308**)	*Micrococcus luteus* QDIO-3		

Table 1. *Cont.*

Sr. No.	Fungus	Source	Local-ity	Compounds Isolated	Biological Target	Biological Activity (MIC/IC$_{50}$/ID$_{50}$)	Reference
120	*Trichoderma koningiopsis* QA-3	*Artemisia argyi.*		Trichodermaketone E (**314**), trichopyranone A (**316**), 3-hydroxyharziandione (**317**) and 10,11-dihydro-11-hydroxycyclonerodiol (**318**), harziandione (**321**)	*E. coli*	MIC values ranging from 0.5 to 64 µg/mL	[139]
				Trichopyranone A (**316**), 3-hydroxyharziandione (**317**), 10,11-dihydro-11-hydroxycyclonerodiol (**318**), cyclonerodiol (**319**), 6-(3-hydroxypent-1-en-1-yl)-2H-pyran-2-one (**320**), harziandione (**321**)	*M. luteus*	MIC values ranging from 1 to 16 µg/mL	
				Trichodermaketone E (**314**), 4-epi-7-O-methylkoninginin D (**315**), 3-hydroxyharziandione (**317**), 10,11-dihydro-11-hydroxycyclonerodiol (**318**), cyclonerodiol (**319**), 6-(3-hydroxypent-1-en-1-yl)-2H-pyran -2-one (**320**), harziandione (**321**)	*P. aeruginosa*	with MIC values ranging from 4 to 16 µg/mL	
				Trichodermaketone E (**314**), 10,11-dihydro-11-hydroxycyclonerodiol (**318**), cyclonerodiol (**319**), 6-(3-hydroxypent-1-en-1-yl)-2H-pyran-2-one (**320**), harziandione (**321**)	*V. parahaemolyticus*	MIC values ranging from 4 to 16 µg/mL.	
				3-Hydroxyharziandione (**317**)	*E. coli*	MIC value of 0.5 µg/mL	
				6-(3-Hydroxypent-1-en-1-yl)-2*H*-pyran-2-one (**320**)	*M. luteus*	MIC value of 1 µg/mL	

Table 1. *Cont.*

Sr. No.	Fungus	Source	Locality	Compounds Isolated	Biological Target	Biological Activity (MIC/IC$_{50}$/ID$_{50}$)	Reference
121	*Trichoderma koningiopsis* QA-3	*Artemisia argyi*	Qichun of the Hubei Province, China	15-Hydroxy-1,4,5,6-tetra-*epi*-koninginin G (**322**)	*Vibrio alginolyticus*	MIC, 1 µg/mL	[140]
				Koninginin U (**323**), 14-ketokoninginin B (**324**)	*Vibrio harveyi* and *Edwardsiella tarda*	MICs 4 and 2 µg/mL	
122	*Trichoderma atroviride* B7	*Colquhounia coccinea* var. *mollis*	Kunming Botanical Garden, Yunnan, China	Harzianol I (**325**)	*S. aureus*, *B. subtilis*, and *M. luteus*	EC$_{50}$ 7.7, 7.7, and 9.9 µg/mL	[141]
123	*Trichoderma longibrachiatum* MD33	*Dendrobium nobile*	Jinshishi, Chishui, China	Dendrobine (**326**)	*Bacillus mycoides*, *B. subtilis*, and *Staphylococcus*	Zone of inhibition of 9, 12 and 8 mm	[142]
124	*Trichoderma virens* QA-8,	*Artemisia argyi*	Qichun of Hubei Province in central China	Trichocadinins B-D and G (**327–330**)	*E. coli* EMBLC-1, *Aeromonas hydrophilia* QDIO-1, *Edwardsiella tarda* QDIO-2, *E. ictarda* QDIO-10, *Micrococcus luteus* QDIO-3, *P. aeruginosa* QDIO-4, *Vibrio alginolyticus* QDIO-5, *V. anguillarum* QDIO-6, *V. harveyi* QDIO-7, *V. parahemolyticus* QDIO-8, and *V. vulnificus* QDIO-9	MIC in the range of 8–64 µg/mL	[143]
				Trichocadinin G (**330**)	*Ed. tarda* and *V. anguillarum*	MIC values of 1 and 2 µg/mL	
125	*Trichoderma koningiopsis* A729	*Morinda officinalis*		Koninginols A-B (**331–332**)	*B. subtilis*	MIC values of 10 and 2 µg/mL	[144]

Table 1. *Cont.*

Sr. No.	Fungus	Source	Locality	Compounds Isolated	Biological Target	Biological Activity (MIC/IC$_{50}$/ID$_{50}$)	Reference
126	*Trichoderma koningiopsis* QA-3	*Artemisia argyi*	Qichun	Ent-koninginin A (**333**)	*V. vulnificus*	MIC, 4 µg/mL	[145]
				Ent-koninginin A (**333**), trichoketide A (**339**)	*E. coli, E. tarda, V. anguillarum,* and *V. parahemolyticus*	MICs ranging from 8 to 64 µg/mL	
				Ent-koninginin A (**333**), 1,6-di-*epi*-koninginin A (**334**), 15-hydroxykoninginin A (**335**), 10-deacetylkoningiopisin D (**336**), koninginin T (**337**), koninginin L (**338**), trichoketide A (**339**)	*E. coli*	MIC, 64 µg/mL each	
					E. tarda, V. alginolyticus, and *V. anguillarum*	MIC values ranging from 4 to 64 µg/mL	
	Alternaria						
127	*Alternaria alternata* ZHJG5	*Cercis chinensis*		Isotalaroflavone (**340**), 4-hydroxyalternariol-9-methyl ether (**341**), verrulactone A (**342**)	*Xanthomonas oryzae* pv. Oryzae, *Xanthomonas oryzae* pv. oryzicola and *Ralstonia solanacearum* (Rs)	MIC ranging from 0.5 to 64 µg/mL.	[146]
128	*Alternaria* sp. PfuH1	*Pogostemon cablin* (Pacholi).		Alternariol (**44**), altertoxin VII (**343**), altenuisol (**344**)	*S. agalactiae*	MIC, 9.3, 17.3 and 85.3 µg/mL	[147]
				Altenuisol (**344**)	*E. coli*	MIC, 128 µg/mL	
129	*Alternaria alternata* ZHJG5	*Cercis chinensis*		Alternariol (**44**), altenuisol (**344**), alterlactone (**345**), Dehydroaltenusin (**346**)	FabH of *Xanthomonas oryzae* pv. *oryzae* (Xoo)	IC$_{50}$ values from 29.5 to 74.1 µM	[148]
					Xanthomonas oryzae pv. Oryzae	MIC values from 4 to 64 µg/mL.	
				Alternariol (**44**), alterlactone (**345**)	Rice bacterial leaf blight	a protective efficiency of 66.2 and 82.5% at the concentration of 200 µg/mL	

Table 1. *Cont.*

Sr. No.	Fungus	Source	Locality	Compounds Isolated	Biological Target	Biological Activity (MIC/IC$_{50}$/ID$_{50}$)	Reference
130	*Alternaria alternata* MGTMMP031	*Vitex negundo*	Madurai, Tamil Nadu, India	Alternariol Me ether (**347**)	*B. cereus, Klebsiella pneumoniae*	MIC, 30 µM/L	[149]
					E. coli, Salmonella typhi, Proteus mirabilis, S. aureus and *S. epidermidis*	MIC, 35 µM/L	
131	*Alternaria alternata*	*Grewia asiatica*		3,7-Dihydroxy-9-methoxy-2-methyl-6*H*-benzo[c]chromen-6-one (**348**)	*S. aureus* (ATCC 29213), VRE, and MRSA	MIC, 32, 32 and 8 µg/mL	[150]
				Alternariol (**44**)	*S. aureus* (ATCC 29213), VRE, and MRSA	MIC, 128, 128, and 64 µg/mL	
132	*Alternaria* sp. Samif01	*Salvia miltiorrhiza*	Beijing Medicinal Plant Garden, Beijing, China	Altenuisol (**344**), 4-hydroxyalternariol-9-methyl ether (**349**) and alternariol (**44**)	*A. tumefaciens, B. subtilis, Pseudomonas lachrymans, Ralstonia solanacearum, Staphylococcus hemolyticus* and *Xanthomonas vesicatorya*	MIC values in the range of 86.7–364.7 µM	[151]
133	*Alternaria* sp. Samif01	*Salvia miltiorrhiza*	Beijing, China	Alternariol 9-Me ether (**347**)	*Bacillus subtilis* ATCC 11562 and *Staphylococcus haemolyticus* ATCC 29970, *A. tumefaciens* ATCC 11158, *Pseudomonas lachrymans* ATCC 11921, *Ralstonia solanacearum* ATCC 11696, and *Xanthomonas vesicatoria* ATCC 11633	IC$_{50}$ values varying from 16.00 to 38.27 g/mL	[152]

Table 1. *Cont.*

Sr. No.	Fungus	Source	Locality	Compounds Isolated	Biological Target	Biological Activity (MIC/IC$_{50}$/ID$_{50}$)	Reference
134	*Alternaria* sp. and *Pyrenochaeta* sp.,	*Hydrastis canadensis*	William Burch in Hendersonville, North Carolina	Altersetin (**350**), macrosphelide A (**351**)	*S. aureus*	MIC, 0.23, and 75 µg/mL	[153]
135	*Simplicillium lanosoniveum*	*Hevea brasiliensis*	Songkhla Province, Thailand	Simplicildones K (**352**)	*S. aureus* ATCC25923, MRSA	MIC, 128µg/mL	[154]
				Botryorhodine C (**353**), simplicildones A (**354**)	*S. aureus* ATCC25923, MRSA	MIC, 32 µg/mL each	
136	*Simplicillium* sp. PSU-H41	*Hevea brasiliensis*	Songkhla Province, Thailand	Botryorhodine C (**353**), simplicildone A (**354**)	*S. aureus*	MIC, 32 µg/mL each	[155]
				Botryorhodine C (**353**)	MRSA	MIC, 32 µg/mL	
	Cladosporium						
137	*Cladosporium cladosporioides*	*Zygophyllum mandavillei*	Al-Ahsa, Saudi Arabia	Isocladosporin (**355**), 5′- hydro xyasperentin (**356**), 1-acetyl-17-methoxyaspidospermidin-20-ol (**357**), and 3-phenylpropionic acid (**358**)	*Xanthomonas oryzae* and *Pseudomonas syringae*	MIC values in the range of 7.81 to 125 µg/mL	[156]
138	*Cladosporium sphaerospermum* WBS017	*Fritillaria unibracteata* var. *wabuensis*	Western Sichuan Plateau of China	Cladosin L (**359**)	*S. aureus* ATCC 29213 and *S. aureus* ATCC 700699	MICs, 50 and 25 mM,	[157]
139	*Cladosporium* sp.	*Rauwolfia serpentina*		Me ether of fusarubin (**360**)	*S. aureus, E. coli, P. aeruginosa* and *B. megaterium*	Zone of inhibition of 27, 25, 24 and 22 mm (40µg/disk)	[158]
	Pestalotiopsis						
140	*Pestalotiopsis* sp. M-23	*Leucosceptrum canum*	Kunming Botanical Garden, China	11-Dehydro-3a-hydro xyisodrimeninol (**361**)	*B. subtilis*	IC$_{50}$, 280.27 µM	[159]

Table 1. *Cont.*

Sr. No.	Fungus	Source	Locality	Compounds Isolated	Biological Target	Biological Activity (MIC/IC$_{50}$/ID$_{50}$)	Reference
141	*Pestalotiopsis* sp.	*Melaleuca quinquenervia*	Toohey Forest, Queensland, Australia	(1*S*,3*R*)-austrocortirubin (**362**), (1*S*,3*S*)-austrocortirubin (**363**), 1-deoxyaustrocortirubin (**364**)	Gram-pos.	100 µM	[160]
142	*Neopestalotiopsis* sp.			Neopestalotins B (**365**)	*B. subtilis, S. aureus, S. pneumoniae*	MIC, 10, 20, and 20 µg/mL	[161]
Phoma							
143	*Phoma cucurbitacearum*	*Glycyrrhiza glabra*	Jammu (J&K).	Thiodiketopiperazine derivatives (**366**) and (**367**)	*S. aureus* and *Streptococcus pyogenes*	IC$_{50}$, 10 µM	[162]
144	*Phoma* sp. JS752	*Phragmites communis*	Seochun, South Korea	Barceloneic acid C (**368**)	*Listeria monocytogenes* and *Staphylococcus pseudintermedius*	MIC, 1.02 µg/mL each	[163]
145	*Setophoma* sp.,	*Psidium guajava* fruits		Thielavins T (**369**), U (**370**) and V (**371**)	*S. aureus* ATCC 25923	MIC, 6.25, 50, and 25 µg/mL	[164]
Colletotrichum							
146	*Colletotrichum gloeosporioides* B12	*Illigera rhodantha*	Qionghai City, Hainan Province, China	Colletolides A (**372**) and B (**373**), and 3-methyleneis oindolinon (**374**)	*Xanthomonas oryzae* pv. *oryzae,*	MIC, 128 µg/mL each	[165]
				Sclerone (**375**)	*X. oryzae* pv. *oryzae*	MIC, 64 µg/mL	
147	*Colletotrichum* sp. BS4	*Buxus sinica*	Guangzhou, Guangdong Province, China	Colletotrichones A (**376**)	*E. coli* and *B. subtilis*	MIC, 1.0 and 0.1 µg/mL	[166]
				Colletotrichone B (**377**)	*S. aureus* (DSM 799)	MIC, 5.0 µg/mL	
				Colletotrichone C (**378**)	*E. coli*	MIC, 5.0 µg/mL	

Table 1. *Cont.*

Sr. No.	Fungus	Source	Locality	Compounds Isolated	Biological Target	Biological Activity (MIC/IC$_{50}$/ID$_{50}$)	Reference
	Minor Taxa of Anamorphic Ascomycetes						
148	*Rhizopycnis vagum* Nitaf22 (synonym *Acrocalymma vagum*)	*Nicotiana tabacum*	Agricultural University Beijing China	Rhizopycnolide A (**379**)	*A. tumefaciens, B. subtilis,* and *P. lachrymans*	MICs 100, 75, and 100 µg/mL	[167]
				Rhizopycnin C (**380**), penicilliumolide D (**384**), alternariol (**44**)	*A. tumefaciens, B. subtilis, Pseudomonas lachrymans, Ralstonia solanacearum, Staphylococcus hemolyticus,* and *Xanthomonas vesicatoria,*	MICs in the range 25–100 µg/mL	
				Rhizopycnin D (**381**)	*A. tumefaciens, B. subtilis,* and *R. solanacearum,*	MIC 50 µg/mL each,	
					X. vesicatoria	MIC, 75 µg/mL.	
				Palmariol B (**383**), Alternariol 9-methyl ether (**347**)	*A. tumefaciens, B. subtilis, P. lachrymans, R. solanacearum,* and *X. vesicatoria,*	IC$_{50}$ values in the range 16.7−34.3 µg/mL	
				TMC-264 (**382**)	*B. subtilis*	MIC 50 µg/mL	
149	*Rhizopycnis vagum* Nitaf22 (synonym *Acrocalymma vagum*)	*Nicotiana tabacum*	China Agricultural University, Beijing	Rhizoperemophilane K (**385**), 1α-hydroxyhydroisofukinon (**386**), 2-oxo-3-hydroxyeremophila-1(10),3,7(11), 8-tetraen-8,12-olide (**387**)	*A. tumefaciens, B. subtilis, P. lachrymans, R. solanacearum, S. haemolyticus,* and *X. vesicatoria,*	MIC, 32~128 µg/mL	[168]

Table 1. *Cont.*

Sr. No.	Fungus	Source	Locality	Compounds Isolated	Biological Target	Biological Activity (MIC/IC$_{50}$/ID$_{50}$)	Reference
150	*Rhizopycnis vagum* Nitaf22 (synonym *Acrocalymma vagum*)	*Nicotiana tabacum*	China Agricultural University (CAU), Beijing 100101, China	Rhizopycnis acid A (**388**)	*A. tumefaciens, B. subtilis, P. lachrymans, R. solanacearum, S. hemolyticus* and *X. vesicatoria*	MICs, 20.82, 16.11, 23.48, 29.46, 21.11, and 24.31 µg/mL	[169]
				Rhizopycnis acid B (**389**)		MICs, 70.89, 81.28, 21.23, 43.40, 67.61, and 34.86 µg/mL	
151	*Leptosphaeria* sp. XL026	*Panax notoginseng*	Shijiazhuang, Hebei province, China	Leptosphin B (**390**), conidiogenone C (**391**), conidiogenone D (**392**), conidiogenone G (**393**)	*B. cereus*	MICs 12.5–6.25 µg/mL	[170]
				Conidiogenone D (**392**)	*P. aeruginosa*	MIC, 12.5 µg/mL	
152	*Lophiostoma* sp. Eef-7	*Eucalyptus exserta.*		Scorpinone (**394**), 5-deoxybostrycoidin (**395**)	*Ralstonia solanacearum*	Zone of inhibition of 9.86 and 9.58 mm at 64 µg concentration	[171]
	Lophiostoma sp. Sigrf10	*Siraitia grosvenorii*	Guangxi Province of China	(8*R*,9*S*)-dihydroisoflavipucine (**396**), (8*S*,9*S*)-dihydroisoflavipucine (**397**)	*B. subtilis, A. tumefaciens, Ralstonia solanacearum,* and *Xanthomonas vesicatoria*	IC$_{50}$ in the range of 35.68–44.85 µM	[172]
153	*Cytospora chrysosperma*	Hippophae rhamnoides		Cytochrysin A (**398**)	Enterococcus faecium	MIC, 25 µg/mL	[173]
				Cytochrysin C (**399**)	MRSA	MIC, 25 µg/mL	
154	*Microsphaeropsis* sp. *Seimatosporium* sp.	*Salsola oppositifolia*	Gomera, Spain	Microsphaerol (**400**)	*B. megaterium* and *E. coli,*	Zone of inhibition 8 and 9 mm at 0.05 mg concentration	[174]
				Seimatorone (**401**)	*B. megaterium* and *E. coli,*	Zone of inhibition 3 and 7 (partial) mm at a 0.05 mg concentration	

Table 1. *Cont.*

Sr. No.	Fungus	Source	Locality	Compounds Isolated	Biological Target	Biological Activity (MIC/IC$_{50}$/ID$_{50}$)	Reference
155	*Epicoccum nigrum* MK214079	*Salix* sp.	*Caucasus mountains* Lago-Naki, Russia	Epicocconigrone A (**402**), epipyrone A (**403**), and epicoccolide B (**404**)	*S. aureus* ATCC 29213	MIC values ranging from 25 to 50 μM	[175]
156	*Epicoccum nigrum*	*Entada abyssinica*	Balatchi (Mbouda), in the West region of Cameroon	p-Hydroxybenzaldehyde (**223**)	*S. aureus, B. cereus, P. aeruginosa,* and *E. coli*	MICs 50, 25, 50, and 25 μg/mL	[176]
				Beauvericin (**267**)	*S. aureus, B. cereus,* and *Salmonella typhimurium*	MICs 3.12, 12.5, and 12.5 μg/mL	
				Indole-3-carboxylic acid (**405**)	*S. aureus* and *E. faecalis*	MIC values of 6.25 and 50 μg/mL	
				Quinizarin (**406**)	*S. aureus, B. cereus* St	MIC values of 50 μg/mL each	
157	*Stemphylium lycopersici*	*S. tonkinensis*		Xylapeptide B (**407**)	*B. subtilis, S. aureus* and *E. coli*	MIC, 12.5, 25 and 25 μg/mL	[177]
				Cytochalasin E (**408**)	*B. subtilis, S. aureus, B. anthracis, S. dysenteriae,* and *E. coli*	MIC 12.5 to 25 μg/mL	
				6-Heptanoyl-4-methoxy-2H-pyran2-one (**409**)	*S. paratyphi* B	MIC, 12.5 μg/mL	
				(–)-5-Carboxymellein (**410**)	*B. subtilis, S. aureus, B. anthracis, S. dysenteriae, S. paratyphi, E. coli* and *S. paratyphi* B	MIC values from 12.5 to 25 μg/mL	
158	*Stemphylium globuliferum,*	*Juncus acutus*	Egypt	Dihydroaltersolanol C (**411**)	*S. aureus*	MICs of 49.7 μM	[178]

Table 1. *Cont.*

Sr. No.	Fungus	Source	Locality	Compounds Isolated	Biological Target	Biological Activity (MIC/IC$_{50}$/ID$_{50}$)	Reference
159	*Lecanicillium* sp. (BSNB-SG3.7 Strain)	*Sandwithia guyanensis*	St Elie, France.	Stephensiolides I (**412**), D (**413**), G (**414**), stephensiolide F (**415**)	MRSA	MICs 4, 32, 16 and 32 µg/mL	[179]
160	*Nigrospora sphaerica*	*Adiantum philippense*	Western Ghats region near Virajpete, India	Phomalactone (**416**)	*E. coli* and *X. campestris*	MIC 3.12 µg/mL	[180]
					S. typhi, B. subtilis, B. cereus, and *K. pneumonia*	MIC value of 6.25 µg/mL	
					S. aureus, S. epidermidis, and *C. albicans*	MIC of 12.5 µg/mL	
161	*Nigrospora* sp. BCC 47789	*Choerospondias axillaris*	Khao Yai National Park, Nakhon Ratchasima Province, Thailand	Nigrosporone B (**417**)	*M. tuberculosis, B. cereus* and *E. faecium*	MICs 172.25, 21.53 and 10.78 µM	[181]
162	*Curvularia sorghina* BRIP 15900	*Rauwolfia macrophylla*	Mount Kalla in Cameroon	2′-Deoxyribolactone (**419**), hexylitaconic acid (**419**)	*E. coli, Micrococcus luteus, Pseudomonas agarici* and *Staphylococcus warneri*	MIC ranging between 0.17 µg/mL and 0.58 µg/mL	[182]
163	*Curvularia lunata*	*Paepalanthus chiquitensis*	Serra do Cipó, in Minas Gerais State, Brazil	Triticones E (**420**), F (**421**)	*E. coli,*	MIC 62.5 µg/mL	[183]
164	*Bipolaris* sp. L1-2	*Lycium barbarum*	Ningxia Province, China	Cochlioquinones B (**422**), C (**423**), isocochlioquinones (**424**)	*B. subtilis, C. perfringens,* and *P. viridiflava*	MICs 26 µM	[184]

Table 1. *Cont.*

Sr. No.	Fungus	Source	Locality	Compounds Isolated	Biological Target	Biological Activity (MIC/IC$_{50}$/ID$_{50}$)	Reference
165	*Bipolaris eleusines*	Potatoes	nursery of Yunnan Agricultural University, Kunming, Yunnan China	(S)-5-Hydroxy-2-(1-hydroxyethyl)-7-methylchromone (**425**), 5,7-dihydroxyl-2,6,8-trimethylchromone (**426**)	*Staphylococcus aureus* subsp. *Aureus*	inhibition rates of 56.3 and 32 %, at the concentration of 128 µg/mL	[185]
166	*Bionectria* sp. Y1085,	*Huperzia serrata*	Xichou County, Yunnan Province, China	Bionectin D (**427**), bionectin E (**428**), verticillin A (**430**), sch 52901 (**429**), gliocladicillin C (**431**)	*E. coli, S. aureus,* and *S. typhimurium* ATCC 6539,	MIC values ranging from 6.25–25 µg/mL	[186]
167	*Cylindrocarpon* sp.,	*Sapium ellipticum*	Haut Plateaux region, Cameroon	Pyrrocidine A (**432**)	*S. aureus,* ATCC 25923, *S. aueus* ATCC 700699, *S. aueus* ATCC 700699, *E. faecalis* ATCC 29212, *E. faecalis* ATCC 51299, *E. faecium* ATCC 35667, *E. faecium* ATCC 700221	MIC values ranging from 0.78 to 25 µM	[187]
				19-O-Methylpyrrocidine B (**433**)	*S. aureus* ATCC25923 and ATCC700699	MIC, 50 and 25 µM,	
168	*Eupenicillium* sp. LG41.9 treated with HDAC inhibitor, nicotinamide (15 mg/100 mL)	*Xanthium sibiricum*	Taian, Shandong Province, China	Eupenicinicol C (**434**)			[188]
				Eupenicinicol D (**435**),	*S. aureus*	MIC 0.1 µg/mL,	
				Eujavanicol A (**436**)	*E. coli*	MIC 5.0 µg/mL	
				Eupenicinicol A (**437**)			
169	*Dendrothyrium variisporum*	*Globularia alypum*	Ain Touta, Batna 05000, Algeria	2-Phenylethyl 3-hydroxyanthranilate (**438**)	*B. subtilis* and *M. luteus*	MICs 8.33 and 16.66 µg/mL	[189]
				2-Phenylethyl anthranilate (**439**)	*B. subtilis* and *M. luteus*	66.67 µg/mL each	

Table 1. *Cont.*

Sr. No.	Fungus	Source	Locality	Compounds Isolated	Biological Target	Biological Activity (MIC/IC$_{50}$/ID$_{50}$)	Reference
170	*Exserohilum rostratum*	*Phanera splendens* (Kunth) Vaz		Ravenelin (**440**)	*Bacillus subtilis* and *Staphylococcus aureus*	MICs, 7.5 and 484 µM	[190]
171	*Exserohilum rostratum*	*Bauhinia guianensis*		Monocerin (**441**)	*P. aeruginosa*	MIC, 62.5 µg/mL	[191]
				Annularin I (**442**)	*E. coli* and *B. subtilis*	MIC, 62.50 and 31.25 µg/mL	
				Annularin J (**443**)	*E. coli* and *B. subtilis*	MIC, 62.50 µg/mL each	
Basidiomycete							
172	*Psathyrella candolleana*	*Ginkgo biloba*		Quercetin (**444**), carboxybenzene (**445**), and nicotinamide (**446**)	*S. aureus*	MIC 0.3906, 0.7812 and 6.25 µg/mL	[192]
173	*Irpex lacteus* DR10-1	*Distylium chinense*	Banan district of Chongqing in the TGR area, China	Irpexlacte A (**447**), irpexlacte B-D (**448–450**)	*P. aeruginosa*	MIC values ranging from 23.8 to 35.4 µM	[193]
Zygomycetes							
174	*Mucor irregularis*			Chlorflavonin (**451**)			[194]

4. Methods Used for Activation of Silent Biosynthetic Genes

It has been reported that fungi have various unexpressed gene clusters related to bioactive secondary metabolites, which do not express in mass multiplications of the axenic form [213,214]. The expression of such gene clusters directly or indirectly depends on the surrounding environment of the microorganism. In axenic form, various induction or activation signals are or may be absent for some bioactive molecule production in the culture, which are usually present in natural habitats [215]. Such biosynthetic gene clusters (BGC) are part of the heterochromatin of fungal chromosomes, which do not express at laboratory conditions [216].

To induce such silent biosynthetic gene clusters two major approaches have been reported, including pleiotropic- and pathway-specific approaches, which include various techniques like knocking down, mutation induction [217], co-culture methods [218], heterologous expression [219,220], interspecies crosstalk [221], one strain many compounds (OSMAC) [222] and epigenetic manipulation [223]. Changes in media composition and physical factors like pH, temperature, light, salt concentration, metal and elicitor also support the induction of silent BGC and improve production of secondary metabolites in microbes. The generation of various types of stresses significantly affects the metabolic activities of growing culture and microbes to release compounds for their survival under stress conditions. Changes in physical conditions or stresses impacted gene regulation by upregulating or downregulating the gene expression [126,224]. Nowadays, high throughput elicitor screening technique (HiTES) is also employed to save time in exposing culture against various types of elicitors. In this technique selected culture is grown in 96 well plates with various elicitors in each well and after the incubation period metabolites are identified by mass spectrometry or assay system.

The mutation is one of the other approaches to induce silent biosynthetic gene clusters (BGC). Mutation in RNA polymerase genes and ribosomal proteins changes the transcription and translational process and upregulates the expression of biosynthetic gene clusters. Some of the genes related to biosynthetic gene clusters are silent from decades and overexpression of *adpA*, a global regulatory gene, induced the expression of silent lucensomycin in *Streptomyces cyanogenus* S136 [225]. Cloning is another type of molecular technique used to express the silent BGC incompatible strains. In the cloning method, isolation of high-quality DNA, fragmentation, library construction and development of suitable expression vectors for large sequences of BGC is a challenging task and many groups are working on this aspect [226]. In addition to this, use of bioinformatics also helps in direct cloning of silent BGCs and their expression for secondary metabolites production. Development of various bioinformatics tools such as PRISM3, BiG-SCAPE and anti-SMASH etc facilitated the scientist to identify bioactive gene clusters in unknown strains without time consumption used in identification of active BGC sites [227]. The CRISPR-Cas system is also a excellent tool for cloning system or genome editing that provides better expression of silent BGC in comparison to conventional molecular techniques [228]. Similarly, promoter engineering, transcriptional regulation engineering and ribosome engineering also support the activation of silent BGC through molecular approaches [229]. Recent use of Cpf1 nuclease in genome editing was also found to be a suitable tool for induction of silent BGC [230].

4.1. Epigenetic Modification

On the other hand, epigenetic modification played a great role to induce the silent genes related to bioactive molecules, which are actively produced under symbiotic interactions. Epigenetics refers to the study of DNA sequences that do not changes in mutation but change in gene function [231]. The epigenetic regulations such as methylation, demethylation, acetylation, deacetylation and phosphorylation of histones also regulate the transcription of biosynthetic genes of fungi and are helpful in silencing or expression of such genes related to the production of secondary metabolites [232]. The importance of epigenetic regulation in secondary metabolite production by fungi has been shown in a few reports published [231,233–236]. Modification or alteration in DNA or chromatin changes the expression level of the selected genes, which directly impacted the biosynthesis of the metabolites in the strain.

4.2. The Co-Culture Strategy

The co-culture is another method to induce the silent biosynthetic gene clusters by interspecies cross-talking of microorganisms. In this method, various combinations of inducers with producer microbial strains are screened for the production of novel molecules. In co-culture technique real-time bioactivity screening can also be measured by the growth of pathogen as co-culture [218]. Recently, Kim et al. [237] reviewed the co-culture interactions of fungi with various actinomycetes for induction of silent biosynthetic gene clusters and reported upregulation and production of novel antibiotics and bioactive compounds. Co-culturing of microbes provides the habitat type environment to producers and helps to promote silent BGCs by producing signal molecules. Exchange of chemical signals of growing organisms is helpful in the induction of defense molecules and other silent BGC, and usually results in the production of new natural products or secondary metabolites in the culture [238].

Another concept has also been introduced to elicit the production of silent secondary metabolites by scaffold technique. In this technique, two types of scaffold named cotton and talc powder are introduced in the medium which physically interacts with the grown culture and elicit chemical signaling of the culture and activate the production of silent BGC. The addition of scaffold in the medium supports the grown culture in formation of biofilm and provides a mimic architecture of natural habitat [239,240]. The addition of scaffold in medium affects the morphology of growing culture and sporulation pattern like an agglomeration of spores, oxygen diffusion in comparison to non-scaffold containing medium and then facilitates more metabolites production [241].

4.3. OSMAC

In the OSMAC technique different cultivation approaches are applied to induce silent bioactive gene clusters to promote more production of secondary metabolites including media variations, variation in media composition, co-cultivation with other strains and variations in cultivations strategy [222,242]. Variation in growth conditions also supports the induction of silent biosynthetic gene clusters and the production of novel compounds. Scherlach and Hertweck [243] and Scherlach et al. [244] reported the production of novel aspoquinolone and aspernidine alkaloid compounds from *Aspergillus nidulans* by variation in growth conditions.

5. Conclusions

Increasing resistance among microbial pathogens against existing antibiotics has been a major concern during the past several decades. Scientists are exploring new sources of novel antibiotics and other bioactive compounds that can curb pathogenic infections and overcome antimicrobial resistance. Endophytic fungi have been reported to secrete a wide spectrum of bioactive compounds to counter pathogens. In the current review, we have reported 453 new bioactive compounds, including volatile compounds, isolated during the period of 2015-21 from various endophytic fungi belonging to the Ascomycetes, Basidiomycetes, and Zygomycetes classes. Newly reported bioactive compounds have shown activity against various pathogenic bacteria and shown scaffold similarity with alkaloids, benzopyranones, chinones, cytochalasins, mullein, peptides, phenols, quinones, flavonoids, steroids, terpenoids, sesquiterpene, tetralones, xanthones, and others. The lowest in vitro activity in terms of minimum inhibitory concentrations (MICs) in the 0.1–1 µg/mL range against various pathogens was reported for the compounds vochysiamides A (**23**) and B (**24**), colletotrichone A (**376**), 15-hydroxy-1,4,5,6-tetra-*epi*-koninginin G (**322**), trichocadinin G (**330**) and eupenicinicol D (**435**). Compounds like fusarubin (**287**), chetomin (**62**), chaetocochin C (**63**), and dethiotetra(methylthio)chetomin (**64**), pretrichodermamide A (**296**), terpestacin (**105**), fusaproliferin (**106**), mutolide (**108**), isoeugenitol (**120**) and nigrosporone B (**417**) were reported to have significant in vitro anti-mycobacterial activity and could be developed as potential drugs against resistant mycobacterial infections. The production of such bioactive compounds and their activity is also affected by the surrounding environment and conditions. Various techniques related to induction of silent gene clusters such as epigenetic modifications, co-culture, OSMAC and mutation have been reported

In most of cases only in vitro data against a limited number of bacteria is reported and there is a great need for extensive in vitro studies including their mode of action, kill curve studies, mutation induction frequency, resistance occurrence frequency studies, in vitro cytotoxicity and initial in vivo evaluation followed by formulation studies. Moreover, there is also a need to perform extensive in vitro efficacy testing studies using panels of references strains and clinical strains to establish MIC_{90} and MIC_{50} values. Generation of comparative efficacy data with benchmark clinical compounds is very important from a further development perspective. These extensive studies also help to generate data for understanding the scope of work when we consider such potent molecules for semisynthetic work. The exact studies to be performed during screening and further shortlisting of semi-synthetic molecules can be extracted from this initial extensive work.

Still, more research is required to investigate a new generation of antibiotics which can control the increasing resistance of infectious microorganisms in a sustainable manner. The success of this exploration depends upon screening more and more endophytic fungi and ways of their isolation, fermentation and scale-up.

Author Contributions: Conceptualization: (S.K.D., L.D.), Literature search and compilation: (S.K.D., H.C., S.S.); Writing abstract, introduction, conclusion, proof reading: (S.K.D., G.B.M., S.S., H.C., M.K.G.). Preparation of data tables: (M.K.G., S.K.D.). Generating structures: M.K.G., S.K.D. Overall compilation and coordination: (S.K.D., L.D.). All authors have read and agreed to the published version of the manuscript.

Funding: This research received no external funding.

Institutional Review Board Statement: Not Applicable.

Informed Consent Statement: Not Applicable.

Data Availability Statement: Not Applicable.

Conflicts of Interest: The authors declare no conflict of interest.

References

1. Deshmukh, S.K.; Verekar, S.A.; Bhave, S. Endophytic fungi: An untapped source for antibacterials. *Front. Microbiol.* **2015**, *5*, 715. [CrossRef] [PubMed]
2. Jakubczyk, D.; Dussart, F. Selected Fungal Natural Products with Antimicrobial Properties. *Molecules* **2020**, *25*, 911. [CrossRef] [PubMed]
3. Xu, T.C.; Lu, Y.H.; Wang, J.F.; Song, Z.Q.; Hou, Y.G.; Liu, S.S.; Liu, C.S.; Wu, S.H. Bioactive secondary metabolites of the genus Diaporthe and anamorph *Phomopsis* from terrestrial and marine habitats and endophytes: 2010–2019. *Microorganisms* **2021**, *9*, 217. [CrossRef] [PubMed]
4. Kim, J.W.; Choi, H.G.; Song, J.H.; Kang, K.S.; Shim, S.H. Bioactive secondary metabolites from an endophytic fungus *Phoma* sp. PF2 derived from *Artemisia princeps* Pamp. *J. Antibiot.* **2019**, *72*, 174–177. [CrossRef]
5. El-hawary, S.S.; Moawad, A.S.; Bahr, H.S.; Abdelmohsen, U.R.; Mohammed, R. Natural product diversity from the endophytic fungi of the genus *Aspergillus*. *RSC Adv.* **2020**, *10*, 22058–22079. [CrossRef]
6. Deshmukh, S.K.; Mishra, P.D.; Kulkarni-Almeida, A.; Verekar, S.A.; Sahoo, M.R.; Periyasamy, G.; Goswami, H.; Khanna, A.; Balakrishnan, A.; Vishwakarma, R. Anti-inflammatory and anti-cancer activity of ergoflavin isolated from an endophytic fungus. *Chem. Biodivers.* **2009**, *6*, 784–789. [CrossRef]
7. Martínez-Luis, S.; Cherigo, L.; Arnold, E.; Spadafora, C.; Gerwick, W.H.; Cubilla-Rios, L. Antiparasitic and anticancer constituents of the endophytic fungus *Aspergillus* sp. strain F1544. *Nat. Prod. Commun.* **2012**, *7*, 165–168. [CrossRef]
8. Deshmukh, S.K.; Verekar, S.A.; Ganguli, B.N. Fungi: An Amazing and Hidden Source of Antimycobacterial compounds. In *Fungi: Applications and Management Strategies*; Deshmukh, S.K., Misra, J.K., Tiwari, J.P., Papp, T., Eds.; CRC Press: Boca Raton, FL, USA, 2016; pp. 32–60.
9. Deshmukh, S.K.; Gupta, M.K.; Prakash, V.; Saxena, S. Endophytic Fungi: A Source of Potential Antifungal Compounds. *J. Fungi* **2018**, *4*, 77. [CrossRef]

10. Deshmukh, S.K.; Gupta, M.K.; Prakash, V.; Reddy, M.S. Mangrove-associated fungi a novel source of potential anticancer Compounds. *J. Fungi* **2018**, *4*, 101. [CrossRef]
11. Deshmukh, S.K.; Agrawala, S.; Gupta, M.K.; Patidar, R.K.; Ranjan, N. Recent advances in the discovery of antiviral metabolites from fungi. *Curr. Pharm. Biotechnol.* **2022**, *23*, 495–537. [CrossRef]
12. Wang, W.X.; Cheng, G.G.; Li, Z.H.; Ai, H.L.; He, J.; Li, J.; Feng, T.; Liu, J.K. Curtachalasins, immunosuppressive agents from the endophytic fungus *Xylaria* cf. *curta*. *Org. Biomol. Chem.* **2019**, *17*, 7985–7994. [CrossRef]
13. Bedi, A.; Gupta, M.K.; Conlan, X.A.; Cahill, D.M.; Deshmukh, S.K. Endophytic and marine fungi are potential source of antioxidants. In *Fungi Bio-Prospects in Sustainable Agriculture, Environment and Nano-Technology*; Sharma, V.K., Shah, M.P., Parmar, S., Kumar, A., Eds.; Elsevier: San Diego, CA, USA, 2021; pp. 23–89.
14. Toghueo, R.M.K.; Boyom, F.F. Endophytic Penicillium species and their agricultural, biotechnological, and pharmaceutical applications. *3 Biotech* **2020**, *10*, 1–35. [CrossRef]
15. Toghueo, R.M.K. Bioprospecting endophytic fungi from Fusarium genus as sources of bioactive metabolites. *Mycology* **2020**, *11*, 1–21. [CrossRef]
16. Selvakumar, V.; Panneerselvam, A. Bioactive compounds from endophytic fungi. In *Fungi and Their Role in Sustainable Development: Current Perspectives*; Gehlot, P., Singh, J., Eds.; Springer: Singapore, 2018; pp. 699–717.
17. Preethi, K.; Manon Mani, V.; Lavanya, N. Endophytic fungi: A potential source of bioactive compounds for commercial and therapeutic applications. In *Endophytes*; Patil, R.H., Maheshwari, V.L., Eds.; Springer: Singapore, 2021; pp. 247–272.
18. Oktavia, L.; Krishna, V.S.; Rekha, E.M.; Fathoni, A.; Sriram, D.; Agusta, A. Anti-mycobacterial activity of two natural Bisanthraquinones:(+)-1, 1′-Bislunatin and (+)-2, 2′-Epicytoskyrin A. In IOP Conference Series: Earth and Environmental Science. *IOP Publ.* **2020**, *591*, 12025.
19. Huang, X.; Zhou, D.; Liang, Y.; Liu, X.; Cao, F.; Qin, Y.; Mo, T.; Xu, Z.; Li, J.; Yang, R. Cytochalasins from endophytic Diaporthe sp. GDG-118. *Nat. Prod. Res.* **2021**, *35*, 3396–3403. [CrossRef]
20. Li, X.M.; Mi, Q.L.; Gao, Q.; Li, J.; Song, C.M.; Zeng, W.L.; Xiang, H.Y.; Liu, X.; Chen, J.H.; Zhang, C.M.; et al. Antibacterial naphthalene derivatives from the fermentation products of the endophytic fungus *Phomopsis fukushii*. *Chem. Nat. Compd.* **2021**, *57*, 293–296. [CrossRef]
21. Yang, H.Y.; Duan, Y.Q.; Yang, Y.K.; Li, J.; Liu, X.; Ye, L.; Mi, Q.L.; Kong, W.S.; Zhou, M.; Yang, G.Y.; et al. Three new naphthalene derivatives from the endophytic fungus *Phomopsis fukushii*. *Phytochem. Lett.* **2017**, *22*, 266–269. [CrossRef]
22. Yang, H.Y.; Duan, Y.Q.; Yang, Y.K.; Liu, X.; Ye, L.; Mi, Q.L.; Kong, W.S.; Zhou, M.; Yang, G.Y.; Hu, Q.F.; et al. Two new diphenyl ether derivatives from the fermentation products of an endophytic fungus *Phomopsis fukushii*. *Chem. Nat. Compd.* **2019**, *55*, 428–431. [CrossRef]
23. Gao, Y.H.; Zheng, R.; Li, J.; Kong, W.S.; Liu, X.; Ye, L.; Mi, Q.L.; Kong, W.S.; Zhou, M.; Yang, G.Y.; et al. Three new diphenyl ether derivatives from the fermentation products of an endophytic fungus *Phomopsis fukushii*. *J. Asian Nat. Prod. Res.* **2019**, *21*, 316–322. [CrossRef]
24. Li, Z.J.; Yang, H.Y.; Li, J.; Liu, X.; Ye, L.; Kong, W.S.; Tang, S.Y.; Du, G.; Liu, Z.H.; Zhou, M.; et al. Isopentylated diphenyl ether derivatives from the fermentation products of an endophytic fungus *Phomopsis fukushii*. *J. Antibiot.* **2018**, *71*, 359–362. [CrossRef]
25. Wu, F.; Zhu, Y.N.; Hou, Y.T.; Mi, Q.L.; Chen, J.H.; Zhang, C.M.; Miao, D.; Zhou, M.; Wang, W.G.; Hu, Q.F.; et al. Two new antibacterial anthraquinones from cultures of an endophytic fungus *Phomopsis* sp. *Chem. Nat. Compd.* **2021**, *57*, 823–827. [CrossRef]
26. Guo, L.; Niu, S.; Chen, S.; Liu, L. Diaporone A, a new antibacterial secondary metabolite from the plant endophytic fungus *Diaporthe* sp. *J. Antibiot.* **2020**, *73*, 116–119. [CrossRef] [PubMed]
27. Qu, H.R.; Yang, W.W.; Zhang, X.Q.; Lu, Z.H.; Deng, Z.S.; Guo, Z.Y.; Cao, F.; Zou, K.; Proksch, P. Antibacterial bisabolane sesquiterpenoids and isocoumarin derivatives from the endophytic fungus *Phomopsis prunorum*. *Phytochem. Lett.* **2020**, *37*, 1–4. [CrossRef]
28. Noriler, S.A.; Savi, D.C.; Ponomareva, L.V.; Rodrigues, R.; Rohr, J.; Thorson, J.S.; Glienke, C.; Shaaban, K.A. Vochysiamides A and B: Two new bioactive carboxamides produced by the new species Diaporthe vochysiae. *Fitoterapia* **2019**, *138*, 104273. [CrossRef] [PubMed]
29. Hu, S.S.; Liang, M.J.; Mi, Q.L.; Chen, W.; Ling, J.; Chen, X.; Li, J.; Yang, G.Y.; Hu, Q.F.; Wang, W.G.; et al. Two new diphenyl ether derivatives from the fermentation products of the endophytic fungus *Phomopsis asparagi*. *Chem. Nat. Compd.* **2019**, *55*, 843–846. [CrossRef]
30. Li, X.M.; Zeng, Y.C.; Chen, J.H.; Yang, Y.K.; Li, J.; Ye, L.; Du, G.; Zhou, M.; Hu, Q.F.; Yang, H.Y.; et al. Two new naphthalene derivatives from the fermentation products of an endophytic fungus *Phomopsis* sp. *Chem. Nat. Compd.* **2019**, *55*, 618–621. [CrossRef]
31. De Medeiros, A.G.; Savi, D.C.; Mitra, P.; Shaaban, K.A.; Jha, A.K.; Thorson, J.S.; Rohr, J.; Glienke, C. Bioprospecting of *Diaporthe terebinthifolii* LGMF907 for antimicrobial compounds. *Folia Microbiol.* **2018**, *63*, 499–505. [CrossRef]
32. Jayanthi, G.; Arun Babu, R.; Ramachandran, R.; Karthikeyan, K.; Muthumary, J. Production, isolation and structural elucidation of a novel antimicrobial metabolite from the endophytic fungus, *Phomopsis/Diaporthe theae*. *Int. J. Pharm. Biol. Sci.* **2018**, *8*, 8–26.
33. Kongprapan, T.; Xu, X.; Rukachaisirikul, V.; Phongpaichit, S.; Sakayaroj, J.; Chen, J.; Shen, X. Cytosporone derivatives from the endophytic fungus *Phomopsis* sp. PSU-H188. *Phytochem. Lett.* **2017**, *22*, 219–223. [CrossRef]

34. Yedukondalu, N.; Arora, P.; Wadhwa, B.; Malik, F.A.; Vishwakarma, R.A.; Gupta, V.K.; Riyaz-Ul-Hassan, S.; Ali, A. Diapolic acid A-B from an endophytic fungus, *Diaporthe terebinthifolii* depicting antimicrobial and cytotoxic activity. *J. Antibiot.* **2017**, *70*, 212–215. [CrossRef]

35. Sousa, J.P.B.; Aguilar-Pérez, M.M.; Arnold, A.E.; Rios, N.; Coley, P.D.; Kursar, T.A.; Cubilla-Rios, L. Chemical constituents and their antibacterial activity from the tropical endophytic fungus *Diaporthe* sp. F2934. *J. Appl. Microbiol.* **2016**, *120*, 1501–1508. [CrossRef]

36. Jouda, J.B.; Mbazoa, C.D.; Douala-Meli, C.; Sarkar, P.; Bag, P.K.; Wandji, J. Antibacterial and cytotoxic cytochalasins from the endophytic fungus *Phomopsis* sp. harbored in *Garcinia kola* (Heckel) nut. *BMC Complement Altern. Med.* **2016**, *16*, 1–9. [CrossRef]

37. Li, G.; Kusari, S.; Kusari, P.; Kayser, O.; Spiteller, M. Endophytic *Diaporthe* sp. LG23 produces a potent antibacterial tetracyclic triterpenoid. *J. Nat. Prod.* **2015**, *78*, 2128–2132. [CrossRef]

38. Patridge, E.V.; Darnell, A.; Kucera, K.; Phillips, G.M.; Bokesch, H.R.; Gustafson, K.R.; Spakowicz, D.J.; Zhou, L.; Hungerford, W.M.; Plummer, M.; et al. Pyrrolocin a, a 3-decalinoyltetramic acid with selective biological activity, isolated from Amazonian cultures of the novel endophyte *Diaporthales* sp. E6927E. *Nat. Prod. Commun.* **2015**, *10*, 1649–1654. [CrossRef]

39. Ibrahim, A.; Tanney, J.B.; Fei, F.; Seifert, K.A.; Cutler, G.C.; Capretta, A.; Miller, J.D.; Sumarah, M.W. Metabolomic-guided discovery of cyclic nonribosomal peptides from *Xylaria ellisii* sp. nov., a leaf and stem endophyte of *Vaccinium angustifolium*. *Sci. Rep.* **2020**, *10*, 1–17. [CrossRef]

40. Liang, Y.; Xu, W.; Liu, C.; Zhou, D.; Liu, X.; Qin, Y.; Cao, F.; Li, J.; Yang, R.; Qin, J. Eremophilane sesquiterpenes from the endophytic fungus *Xylaria* sp. GDG-102. *Nat. Prod. Res.* **2019**, *33*, 1304–1309. [CrossRef]

41. Zheng, N.; Yao, F.; Liang, X.; Liu, Q.; Xu, W.; Liang, Y.; Liu, X.; Li, J.; Yang, R. A new phthalide from the endophytic fungus *Xylaria* sp. GDG-102. *Nat. Prod. Res.* **2018**, *32*, 755–760. [CrossRef]

42. Zheng, N.; Liu, Q.; He, D.L.; Liang, Y.; Li, J.; Yang, R.Y. A New compound from the endophytic fungus *Xylaria* sp. from *Sophora tonkinensis*. *Chem. Nat. Compd.* **2018**, *54*, 447–449. [CrossRef]

43. Lin, X.; Yu, M.; Lin, T.; Zhang, L. Secondary metabolites of *Xylaria* sp., an endophytic fungus from Taxus mairei. *Nat. Prod. Res.* **2016**, *30*, 2442–2447. [CrossRef]

44. Zhang, Q.; Li, H.Q.; Zong, S.C.; Gao, J.M.; Zhang, A.L. Chemical and bioactive diversities of the genus Chaetomium secondary metabolites. *Mini Rev. Med. Chem.* **2012**, *12*, 127–148. [CrossRef]

45. Tantapakul, C.; Promgool, T.; Kanokmedhakul, K.; Soytong, K.; Song, J.; Hadsadee, S.; Jungsuttiwong, S.; Kanokmedhakul, S. Bioactive xanthoquinodins and epipolythiodioxopiperazines from *Chaetomium globosum* 7s-1, an endophytic fungus isolated from *Rhapis cochinchinensis* (Lour.) Mart. *Nat. Prod. Res.* **2020**, *34*, 494–502. [CrossRef] [PubMed]

46. Peng, F.; Hou, S.Y.; Zhang, T.Y.; Wu, Y.Y.; Zhang, M.Y.; Yan, X.M.; Xia, M.Y.; Zhang, Y.X. Cytotoxic and antimicrobial indole alkaloids from an endophytic fungus *Chaetomium* sp. SYP-F7950 of *Panax notoginseng*. *RSC Adv.* **2019**, *9*, 28754–28763. [CrossRef]

47. Liu, P.; Zhang, D.; Shi, R.; Yang, Z.; Zhao, F.; Tian, Y. Antimicrobial potential of endophytic fungi from *Astragalus chinensis*. *3 Biotech* **2019**, *9*, 1–9. [CrossRef] [PubMed]

48. Wang, H.H.; Li, G.; Qiao, Y.N.; Sun, Y.; Peng, X.P.; Lou, H.X. Chamiside A, a cytochalasan with a tricyclic core skeleton from the endophytic fungus *Chaetomium nigricolor* F5. *Org. Lett.* **2019**, *21*, 3319–3322. [CrossRef]

49. Yang, S.X.; Zhao, W.T.; Chen, H.Y.; Zhang, L.; Liu, T.K.; Chen, H.P.; Yang, J.; Yang, X.L. Aureonitols A and B, Two New C13-Polyketides from *Chaetomium globosum*, an endophytic fungus in *Salvia miltiorrhiza*. *Chem. Biodivers.* **2019**, *16*, e1900364. [CrossRef]

50. Ouyang, J.; Mao, Z.; Guo, H.; Xie, Y.; Cui, Z.; Sun, J.; Wu, H.; Wen, X.; Wang, J.; Shan, T. Mollicellins O–R, Four new depsidones isolated from the endophytic fungus *Chaetomium* sp. Eef-10. *Molecules* **2018**, *23*, 3218. [CrossRef]

51. Yu, F.X.; Chen, Y.; Yang, Y.H.; Li, G.H.; Zhao, P.J. A new epipolythiodioxopiperazine with antibacterial and cytotoxic activities from the endophytic fungus *Chaetomium* sp. M336. *Nat. Prod Res.* **2018**, *32*, 689–694. [CrossRef]

52. Dissanayake, R.K.; Ratnaweera, P.B.; Williams, D.E.; Wijayarathne, C.D.; Wijesundera, R.L.; Andersen, R.J.; de Silva, E.D. Antimicrobial activities of endophytic fungi of the Sri Lankan aquatic plant *Nymphaea nouchali* and chaetoglobosin A and C, produced by the endophytic fungus *Chaetomium globosum*. *Mycology* **2016**, *7*, 1–8. [CrossRef]

53. Zhao, W.T.; Shi, X.; Xian, P.J.; Feng, Z.; Yang, J.; Yang, X.L. A new fusicoccane diterpene and a new polyene from the plant endophytic fungus *Talaromyces pinophilus* and their antimicrobial activities. *Nat. Prod. Res.* **2021**, *35*, 124–130. [CrossRef]

54. Feng, L.X.; Zhang, B.Y.; Zhu, H.J.; Pan, L.; Cao, F. Bioactive metabolites from *Talaromyces purpureogenus*, an endophytic fungus from *Panax notoginseng*. *Chem. Nat. Compd.* **2020**, *56*, 974–976. [CrossRef]

55. Bingyang, Z.; Yangyang, M.; Hua, G.; Huajie, Z.; Wan, L. Absolute configuration determination of two drimane sesquiterpenoids from the endophytic fungi Talaromyces Purpureogenus of *Panax notoginseng*. *Chem. J. Chin. Univ.-Chin.* **2017**, *38*, 1046–1051.

56. Guo, J.; Ran, H.; Zeng, J.; Liu, D.; Xin, Z. Tafuketide, a phylogeny-guided discovery of a new polyketide from Talaromyces funiculosus Salicorn 58. *Appl. Microbiol. Biotechnol.* **2016**, *100*, 5323–5338. [CrossRef]

57. Zhao, Q.H.; Yang, Z.D.; Shu, Z.M.; Wang, Y.G.; Wang, M.G. Secondary metabolites and biological activities of *Talaromyces* sp. LGT-2, an endophytic fungus from *Tripterygium wilfordii*. *Iran. J. Pharm Res.* **2016**, *15*, 453–457.

58. Zhang, S.; Chen, D.; Kuang, M.; Peng, W.; Chen, Y.; Tan, J.; Kang, F.; Xu, K.; Zou, Z. Rhytidhylides A and B, two new phthalide derivatives from the endophytic fungus *Rhytidhysteron* sp. BZM -9. *Molecules* **2021**, *26*, 6092. [CrossRef]

59. Wang, J.T.; Li, H.Y.; Rao, R.; Yue, J.Y.; Wang, G.K.; Yu, Y. (±)-Stagonosporopsin A, stagonosporopsin B and stagonosporopsin C, antibacterial metabolites produced by endophytic fungus *Stagonosporopsis oculihominis*. *Phytochem. Lett.* **2021**, *45*, 157–160. [CrossRef]

60. Zhang, W.; Lu, X.; Wang, H.; Chen, Y.; Zhang, J.; Zou, Z.; Tan, H. Antibacterial secondary metabolites from the endophytic fungus Eutypella scoparia SCBG-8. *Tetrahedron Lett.* **2021**, *79*, 153314. [CrossRef]

61. Zhang, W.; Lu, X.; Huo, L.; Zhang, S.; Chen, Y.; Zou, Z.; Tan, H. Sesquiterpenes and steroids from an endophytic *Eutypella scoparia*. *J. Nat. Prod.* **2021**, *84*, 1715–1724. [CrossRef]

62. Zhang, Z.B.; Du, S.Y.; Ji, B.; Ji, C.J.; Xiao, Y.W.; Yan, R.M.; Zhu, D. New Helvolic Acid derivatives with antibacterial activities from *Sarocladium oryzae* DX-THL3, an endophytic fungus from Dongxiang wild rice (*Oryza rufipogon* Griff.). *Molecules* **2021**, *26*, 1828. [CrossRef]

63. Carrieri, R.; Borriello, G.; Piccirillo, G.; Lahoz, E.; Sorrentino, R.; Cermola, M.; Bolletti Censi, S.; Grauso, L.; Mangoni, A.; Vinale, F. Antibiotic Activity of a Paraphaeosphaeria sporulosa -Produced diketopiperazine against *Salmonella enterica*. *J. Fungi* **2020**, *6*, 83. [CrossRef]

64. Gao, Y.; Stuhldreier, F.; Schmitt, L.; Wesselborg, S.; Wang, L.; Müller, W.E.; Kalscheuer, R.; Guo, Z.; Zou, K.; Liu, Z.; et al. Sesterterpenes and macrolide derivatives from the endophytic fungus *Aplosporella javeedii*. *Fitoterapia* **2020**, *146*, 104652. [CrossRef]

65. Lai, D.; Mao, Z.; Zhou, Z.; Zhao, S.; Xue, M.; Dai, J.; Zhou, L.; Li, D. New chlamydosporol derivatives from the endophytic fungus *Pleosporales* sp. Sigrf05 and their cytotoxic and antimicrobial activities. *Sci. Rep.* **2020**, *10*, 1–9. [CrossRef]

66. Gao, Y.; Wang, L.; Kalscheuer, R.; Liu, Z.; Proksch, P. Antifungal polyketide derivatives from the endophytic fungus *Aplosporella javeedii*. *Bioorg. Med. Chem.* **2020**, *28*, 115456. [CrossRef]

67. Abbas, Z.; Siddiqui, B.S.; Shahzad, S.; Sattar, S.; Begum, S.; Batool, A.; Choudhary, M.I. Lawsozaheer, a new chromone produced by an endophytic fungus *Paecilomyces variotii* isolated from Lawsonia Alba Lam. inhibits the growth of *Staphylococcus aureus*. *Nat. Prod. Res.* **2021**, *35*, 4448–4453. [CrossRef]

68. Chen, H.L.; Zhao, W.T.; Liu, Q.P.; Chen, H.Y.; Zhao, W.; Yang, D.F.; Yang, X.L. (±)-Preisomide: A new alkaloid featuring a rare naturally occurring tetrahydro-2H-1, 2-oxazin skeleton from an endophytic fungus Preussia isomera by using OSMAC strategy. *Fitoterapia* **2020**, *141*, 104475. [CrossRef]

69. Xu, L.L.; Chen, H.L.; Hai, P.; Gao, Y.; Xie, C.D.; Yang, X.L.; Abe, I. (+)-and (−)-Preuisolactone A: A pair of caged norsesquiterpenoidal enantiomers with a tricyclo.4.4. 01, 6.02, 8. decane carbon skeleton from the endophytic fungus Preussia isomera. *Org. Lett.* **2019**, *21*, 1078–1081. [CrossRef]

70. Macabeo, A.P.G.; Cruz, A.J.C.; Narmani, A.; Arzanlou, M.; Babai-Ahari, A.; Pilapil, L.A.E.; Garcia, K.Y.M.; Huch, V.; Stadler, M. Tetrasubstituted α-pyrone derivatives from the endophytic fungus, *Neurospora udagawae*. *Phytochem. Lett.* **2020**, *35*, 147–151. [CrossRef]

71. Lai, D.; Li, J.; Zhao, S.; Gu, G.; Gong, X.; Proksch, P.; Zhou, L. Chromone and isocoumarin derivatives from the endophytic fungus *Xylomelasma* sp. Samif07, and their antibacterial and antioxidant activities. *Nat. Prod. Res.* **2021**, *35*, 4616–4620. [CrossRef]

72. Nguyen, H.T.; Kim, S.; Yu, N.H.; Park, A.R.; Yoon, H.; Bae, C.H.; Yeo, J.H.; Kim, I.S.; Kim, J.C. Antimicrobial activities of an oxygenated cyclohexanone derivative isolated from *Amphirosellinia nigrospora* JS-1675 against various plant pathogenic bacteria and fungi. *J. Appl. Microbiol.* **2019**, *126*, 894–904. [CrossRef]

73. Wu, X.; Pang, X.J.; Xu, L.L.; Zhao, T.; Long, X.Y.; Zhang, Q.Y.; Qin, H.L.; Yang, D.F.; Yang, X.L. Two new alkylated furan derivatives with antifungal and antibacterial activities from the plant endophytic fungus *Emericella* sp. XL029. *Nat. Prod. Res.* **2018**, *32*, 2625–2631. [CrossRef]

74. Wu, X.; Fang, L.Z.; Liu, F.L.; Pang, X.J.; Qin, H.L.; Zhao, T.; Xu, L.L.; Yang, D.F.; Yang, X.L. New prenylxanthones, polyketide hemiterpenoid pigments from the endophytic fungus *Emericella* sp. XL029 and their anti-agricultural pathogenic fungal and antibacterial activities. *RSC Adv.* **2017**, *7*, 31115–31122. [CrossRef]

75. Wu, Y.Z.; Zhang, H.W.; Sun, Z.H.; Dai, J.G.; Hu, Y.C.; Li, R.; Lin, P.C.; Xia, G.Y.; Wang, L.Y.; Qiu, B.L.; et al. Bysspectin A, an unusual octaketide dimer and the precursor derivatives from the endophytic fungus *Byssochlamys spectabilis* IMM0002 and their biological activities. *Eur. J. Med. Chem.* **2018**, *145*, 717–725. [CrossRef] [PubMed]

76. Kawashima, D.; Hosoya, T.; Tomoda, H.; Kita, M.; Shigemori, H. Sydowianumols A, B, and C, Three new compounds from discomycete *Poculum pseudosydowianum*. *Chem. Pharm. Bull.* **2018**, *66*, 826–829. [CrossRef] [PubMed]

77. Zhao, M.; Yuan, L.Y.; Guo, D.L.; Ye, Y.; Da-Wa, Z.M.; Wang, X.L.; Ma, F.W.; Chen, L.; Gu, Y.C.; Ding, L.S.; et al. Bioactive halogenated dihydroisocoumarins produced by the endophytic fungus Lachnum palmae isolated from *Przewalskia tangutica*. *Phytochemistry* **2018**, *148*, 97–103. [CrossRef] [PubMed]

78. Ibrahim, A.; Sørensen, D.; Jenkins, H.A.; Ejim, L.; Capretta, A.; Sumarah, M.W. Epoxynemanione A, nemanifuranones A–F, and nemanilactones A–C, from *Nemania serpens*, an endophytic fungus isolated from Riesling grapevines. *Phytochemistry* **2017**, *140*, 16–26. [CrossRef] [PubMed]

79. Amand, S.; Vallet, M.; Guedon, L.; Genta-Jouve, G.; Wien, F.; Mann, S.; Dupont, J.; Prado, S.; Nay, B. A reactive eremophilane and its antibacterial 2 (1 H)-naphthalenone rearrangement product, witnesses of a microbial chemical warfare. *Org. Lett.* **2017**, *19*, 4038–4041. [CrossRef] [PubMed]

80. Deng, Z.; Li, C.; Luo, D.; Teng, P.; Guo, Z.; Tu, X.; Zou, K.; Gong, D. A new cinnamic acid derivative from plant-derived endophytic fungus *Pyronema* sp. *Nat. Prod. Res.* **2017**, *31*, 2413–2419. [CrossRef] [PubMed]

81. Wijeratne, E.K.; Xu, Y.; Arnold, A.E.; Gunatilaka, A.L. Pulvinulin A, graminin C, and cis-gregatin B–new natural furanones from *Pulvinula* sp. 11120, a fungal endophyte of *Cupressus arizonica*. *Nat. Prod. Commun.* **2015**, *10*, 107–111. [CrossRef]

82. Forcina, G.C.; Castro, A.; Bokesch, H.R.; Spakowicz, D.J.; Legaspi, M.E.; Kucera, K.; Villota, S.; Narva'ez-Trujillo, A.; McMahon, J.B.; Gustafson, K.R.; et al. Stelliosphaerols A and B, sesquiterpene–polyol conjugates from an ecuadorian fungal endophyte. *J. Nat. Prod.* **2015**, *78*, 3005–3010. [CrossRef]

83. Hussain, H.; Jabeen, F.; Krohn, K.; Al-Harrasi, A.; Ahmad, M.; Mabood, F.; Shah, A.; Badshah, A.; Rehman, N.U.; Green, I.R.; et al. Antimicrobial activity of two mellein derivatives isolated from an endophytic fungus. *Med. Chem. Res.* **2015**, *24*, 2111–2114. [CrossRef]

84. Qader, M.; Zaman, K.H.; Hu, Z.; Wang, C.; Wu, X.; Cao, S. Aspochalasin H1: A New Cyclic Aspochalasin from Hawaiian Plant-Associated Endophytic Fungus *Aspergillus* sp. T1307. *Molecules* **2021**, *26*, 4239. [CrossRef]

85. Wang, M.L.; Chen, R.; Sun, F.J.; Cao, P.R.; Chen, X.R.; Yang, M.H. Three alkaloids and one polyketide from *Aspergillus cristatus* harbored in *Pinellia ternate* tubers. *Tetrahedron Lett.* **2021**, *68*, 152914. [CrossRef]

86. Ebada, S.S.; Ebrahim, W. A new antibacterial quinolone derivative from the endophytic fungus *Aspergillus versicolor* strain Eich. 5.2.2. *S. Afr. J. Bot.* **2020**, *134*, 151–155. [CrossRef]

87. Mohamed, G.A.; Ibrahim, S.R.; Asfour, H.Z. Antimicrobial metabolites from the endophytic fungus *Aspergillus versicolor*. *Phytochem. Lett.* **2020**, *35*, 152–155. [CrossRef]

88. Luo, P.; Shao, G.; Zhang, S.Q.; Zhu, L.; Ding, Z.T.; Cai, L. Secondary metabolites of endophytic fungus *Aspergillus ochraceus* SX-C7 from *Selaginella stauntoniana*. *Zhongcaoyao* **2020**, *51*, 17–23.

89. Fathallah, N.; Raafat, M.M.; Issa, M.Y.; Abdel-Aziz, M.M.; Bishr, M.; Abdelkawy, M.A.; Salama, O. Bio-guided fractionation of prenylated benzaldehyde derivatives as potent antimicrobial and antibiofilm from *Ammi majus* L. fruits-associated *Aspergillus amstelodami*. *Molecules* **2019**, *24*, 4118. [CrossRef]

90. Wu, Z.; Zhang, X.; Anbari, W.H.A.; Zhou, Q.; Zhou, P.; Zhang, M.; Zeng, F.; Chen, C.; Tong, Q.; Wang, J.; et al. Cysteine Residue Containing Merocytochalasans and 17, 18-seco-Aspochalasins from *Aspergillus micronesiensis*. *J. Nat. Prod.* **2019**, *82*, 2653–2658. [CrossRef]

91. Mawabo, I.K.; Nkenfou, C.; Notedji, A.; Jouda, J.B.; Lunga, P.K.; Eke, P.; Fokou, V.T.; Kuiate, J.R. Antimicrobial activities of two secondary metabolites isolated from *Aspergillus niger*, endophytic fungus harbouring stems of *Acanthus montanus*. *Issues Biol. Sci. Pharm. Res.* **2019**, *7*, 7–15.

92. Yang, X.F.; Wang, N.N.; Kang, Y.F.; Ma, Y.M. A new furan derivative from an endophytic *Aspergillus tubingensis* of Decaisnea insignis (Griff.) Hook. f. & Thomson. *Nat. Prod. Res.* **2019**, *33*, 2777–2783.

93. Akhter, N.; Pan, C.; Liu, Y.; Shi, Y.; Wu, B. Isolation and structure determination of a new indene derivative from endophytic fungus *Aspergillus flavipes* Y-62. *Nat. Prod. Res.* **2019**, *33*, 2939–2944. [CrossRef]

94. Zhang, H.; Ruan, C.; Bai, X.; Chen, J.; Wang, H. Heterocyclic alkaloids as antimicrobial agents of *Aspergillus fumigatus* D endophytic on *Edgeworthia chrysantha*. *Chem. Nat. Compd.* **2018**, *54*, 411–414. [CrossRef]

95. Liu, R.; Li, H.; Yang, J.; An, Z. Quinazolinones isolated from *Aspergillus* sp., an endophytic fungus of *Astragalus membranaceus*. *Chem. Nat. Compd.* **2018**, *54*, 808–810. [CrossRef]

96. Ola, A.R.; Tawo, B.D.; Belli, H.L.L.; Proksch, P.; Tommy, D.; Hakim, E.H. A new antibacterial polyketide from the endophytic fungi *Aspergillus fumigatiaffinis*. *Nat. Prod. Commun.* **2018**, *13*, 1573–1574. [CrossRef]

97. Qiao, Y.; Zhang, X.; He, Y.; Sun, W.; Feng, W.; Liu, J.; Hu, Z.; Xu, Q.; Zhu, H.; Zhang, J.; et al. Aspermerodione, a novel fungal metabolite with an unusual 2, 6-dioxabicyclo.2.2. 1. heptane skeleton, as an inhibitor of penicillin-binding protein 2a. *Sci. Rep.* **2018**, *8*, 1–11.

98. Yan, W.; Li, S.J.; Guo, Z.K.; Zhang, W.J.; Wei, W.; Tan, R.X.; Jiao, R.H. New p-terphenyls from the endophytic fungus *Aspergillus* sp. YXf3. *Bioorg. Med. Chem. Lett.* **2017**, *27*, 51–54. [CrossRef]

99. Gombodorj, S.; Yang, M.H.; Shang, Z.C.; Liu, R.H.; Li, T.X.; Yin, G.P.; Kong, L.Y. New phenalenone derivatives from *Pinellia ternata* tubers derived *Aspergillus* sp. *Fitoterapia* **2017**, *120*, 72–78. [CrossRef]

100. Goutam, J.; Sharma, G.; Tiwari, V.K.; Mishra, A.; Kharwar, R.N.; Ramaraj, V.; Koch, B. Isolation and characterization of "terrein" an antimicrobial and antitumor compound from endophytic fungus *Aspergillus terreus* (JAS-2) associated from *Achyranthus aspera* Varanasi, India. *Front. Microbiol.* **2017**, *8*, 1334. [CrossRef]

101. Elkhayat, E.S.; Ibrahim, S.R.; Mohamed, G.A.; Ross, S.A. Terrenolide S, a new antileishmanial butenolide from the endophytic fungus *Aspergillus terreus*. *Nat. Prod. Res.* **2016**, *30*, 814–820. [CrossRef]

102. Ma, Y.M.; Ma, C.C.; Li, T.; Wang, J. A new furan derivative from an endophytic *Aspergillus flavus* of *Cephalotaxus fortunei*. *Nat. Prod. Res.* **2016**, *30*, 79–84. [CrossRef]

103. Sadorn, K.; Saepua, S.; Boonyuen, N.; Laksanacharoen, P.; Rachtawee, P.; Prabpai, S.; Kongsaeree, P.; Pittayakhajonwut, P. Allahabadolactones A and B from the endophytic fungus, *Aspergillus allahabadii* BCC45335. *Tetrahedron* **2016**, *72*, 489–495. [CrossRef]

104. Ma, Y.M.; Li, T.; Ma, C.C. A new pyrone derivative from an endophytic *Aspergillus tubingensis* of *Lycium ruthenicum*. *Nat. Prod. Res.* **2016**, *30*, 1499–1503. [CrossRef]

105. Ma, Y.M.; Liang, X.A.; Zhang, H.C.; Liu, R. Cytotoxic and antibiotic cyclic pentapeptide from an endophytic *Aspergillus tamarii* of *Ficus carica*. *J. Agric. Food Chem.* **2016**, *64*, 3789–3793. [CrossRef] [PubMed]

106. Ibrahim, S.R.M.; Elkhayat, E.S.; Mohamed, G.A.; Khedr, A.I.M.; Fouad, M.A.; Kotb, M.H.R.; Ross, S.A. Aspernolides F and G, new butyrolactones from the endophytic fungus *Aspergillus terreus*. *Phytochem. Lett.* **2015**, *14*, 84–90. [CrossRef]
107. Elfita, E.; Munawar, M.; Muharni, M.; Ivantri, I. Chemical constituen from an endophytic fungus *Aspergillus* sp. (SbD5) isolated from Sambiloto (*Andrographis paniculata* Nees). *Microbiol. Indones.* **2015**, *9*, 6.
108. Zhang, W.; Wei, W.; Shi, J.; Chen, C.; Zhao, G.; Jiao, R.; Tan, R. Natural phenolic metabolites from endophytic *Aspergillus* sp. IFB-YXS with antimicrobial activity. *Bioorg. Med. Chem. Lett.* **2015**, *25*, 2698–2701. [CrossRef]
109. Song, H.C.; Qin, D.; Liu, H.Y.; Dong, J.Y.; You, C.; Wang, Y.M. Resorcylic acid lactones produced by an endophytic *Penicillium ochrochloron* strain from *Kadsura angustifolia*. *Planta Med.* **2021**, *87*, 225–235. [CrossRef]
110. Syarifah, S.; Elfita, E.; Widjajanti, H.; Setiawan, A.; Kurniawati, A.R. Diversity of endophytic fungi from the root bark of *Syzygium zeylanicum*, and the antibacterial activity of fungal extracts, and secondary metabolite. *Biodivers. J.* **2021**, *22*, 4572–4582. [CrossRef]
111. Qin, Y.Y.; Huang, X.S.; Liu, X.B.; Mo, T.X.; Xu, Z.L.; Li, B.C.; Qin, X.Y.; Li, J.; Sch berle, T.F.; Yang, R.Y. Three new andrastin derivatives from the endophytic fungus *Penicillium vulpinum*. *Nat. Prod. Res.* **2020**, 1–9. [CrossRef]
112. Zhu, Y.X.; Peng, C.; Ding, W.; Hu, J.F.; Li, J. Chromenopyridin A, a new N-methoxy-1-pyridone alkaloid from the endophytic fungus *Penicillium nothofagi* P-6 isolated from the critically endangered conifer *Abies beshanzuensis*. *Nat. Prod. Res.* **2020**, 1–7. [CrossRef]
113. Graf, T.N.; Kao, D.; Rivera-Chávez, J.; Gallagher, J.M.; Raja, H.A.; Oberlies, N.H. Drug leads from endophytic fungi: Lessons learned via scaled production. *Planta Med.* **2020**, *86*, 988–996. [CrossRef]
114. Qin, Y.; Liu, X.; Lin, J.; Huang, J.; Jiang, X.; Mo, T.; Xu, Z.; Li, J.; Yang, R. Two new phthalide derivatives from the endophytic fungus *Penicillium vulpinum* isolated from *Sophora tonkinensis*. *Nat. Prod. Res.* **2021**, *35*, 421–427. [CrossRef]
115. Xu, Y.; Wang, L.; Zhu, G.; Zuo, M.; Gong, Q.; He, W.; Li, M.; Yuan, C.; Hao, X.; Zhu, W. New phenylpyridone derivatives from the *Penicillium sumatrense* GZWMJZ-313, a fungal endophyte of *Garcinia multiflora*. *Chin. Chem. Lett.* **2019**, *30*, 431–434. [CrossRef]
116. Zhao, T.; Xu, L.L.; Zhang, Y.; Lin, Z.H.; Xia, T.; Yang, D.F.; Chen, Y.M.; Yang, X.L. Three new α-pyrone derivatives from the plant endophytic fungus *Penicillium ochrochloronthe* and their antibacterial, antifungal, and cytotoxic activities. *J. Asian Nat. Prod. Res.* **2019**, *21*, 851–858. [CrossRef]
117. Xie, J.; Wu, Y.Y.; Zhang, T.Y.; Zhang, M.Y.; Peng, F.; Lin, B.; Zhang, Y.X. New antimicrobial compounds produced by endophytic Penicillium janthinellum isolated from *Panax notoginseng* as potential inhibitors of FtsZ. *Fitoterapia* **2018**, *131*, 35–43. [CrossRef]
118. Wu, Y.Y.; Zhang, T.Y.; Zhang, M.Y.; Cheng, J.; Zhang, Y.X. An endophytic Fungi of *Ginkgo biloba* L. produces antimicrobial metabolites as potential inhibitors of FtsZ of *Staphylococcus aureus*. *Fitoterapia* **2018**, *128*, 265–271. [CrossRef]
119. Yang, M.H.; Li, T.X.; Wang, Y.; Liu, R.H.; Luo, J.; Kong, L.Y. Antimicrobial metabolites from the plant endophytic fungus *Penicillium* sp. *Fitoterapia* **2017**, *116*, 72–76. [CrossRef]
120. Ma, Y.M.; Qiao, K.; Kong, Y.; Li, M.Y.; Guo, L.X.; Miao, Z.; Fan, C. A new isoquinolone alkaloid from an endophytic fungus R22 of *Nerium indicum*. *Nat. Prod. Res.* **2017**, *31*, 951–958. [CrossRef]
121. Feng, Z.W.; Lv, M.M.; Li, X.S.; Zhang, L.; Liu, C.X.; Guo, Z.Y.; Deng, Z.S.; Zou, K.; Proksch, P. Penicitroamide, an antimicrobial metabolite with high carbonylization from the endophytic fungus *Penicillium* sp. (No. 24). *Molecules* **2016**, *21*, 1438. [CrossRef]
122. Jouda, J.B.; Mbazoa, C.D.; Sarkar, P.; Bag, P.K.; Wandji, J. Anticancer and antibacterial secondary metabolites from the endophytic fungus *Penicillium* sp. CAM64 against multi-drug resistant Gram-negative bacteria. *Afr. Health Sci.* **2016**, *16*, 734–743. [CrossRef]
123. Lenta, B.N.; Ngatchou, J.; Frese, M.; Ladoh-Yemeda, F.; Voundi, S.; Nardella, F.; Michalek, C.; Wibberg, D.; Ngouela, S.; Tsamo, E.; et al. Purpureone, an antileishmanial ergochrome from the endophytic fungus *Purpureocillium lilacinum*. *Z. Naturforsch. B.* **2016**, *71*, 1159–1167. [CrossRef]
124. Klomchit, A.; Calderin, J.D.; Jaidee, W.; Watla-Iad, K.; Brooks, S. Napthoquinones from *Neocosmospora* sp.—Antibiotic Activity against *Acidovorax citrulli*, the Causative Agent of Bacterial Fruit Blotch in Watermelon and Melon. *J. Fungi* **2021**, *7*, 370. [CrossRef]
125. Ibrahim, S.R.M.; Mohamed, G.A.; Khayat, M.T.; Al Haidari, R.A.; El-Kholy, A.A.; Zayed, M.F. A new antifungal aminobenzamide derivative from the endophytic fungus *Fusarium* sp. *Pharmacogn. Mag.* **2019**, *15*, 204–207. [CrossRef]
126. Jiang, C.X.; Li, J.; Zhang, J.M.; Jin, X.J.; Yu, B.; Fang, J.G.; Wu, Q.X. Isolation, identification, and activity evaluation of chemical constituents from soil fungus *Fusarium avenaceum* SF-1502 and endophytic fungus *Fusarium proliferatum* AF-04. *J. Agric. Food Chem.* **2019**, *67*, 1839–1846. [CrossRef]
127. Shi, S.; Li, Y.; Ming, Y.; Li, C.; Li, Z.; Chen, J.; Luo, M. Biological activity and chemical composition of the endophytic fungus *Fusarium* sp. TP-G1 obtained from the root of *Dendrobium officinale* Kimura et Migo. *Rec. Nat. Prod.* **2018**, *12*, 549–556. [CrossRef]
128. Yan, C.; Liu, W.; Li, J.; Deng, Y.; Chen, S.; Liu, H. Bioactive terpenoids from Santalum album derived endophytic fungus *Fusarium* sp. YD-2. *RSC Adv.* **2018**, *8*, 14823–14828. [CrossRef]
129. Ibrahim, S.R.; Mohamed, G.A.; Al Haidari, R.A.; Zayed, M.F.; El-Kholy, A.A.; Elkhayat, E.S.; Ross, S.A. Fusarithioamide B, a new benzamide derivative from the endophytic fungus *Fusarium chlamydosporium* with potent cytotoxic and antimicrobial activities. *Bioorg. Med. Chem.* **2018**, *26*, 786–790. [CrossRef]
130. Shah, A.; Rather, M.A.; Hassan, Q.P.; Aga, M.A.; Mushtaq, S.; Shah, A.M.; Hussain, A.; Baba, S.A.; Ahmad, Z. Discovery of anti-microbial and anti-tubercular molecules from Fusarium solani: An endophyte of *Glycyrrhiza glabra*. *J. Appl. Microbiol.* **2017**, *122*, 1168–1176. [CrossRef]
131. Ibrahim, S.R.M.; Elkhayat, E.S.; Mohamed, G.A.A.; Fat'hi, S.M.; Ross, S.A. Fusarithioamide A, a new antimicrobial and cytotoxic benzamide derivative from the endophytic fungus *Fusarium chlamydosporium*. *Biochem. Biophys Res. Commun.* **2016**, *479*, 211–216. [CrossRef] [PubMed]

132. Alvin, A.; Kalaitzis, J.A.; Sasia, B.; Neilan, B.A. Combined genetic and bioactivity-based prioritization leads to the isolation of an endophyte-derived antimycobacterial compound. *J. Appl. Microbiol.* **2016**, *120*, 1229–1239. [CrossRef]

133. Liang, X.A.; Ma, Y.M.; Zhang, H.C.; Liu, R. A new helvolic acid derivative from an endophytic *Fusarium* sp. of *Ficus carica*. *Nat. Prod. Res.* **2016**, *30*, 2407–2412. [CrossRef]

134. Hussain, H.; Drogies, K.H.; Al-Harrasi, A.; Hassan, Z.; Shah, A.; Rana, U.A.; Green, I.R.; Draeger, S.; Schulz, B.; Krohn, K. Antimicrobial constituents from endophytic fungus *Fusarium* sp. *Asian Pac. J. Trop. Dis.* **2015**, *5*, 186–189. [CrossRef]

135. Ratnaweera, P.B.; de Silva, E.D.; Williams, D.E.; Andersen, R.J. Antimicrobial activities of endophytic fungi obtained from the arid zone invasive plant *Opuntia dillenii* and the isolation of equisetin, from endophytic *Fusarium* sp. *BMC Complement. Altern. Med.* **2015**, *15*, 1–7. [CrossRef]

136. Harwoko, H.; Daletos, G.; Stuhldreier, F.; Lee, J.; Wesselborg, S.; Feldbrügge, M.; Müller, W.E.; Kalscheuer, R.; Ancheeva, E.; Proksch, P. Dithiodiketopiperazine derivatives from endophytic fungi *Trichoderma harzianum* and *Epicoccum nigrum*. *Nat. Prod. Res.* **2021**, *35*, 257–265. [CrossRef] [PubMed]

137. Wang, Y.L.; Hu, B.Y.; Qian, M.A.; Wang, Z.H.; Zou, J.M.; Sang, X.Y.; Li, L.; Luo, X.D.; Zhao, L.X. Koninginin W, a new polyketide from the endophytic fungus *Trichoderma koningiopsis* YIM PH30002. *Chem. Biodivers.* **2021**, *18*, e2100460. [CrossRef]

138. Shi, X.S.; Song, Y.P.; Meng, L.H.; Yang, S.Q.; Wang, D.J.; Zhou, X.W.; Ji, N.Y.; Wang, B.G.; Li, X.M. Isolation and Characterization of antibacterial carotane sesquiterpenes from *Artemisia argyi* associated endophytic *Trichoderma virens* QA-8. *Antibiotics* **2021**, *10*, 213. [CrossRef]

139. Shi, X.S.; Meng, L.H.; Li, X.; Wang, D.J.; Zhou, X.W.; Du, F.Y.; Wang, B.G.; Li, X.M. Polyketides and Terpenoids with Potent Antibacterial Activities from the *Artemisia argyi*-Derived Fungus *Trichoderma koningiopsis* QA-3. *Chem. Biodivers.* **2020**, *17*, e2000566. [CrossRef]

140. Shi, X.S.; Li, H.L.; Li, X.M.; Wang, D.J.; Li, X.; Meng, L.H.; Zhou, X.W.; Wang, B.G. Highly oxygenated polyketides produced by *Trichoderma koningiopsis* QA-3, an endophytic fungus obtained from the fresh roots of the medicinal plant *Artemisia argyi*. *Bioorg. Chem.* **2020**, *94*, 103448. [CrossRef]

141. Li, W.Y.; Liu, Y.; Lin, Y.T.; Liu, Y.C.; Guo, K.; Li, X.N.; Luo, S.H.; Li, S.H. Antibacterial harziane diterpenoids from a fungal symbiont *Trichoderma atroviride* isolated from *Colquhounia coccinea* var. *mollis*. *Phytochemistry* **2020**, *170*, 112198. [CrossRef]

142. Sarsaiya, S.; Jain, A.; Fan, X.; Jia, Q.; Xu, Q.; Shu, F.; Zhou, Q.; Shi, J.; Chen, J. New insights into detection of Front Microbiol, 11a dendrobine compound from a novel endophytic *Trichoderma longibrachiatum* strain and its toxicity against phytopathogenic bacteria. *Front. Microbiol.* **2020**, *11*, 337. [CrossRef]

143. Shi, X.S.; Meng, L.H.; Li, X.M.; Li, X.; Wang, D.J.; Li, H.L.; Zhou, X.W.; Wang, B.G. Trichocadinins B–G: Antimicrobial cadinane sesquiterpenes from *Trichoderma virens* QA-8, an endophytic fungus obtained from the medicinal plant *Artemisia argyi*. *J. Nat. Prod.* **2019**, *82*, 2470–2476. [CrossRef]

144. Chen, S.; Li, H.; Chen, Y.; Li, S.; Xu, J.; Guo, H.; Liu, Z.; Zhu, S.; Liu, H.; Zhang, W. Three new diterpenes and two new sesquiterpenoids from the endophytic fungus *Trichoderma koningiopsis* A729. *Bioorg. Chem.* **2019**, *86*, 368–374. [CrossRef]

145. Shi, X.S.; Wang, D.J.; Li, X.M.; Li, H.L.; Meng, L.H.; Li, X.; Pi, Y.; Zhou, X.W.; Wang, B.G. Antimicrobial polyketides from *Trichoderma koningiopsis* QA-3, an endophytic fungus obtained from the medicinal plant *Artemisia argyi*. *Rsc. Adv.* **2017**, *7*, 51335–51342. [CrossRef]

146. Zhao, S.; Wang, B.; Tian, K.; Ji, W.; Zhang, T.; Ping, C.; Yan, W.; Ye, Y. Novel metabolites from the *Cercis chinensis* derived endophytic fungus *Alternaria alternata* ZHJG5 and their antibacterial activities. *Pest Manag. Sci.* **2021**, *77*, 2264–2271. [CrossRef] [PubMed]

147. Kong, F.D.; Yi, T.F.; Ma, Q.Y.; Xie, Q.Y.; Zhou, L.M.; Chen, J.P.; Dai, H.F.; Wu, Y.G.; Zhao, Y.X. Biphenyl metabolites from the patchouli endophytic fungus *Alternaria* sp. PfuH1. *Fitoterapia* **2020**, *146*, 104708. [CrossRef] [PubMed]

148. Zhao, S.; Xiao, C.; Wang, J.; Tian, K.; Ji, W.; Yang, T.; Khan, B.; Qian, G.; Yan, W.; Ye, Y. Discovery of natural FabH inhibitors using an immobilized enzyme column and their antibacterial activity against *Xanthomonas oryzae* pv. *oryzae*. *J. Agric. Food Chem.* **2020**, *68*, 14204–14211. [CrossRef]

149. Palanichamy, P.; Kannan, S.; Murugan, D.; Alagusundaram, P.; Marudhamuthu, M. Purification, crystallization and anticancer activity evaluation of the compound alternariol methyl ether from endophytic fungi *Alternaria alternata*. *J. Appl. Microbiol.* **2019**, *127*, 1468–1478. [CrossRef]

150. Deshidi, R.; Devari, S.; Kushwaha, M.; Gupta, A.P.; Sharma, R.; Chib, R.; Khan, I.A.; Jaglan, S.; Shah, B.A. Isolation and quantification of alternariols from endophytic fungus, *Alternaria alternata*: LC-ESI-MS/MS analysis. *ChemistrySelect* **2017**, *2*, 364–368. [CrossRef]

151. Tian, J.; Fu, L.; Zhang, Z.; Dong, X.; Xu, D.; Mao, Z.; Liu, Y.; Lai, D.; Zhou, L. Dibenzo-α-pyrones from the endophytic fungus *Alternaria* sp. Samif01: Isolation, structure elucidation, and their antibacterial and antioxidant activities. *Nat. Prod. Res.* **2017**, *31*, 387–396. [CrossRef]

152. Lou, J.; Yu, R.; Wang, X.; Mao, Z.; Fu, L.; Liu, Y.; Zhou, L. Alternariol 9-methyl ether from the endophytic fungus *Alternaria* sp. Samif01 and its bioactivities. *Braz. J. Microbiol.* **2016**, *47*, 96–101. [CrossRef]

153. Kellogg, J.J.; Todd, D.A.; Egan, J.M.; Raja, H.A.; Oberlies, N.H.; Kvalheim, O.M.; Cech, N.B. Biochemometrics for natural products research: Comparison of data analysis approaches and application to identification of bioactive compounds. *J. Nat. Prod.* **2016**, *79*, 376–386. [CrossRef]

154. Rukachaisirikul, V.; Chinpha, S.; Saetang, P.; Phongpaichit, S.; Jungsuttiwong, S.; Hadsadee, S.; Sakayaroj, J.; Preedanon, S.; Temkitthawon, P.; Ingkaninan, K. Depsidones and a dihydroxanthenone from the endophytic fungi *Simplicillium lanosoniveum* (JFH Beyma) Zare & W. Gams PSU-H168 and PSU-H261. *Fitoterapia* **2019**, *138*, 104286.

155. Saetang, P.; Rukachaisirikul, V.; Phongpaichit, S.; Preedanon, S.; Sakayaroj, J.; Borwornpinyo, S.; Seemakhan, S.; Muanprasat, C. Depsidones and an α-pyrone derivative from *Simpilcillium* sp. PSU-H41, an endophytic fungus from *Hevea brasiliensis* leaf. *Phytochemistry* **2017**, *143*, 115–123. [CrossRef]

156. Yehia, R.S.; Osman, G.H.; Assaggaf, H.; Salem, R.; Mohamed, M.S. Isolation of potential antimicrobial metabolites from endophytic fungus *Cladosporium cladosporioides* from endemic plant *Zygophyllum mandavillei*. *S. Afr. J. Bot.* **2020**, *134*, 296–302. [CrossRef]

157. Pan, F.; El-Kashef, D.H.; Kalscheuer, R.; Müller, W.E.; Lee, J.; Feldbrügge, M.; Mándi, A.; Kurtán, T.; Liu, Z.; Wu, W.; et al. Cladosins LO, new hybrid polyketides from the endophytic fungus *Cladosporium sphaerospermum* WBS017. *Eur. J. Med. Chem.* **2020**, *191*, 112159. [CrossRef]

158. Khan, M.I.H.; Sohrab, M.H.; Rony, S.R.; Tareq, F.S.; Hasan, C.M.; Mazid, M.A. Cytotoxic and antibacterial naphthoquinones from an endophytic fungus, *Cladosporium* sp. *Toxicol. Rep.* **2016**, *3*, 861–865. [CrossRef]

159. Kuang, C.; Jing, S.X.; Liu, Y.; Luo, S.H.; Li, S.H. Drimane sesquiterpenoids and isochromone derivative from the endophytic fungus *Pestalotiopsis* sp. M-23. *Nat. Prod. Bioprospect.* **2016**, *6*, 155–160. [CrossRef]

160. Beattie, K.D.; Ellwood, N.; Kumar, R.; Yang, X.; Healy, P.C.; Choomuenwai, V.; Quinn, R.J.; Elliott, A.G.; Huang, J.X.; Chitty, J.L.; et al. Antibacterial and antifungal screening of natural products sourced from Australian fungi and characterisation of pestalactams D–F. *Phytochemistry* **2016**, *124*, 79–85. [CrossRef]

161. Zhao, S.; Chen, S.; Wang, B.; Niu, S.; Wu, W.; Guo, L.; Che, Y. Four new tetramic acid and one new furanone derivatives from the plant endophytic fungus *Neopestalotiopsis* sp. *Fitoterapia* **2015**, *103*, 106–112. [CrossRef]

162. Arora, P.; Wani, Z.A.; Nalli, Y.; Ali, A.; Riyaz-Ul-Hassan, S. Antimicrobial potential of thiodiketopiperazine derivatives produced by *Phoma* sp., an endophyte of *Glycyrrhiza glabra* Linn. *Microb. Ecol.* **2016**, *72*, 802–812. [CrossRef]

163. Xia, X.; Kim, S.; Bang, S.; Lee, H.J.; Liu, C.; Park, C.I.; Shim, S.H. Barceloneic acid C, a new polyketide from an endophytic fungus *Phoma* sp. JS752 and its antibacterial activities. *J. Antibiot.* **2015**, *68*, 139–141. [CrossRef]

164. De Medeiros, L.S.; Abreu, L.M.; Nielsen, A.; Ingmer, H.; Larsen, T.O.; Nielsen, K.F.; Rodrigues-Filho, E. Dereplication-guided isolation of depsides thielavins S–T and lecanorins D–F from the endophytic fungus *Setophoma* sp. *Phytochemistry* **2015**, *111*, 154–162. [CrossRef]

165. Li, Y.; Wei, W.; Wang, R.L.; Liu, F.; Wang, Y.K.; Li, R.; Khan, B.; Lin, J.; Yan, W.; Ye, Y.H. Colletolides A and B, two new γ-butyrolactone derivatives from the endophytic fungus *Colletotrichum gloeosporioides*. *Phytochem. Lett.* **2019**, *33*, 90–93. [CrossRef]

166. Wang, W.X.; Kusari, S.; Laatsch, H.; Golz, C.; Kusari, P.; Strohmann, C.; Kayser, O.; Spiteller, M. Antibacterial azaphilones from an endophytic fungus, *Colletotrichum* sp. BS4. *J. Nat. Prod.* **2016**, *79*, 704–710. [CrossRef]

167. Wang, A.; Yin, R.; Zhou, Z.; Gu, G.; Dai, J.; Lai, D.; Zhou, L. Eremophilane-type sesquiterpenoids from the endophytic fungus *Rhizopycnis vagum* and their antibacterial, cytotoxic, and phytotoxic activities. *Front. Chem.* **2020**, *8*, 980. [CrossRef]

168. Wang, A.; Li, P.; Zhang, X.; Han, P.; Lai, D.; Zhou, L. Two new anisic acid derivatives from endophytic fungus *Rhizopycnis vagum* Nitaf22 and their antibacterial activity. *Molecules* **2018**, *23*, 591. [CrossRef] [PubMed]

169. Lai, D.; Wang, A.; Cao, Y.; Zhou, K.; Mao, Z.; Dong, X.; Tian, J.; Xu, D.; Dai, J.; Peng, Y.; et al. Bioactive dibenzo-α-pyrone derivatives from the endophytic fungus *Rhizopycnis vagum* Nitaf22. *J. Nat. Prod.* **2016**, *79*, 2022–2031. [CrossRef] [PubMed]

170. Chen, H.Y.; Liu, T.K.; Shi, Q.; Yang, X.L. Sesquiterpenoids and diterpenes with antimicrobial activity from *Leptosphaeria* sp. XL026, an endophytic fungus in *Panax notoginseng*. *Fitoterapia* **2019**, *137*, 104243. [CrossRef]

171. Mao, Z.; Zhang, W.; Wu, C.; Feng, H.; Peng, Y.; Shahid, H.; Cui, Z.; Ding, P.; Shan, T. Diversity and antibacterial activity of fungal endophytes from *Eucalyptus exserta*. *BMC Microbiol.* **2021**, *21*, 1–12. [CrossRef] [PubMed]

172. Mou, Q.L.; Yang, S.X.; Xiang, T.; Liu, W.W.; Yang, J.; Guo, L.P.; Wang, W.J.; Yang, X.L. New cytochalasan alkaloids and cyclobutane dimer from an endophytic fungus *Cytospora chrysosperma* in *Hippophae rhamnoides* and their antimicrobial activities. *Tetrahedron Lett.* **2021**, *87*, 153207. [CrossRef]

173. Mao, Z.; Xue, M.; Gu, G.; Wang, W.; Li, D.; Lai, D.; Zhou, L. Lophiostomin A–D: New 3, 4-dihydroisocoumarin derivatives from the endophytic fungus *Lophiostoma* sp. Sigrf10. *RSC Adv.* **2020**, *10*, 6985–6991. [CrossRef]

174. Hussain, H.; Root, N.; Jabeen, F.; Al-Harrasi, A.; Ahmad, M.; Mabood, F.; Hassan, Z.; Shah, A.; Green, I.R.; Schulz, B.; et al. Microsphaerol and seimatorone: Two new compounds isolated from the endophytic fungi, *Microsphaeropsis* sp. and *Seimatosporium* sp. *Chem. Biodivers.* **2015**, *12*, 289–294. [CrossRef]

175. Harwoko, H.; Lee, J.; Hartmann, R.; Mándi, A.; Kurtán, T.; Müller, W.E.; Feldbrügge, M.; Kalscheuer, R.; Ancheeva, E.; Daletos, G.; et al. Azacoccones FH, new flavipin-derived alkaloids from an endophytic fungus *Epicoccum nigrum* MK214079. *Fitoterapia* **2020**, *146*, 104698. [CrossRef]

176. Dzoyem, J.P.; Melong, R.; Tsamo, A.T.; Maffo, T.; Kapche, D.G.; Ngadjui, B.T.; McGaw, L.J.; Eloff, J.N. Cytotoxicity, antioxidant and antibacterial activity of four compounds produced by an endophytic fungus *Epicoccum nigrum* associated with *Entada abyssinica*. *Rev. Bras. Farmacogn.* **2017**, *27*, 251–253. [CrossRef]

177. Xu, Z.L.; Zheng, N.; Cao, S.M.; Li, S.T.; Mo, T.X.; Qin, Y.Y.; Li, J.; Yang, R.Y. Secondary metabolites from the endophytic fungus *Stemphylium lycopersici* and their antibacterial activities. *Chem. Nat. Compd.* **2020**, *56*, 1162–1165. [CrossRef]

178. Liu, Y.; Marmann, A.; Abdel-Aziz, M.S.; Wang, C.Y.; Müller, W.E.; Lin, W.H.; Mándi, A.; Kurtán, T.; Daletos, G.; Proksch, P. Tetrahydroanthraquinone derivatives from the endophytic fungus *Stemphylium globuliferum*. *Eur. J. Org. Chem.* **2015**, *2015*, 2646–2653. [CrossRef]

179. Mai, P.Y.; Levasseur, M.; Buisson, D.; Touboul, D.; Eparvier, V. Identification of antimicrobial compounds from *Sandwithia guyanensis*-associated endophyte using molecular network approach. *Plants* **2020**, *9*, 47. [CrossRef]

180. Ramesha, K.P.; Mohana, N.C.; Nuthan, B.R.; Rakshith, D.; Satish, S. Antimicrobial metabolite profiling of *Nigrospora sphaerica* from *Adiantum philippense* L. *J. Genet. Eng. Biotechnol.* **2020**, *18*, 1–9. [CrossRef]

181. Kornsakulkarn, J.; Choowong, W.; Rachtawee, P.; Boonyuen, N.; Kongthong, S.; Isaka, M.; Thongpanchang, C. Bioactive hydroanthraquinones from endophytic fungus *Nigrospora* sp. BCC 47789. *Phytochem. Lett.* **2018**, *24*, 46–50. [CrossRef]

182. Kaaniche, F.; Hamed, A.; Abdel-Razek, A.S.; Wibberg, D.; Abdissa, N.; El Euch, I.Z.; Allouche, N.; Mellouli, L.; Shaaban, M.; Sewald, N. Bioactive secondary metabolites from new endophytic fungus *Curvularia* sp. isolated from *Rauwolfia macrophylla*. *PLoS ONE* **2019**, *14*, e0217627. [CrossRef]

183. Hilario, F.; Polinário, G.; de Amorim, M.R.; de Sousa Batista, V.; do Nascimento Júnior, N.M.; Araújo, A.R.; Bauab, T.M.; Dos Santos, L.C. Spirocyclic lactams and curvulinic acid derivatives from the endophytic fungus *Curvularia lunata* and their antibacterial and antifungal activities. *Fitoterapia* **2020**, *141*, 104466. [CrossRef]

184. Long, Y.; Tang, T.; Wang, L.Y.; He, B.; Gao, K. Absolute configuration and biological activities of meroterpenoids from an endophytic fungus of *Lycium barbarum*. *J. Nat. Prod.* **2019**, *82*, 2229–2237. [CrossRef]

185. He, J.; Li, Z.H.; Ai, H.L.; Feng, T.; Liu, J.K. Anti-bacterial chromones from cultures of the endophytic fungus *Bipolaris eleusines*. *Nat. Prod. Res.* **2019**, *33*, 3515–3520. [CrossRef]

186. Yang, Y.H.; Yang, D.S.; Li, G.H.; Pu, X.J.; Mo, M.H.; Zhao, P.J. Antibacterial diketopiperazines from an endophytic fungus *Bionectria* sp. Y1085. *J. Antibiot.* **2019**, *72*, 752–758. [CrossRef]

187. Kamdem, R.S.; Pascal, W.; Rehberg, N.; van Geelen, L.; Höfert, S.P.; Knedel, T.O.; Janiak, C.; Sureechatchaiyan, P.; Kassack, M.U.; Lin, W.; et al. Metabolites from the endophytic fungus *Cylindrocarpon* sp. isolated from tropical plant *Sapium ellipticum*. *Fitoterapia* **2018**, *128*, 175–179. [CrossRef]

188. Li, G.; Kusari, S.; Golz, C.; Laatsch, H.; Strohmann, C.; Spiteller, M. Epigenetic modulation of endophytic *Eupenicillium* sp. LG41 by a histone deacetylase inhibitor for production of decalin-containing compounds. *J. Nat. Prod.* **2017**, *80*, 983–988. [CrossRef]

189. Teponno, R.B.; Noumeur, S.R.; Helaly, S.E.; Hüttel, S.; Harzallah, D.; Stadler, M. Furanones and anthranilic acid derivatives from the endophytic fungus *Dendrothyrium variisporum*. *Molecules* **2017**, *22*, 1674. [CrossRef]

190. Pina, J.R.S.; Silva-Silva, J.V.; Carvalho, J.M.; Bitencourt, H.R.; Watanabe, L.A.; Fernandes, J.M.P.; Souza, G.E.D.; Aguiar, A.C.C.; Guido, R.V.C.; Almeida-Souza, F.; et al. Antiprotozoal and antibacterial activity of ravenelin, a xanthone isolated from the endophytic fungus *Exserohilum rostratum*. *Molecules* **2021**, *26*, 3339. [CrossRef]

191. Pinheiro, E.A.; Borges, F.C.; Pina, J.R.; Ferreira, L.R.; Cordeiro, J.S.; Carvalho, J.M.; Feitosa, A.O.; Campos, F.R.; Barison, A.; Souza, A.D.; et al. Annularins I and J: New metabolites isolated from endophytic fungus *Exserohilum rostratum*. *J. Braz. Chem. Soc.* **2016**, *27*, 1432–1436.

192. Pan, Y.; Zheng, W.; Yang, S. Chemical and activity investigation on metabolites produced by an endophytic fungi *Psathyrella candolleana* from the seed of *Ginkgo biloba*. *Nat. Prod. Res.* **2020**, *34*, 3130–3133. [CrossRef]

193. Duan, X.X.; Qin, D.; Song, H.C.; Gao, T.C.; Zuo, S.H.; Yan, X.; Wang, J.Q.; Ding, X.; Di, Y.T.; Dong, J.Y. Irpexlacte A-D, four new bioactive metabolites of endophytic fungus Irpex lacteus DR10-1 from the waterlogging tolerant plant *Distylium chinense*. *Phytochem. Lett.* **2019**, *32*, 151–156. [CrossRef]

194. Rehberg, N.; Akone, H.S.; Ioerger, T.R.; Erlenkamp, G.; Daletos, G.; Gohlke, H.; Proksch, P.; Kalscheuer, R. Chlorflavonin targets acetohydroxyacid synthase catalytic subunit IlvB1 for synergistic killing of *Mycobacterium tuberculosis*. *ACS Infect. Dis.* **2018**, *4*, 123–134. [CrossRef] [PubMed]

195. Schulz, S.; Dickschat, J.S. Bacterial volatiles: The smell of small organisms. *Nat. Prod. Rep.* **2007**, *24*, 814–842. [CrossRef] [PubMed]

196. Morath, S.U.; Hung, R.; Bennett, J.W. Fungal volatile organic compounds: A review with emphasis on their biotechnological potential. *Fungal Biol. Rev.* **2012**, *26*, 73–83. [CrossRef]

197. Guo, Y.; Jud, W.; Weikl, F.; Ghirardo, A.; Junker, R.R.; Polle, A.; Benz, J.P.; Pritsch, K.; Schnitzler, J.P.; Rosenkranz, M. Volatile organic compound patterns predict fungal trophic mode and lifestyle. *Commun. Biol.* **2021**, *4*, 1–12. [CrossRef] [PubMed]

198. Weisskopf, L.; Schulz, S.; Garbeva, P. Microbial volatile organic compounds in intra-kingdom and inter-kingdom interactions. *Nat. Rev. Microbiol.* **2021**, *19*, 391–404. [CrossRef]

199. Chen, J.J.; Feng, X.; Xia, C.Y.; Kong, D.; Qi, Z.Y.; Liu, F.; Chen, D.; Lin, F.; Zhang, C. Confirming the phylogenetic position of the genus *Muscodor* and the description of a new *Muscodor* species. *Mycosphere* **2019**, *10*, 187–201. [CrossRef]

200. Saxena, S.; Strobel, G.A. Marvellous *Muscodor* spp.: Update on Their Biology and Applications. *Microb. Ecol.* **2020**, *82*, 5–20. [CrossRef]

201. Ezra, D.; Hess, W.; Strobel, G. Unique wild type endophytic isolates of *Muscodor albus*, a volatile antibiotic producing fungus. *Microbiology* **2004**, *150*, 4023–4031. [CrossRef]

202. Atmosukarto, I.; Castillo, U.; Hess, W.M.; Sears, J.; Strobel, G. Isolation and characterization of *M. albus* I-41.3 s, a volatile antibiotic producing fungus. *Plant Sci.* **2010**, *169*, 854–861. [CrossRef]

203. Mitchell, A.M.; Strobel, G.A.; Moore, E.; Robison, R.; Sears, J. Volatile antimicrobials from *Muscodor crispans*, a novel endophytic fungus. *Microbiology* **2010**, *156*, 270–277. [CrossRef]

204. Zhang, C.L.; Wang, G.P.; Mao, L.J.; Komon-Zelazowska, M.; Yuan, Z.L.; Lin, F.C.; Druzhinina, I.S.; Kubicek, C.P. *Muscodor fengyangensis* sp. nov. from southeast China: Morphology, physiology and production of volatile compounds. *Fungal Biol.* **2010**, *114*, 797–808. [CrossRef]
205. Meshram, V.; Kapoor, N.; Saxena, S. *Muscodor kashayum* sp. nov.–a new volatile anti-microbial producing endophytic fungus. *Mycology* **2013**, *4*, 196–204. [CrossRef]
206. Suwannarach, N.; Kumla, J.; Bussaban, B.; Hyde, K.D.; Matsui, K.; Lumyong, S. Molecular and morphological evidence support four new species in the genus *Muscodor* from northern Thailand. *Ann. Microbiol.* **2013**, *63*, 1341–1351. [CrossRef]
207. Strobel, G.A.; Blatt, B. Volatile Organic Compound Formulations Having Antimicrobial Activity. U.S. Patent Application No. 16/179,370, 5 September 2019.
208. Suwannarach, N.; Kaewyana, C.; Yodmeeklin, A.; Kumla, J.; Matsui, K.; Lumyong, S. Evaluation of *Muscodor cinnamomi* as an egg biofumigant for the reduction of microorganisms on the eggshell surface and its effect on egg quality. *Int. J. Food Microbiol.* **2017**, *244*, 52–61. [CrossRef]
209. Huang, C.B.; Alimova, Y.; Myers, T.M.; Ebersole, J.L. Short- and medium- chain fatty acids exhibit antimicrobial activity for oral microorganisms. *Arch. Oral Biol.* **2011**, *56*, 650–654. [CrossRef]
210. Levison, M. Effect of colon flora and short chain fatty acids on in vitro growth of *Pseudomonas aeruginosa* and Enterobacteriaceae. *Infect. Immun.* **1973**, *8*, 30–35. [CrossRef]
211. Moo, C.L.; Yang, S.K.; Osman, M.A.; Yuswam, M.H.; Loh, J.Y.; Lim, W.M.; Lim, S.H.E.; Lai, K.S. Antibacterial activity and mode of action of β- caryophyllene on *Bacillus cereus*. *Pol. J. Microbiol.* **2020**, *68*, 49–54. [CrossRef]
212. Bakun, P.; Czarczynska-Goslinka, B.; Goslinka, T.; Lijewiski, S. In vitro and in vivo biological activities of azulene derivatives with potential applications in medicine. *Med. Chem. Res.* **2021**, *30*, 834–846. [CrossRef]
213. Khaldi, N.; Seifuddin, F.T.; Turner, G.; Haft, D.; Nierman, W.C.; Wolfe, K.H.; Fedorova, N.D. SMURF: Genomic mapping of fungal secondary metabolite clusters. *Fungal Genet. Biol.* **2010**, *47*, 736–741. [CrossRef]
214. Wasil, Z.; Pahirulzaman, K.A.K.; Butts, C.; Simpson, T.J.; Lazarus, C.M.; Cox, R.J. One pathway, many compounds: Heterologous expression of a fungal biosynthetic pathway reveals its intrinsic potential for diversity. *Chem. Sci.* **2013**, *4*, 3845–3856. [CrossRef]
215. Rutledge, P.J.; Challis, G.L. Discovery of microbial natural products by activation of silent biosynthetic gene clusters. *Nat. Rev. Microbiol.* **2015**, *13*, 509–523. [CrossRef]
216. Bharatiya, P.; Rathod, P.; Hiray, A.; Kate, A.S. Multifarious elicitors: Invoking biosynthesis of various bioactive secondary metabolite in fungi. *Appl. Biochem. Biotechnol.* **2021**, *193*, 668–686. [CrossRef]
217. Schneider, P.; Misiek, M.; Hoffmeister, D. In vivo and in vitro production o ptions for fungal secondary metabolites. *Mol. Pharm.* **2008**, *5*, 234–242. [CrossRef]
218. Yu, M.; Li, Y.; Banakar, S.P.; Liu, L.; Shao, C.; Li, Z.; Wang, C. New metabolites from the co-culture of marine derived actinomycete Streptomyces rochei MB037 and fungus *Rhinocladiella similis* 35. *Front. Microbiol.* **2019**, *10*, 915. [CrossRef]
219. Huo, L.; Hug, J.J.; Fu, C.; Bian, X.; Zhang, Y.; Müller, R. Heterologous expression of bacterial natural product biosynthetic pathways. *Nat. Prod. Rep.* **2019**, *36*, 1412–1436. [CrossRef]
220. Zhang, W.; Shao, C.L.; Chen, M.; Liu, Q.A.; Wang, C.Y. Brominated resorcylic acid lactones from the marine-derived fungus *Cochliobolus lunatus* induced by histone deacetylase inhibitors. *Tetrahedron Lett.* **2014**, *55*, 4888–4891. [CrossRef]
221. Wang, Z.R.; Li, G.; Ji, L.X.; Wang, H.H.; Gao, H.; Peng, X.P.; Lou, H.X. Induced production of steroids by co-cultivation of two endophytes from *Mahonia fortunei*. *Steroids* **2019**, *145*, 1–4. [CrossRef]
222. Pan, R.; Bai, X.; Chen, J.; Zhang, H.; Wang, H. Exploring structural diversity of microbe secondary metabolites using OSMAC strategy: A literature review. *Front. Microbiol.* **2019**, *10*, 294. [CrossRef]
223. Shi, T.; Shao, C.L.; Liu, Y.; Zhao, D.L.; Cao, F.; Fu, X.M.; Yu, J.Y.; Wu, J.S.; Zhang, Z.K.; Wang, C.Y. Terpenoids from the coral-derived fungus *Trichoderma harzianum* (XS-20090075) induced by chemical epigenetic manipulation. *Front. Microbiol.* **2020**, *11*, 572. [CrossRef]
224. Feng, X.; He, C.; Jiao, L.; Liang, X.; Zhao, R.; Guo, Y. Analysis of differential expression proteins reveals the key pathway in response to heat stress in *Alicyclobacillus acidoterrestris* DSM 3922T. *Food Microbiol.* **2019**, *80*, 77–84. [CrossRef]
225. Yushchuk, O.; Ostash, I.; Mösker, E.; Vlasiuk, I.; Deneka, M.; Rückert, C.; Busche, T.; Fedorenko, V.; Kalinowski, J.; Süssmuth, R.D.; et al. Eliciting the silent lucensomycin biosynthetic pathway in *Streptomyces cyanogenus* S136 via manipulation of the global regulatory gene adpA. *Sci. Rep.* **2021**, *11*, 3507. [CrossRef]
226. Libis, V.; Antonovsky, N.; Zhang, M.; Shang, Z.; Montiel, D.; Maniko, J.; Ternei, M.A.; Calle, P.Y.; Lemetre, C.; Owen, J.G.; et al. Uncovering the biosynthetic potential of rare metagenomic DNA using co-occurrence network analysis of targeted sequences. *Nat. Commun.* **2019**, *10*, 3848. [CrossRef] [PubMed]
227. Alberti, F.; Leng, D.J.; Wilkening, I.; Song, L.; Tosin, M.; Corre, C. Triggering the expression of a silent gene cluster from genetically intractable bacteria results in scleric acid discovery. *Chem. Sci.* **2019**, *10*, 453–463. [CrossRef] [PubMed]
228. Tao, W.; Chen, L.; Zhao, C.; Wu, J.; Yan, D.; Deng, Z.; Sun, Y. In vitro packaging mediated one-step targeted cloning of natural product pathway. *ACS Synth. Biol.* **2019**, *8*, 1991–1997. [CrossRef] [PubMed]
229. Liu, Z.; Zhao, Y.; Huang, C.; Luo, Y. Recent Advances in Silent Gene Cluster Activation in Streptomyces. *Front. Bioeng. Biotechnol.* **2021**, *9*, 632230. [CrossRef]
230. Li, L.; Wei, K.; Zheng, G.; Liu, X.; Chen, S.; Jiang, W.; Lu, Y. CRISPR-Cpf1-assisted multiplex genome editing and transcriptional repression in Streptomyces. *Appl. Environ. Microbiol.* **2018**, *84*, e00827-18. [CrossRef]

231. Poças-Fonseca, M.J.; Cabral, C.G.; Manfrão-Netto, J.H.C. Epigenetic manipulation of filamentous fungi for biotechnological applications: A systematic review. *Biotechnol. Lett.* **2020**, *42*, 885–904. [CrossRef]
232. Mao, X.M.; Xu, W.; Li, D.; Yin, W.B.; Chooi, Y.H.; Li, Y.Q.; Tang, Y.; Hu, Y. Epigenetic genome mining of an endophytic fungus leads to the pleiotropic bio- synthesis of natural products. *Angew. Chem. Int. Ed.* **2015**, *54*, 7592–7596. [CrossRef]
233. Strauss, J.; Reyes-Dominguez, Y. Regulation of secondary metabolism by chromatin structure and epigenetic codes. *Fungal Genet. Biol.* **2011**, *48*, 62–69. [CrossRef]
234. Gacek, A.; Strauss, J. The chromatin code of fungal secondary metabolite gene clusters. *Appl. Microbiol. Biotechnol.* **2012**, *95*, 1389–1404. [CrossRef]
235. Aghcheh, R.K.; Kubicek, C.P. Epigenetics as an emerging tool for improvement of fungal strains used in biotechnology. *Appl. Microbiol. Biotechnol.* **2015**, *99*, 6167–6181. [CrossRef]
236. Li, C.Y.; Chung, Y.M.; Wu, Y.C.; Hunyadi, A.; Wang, C.C.; Chang, F.R. Natural products development under epigenetic modulation in fungi. *Phytochem. Rev.* **2020**, *19*, 1323–1340. [CrossRef]
237. Kim, J.H.; Lee, N.; Hwang, S.; Kim, W.; Lee, Y.; Cho, S.; Palsson, B.O.; Cho, B.K. Discovery of novel secondary metabolites encoded in actinomycete genomes through coculture. *J. Ind. Microbiol. Biotechnol.* **2021**, *48*, kuaa001. [CrossRef]
238. Tomm, H.A.; Ucciferri, L.; Ross, A.C. Advances in microbial culturing conditions to activate silent biosynthetic gene clusters for novel metabolite production. *J. Ind. Microbiol. Biotechnol.* **2019**, *46*, 1381–1400. [CrossRef]
239. Gonciarz, J.; Bizukojc, M. Adding talc microparticles to *Aspergillus terreus* ATCC 20542 preculture decreases fungal pellet size and improves lovastatin production. *Eng. Life Sci.* **2014**, *14*, 190–200. [CrossRef]
240. Timmermans, M.L.; Picott, K.J.; Ucciferri, L.; Ross, A.C. Culturing marine bacteria from the genus Pseudoalteromonas on a cotton scaffold alters secondary metabolite production. *Microbiologyopen* **2019**, *8*, e00724. [CrossRef]
241. Boruta, T.; Bizukojc, M. Application of aluminum oxide nanoparticles in *Aspergillus terreus* cultivations: Evaluating the effects on lovastatin production and fungal morphology. *Biomed. Res. Int.* **2019**, *2019*, 1–11. [CrossRef]
242. Bode, H.B.; Bethe, B.; Hofs, R.; Zeeck, A. Big effects from small changes: Possible ways to explore nature's chemical diversity. *Chembiochemistry* **2002**, *3*, 619–627. [CrossRef]
243. Scherlach, K.; Hertweck, C. Discovery of aspoquinolones A–D, prenylated quinoline-2-one alkaloids from *Aspergillus nidulans*, motivated by genome mining. *Org. Biomol. Chem.* **2006**, *4*, 3517–3520. [CrossRef]
244. Scherlach, K.; Schuemann, J.; Dahse, H.M.; Hertweck, C. Aspernidine A and B, prenylated isoindolinone alkaloids from the model fungus *Aspergillus nidulans*. *J. Antibiot.* **2010**, *63*, 375–377. [CrossRef]

Journal of
Fungi

MDPI

Review

Recent Developments in Metabolomics Studies of Endophytic Fungi

Kashvintha Nagarajan [1], Baharudin Ibrahim [2,3], Abdulkader Ahmad Bawadikji [2], Jun-Wei Lim [4], Woei-Yenn Tong [5], Chean-Ring Leong [5], Kooi Yeong Khaw [6] and Wen-Nee Tan [1,*]

1   Chemistry Section, School of Distance Education, Universiti Sains Malaysia, Penang 11800, Malaysia; kashvintha18@gmail.com
2   School of Pharmaceutical Sciences, Universiti Sains Malaysia, Penang 11800, Malaysia; baharudin.ibrahim@usm.my (B.I.); a.bawadkji@yahoo.com (A.A.B.)
3   Department of Clinical Pharmacy & Pharmacy Practice, Faculty of Pharmacy, Universiti Malaya, Kuala Lumpur 50603, Malaysia
4   Department of Fundamental and Applied Sciences, HICoE-Centre for Biofuel and Biochemical Research, Institute of Self-Sustainable Building, Universiti Teknologi PETRONAS, Seri Iskandar 32610, Perak Darul Ridzuan, Malaysia; junwei.lim@utp.edu.my
5   Drug Discovery and Delivery Research Laboratory, Malaysian Institute of Chemical and Bioengineering Technology, Universiti Kuala Lumpur, Alor Gajah, Melaka 78000, Malaysia; wytong@unikl.edu.my (W.-Y.T.); crleong@unikl.edu.my (C.-R.L.)
6   School of Pharmacy, Monash University Malaysia, Jalan Lagoon Selatan, Bandar Sunway 47500, Malaysia; khaw.kooiyeong@monash.edu
*   Correspondence: tanwn@usm.my

Citation: Nagarajan, K.; Ibrahim, B.; Ahmad Bawadikji, A.; Lim, J.-W.; Tong, W.-Y.; Leong, C.-R.; Khaw, K.Y.; Tan, W.-N. Recent Developments in Metabolomics Studies of Endophytic Fungi. *J. Fungi* **2022**, *8*, 28. https://doi.org/10.3390/jof8010028

Academic Editor: Laurent Dufossé

Received: 30 November 2021
Accepted: 24 December 2021
Published: 29 December 2021

Publisher's Note: MDPI stays neutral with regard to jurisdictional claims in published maps and institutional affiliations.

Abstract: Endophytic fungi are microorganisms that colonize living plants' tissues without causing any harm. They are known as a natural source of bioactive metabolites with diverse pharmacological functions. Many structurally different chemical metabolites were isolated from endophytic fungi. Recently, the increasing trends in human health problems and diseases have escalated the search for bioactive metabolites from endophytic fungi. The conventional bioassay-guided study is known as laborious due to chemical complexity. Thus, metabolomics studies have attracted extensive research interest owing to their potential in dealing with a vast number of metabolites. Metabolomics coupled with advanced analytical tools provides a comprehensive insight into systems biology. Despite its wide scientific attention, endophytic fungi metabolomics are relatively unexploited. This review highlights the recent developments in metabolomics studies of endophytic fungi in obtaining the global metabolites picture.

Keywords: metabolomics; endophytic fungi; bioactive metabolites; nuclear magnetic resonance; liquid chromatography–high-resolution mass spectrometry; extraction; analytical tool; systems biology

1. Introduction

Endophytic fungi are highly diverse microorganisms that inhabit intracellular and intercellular plant tissues. According to Rashmi et al. (2019), there are approximately one million endophytic fungi [1]. Their wide appearance and distribution in the natural ecosystem indicate the importance of ecological and evolutionary fungi [1–3]. Endophytic fungi offer significant advantages to the host plants by producing various metabolites to counter the attack by pathogens, insects, and herbivores. They are a promising group of microorganisms that produce plant-associated bioactive metabolites with diverse chemical entities and structural functions [4–6]. Metabolites isolated from endophytic fungi exhibited various pharmacological properties, such as antimicrobial, anticancer, antioxidant, anti-inflammatory, and antidiabetic activities. For instance, Taxol is one of the clinical anticancer drugs produced by endophytic fungi used in the treatment of human cancer diseases [7,8].

Nevertheless, there are challenges in the isolation of bioactive metabolites from endophytic fungi due to the high complexity of the crude extracts. The traditional extraction techniques are laborious and slow. Occasionally, some bioactive metabolites are detected in minor and trace quantities. Thus, metabolomics has emerged as an indispensable approach in the comprehensive analysis of complex crude samples [9,10].

Metabolomics is an emerging research field that utilizes technological advances in analytical chemistry to measure and compare the metabolites present [11,12]. Nonetheless, the vast chemical diversity of metabolites has posed challenges in data analysis and interpretation. The enormous metabolomics data obtained from analytical tools require multivariate analysis in classifying different sample groups and metabolites distribution when subjected to different experimental parameters. Therefore, advances in analytical tools coupled with multivariate analysis allow the overview of the metabolites presented aided with visual representation. Principal component analysis (PCA), partial least square discriminant analysis (PLS-DA), and orthogonal partial least square discriminant analysis (OPLS-DA) are among the primary analyses used for this function [9,13]. Generally, metabolomics studies start with study design, followed by sample preparation, data acquisition using analytical tools, and then proceed with data processing and analysis. Lastly, metabolites are identified using the established databases (Figure 1) [11,14]. To date, the most commonly used analytical tools in metabolomics studies are mass spectrometry (MS) coupled with liquid chromatography (LC) and nuclear magnetic resonance (NMR). In metabolomics studies, analytical reproducibility is considered one of the most important criteria. Multiple replicates are recommended to account for the effects and/or variations during the data acquisition [12,15].

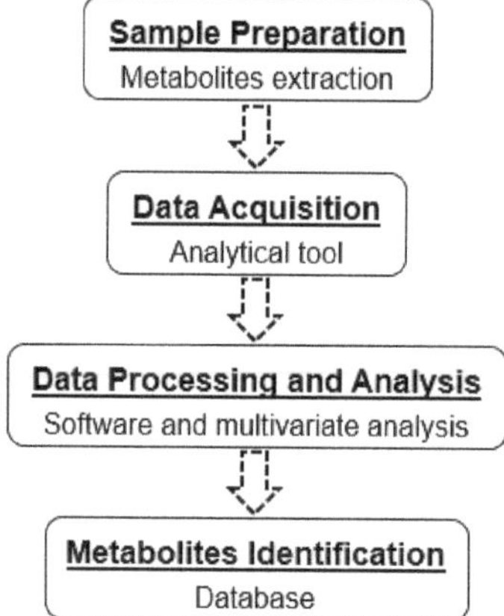

Figure 1. A workflow in metabolomics studies of fungal endophytes.

Endophytic fungi have been regarded as a less exploited niche despite their substantial roles in plants' protection and drug discovery [16,17]. The present review focused on the recent developments in the metabolomics studies of endophytic fungi isolated from host plants. In the following sections, contemporary metabolites extraction from endophytic fungi is described and discussed. Advanced analytical tools, in particular, liquid chromatography–high-resolution mass spectrometry (LC-HRMS) and NMR are reviewed in relation to metabolomics of endophytic fungi (Table 1). In addition, various experimental parameters, such as fungal culture media, time-based incubation, and analysis, as well as bioactivity of fungal extracts that affect the metabolomics data, are also discussed. Perspectives pertaining to the prospects and challenges of metabolomics studies in endophytic fungi are then concluded.

Table 1. Metabolites identification in endophytic fungi via LC-HRMS- and NMR-based metabolomics.

Endophytic Fungi	Host Plant	Metabolite Extraction	Solvent Used	Analytical Tool	Database	Metabolites	Ref.
Aspergillus terreus (AFL, AFSt, AFR)	*Artemisia annua, Medicago sativa*	UAE	Ethyl acetate	LC-HRMS	MarinLit, Dictionary of Natural Products	Paeonol, p-hydroxy benzoic acid, p-coumaric acid, dihydrosinapic acid, osmundacetone, shikimic acid, parvulenone, nidulol, tyrosol, asperpanoid A, maltoryzine, isopestacin, globoscinic acid, 5,7-dihydroxy-4-methylcoumarin, β-methylumbelliferone, hymecromone, 7-hydroxycoumarin, scopoletin, citropten, similanpyrone A, flavipin, gliotoxin, isotryptoquivaline, neoxaline, ochratoxin B, indole-3-acetic acid, phenethylamine, gregatin A, aflatoxin B1, aflatoxin B1 exo-8,9-epoxide, penicillic acid, terrein, physcion	[2]
Aspergillus terreus, A. favus, A. oryzae, Penicillium commune, P. chrysogenum, P. chrysogenum, Talaromyces piophilus, T. piophilus, Fusarium oxysporum, F. nematophilum, Pleosporaceae sp.	*Artemisia annua*	UAE	Ethyl acetate	LC-HRMS	Dictionary of Natural Products	Physcion, emodin, katenarin, norjavanicin, dechlorogriseofulvin, benzyl benzoate, 4-hydroxy benzyl benzoate, benzyl anisate	[9]
Xylaria ellisii sp. nov.	*Vaccinium angustifolium*	LLE	Ethyl acetate; Methanol/acetone (1:1)	LC-HRMS	-	Griseofulvin, dechlorogriseofulvin, cytochalasin D, zygosporin E, epoxycytochalasin D, hirsutatin A, piliformic acid, 2,3-dihydro,2,4-dimethylbenzofuran-7-carboxylic acid, cyclic pentapeptide 1 and 2, xylarotide A, ellisiiamides A-H	[13]
Lasiodiplodia theobromae	*Vitex pinnata*	LLE	Ethyl acetate	LC-HRMS	AntiBase, Dictionary of Natural Products	6,8-Dihydroxy-3-methylisocoumarin, 6-oxo-de-O-methyllasiodiplodin, preussomerins-C and H, palmarumycin CP17, cladospirone B, phomopsin B, desmethyl-lasiodiplodin	[18]
Curvularia sp.	*Terminalia laxiflora*	LLE	Ethyl acetate	LC-HRMS	Natural Product Database	*N*-Acetyl-leucine, afalanine, herbarin A, picroroccellin, dihydroxyisoechinulin A, cyclopiamine B, sengosterone, (E)-11-hydroxyoctadeca-12-enoic acid	[19]

Table 1. *Cont.*

Endophytic Fungi	Host Plant	Metabolite Extraction	Solvent Used	Analytical Tool	Database	Metabolites	Ref.
Aspergillus ochraceus MSEF6	*Medicago sativa*	LLE	Ethyl acetate	LC-HRMS	Dictionary of Natural Products, METLIN	Anisole, 3-hydroxytoluquinone, versicolin, phenoxyacetic acid, terreic acid, terremurin, terredionol, fumigatin, aspyrone, isoaspinonene, 4-hydroxymellein, nidulol, aspyrone	[20]
Aspergillus terreus GMEF1	*Glycine max* L.	UAE	Ethyl acetate	LC-HRMS	Dictionary of Natural Products	Terreic acid, terremutin, (-)-terredionol, terremutin hydrate, 3-methylorsellinic acid, flavipin, astepyrone, reticulol, (3S,6S)-terramide A, emodin, terrelactone A, aspergiketal, 4-hydroxykigelin, 8-hydroxyquadrone, dihydrocitrinone, aspergillide B1, sulochrin, 3α-hydroxy-3,5-dihydromonacolin L	[21]
Penicillium setosum	*Withania somnifera*	UAE/LLE	Dichloromethane: ethyl acetate: methanol (3:2:1)/Ethyl acetate	LC-Q-TOF-MS	METLIN	Kaempferol, quercetin, quercetin acetate, luteolin, dihydroqueretin, dihydromyricetin, quinalizarin, isofraxidin, andrastin D, citromycetin, patulin, 6-deoxyerythronolide B, vanillic acid, 2-dehydro-3-deoxy-darabino-heptonate 7-phosphate (DAHP)	[22]
Coohinforma mamane, Fusarium solani	*C. mamane* (from *Bixa orellana* L.), *F. solani* (from *Plantago lanceolata*)	UAE	Dichloromethane, methanol and water	UHPLC-HRMS	Dictionary of Natural Products, SciFinder, MS Finder, Natural Product Database, KNApSAcK, Chemical Entities of Biological Interest (ChEBI), STOFF, The Toxin and Toxin Target Database (T3DB), Northern African Natural Products Database (NANPDB), Drugbank, FooDB, PlantCyc	Cyclosporins A and E, botryosulfuranol C and B, cyclo-(L-Pro-L-Val), (R)-(-)-mellein, cyclo-(L-Leu-L-Leu-D-Leu-L-Leu-L-Val)	[23]
Colletotrichum sp., *Diaporthe* sp., *Periconia* sp.	*Crescentia alata* Kunth	UAE/LLE	Methanol/Ethyl acetate	^{1}H-NMR	-	Terpenes, phenolics, alkaloids, pigments, steroids, polyketides, glycosides	[24]

Note: UAE = Ultrasonic-assisted extraction; LLE = Liquid–liquid extraction.

2. Multivariate Analysis

Complications in data analysis in metabolomics studies have posed challenges to researchers. Multivariate analysis plays an important part in mining the key metabolites information from the vast raw dataset [25]. PCA is the most widely used multivariate analysis in metabolomics studies. It is employed to observe the distribution pattern of the sample in a general view along with/without the outliers. Meanwhile, partial least squares (PLS) are used to detect the correlations between variables. The quality and effectiveness of the statistical model are then validated using PLS-DA by obtaining the predictable variables and degree. OPLS-DA is then applied to maximize the variations between different groups in the model. In addition to that, statistical significance using analysis of coefficient is performed to summarize the metabolites information [26,27]. In metabolomics studies, the selection of appropriate multivariate analysis is crucial in achieving the experimental goals. The multivariate analysis provides an essential platform to understand the complex metabolites information in a metabolic system [28].

3. Metabolites Extraction

Metabolites extraction from endophytic fungi plays a fundamental role in obtaining a comprehensive metabolites profile [29]. Before metabolites extraction, the isolation of the endophytic fungus from the host takes place. It involves the combination of the sterilized tissue from endophytic fungus with the plant tissue and streaking onto nutrient agar. Occasionally, small sterilized tissues could be plated on nutrient agar [30,31]. Fungal fermentation is followed in either solid-state fermentation (SSF) or submerged fermentation (SMF). Generally, SSF is conducted on a solid substrate. It requires low energy but produces a high concentration of metabolites. SSF has a low water content due to water absorption by the solid substrate. Thus, it enhances the oxygen transfer for the growth of microorganisms. Rice straw, rice hull, and sugarcane bagasse are among the common substrates used in SSF [32,33]. Meanwhile, SMF is a simple fermentation process occurring in excess of water content. Fungi are grown on soluble and/or insoluble substrates, which are submerged in liquid nutrient media. It is widely used due to its better control of fermentation parameters. The pH, temperature, types of culture media, dissolved oxygen, etc., in SMF, could be altered to obtain a wide variety of metabolites [34,35].

Subsequently, metabolites are extracted using appropriate organic solvents such as chloroform, dichloromethane, or ethyl acetate. A suitable and optimized extraction technique is vital to obtain a full profile of metabolites [36]. Liquid–liquid extraction (LLE) and ultrasonic-assisted extraction (UAE) are among the popular techniques used in the preparation of endophytic fungal extract for metabolomics studies. LLE, or known as solvent extraction and partitioning, involves the transferring of metabolites from aqueous to organic solvents by phase separation. This technique separates the metabolites based on solubility in two different immiscible liquids. Separation funnel extraction is the conventional way of performing LLE [37–39]. On the contrary, UAE employs high frequency to produce cavitation, providing better penetration between sample and solvent during the extraction process. The ultrasound waves increase the extraction efficiency by disrupting cells for effective mass transfer. UAE is capable of shortening the extraction duration to achieve the ideal extraction efficiency [37,40].

4. Advanced Analytical Tools for Metabolomics Studies

Analytical tools that are primarily used in metabolomics studies are LC-HRMS and NMR spectroscopy. They possess their own strengths and limitations in metabolomics studies. It is essential to select the right analytical tool to detect metabolites of interest. The choice of the analytical tool is fundamentally relying on the nature of the metabolomics samples as well as the essence of the study [15,41,42]. Developments and challenges of LC-HRMS and NMR in the metabolomics studies of endophytic fungi are highlighted in the following sections.

4.1. LC-HRMS-Based Metabolomics

LC-HRMS is a versatile analytical tool that employs the separation technique of LC with high-resolution mass spectrometry (HRMS) in measuring the mass-to-charge (*m/z*) ratio of metabolites. In order to produce the metabolites' peak signals for identification, ionization occurs within the metabolomics sample when injected into the instrument. Electrospray ionization (ESI) is among the common ionization methods used in LC-HRMS-based metabolomics. Meanwhile, the high resolution of mass enables the separation or differentiation of metabolites with identical nominal mass but distinct elemental compositions [15,41,43].

Over the past decades, LC-HRMS-based metabolomics studies in endophytic fungi have increased rapidly, accounting for ~80% of the published literature. The rise has indicated the unique strengths of LC-HRMS-based metabolomics. Generally, LC-HRMS is highly sensitive and selective. It can detect samples with concentrations up to nanomolar (nM) with extensive coverage of metabolites. It is a superior tool for targeted and non-targeted metabolomics studies. Another merit in employing LC-HRMS-based metabolomics is the large access to the spectral databases, which is the key to metabolites identification [44–46]. For instance, the database MS-Finder has provided more than 290,000 and 35,000 spectral libraries for positive and negative mode MS, respectively [47,48]. METLIN is among the major electronic databases used in LC-HRMS-based metabolomics. It consists of MS/MS data from positive and negative modes collected at three collision energies (10, 20, and 40 V). Comprehensive high-resolution MS/MS spectral data have allowed the researchers to access the spectral resources for metabolomics studies freely. Furthermore, databases such as AntiBase, ChEBI, Dictionary of Natural Products (DNP), Drugbank, FooDB, KNApSaCK, MarineLit, NANPDB, and National Institute of Science and Technology (NIST) have developed rapidly over the past decades [49,50].

Metabolomics studies of endophytic fungi using different culture media and incubation periods are gaining attention in the mining of bioactive metabolites [51,52]. Endophytic fungi *Lasiodiplodia theobromae* grown in liquid and rice media for 7-, 15-, and 30-day incubation were studied in correlation to anti-trypanosomal activity. Based on the results, 7- and 15-day incubation extracts provided a similar chemical profile while a 30-day incubation extract yielded lesser metabolites. However, a bioactivity assay revealed a 30-day rice medium incubated extract exhibited the lowest minimum inhibitory concentration (MIC). Bioactive metabolites 6,8-dihydroxy-3-methylisocoumarin (**1**), 6-oxo-de-*O*-methyllasiodiplodin (**2**), preussomerins-C and H (**3-4**), palmarumycin CP17 (**5**), cladospirone B (**6**), phomopsin B (**7**), and desmethyl-lasiodiplodin (**8**) (Figure 2) were identified in the study using OPLS-DA model [18]. Tawfike and co-workers performed metabolomics profiling on *Curvularia* sp. extracts from liquid and rice media on 7-, 15- and 30-day incubation periods in correspondence to cytotoxicity. The findings revealed that 7- and 30-day incubation extracts produced more metabolites than a 15-day incubation extract. A 15-day fungal incubation was regarded as an intermediate phase where metabolites were consumed for survival to counter environmental stress. Different chemical profiles were noticed over the incubation period as different metabolites were synthesized and depleted for different mechanisms. Mass spectral dereplication showed that *N*-acetyl-leucine (**9**), afalanine (**10**), herbarin A (**11**), picroroccellin (**12**), dihydroxyisoechinulin A (**13**), cyclopiamine B (**14**), sengosterone (**15**), and (*E*)-11-hydroxyoctadeca-12-enoic acid (**16**) were detected as unique bioactive metabolites during the growth phase [19]. In a study conducted by Attia et al. (2020), *Aspergillus ochraceus* MSEF6 isolated from *Medicago sativa* was grown in different media, namely, potato dextrose broth (PDB), Sabouraud broth (SAB), malt extract broth (MEB), and rice extract broth (REB) to discover its most potent antimicrobial activity. Fungal extract cultured in PDB was found as the most active with a minimum inhibitory concentration (MIC) of 15–30 mg/mL. Metabolomics-based chemical profiling revealed anisole (**17**), 3-hydroxytoluquinone (**18**), versicolin (**19**), phenoxyacetic acid (**20**), terreic acid (**21**), terremurin (**22**), terredionol (**23**), fumigatin (**24**), aspyrone (**25**), isoaspinonene (**26**), 4-hydroxymellein (**27**), and nidulol (**28**) were present and may contribute to the observed activity [20]. In another study, endophytic fungi *Aspergillus terreus* isolated from soybean was cultured in PDB, acidified potato dextrose broth (MPDB), SAB, MEB, and REB. The obtained ethyl acetate fungal extracts were then subjected to LC-HRMS to produce 2319 and 1230 peaks in positive and negative modes, respectively. Eighteen metabolites were identified after dereplication using the Dictionary of Natural Products database. The metabolites comprised mainly quinones, isocoumarins, and polyketides. Multivariate data analysis reported that MEB, PDB, and MPDB extracts consisted of characteristic chemical fingerprints compared to other cultured media extracts [21]. Endophytic fungi are stimulated by different growth mediums and culture conditions to produce different metabolites. Optimization of culture parameters, co-culture fermentation, as well as the addition of elicitors may be used to increase the production of bioactive metabolites. Metabolomics plays a crucial role in the search for bioactive metabolites from a specific endophyte at specific conditions and parameters [16,53–55].

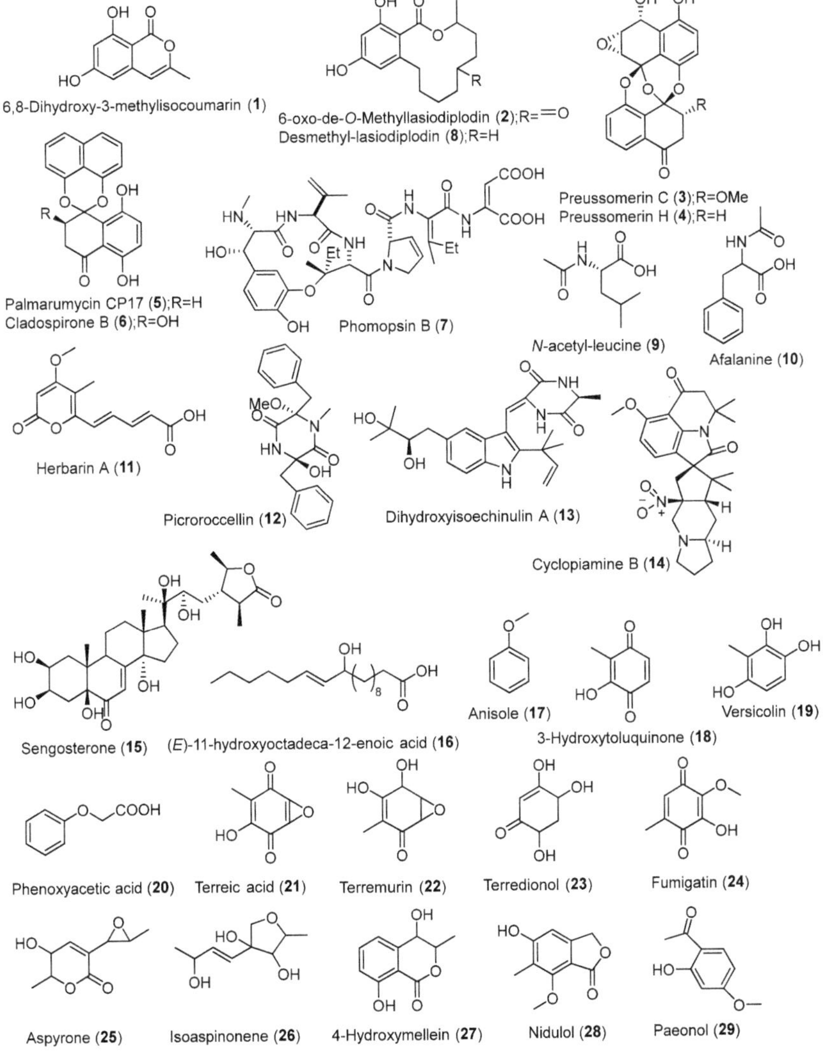

Figure 2. *Cont.*

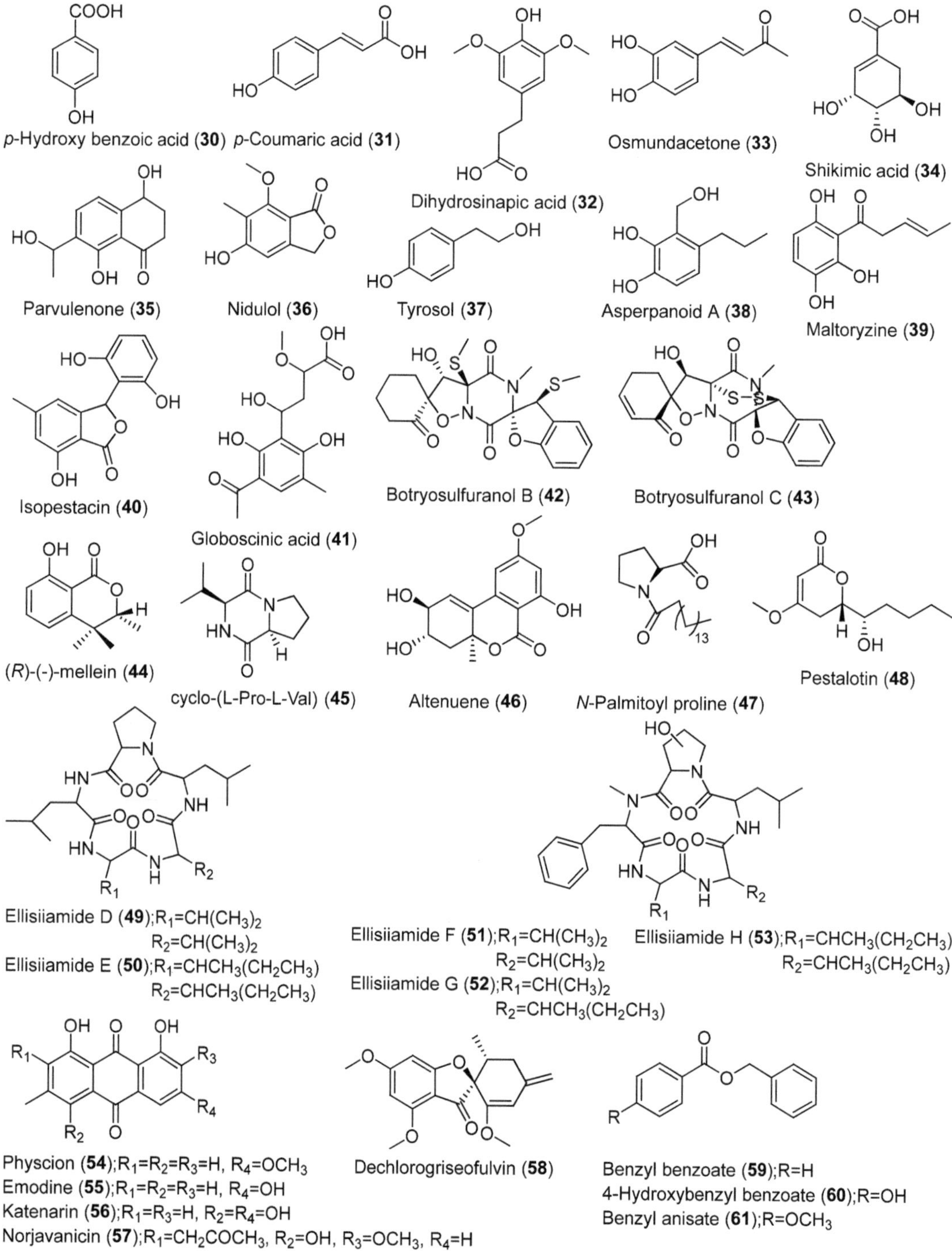

Figure 2. Chemical structures of metabolites (**1**)–(**61**).

Chemical fingerprints of fungal endophytes may affect by metabolites extraction methods [5,56–58]. George et al. (2019) investigated the metabolites produced by *Penicillium setosum* by employing UAE and LLE techniques. Based on the results, endophytic fungal extracts from both extraction techniques gave an almost similar chromatogram when subjected to liquid chromatography–quadrupole time-of-flight mass spectrometry (LC-Q-ToF-MS). Fourteen metabolites were identified from LLE, while eleven metabolites were detected from UAE. They comprised of chemical classes flavonol, flavone, dihydroflavonol, anthraquinone, coumarin, *Penicillium* metabolites, and other organic compounds [22]. The utilization of suitable extraction methods is vital in discriminating the metabolites profiles of endophytic fungi when subjecting to different extraction methods [5,59,60].

Statistical analyses are indispensable in any research study. Multivariate analysis associated with metabolomics serves as a significant approach in dealing with a large number of datasets produced by multiple experiments [25,61–63]. Three endophytic fungi (AFL, AFSt, and AFR) isolated from *Artemisia annua* and *Medicago sativa* were studied for their metabolites profiles using LC-HRMS. It revealed the presence of 682 metabolites in the ethyl acetate extracts of AFL, AFSt, and AFR. Phenolic derivatives paeonol (**29**), *p*-hydroxy benzoic acid (**30**), (Figure 2) *p*-coumaric acid (**31**), dihydrosinapic acid (**32**), osmundacetone (**33**), shikimic acid (**34**), parvulenone (**35**), nidulol (**36**), tyrosol (**37**), asperpanoid A (**38**), maltoryzine (**39**), isopestacin (**40**), and globoscinic acid (**41**) were identified as the major metabolites. Additionally, coumarins, alkaloids, and polyketides were detected, among others. Multivariate analysis employing PCA has discriminated the endophytic fungal extracts into three different clusters, indicating their distinctive chemical fingerprints. The PLS-DA-derived heat map displayed the abundance of phenolics in AFL and coumarins in AFSt. Meanwhile, polyketides and alkaloids were predominant in the extract of AFR. The findings are in accordance with the strongest antioxidant activity shown by AFL, followed by AFSt and AFR [1]. Triastuti and colleagues recently worked on a co-culture endophytic fungi *Cophinforma mamane* and *Fusarium solani* in a time-series metabolomics study. By using ultra-high-performance liquid chromatography–high-resolution mass spectrometry (UHPLC-HRMS), 120 and 108 metabolites were identified from positive and negative ionization modes, respectively. Differences in metabolites profile were observed over time (3, 5, and 10 days) in the endophytic fungal extracts of monoculture as well as co-culture. Multivariate analysis using PLS-DA has revealed 25 metabolites that contributed to the group discrimination. Among them, botryosulfuranols B and C (**42-43**), cyclosporins A and E, (*R*)-(-)-mellein (**44**), cyclo-(L-Pro-L-Val) (**45**), and cyclo-(L-Leu-L-Leu-D-Leu-L-Leu-L-Val) were identified. In detail, the Venn diagram was used to analyze the metabolites variation in the monoculture and co-culture of fungal extracts. Generally, the number of metabolites increased over time in both monoculture and co-culture of fungal extracts. It is worth noting that co-culture endophytic fungal extract has induced five de novo metabolites, of which three were identified as altenuene (**46**), *N*-palmitoyl proline (**47**), and pestalotin (**48**). Hence, the findings indicated that fungal co-culture in the time-series analysis is worth exploring by researchers in an attempt to discover interesting metabolomes as well as the biosynthetic pathways [23]. Ibrahim and co-workers investigated *Xylaria ellisii*, a new endophytic fungus from *Vaccinium angustifolium*. Fungal filtrates and mycelium from *V. angustifolium* grown in wild and highbush were studied. The ethyl acetate fungal filtrate extract and methanol/acetone (1:1) mycelium extract were profiled using LC-MS-based metabolomics. Nineteen metabolites were identified, consisting mainly of nonribosomal peptides and polyketides. Ellisiiamides D–H (**49–53**) were detected as outliers in the extracted filtrates and mycelium employing the OPLS-DA model [13]. In a recent study on endophytic fungi isolated from *Artemisia annua*, antimalarial metabolites were identified using LC-HRMS-based metabolomics. Eleven endophytic fungal isolates from the family of Trichocomaceae, Nectriaceae, and Pleosporaceae were fermented and extracted with ethyl acetate via the UAE technique. Among the 2363 peaks detected in LC-HRMS, eight metabolites were identified. Physcion (**54**), emodine (**55**), katenarin (**56**), norjavanicin (**57**), dechlorogriseofulvin (**58**), benzyl benzoate (**59**), 4-hydroxybenzyl benzoate (**60**), and benzyl anisate (**61**) were found to be positively correlated with the anti-plasmodial activity using multivariate analysis. In further investigation, neural network and deep learning-based software were employed to identify metabolites with the most possible active hits [9]. By applying multivariate analysis in metabolomics, metabolites of interest could be systematically detected and identified from a complex mixture of chemical compounds [64–66].

4.2. NMR-Based Metabolomics

NMR spectroscopy is a non-destructive, strongly quantitative, and reproducible analytical tool. Owing to its robustness, NMR is automatable, has high throughput, and requires simple sample preparation [67–69]. Among the various NMR experiments, one-dimensional proton-NMR (1D ^{1}H-NMR) is broadly employed in metabolomics due to its high intense NMR signals and short collection time. The ^{1}H-NMR spectrum could be obtained within a minute, and it is well suited for large-scale sample analysis. Furthermore, different nucleic experiments (^{13}C, ^{15}N, and ^{31}P) could be used in NMR to study different chemical classes of metabolites [44,45,70]. Despite its numerous advantages, there are challenges in performing NMR-based metabolomics. NMR is 10–100 times less sensitive compared to MS-based metabolomics. More often, overlapping peaks in ^{1}H-NMR spectra have become a great challenge in the characterization of metabolites. Limited NMR databases and software have restricted NMR in metabolomics applications [44,45,71]. Thus, the literature studies employing solely NMR-based metabolomics in endophytic fungi are rather scarce.

Metabolomics investigations using ^{1}H-NMR on endophytic fungal isolates (*Colletotrichum* sp., *Diaporthe* sp. and *Periconia* sp.) from *Crescentia alata* Kunth discriminated the classes of compounds present in each extract. The multivariate PLS-DA revealed the strongest anti-inflammatory activity was exhibited by *Colletotrichum* extract, followed by *Diaporthe* and *Periconia* extracts. Chemical shifts δ_H 0.76, 1.2, 1.44, 1.72, and 1.76 ppm were noticed in *Colletotrichum* extract, suggesting the presence of terpene-type metabolites. These metabolites were believed to contribute to the observed activity in the *Colletotrichum* extract. Meanwhile, olefinic and aromatic protons were detected in the *Periconia* and *Diaporthe* extracts, respectively. The presence of characteristic NMR signals among the extracts has discriminated the chemical profiles of fungal endophytes isolated from the same host [24]. NMR chemical shift databases could potentially target a list of metabolites hits to the peaks generated from metabolomics experiments. Nonetheless, selecting the right metabolite could be challenging when dealing with metabolites of low concentration. Moreover, the peak assignment becomes complicated when some of the NMR peaks are partly and/or entirely overlapped by highly concentrated metabolites signals [72–75].

5. Conclusions and Future Perspectives

Endophytic fungi are promising producers of biologically interesting metabolites. The bioactive metabolites play a vital role in the treatment and enhancement of human health. Occasionally, potent bioactive metabolites are present at low yields. Thus, metabolites isolation could be challenging at the preliminary stage of works. As such, metabolomics has appeared as a robust approach to measuring the global metabolites from a minimal amount of samples. Innovations in analytical tools have advanced the progress of metabolomics, particularly in endophytic fungi. In-depth research on endophytic fungi by integrating both MS and NMR is warranted in order to perform high-throughput metabolomics studies with better detection and metabolites identification. Continuous metabolomics research on endophytic fungi must remain to ensure contribution to the chemical database of fungal metabolites is increasing for biomarker discovery.

Though metabolomics of endophytic fungi is still in its developing stage, it is highly prospective that the next 10 years will be impactful by integrating both LC-HRMS and NMR in the study design for direct comparison and correlation of the metabolites data. In addition, statistical tools and software, as well as machine learning and neural networks, may significantly aid the data analysis and visual representation of the metabolomics experiments. Studies involving the mechanisms between host plant–endophyte interaction, genetic manipulations of endophytic fungi, and the modified biosynthetic pathways for natural bioactive metabolites from fungal endophytes could be focused on.

Author Contributions: Conceptualization, W.-N.T.; data collection; methodology, K.N.; software, W.-N.T. and J.-W.L.; writing—original draft preparation, K.N., A.A.B. and W.-N.T.; writing—review and editing, B.I., W.-Y.T., C.-R.L., K.Y.K. and W.-N.T. All authors have read and agreed to the published version of the manuscript.

Funding: This research was funded by the Ministry of Higher Education Malaysia for Fundamental Research Grant Scheme with Project Code: FRGS/1/2018/STG01/USM/02/3.

Institutional Review Board Statement: Not applicable.

Informed Consent Statement: Not applicable.

Data Availability Statement: Not applicable.

Acknowledgments: The authors would like to acknowledge Ministry of Higher Education Malaysia for Fundamental Research Grant Scheme with Project Code: FRGS/1/2018/STG01/USM/02/3 and Universiti Sains Malaysia for facilities and support.

Conflicts of Interest: The authors declare no conflict of interest.

Abbreviations

ChEBI	Chemical Entities of Biological Interest
ESI	Electrospray ionization
LC	Liquid chromatography
LC-HRMS	Liquid chromatography–high-resolution mass spectrometry
LC-Q-ToF-MS	Liquid chromatography–quadrupole time-of-flight-mass spectrometry
LLE	Liquid–liquid extraction
MEB	Malt extract broth
MPDB	Acidified potato dextrose broth
MS	Mass spectrometry
m/z	Mass-to-charge
MIC	Minimum inhibitory concentration
NIST	National Institute of Science and Technology
NANPDB	Northern African Natural Products Database
NMR	Nuclear magnetic resonance
OPLS-DA	Orthogonal partial least square discriminant analysis
PLS-DA	Partial least square discriminant analysis
PCA	Principal component analysis
PDB	Potato dextrose broth
REB	Rice extract broth
SAB	Sabouraud broth
SSF	Solid-state fermentation
SMF	Submerged fermentation
T3DB	The Toxin and Toxin Target Database
UHPLC-HRMS	Ultra high-performance liquid chromatography–high-resolution mass spectrometry
UAE	Ultrasonic-assisted extraction
1D ^{1}H-NMR	One-dimensional proton-NMR

References

1. Rashmi, M.; Kushveer, J.S.; Sarma, V.V. A worldwide list of endophytic fungi with notes on ecology and diversity. *Mycosphere* **2019**, *10*, 798–1079. [CrossRef]
2. Sayed, A.M.; Sherif, N.H.; El-Gendy, A.O.; Shamikh, Y.I.; Ali, A.T.; Attia, E.Z.; El-Katatny, M.H.; Khalifa, B.A.; Hassan, H.M.; Abdelmohsen, U.R. Metabolomic profiling and antioxidant potential of three fungal endophytes derived from *Artemisia annua* and *Medicago sativa*. *Nat. Prod. Res.* **2020**. *In Press.* [CrossRef]
3. Hartley, S.E.; Eschen, R.; Horwood, J.M.; Gange, A.C.; Hill, E.M. Infection by a foliar endophyte elicits novel arabidopside-based plant defence reactions in its host, *Cirsium arvense*. *New Phytol.* **2015**, *205*, 816–827. [CrossRef] [PubMed]

4. Nagarajan, K.; Tong, W.-Y.; Leong, C.-R.; Tan, W.-N. Potential of endophytic *Diaporthe* sp. as a new source of bioactive compounds. *J. Microbiol. Biotechnol.* **2021**, *31*, 493–500. [CrossRef] [PubMed]
5. Liu, J.; Liu, G. Analysis of secondary metabolites from plant endophytic fungi. *Methods Mol. Biol.* **2018**, *1848*, 25–38. [CrossRef]
6. Vinale, F.; Nicoletti, R.; Lacatena, F.; Marra, R.; Sacco, A.; Lombardi, N.; d'Errico, G.; Digilio, M.C.; Lorito, M.; Woo, S.L. Secondary metabolites from the endophytic fungus *Talaromyces pinophilus*. *Nat. Prod. Res.* **2017**, *31*, 1778–1785. [CrossRef]
7. Uzma, F.; Mohan, C.D.; Hashem, A.; Konappa, N.M.; Rangappa, S.; Kamath, P.V.; Singh, B.P.; Mudili, V.; Gupta, V.K.; Siddaiah, C.N.; et al. Endophytic fungi—Alternative sources of cytotoxic compounds: A review. *Front. Pharmacol.* **2018**, *9*, 309. [CrossRef] [PubMed]
8. Gupta, J.; Sharma, S. Endophytic fungi: A new hope for drug discovery. In *New and Future Developments in Microbial Biotechnology and Bioengineering*; Singh, J., Gehlot, P., Eds.; Elsevier: Amsterdam, The Netherlands, 2020; pp. 39–49.
9. Alhadrami, H.A.; Sayed, A.M.; El-Gendy, A.O.; Shamikh, Y.I.; Gaber, Y.; Bakeer, W.; Sheirf, N.H.; Attia, E.Z.; Shaban, G.M.; Khalifa, B.A.; et al. A metabolomic approach to target antimalarial metabolites in the *Artemisia annua* fungal endophytes. *Sci. Rep.* **2021**, *11*, 2770. [CrossRef]
10. Naik, B.S. Developments in taxol production through endophytic fungal biotechnology: A review. *Orient. Pharm. Exp. Med.* **2019**, *19*, 1–13. [CrossRef]
11. Dayalan, S.; Xia, J.; Spicer, R.A.; Salek, R.; Roessner, U. Metabolome Analysis. In *Encyclopedia of Bioinformatics and Computational Biology*; Ranganathan, S., Gribskov, M., Nakai, K., Schönbach, C., Eds.; Academic Press: Cambridge, MA, USA, 2019; pp. 396–409.
12. Manchester, M.; Anand, A. Metabolomics: Strategies to define the role of metabolism in virus infection and pathogenesis. In *Advances in Virus Research*; Kielian, M., Mettenleiter, T.C., Roossinck, M.J., Eds.; Academic Press: Cambridge, MA, USA, 2017; Volume 98, pp. 57–81.
13. Ibrahim, A.; Tanney, J.B.; Fei, F.; Seifert, K.A.; Cutler, G.C.; Capretta, A.; Miller, J.D.; Sumarah, M.W. Metabolomic-guided discovery of cyclic nonribosomal peptides from *Xylaria ellisii* sp. nov., a leaf and stem endophyte of *Vaccinium angustifolium*. *Sci. Rep.* **2020**, *10*, 4599. [CrossRef]
14. Tawfik, N.F.; Tawfike, A.F.; Abdou, R.; Abbott, G.; Abdelmohsen, U.R.; Edrada-Ebelm, R.; Haggag, E.G. Metabolomics and bioactivity guided isolation of secondary metabolites from the endophytic fungus *Chaetomium* sp. *J. Adv. Pharm. Res.* **2017**, *1*, 66–74. [CrossRef]
15. Alonso, A.; Marsal, S.; Julia, A. Analytical methods in untargeted metabolomics: State of the art in 2015. *Front. Bioeng. Biotechnol.* **2015**, *3*, 23. [CrossRef]
16. Fadiji, A.E.; Babalola, O.O. Elucidating mechanisms of endophytes used in plant protection and other bioactivities with multifunctional prospects. *Front. Bioeng. Biotechnol.* **2020**, *8*, 467. [CrossRef]
17. Manganyi, M.C.; Ateba, C.N. Untapped potentials of endophytic fungi: A review of novel bioactive compounds with biological applications. *Microorganisms* **2020**, *8*, 1934. [CrossRef]
18. Kamal, N.; Viegelmann, C.V.; Clements, C.J.; Edrada-Ebel, R. Metabolomics-guided isolation of anti-trypanosomal metabolites from the endophytic fungus *Lasiodiplodia theobromae*. *Planta Med.* **2017**, *83*, 565–573. [CrossRef]
19. Tawfike, A.F.; Abbott, G.; Young, L.; Edrada-Ebel, R. Metabolomic-guided isolation of bioactive natural products from *Curvularia* sp., an endophytic fungus of *Terminalia laxiflora*. *Planta Med.* **2018**, *84*, 182–190. [CrossRef] [PubMed]
20. Attia, E.Z.; Farouk, H.M.; Abdelmohsen, U.R.; El-Katatny, M.H. Antimicrobial and extracellular oxidative enzyme activities of endophytic fungi isolated from alfalfa (*Medicago sativa*) assisted by metabolic profiling. *S. Afr. J. Bot.* **2020**, *134*, 156–162. [CrossRef]
21. El-Hawary, S.S.; Mohammed, R.; Bahr, H.S.; Attia, E.Z.; El-Katatny, M.H.; Abelyan, N.; Al-Sanea, M.M.; Moawad, A.S.; Abdelmohsen, U.R. Soybean-associated endophytic fungi as potential source for anti-COVID-19 metabolites supported by docking analysis. *J. Appl. Microbiol.* **2021**, *131*, 1193–1211. [CrossRef]
22. George, T.K.; Devadasan, D.; Jisha, M.S. Chemotaxonomic profiling of *Penicillium setosum* using high-resolution mass spectrometry (LC-Q-ToF-MS). *Heliyon* **2019**, *5*, e02484. [CrossRef] [PubMed]
23. Triastuti, A.; Haddad, M.; Barakat, F.; Mejia, K.; Rabouille, G.; Fabre, N.; Amasifuen, C.; Jargeat, P.; Vansteelandt, M. Dynamics of chemical diversity during co-cultures: An integrative time-scale metabolomics study of fungal endophytes *Cophinforma mamane* and *Fusarium solani*. *Chem. Biodivers.* **2021**, *18*, e2000672. [CrossRef] [PubMed]
24. Flores-Vallejo, R.C.; Folch-Mallol, J.L.; Sharma, A.; Cardoso-Taketa, A.; Alvarez-Berber, L.; Villarreal, M.L. ITS2 ribotyping, in vitro anti-inflammatory screening, and metabolic profiling of fungal endophytes from the Mexican species *Crescentia alata* Kunth. *S. Afr. J. Bot.* **2020**, *134*, 213–224. [CrossRef]
25. Percival, B.; Gibson, M.; Leenders, J.; Wilson, P.B.; Grootveld, M. Univariate and multivariate statistical approaches to the analysis and interpretation of NMR-based metabolomics datasets of increasing complexity. In *Computational Techniques for Analytical Chemistry and Bioanalysis*; Wilson, P.B., Grootveld, M., Eds.; Royal Society of Chemistry: London, UK, 2020.
26. Worley, B.; Powers, R. Multivariate Analysis in Metabolomics. *Curr. Metab.* **2013**, *1*, 92–107. [CrossRef]
27. Wu, J.F.; Wang, Y. Multivariate analysis of metabolomics data. In *Plant Metabolomics*; Qi, X., Chen, X., Wang, Y., Eds.; Springer: Dordrecht, Germany, 2015.
28. Ramana, P.; Adams, E.; Augustijns, P.; Van Schepdael, A. Metabonomics and drug development. In *Metabonomics. Methods in Molecular Biology*; Bjerrum, J., Ed.; Humana Press: New York, NY, USA, 2015; Volume 1277.
29. Synytsya, A.; Monkai, J.; Bleha, R.; Macurkova, A.; Ruml, T.; Ahn, J.; Chukeatirote, E. Antimicrobial activity of crude extracts prepared from fungal mycelia. *Asian Pac. J. Trop. Biomed.* **2017**, *7*, 257–261. [CrossRef]

30. Ezeobiora, C.E.; Igbokwe, N.H.; Amin, D.H.; Mendie, U.E. Endophytic microbes from Nigerian ethnomedicinal plants: A potential source for bioactive secondary metabolites—A review. *Bull. Natl. Res. Cent.* **2021**, *45*, 103. [CrossRef]

31. de Carvalho, C.C.C.R. Fungi in Fermentation and Biotransformation Systems. In *Biology of Microfungi. Fungal Biology*; Li, D.W., Ed.; Springer: Cham, Switzerland, 2016; pp. 525–541.

32. Srivastava, N.; Srivastava, M.; Ramteke, P.W.; Mishra, P.K. Solid-state fermentation strategy for microbial metabolites production: An overview. In *New and Future Developments in Microbial Biotechnology and Bioengineering*; Gupta, V.J., Pandey, A., Eds.; Elsevier: Amsterdam, The Netherlands, 2019; pp. 345–354.

33. Costa, J.A.V.; Treichel, H.; Kumar, V.; Pandey, A. Advances in solid-state fermentation. In *Current Developments in Biotechnology and Bioengineering*; Pandey, A., Larroche, C., Soccol, C.R., Eds.; Elsevier: Amsterdam, The Netherlands, 2018; pp. 1–17.

34. Martínez-Medina, G.A.; Barragán, A.P.; Ruiz, H.A.; Ilyina, A.; Hernández, J.L.M.; Rodríguez-Jasso, R.M.; Hoyos-Concha, J.L.; Aguilar-González, C.N. Fungal proteases and production of bioactive peptides for the food industry. In *Enzymes in Food Biotechnology*; Kuddus, M., Ed.; Academic Press: Cambridge, MA, USA, 2019; pp. 221–246.

35. Kapoor, M.; Panwar, D.; Kaira, G.S. Bioprocesses for enzyme production using agro-industrial wastes: Technical challenges and commercialization potential. In *Agro-Industrial Wastes as Feedstock for Enzyme Production*; Dhillon, G.S., Kaur, S., Eds.; Academic Press: Cambridge, MA, USA, 2016; pp. 61–93.

36. Song, R.; Wang, J.; Sun, L.; Zhang, Y.; Ren, Z.; Zhao, B.; Lu, H. The study of metabolites from fermentation culture of *Alternaria oxytropis*. *BMC Microbiol.* **2019**, *19*, 35. [CrossRef] [PubMed]

37. Fierascu, R.C.; Fierascu, I.; Ortan, A.; Georgiev, M.I.; Sieniawska, E. Innovative approaches for recovery of phytoconstituents from medicinal/aromatic plants and biotechnological production. *Molecules* **2020**, *25*, 309. [CrossRef]

38. Urkude, R.; Dhurvey, V.; Kochhar, S. Pesticide residues in beverages. In *Quality Control in the Beverage Industry*; Grumezescu, A.M., Holban, A.M., Eds.; Academic Press: Cambridge, MA, USA, 2019; pp. 529–560.

39. Kyle, P.B. Toxicology: GCMS. In *Mass Spectrometry for the Clinical Laboratory*; Nair, H., Clarke, W., Eds.; Academic Press: Cambridge, MA, USA, 2017; pp. 131–163.

40. Zahari, N.A.A.R.; Chong, G.H.; Abdullah, L.C.; Chua, B.L. Ultrasonic-assisted extraction (UAE) process on thymol concentration from *Plectranthus Amboinicus* leaves: Kinetic modeling and optimization. *Processes* **2020**, *8*, 322. [CrossRef]

41. David, A.; Rostkowski, P. *Environmental Metabolomics*; Elsevier Inc.: Amsterdam, The Netherlands, 2020; pp. 35–64.

42. Zhang, T.; Chen, C.; Xie, K.; Wang, J.; Pan, Z. Current state of metabolomics research in meat quality analysis and authentication. *Foods* **2021**, *10*, 2388. [CrossRef] [PubMed]

43. Lei, Z.; Huhman, D.V.; Sumner, L.W. Mass spectrometry strategies in metabolomics. *J. Biol. Chem.* **2011**, *286*, 25435–25442. [CrossRef] [PubMed]

44. Emwas, A.-H.; Roy, R.; McKay, R.T.; Tenori, L.; Saccenti, E.; Gowda, G.A.N.; Raftery, D.; Alahmari, F.; Jaremko, L.; Jaremko, M.; et al. NMR spectroscopy for metabolomics research. *Metabolites* **2019**, *9*, 123. [CrossRef]

45. Wishart, D.S. NMR metabolomics: A look ahead. *J. Magn. Reson.* **2019**, *306*, 155–161. [CrossRef] [PubMed]

46. Sands, C.J.; Gómez-Romero, M.; Correia, G.; Chekmeneva, E.; Camuzeaux, S.; Izzi-Engbeaya, C.; Dhillo, W.S.; Takats, Z.; Lewis, M.R. Representing the metabolome with high fidelity: Range and response as quality control factors in LC-MS-based global profiling. *Anal. Chem.* **2021**, *93*, 1924–1933. [CrossRef]

47. Tsugawa, H.; Kind, T.; Nakabayashi, R.; Yukihira, D.; Tanaka, W.; Cajka, T.; Saito, K.; Fiehn, O.; Arita, M. Hydrogen rearrangement rules: Computational MS/MS fragmentation and structure elucidation using MS-FINDER software. *Anal. Chem.* **2016**, *88*, 7946–7958. [CrossRef]

48. Lai, Z.; Tsugawa, H.; Wohlgemuth, G.; Mehta, S.; Mueller, M.; Zheng, Y.; Ogiwara, A.; Meissen, J.; Showalter, M.; Takeuchi, K.; et al. Identifying metabolites by integrating metabolome databases with mass spectrometry cheminformatics. *Nat. Methods* **2017**, *15*, 53–56. [CrossRef]

49. Sorokina, M.; Steinbeck, C. Review on natural products databases: Where to find data in 2020. *J. Cheminf.* **2020**, *12*, 1–51. [CrossRef]

50. Vinaixa, M.; Schymanski, E.L.; Neumann, S.; Navarro, M.; Salek, R.M.; Yanes, O. Mass spectral databases for LC/MS- and GC/MS-based metabolomics: State of the field and future prospects. *Trends Anal. Chem.* **2016**, *78*, 23–35. [CrossRef]

51. Hyde, K.D.; Xu, J.; Rapior, S.; Jeewon, R.; Lumyong, S.; Niego, A.G.T.; Abeywickrama, P.D.; Aluthmuhandiram, J.V.S.; Brahamanage, R.S.; Brooks, S.; et al. The amazing potential of fungi: 50 ways we can exploit fungi industrially. *Fungal Divers.* **2019**, *97*, 1–136. [CrossRef]

52. Zimowska, B.; Bielecka, M.; Abramczyk, B.; Nicoletti, R. Bioactive products from endophytic fungi of Sages (*Salvia* spp.). *Agriculture* **2020**, *10*, 543. [CrossRef]

53. Gakuubi, M.M.; Munusamy, M.; Liang, Z.X.; Ng, S.B. Fungal endophytes: A promising frontier for discovery of novel bioactive compounds. *J. Fungi* **2021**, *7*, 786. [CrossRef]

54. Vinale, F.; Nicoletti, R.; Borrelli, F.; Mangoni, A.; Parisi, O.A.; Marra, R.; Lombardi, N.; Lacatena, F.; Grauso, L.; Finizio, S.; et al. Co-Culture of Plant Beneficial Microbes as Source of Bioactive Metabolites. *Sci. Rep.* **2017**, *7*, 14330. [CrossRef]

55. Rai, N.; Keshri, P.K.; Verma, A.; Kamble, S.C.; Mishra, P.; Barik, S.; Singh, S.K.; Gautam, V. Plant associated fungal endophytes as a source of natural bioactive compounds. *Mycology* **2021**, *12*, 139–159. [CrossRef] [PubMed]

56. Ramdani, D.; Chaudhry, A.S.; Seal, C.J. Chemical composition, plant secondary metabolites, and minerals of green and black teas and the effect of different tea-to-water ratios during their extraction on the composition of their spent leaves as potential additives for ruminants. *J. Agric. Food Chem.* **2013**, *61*, 4961. [CrossRef]

57. Ser, Z.; Liu, X.; Tang, N.N.; Locasale, J.W. Extraction parameters for metabolomics from cultured cells. *Anal Biochem.* **2015**, *475*, 22–28. [CrossRef]

58. Sitnikov, D.; Monnin, C.; Vuckovic, D. Systematic assessment of seven solvent and solid-phase extraction methods for metabolomics analysis of human plasma by LC-MS. *Sci. Rep.* **2016**, *6*, 38885. [CrossRef] [PubMed]

59. Mahmud, I.; Sternberg, S.; Williams, M.; Garrett, T.J. Comparison of global metabolite extraction strategies for soybeans using UHPLC-HRMS. *Anal. Bioanal. Chem.* **2017**, *409*, 6173–6180. [CrossRef] [PubMed]

60. Tokuoka, M.; Sawamura, N.; Kobayashi, K.; Mizuno, A. Simple metabolite extraction method for metabolic profiling of the solid-state fermentation of *Aspergillus oryzae*. *J. Biosci. Bioeng.* **2010**, *110*, 665–669. [CrossRef]

61. Vinaixa, M.; Samino, S.; Saez, I.; Duran, J.; Guinovart, J.J.; Yanes, O. A guideline to univariate statistical analysis for LC/MS-based untargeted metabolomics-derived data. *Metabolites* **2012**, *2*, 775–795. [CrossRef]

62. Berg, M.; Vanaerschot, M.; Jankevics, A.; Cuypers, B.; Breitling, R.; Dujardin, J.C. LC-MS metabolomics from study design to data-analysis–Using a versatile pathogen as a test case. *Comput. Struct. Biotechnol. J.* **2013**, *4*, e201301002. [CrossRef]

63. De Souza, L.P.; Alseekh, S.; Brotman, Y.; Fernie, A.R. Network-based strategies in metabolomics data analysis and interpretation: From molecular networking to biological interpretation. *Expert Rev. Proteom.* **2020**, *17*, 243–255. [CrossRef]

64. Beniddir, M.A.; Bin Kang, K.; Genta-Jouve, G.; Huber, F.; Rogers, S.; van der Hooft, J.J.J. Advances in decomposing complex metabolite mixtures using substructure- and network-based computational metabolomics approaches. *Nat. Prod. Rep.* **2021**, *38*, 1967–1993. [CrossRef] [PubMed]

65. Nalbantoglu, S. Metabolomics: Basic principles and strategies. In *Molecular Medicine*; Nalbantoglu, S., Amri, H., Eds.; IntechOpen: London, UK, 2019.

66. Kellogg, J.J.; Graf, T.N.; Paine, M.F.; McCune, J.S.; Kvalheim, O.M.; Oberlies, N.H.; Cech, N.B. Comparison of metabolomics approaches for evaluating the variability of complex botanical preparations: Green tea (*Camellia sinensis*) as a case study. *J. Nat. Prod.* **2017**, *80*, 1457–1466. [CrossRef] [PubMed]

67. Crowley, T.E. Nuclear magnetic resonance spectroscopy. In *Purification and Characterization of Secondary Metabolites*; Crowley, T.E., Ed.; Academic Press: Cambridge, MA, USA, 2020; pp. 67–78.

68. Tampieri, A.; Szabó, M.; Medina, F.; Gulyás, H. A brief introduction to the basics of NMR spectroscopy and selected examples of its applications to materials characterization. *Phys. Sci. Rev.* **2020**, *6*, 1–41. [CrossRef]

69. Decker, S.R.; Harman-Ware, A.E.; Happs, R.M.; Wolfrum, E.J.; Tuskan, G.A.; Kainer, D.; Oguntimein, G.B.; Rodriguez, M.; Weighill, D.; Jones, P.; et al. High throughput screening technologies in biomass characterization. *Front. Energy Res.* **2018**, *6*, 120. [CrossRef]

70. Vögele, J.; Ferner, J.-P.; Altincekic, N.; Bains, J.K.; Ceylan, B.; Fürtig, B.; Grün, J.T.; Hengesbach, M.; Hohmann, K.F.; Hymon, D.; et al. ^{1}H, ^{13}C, ^{15}N and ^{31}P chemical shift assignment for stem-loop 4 from the 5′-UTR of SARS-CoV-2. *Biomol. NMR Assign.* **2021**, *15*, 335–340. [CrossRef] [PubMed]

71. Liu, R.; Bao, Z.-X.; Zhao, P.-J.; Li, G.-H. Advances in the study of metabolomics and metabolites in some species interactions. *Molecules* **2021**, *26*, 3311. [CrossRef]

72. Gowda, G.A.N.; Raftery, D. Can NMR solve some significant challenges in metabolomics? *J. Magn. Reson.* **2015**, *260*, 144–160. [CrossRef]

73. Bingol, K. Recent advances in targeted and untargeted metabolomics by NMR and MS/NMR Methods. *High-Throughput* **2018**, *7*, 9. [CrossRef]

74. Dona, A.C.; Kyriakides, M.; Scott, F.; Shephard, E.A.; Varshavi, D.; Veselkov, K.; Everett, J.R. A guide to the identification of metabolites in NMR-based metabonomics/metabolomics experiments. *Comput. Struct. Biotechnol. J.* **2016**, *14*, 135–153. [CrossRef] [PubMed]

75. Garcia-Perez, I.; Posma, J.M.; Serrano-Contreras, J.I.; Boulangé, C.L.; Chan, Q.; Frost, G.; Stamler, J.; Elliott, P.; Lindon, J.; Holmes, E.; et al. Identifying unknown metabolites using NMR-based metabolic profiling techniques. *Nat. Protoc.* **2020**, *15*, 2538–2567. [CrossRef]

Journal of
Fungi

Review

Mycotherapy: Potential of Fungal Bioactives for the Treatment of Mental Health Disorders and Morbidities of Chronic Pain

Elaine Meade [1], Sarah Hehir [1,2], Neil Rowan [3] and Mary Garvey [1,2,*]

1   Department of Life Science, Sligo Institute of Technology, F91 YW50 Sligo, Ireland; elaine.meade@mail.itsligo.ie (E.M.); hehir.sarah@itsligo.ie (S.H.)

2   Centre for Precision Engineering, Materials and Manufacturing Research (PEM), Institute of Technology, F91 YW50 Sligo, Ireland

3   Bioscience Research Institute, Technical University Shannon Midlands Midwest, N37 HD68 Athlone, Ireland; neil.rowan@tus.ie

*   Correspondence: garvey.mary@itsligo.ie; Tel.: +353-071-9305529

Abstract: Mushrooms have been used as traditional medicine for millennia, fungi are the main natural source of psychedelic compounds. There is now increasing interest in using fungal active compounds such as psychedelics for alleviating symptoms of mental health disorders including major depressive disorder, anxiety, and addiction. The anxiolytic, antidepressant and anti-addictive effect of these compounds has raised awareness stimulating neuropharmacological investigations. Micro-dosing or acute dosing with psychedelics including Lysergic acid diethylamide (LSD) and psilocybin may offer patients treatment options which are unmet by current therapeutic options. Studies suggest that either dosing regimen produces a rapid and long-lasting effect on the patient post administration with a good safety profile. Psychedelics can also modulate immune systems including pro-inflammatory cytokines suggesting a potential in the treatment of auto-immune and other chronic pain conditions. This literature review aims to explore recent evidence relating to the application of fungal bioactives in treating chronic mental health and chronic pain morbidities.

Keywords: fungi; biologics; mushrooms; chronic pain; mycotherapy; mental health

check for
updates

Citation: Meade, E.; Hehir, S.; Rowan, N.; Garvey, M. Mycotherapy: Potential of Fungal Bioactives for the Treatment of Mental Health Disorders and Morbidities of Chronic Pain. *J. Fungi* **2022**, *8*, 290. https://doi.org/10.3390/jof8030290

Academic Editor: Laurent Dufossé

Received: 24 February 2022
Accepted: 10 March 2022
Published: 11 March 2022

Publisher's Note: MDPI stays neutral with regard to jurisdictional claims in published maps and institutional affiliations.

1. Introduction

There are many recognised mental health disorders or mental illnesses including (but not limited to) anxiety disorders, mood disorders (depression, bipolar disorder, and cyclothymic disorder) psychotic disorders (schizophrenia), eating disorders (anorexia nervosa, bulimia nervosa), impulse control and addiction disorders, obsessive-compulsive disorder (OCD), post-traumatic stress disorder (PTSD) and personality disorders [1]. Each mental illness manifests with a variable array of symptoms that differ depending on the illness present; it is accepted, however, that a person's mood, thinking, perceptions, anhedonia (pleasure sensation) and behaviours are affected. According to the 2013 Diagnostic and Statistical Manual of Mental Disorders 5 (DSM-5), "A mental disorder is a syndrome characterized by clinically significant disturbance in an individual's cognition, emotion regulation, or behaviour that reflects a dysfunction in the psychological, biological, or developmental processes underlying mental functioning" [1]. These issues are undoubtedly proliferated by the stigma toward mental disorders such as anxiety, clinical depression (major depressive disorder), and OCD where a lack of understanding in society prohibits afflicted persons from seeking help or speaking out [2]. Stigma where mentally ill persons are perceived as dangerous, unpredictable, or as having a weakness of character has excluded them from society to an extent, further exasperating the issue [2]. The use of the DSM-5 in the diagnosis of mental illness, however, remains under scrutiny as understanding and expansion of mental illness progresses [3]. The World Health Organisation (WHO, Geneva, Switzerland) created the WHO International Consortium in Psychiatric Epidemiology

(ICPE) in 1998 with the aim of gathering data on mental illness via a diagnostic instrument, i.e., the WHO Composite International Diagnostic Interview (CIDI) surveys [4]. The WHO states that mental health illness is increasing yearly across the globe with approximately 20% of children and adolescents being diagnosed with a mental illness where suicide is the second leading cause of death in 15- to 29-year-olds [5]. In Europe, mental health disorders affect approximately 165 million people yearly with anxiety, mood and addictive disorders being most common [6]. Suffering a mental health disorder is devasting for the individual afflicted but also impacts the family unit as substance abuse and suicidal idealisation often manifests. Considering the wider burden of mental health disease and its associated functional impairment, the socio-economic impact must also be considered due to their increasing prevalence and high disability rate where treatment options often remain unmet [7].

The direct economic burden mental illness relates to diagnosis, treatment, or hospitalisation costs with indirect costs associated with the impact on economic growth including loss of income due to disability, absence from work and cost of additional care [6].

The impact of chronic pain co-morbidities associated with mental health must also be considered including functional disorders such as fibromyalgia and other functional somatic syndromes (FSSs). The treatment of mental health disorders is typically of combination of drug therapy based on psychoactive drugs (Table 1), mood stabilisers or antipsychotic drugs, and psychotherapeutic treatments. In many cases, however, mentally ill persons are non-responsive to therapy, termed treatment resistant [8]. Currently, 30% of clinically depressed are treatment resistant with this cohort having increased issues with social interactions, occupational difficulties, declining physical health and often suicidal thoughts [9]. Additionally, therapeutics currently in use are prone to numerous side effects where treatment is ultimately discontinued in patients (Table 1). To provide therapeutic options for such persons, and to reduce the burden of disease, it is imperative that we seek effective alternatives. The use of psychedelics, such as psilocybin (magic mushrooms), Lysergic acid diethylamide (LSD), and ayahuasca have drawn attention due to their noticeable ability to alter consciousness in personally meaningful, therapeutic, and spiritual ways [10]. Indeed, clinical trials on psilocybin demonstrated its efficacy in reducing cancer-related anxiety and depression, treatment-resistant depression, major depressive disorder (MDD), and substance misuse [11]. This review aims to highlight and discuss the potential of fungal biologics including psilocybin, LSD and others as therapeutic alternatives (mycotherapy) for alleviating the symptoms of mental health disorders. The current knowledge underlining the potential of fungal biologics is highlighted and discussed in the context of common mental health disorders including addiction, MDD and anxiety where current treatment is ineffective or causes significant undesirable side effects. Additionally, this review will review the potential of said biologics to also alleviate the symptoms of chronic pain conditions commonly associated with mental illness including fibromyalgia and irritable bowel syndrome.

Table 1. Treatment of mental health disorders.

Disorder	Treatment	Mode of Action	Efficacy	Side Effects
Anxiety disorders *	SSRIs e.g., sertraline, escitalopram	Inhibit the reuptake of 5-HT	First-line treatments for PD, GAD, SAD, and PTSD [12]	GI problems (nausea, diarrhoea, dyspepsia, bleeding), dry mouth, headaches, dizziness, anxiety, insomnia, and sexual dysfunction [13]
	SNRIs e.g., venlafaxine, duloxetine	Inhibit the reuptake of NE and 5-HT (and/or DA)		
	TCAs e.g., amitriptyline, imipramine	Inhibit the reuptake of NE and 5-HT	Equivocal efficacy with SSRIs; however, cause more adverse side effects due to their anticholinergic activity [14]	Nausea, dry mouth, constipation, weight gain, blurred vision, light-headedness, confusion, sedation, urine retention and tachycardia [15]
	MAOIs e.g., moclobemide, phenelzine	Inhibit the mitochondrial enzyme monoamine oxidase	Third-line treatment for refractory SAD and PD, i.e., considered for patients who are non-responsive to other treatments [16]	Dry mouth, nausea, diarrhoea, constipation, drowsiness, insomnia, dizziness/or light-headedness, fatigue, urinary problems, sexual dysfunction, hypertensive crisis reaction, and serotonin syndrome [17]
	Benzodiazepines e.g., alprazolam, diazepam, clonazepam	Positive allosteric modulators of GABA-A, resulting in increased frequency of chloride channel opening	Effective and fast acting in the treatment of GAD. Recommended as second-line therapy and for short duration use due potential risks of tolerance, dependence, abuse, or misuse [13]	Drowsiness, lethargy, fatigue, and potential for dependence. Higher doses can cause impaired motor coordination, dizziness, vertigo, slurred speech, blurry vision, mood swings, euphoria and hostile or erratic behaviour
	Pregabalin	Calcium channel modulator [18]	Effective as a monotherapy for GAD, or as an adjunct to SSRIs/SNRIs in treatment-resistant GAD [19,20]	Sedation, dizziness, somnolence, dry mouth, amblyopia, diarrhoea, weight gain and potential for dependence [12]
	Buspirone	High affinity for 5-HT$_{1A}$ receptors [21]		Nausea, headaches, dizziness, and fatigue [13]

Table 1. *Cont.*

Disorder	Treatment	Mode of Action	Efficacy	Side Effects
Mood disorders *	Lithium	Multiple mechanisms including modulation of (GABA)-ergic and glutamatergic neurotransmission, and alteration of voltage-gated ion channels or intracellular signalling pathways [22,23]	First-line treatment for prevention of manic and depressive episodes of bipolar disorder (BD) [24]	Cardiac problems, cognitive problems, acne, psoriasis, thyroid problems, nausea, vomiting, weight gain, hyponatremia, sedation, decreased libido, and teratogenic [25]
	valproic acid		First-line treatment for acute mania and maintenance of BD [26]	Cardiac problems, cognitive problems, hair loss, hypothyroidism, aplastic anaemia, Leukopenia, increased transaminases, hepatitis, SLE-like syndrome, hyponatremia, tremor, decreased libido, infertility and teratogenic [25]
	Carbamazepine		Effective as a monotherapy to treat manic symptoms of bipolar or as adjunct to lithium or valproic acid [27]	Cardiac problems, cognitive problems, acne, hair loss, hypothyroidism, PCOS. diarrhoea, nausea, vomiting, pancreatitis, increased transaminases, metabolic syndrome, weight gain, sedation, tremor, decreased sexual function, infertility and teratogenic [25]
Psychotic disorders	First-generation antipsychotics (FGA) e.g., Chlorpromazine, haloperidol	D2 antagonists: work by inhibiting dopaminergic neurotransmission [28]	Effective in the treatment and maintenance of schizophrenia, acute mania with psychotic symptoms, major depressive order with psychotic features, and delusional disorder [28]	Adverse effects are drug specific and include anticholinergic effects (dry mouth, blurry vision, tachycardia, constipation), sedation, distonias, weight gain, increased lipids, parkinsonism (tremor, rigidity, bradykinesia), akathisia tardive dyskinesia, sialorrhea, orthostatic hypotension, neuroleptic malignant syndrome, sexual disfunction, neutropenia/agranulocytosis, and myocarditis [29]
	Second-generation antipsychotics (SGA) e.g., quetiapine, aripiprazole	Serotonin-dopamine antagonists: work by blocking D2 dopamine receptors as well as serotonin receptor antagonist action [28]	Same clinical efficacy as FGA, with the exception of clozapine, which has unique efficacy against treatment resistant schizophrenia [30]	

Table 1. *Cont.*

Disorder	Treatment	Mode of Action	Efficacy	Side Effects
Eating disorder	Olanzapine (SGA)	Block dopaminergic (D1-4 antagonism) and serotonergic (5-HT$_{2A/2C}$ antagonism) receptors [31]	Effective as an adjunctive therapy in treatment of AN, increasing appetite and decreasing anxiety and ruminating thoughts involving body image and food [32]	Dizziness, orthostatic hypotension, hypercholesterolemia, hypertriglyceridemia, hyperglycaemia, weight gain, extra-pyramidal symptoms, dry mouth, hyperprolactinemia, and insomnia [32]
	Antidepressants(SSRIs, SNRIs, TCAs, MAOIs)	Defined above	Effective as an adjunctive therapy in treatment of BN and BED, reducing the crisis of binge eating, purging phenomena and improving mood and anxiety [33]	Listed above
	Mood stabilizers e.g., topiramate	Blocks voltage gated sodium channels, enhances GABA-A receptor activity, reduces membrane depolarization by AMPA/Kainate receptors and is a weak inhibitor of carbonic anhydrase [34]	Shown efficacy in treatment of BN and BED, reducing the crisis of binge eating, purging phenomena and promoting weight loss (in overweight or obese patients) [33]	Paraesthesia, fatigue, cognitive problems, dizziness, somnolence, psychomotor slowing, memory/concentration difficulties, nervousness, confusion, weight loss [34]
Impulse control, addiction, and obsessive-compulsive disorders	Antidepressants e.g., SSRIs and clomipramine (TCA)	Potently inhibit the reuptake of 5-HT	Effective as a monotherapy or as an augmentation agent in the treatment of impulsive (PG, KM, TTM, IED and pyromania), addiction and compulsive disorders [35–39]	Listed above
	Mood stabilisers e.g., olanzapine, carbamazepine	Defined above		
	Naltrexone	Non-specific competitive opioid antagonist with highest affinity for the mu-opioid receptors in the CNS [39]		Nausea, vomiting, abdominal pain, decreased appetite, dizziness, lethargy, headaches and sleep disorders [40]
Personality disorders	Antidepressants (SSRIs, SNRIs) Quetiapine Naltrexone	Defined above	Shown efficacy in the treatment of BPD [41,42]	Listed above

Abbreviations: SSRI = selective serotonin reuptake inhibitors, SNRI = serotonin and norepinephrine reuptake inhibitor, TCA = Tricyclic antidepressants, MAOIs = Monoamine oxidase inhibitors, FGA = first generation antipsychotics, SGA = second generation antipsychotics, 5-HT = 5-hydroxytryptamine, NE = norepinephrine, DA = dopamine, GABA = γ-AMPA = aminobutyric acid, α-Amino-3-hydroxy-5-methyl-4-isoxazolepropionic acid, PD = panic disorder, GAD = generalised anxiety disorder, SAD = social anxiety disorder, PTSD = post-traumatic stress disorder, BD = bipolar disorder, AN = anorexia nervosa, BN = bulimia nervosa, BED = binge eating disorder, PG = pathological gambling, KM = kleptomania, TTM = trichotillomania, IED = intermittent explosive disorder, BPD = Borderline Personality Disorder, SLE = systemic lupus erythematosus, PCOS = Polycystic ovary syndrome, CNS = central nervous system. * Many drugs listed for the treatment of anxiety are also employed for the treatment of mood disorders including SSRIs, NSRIs and TCAs.

2. Mental Health Disorders

The incidence of mental health disorders is increasing globally, yet a fundamental understanding of the aetiology of mental health disease remains elusive. Assumptions relate severe mental illness to a small number of genes with a specific relationship between genotype and mental illness where biological pathways and environmental factors impact on illness [43]. Studies have demonstrated a strong relationship between severe trauma such as childhood abuse and/or abandonment with adult mental health disorders and a diminished quality of life [44]. The most widely discussed theories of depression are based on the regulation of the monoamines, norepinephrine (NE), dopamine and serotonin. The catecholamine hypothesis of depression developed in the 1960s remains the basis of treatment of mental health disorders such as MDD and anxiety disorder. Based on the theory that such mental illness is a result of a NE or dopamine deficiency at central nervous system (CNS) synapses, treatment aims to increase dopamine or NE levels to alleviate symptoms [45]. The serotonin hypothesis of depression also from the 1960s relates mood disorders to a deficiency in serotonin (5-hydroxytryptamine, 5-HT) a neurotransmitter involved in regulating emotion and mood [46]. There are currently 14 5-HT receptor subtypes identified having varying affinities for serotonin, present presynaptically, postsynaptically in the CNS and throughout the brain being predominately G protein coupled receptors (except 5-HT3 receptor) [47]. Agonism of the 5-HT_{2A} receptors is involved in increasing cortical glutamate levels [48]. The neuroendocrine hypothalamic-pituitary-adrenal (HPA) axis a natural pathway activated in times of stress also becomes hyperactive in states of depression. The HPA axis is involved in maintaining blood glucocorticoid levels, where elevated levels can disrupt HPA function [49]. Studies have shown that cortisol elevation long term results in cognitive and other medical issues in patients and exacerbate depressive symptoms as shown in mice [50]. Thus, it is generally accepted that neuropsychiatric diseases such as PTSD, MDD, anxiety etc are a result of a chemical imbalance in neural circuits distributed across the brain and CNS related to a loss of neuro plasticity [51]. Consequently, treatment of mental health disorders is based on prescribing anti-depressants namely selective serotonin reuptake inhibitors (SSRIs), serotonin-norepinephrine reuptake inhibitors (SNRIs), dopamine reuptake inhibitors (DRIs) and tricyclic antidepressants (TCAs) which aim to increase neurotransmitter levels at the synapses and increase neuroplasticity. While the efficacy of this approach remains under question the side effects associated with the varying therapeutics in these categories are undeniable. Serotonin syndrome for example which is a potentially life-threatening condition can be caused by using serotonergic drugs and overactivation of 5-HT_{1A} and 5-HT_{2A} receptors [52]. Additionally, antidepressants such as TCAs and SSRIs are known to increase the concentration of neurotransmitters at synapses within minutes, yet it takes weeks to months for mental illness symptoms to regress in patients [53]. Additional theories of depression include the neurogenic hypothesis of depression and inflammatory hypothesis based on an increase in proinflammatory cytokines in the nervous system [7]. The 5-HT receptors are known to regulate the release of cytokines interleukin (IL) and tumour necrosis factor (TNF) and modulate immune macrophage and dendritic cell function and aid in maintaining an anti-inflammatory state in vivo [54]. With increasing rates of mental illness and suicide ideation and tendencies, it is imperative that a more rapid acting treatment option is established.

Mental Health and Co-Morbidities of Chronic Pain

Functional somatic syndrome is the term used for conditions that do not appear to have an underlying biological pathology [55]. These debilitating conditions are becoming increasingly frequent imposing a significant economic burden. Chronic fatigue syndrome (CFS), irritable bowel syndrome (IBS) and fibromyalgia (FM) amongst others are FSSs where pain, fatigue, muscle, and joint aches often present without an identifiable cause. Theories have arisen (the 'Lumpers' vs. 'Splitters' debate) suggesting that certain FSSs have one underlying aetiology and are associated with psychiatric disorder's including anxiety and depression. While the pathophysiology of FM remains unknown it has been postulated that thinly myelinated and unmyelinated C-nerve fibres have been reported in FMS patients [56] suggesting a link between loss of nerve myelin and chronic pain. Additionally, studies suggest that a dysfunction of dopamine is present in FM patients where dopamine agonists appear effective in treating symptoms [57]. Studies have demonstrated decreased cortical dopamine receptor binding in FM patients [57]. FM characterised by extreme hyperalgesia is highly prevalent (3 times more prevalent) in persons with major depressive disorder than in patients without, studies also show high rates of MDD amongst family members of FM patients suggesting a genetic aspect [58]. Additionally, sleep abnormalities in FM are similar to those present with a 5-HT dysfunction in depressed individuals [58]. CFS appears less linked to mental health disorders where psychiatric status is not considered an important causal contributor to CFS [59]. IBS is a painful syndrome characterized by chronic abdominal pain, altered bowel habits, correlated to significant psychological distress and psychiatric comorbidities anxiety, panic disorder, MDD and suicidal ideation [60]. Polymorphisms in the serotonin transporter gene (SERT) has been associated with higher risk of depression in IBS patients with a dysfunction of serotonin transporters impacting the emotion regulating regions of the brain [61,62]. At present, treatment of FM and IBS is also reliant on the use of antidepressant drugs TCAs and SNRIs where successful therapy has supported the belief that a dysfunction of serotonin and NE is present in FSSs similarly to MDD [58].

The relationship between mental health disorders and autoimmune conditions (including multiple sclerosis, rheumatoid arthritis, colitis, and lupus erythematosus) has been well established where there is increased risk of MDD, anxiety and schizophrenia in autoimmune patients. Theories suggest the excessive inflammation characteristic of autoimmune diseases may be a casual factor impacting the CNS or via cytokine interactions between nerve and immune cells of the patient [54]. The relationship between increased levels of cytokines including tumour necrosis factor-α (TNF-α) and IL-6 and major depression disorder is now recognised [47]. This raises the question if fungal extracts showing promise in treating mental health disorders may also alleviate the symptoms of certain FSSs namely FM and IBS and auto immune conditions.

3. Fungal Biologics: Unlocking the Potential of Eastern Practice into Western Medicine

There has been an increasing interest in translating potential medicinal cures derived from medicinal mushroom used for centuries in Eastern practice into Western medicine [63,64]. Modern medical practice relies heavily on the use of highly purified pharmaceutical compounds whose purity can be easily assessed and whose pharmaceutical activity and toxicity show clear structure-function relationships. In contrast, for decades, many mushroom-derived medicines contain mixtures of natural compounds had not undergone detailed chemical analyses and whose mechanism of action had not been elucidated [65]. With the decline in the number of new therapeutics produced from the pharmaceutical industry, novel biologic agents are being sought from traditional medicine. Translating traditional Eastern practices into acceptable evidence-based Western therapies is challenging given different manufacturing standards, criteria of purity, and under-powered clinical trials making assessment of efficacy and toxicity by Western standards of clinical evidence difficult. Purified bioactive compounds derived from medicinal mushrooms using appropriate drug discovery programs are a potentially important new source of psychoactive agents. For example, the therapeutic potential of medicinal mushrooms is evidenced by the fact that two glucan isolates were licensed as drugs in Japan as immune-adjuvant therapy for cancer in 1980. Moreover, Murphy et al. (2021) noted that approximately 200 registered clinical trials have appeared in the literature focusing on use of beta-glucans extracted from a diversity of medicinal mushrooms with a focus on cancer treatments using mainly oral administration [65]. These authors observed significant challenges exist to further clinical testing and translation of mushroom-derived biologics. The diverse range of conditions for which biologics from medicinal fungi are in clinical testing underlines the incomplete understanding of the diverse mechanisms of action that is a key knowledge gap.

By far the greatest mechanistic information relates to elucidating immunomodulatory potential of medicinal mushrooms [66,67]. Furthermore, important differences appear to exist in the effects of apparently similar mushroom-derived preparations, which may be due to differences in sources and extraction procedures, another poorly understood issue [65]. In general, there is a dearth in evidence-based knowledge from registered clinical trials on potential use of biologics derived from medicinal mushrooms for alleviating mental health conditions. Therefore, there is growing interest in adopting a Quadruple Helix approach (academia–industry–policy–society) to support emerging industries, including medicinal fungi companies, in the 'bioprospecting' and development of new therapeutic products informed by compliance with appropriate regulatory framework and clinical trials; moreover, this multi-actor framework also enables greater societal awareness of the potential for higher fungi to meet a diversity of emerging needs [68–70].

Outside of the use of extracts and along with the beta-glucans, a plethora of fungal bioactives have been isolated and fully purified including further polysaccharides, peptides and polyphenols, medium-sized macromolecules such as triterpenoids and the small molecule type structures of the indole alkaloids including tryptamines. Yildiz et al. (2015), for example, reports on the potent antioxidant action of phenol compounds in many medicinal mushrooms [71]. The focus here will be on the bioactives with low molecular weights which can cross the BBB and have structural features and effects similar to the common neurotransmitters. The neurotransmitters, serotonin, dopamine and NE are monoamines derived from either tyrosine or tryptophan. Some of the most interesting and potent compounds discussed herein are derived from the same precursors. Indeed, serotonin itself and psilocybin are in fact both substituted tryptamines (Figure 1).

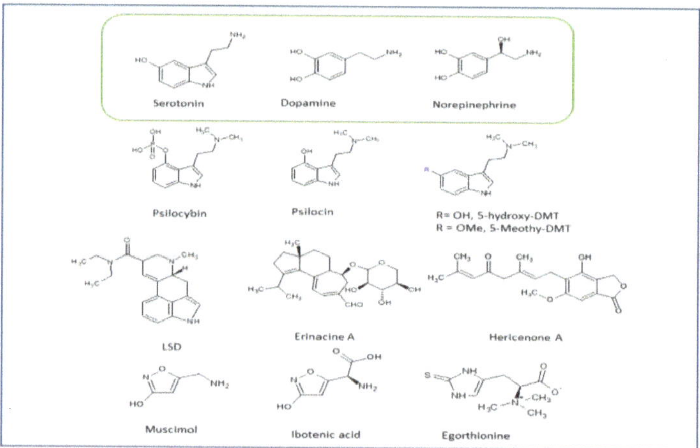

Figure 1. Monoamine neurotransmitters and fungal bioactives.

Psychedelics or serotonergic hallucinogens are natural substances which have agonism for the 5-HT receptors and include psilocybin, dimethyltryptamine (DMT) from the chacruna (ayahuasca brew) and jurema plants and mescaline from the peyote and San Pedro cacti [72]. Psychedelic compounds produce hallucinogenic experiences via activity in brain regions regulating cognition, emotions, self-awareness and perception of pain [73]. Psychedelics are included in the psychoplastogens class of compounds which have effect on neural plasticity in key circuits involved in brain health [53]. By definition psychoplastogens produce a quantifiable change in plasticity within a short time frame (>72 h) post administration [74]. Agonism of the 5-HT receptor group is associated with neuroplastic changes potentially reducing depression and anxiety symptoms [75]. Studies show that psychoplastogens have a range of therapeutic potential in treating neuropsychiatric diseases including anxiety disorder, mood disorders, PTSD, and addiction [74].

The 5-HT receptor group also has an important role in the gut-brain axis and is linked to the pathophysiology of IBS via dysregulated serotonin signalling [76]. Mushrooms containing psychoplastogens have been in use as traditional medicine in numerous cultures as healing rituals for millennia. Fungal biologics of therapeutic potential include lysergic acid diethylamide (LSD) a synthetic derivative of the precursor LSA (d-lysergic acid) or ergine from the rye ergot fungus *Claviceps purpurea*, muscimol and a DMT derivative from *Amanita muscaria* mushroom (amongst others) and psilocybin, a hallucinogenic pro-drug of psilocin found in Psilocybe mushrooms (magic mushrooms) (Table 2). Psilocybe alkaloids are more recently obtained from magic truffles (fungis sclerotia) as this species is currently overlooked in the prohibition laws banning sale [48]. Non hallucinogenic fungal biologics showing potential for mental health treatment include compounds from the edible mushroom *Hericium erinaceus* and ergothioneine from *Pleurotus cornucopiae*.

Table 2. Fungal biologics of therapeutic potential.

	Mescaline	Psilocybin/Psilocin	LSD
Pharmacodynamics	Naturally occurring substituted phenethylamine extracted from the peyote cactus 5-HT releasing agent, catecholamine-like structure [77] Primarily interacts at 5-HT$_{2A/2C}$ and α_2-adrenergic receptors [78] Low binding affinity at dopaminergic and histaminergic receptors [78]	Naturally occurring indole-alkylamine (tryptamine) extracted from Psilocybe mushrooms Close structural analogue of 5-HT Primarily interacts at 5-HT$_{2A/2C}$ and 5-HT$_{1A}$, 5-HT$_{2C}$ [79] Indirectly increases DA concentration but has no affinity for D2 receptors [81]	Semisynthetic indole-alkylamine (ergoline) derived from lysergic acid found in *Claviceps purpurea* Close structural analogue of 5-HT Mixed 5-HT$_2$/5-HT$_1$ receptor partial agonist [80] High affinity dopaminergic, adrenergic, and histaminergic receptors [82,83]
Pharmacokinetics	Can be ingested orally, smoked, or insufflated Relatively low-potency: active doses in the 200–400 mg range) [84] Rapidly absorbed in GI and distributed to the kidneys and liver Low lipid solubility, weak ability to penetrate BBB [77] Detoxification via oxidative deamination Long-lasting, half-life of 6 hrs Eliminated in urine mainly in the unchanged form (81.4%) and the remaining as the metabolite TMPA [88]	Can be ingested orally or intravenously 20× more potent than mescaline: active doses in 10–30 mg range [81] Rapidly absorbed and dephosphorylated to psilocin (bioactive form) [81] Lipid soluble, can easily cross BBB [86] Detoxification via demethylation and oxidative deamination Half-life of 3 hrs Eliminated in urine mainly as glucuronidated metabolites (80%) as well as unaltered psilocybin (3–10%) [89]	Can be ingested orally, smoked, injected, or snorted 2000× more potent mescaline: active doses in 25–200 µg range [85] Rapidly absorbed in GI and distributed to tissues and organs Can easily cross BBB [87] Detoxification via N-dealkylation and/or oxidation processes Half-life of 3.6 hrs Eliminated in urine mainly as metabolites, only 1% of the dose is excreted unchanged) [87]
Efficacy	Acute experiences of psychological insight during mescaline use are associated with self-reported improvements in anxiety disorders, depression, and substance abuse [78,84]	Therapeutic efficacy in treating mood and anxiety disorders, depression, cluster headaches, chronic pain, intractable phantom pain, obsessive-compulsive disorder, and substance abuse [79,81,90]	Therapeutic efficacy in treating anxiety disorders, depression, cluster headaches, obsessive-compulsive disorder, substance abuse, psychosomatic illnesses, and anxiety in relation to life-threatening diseases [91,92]

3.1. Psilocybe Mushrooms

Undoubtedly psychedelics from the Psilocybe species are currently the more commonly investigated fungal compounds for their biological and therapeutic potential. Psychedelic compounds are known to exert their effects via activation of serotonergic 5-HT receptors dispersed throughout the central and peripheral nervous system impacting on cognition, mood and emotion [93]. Psilocybin is a potent agonist of 5-HT$_{2A}$ receptors and moderate agonist of 5-HT$_{1A}$ and 5-HT$_{2C3}$ receptors present in the thalamus and cortex of the brain [79]. These receptors are also associated with peripheral and central nervous system pain perception [73] justifying the use of SSRIs in the treatment of chronic pain conditions such as FM. Psilocybin has also demonstrated affinity for dopamine receptors and serotonin transporter protein [94]. Both psilocybin and its metabolite psilocin, pass through the blood–brain barrier (BBB) where psilocin is 1.5 times more potent. Indeed, between the 1950s and 1970s, researchers actively investigated the potential of psilocybin in the treatment of OCD, addiction, and neurotic behaviours [72] but studies were inhibited by prohibition until the early 1990s. Psychedelics are now believed to have additional beneficial biological activities including promoting neural plasticity and immune modulation [95]. Additionally, psilocybin positively impacted symptoms of MDD, treatment resistant depression, substance addiction and anxiety (anxiolytic) in afflicted persons and terminal cancer patients [96]. Comparative studies where one dose of ca. 30 mg/70 kg psilocybin was administered to terminally ill patients reduced symptoms of depression and anxiety and improved optimism in 51 cancer patients for up to 6 months post administration [97]. In one trial study, psilocybin reduced symptoms of MDD, anxiety and anhedonia as early as 1 week post administration of 10 mg and 25 mg (seven days apart) with no observed side effects to the patient [75]. Studies assessing the impact of psilocybin on alcohol addiction showed patients had significantly fewer drinking days in the treatment period of 8 weeks at 300 µg/kg 4 weeks apart [79]. Promising results were also observed with the use of psilocybin in conjunction with cognitive behavioural therapy (CBT) to aid in smoking cessation [98]. A long term follow up study reported 60% abstinence rates after more than 12 months compared with a 31% anstinence rate at 12 months using conventional therapies. Small scale clinical studies also demonstrate the efficacy of psilocybin at reducing the symptoms of OCD where SSRI treatment failed [99]. Similar to the action of SSRIs on 5-HT receptors alleviating chronic pain in patients with neuropathic, musculoskeletal pain and fibromyalgia, psilocybin may reduce chronic pain symptoms [73]. Clinical trials are warranted on the administration of psilocybin to chronic pain patients including FSSs comparative to current treatment options such as SSRIs and opioids.

3.2. Claviceps purpurea

The fungus *Claviceps purpurea* is extensively known for its production of ergot alkaloids having activity on the nervous system and smooth muscles where therapeutic application includes treating Parkinson's disease, cluster headaches and migraine [100]. LSD is a semi-synthetic derivative of lysergic acid found in fungus *Claviceps purpurea* [73]. LSD was also a psychedelic compound highly researched in the 1950s until the Controlled Substances Act was introduced in 1970 [72]. Like psilocybin, LSD has activity on brain processes involved in cognition, emotion, and perception via affinity for serotonin receptors [101]. LSD is an agonist of 5-HT with high affinity for 5-HT2A and 5-HT2C and is small enough to pass the BBB being approximately 100 times more potent than psilocybin [73]. Interestingly, LSD is also an agonist of the dopaminergic receptors with relatively high affinity for dopamine 2 receptors [102] and an agonist of the trace amine associate receptor 1 (TAAR1) [103]. Currently, in the treatment of psychotic illness the therapeutic efficacy of active pharmaceutical ingredients (APIs) is required to be greater than 70% occupancy of D2 receptors [103]. TAAR1 is essential in the regulation of monoamines (dopamine, NE, serotonin) in neurons of the CNS and neuro immune modulation [104] via agonism of trace amines (β-phenylethylamine, p-tyramine, and tryptamine). Dysregulation of trace amines and TAAR1 receptor dysfunction has been identified in psychotic disorders including MDD, bipolar disorder and schizophrenia [103]. Additionally, the TAAR1 receptors are associated with the pathophysiology of IBS [76]. Studies have demonstrated the positive effects of LSD on anxiety relating to terminal illness [93] with positive affects also observed for PTSD and addiction. The analgesic activity of LSD has also been described in terminally ill patients lasting up to 12 h post administration [73]. While reports on pain management of headaches are self-reported, patients state that sub hallucinogenic concentrations (micro dosing) of LSD prophylactically and metaphylactically alleviated migraine and cluster headaches [105]. One clinical study determined that 20 µg LSD increased pain tolerance and reduced pain perception in patients compared to placebo with similar outcomes to oxycodone and morphine [106]. Such analgesic effects may be attributed to the dopaminergic activity of LSD suggesting positive effects on neurological disease such as FM. LSD's mechanism of action is pleiotropic affecting serotonin (5-HT), dopamine and TAAR1 receptor pathways potentially alleviating symptoms of mental illness and chronic pain conditions FM and IBS. It is important to note that LSD is a complex molecule, effecting many receptor pathways, where long-term administration may result in psychotic-like symptoms or Hallucinogen Persisting Perception Disorder (HPPD) or flashbacks in the user [103]. Such effects, however, are deemed low risk, uncommon and mostly associated with recreational use [107]. Additionally, secalonic acid A and its isomers are also found in Claviceps purpurea which has anti-cancer activity via topoisomerase I and II inhibition [108].

3.3. Amanita muscaria

A. muscaria contains many biologically active compounds including the psychoactive alkaloids: muscarine, ibotenic acid and muscimol [109] where ibotenic and muscimol are structurally similar to gamma-aminobutyric acid (GABA) having effect on glutamate receptors in the CNS [110] and can cross the BBB. Ibotenic acid is a potent neurotoxin via activation of N-methyl-d-aspartate (NMDA) receptors where muscimol has a strong psychoactivating action [111]. The alkaloid muscarine is an acetylcholine agonist of the parasympathetic nervous system typically negatively impacting the functioning of numerous organs but cannot pass the BBB [109]. Over stimulation of muscarinic cholinergic system and acetylcholine is associated with the aetiology of depression suggesting that the agonist muscarine will elevate depressive symptoms in patients. Studies by Corbett 1991, demonstrate that muscimol improved social behaviour and had anxiolytic-like effects post systemic administration to test rats, similar to diazepam [112]. Additionally, muscimol when used as a combination therapy with endomorphin-1 reduced the symptoms of neuropathic pain due to spinal cord injury when administered for 7 consecutive days [113]. Bufotenine (5-HO-DMT) is an alkaloid and a substituted tryptamine similar to serotonin and psilocin, also present in the Amanita mushroom. As such, it has the ability to cross the BBB having agonistic effects on 5-HT1A and 5-HT1B receptors [114]. Another substituted tryptamine,5-MeO-DMT, requires enzymatic activity of a monoamine oxidase (MAO) inhibitor to induce its psychedelic effects orally [48]. Recently, β-carbolines, which are well-known MAO inhibitors have been isolated directly from psilocybe mushrooms. β-carbolines, as well as psilocybin itself, are derivatives of tryptophan. This represents an interesting natural product pathway where two different types of products from the same precursor contribute to the same pharmacological effect [115]. Tryptamines have been used to treat symptoms of PTSD, MDD, addiction and anxiety [116] and show potential as treatment with low risks for addiction.

3.4. Hericium erinaceus (H. erinaceus)

The medical benefits of *H. erinaceus* have been well documented with anticancer, antioxidative, anti-inflammatory and antimicrobial activity amongst others [7] where neurotropic compounds (Figure 1) present in the mushroom are more recently being considered for neuropsychiatric disorders. Hericenones and erinacines (cyathane derivatives) are biological neurotropic compounds found in the fruiting body and mycelium of *H. erinaceus*. These compounds can pass the BBB due to their low molecular weight [117] which may beneficially impact on Alzheimer's and Parkinson's disease due to their impact on nerve growth factor (NGF) [118]. Indeed, many biological active compounds are present in the fruiting bodies of *H. erinaceus* including aromatic compounds, diterpenoids, steroids, and polysaccharides [119]. NGF is essential for protecting nerve tissue and maintaining neuron functionality where studies have shown NGF levels in MDD patients is also significantly reduced [120]. Interestingly, NGF itself is unable to pass the BBB highlighting the benefits of these compounds which stimulate NGF production within the brain and encourage nerve myelination throughout the CNS [121]. Perhaps such myelination effects may also aid in alleviating the symptoms of FM in patients where improper myelination of nerves is a possible issue [56]. Initial studies suggest that the NGF enhancing activity of these extracts may reduce MDD and anxiety [122]. Elevated levels of NGF are associated with neuroplasticity and neurogenesis which may improve mood in MDD patients based on the neurogenic hypothesis of depression [7]. Additionally, in vivo studies on depressed animals demonstrated that when fed with *H. erinaceus* levels of the monoamine's serotonin, dopamine and NE were elevated [123]. In vivo studies report that erinacine can increase monamine expression and modulate anti-inflammatory brain-derived neurotrophic factor (BDNF) signalling in depressed animals [124]. Amycenone, a proprietary mixture of compounds which are present in the fruiting body of *H. erinaceus*, demonstrated anti-inflammatory activity against cytokines TNF and IL and reduced inflammatory induced depression in test animals [119] a similar activity observed with SSRIs and SNRIs. Such anti-inflammatory activity may also prove beneficial in alleviating symptoms of autoimmune co-morbidities of mental illness. Studies show when consuming H. erinaceus adults with mild cognitive impairment and menopausal women were less depressed, anxious with improved cognitive abilities compared to placebo control groups [121]. Studies are warranted however, confirming the efficacy of biological extracts from *H. erinaceus* mushrooms comparative to consuming the entire fruiting body of the mushroom at a therapeutic level. Additionally, while these compounds have been shown to affect the levels of monoamines, the mechanism of modulation requires further investigation.

3.5. Pleurotus cornucopiae

One of the fungal bioactives with the most diverse effects and potential uses currently under widespread investigation is Ergothionine. Found in large amounts in oyster mushrooms (*Pleurotus cornucopiae* var. *citrinopileatus*), Ergothione is a metabolite of the aromatic amino acid histidine. It has been shown to have neuroprotective and anti-inflammatory effects as well as being potentially involved in neurogenesis, all of which are characteristics of compounds which may be useful in the treatment of the aforementioned disorders [125]. Dietary ergothionine was shown to significantly reduce immobility time in forced swim test (FST) and tail spin test (TST) which are used as models of depression in mice [126]. Additionally, oral administration of ergothioneine prior to and during another mouse model of depression (Social Defeat Stress (SDS) paradigm) had a preventative effect on depressive behaviours such as social avoidance and sleep abnormalities [127]. Ergothioneine is a substrate of carnitine/organic cation transporter OCTN1 and this has led to investigation of the potential targeting of such atypical monoamine transporters as a strategy for the development of new anti-depressants [128].

4. Pharmacological Consideration of Mycotherapy

Bioactive compounds which promote neuro plasticity and induce long term changes in mood, emotion and cognition may offer therapeutic options to chronically mentally ill patients [95]. When considering the administration of psychedelic compounds as therapeutics the efficacy of single dosing versus micro-dosing must be considered as current studies vary between self-medicating/micro dosing patients and single dose clinical trials. The time frame required for onset of action is an important consideration as current drug therapy (SSRI, NSRIs) for mental illness requires weeks to months for therapeutic effect. Studies indicate the effects of the fungal bioactives described occur rapidly and have a prolonged lasting effect on the symptoms of the diseases investigated [93]. Additionally, there appears to be no, or limited side effects or withdrawal symptoms associated with acute therapy or micro-dosing. Tolerance can develop however, relating to down regulation or de-sensitization of the 5-HT receptors and a cross tolerance with other psychedelic compounds may occur [47]. Psilocybin has low toxicity with chronic exposure and moderate toxicity in cases of acute exposure and has low addiction and dependence issues [75]. LSD which is more potent than psilocybin induced some dose dependent physical and psychological symptoms including derealization, depersonalization and dissociation [106]. Psilocybin has pharmacodynamic effect lasting between 4–6 h with LSD lasting up to 12 h, an elimination half-life of 3.5 h and 3 h for LSD [129] and psilocybin [130] respectively, has been established. Route of administration is an important formulation consideration for psychotherapy as tryptamine and its derivates are prone to first pass metabolism. The addition of a monoamine oxidase inhibitor may be needed for oral delivery formulations [48]. Such an inhibitor would increase bioavailability and pharmacological efficacy of bioactive amines including psilocybin and LSD. The administration of such compounds is not straight forward as each has specific receptor affinities, duration of action and potency. The therapeutic index for each compound must be established to fully determine the safety profile as long-term treatment may be required for chronic mental health illness or chronic pain conditions. Currently, mood and behavioural studies are conducted in animals where analysis of symptoms and extrapolation to humans is difficult. Consuming such compounds may affect the patient's consciousness and provide insights and existential and spiritual questions [99]. For example, at 12 month follow up in the smoking cessation work by Johnson et al., 86.7% of the participants rated their psilocybin experience among the 5 most personally meaningful and spiritually significant of their lives. Administration of these compounds, therefore, must be in a controlled manner to avoid the risk to persons prone to psychosis and to determine the patient's mindset prior to administration. To this end, many of the studies involving the psychedelics and compounds which alter consciousness have been carried out in the presence of psychiatrists and counselling professionals specifically trained in psychedelic psychotherapy. Some alternative methods of determining activity have been utilized based on gene ontology, predictive analysis and Kyoto Encyclopedia of Genes and Genomes (KEGG) pathway analysis to investigate fungal biologics [124].

There is a pressing need to reach international consensus and agreement on the defined production methods for bioactives produced from medicinal fungi for this potential purpose as batch-to-batch variation in production methods and types of mushrooms may generated different structural-bioactives that may lead to vary functional activities [67,131,132]. This situation will be addressed by promoting international collaborations between academia and industry partners in the areas of biotechnology, mycology, toxicology, bioinformatics and biopharma.

5. Conclusions

Bioactives found in higher fungi such as mushrooms have chemical structures similar to neurotransmitters and can act as agonists of receptor pathways involved in psychiatric conditions. Harnessing this activity as therapy for chronic mental health and pain diseases may offer benefits where therapeutic needs are currently unmet. Considering the high morbidity and mortality rates associated with these disorders and the economic burden placed on health systems, it is imperative competent treatment options are investigated. Bioactives such as psychedelics effect a person's cognition, emotion and mood often having a long-lasting effect on symptoms of mental illness. Interest has arisen in tryptamine derivatives LSD and psilocybin and actives including hericenones and erinacines for the treatment of mood disorders, anxiety disorders, PTSD and addiction. Additionally, compounds acting as agonists of serotonin receptors may aid in alleviating symptoms of inflammatory conditions by regulating pro-inflammatory cytokines TNF and IL. While the potential benefits of such mycotherapy emerge, studies are warranted to determine their full pharmacokinetic and pharmacodynamic profiles before implementation as alternative drug therapy can be considered. While addiction and dependence do not appear to be an issue with such compounds, tolerance and cumulative effects must be considered. The complete aetiology of mental health disorders remains undetermined with the relationship between the mind, the brain and CNS a multifaceted interaction. As knowledge emerges giving a better understanding of mental illness, advances in neuropharmacology will undoubtedly follow. Perhaps such advances will include mycotherapy. Currently, the biggest challenge associated with psychedelic pharmacology studies is prohibition where psychedelics are classed as a Schedule 1 drug and the social stigma associated with their use. Chong et al., 2021 implied a predictive analysis method to investigate the molecular mechanisms of *H. erinaceus* [124], perhaps this offers a novel means of determining efficacy prior to clinical trials in line with prohibition. Other methods including ontology and Kyoto Encyclopedia of Genes and Genomes pathway analysis to investigate fungal biologics may also be implemented. Using animal species to model depression system in humans is currently standard protocol. Such assays allow researchers to investigate neural circuitry and molecular and cellular pathways in acute and chronic states. While such models off advantages, perhaps going forward inclusive studies of depression may employ several strains and multiple testing models. Where studies must compare genetic and epigenetic expressions, and activity to better extrapolate to human patients. Until a more comprehensive understanding of the aetiology of depression is established critical analysis of current drug therapy and fungal bioactives is warranted on existing animal and clinical studies to accurately determine the exact mode of action of fungal biologicals. Adopting 'Quadruple Helix' framework that unites academia, industry, policy, and society will further enable and support bioprospecting, development, testing and validation of bioactives of interest extracted from higher fungi (mushrooms) such as to advance the field of chronic pain and mental health disorders and conditions. Currently, international drug policies are slowly evolving as the potential benefits of these compounds are being revealed; however, care must be taken at clinical therapy stage.

Author Contributions: Conceptualization of this article was by M.G. and S.H. Design, research, and writing of this article was conducted by E.M., S.H., N.R. and M.G. All authors have read and agreed to the published version of the manuscript.

Funding: This research received no external funding.

Institutional Review Board Statement: Not applicable.

Informed Consent Statement: Not applicable.

Acknowledgments: M.G., S.H. and E.M. would like to acknowledge the PEM centre and IT Sligo.

Conflicts of Interest: The author declares no conflict of interest.

References

1. American Psychiatric Association. *DSM 5 Diagnostic and Statistical Manual of Mental Disorders*; American Psychiatric Association: Washington, DC, USA, 2013; 947p.
2. Ebneter, D.S.; Latner, J.D. Stigmatizing attitudes differ across mental health disorders: A comparison of stigma across eating disorders, obesity, and major depressive disorder. *J. Nerv. Ment. Dis.* **2013**, *201*, 281–285. [CrossRef]
3. Adam, D. Mental health: On the spectrum. *Nature* **2013**, *496*, 416–418. [CrossRef]
4. Kessler, R.C.; Angermeyer, M.; Anthony, J.C.; De Graaf, R.; Demyttenaere, K.; Gasquet, I.; De Girolamo, G.; Gluzman, S.; Gureje, O.; Haro, J.M.; et al. Lifetime prevalence and age-of-onset distributions of mental disorders in the World Health Organization's World Mental Health Survey Initiative. *World Psychiatry* **2007**, *6*, 168–176. [PubMed]
5. World Health Organisation. Mental Health. Available online: https://www.who.int/health-topics/mental-health#tab=tab_12021 (accessed on 23 February 2022).
6. Trautmann, S.; Rehm, J.; Wittchen, H.-U. The economic costs of mental disorders: Do our societies react appropriately to the burden of mental disorders? *EMBO Rep.* **2016**, *17*, 1245–1249. [CrossRef] [PubMed]
7. Chong, P.S.; Fung, M.-L.; Wong, K.H.; Lim, L.W. Therapeutic Potential of Hericium erinaceus for Depressive Disorder. *Int. J. Mol. Sci.* **2019**, *21*, 163. [CrossRef] [PubMed]
8. Keller, M.B. Issues in treatment-resistant depression. *J. Clin. Psychiatry* **2005**, *66* (Suppl. 8), 5–12.
9. Al-Harbi, K.S. Treatment-resistant depression: Therapeutic trends, challenges, and future directions. *Patient Prefer. Adherence* **2012**, *6*, 369–388. [CrossRef]
10. Johansen, P.; Krebs, T.S. Psychedelics not linked to mental health problems or suicidal behavior: A population study. *J. Psychopharmacol.* **2015**, *29*, 270–279. [CrossRef]
11. Davis, A.K.; Agin-Liebes, G.; España, M.; Pilecki, B.; Luoma, J. Attitudes and Beliefs about the Therapeutic Use of Psychedelic Drugs among Psychologists in the United States. *J. Psychoact. Drugs* **2021**, *53*, 1–10. [CrossRef]
12. Strawn, J.R.; Geracioti, L.; Rajdev, N.; Clemenza, K.; Levine, A. Pharmacotherapy for generalized anxiety disorder in adult and pediatric patients: An evidence-based treatment review. *Expert Opin. Pharmacother.* **2018**, *19*, 1057–1070. [CrossRef]

13. Garakani, A.; Murrough, J.W.; Freire, R.C.; Thom, R.P.; Larkin, K.; Buono, F.D.; Iosifescu, D.V. Pharmacotherapy of Anxiety Disorders: Current and Emerging Treatment Options. Review. *Front. Psychiatry* **2020**, *11*, 1412. [CrossRef]
14. Moraczewski, J.; Aedma, K.K. *Tricyclic Antidepressants*; StatPearls Publishing: Treasure Island, FL, USA, 2022.
15. Schneider, J.; Patterson, M.; Jimenez, X.F. Beyond depression: Other uses for tricyclic antidepressants. *Clevel. Clin. J. Med.* **2019**, *86*, 807–814. [CrossRef]
16. Chamberlain, S.R.; Baldwin, D.S. Monoamine oxidase inhibitors (MAOIs) in psychiatric practice: How to use them safely and effectively. *CNS Drugs* **2021**, *35*, 703–716. [CrossRef]
17. Sub Laban, T.; Saadabadi, A. *Monoamine Oxidase Inhibitors (MAOI)*; StatPearls Publishing: Treasure Island, FL, USA, 2022.
18. Ansara, E.D. Management of treatment-resistant generalized anxiety disorder. *Ment. Health Clin.* **2020**, *10*, 326–334. [CrossRef]
19. Sartori, S.B.; Singewald, N. Novel pharmacological targets in drug development for the treatment of anxiety and anxiety-related disorders. *Pharmacol. Ther.* **2019**, *204*, 107402. [CrossRef]
20. Shmuts, R.; Kay, A.; Beck, M. Buspirone: A forgotten friend. *Curr. Psychiatry* **2020**, *19*, 20.
21. Wilson, T.K.; Tripp, J. *Buspirone*; StatPearls Publishing: Treasure Island, FL, USA, 2022.
22. Schloesser, R.J.; Martinowich, K.; Manji, H.K. Mood-stabilizing drugs: Mechanisms of action. *Trends Neurosci.* **2012**, *35*, 36–46. [CrossRef]
23. Iannaccone, T.; Sellitto, C.; Manzo, V.; Colucci, F.; Giudice, V.; Stefanelli, B.; Iuliano, A.; Corrivetti, G.; Filippelli, A. Pharmacogenetics of carbamazepine and valproate: Focus on polymorphisms of drug metabolizing enzymes and transporters. *Pharmaceuticals* **2021**, *14*, 204. [CrossRef]
24. Pérez de Mendiola, X.; Hidalgo-Mazzei, D.; Vieta, E.; González-Pinto, A. Overview of lithium's use: A nationwide survey. *Int. J. Bipolar Disord.* **2021**, *9*, 10. [CrossRef]
25. Murru, A.; Popovic, D.; Pacchiarotti, I.; Hidalgo, D.; León-Caballero, J.; Vieta, E. Management of Adverse Effects of Mood Stabilizers. *Curr. Psychiatry Rep.* **2015**, *17*, 66. [CrossRef]
26. Bourin, M. Mechanism of action of valproic acid and its derivatives. *SOJ Pharm. Sci.* **2020**, *7*, 1–4.
27. Grunze, A.; Amann, B.L.; Grunze, H. Efficacy of Carbamazepine and Its Derivatives in the Treatment of Bipolar Disorder. *Medicina* **2021**, *57*, 433. [CrossRef]
28. Chokhawala, K.; Stevens, L. *Antipsychotic Medications*; StatPearls Publishing: Treasure Island, FL, USA, 2022.
29. Stroup, T.S.; Gray, N. Management of common adverse effects of antipsychotic medications. *World Psychiatry Off. J. World Psychiatr. Assoc.* **2018**, *17*, 341–356. [CrossRef]
30. Lally, J.; MacCabe, J.H. Antipsychotic medication in schizophrenia: A review. *Br. Med. Bull.* **2015**, *114*, 169–179. [CrossRef]
31. Norris, M.L.; Spettigue, W.; Buchholz, A.; Henderson, K.A.; Gomez, R.; Maras, D.; Gaboury, I.; Ni, A. Olanzapine use for the adjunctive treatment of adolescents with anorexia nervosa. *J. Child Adolesc. Psychopharmacol.* **2011**, *21*, 213–220. [CrossRef]
32. Çöpür, S.; Çöpür, M. Olanzapine in the treatment of anorexia nervosa: A systematic review. *Egypt. J. Neurol. Psychiatry Neurosurg.* **2020**, *56*, 60. [CrossRef]
33. Milano, W.; De Rosa, M.; Milano, L.; Riccio, A.; Sanseverino, B.; Capasso, A. The Pharmacological Options in the Treatment of Eating Disorders. *ISRN Pharmacol.* **2013**, *2013*, 352865. [CrossRef]
34. Fariba, K.; Saadabadi, A. *Topiramate*; StatPearls Publishing: Treasure Island, FL, USA, 2020.
35. Grant, J.E.; Schreiber, L.; Odlaug, B.L. Impulse control disorders: Updated review of clinical characteristics and pharmacological management. *Front. Psychiatry* **2011**, *2*, 1.
36. Douaihy, A.B.; Kelly, T.M.; Sullivan, C. Medications for substance use disorders. *Soc. Work Public Health* **2013**, *28*, 264–278. [CrossRef]
37. Fontenelle, L.F.; Oostermeijer, S.; Harrison, B.J.; Pantelis, C.; Yücel, M. Obsessive-compulsive disorder, impulse control disorders and drug addiction. *Drugs* **2011**, *71*, 827–840. [CrossRef]
38. Kayser, R.R. Pharmacotherapy for treatment-resistant obsessive-compulsive disorder. *J. Clin. Psychiatry* **2020**, *81*, 14428. [CrossRef] [PubMed]
39. Mouaffak, F.; Leite, C.; Hamzaoui, S.; Benyamina, A.; Laqueille, X.; Kebir, O. Naltrexone in the treatment of broadly defined behavioral addictions: A review and meta-analysis of randomized controlled trials. *Eur. Addict. Res.* **2017**, *23*, 204–210. [CrossRef] [PubMed]
40. Bolton, M.; Hodkinson, A.; Boda, S.; Mould, A.; Panagioti, M.; Rhodes, S.; Riste, L.; van Marwijk, H. Serious adverse events reported in placebo randomised controlled trials of oral naltrexone: A systematic review and meta-analysis. *BMC Med.* **2019**, *17*, 10. [CrossRef] [PubMed]
41. Timäus, C.; Meiser, M.; Bandelow, B.; Engel, K.R.; Paschke, A.M.; Wiltfang, J.; Wedekind, D. Pharmacotherapy of borderline personality disorder: What has changed over two decades? A retrospective evaluation of clinical practice. *BMC Psychiatry* **2019**, *19*, 393. [CrossRef]
42. Bozzatello, P.; Ghirardini, C.; Uscinska, M.; Rocca, P.; Bellino, S. Pharmacotherapy of personality disorders: What we know and what we have to search for. *Future Neurol.* **2017**, *12*, 199–222. [CrossRef]
43. Uher, R.; Zwicker, A. Etiology in psychiatry: Embracing the reality of poly-gene-environmental causation of mental illness. *World Psychiatry Off. J. World Psychiatr. Assoc.* **2017**, *16*, 121–129. [CrossRef]
44. Wu, N.S.; Schairer, L.C.; Dellor, E.; Grella, C. Childhood trauma and health outcomes in adults with comorbid substance abuse and mental health disorders. *Addict. Behav.* **2010**, *35*, 68–71. [CrossRef]

45. Gold, P.W.; Wong, M.-L. Re-assessing the catecholamine hypothesis of depression: The case of melancholic depression. *Mol. Psychiatry* **2021**, *26*, 6121–6124. [CrossRef]
46. Albert, P.R.; Benkelfat, C.; Descarries, L. The neurobiology of depression—Revisiting the serotonin hypothesis. I. Cellular and molecular mechanisms. *Philos. Trans. R. Soc. Lond. Ser. B Biol. Sci.* **2012**, *367*, 2378–2381. [CrossRef]
47. Baumeister, D.; Barnes, G.; Giaroli, G.; Tracy, D. Classical hallucinogens as antidepressants? A review of pharmacodynamics and putative clinical roles. *Ther. Adv. Psychopharmacol.* **2014**, *4*, 156–169. [CrossRef]
48. Malaca, S.; Lo Faro, A.F.; Tamborra, A.; Pichini, S.; Busardò, F.P.; Huestis, M.A. Toxicology and Analysis of Psychoactive Tryptamines. *Int. J. Mol. Sci.* **2020**, *21*, 9279. [CrossRef]
49. Lew, S.Y.; Lim, S.H.; Lim, L.W.; Wong, K.H. Neuroprotective effects of *Hericium erinaceus* (Bull.: Fr.) Pers. against high-dose corticosterone-induced oxidative stress in PC-12 cells. *BMC Complement. Med. Ther.* **2020**, *20*, 340. [CrossRef]
50. Conrad, C.D. Chronic stress-induced hippocampal vulnerability: The glucocorticoid vulnerability hypothesis. *Rev. Neurosci.* **2008**, *19*, 395–412. [CrossRef]
51. Crocker, L.; Heller, W.; Warren, S.; O'Hare, A.; Infantolino, Z.; Miller, G. Relationships among cognition, emotion, and motivation: Implications for intervention and neuroplasticity in psychopathology. Review. *Front. Hum. Neurosci.* **2013**, *7*, 261. [CrossRef]
52. Volpi-Abadie, J.; Kaye, A.M.; Kaye, A.D. Serotonin syndrome. *Ochsner J.* **2013**, *13*, 533–540.
53. Vargas, M.V.; Meyer, R.; Avanes, A.A.; Rus, M.; Olson, D.E. Psychedelics and Other Psychoplastogens for Treating Mental Illness. Review. *Front. Psychiatry* **2021**, *12*, 1691. [CrossRef]
54. Thompson, C.; Szabo, A. Psychedelics as a novel approach to treating autoimmune conditions. *Immunol. Lett.* **2020**, *228*, 45–54. [CrossRef]
55. Janssens, K.A.; Zijlema, W.L.; Joustra, M.L.; Rosmalen, J.G. Mood and Anxiety Disorders in Chronic Fatigue Syndrome, Fibromyalgia, and Irritable Bowel Syndrome: Results from the LifeLines Cohort Study. *Psychosom. Med.* **2015**, *77*, 449–457. [CrossRef]
56. Üçeyler, N.; Sommer, C. Fibromyalgiesyndrom. *Z. Rheumatol.* **2015**, *74*, 490–495. [CrossRef]
57. Albrecht, D.S.; MacKie, P.J.; Kareken, D.A.; Hutchins, G.D.; Chumin, E.J.; Christian, B.T.; Yoder, K.K. Differential dopamine function in fibromyalgia. *Brain Imaging Behav.* **2016**, *10*, 829–839. [CrossRef]
58. Moret, C.; Briley, M. Antidepressants in the treatment of fibromyalgia. *Neuropsychiatr. Dis. Treat.* **2006**, *2*, 537–548. [CrossRef] [PubMed]
59. Natelson, B.H.; Lin, J.S.; Lange, G.; Khan, S.; Stegner, A.; Unger, E.R. The effect of comorbid medical and psychiatric diagnoses on chronic fatigue syndrome. *Ann. Med.* **2019**, *51*, 371–378. [CrossRef] [PubMed]
60. Fadgyas-Stanculete, M.; Buga, A.-M.; Popa-Wagner, A.; Dumitrascu, D.L. The relationship between irritable bowel syndrome and psychiatric disorders: From molecular changes to clinical manifestations. *J. Mol. Psychiatry* **2014**, *2*, 4. [CrossRef] [PubMed]
61. Jarrett, M.E.; Kohen, R.; Cain, K.C.; Burr, R.L.; Poppe, A.; Navaja, G.P.; Heitkemper, M.M. Relationship of SERT polymorphisms to depressive and anxiety symptoms in irritable bowel syndrome. *Biol. Res. Nurs.* **2007**, *9*, 161–169. [CrossRef]
62. Fukudo, S.; Kanazawa, M.; Mizuno, T.; Hamaguchi, T.; Kano, M.; Watanabe, S.; Sagami, Y.; Shoji, T.; Endo, Y.; Hongo, M.; et al. Impact of serotonin transporter gene polymorphism on brain activation by colorectal distention. *Neuroimage* **2009**, *47*, 946–951. [CrossRef]
63. Smith, J.E.; Rowan, N.J.; Sullivan, R. Medicinal mushrooms: A rapidly developing area of biotechnology for cancer therapy and other bioactivities. *Biotechnol. Lett.* **2002**, *24*, 1839–1845. [CrossRef]
64. Sullivan, R.; Smith, J.E.; Rowan, N.J. Medicinal mushrooms and cancer therapy: Translating a traditional practice into Western medicine. *Perspect. Biol. Med.* **2006**, *49*, 159–170. [CrossRef]
65. Murphy, E.J.; Rezoagli, E.; Major, I.; Rowan, N.J.; Laffey, J.G. β-glucan metabolic and immunomodulatory properties and potential for clinical application. *J. Fungi* **2020**, *6*, 356. [CrossRef]
66. Murphy, E.J.; Masterson, C.; Rezoagli, E.; O'Toole, D.; Major, I.; Stack, G.D.; Lynch, M.; Laffey, J.G.; Rowan, N.J. β-Glucan extracts from the same edible shiitake mushroom Lentinus edodes produce differential in-vitro immunomodulatory and pulmonary cytoprotective effects—Implications for coronavirus disease (COVID-19) immunotherapies. *Sci. Total Environ.* **2020**, *732*, 139330. [CrossRef]
67. Murphy, E.J.; Rezoagli, E.; Pogue, R.; Simonassi-Paiva, B.; Izwani, I.; Abidin, Z.; Waltzer Fehrenbach, G.; O'Neil, E.; Major, I.; Laffey, J.G.; et al. Immunomodulatory activity of β-glucan polysaccharides isolated from different species of mushroom—A potential treatment for inflammatory lung conditions. *Sci. Total Environ.* **2022**, *809*, 152177. [CrossRef]
68. Rowan, N.J.; Galanakis, C.M. Unlocking challenges and opportunities presented by COVID-19 pandemic for cross-cutting disruption in agri-food and green deal innovations: Quo Vadis? *Sci. Total Environ.* **2020**, *748*, 141362. [CrossRef] [PubMed]
69. Rowan, N.J.; Casey, O. Empower Eco multiactor HUB: A triple helix 'academia-industry-authority' approach to creating and sharing potentially disruptive tools for addressing novel and emerging new Green Deal opportunities under a United Nations Sustainable Development Goals framework. *Curr. Opin. Environ. Sci. Health* **2021**, *21*, 100254. [CrossRef]
70. Allam, Z.; Sharifi, A.; Giurco, D.; Sharpe, S.A. On the theoretical conceptualisations, knowledge structures and trends of green new deals. *Sustainability* **2021**, *13*, 12529. [CrossRef]
71. Yildiz, O.; Can, Z.; Laghari, A.Q.; Şahin, H.; Malkoç, M. Wild edible mushrooms as a natural source of phenolics and antioxidants. *J. Food Biochem.* **2015**, *39*, 148–154. [CrossRef]

72. Dos Santos, R.G.; Bouso, J.C.; Rocha, J.M.; Rossi, G.N.; Hallak, J.E. The use of classic hallucinogens/psychedelics in a therapeutic context: Healthcare policy opportunities and challenges. *Risk Manag. Healthc. Policy* **2021**, *14*, 901. [CrossRef]
73. Whelan, A.; Johnson, M.I. Lysergic acid diethylamide and psilocybin for the management of patients with persistent pain: A potential role? *Pain Manag.* **2018**, *8*, 217–229. [CrossRef]
74. Olson, D.E. Psychoplastogens: A Promising Class of Plasticity-Promoting Neurotherapeutics. *J. Exp. Neurosci.* **2018**, *12*, 1–4. [CrossRef]
75. Gill, H.; Gill, B.; Chen-Li, D.; El-Halabi, S.; Rodrigues, N.B.; Cha, D.S.; Lipsitz, O.; Lee, Y.; Rosenblat, J.D.; Majeed, A.; et al. The emerging role of psilocybin and MDMA in the treatment of mental illness. *Expert Rev. Neurother.* **2020**, *20*, 1263–1273. [CrossRef]
76. Pretorius, L.; Smith, C. The trace aminergic system: A gender-sensitive therapeutic target for IBS? *J. Biomed. Sci.* **2020**, *27*, 95. [CrossRef]
77. Mayet, S. A review of common psychedelic drugs. *S. Afr. J. Anaesth. Analg.* **2020**, *26*, S113–S117. [CrossRef]
78. Uthaug, M.V.; Davis, A.K.; Haas, T.F.; Dawis, D.; Dolan, S.B.; Lancelotta, R.; Timmermann, C.; Ramaekers, J.G. The epidemiology of mescaline use: Pattern of use, motivations for consumption, and perceived consequences, benefits, and acute and enduring subjective effects. *J. Psychopharmacol.* **2021**, *36*, 309–320. [CrossRef]
79. Daniel, J.; Haberman, M. Clinical potential of psilocybin as a treatment for mental health conditions. *Ment. Health Clin.* **2017**, *7*, 24–28. [CrossRef]
80. Passie, T.; Halpern, J.H.; Stichtenoth, D.O.; Emrich, H.M.; Hintzen, A. The pharmacology of lysergic acid diethylamide: A review. *CNS Neurosci. Ther.* **2008**, *14*, 295–314. [CrossRef]
81. Lowe, H.; Toyang, N.; Steele, B.; Valentine, H.; Grant, J.; Ali, A.; Ngwa, W.; Gordon, L. The Therapeutic Potential of Psilocybin. *Molecules* **2021**, *26*, 2948. [CrossRef]
82. Halberstadt, A.L.; Geyer, M.A. Multiple receptors contribute to the behavioral effects of indoleamine hallucinogens. *Neuropharmacology* **2011**, *61*, 364–381. [CrossRef]
83. Cumming, P.; Scheidegger, M.; Dornbierer, D.; Palner, M.; Quednow, B.B.; Martin-Soelch, C. Molecular and functional imaging studies of psychedelic drug action in animals and humans. *Molecules* **2021**, *26*, 2451. [CrossRef]
84. Agin-Liebes, G.; Haas, T.F.; Lancelotta, R.; Uthaug, M.V.; Ramaekers, J.G.; Davis, A.K. Naturalistic use of mescaline is associated with self-reported psychiatric improvements and enduring positive life changes. *ACS Pharmacol. Transl. Sci.* **2021**, *4*, 543–552. [CrossRef]
85. Holze, F.; Vizeli, P.; Ley, L.; Müller, F.; Dolder, P.; Stocker, M.; Duthaler, U.; Varghese, N.; Eckert, A.; Borgwardt, S.; et al. Acute dose-dependent effects of lysergic acid diethylamide in a double-blind placebo-controlled study in healthy subjects. *Neuropsychopharmacology* **2021**, *46*, 537–544. [CrossRef]
86. Dinis-Oliveira, R.J. Metabolism of psilocybin and psilocin: Clinical and forensic toxicological relevance. *Drug Metab. Rev.* **2017**, *49*, 84–91. [CrossRef]
87. Libânio Osório Marta, R.F. Metabolism of lysergic acid diethylamide (LSD): An update. *Drug Metab. Rev.* **2019**, *51*, 378–387. [CrossRef]
88. Dinis-Oliveira, R.J.; Pereira, C.L.; da Silva, D.D. Pharmacokinetic and Pharmacodynamic Aspects of Peyote and Mescaline: Clinical and Forensic Repercussions. *Curr. Mol. Pharmacol.* **2019**, *12*, 184–194. [CrossRef]
89. Passie, T.; Seifert, J.; Schneider, U.; Emrich, H.M. The pharmacology of psilocybin. *Addict. Biol.* **2002**, *7*, 357–364. [CrossRef]
90. Carhart-Harris, R.L.; Bolstridge, M.; Rucker, J.; Day, C.M.J.; Erritzoe, D.; Kaelen, M.; Bloomfield, M.; Rickard, J.A.; Forbs, B.; Feilding, A.; et al. Psilocybin with psychological support for treatment-resistant depression: An open-label feasibility study. *Lancet Psychiatry* **2016**, *3*, 619–627. [CrossRef]
91. Fuentes, J.J.; Fonseca, F.; Elices, M.; Farré, M.; Torrens, M. Therapeutic use of LSD in psychiatry: A systematic review of randomized-controlled clinical trials. *Front. Psychiatry* **2020**, *10*, 943. [CrossRef]
92. Nichols, D.E. Psychedelics. *Pharmacol. Rev.* **2016**, *68*, 264–355. [CrossRef]
93. Sherwood, A.M.; Prisinzano, T.E. Novel psychotherapeutics—A cautiously optimistic focus on hallucinogens. *Expert Rev. Clin. Pharmacol.* **2018**, *11*, 1–3. [CrossRef]
94. Carlini, E.A.; Maia, L.O. Plant and fungal hallucinogens as toxic and therapeutic agents. In *Plant Toxins Toxinology*; Gopalakrishnakone, P., Carlini, C., Ligabue-Braun, R., Eds.; Springer: Berlin, Germany, 2017; pp. 37–80.
95. Olson, D.E. The Promise of Psychedelic Science. *ACS Pharmacol. Transl. Sci.* **2021**, *4*, 413–415. [CrossRef]
96. Ross, S.; Bossis, A.; Guss, J.; Agin-Liebes, G.; Malone, T.; Cohen, B.; Mennenga, S.E.; Belser, A.; Kalliontzi, K.; Babb, J.; et al. Rapid and sustained symptom reduction following psilocybin treatment for anxiety and depression in patients with life-threatening cancer: A randomized controlled trial. *J. Psychopharmacol.* **2016**, *30*, 1165–1180. [CrossRef]
97. Griffiths, R.R.; Johnson, M.W.; Carducci, M.A.; Umbrich, A.; Richards, W.A.; Richards, B.D.; Cosimano, M.P.; Klinedinst, M.A. Psilocybin produces substantial and sustained decreases in depression and anxiety in patients with life-threatening cancer: A randomized double-blind trial. *J. Psychopharmacol.* **2016**, *30*, 1181–1197. [CrossRef]
98. Johnson, M.W.; Garcia-Romeu, A.; Griffiths, R.R. Long-term follow-up of psilocybin-facilitated smoking cessation. *Am. J. Drug Alcohol Abus.* **2017**, *43*, 55–60. [CrossRef]
99. Moreno, F.A.; Wiegand, C.B.; Taitano, E.K.; Delgado, P.L. Safety, tolerability, and efficacy of psilocybin in 9 patients with obsessive-compulsive disorder. *J. Clin. Psychiatry* **2006**, *67*, 1735–1740. [CrossRef] [PubMed]

100. Negård, M.; Uhlig, S.; Kauserud, H.; Andersen, T.; Høiland, K.; Vrålstad, T. Links between genetic groups, indole alkaloid profiles and ecology within the grass-parasitic *Claviceps purpurea* species complex. *Toxins* **2015**, *7*, 1431–1456. [CrossRef] [PubMed]
101. Rodríguez Arce, J.M.; Winkelman, M.J. Psychedelics, sociality, and human evolution. *Front. Psychol.* **2021**, *12*, 729425. [CrossRef] [PubMed]
102. Halberstadt, A.L. Hallucinogenic drugs: A new study answers old questions about LSD. *Curr. Biol.* **2017**, *27*, R156–R158. [CrossRef]
103. De Gregorio, D.; Comai, S.; Posa, L.; Gobbi, G. d-Lysergic acid diethylamide (LSD) as a model of psychosis: Mechanism of action and pharmacology. *Int. J. Mol. Sci.* **2016**, *17*, 1953. [CrossRef]
104. Rogers, T.J. The molecular basis for neuroimmune receptor signaling. *J. Neuroimmune Pharmacol.* **2012**, *7*, 722–724. [CrossRef]
105. Hutten, N.R.P.W.; Mason, N.L.; Dolder, P.C.; Kuypers, K.P.C. Self-rated effectiveness of microdosing with psychedelics for mental and physical health problems among microdosers. *Front. Psychiatry* **2019**, *10*, 672. [CrossRef]
106. Ramaekers, J.G.; Hutten, N.; Mason, N.L.; Dolder, P.; Theunissen, E.L.; Holze, F.; Liechti, M.E.; Feilding, A.; Kuypers, K.P.C. A low dose of lysergic acid diethylamide decreases pain perception in healthy volunteers. *J. Psychopharmacol.* **2021**, *35*, 398–405. [CrossRef]
107. Tupper, K.W.; Wood, E.; Yensen, R.; Johnson, M.W. Psychedelic medicine: A re-emerging therapeutic paradigm. *Can. Med. Assoc. J.* **2015**, *187*, 1054–1059. [CrossRef]
108. Lünne, F.; Köhler, J.; Stroh, C.; Müller, L.; Daniliuc, C.G.; Mück-Lichtenfeld, C.; Würthwein, E.-U.; Esselen, M.; Humpf, H.-U.; Kalinina, S.A. Insights into Ergochromes of the Plant Pathogen *Claviceps purpurea*. *J. Nat. Prod.* **2021**, *84*, 2630–2643. [CrossRef]
109. Carboué, Q.; Lopez, M. *Amanita muscaria*: Ecology, Chemistry, Myths. *Encyclopedia* **2021**, *1*, 905–914. [CrossRef]
110. Michelot, D.; Melendez-Howell, L.M. *Amanita muscaria*: Chemistry, biology, toxicology, and ethnomycology. *Mycol. Res.* **2003**, *107*, 131–146. [CrossRef]
111. Mikaszewska-Sokolewicz, M.A.; Pankowska, S.; Janiak, M.; Pruszczyk, P.; Łazowski, T.; Jankowski, K. Coma in the course of severe poisoning after consumption of red fly agaric (*Amanita muscaria*). *Acta Biochim. Pol.* **2016**, *63*, 181–182. [CrossRef]
112. Corbett, R.; Fielding, S.; Cornfeldt, M.; Dunn, R.W. GABAmimetic agents display anxiolytic-like effects in the social interaction and elevated plus maze procedures. *Psychopharmacology* **1991**, *104*, 312–316. [CrossRef]
113. Hosseini, M.; Karami, Z.; Yousefifard, M.; Janzadeh, A.; Zamani, E.; Nasirinezhad, F. Simultaneous intrathecal injection of muscimol and endomorphin-1 alleviates neuropathic pain in rat model of spinal cord injury. *Brain Behav.* **2020**, *10*, e01576. [CrossRef]
114. Mishraki-Berkowitz, T.; Kochelski, E.; Kavanagh, P.; O'Brien, J.; Dunne, C.; Talbot, B.; Ennis, P.; Wolf, U. The Psilocin (4-hydroxy-N, N-dimethyltryptamine) and Bufotenine (5-hydroxy-N, N-dimethyltryptamine) Case: Ensuring the Correct Isomer has Been Identified. *J. Forensic Sci.* **2020**, *65*, 1450–1457. [CrossRef]
115. Blei, F.; Dörner, S.; Fricke, J.; Baldeweg, F.; Trottmann, F.; Komor, A.; Meyer, F.; Hertweck, C.; Hoffmeister, D. Simultaneous production of psilocybin and a cocktail of β–carboline monoamine oxidase inhibitors in "magic" mushrooms. *Chem.–A Eur. J.* **2020**, *26*, 729–734. [CrossRef]
116. Davis, A.K.; Barsuglia, J.P.; Lancelotta, R.; Grant, R.M.; Renn, E. The epidemiology of 5-methoxy- N, N-dimethyltryptamine (5-MeO-DMT) use: Benefits, consequences, patterns of use, subjective effects, and reasons for consumption. *J. Psychopharmacol.* **2018**, *32*, 779–792. [CrossRef]
117. Ma, B.-J.; Shen, J.-W.; Yu, H.-Y.; Ruan, Y.; Wu, T.-T.; Zhao, X. Hericenones and erinacines: Stimulators of nerve growth factor (NGF) biosynthesis in *Hericium erinaceus*. *Mycology* **2010**, *1*, 92–98. [CrossRef]
118. Rai, S.N.; Mishra, D.; Singh, P.; Vamanu, E.; Singh, M.P. Therapeutic applications of mushrooms and their biomolecules along with a glimpse of in silico approach in neurodegenerative diseases. *Biomed. Pharmacother.* **2021**, *137*, 111377. [CrossRef]
119. Yao, W.; Zhang, J.-c.; Dong, C.; Zhuang, C.; Hirota, S.; Inanaga, K.; Hashimoto, K. Effects of amycenone on serum levels of tumor necrosis factor-α, interleukin-10, and depression-like behavior in mice after lipopolysaccharide administration. *Pharmacol. Biochem. Behav.* **2015**, *136*, 7–12. [CrossRef] [PubMed]
120. Li, I.; Lee, L.-Y.; Tzeng, T.-T.; Chen, W.-P.; Chen, Y.-P.; Shiao, Y.-J.; Chen, C.-C. Neurohealth properties of *Hericium erinaceus* mycelia enriched with erinacines. *Behav. Neurol.* **2018**, *2018*, 5802634. [CrossRef] [PubMed]
121. Kim, Y.O.; Lee, S.W.; Kim, J.S. A comprehensive review of the therapeutic effects of *Hericium erinaceus* in neurodegenerative disease. *J. Mushroom* **2014**, *12*, 77–81. [CrossRef]
122. Nagano, M.; Shimizu, K.; Kondo, R.; Hayashi, C.; Sato, D.; Kitagawa, K.; Ohnuki, K. Reduction of depression and anxiety by 4 weeks *Hericium erinaceus* intake. *Biomed. Res.* **2010**, *31*, 231–237. [CrossRef]
123. Chiu, C.-H.; Chyau, C.-C.; Chen, C.-C.; Lee, L.-Y.; Chen, W.-P.; Liu, Y.-L.; Lin, W.-H.; Mong, M.-C. Erinacine A-enriched *Hericium erinaceus* mycelium produces antidepressant-like effects through modulating BDNF/PI3K/Akt/GSK-3β signaling in mice. *Int. J. Mol. Sci.* **2018**, *19*, 341. [CrossRef]
124. Chong, P.S.; Poon, C.H.; Roy, J.; Tsui, K.C.; Lew, S.Y.; Lok Phang, M.W.; Yuenyinn Tan, R.J.; Cheng, P.G.; Fung, M.-L.; Wong, K.H.; et al. Neurogenesis-dependent antidepressant-like activity of *Hericium erinaceus* in an animal model of depression. *Chin. Med.* **2021**, *16*, 132. [CrossRef]
125. Cheah, I.K.; Halliwell, B. Ergothioneine, recent developments. *Redox Biol.* **2021**, *42*, 101868. [CrossRef]

126. Nakamichi, N.; Nakayama, K.; Ishimoto, T.; Masuo, Y.; Wakayama, T.; Sekiguchi, H.; Sutoh, K.; Usumi, K.; Iseki, S.; Kato, Y. Food-derived hydrophilic antioxidant ergothioneine is distributed to the brain and exerts antidepressant effect in mice. *Brain Behav.* **2016**, *6*, e00477. [CrossRef]
127. Matsuda, Y.; Ozawa, N.; Shinozaki, T.; Wakabayashi, K.-i.; Suzuki, K.; Kawano, Y.; Ohtsu, I.; Tatebayashi, Y. Ergothioneine, a metabolite of the gut bacterium *Lactobacillus reuteri*, protects against stress-induced sleep disturbances. *Transl. Psychiatry* **2020**, *10*, 170. [CrossRef]
128. Orrico-Sanchez, A.; Chausset-Boissarie, L.; Alves de Sousa, R.; Coutens, B.; Rezai Amin, S.; Vialou, V.; Louis, F.; Hessani, A.; Dansette, P.M.; Zornoza, T.; et al. Antidepressant efficacy of a selective organic cation transporter blocker in a mouse model of depression. *Mol. Psychiatry* **2020**, *25*, 1245–1259. [CrossRef]
129. Dolder, P.C.; Schmid, Y.; Steuer, A.E.; Kraemer, T.; Rentsch, K.M.; Hammann, F.; Liechti, M.E. Pharmacokinetics and pharmacodynamics of lysergic acid diethylamide in healthy subjects. *Clin. Pharmacokinet.* **2017**, *56*, 1219–1230. [CrossRef]
130. Brown, R.T.; Nicholas, C.R.; Cozzi, N.V.; Gassman, M.C.; Cooper, K.M.; Muller, D.; Thomas, C.D.; Hetzel, S.J.; Henriquez, K.M.; Ribaudo, A.S.; et al. Pharmacokinetics of escalating doses of oral psilocybin in healthy adults. *Clin. Pharmacokinet.* **2017**, *56*, 1543–1554. [CrossRef]
131. Masterson, C.H.; Murphy, E.J.; Gonzalez, H.; Major, I.; McCarthy, S.D.; O'Toole, D.; Laffey, J.G.; Rowan, N.J. Purified β-glucans from the Shiitake mushroom ameliorates antibiotic-resistant *Klebsiella pneumoniae*-induced pulmonary sepsis. *Lett. Appl. Microbiol.* **2020**, *71*, 405–412. [CrossRef]
132. Usuldin, S.R.A.; Wan-Mohtar, W.A.A.Q.I.; Ilham, Z.; Jamaludin, A.A.; Abdullah, N.R.; Rowan, N. In vivo toxicity of bioreactor-grown biomass and exopolysaccharides from Malaysian tiger milk mushroom mycelium for potential future health applications. *Sci. Rep.* **2021**, *11*, 23079. [CrossRef]

Journal of *Fungi*

Article

Nephroprotective Effects of Two Ganoderma Species Methanolic Extracts in an In Vitro Model of Cisplatin Induced Tubulotoxicity

Sébastien Sinaeve [1,*], Cécile Husson [2], Marie-Hélène Antoine [2], Stéphane Welti [3], Caroline Stévigny [1] and Joëlle Nortier [2]

1 RD3-Pharmacognosy, Bioanalysis and Drug Discovery Unit, Faculty of Pharmacy, Université Libre de Bruxelles, 1050 Brussels, Belgium
2 Laboratory of Experimental Nephrology, Faculty of Medicine, Université Libre de Bruxelles, 1050 Brussels, Belgium
3 Laboratoire de Génie Civil et Géo-Environnement- EA 4515, Université Lille, 59000 Lille, France
* Correspondence: sebastien.sinaeve@ulb.be

Abstract: Although cisplatin is used as a first-line therapy in many cancers, its nephrotoxicity remains a real problem. Acute kidney injuries induced by cisplatin can cause proximal tubular necrosis, possibly leading to interstitial fibrosis, chronic dysfunction, and finally to a cessation of chemotherapy. There are only a few nephroprotective actions that can help reduce cisplatin nephrotoxicity. This study aims to identify new prophylactic properties with respect to medicinal mushrooms. Among five *Ganoderma* species, the methanolic extracts of *Ganoderma tuberculosum* Murill., *Ganoderma parvigibbosum* Welti & Courtec. (10 µg/mL), and their association (5 + 5 µg/mL) were selected to study respective in vitro effects on human proximal tubular cells (HK-2) intoxicated by cisplatin. Measurements were performed after a pretreatment of 1 h with the extracts before adding cisplatin (20 µM). A viability assay, antioxidant activity, intracytoplasmic β-catenin, calcium, caspase-3, p53, cytochrome C, IL-6, NFκB, membranous KIM-1, and ROS overproduction were studied. Tests showed that both methanolic extracts and their association prevented a loss of viability, apoptosis, and its signaling pathway. *G. parvigibbosum* and the association prevented an increase in intracytoplasmic β-catenin. *G. parvigibbosum* prevented ROS overproduction and exhibited scavenger activity. None of the extracts could interfere with pro-inflammatory markers or calcium homeostasis. Our in vitro data demonstrate that these mushroom extracts have interesting nephroprotective properties. Finally, the chemical content was investigated through a phytochemical screening, and the determination of the total phenolic and triterpenoid content. Further studies about the chemical composition need to be conducted.

Keywords: ganoderma; cisplatin; nephrotoxicants; nephroprotection; apoptosis; anti-inflammatory; antioxidant; β-catenin; calcium; oxidative stress

Citation: Sinaeve, S.; Husson, C.; Antoine, M.-H.; Welti, S.; Stévigny, C.; Nortier, J. Nephroprotective Effects of Two Ganoderma Species Methanolic Extracts in an In Vitro Model of Cisplatin Induced Tubulotoxicity. *J. Fungi* **2022**, *8*, 1002. https://doi.org/ 10.3390/jof8101002

Academic Editor: Laurent Dufossé

Received: 26 August 2022
Accepted: 19 September 2022
Published: 24 September 2022

Publisher's Note: MDPI stays neutral with regard to jurisdictional claims in published maps and institutional affiliations.

1. Introduction

Cisplatin (cis-diamminedichloroplatinum or CisPt) is commonly used as a chemotherapeutic drug to treat numerous cancers, such as lung, testicular, or ovarian cancer [1]. Studies on CisPt have shown that its primary target is the DNA by being an alkylating agent, although the entire mechanism is not completely understood. The CisPt's uptake leads to DNA damage, affecting RNA transcription, the cell cycle, and therefore inducing the process of apoptosis [2,3]. CisPt has not only demonstrated good efficacy but also significant toxicity, including nephrotoxicity, ototoxicity, neurotoxicity, and hepatotoxicity [4]. Among these, nephrotoxicity has been widely studied as it can lead to a discontinuation of the treatment.

This toxicity prevalence remains high, occurring in about one-third of treated patients [4]. The clinical manifestations observed are a reduced glomerular filtration rate, higher serum creatinine level, and hypomagnesemia associated with hypokalemia. CisPt accumulates mainly in the proximal tubular cells where organic cation transporters (OCTs) are expressed [5]. CisPt is mainly transported into kidney cells by OCT-2 and to a lower level by the copper transporter 1 (Ctr1), which are mostly expressed in the basolateral membrane of proximal tubules [6]. Due to this uptake, CisPt accumulates in kidney cells and can interact with various reactive groups and induces DNA damages, activating the pro-apoptotic and pro-inflammatory signaling pathways [4–6].

In order to counteract this nephrotoxicity and allow a full and adequate treatment in the patient, there are only some recommendations for the clinician, such as increasing the patient's hydration, regularly checking renal functional parameters, and using lower CisPt doses [7,8]. CisPt analogs have also been synthesized, such as carboplatin and oxaliplatin, to reduce this toxicity and can be used as a combination therapy with CisPt [6,7].

In addition, there is currently a lack in new potential adjuvant treatments for preventing the tubulotoxicity induced by CisPt. Natural products have been studied to supplement this paucity and can potentially increase the patient's treatment adherence. Medicinal mushrooms are, therefore, an interesting path to study and to obtain new potential adjuvant drugs [6–8]. Indeed, various species have been studied as potential nephroprotective agents in the CisPt-induced model either in vitro [9–11] or in vivo [12,13].

Genus *Ganoderma* P. Karst has been widely studied for its potential medicinal properties. It is estimated to comprise nearly 300 species from which 20 are intensively studied for their biological activities [14]. The Ling Zhi, a member of the genus *Ganoderma*, is well-known in Traditional Chinese Medicine and is represented in the *Chinese Pharmacopoeia*. It consists of the dried sporophores of two species: *G. lucidum* and *G. sinense* [15]. *G. lucidum* mainly and other species, such as *G. applanatum* or *G. resinaceum*, have been studied for various activities such as for their anticancer, antidiabetic, and antiviral properties [16–21]. Following the increase in interest with respect to *G. lucidum*, it was added to the *European Pharmacopoeia* (Version 10.8). *Ganoderma*'s have been studied for their rich chemical content, including mainly polysaccharides and terpenes [14]. Among these, *G. lucidum* has been already studied as a nephroprotective agent in CisPt-induced tubulotoxicity [13].

In this study, methanolic extracts of five *Ganoderma* species have been screened in vitro on Human Kidney (HK-2) cells in order to test their possibly protective effects against CisPt tubulotoxicity. It was interesting to identify if other species from the same genus could also have a nephroprotective effect. The species studied were the following: *Ganoderma tuberculosum* Murill., *G. applanatum* Pat., *G. parvigibbosum* Welti & Courtec., *G. martinicense* Welti & Courtec., and *G. resinaceum* Boud. Among them, two belong to the list of the twenty most-studied [14], while the others are less known for their potential bioactivity. On top of that, to the best of our knowledge, *G. parvigibbosum* has not been studied yet and is, therefore, interesting for identifying new medicinal species. Targeted tests on the most promising species have then been performed to characterize the mechanisms of observed protection by evaluating, in particular, antioxidant, anti-inflammatory, and antiapoptotic potentials.

2. Materials and Methods

2.1. Reagents and Culture Media

Ascorbic acid, anhydrous sodium carbonate, sodium citrate, copper (II) sulfate pentahydrate, iron (III) chloride, magnesium shreds, bismuth nitrate, mercuric chloride, iodine, potassium iodide, sulfuric acid, chlorohydric acid, chloroform, DMSO, Fluo-3 AM solution, Folin Ciocalteu's reagent, glacial acetic acid, methanol, vanillin, $2',7'$-dichlorodihydrofluorescein diacetate (H_2DCF-DA), and $2,2'$-diphenyl-1-picrylhydrazyl (DPPH) were purchased from Sigma-Aldrich, Overijse, Belgium.

Quercetin dihydrate was obtained from Riedel de Haën, Selze, Germany.

Accutase, Dulbecco's Modified Eagle Medium (DMEM) low glucose, Dulbecco's PBS, Fetal bovine serum (FBS), L-glutamine, and Penicillin/Streptomycin solution were acquired from Capricorn Scientific, Ebsdorfergrund, Germany.

Anti-human β-catenin-phycoerythrin monoclonal antibody was purchased from R&D Systems, Abingdon, UK.

Pluronic F127 was obtained from Thermo Fisher, Geel, Belgium.

FITC Annexin V Apoptosis Detection Kit I, Cytofix/Cytoperm solution, anti-human IL-6-APC, anti-human KIM-1-phycoerythrin, and anti-human NFκB-Alexa Fluor 488 monoclonal antibodies were acquired from BD Biosciences, Franklin Lakes, NJ, USA.

Cell Counting Kit-8 (CCK-8) assay was purchased from the Dojindo Laboratories, Kumamoto, Japan.

CisPt solution was obtained from TEVA Pharmaceutical, Antwerpen, Belgium.

Anti-human p53-Alexa Fluor 647, anti-human caspase-3-phycoerythrin, and anti-human cytochrome C-FITC monoclonal antibodies were acquired from Santa Cruz Biotechnology, Dallas, TX, USA.

2.2. Sample Preparation

Seven specimens belonging to five *Ganoderma* species were collected. The species are namely two *G. tuberculosum* (G. tub.), two *G. applanatum* (G. app.), one *G. parvigibbosum* (G. par.), one *G. martinicense* (G. mar.), and finally one *G. resinaceum* (G. res.). They were collected as follows: G. tub. in Martinique in 2006 and 2008; G. app. in Martinique in 2006 and Belgium in 2018; G. par. in Martinique in 2007; G. mar. in Martinique in 2008; and G. res. in Belgium in 2009. They were directly dried after collection and then stored and identified at the Belgian Coordinated Collection of Microorganisms/Mycotheque of the Université Catholique de Louvain (BCCM/MUCL), Belgium. Prior to tests, they were ground into a powder using a ZM 100 grinder, Retsch, Germany. An extraction of 5 g of the powdered mushrooms with 200 mL of methanol was performed under agitation thrice for 24 h. Extracts were then dried using a rotary vacuum evaporator (Büchi, The Netherlands). Stock solutions were prepared at 100 mg/mL in DMSO and stored at −20 °C. Sample solutions were obtained by dilution in Serum Free DMEM low glucose at 1 mg/mL and filtered through a 0.2 μm cellulose acetate membrane (VWR, Radnor, PA, USA) before being diluted to reach a working concentration.

2.3. Cell Culture and Treatments

Human proximal tubular epithelial (HK-2) cells were obtained from the American Type Culture Collection (ATCC, Virginia, USA). They were grown in low glucose DMEM containing 10% FBS, 1% L-glutamine, and 1% penicillin–streptomycin. Cells were subcultured and harvested for experiments every week. Cells were used between passages 8 and 15 for experimental purposes, harvested using accutase, and seeded in 12-well plates (1 or 2×10^5 cells) or 96-well plates (1×10^4 cells). Cells were then incubated for 24 h in complete medium, rinsed twice with DMEM, and treated with test substances for 24 to 48 h in FBS-free medium.

Four groups were compared in each experiment: control (untreated), negative control (10 μg/mL mushroom extract treated), positive control (20 μM of CisPt), and tested group (pre-treatment with the mushroom extract at 10 μg/mL 1 h prior to the addition of CisPt 20 μM). In the case of associations, methanolic extracts were added each at 5 μg/mL. CisPt's concentration was chosen to reach about 85% of cell survival.

2.4. Cell Viability Assay (CCK-8)

The cell's viability was determined with an extract of the 7 specimens individually as well as with the association of G. tub. and G. par. using Cell Counting Kit-8 (CCK-8). CCK-8 is a tetrazolium salt that will be reduced by metabolically active cells in a formazan, allowing an estimation of cell viability [22]. The tests took place in a 96-well for 24 h and after treatment, 10 μL of CCK-8 solution was added to each well, and the 96-well plate was continuously incubated at 37 °C for 2 h. The OD value for each well was read at a wavelength of 450 nm to determine the cell's viability with a Labsystems iEMS reader/dispenser MF (Labsystems, Vantaa, Finland). The assay was repeated six times.

2.5. Scavenger Activity

The scavenger activity was assessed by the measurement of the scavenging ability of extracts towards the stable free radical 2,2′-diphenyl-1-picrylhydrazyl (DPPH). Serial dilutions (2-fold) of extracts (1 mg/mL) were mixed with 225 µL of methanolic 0.04% DPPH in 96-wells plates and left for 30 min in the dark; absorbances were measured with a Labsystems iEMS reader/dispenser MF (Labsystems, Finland) at 540 nm and 620 nm, using methanol as a blank and a 0.04% methanolic DPPH solution as the control. The assay was repeated three times, and for each, two serial dilutions of standards were added, namely quercetin dihydrate and ascorbic acid. The assay was repeated three times. The scavenger activity was calculated as described in Equation (1), the IC_{50} was calculated, and the residual DPPH free radical at working concentrations was graphically determined.

$$\% \text{ Scavenger activity} = [(OD_{540} - OD_{620})_{tested}/(OD_{540} - OD_{620})_{control}] \times 100 \qquad (1)$$

2.6. Flux Cytometry Analyses

Cells were seeded in a 12-well plate (1 or 2×10^5 cells). The analyses were performed on a BD FACSCanto II flow cytometer (BD Pharmingen, San Diego, CA); a minimum of 10^4 cells was recorded. The mean fluorescence intensities were estimated, compared to controls, and expressed as proportions. All antibodies were used in accordance to the working concentration specified by the manufacturer. The assays were repeated three times.

2.6.1. Apoptosis Detection: Annexin V/PI Assay and Estimation of the p53, Caspase 3 and Cytochrome C Pro-Apoptotic Markers

Firstly, the Annexin V/PI assay was performed. Cells were harvested after 24 h and centrifuged at $1400\times g$. Cells were then incubated with a solution of binding buffer containing the Annexin V and PI for 15 min at 4 °C. Cells were washed and analyzed. The total apoptosis was estimated by the sum of the early apoptotic cells (Annexin V positive and PI negative) and the late apoptotic cells (Annexin V positive and PI positive).

Secondly, p53, caspase-3, and cytochrome C, three markers of the pro-apoptotic signaling pathway, were estimated. For this, cells were harvested after 24 h, centrifuged at $500\times g$, and fixed in Cytofix/Cytoperm solution for 20 min at 4 °C. The suspension was washed and incubated with a solution of anti-human p53, anti-human caspase-3 and anti-human cytochrome C monoclonal antibodies for 30 min in the dark at 4 °C. Cells were then washed and analyzed.

2.6.2. Oxidative Stress Measurement

The reactive oxygen species (ROS) overexpression was determined using the H_2DCF-DA reagent, a reduced form of fluorescein, which, when cleaved and oxidized, especially by ROS, is converted into a highly fluorescent form. After 24 h of treatment conditions, cells were incubated for 30 min with a 10 µM H_2DCF-DA solution. They were then rinsed, harvested, centrifuged at $1400\times g$, and analyzed.

2.6.3. Anti-Inflammatory Potential Measurements

The anti-inflammatory potential was assessed by estimating the proportion of three pro-inflammatory markers, namely NFκB, IL-6 and KIM-1. Cells were harvested after 24 h, centrifuged at $500\times g$, and fixed in Cytofix/Cytoperm solution for 20 min at 4 °C. The suspension was washed and incubated with a solution of anti-human NFκB, anti-human IL-6, and anti-human KIM-1 monoclonal antibodies for 30 min in the dark at 4 °C. Cells were then washed and analyzed.

2.6.4. Intracytoplasmic Calcium Estimation

The intracytoplasmic calcium proportion was assessed using Fluo-3 AM in order to observe the late dysregulation in its homeostasis. Cells were harvested after 48 h and centrifugated at $1400\times g$; then, the cells were incubated in 200 µL of a 1 µM Fluo-3 and 1 mg/mL Pluronic mix solution at room temperature for 30 min. Cells were washed twice and then analyzed.

2.6.5. Intracytoplasmic β-catenin Estimation

The intracytoplasmic β-catenin estimation was assessed in order to estimate the cell adherence and its decrease in stress conditions. Cells were harvested after 48 h, centrifuged at $500\times g$ and fixed in Cytofix/Cytoperm solution for 20 min at 4 °C. The suspension was washed twice and incubated with an anti-human β-catenin-PE monoclonal antibody in the dark for 30 min. Cells were washed twice and then analyzed.

2.7. Chemical Content

2.7.1. Phytochemical Screening

The phytochemical screening was performed in order to identify the presence or absence of 6 types of secondary metabolites. They are namely the following: simple carbohydrates, terpenes, saponins, alkaloids, flavonoids, and tannins. For these tests, 0.5 g of the mushroom was extracted in 10 mL of methanol for one hour in an ultrasonic bath. Extracts were then filtered using Büchner filtration, and volumes were then completed at 10 mL. All tests were performed by adapting methods from the literature [23–25].

The presence of simple reducing carbohydrates was assessed using the Benedict's test. For this, 2 mL of the Benedict's reagent (1 g anhydrous sodium carbonate, 1.73 g of sodium citrate, and 0.173 g of copper (II) sulfate pentahydrate in 10 mL of water) was added to 1 mL of the extract. The mixture was then heated in a boiling bath for 3 min. The color obtained determined the presence or absence of simple reducing carbohydrates (blue: none; green: traces; orange: moderate; red: large amount)

The terpenes' presence was determined by the Salkowski's test. To 1 mL of extract, 3 mL of chloroform were added, followed by few drops of sulfuric acid. A red–brown color appearing between the two phases confirms the presence of terpenes.

The presence of saponins was permitted by the foam test. Firstly, 1 mL of extract was evaporated. Then, the residue was then dissolved in 10 mL of water and vigorously agitated. If a minimum of 1 cm of persistent foam (>10 s) formed, the test was considered positive with respect to the presence of saponins.

The alkaloids' presence was estimated by three reagents: Dragendorff, Mayer, and Bouchardat. The apparition of a precipitate after the addition of a few drops of reagent to 1 mL of extract allowed the confirmation of the presence of alkaloids.

The presence of flavonoids was assessed by the Shinoda test. To 1 mL of extract, 1 mL of hydrochloric acid was added in addition to 1 mL of water and a few magnesium shreds. Hydrogen gas was produced and a red coloration of the solution indicates the presence of flavonoids.

The tannins' presence was determined by adding a few drops of 5% aqueous ferric chloride solution (m/V). The apparition of a brown–black or blue–green precipitate indicates the presence of tannins.

2.7.2. Total Phenolic Content

The total phenolic content (TPC) was determined following the Folin Ciocalteu's method. The method was adapted from Rojo-Poveda et al. [26]. Briefly, 100 µL of Folin 10% solution was added to 20 µL of standard or to a 1 mg/mL methanolic extract solution in a 96-well plate. It was then incubated at room temperature in the dark for 3 min. 75 µL of a sodium carbonate 7.5% aqueous solution was then added before incubation at room temperature in the dark for one hour. Absorbance was then measured at 740 nm using a BioTek Synergy HT spectrophotometric multi-detection microplate reader (BioTek

Instruments, Milan, Italy). Gallic acid was used as a standard from 0 to 700 µM. All measurements were repeated thrice.

2.7.3. Total Triterpenoid Content

The total triterpenoid content (TTC) was determined by adapting the method used by Wei et al. [27]. Briefly, 200 µL of standard or 1 mg/mL methanolic extract solution was evaporated in test tubes. 1 mL of a five percent vanillin in a glacial acetic acid solution was added prior to 1.8 mL of sulfuric acid. Tubes were then heated at 70 °C for 30 min. 7.2 mL of glacial acetic acid was added before the absorbance was measured at 573 nm using a Genesys 10 spectrophotometer (Spectronic Unicam, Gent, Belgium). Oleanolic acid was used as a standard from 0 to 1000 µM. All measurements were repeated thrice.

2.8. Statistical Analyses

Statistical analyses were performed on results normalized and compared to the respective controls. The data were compared by means of a one-way ANOVA using GraphPad Prism 8 software (San Diego, CA, USA), and p values < 0.05 were considered significant.

3. Results

3.1. Cell Viability Assay (CCK-8)

The effect on the cell viability after CisPt incubation with or without the mushroom extracts is shown in Figure 1. Results were obtained after 24 h incubation with CisPt 20 µM by the use of the CCK-8 assay. Six independent experiments were performed for each experimental condition and *Ganoderma* species.

The two G. tub. extracts did not affect the cell survival rate (Figure 1a), whereas the exposure to CisPt alone led to a significant decrease in cell survival (87 ± 6%). Both G. tub. extracts were able to significantly prevent the cell mortality (respectively 99 ± 8% and 96 ± 3%).

The two G. app., G. mar. and G. res., did not affect cell survival rate (Figure 1b,d,e), whereas the exposure to CisPt alone led to a significant decrease in cell survival (respectively 88 ± 4%; 88 ± 3% and 87 ± 5%). None of these extracts were able to significantly prevent the cell mortality (respectively 90 ± 4% and 91 ± 6 %; 89 ± 4%; 88 ± 6%).

G. par. positively affected cell survival (104 ± 4%) (Figure 1f), while the addition of CisPt led to a significant decrease (89 ± 4%). G. par. extract allowed a significant prevention of cell mortality (103 ± 5%).

These screening results led to the identification of two interesting species: *Ganoderma tuberculosum* and *Ganoderma parvigibbosum*. Following these results, G. tub. 1 and G. par were further studied and a synergetic potential was also tested between both active extracts by reaching a total concentration of 10 µg/mL with an equivalent proportion of both extracts (5 µg/mL for both). The association of G. tub. and G. par. did not affect cell survival (Figure 1c), whereas the exposure to CisPt alone led to a significant decrease in cell survival (88 ± 6%). The association allowed a significant prevention in cell mortality (96 ± 5%).

Considering these screening results, a decision was made to investigate the mechanism of action of *G. tuberculosum 1*, *G. parvigibbosum*, and their associations.

3.2. Apoptosis Detection: Annexin V/PI Assay and Estimation of the p53, Caspase 3, and Cytochrome C pro-Apoptotic Markers

The total apoptosis assessment is shown in Figure 2a. The G. tub., G. par. extracts, and their association had no significant impact on the total apoptosis process compared to the controls. CisPt significantly increased apoptosis (14.9 ± 3.6%), and this increase was significantly prevented by a pre-treatment with G. tub., G. par, and their association (11.4 ± 3.8%, 11.8 ± 2.5%, and 11.3 ± 2.2%, respectively).

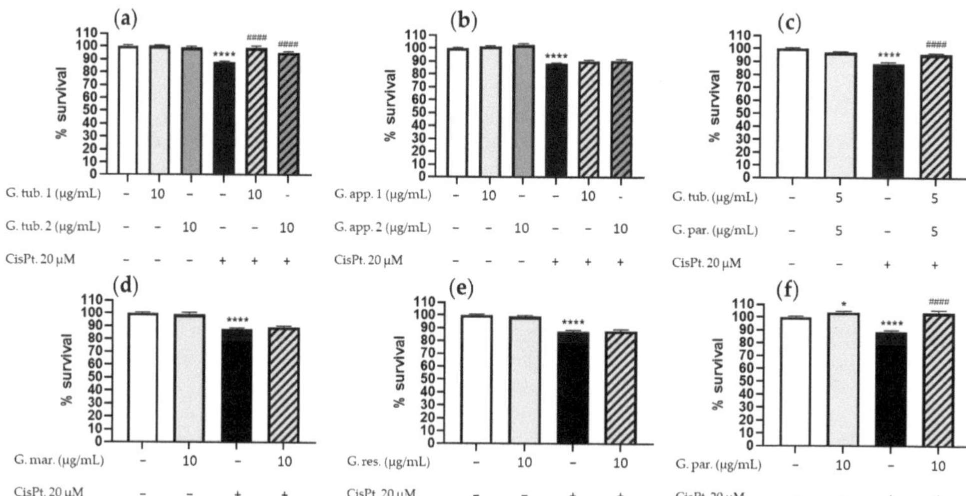

Figure 1. Cell survival rate determined using the CCK-8 assay after a 24 h treatment with CisPt 20 µM and/or: (**a**) one of the G. tub. extract (10 µg/mL); (**b**) one of the G. app. extract (10 µg/mL); (**c**) the G. tub. and G. par. extract (each at 5 µg/mL); (**d**) the G. par. extract (10 µg/mL); (**e**) the G. mar. extract (10 µg/mL); (**f**) the G. res. extract (10 µg/mL). Results are shown as mean ± SD of six independent experiments (**** $p < 0.0001$ and * $p < 0.05$ compared to controls; #### $p < 0.0001$ compared to CisPt. 20µM). The signs + and − indicate the presence or the absence of the corresponding treatment, respectively.

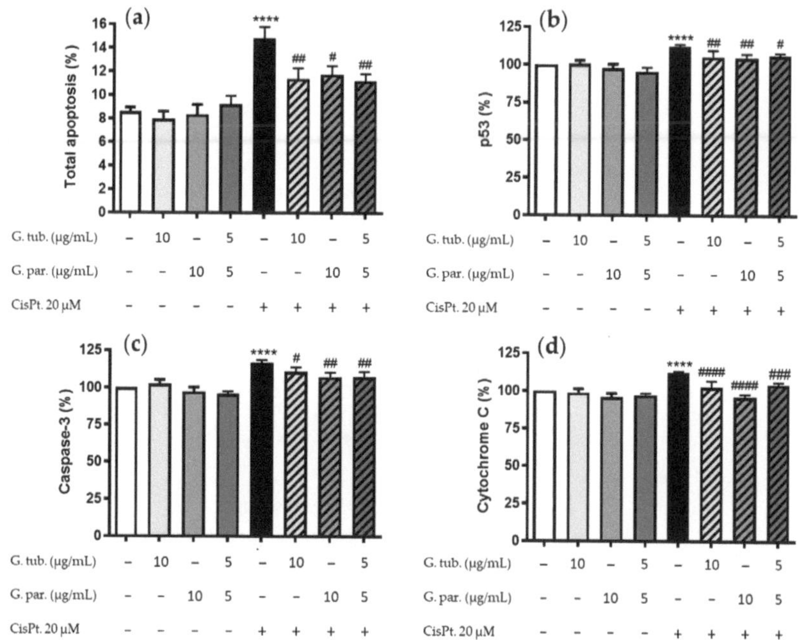

Figure 2. Apoptosis determinations assessed using flow cytometry of cells exposed during 24 h to CisPt 20 µM and/or the G. tub. extract (10 µg/mL), the G. par. extract (10 µg/mL), or the G. tub. and

G. par. association extract (each at 5 µg/mL). (**a**) The total apoptosis process was determined following Annexin V/PI staining, and results are shown as mean ± SD of three independent experiments (**** $p < 0.0001$ compared to controls; ## $p < 0.01$ and # $p < 0.05$ compared to CisPt. 20 µM). Three markers were analyzed, and their respective normalized fluorescence compared to the controls is shown as follows: (**b**) intracytoplasmic p53 proportion, (**c**) membranous caspase-3 proportion, and (**d**) intracytoplasmic cytochrome C proportion. Results are shown as mean ± SD of three independent experiments (**** $p < 0.0001$ compared to controls; #### $p < 0.0001$, ### $p < 0.001$ ## $p < 0.01$, and # $p < 0.05$ compared to CisPt. 20 µM). The signs + and − indicate the presence or the absence of the corresponding treatment, respectively.

Among the three markers studied, namely p53, caspase-3, and cytochrome C, the mushroom extracts did not affect their intracytoplasmic proportions compared to the controls. The p53 proportion increased up to 112 ± 5% by CisPt (Figure 2b). This increase was prevented by G. tub. (105 ± 3%), G. par. (105 ± 6%), and their association (107 ± 4%). The caspase-3 proportion increased up to 117 ± 7% by CisPt (Figure 2c). This increase was prevented by G. tub. (110 ± 7%), G. par. (107 ± 8%), and their association (107 ± 8%). The cytochrome C proportion increased up to 112 ± 4 % by CisPt (Figure 2d). This increase was prevented by G. tub. (103 ± 9%), G. par. (96 ± 4%), and their association (104 ± 3%).

3.3. Antioxidant Potential: Scavenger Activity and ROS Determination (H₂DCFDA Assay)

3.3. Antioxidant Potential: Scavenger Activity and ROS Determination (H$_2$DCFDA Assay)

Scavenging activities at the working concentration (10 µg/mL) as well as IC$_{50}$ have been determined using the DPPH assay. The results obtained are shown in Figure 3a. IC$_{50}$ scavenging activities for G. tub., G. par., and their association were 124 ± 4 µg/mL, 13.8 ± 0.8 µg/mL, and 21.6 ± 0.8 µg/mL, respectively. Residual DPPH at the working concentration were 95.7%, 59.9%, and 72.9%, respectively.

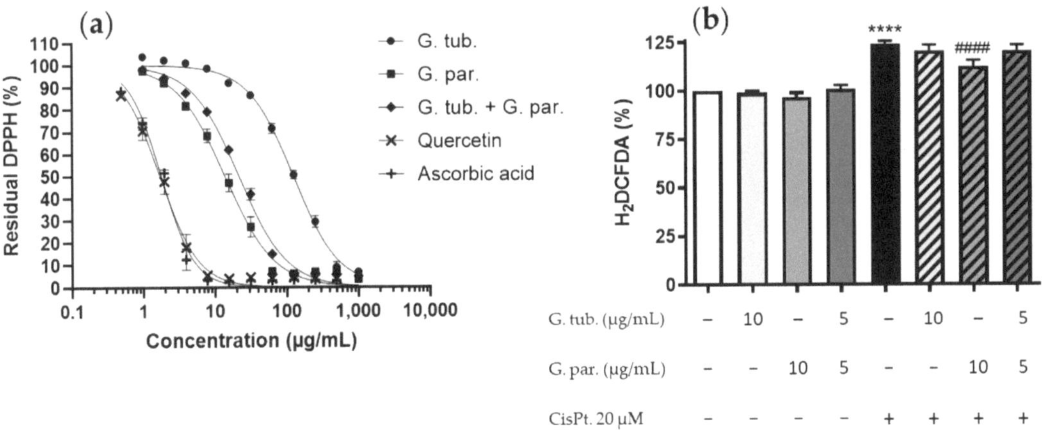

Figure 3. Antioxidant potential of G. tub., G. par. and their association represented by (**a**) their scavenger activity based on residual DPPH percentages on serial 2-fold dilutions (starting concentration from 1000 µg/mL; while standards were added at 500 µg/mL); (**b**) normalized fluorescence of H₂DCFDA, expressed as percentages compared to controls after a 24 h treatment with CisPt 20 µM and/or the G. tub. extract (10 µg/mL); the G. par. extract (10 µg/mL) or the G. tub. and G. par. association extract (each at 5 µg/mL). Results are shown as mean ± SD of three independent experiments (**** $p < 0.0001$ compared to controls; #### $p < 0.0001$ compared to CisPt 20 µM). The signs + and − indicate the presence or the absence of the corresponding treatment respectively.

The effect on the ROS's increase due to CisPt incubation with or without the extracts is shown in Figure 3b. ROS production was not significantly affected by the extracts alone, whereas it was significantly increased by CisPt, up to 124 ± 8%. This ROS overproduction was significantly prevented by the G. par. methanolic extract, as reflected by a significant reduction to $112.4 \pm 8\%$. The G. tub. and the association did not significantly modify the ROS's production compared to the effect obtained with CisPt alone.

3.4. Anti-Inflammatory Measurement: Impact on the NFkB, KIM-1, IL-6 Markers

The anti-inflammatory potential effect of the two mushroom extracts and their association was tested using three markers, namely NFκB, KIM-1 and IL-6. The results obtained are shown in Figure 4. None of the extracts, used alone, significantly modified the expression of these three markers. By contrast, CisPt exposure led to an increased expression of NFκB, KIM-1, and IL-6 in proportions to $119 \pm 7\%$, $115 \pm 6\%$, and $114 \pm 5\%$ compared to the controls, respectively (Figure 4a–c). In the test conditions, neither G. tub., G. par. extracts, nor their association could significantly prevent the increase in NFκB, IL-6, and KIM-1 proportions.

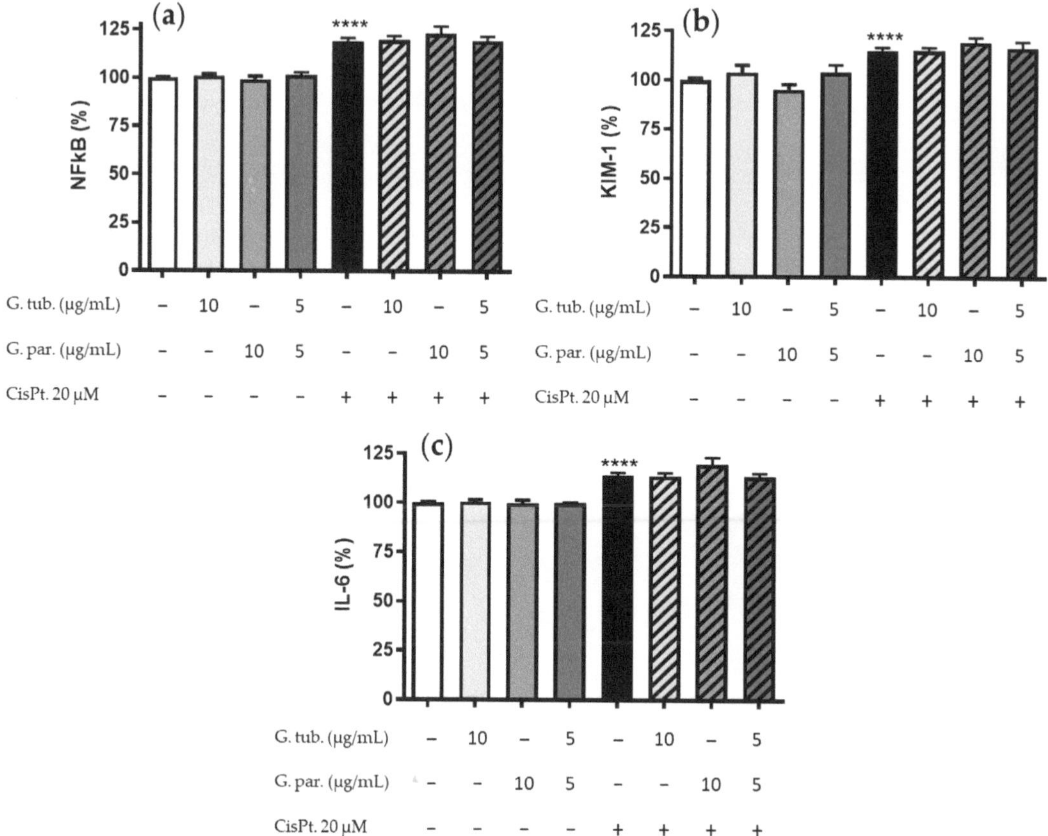

Figure 4. Anti-inflammatory measurements of immunostained cells after a 24 h treatment with CisPt 20 μM and/or the G. tub. extract (10 μg/mL); the G. par. extract (10 μg/mL) or the G. tub. and G. par. association extract (each at 5 μg/mL). Three markers were analyzed and their normalized fluorescences compared to controls are shown as follows: (**a**) intracytoplasmic NFκB proportion, (**b**) membranous KIM-1 proportion, and (**c**) intracytoplasmic IL-6 proportion. Results are shown as mean $\pm$ SD of three independent experiments (**** $p < 0.0001$ compared to controls). The signs + and − indicate the presence or the absence of the corresponding treatment, respectively.

3.5. Intracytoplasmic Calcium Estimation

The estimation of intracytoplasmic calcium after 48 h was determined in order to evaluate the dysregulation in late cell homeostasis. The results obtained are shown in Figure 5. The mushroom extracts did not significantly affect the intracytoplasmic calcium compared to the controls. CisPt, on the other hand, increased the proportion of intracytoplasmic calcium up to $166 \pm 22\%$. A pre-treatment with the mushroom extracts could not prevent the CisPt-induced increase.

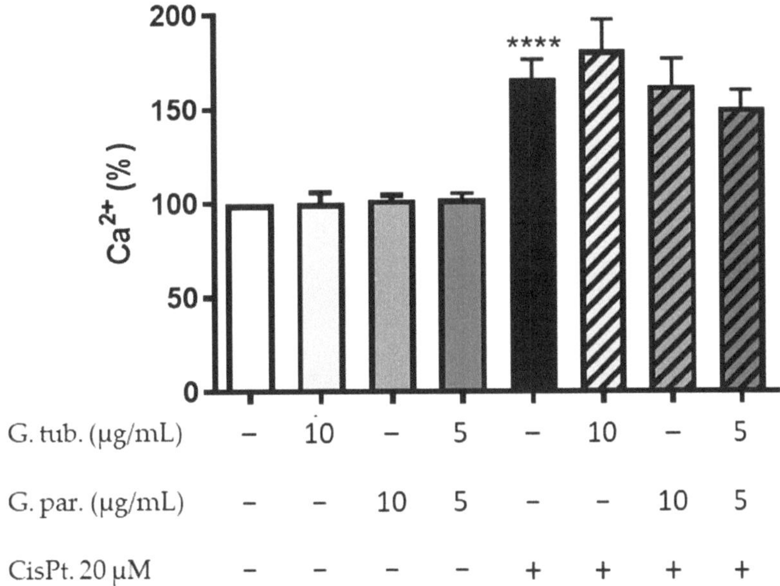

Figure 5. Intracytoplasmic calcium proportion of Fluo-3 AM dyed cells after a 48 h treatment with CisPt 20 µM and/or the G. tub. extract (10 µg/mL), the G. par. extract (10 µg/mL) or the G. tub., and G. par. association extract (each at 5 µg/mL). Cells were analyzed by flux cytometry, and results are shown as mean of normalized fluorescence compared to controls $\pm$ SD of three independent experiments (**** $p < 0.0001$ compared to the control). The signs + and − indicate the presence or the absence of the corresponding treatment, respectively.

3.6. Intracytoplasmic β-Catenin Determination

The intracytoplasmic proportion of β-catenin measured in all conditions is shown in Figure 6. The mushroom extracts did not significantly affect its proportion. However, CisPt exposure increased it up to $131 \pm 11\%$ compared to the controls. This increase was not prevented by pre-treating cells with G. tub. ($127 \pm 9\%$), but it was significantly prevented by G. par. ($113 \pm 4\%$) and by the association of both extracts ($109 \pm 8\%$).

3.7. Chemical Content

3.7.1. Phytochemical Screening

The phytochemical screening results are shown in Table 1. All methanolic extracts showed the presence of terpenes. G. tub. 1, G. app. 1, G. tub. 2, G. par., and G. app. 2 showed the presence of simple reducing carbohydrates. Moreover, saponins were detected in G. par. and G. app. 1 methanolic extracts. Finally, the presence of tannins was suspected in G. app. 1, G. par., and G. app. 2 (change in color of the solution but no precipitate was observed).

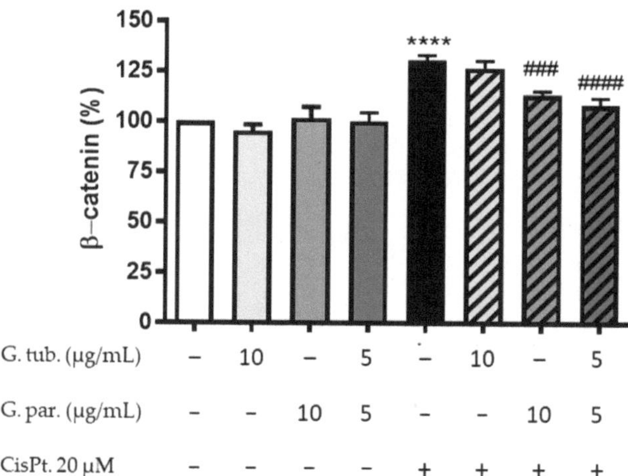

Figure 6. Intracytoplasmic β-catenin immunostained cells after a 24 h treatment with CisPt 20 µM and/or the G. tub. extract (10 µg/mL), the G. par. extract (10 µg/mL) or the G. tub., and G. par. association extract (each at 5 µg/mL). Cells were analyzed by flux cytometry and results are shown as mean of normalized fluorescence compared to controls $\pm$ SD of three independent experiments (**** $p < 0.0001$ compared to controls; #### $p < 0.0001$ and ### $p < 0.001$ compared to CisPt. 20 µM). The signs + and − indicate the presence or the absence of the corresponding treatment, respectively.

Table 1. Phytochemical screening results of methanolic extracts of the 7 studied specimens. The signs +, $\pm$ and − indicate the presence, the suspected presence or the absence of the corresponding secondary metabolite, respectively.

Secondary Metabolite	G. tub. 1	G. app. 1	G. tub. 2	G. par.	G. mar.	G. res.	G. app. 2
Simple carbohydrates	+	++	+	++	−	−	++
Terpenes	+	+	+	+	+	+	+
Saponins	−	+	−	+	−	−	−
Alkaloids	−	−	−	−	−	−	−
Flavonoids	−	−	−	−	−	−	−
Tannins	−	$\pm$	−	$\pm$	−	−	$\pm$

3.7.2. Total Phenolic and Triterpenoid Contents

Both TPC and TTC are shown in Table 2. Results have shown that G. app. 1 and G. par. have a higher TPC, with 121 ± 14 and 201 ± 12 mg of Gallic Acid Equivalent (GAE) per gram of extract, respectively. Both G. tub. and G. par., have a TTC over 200 mg Oleanolic Acid Equivalent (OAE) per gram of extract.

Table 2. Total phenolic and triterpenoid contents in the methanolic extracts of the 7 specimens studied.

	Total Phenolic Content [1]	Total Triterpenoid Content [2]
G. tub. 1	47 ± 4	212 ± 12
G. app. 1	121 ± 14	110 ± 13
G. tub. 2	40 ± 3	256 ± 13
G. par.	201 ± 12	261 ± 16
G. mar.	26 ± 2	138 ± 12
G. res.	30 ± 2	169 ± 14
G. app. 2	77 ± 7	157 ± 15

[1] mg GAE/g extract. [2] mg OAE/g extract.

4. Discussion

Cisplatin is a currently and frequently used cancer treatment and is well known in its induction of tubulotoxicity [5]. As previously described by others, the present study confirms that CisPt has various deleterious effects on proximal tubular epithelial cells [3,5,28]. It activates the apoptosis pathway, dysregulates the late calcium homeostasis, increases the oxidative stress as well as pro-inflammatory and stress markers, and finally increases β-catenin translocation. Previous studies have shown the impact on various biomarkers in vitro in CisPt-induced nephrotoxicity on HK-2 cells. The results demonstrated that the apoptosis was induced, including a decrease in the Bcl-2 family and an increase on the activated caspase-3. Finally, the protein levels of KIM-1, calbindin, and TIMP-1 were also increased [28].

Numerous natural products have been tested for their potential positive impact on human health and among these, medicinal mushrooms are increasingly investigated. The *Ganoderma* genus and more specifically the *Ganoderma lucidum*, used in traditional Chinese medicine, have been studied in the case of CisPt nephrotoxicity in vivo as well as on cultured mice tubular epithelial cells. Results have introduced an interest in small molecules extracted from *G. lucidum*, as well as its methanolic and chloroformic extracts. The *G. lucidum* methanolic extract was used at 500 mg/kg, while the chloroformic extract was used at 100 mg/kg. In all studies, the *G. lucidum* extract has been given prior to CisPt injection. Interestingly, this extract exhibited antioxidant and anti-inflammatory activities and reduced apoptosis [13,15,29–31]. Finally, complex Chinese medicinal herb mixes (CMHs), containing *G. lucidum*, have been studied in CisPt-treated patients for non-small-cell lung cancer, and an improved median survival rate was reported. The authors related this result to a protective effect on the bone marrow hematopoietic system. Renal function parameters (serum levels of urea and creatinine) were not improved. The CMH was used as a decoction and results were observed [32].

Starting from these quite limited results, the present study aimed to increase the knowledge with respect to this genus by examining various *Ganoderma* species in vitro. Working concentrations were chosen based on a non-toxic concentration of the methanolic extracts, which was determined to be 10 µg/mL and 5 + 5 µg/mL when an association was considered. CisPt concentrations were chosen to reach about 85% of cell survival. The cell viability assay results confirmed the interest on *Ganoderma tuberculosum*, *Ganoderma parvigibbosum*, and their association as they prevented decreases in cell viability compared to cells exposed to CisPt alone. Following this, the investigations focused on the mechanisms of tubulotoxicity prevention by using the corresponding extracts.

Apoptosis prevention.

CisPt is a known pro-apoptotic xenobiotic drug. It induces apoptosis at micromolar concentrations and necrosis at millimolar concentrations using different mechanisms [13,33]. Previous studies have shown the apoptosis induction of CisPt on HK-2 cells. Indeed, their results showed an increase of up to 14.2% of apoptotic cells, which are comparable to our results (14.9%) [28]. CisPt is known to induce DNA damages as alkylating agent and to induce cellular stress by increasing ROS production [4]. The results of the present study confirmed the significant CisPt-induced apoptosis of cultured HK-2 cells. They also showed that G. tub., G. par., and the extract association prevented this apoptosis induction, using the annexin V/PI assay to determine and quantify the total apoptosis.

This prevention was observed at different steps of the apoptosis induction: p53, a transcriptional protein, can be early activated by a stimulus, such as a DNA damage or cellular stress. It is followed by a cytochrome C release from the mitochondria. Both result in the activation of the caspase pathway [33,34] and cell death [35–37]. Finally, caspase-3, a protease involved at the end of the proapoptotic signaling pathway is activated, leading to an enhanced apoptosis. It is, therefore, interesting for new potential cancer treatments [38,39]. In our in vitro study, G. par., G. tub., and their association were able to mitigate CisPt-induced apoptosis, including the increase in caspase 3.

Our in vitro results confirm prior in vivo observations showing the crucial impact of CisPt in apoptosis regulation. Mahran et al. demonstrated that the caspase 3 increase was significantly reduced by *G. lucidum* (about 0.5 fold compared to the CisPt group) [13]. Other authors reported the potential interest of *G. lucidum* small molecules through molecular docking. They have shown their capacity to inhibit the activation of the caspase-1 [31]. Finally, an ethanolic extract of *G. lucidum* has also shown the prevention of apoptosis in cadmium intoxicated chicken. Data of this in vivo study have demonstrated its ability to prevent the increase in caspase-3, Bax, and Bcl-2, all markers of the pro-apoptotic pathway in chickens' spleen [40].

Antioxidant and anti-inflammatory activities.

As the production of ROS induces apoptosis signal transduction and activates apoptotic proteins, such as caspase 3, the antioxidant property of the mushroom extracts was studied [41]. Our results have shown a partial scavenger activity of the *G. parvigibbosum* extract at working concentrations. These results can be linked to its total phenolic content as polyphenols are known to show antioxidant activity [26]. Comparable results were observed with the association of the two extracts while *G. tuberculosum* did not exhibit this activity. Interestingly and logically, the ROS overproduction due to CisPt was prevented by the *G. par.* extract, linkable to its scavenger activity. Indeed, a scavenger activity can prevent the harmful effect of ROS increase measured in intoxicated cells and, therefore, decrease the induced toxicity [42]. Previous studies have shown that *G. lucidum* extracts are antioxidants in vivo. They have shown the ability to prevent superoxide dismutase, catalase, and glutathione peroxidase decreases as well as to restore glutathione and malondialdehyde levels [29,30]. It was also shown that the H_2O_2 renal level was reduced by 1.8 fold in pre-treated animals prior to CisPt administration [13].

CisPt is also known to induce a pro-inflammatory response in intoxicated cells [5,6], including in HK-2 cells. To further examine this, three markers of the pro-inflammation cell response were studied: the intracytoplasmic increase in NFκB, IL-6, and the membranous overexpression of KIM-1. NFκB is a transcription factor that has a key role in immune and inflammatory responses [43,44]. IL-6 is an inflammatory cytokine that is overexpressed in response to stimuli, allowing an immune reaction of cells [45]. KIM-1 is a specific glycoprotein mainly expressed by proximal tubule cells, and it is upregulated when cells are exposed to a nephrotoxin, allowing a response to the injury [46,47]. As *G. parvigibbosum* prevented ROS overproduction, it could have decreased pro-inflammatory responses. In our hands, despite observing antioxidant activities, no anti-inflammatory response due to the pre-treatment of the cells with the mushroom's extracts was detected. These results actually contrast with previous studies that report the capacity to prevent the CisPt-mediated inflammatory signal through NFκB [13,31], interleukin 1 [31], and the high-mobility group box-1 [13]. These studies were conducted in vitro with a purified small molecule from *G. lucidum* [31] or in vivo in mice receiving *G. lucidum* [13]. However, it must be noted that the results from Xue et al. are preliminary results, and further information on the results are needed.

Effect on the late calcium homeostasis.

The late calcium homeostasis dysregulation observed by the addition of CisPt can be an important cause of the ROS overproduction and can lead to cell damages [48,49]. It is known that CisPt treatments lead to mitochondrial and endoplasmic reticulum dysfunctions, including a decrease in calcium uptake; thus, an increase in intracytoplasmic calcium concentration [6]. Our study contributed to demonstrating that CisPt was able to interfere with the long-term homeostasis of calcium in vitro but the mushroom extracts were not able to prevent it.

Effect on the β-Catenin translocation.

β-Catenin is a protein that is known to play a role in cell adherence and proliferation, apoptosis, and pro-inflammatory-associated cancer through the Wnt/β-catenin signaling pathway. Inhibiting this pathway can mitigate renal fibrosis [50]. β-catenin is mostly located on the cytoplasmic side of the membrane before being translocated to the nucleus in the presence of a stimulus [51–53]. While these stimuli can be numerous, external stimuli, such as xenobiotics, and more specifically CisPt can lead to the delocalization of the β-catenin, hence increasing its intracytoplasmic concentration and resulting in higher fibrosis-occurring risks [54]. Interestingly, in our in vitro study, G. par. and the association of G. par. and G. tub. were able to prevent this internalization. The inhibition of β-catenin translocation by the mushrooms extracts may be due to their antioxidant potential since ROS can activate the Wnt/β-Catenin signaling pathway [55]. Their antiapoptotic potential could also favor cell adhesion, thereby preventing β-cat translocation following loss of E-cadherin. That type of protection may therefore lead to a lower risk of cell inadherence and fibrosis [51–53].

Chemical content

The *Ganoderma* genus is known to be rich in acid triterpenes and polysaccharides [56,57]. Triterpenes have been widely studied for various applications such as antidiabetic, antiviral and anticancer properties [13,16,17]. The species itself as well as the extracts tested lead to a better understanding of the observed activity. Indeed, the polysaccharides of *G. lucidum*, *G. sinense*, and *G. microsporum* exhibit an enhanced apoptosis in cancer cells in association with CisPt [58–60]. Our study has shown that the five species studied contain terpenes when performing the phytochemical screening. Most of the mushrooms also contain simple reducing carbohydrates. TTC determination indicates that a higher concentration of triterpenoids increases the protective effect on healthy HK-2 cells exposed to a nephrotoxicant. Interestingly, although CisPt is well-known to induce a pro-inflammatory and a pro-oxidant response [5], the total phenolic content, linkable to the antioxidant activity, does not allow a decrease in cell death. Indeed, while G. app. 1 has thrice the amount of polyphenols than G. tub. 1 and 2, it was unable to prevent CisPt-induced tubulotoxicity. The combination of a higher TTC and TPC, as for G. par., is more interesting. Indeed, it not only significantly decreased the total apoptosis and its signaling pathway, likewise observed for G. tub., but also decreased the internalization of β-catenin and prevented the ROS overproduction. These promising results highlight the importance of the triterpenoid content of the *Ganoderma*'s genus and their potential activity. Metabolomic and analytical studies are currently conducted in order to further identify the chemical composition of the extracts and to understand the link between the chemical compounds and the observed cell's protective activities.

5. Conclusions

All together, these results confirm the promising nephroprotective effect of *Ganoderma parvigibbosum* and *Ganoderma tuberculosum* against in vitro CisPt tubulotoxicity. The results show the importance of continuing the study of new species. The prevention of this toxicity can be mainly explained by the capacity to downregulate the pro-apoptotic pathway. Both methanolic extracts can therefore be useful as adjuvant agents. In order to confirm this, more studies will be required to analyze the chemical composition and the cell selectivity of the protective properties observed. Tests on cancer cells should be performed.

Author Contributions: Conceptualization, S.S., C.S. and J.N.; methodology, S.S., M.-H.A. and C.H.; validation, M.-H.A., C.S. and J.N.; formal analysis, S.S. and C.H.; investigation, S.S., M.-H.A. and C.H.; resources, S.S. and S.W.; data curation, S.S.; writing—original draft preparation, S.S.; writing—review and editing, C.H., M.-H.A., S.W., C.S. and J.N.; visualization, S.S.; supervision, C.S. and J.N.; project administration, C.S. and J.N. All authors have read and agreed to the published version of the manuscript.

Funding: This research received no external funding.

Institutional Review Board Statement: Not applicable.

Informed Consent Statement: Not applicable.

Data Availability Statement: Not applicable.

Acknowledgments: The authors thank Cony Decock, Mario Amalfi, the Meise Botanic Garden and the BCCM/MUCL for their help and collaboration. We would like to thank the Fondation Universitaire de Belgique for its contribution to publication fees.

Conflicts of Interest: The authors declare no conflict of interest.

References

1. Ghosh, S. Cisplatin: The First Metal Based Anticancer Drug. *Bioorganic Chem.* **2019**, *88*, 102925. [CrossRef] [PubMed]
2. Galluzzi, L.; Senovilla, L.; Vitale, I.; Michels, J.; Martins, I.; Kepp, O.; Castedo, M.; Kroemer, G. Molecular Mechanisms of Cisplatin Resistance. *Oncogene* **2012**, *31*, 1869–1883. [CrossRef] [PubMed]
3. Dasari, S.; Tchounwou, P.B. Cisplatin in Cancer Therapy: Molecular Mechanisms of Action. *Eur. J. Pharmacol.* **2014**, *740*, 364–378. [CrossRef] [PubMed]
4. Aldossary, S.A. Review on Pharmacology of Cisplatin: Clinical Use, Toxicity and Mechanism of Resistance of Cisplatin. *Biomed. Pharmacol. J.* **2019**, *12*, 7–15. [CrossRef]
5. Manohar, S.; Leung, N. Cisplatin Nephrotoxicity: A Review of the Literature. *J. Nephrol.* **2018**, *31*, 15–25. [CrossRef] [PubMed]
6. Miller, R.P.; Tadagavadi, R.K.; Ramesh, G.; Reeves, W.B. Mechanisms of Cisplatin Nephrotoxicity. *Toxins* **2010**, *2*, 2490–2518. [CrossRef] [PubMed]
7. Crona, D.J.; Faso, A.; Nishijima, T.F.; McGraw, K.A.; Galsky, M.D.; Milowsky, M.I. A Systematic Review of Strategies to Prevent Cisplatin-Induced Nephrotoxicity. *Oncologist* **2017**, *22*, 609–619. [CrossRef]
8. Hayati, F.; Hossainzadeh, M.; Shayanpour, S.; Abedi-Gheshlaghi, Z.; Beladi Mousavi, S.S. Prevention of Cisplatin Nephrotoxicity. *J. Nephropharmacology* **2016**, *5*, 57–60.
9. Hwang, B.S.; Lee, D.; Choi, P.; Kim, K.S.; Choi, S.-J.; Song, B.G.; Kim, T.; Song, J.H.; Kang, K.S.; Ham, J. Renoprotective Effects of Hypoxylonol C and F Isolated from Hypoxylon Truncatum against Cisplatin-Induced Cytotoxicity in LLC-PK1 Cells. *Int. J. Mol. Sci.* **2018**, *19*, 948. [CrossRef]
10. Lee, S.R.; Lee, D.; Lee, H.-J.; Noh, H.J.; Jung, K.; Kang, K.S.; Kim, K.H. Renoprotective Chemical Constituents from an Edible Mushroom, Pleurotus Cornucopiae in Cisplatin-Induced Nephrotoxicity. *Bioorganic Chem.* **2017**, *71*, 67–73. [CrossRef]
11. Zhou, S.; Zhou, Y.; Yu, J.; Jiang, L.; Xiang, Y.; Wang, J.; Du, Y.; Cui, X.; Ge, F. A Neutral Polysaccharide from Ophiocordyceps Lanpingensis Restrains Cisplatin-Induced Nephrotoxicity. *Food Sci. Nutr.* **2021**, *9*, 3602–3616. [CrossRef] [PubMed]
12. Nitha, B.; Janardhanan, K.K. Aqueous-Ethanolic Extract of Morel Mushroom Mycelium Morchella Esculenta, Protects Cisplatin and Gentamicin Induced Nephrotoxicity in Mice. *Food Chem. Toxicol.* **2008**, *46*, 3193–3199. [CrossRef] [PubMed]
13. Mahran, Y.F.; Hassan, H.M. Ganoderma Lucidum Prevents Cisplatin-Induced Nephrotoxicity through Inhibition of Epidermal Growth Factor Receptor Signaling and Autophagy-Mediated Apoptosis. *Oxid. Med. Cell. Longev.* **2020**, *2020*, 4932587. [CrossRef] [PubMed]
14. Gong, T.; Yan, R.; Kang, J.; Chen, R. Chemical Components of Ganoderma. *Adv. Exp. Med. Biol.* **2019**, *1181*, 59–106. [CrossRef]
15. Da, J.; Wu, W.-Y.; Hou, J.-J.; Long, H.-L.; Yao, S.; Yang, Z.; Cai, L.-Y.; Yang, M.; Jiang, B.-H.; Liu, X.; et al. Comparison of Two Officinal Chinese Pharmacopoeia Species of Ganoderma Based on Chemical Research with Multiple Technologies and Chemometrics Analysis. *J. Chromatogr. A* **2012**, *1222*, 59–70. [CrossRef]
16. Ahmad, M.F.; Ahmad, F.A.; Khan, M.I.; Alsayegh, A.A.; Wahab, S.; Alam, M.I.; Ahmed, F. Ganoderma Lucidum: A Potential Source to Surmount Viral Infections through β-Glucans Immunomodulatory and Triterpenoids Antiviral Properties. *Int. J. Biol. Macromol.* **2021**, *187*, 769–779. [CrossRef]
17. Wińska, K.; Mączka, W.; Gabryelska, K.; Grabarczyk, M. Mushrooms of the Genus Ganoderma Used to Treat Diabetes and Insulin Resistance. *Molecules* **2019**, *24*, 4075. [CrossRef]
18. Elkhateeb, W.A.; Zaghlol, G.M.; El-Garawani, I.M.; Ahmed, E.F.; Rateb, M.E.; Abdel Moneim, A.E. Ganoderma Applanatum Secondary Metabolites Induced Apoptosis through Different Pathways: In Vivo and in Vitro Anticancer Studies. *Biomed. Pharmacother.* **2018**, *101*, 264–277. [CrossRef]
19. Hanyu, X.; Lanyue, L.; Miao, D.; Wentao, F.; Cangran, C.; Hui, S. Effect of Ganoderma Applanatum Polysaccharides on MAPK/ERK Pathway Affecting Autophagy in Breast Cancer MCF-7 Cells. *Int. J. Biol. Macromol.* **2020**, *146*, 353–362. [CrossRef]
20. Kou, R.-W.; Xia, B.; Wang, Z.-J.; Li, J.-N.; Yang, J.-R.; Gao, Y.-Q.; Yin, X.; Gao, J.-M. Triterpenoids and Meroterpenoids from the Edible Ganoderma Resinaceum and Their Potential Anti-Inflammatory, Antioxidant and Anti-Apoptosis Activities. *Bioorganic Chem.* **2022**, *121*, 105689. [CrossRef]
21. El-Sherif, N.F.; Ahmed, S.A.; Ibrahim, A.K.; Habib, E.S.; El-Fallal, A.A.; El-Sayed, A.K.; Wahba, A.E. Ergosterol Peroxide from the Egyptian Red Lingzhi or Reishi Mushroom, Ganoderma Resinaceum (Agaricomycetes), Showed Preferred Inhibition of MCF-7 over MDA-MB-231 Breast Cancer Cell Lines. *Int. J. Med. Mushrooms* **2020**, *22*, 389–396. [CrossRef] [PubMed]
22. Cai, L.; Qin, X.; Xu, Z.; Song, Y.; Jiang, H.; Wu, Y.; Ruan, H.; Chen, J. Comparison of Cytotoxicity Evaluation of Anticancer Drugs between Real-Time Cell Analysis and CCK-8 Method. *ACS Omega* **2019**, *4*, 12036–12042. [CrossRef] [PubMed]
23. Hossen, S.M.M.; Hossain, M.S.; Yusuf, A.T.M.; Chaudhary, P.; Emon, N.U.; Janmeda, P. Profiling of Phytochemical and Antioxidant Activity of Wild Mushrooms: Evidence from the in Vitro Study and Phytoconstituent's Binding Affinity to the Human Erythrocyte Catalase and Human Glutathione Reductase. *Food Sci. Nutr.* **2022**, *10*, 88–102. [CrossRef] [PubMed]

24. Ogidi, O.C.; Oyetayo, V.O. Phytochemical Property and Assessment of Antidermatophytic Activity of Some Selected Wild Macrofungi against Pathogenic Dermatophytes. *Mycology* **2016**, *7*, 9–14. [CrossRef] [PubMed]
25. Evans, W.C. *Trease and Evan's Pharmacognosy*, 13th ed.; Baillière Tendall: London, UK, 1989; ISBN 978-0-7020-1357-7.
26. Rojo-Poveda, O.; Barbosa-Pereira, L.; El Khattabi, C.; Youl, E.N.H.; Bertolino, M.; Delporte, C.; Pochet, S.; Stévigny, C. Polyphenolic and Methylxanthine Bioaccessibility of Cocoa Bean Shell Functional Biscuits: Metabolomics Approach and Intestinal Permeability through Caco-2 Cell Models. *Antioxidants* **2020**, *9*, 1164. [CrossRef]
27. Wei, L.; Zhang, W.; Yin, L.; Yan, F.; Xu, Y.; Chen, F. Extraction Optimization of Total Triterpenoids from Jatropha Curcas Leaves Using Response Surface Methodology and Evaluations of Their Antimicrobial and Antioxidant Capacities. *Electron. J. Biotechnol.* **2015**, *18*, 88–95. [CrossRef]
28. Sohn, S.-J.; Kim, S.Y.; Kim, H.S.; Chun, Y.-J.; Han, S.Y.; Kim, S.H.; Moon, A. In Vitro Evaluation of Biomarkers for Cisplatin-Induced Nephrotoxicity Using HK-2 Human Kidney Epithelial Cells. *Toxicol. Lett.* **2013**, *217*, 235–242. [CrossRef]
29. Pillai, T.G.; John, M.; Sara Thomas, G. Prevention of Cisplatin Induced Nephrotoxicity by Terpenes Isolated from Ganoderma Lucidum Occurring in Southern Parts of India. *Exp. Toxicol. Pathol.* **2011**, *63*, 157–160. [CrossRef]
30. Sheena, N.; Ajith, T.; Janardhanan, K. Prevention of Nephrotoxicity Induced by the Anticancer Drug Cisplatin, Using Ganoderma Lucidum, a Medicinal Mushroom Occurring in South India. *Curr. Sci.* **2003**, *85*, 478–482, *undefined*.
31. Xue, L.; Wu, W.; You, Y.; Chen, H. A Small Molecule from Ganoderma lucidum Protects against Cisplatin-induced Kidney Injury via Suppressing NLRP3/Caspase-1 Related Pyroptosis. Congress Abstract. *Kidney Int. Rep.* **2020**, *5*, S217. [CrossRef]
32. Xu, Z.Y.; Jin, C.J.; Zhou, C.C.; Wang, Z.Q.; Zhou, W.D.; Deng, H.B.; Zhang, M.; Su, W.; Cai, X.Y. Treatment of Advanced Non-Small-Cell Lung Cancer with Chinese Herbal Medicine by Stages Combined with Chemotherapy. *J. Cancer Res. Clin. Oncol.* **2011**, *137*, 1117–1122. [CrossRef] [PubMed]
33. Pfeffer, C.; Singh, A. Apoptosis: A Target for Anticancer Therapy. *Int. J. Mol. Sci.* **2018**, *19*, 448. [CrossRef] [PubMed]
34. Elmore, S. Apoptosis: A Review of Programmed Cell Death. *Toxicol. Pathol.* **2007**, *35*, 495–516. [CrossRef] [PubMed]
35. Jiang, X.; Wang, X. Cytochrome C-Mediated Apoptosis. *Annu. Rev. Biochem.* **2004**, *73*, 87–106. [CrossRef]
36. Ow, Y.-L.P.; Green, D.R.; Hao, Z.; Mak, T.W. Cytochrome c: Functions beyond Respiration. *Nat. Rev. Mol. Cell Biol.* **2008**, *9*, 532–542. [CrossRef]
37. Tuppy, H.; Kreil, G. Cytochrome c. In *Encyclopedia of Biological Chemistry (Second Edition)*; Lennarz, W.J., Lane, M.D., Eds.; Academic Press: Waltham, MA, USA, 2013; pp. 599–601. ISBN 978-0-12-378631-9.
38. Asadi, M.; Taghizadeh, S.; Kaviani, E.; Vakili, O.; Taheri-Anganeh, M.; Tahamtan, M.; Savardashtaki, A. Caspase-3: Structure, Function, and Biotechnological Aspects. *Biotechnol. Appl. Biochem.* **2021**, *69*, 1633–1645. [CrossRef]
39. Jiang, M.; Qi, L.; Li, L.; Li, Y. The Caspase-3/GSDME Signal Pathway as a Switch between Apoptosis and Pyroptosis in Cancer. *Cell Death Discov.* **2020**, *6*, 112. [CrossRef]
40. Teng, X.; Zhang, W.; Song, Y.; Wang, H.; Ge, M.; Zhang, R. Protective Effects of Ganoderma Lucidum Triterpenoids on Oxidative Stress and Apoptosis in the Spleen of Chickens Induced by Cadmium. *Environ. Sci. Pollut. Res.* **2019**, *26*, 23967–23980. [CrossRef]
41. Cummings, B.S.; Schnellmann, R.G. Cisplatin-Induced Renal Cell Apoptosis: Caspase 3-Dependent and -Independent Pathways. *J. Pharmacol. Exp. Ther.* **2002**, *302*, 8–17. [CrossRef]
42. Somogyi, A.; Rosta, K.; Pusztai, P.; Tulassay, Z.; Nagy, G. Antioxidant Measurements. *Physiol. Meas.* **2007**, *28*, R41–R55. [CrossRef]
43. Mitchell, S.; Vargas, J.; Hoffmann, A. Signaling via the NFκB System. *Wiley Interdiscip. Rev. Syst. Biol. Med.* **2016**, *8*, 227–241. [CrossRef] [PubMed]
44. Nennig, S.E.; Schank, J.R. The Role of NFkB in Drug Addiction: Beyond Inflammation. *Alcohol Alcohol.* **2017**, *52*, 172–179. [CrossRef] [PubMed]
45. Tanaka, T.; Narazaki, M.; Kishimoto, T. IL-6 in Inflammation, Immunity, and Disease. *Cold Spring Harb. Perspect. Biol.* **2014**, *6*, a016295. [CrossRef] [PubMed]
46. Bonventre, J.V. Kidney Injury Molecule-1 (KIM-1): A Urinary Biomarker and Much More. *Nephrol. Dial. Transplant.* **2009**, *24*, 3265–3268. [CrossRef]
47. Han, W.K.; Bailly, V.; Abichandani, R.; Thadhani, R.; Bonventre, J.V. Kidney Injury Molecule-1 (KIM-1): A Novel Biomarker for Human Renal Proximal Tubule Injury. *Kidney Int.* **2002**, *62*, 237–244. [CrossRef]
48. Gleichmann, M.; Mattson, M.P. Neuronal Calcium Homeostasis and Dysregulation. *Antioxid. Redox Signal.* **2011**, *14*, 1261–1273. [CrossRef]
49. Carlström, M.; Wilcox, C.S.; Arendshorst, W.J. Renal Autoregulation in Health and Disease. *Physiol. Rev.* **2015**, *95*, 405–511. [CrossRef]
50. Katoh, M. Multi-Layered Prevention and Treatment of Chronic Inflammation, Organ Fibrosis and Cancer Associated with Canonical WNT/β-Catenin Signaling Activation (Review). *Int. J. Mol. Med.* **2018**, *42*, 713–725. [CrossRef]
51. Valenta, T.; Hausmann, G.; Basler, K. The Many Faces and Functions of β-Catenin. *EMBO J.* **2012**, *31*, 2714–2736. [CrossRef]
52. Silva-García, O.; Valdez-Alarcón, J.J.; Baizabal-Aguirre, V.M. Wnt/β-Catenin Signaling as a Molecular Target by Pathogenic Bacteria. *Front. Immunol.* **2019**, *10*, 2135. [CrossRef]
53. Wang, Z.; Li, Z.; Ji, H. Direct Targeting of β-Catenin in the Wnt Signaling Pathway: Current Progress and Perspectives. *Med. Res. Rev.* **2021**, *41*, 2109–2129. [CrossRef] [PubMed]
54. Guo, Y.; Xiao, L.; Sun, L.; Liu, F. Wnt/Beta-Catenin Signaling: A Promising New Target for Fibrosis Diseases. *Physiol. Res.* **2012**, *61*, 337–346. [CrossRef]

55. Zhang, D.; Pan, Y.; Zhang, C.; Yan, B.; Yu, S.; Wu, D.; Shi, M.; Shi, K.; Cai, X.; Zhou, S.; et al. Wnt/β-Catenin Signaling Induces the Aging of Mesenchymal Stem Cells through Promoting the ROS Production. *Mol. Cell. Biochem.* **2013**, *374*, 13–20. [CrossRef] [PubMed]

56. Zhou, X.; Lin, J.; Yin, Y.; Zhao, J.; Sun, X.; Tang, K. Ganodermataceae: Natural Products and Their Related Pharmacological Functions. *Am. J. Chin. Med.* **2007**, *35*, 559–574. [CrossRef] [PubMed]

57. Sharma, D.; Singh, V.P.; Singh, N.K. A Review on Phytochemistry and Pharmacology of Medicinal as Well as Poisonous Mushrooms. *Mini Rev. Med. Chem.* **2018**, *18*, 1095–1109. [CrossRef]

58. Hsin, I.-L.; Ou, C.-C.; Wu, M.-F.; Jan, M.-S.; Hsiao, Y.-M.; Lin, C.-H.; Ko, J.-L. GMI, an Immunomodulatory Protein from Ganoderma Microsporum, Potentiates Cisplatin-Induced Apoptosis via Autophagy in Lung Cancer Cells. *Mol. Pharm.* **2015**, *12*, 1534–1543. [CrossRef] [PubMed]

59. Zhu, J.; Xu, J.; Jiang, L.-L.; Huang, J.-Q.; Yan, J.-Y.; Chen, Y.-W.; Yang, Q. Improved Antitumor Activity of Cisplatin Combined with Ganoderma Lucidum Polysaccharides in U14 Cervical Carcinoma-Bearing Mice. *Kaohsiung J. Med. Sci.* **2019**, *35*, 222–229. [CrossRef]

60. Jiang, Y.; Chang, Y.; Liu, Y.; Zhang, M.; Luo, H.; Hao, C.; Zeng, P.; Sun, Y.; Wang, H.; Zhang, L. Overview of Ganoderma Sinense Polysaccharide–an Adjunctive Drug Used during Concurrent Chemo/Radiation Therapy for Cancer Treatment in China. *Biomed. Pharmacother.* **2017**, *96*, 865–870. [CrossRef]

Journal of
Fungi

Article

Multigene Phylogeny, Beauvericin Production and Bioactive Potential of *Fusarium* Strains Isolated in India

Shiwali Rana [1,2], Sanjay Kumar Singh [1,2,*] and Laurent Dufossé [3,*]

1   National Fungal Culture Collection of India, Biodiversity and Palaeobiology Group, MACS' Agharkar Research Institute, G.G. Agarkar Road, Pune 411004, India; shiwalirana@aripune.org
2   Faculty of Science, Savitribai Phule Pune University, Ganeshkhind Road, Ganeshkhind, Pune 411007, India
3   Chembiopro Chimie et Biotechnologie des Produits Naturels, ESIROI Département Agroalimentaire, Université de la Réunion, F-97490 Sainte-Clotilde, Ile de La Réunion, France
*   Correspondence: sksingh@aripune.org or singhsksingh@gmail.com (S.K.S.); laurent.dufosse@univ-reunion.fr (L.D.); Tel.: +91-20-2532-5103 (S.K.S.); +33-66-873-1906 (L.D.)

Abstract: The taxonomy of the genus *Fusarium* has been in a *flux* because of ambiguous circumscription of species-level identification based on morphotaxonomic criteria. In this study, multigene phylogeny was conducted to resolve the evolutionary relationships of 88 Indian *Fusarium* isolates based on the internal transcribed spacer region, 28S large subunit, translation elongation factor 1-alpha, RNA polymerase second largest subunit, beta-tubulin and calmodulin gene regions. *Fusarium* species are well known to produce metabolites such as beauvericin (BEA) and enniatins. These identified isolates were subjected to fermentation in *Fusarium*-defined media for BEA production and tested using TLC, HPLC and HRMS. Among 88 isolates studied, 50 were capable of producing BEA, which varied from 0.01 to 15.82 mg/g of biomass. *Fusarium tardicrescens* NFCCI 5201 showed maximum BEA production (15.82 mg/g of biomass). The extract of *F. tardicrescens* NFCCI 5201 showed promising antibacterial activity against *Staphylococcus aureus* MLS16 MTCC 2940 and *Micrococcus luteus* MTCC 2470 with MIC of 62.5 and 15.63 µg/mL, respectively. Similarly, the *F. tardicrescens* NFCCI 5201 extract in potato dextrose agar (40 µg/mL) exhibited antifungal activity in the food poison technique against plant pathogenic and other fungi, *Rhizoctonia solani* NFCCI 4327, *Sclerotium rolfsii* NFCCI 4263, *Geotrichum candidum* NFCCI 3744 and *Pythium* sp. NFCCI 3482, showing % inhibition of 84.31, 49.76, 38.22 and 35.13, respectively. The antibiotic effect was found to synergize when *Fusarium* extract and amphotericin B (20 µg/mL each in potato dextrose agar) were used in combination against *Rhizopus* sp. NFCCI 2108, *Sclerotium rolfsii* NFCCI 4263, *Bipolaris sorokiniana* NFCCI 4690 and *Absidia* sp. NFCCI 2716, showing % inhibition of 50.35, 79.37, 48.07 and 76.72, respectively. The extract also showed satisfactory dose-dependent DPPH radical scavenging activity with an IC_{50} value of 0.675 mg/mL. This study reveals the correct identity of the Indian *Fusarium* isolates based on multigene phylogeny and also throws light on BEA production potential, suggesting their possible applicability in the medicine, agriculture and industry.

Keywords: antimicrobial; fermentation; *Fusarium*; multilocus DNA sequencing; synergistic effect

Citation: Rana, S.; Singh, S.K.; Dufossé, L. Multigene Phylogeny, Beauvericin Production and Bioactive Potential of *Fusarium* Strains Isolated in India. *J. Fungi* 2022, 8, 662. https://doi.org/10.3390/jof8070662

Academic Editor: Ivan M. Dubovskiy

Received: 17 May 2022
Accepted: 21 June 2022
Published: 24 June 2022

Publisher's Note: MDPI stays neutral with regard to jurisdictional claims in published maps and institutional affiliations.

1. Introduction

Members of the genus *Fusarium* are ubiquitous and found in soil, air and on plants [1]. *Fusarium* is an important group of plant-pathogenic fungi which affect a vast diversity of crops in all climatic zones across the globe [2]. Being a genus of great concern, accurate and correct identification of its species becomes a necessity. However, its taxonomy has been in flux due to ambiguous circumscription based on morphotaxonomic criteria. Over the past two decades, many phylogenetic studies have established that morphological species recognition frequently fails to distinguish many *Fusarium* species [3]. Phylogenetic studies have revealed that nuclear ribosomal DNA (nrDNA) is nearly useless for species-level recognition in the case of *Fusarium* and related genera; these markers are useful only in the discrimination of *Fusarium* species complexes [4]. Discordance on using the nuclear rDNA internal transcribed spacer 2 (ITS2) gene region is because of the highly divergent ITS2 rDNA xenologs or paralogs discovered in many species' complexes [5,6]. Many protein-coding genes have been explored for identification and taxonomic purposes in *Fusarium* and related genera. The two primary genes commonly used nowadays for identification that offer high discriminatory power and are well represented in public databases are translation elongation factor 1-alpha (*tef-1α*) and RNA polymerase second largest subunit (*rpb2*). *Tef1-α* is the most common marker, which has very good resolution power for most species [7], and *Rpb2* provides enhanced discrimination between closely related species [8–10]; however, PCR amplification and sequencing success rate is better for *Tef1-α* than for *Rpb2*. In multigene phylogenetic analyses, additional genetic markers are often used with the previously mentioned genes, including *β-tub*, *CaM* and *Rpb1*; these markers vary in resolution depending on species complexes [11].

Fusarium spp. are known to produce several secondary metabolites including beauvericin (BEA) and enniatins. BEA is a cyclic hexadepsipeptide that consists of alternating D-2-hydroxyisovaleric acid and N-methylphenylalanine [12]. BEA is a potent bioactive compound that exhibits very effective anticancer, cytotoxic, antiplatelet aggregation, antimicrobial, leishmanicidal and insecticidal activities [13]. These activities are primarily due to its ionophoric properties that generally disrupt the normal physiological concentration of cations across the plasma membrane [14–16]. It releases Ca^{2+} from internal Ca^{2+} stores of the endoplasmic reticulum [17].

Studies prove that BEA can exhibit promising antibacterial activity against grampositive or gram-negative bacteria, for example, *Agrobacterium tumefaciens*, *Bacillus cereus*, *B. pumilus*, *B. sphaericus*, *B. subtilis*, *B. mycoides*, *Bifidobacterium adolescentis*, *Clostridium perfringens*, *Escherichia coli*, *Eubacterium biforme*, *Mycobacterium tuberculosis*, *Paenibacillus azotofixans*, *P. alvei*, *P. pulvifaciens*, *P. validus P. macquariensis*, *Peptostreptococcus productus*, *P. anaerobius*, *Pseudomonas lachrymans*, *Salmonella typhimurium*, *Staphylococcus haemolyticus*, *Vibrio fisheri* and *Xanthomonas vesicatoria* are inhibited by BEA [18–23].

There has been public concern about certain opportunistic fungi that can cause infections; particularly, cases of *Candida albicans* are greater than before. Patients with compromised immune systems, such as those receiving organ transplants and cancer chemotherapy or those infected by human immunodeficiency virus (HIV), are more prone to such opportunistic infections [24]. BEA alone does not exhibit any antifungal activity; however, it has been found to enhance miconazole activity against *C. albicans* (wild and fluconazole-resistant) [25]. The activity of azole antifungals becomes enhanced against azole-resistant *Candida* isolates by BEA via inhibition of multidrug efflux [26]. The BEA with ketoconazole is known to synergize the antifungal effect against *Candida albicans* and *C. tropicalis*, signifying BEA's possible use as a co-drug for antifungal infections [23,27,28]. A promising strategy using BEA, which involves simultaneous targeting drug resistance and morphogenesis, can boost antifungal efficacy against human fungal pathogens to combat life-threatening fungal infections.

In this study, 88 isolates of *Fusarium* and related genera from various substrates were identified by multigene phylogeny based on the internal transcribed spacer region, translation elongation factor 1-alpha, 28S large subunit, RNA polymerase second largest subunit, beta-tubulin and calmodulin gene regions. Further, these isolates were screened for beauvericin production. The isolate which showed maximum BEA production was subjected to large-scale fermentation, and its extract was used to study antibacterial, antifungal and antioxidant potential.

2. Materials and Methods

2.1. Isolates and Fungarium Specimens

First, 64 fungal strains identified as *Fusarium* based on morphology were procured from the National Fungal Culture Collection of India (NFCCI-WDCM 932) Table 1. The bacterial isolates used in this study, *Escherichia coli* MTCC 739, *Bacillus subtilis* MTCC 121,

Micrococcus luteus MTCC 2470, *Raoultella planticola* MTCC 530, *Pseudomonas aeruginosa* MTCC 2453, *Staphylococcus aureus* MLS16 MTCC 2940 and *Staphylococcus aureus* MTCC 96, were procured from Microbial Type Culture Collection (MTCC). Additionally, fungal pathogens *Pythium* sp. NFCCI 3482, *Geotrichum candidum* NFCCI 3744, *Rhizoctonia solani* NFCCI 4327, *Rhizopus* sp. NFCCI 2108, *Sclerotium rolfsii* NFCCI 4263, *Bipolaris sorokiniana* NFCCI 4690 and *Absidia* sp. NFCCI 2716 were also procured from the National Fungal Culture Collection of India (NFCCI) to study the antifungal activity.

Table 1. Sixty-four *Fusarium* isolates procured from the National Fungal Culture Collection of India (NFCCI), along with other details such as host, place of collection and date of isolation, are listed.

NFCCI No.	Host	Place of Collection	Date of Isolation
NFCCI 680	*Azadirachta indica* (Endophyte)	Maharashtra	25 September 2006
NFCCI 681	*Azadirachta indica* (Endophyte)	Maharashtra	20 December 2006
NFCCI 1127	*Jatropha curcus* (Phyloplane)	Durg, Chhattisgarh	23 May 2007
NFCCI 1788	Soil	Imphal, Manipur	23 June 2009
NFCCI 1895	Soil	Mahabaleshwar, Maharashtra	25 August 2009
NFCCI 2053	Wilted Guava plant (Root)	Rajasthan	9 February 2010
NFCCI 2150	Soil	Arctic Tundra	25 June 2010
NFCCI 2315	Potato	Goa	15 June 2009
NFCCI 2460	*Zanthooxylum armatum*	Udhampur, Jammu and Kashmir	13 June 2011
NFCCI 2467	Bottle Gourd	Dapoli, Maharashtra	16 June 2011
NFCCI 2470	Coconut leaves	Vellayani, Thiruvananthapuram, Kerala	22 June 2011
NFCCI 2491	Soil	Bangalore, Karnataka	8 July 2012
NFCCI 2696	*Coleus* sp.	Puthenthope, Kerala	19 March 2012
NFCCI 2871	Wilted tomato plant (Root)	Manipur	25 September 2012
NFCCI 2872	Wilted tomato plant (Root)	Manipur	25 September 2012
NFCCI 2885	*Azadirachta indica* (Endophyte)	Chandravati, Rajasthan	9 April 2012
NFCCI 2886	*Azadirachta indica* (Endophyte)	Chikkamagaluru, Karnataka	15 July 2012
NFCCI 2904	*Aegle marmelos*	Karnataka	28 November 2009
NFCCI 2945	Soil	Tiruchirappalli, Tamil Nadu	7 January 2012
NFCCI 2946	*Aegle marmelos*	Karnataka	25 November 2010
NFCCI 2949	Cotton field	Shirpur, Maharashtra	9 October 2012
NFCCI 2953	Cotton field	Shirpur, Maharashtra	6 August 2015
NFCCI 2956	Cotton field	Shirpur, Maharashtra	19 October 2012
NFCCI 2959	Pigeon pea (Root)	Amravati, Maharashtra	30 October 2012
NFCCI 2960	Pigeon pea (Root)	Amravati, Maharashtra	30 October 2012
NFCCI 2961	Pigeon pea (Root)	Beed, Maharashtra	26 October 2012
NFCCI 2962	Pigeon pea (Root)	Kalamb, Osmanabad, Maharashtra	26 October 2011
NFCCI 2964	Pigeon pea (Root)	Latur, Maharashtra	30 August 2012
NFCCI 2972	Pigeon pea (Root)	Ahmednagar, Maharashtra	30 October 2012
NFCCI 3020	Soil	Karnataka	24 December 2010
NFCCI 3031	Piper betle (Leaf-endophyte)	Tamil Nadu	26 December 2011
NFCCI 3038	Wilted cumin plant	Gujarat	24 February 2013
NFCCI 3044	Wilted cumin plant	Gujarat	24 February 2013
NFCCI 3048	Wilted cumin plant	Gujarat	24 February 2013
NFCCI 3049	Wilted cumin plant	Gujarat	24 February 2013
NFCCI 3051	Wilted cumin plant	Gujarat	24 February 2013
NFCCI 3065	*Ocimum sanctum*	Maharashtra	20 August 2012
NFCCI 3072	*Barleria prionitis* (Stem)	Maharashtra	24 September 2012
NFCCI 3074	*Barleria prionitis*	Maharashtra	20 September 2012

Table 1. *Cont.*

NFCCI No.	Host	Place of Collection	Date of Isolation
NFCCI 3091	Textile sludge	Tamil Nadu	14 September 2012
NFCCI 3092	Textile sludge	Tamil Nadu	8 September 2012
NFCCI 3093	Textile sludge	Tamil Nadu	15 September 2012
NFCCI 3147	Rotten sugarcane	Dapoli, Maharashtra	5 January 2013
NFCCI 3239	Marigold (Seeds)	Mumbai, Maharashtra	8 February 2013
NFCCI 3243	Cow pea	Thiruvananthapuram, Kerala	20 June 2012
NFCCI 3264	Cow pea	Thiruvananthapuram, Kerala	12 September 2012
NFCCI 3270	Dead wood	Mumbai, Maharashtra	29 March 2013
NFCCI 3282	Soil	Amarkantak, Madhya Pradesh	26 September 2012
NFCCI 3300	*Cathranthus roseus*	Raipur, Chhattisgarh	12 August 2012
NFCCI 3475	*Ocimum sanctum* (Endophyte)	Amaravati, Maharashtra	21 July 2013
NFCCI 3703	Onion	Nanded, Maharashtra	14 September 2014
NFCCI 3706	*Castor* sp. (Root)	Aanand, Gujarat	20 December 2014
NFCCI 4095	Soil	Pune, Maharashtra	2 April 2014
NFCCI 4180	Soil	Chandigarh	19 June 2018
NFCCI 4759	*Anaphalis contorta* (Root-endophyte)	Imphal, Manipur	18 February 2019
NFCCI 4792	Chickpea (Root)	Pratapgarh, Uttar Pradesh	21 January 2018
NFCCI 4830	Cotton (Root)	Madurai, Tamil Nadu	5 September 2018
NFCCI 4859	*Dillenia indica*	Arunachal Pradesh	15 October 2016
NFCCI 4885	Cucumber (Twig and leaf)	Jobner, Jaipur, Rajasthan	3 May 2020
NFCCI 4888	Cucumber (Twig)	Jobner, Jaipur, Rajasthan	2 June 2020
NFCCI 4889	Unknown insect	Vishakhapatnum Andhra Pradesh	30 October 2020
NFCCI 4919	Soil	Pathanamthitta, Kerala	15 September 2020
NFCCI 4963	Plant litter	Raiganj, Uttar dinajpur, West Bengal	25 March 2019
NFCCI 4972	Sugarcane bagasse	Dharur, Beed, Maharashtra	22 December 2020

2.2. Collection, Isolation and Morphological Identification

During the course of this study, *Fusarium* and other related fungi were isolated from various substrates collected from different regions of India. Table 2 enlists the 24 isolates which were isolated afresh in this study along with details of their host, place of collection, date of collection and date of isolation. Pure cultures of all 24 isolates were raised using a single spore isolation technique.

Colony characteristics of all *Fusarium* isolates were studied on multiple media, such as potato dextrose agar (PDA), malt extract agar (MEA), potato carrot agar (PCA), oatmeal agar (OMA), cornmeal agar (CMA) and synthetic nutrient agar (SNA). Microscopic structures of these isolates were observed using lactophenol-cotton blue under a Carl Zeiss Image Analyzer 2 (Germany) microscope. Measurements and photomicrographs of the fungal structures were recorded using Axiovision Rel 4.8 software and Digi-Cam with an attached Carl Zeiss Image Analyzer 2 microscope (Figures 1 and 2). The pure cultures are deposited and accessioned in the National Fungal Culture Collection of India (NFCCI 5189-5212) (Table 2).

Table 2. Twenty-four isolates isolated in this study along with their accession nos., host, place of collection, date of collection and date of isolation details.

NFCCI No.	Host	Place of Collection	Date of Collection	Date of Isolation
NFCCI 5189	Soil	Haridwar, Uttarakhand	02 July 2018	12 July 2018
NFCCI 5190	Ginger	Pune, Maharashtra	27 August 2017	8 September 2017
NFCCI 5191	Pomegranate	Pune, Maharashtra	07 September 2019	15 September 2019
NFCCI 5192	*Zingiber officinale*	Aurangabad, Maharashtra	9 March 2017	10 March 2017
NFCCI 5193	Soil	Jalgaon, Maharashtra	10 October 2017	14 October 2017
NFCCI 5194	Forest litter	Mahabaleshwar, Maharashtra	12 October 2017	14 October 2017
NFCCI 5195	Soil	Chandigarh	1 April 2019	19 April 2019
NFCCI 5196	Wilted sugarcane plant	Gorakhpur, Uttar Pradesh	2 July 2021	17 July 2021
NFCCI 5197	*Chrysanthemum roseum*	Mahabaleshwar, Maharashtra	25 September 2018	27 September 2018
NFCCI 5198	Soil	Pune, Maharashtra	5 March 2020	16 March 2020
NFCCI 5199	Soil	Jalgaon, Maharashtra	8 October 2017	14 October 2017
NFCCI 5200	Pea	Pune, Maharashtra	15 April 2019	21 April 2019
NFCCI 5201	Dead bark	Simbal, Himachal Pradesh	14 December 2017	25 December 2017
NFCCI 5202	Wilted Cashew plant	Pilcode, Kerala	6 August 2019	20 August 2019
NFCCI 5203	Rotten mushroom	Simbal, Himachal Pradesh	19 May 2019	25 May 2019
NFCCI 5204	Potato	Pune, Maharashtra	28 December 2018	12 January 2019
NFCCI 5205	*Aloe vera* (Leaf)	Pune, Maharashtra	5 October 2019	16 October 2019
NFCCI 5206	Soil	Vadodara, Gujarat	3 October 2019	15 October 2019
NFCCI 5207	Coconut	Goa	15 October 2018	29 October 2018
NFCCI 5208	*Zingiber officinale*	Aurangabad, Maharashtra	5 October 2018	12 March 2018
NFCCI 5209	Carrot rhizosphere	Pithoragarh, Uttarakhand	3 October 2019	7 October 2019
NFCCI 5210	Chilli (Fruit)	Simbal, Himachal Pradesh	5 February 2019	12 February 2019
NFCCI 5211	Damping-off Onion	Udaipur, Rajasthan	19 September 2019	27 September 2019
NFCCI 5212	Grape (Leaf)	Nashik, Maharashtra	16 March 2020	17 April 2020

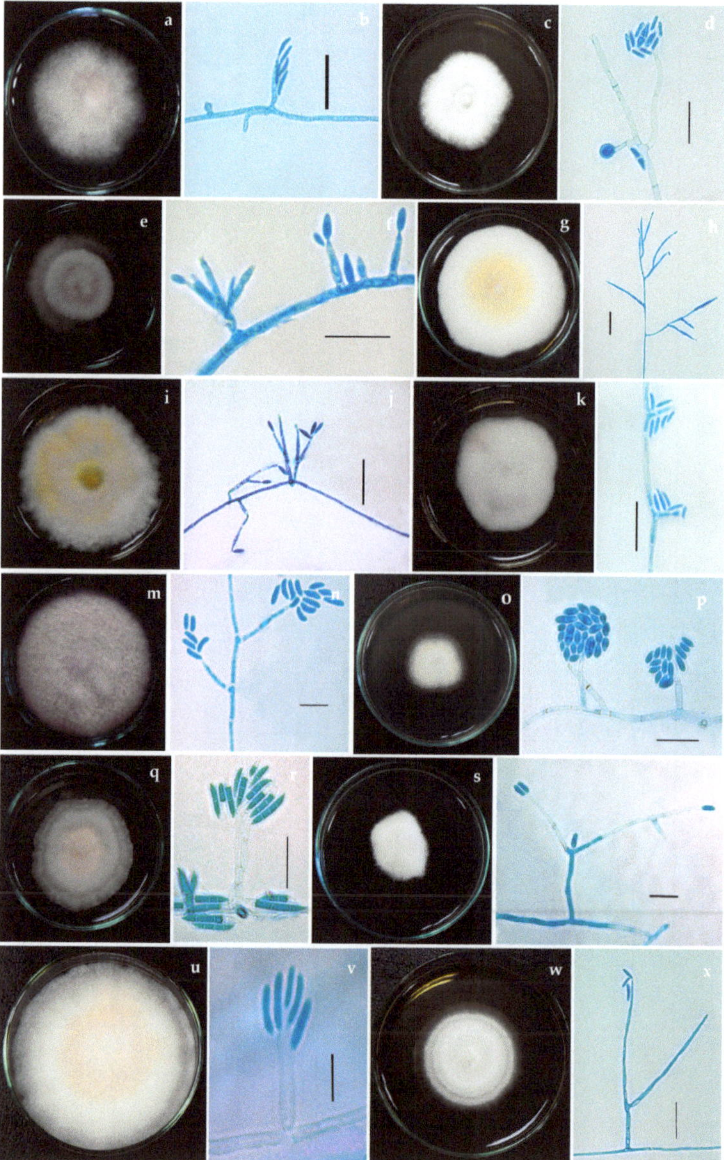

Figure 1. *Fusarium* isolates with phialides bearing conidia (**a,b**) *Fusarium nirenbergiae* NFCCI 5189 (**c,d**) *Neocosmospora solani* NFCCI 2315 (**e,f**) *F. mangiferae* NFCCI 2885 (**g,h**) *F. annulatum* NFCCI 2962 (**i,j**) *F. microconidium* NFCCI 3020 (**k,l**) *F. sacchari* NFCCI 3147 (**m,n**) *N. oblonga* NFCCI 2150 (**o,p**) *Albonectria rigidiuscula* NFCCI 5202 (**q,r**) *F. irregulare* NFCCI 2460 (**s,t**) *Neocosmospora suttoniana* NFCCI 2961 (**u,v**) *F. annulatum* NFCCI 2964 (**w,x**) *Neocosmospora vasinfecta* NFCCI 2972. Scale bars: b, d, f, h, j, l, p, r, t, v, x = 20 µm; n = 10 µm.

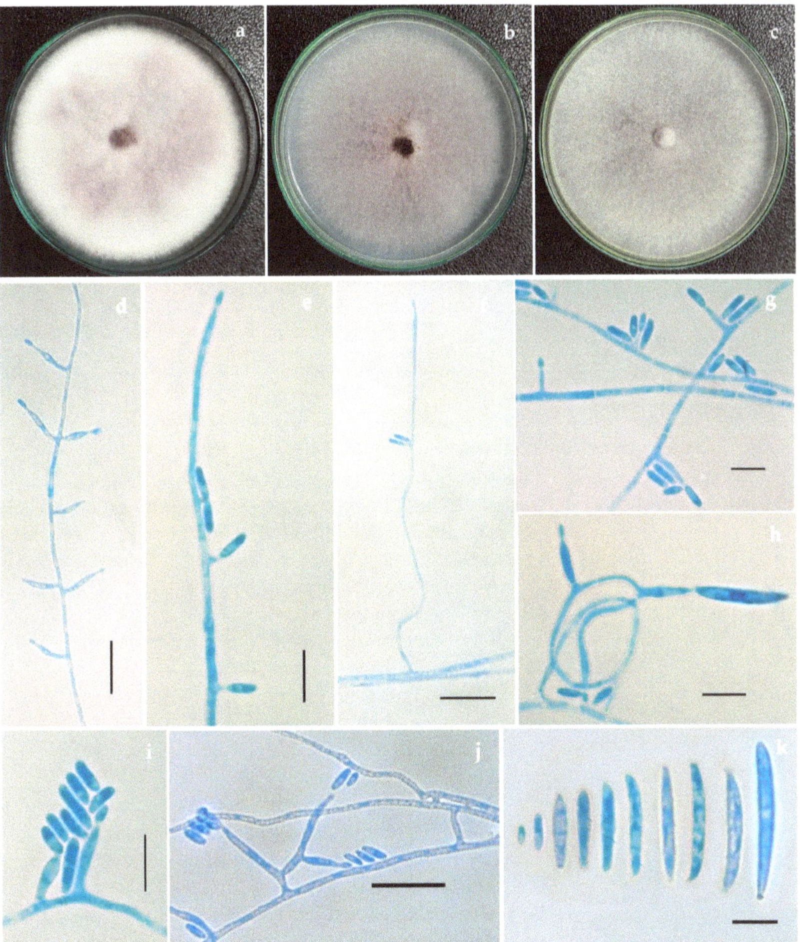

Figure 2. *Fusarium tardicrescens* (NFCCI 5201) culture grown on (**a**) Potato dextrose agar (**b**) Malt extract agar (**c**) Cornmeal agar (**d–j**) Phialides arising from mycelial hyphae producing micro and macroconidia (**k**) Micro and macroconidia. Scale bars: d, f, j = 20 µm; e, g, h, i, k = 10 µm.

2.3. DNA Extraction, PCR Amplification and DNA Sequencing

Genomic DNA was isolated from pure colonies of all 88 isolates growing on Potato Dextrose Agar (PDA) petri-plates, incubated for a week's time using a simple and rapid DNA extraction protocol using FastPrep®24 tissue homogenizer (MP Biomedicals GmbH, Eschwege, Germany) [29]. Partial gene sequences of isolates were determined for six gene markers, i.e., ITS, *tef1-α*, LSU, *rpb2*, *β-tub* and *CaM*. The primer sets used to amplify particular gene region are summarized in Table 3.

Table 3. PCR primer sets used in this study for amplification as well as sequencing.

Name of the Gene Region	Primer	Direction	Sequence (5′-3′)	Reference
Internal transcribed spacer	ITS-5	Forward	GGAAGTAAAAGTCGTAACAAGG	[30]
region of the nrDNA (ITS)	ITS-4	Reverse	TCCTCCGCTTATTGATATGC	[30]
Translation elongation factor	EF-1	Forward	ATGGGTAAGGARGACAAGAC	[31]
1-alpha (*tef1-α*)	EF-2	Reverse	GGARGTACCAGTSATCATG	[31]
28S large subunit of the	LR-OR	Forward	ACCCGCTGAACTTAAGC	[32]
nrDNA (LSU)	LR-7	Reverse	TACTACCACCAAGATCT	[33]
RNA polymerase second largest	fRPB2-5f	Forward	GAYGAYMGWGATCAYTTYGG	[34]
subunit (*rpb2*)	fRPB2-7cR	Reverse	CCCATRGCTTGYTTRCCCAT	[34]
Beta-tubulin (*β-tub*)	βtuFFo1	Forward	CAGACCGGTCAGTGCGTAA	[35]
	βtuFRo1	Reverse	TTGGGGTCGAACATCTGCT	[35]
Calmodulin (*CaM*)	CF-1	Forward	GCCGACTCTTTGACYGARGAR	[36]
	CF-4	Reverse	TTTYTGCATCATRAGYTGGAC	[36]

PCR was carried out in a 25 μL reaction using 12.5 μL 2X Invitrogen Platinum SuperFi PCR Mastermix, 2 μL template DNA (10–20 ng), 1.5 μL 10 pmol primer, 5 μL 5X GC enhancer and H_2O (Sterile Ultra-Pure Water, Sigma, St. Louis, MO, USA), with the volume made to 25 μL. The conditions of the thermo-cycling involved:

An initial denaturation at 94 °C for 5 min, 35 cycles of 1 min at 94 °C, 30 s at 52 °C, 1 min at 72 °C and a final extension at 72 °C for 8 min for ITS gene region;

An initial denaturation of 5 min at 94 °C, 30 cycles of 45 s at 94 °C, 30 s at 57 °C and 1 min at 72 °C followed by a final 7 min extension at 72 °C for *Tef1-α*;

5 min denaturation at 94 °C, 35 cycles of 1 min at 94 °C, 50 s at 52 °C and 1.2 min at 72 °C with a final 8 min extension at 72 °C for LSU;

5 min denaturation at 95 °C, 35 cycles of 45 s at 95 °C, 1 min at 52 °C and 1.5 min at 72 °C with a final 10 min extension at 72 °C for *Rpb2*;

2 min denaturation at 94 °C, 40 cycles of 35 s at 94 °C, 55 s at 52 °C and 2 min at 72 °C with a final 10 min extension at 72 °C for *β-tub*;

An initial denaturation of 5 min at 94 °C, 30 cycles of 1 min at 94 °C, 30 s at 55 °C and 2 min at 72 °C followed by a final 7 min extension at 72 °C for *CaM*.

The PCR amplicons were purified with a FavorPrep™ PCR purification kit as per the manufacturer's instructions. Purified PCR products of all marker genes were checked on 1.2% agarose electrophoresis gels stained with ethidium bromide and were further subjected to a sequencing PCR reaction using a BigDye®Terminator v3.1 Cycle Sequencing Kit, as per manufacturer's instructions.

The sequencing PCR reaction of 20 μL included 4 μL of 5× sequencing buffer, 2 μL of BigDye™ Terminator premix, 4 μL of primer (5 pmol), 4 μL of the purified amplicon and H_2O (Sterile Ultra-Pure Water, Sigma), with the volume made to 20 μL. Thermal cycling conditions consisted of an initial denaturing at 96 °C for 3 min, followed by 30 cycles of 94 °C for 10 s, 50 °C for 40 s and 60 °C for 4 min. The BigDye® terminators and salts were removed from using The BigDye Xterminator® Purification Kit (Thermo Fisher Scientific, Waltham, MA, USA) as per the manufacturer's instructions. The purified sequencing products were transferred into a 96-well microplate. The sequence was elucidated on Applied Biosystems SeqStudio Genetic Analyzer (Applied Biosystems, Foster City, CA, USA). Sequences obtained were submitted in the NCBI GenBank (accession numbers in Table S1).

2.4. Phylogenetic Analysis

To determine the phylogenetic status of all 88 isolates, the Internal transcribed spacer region of the nrDNA (ITS), translation elongation factor 1-alpha (*tef1-α*), 28S large subunit of the nrDNA (LSU), RNA polymerase second largest subunit (*rpb2*), beta-tubulin (*β-tub*) and calmodulin (*CaM*) loci were used to compare the isolates used in this study with already known authentic strains in the genus *Fusarium*. The sequences of the type/reference strains were retrieved from NCBI. A total of 186 isolates of *Fusarium* and related genera were used in the phylogenetic analysis (Table S1). *Atractium stilbaster* CBS 410.67 was selected as the outgroup taxon. The chosen strains used in the construction of the phylogenetic tree, along with their accession numbers and other related details, are listed in Table S1. Each gene region was individually aligned using MAFFT v. 6.864b [37]. The alignments were checked and adjusted manually using Aliview [38]. Further, alignments were concatenated and subjected to phylogenetic analyses. The best substitution model out of 286 DNA models was figured using ModelFinder [39]. Additionally, windows version IQ-tree tool v.1.6.11 [40] was used to construct the phylogenetic tree. The reliability of the tree branches was assessed and tested based on 1000 ultrafast bootstrap (UFBoot) support replicates and a SH-like approximate likelihood ratio test (SH-like aLRT) with 1000 replicates. The constructed phylogenetic tree was visualized in FigTree v.1.4.4 (http://tree.bio.ed.ac.uk/software/figtree/, accessed on 20 February 2022).

2.5. Fermentation for Beauvericin Production

Fusarium isolates were subcultured onto PDA and incubated at 25 °C for a week. Once the cultures were mature enough, spore suspensions were prepared separately for all isolates. Then, 10 µL of Tween 20 was added to 10 mL of normal saline (0.89%) and mixed adequately, saline solution was added to the culture and the spores were harvested carefully with the help of a sterile cotton swab and vortexed. The spore suspension's optical density (OD) was adjusted between 0.08 and 0.1 at 530 nm [41]. Next, 1 mL of this spore suspension was added to 100 mL of *Fusarium* defined media (FDM) (Sucrose 25 g, Sodium nitrate ($NaNO_3$) 4.25 g, Sodium chloride (NaCl) 5 g, Magnesium sulfate heptahydrate ($MgSO_4 \cdot 7H_2O$) 2.5 g, Potassium dihydrogen phosphate (KH_2PO_4) 1.36 g, Ferrous sulphate heptahydrate ($FeSO_4 \cdot 7H_2O$) 0.01 g, Zinc sulfate heptahydrate ($ZnSO_4 \cdot 7H_2O$) 0.0029 g/L; pH 5.5) [42] in a 250 mL Erlenmeyer flask and was incubated in a shaking incubator at 25 °C at 150 rpm for 7 days (Figure S1).

2.6. Extraction of Beauvericin

Beauvericin was extracted from the fermented broth as per the protocol described earlier, with slight modification [43]. The fermented broth was filtered using Whatman filter paper no. 4. Biomass was collected, washed twice with distilled water and was allowed to dry at 50 °C. Dried biomass was extracted overnight in 25 mL of acetonitrile:water 90:10 (*v/v*). Further, the mixture was sonicated twice for 15 min. The mixture was filtered; filtrate was extracted with 25 mL of heptane and separated using a separating funnel. The bottom layer was evaporated to dryness in a rotary evaporator (Heidolph, Schwabach, Germany). The dry residue was dissolved in 25 mL of methanol:water, 60:40 (*v/v*) and extracted twice with 25 mL of dichloromethane. The dichloromethane phase containing beauvericin (BEA) was collected and evaporated to dryness in a rotary evaporator. The evaporated extract containing beauvericin was dissolved in 1 mL of Acetonitrile.

2.7. Detection of Beauvericin

The presence of beauvericin was initially qualitatively analyzed by thin-layer chromatography (TLC) and quantitatively by high-performance liquid chromatography (HPLC). In addition, high-resolution mass spectrometry (HRMS) was carried out for a few samples for further confirmation.

2.7.1. Thin-Layer Chromatography (TLC)

About 5 to 10 µL of the extracts were spotted onto TLC Silica gel plates (TLC Silica gel 60 F254; Merck life science Pvt. Ltd., Bengaluru, Karnataka, India) along with standard Beauvericin (BEA) (500 mg/mL) (Sigma) with the help of glass capillary tube. TLC plates were developed in petroleum ether:ethyl acetate (1:1) as a solvent system. The plates were air-dried, and consequently, the spots formed on the TLC plates were detected using iodine vapours. Retention factor (Rf) values for the standard and extracts were measured as described earlier [44,45]. The extracts were further subjected to high-performance liquid chromatography (HPLC).

2.7.2. High-Performance Liquid Chromatography (HPLC)

The amount of BEA in the extracts was determined by HPLC (Waters Corporation, Milford, MA, USA) using a C18 column and acetonitrile: H_2O (85:15 v/v) as the mobile phase at a flow rate of 1 mL/min under isocratic conditions and U.V. detection at 210 nm [46]. A stock solution of standard beauvericin with a concentration of 1 mg/mL was prepared in acetonitrile and was further diluted to 15.6 µg/mL, 31.25 µg/mL, 62.5 µg/mL, 125 µg/mL, 250 µg/mL and 500 µg/mL. The retention time for BEA was found to be 9.1 min. The injection volume was 20 µL. The peak area was calibrated to the amount of BEA standard (Sigma). The amount of BEA produced was calculated with respect to the area under the peak.

2.7.3. High-Resolution Mass Spectrometry (HRMS)

The mass of beauvericin was determined in selected extracts; high-resolution product ion spectra were acquired using ESI-UHR-Q-TOF (Bruker Impact II Ultra-High Resolution-TOF) to confirm the presence of Beauvericin. The system ionization source type was ESI with positive ion polarity. The instrument was run in the full-scan mode (m/z 150–1200). Nitrogen was used for drying (200 °C; 7 L/min), and the nebulizer gas pressure was 1.7 Bar. Other typical operating parameters were set as capillary voltage—4500 V, end plate offset—500 V, and charging voltage—2000 V.

2.8. Large Scale Fermentation and Extraction of Beauvericin from Fusarium tardicrescens NFCCI 5201

Among the 88 tested isolates, the highest beauvericin producer, *F. tardicrescens* NFCCI 5201, was subjected to flask scale fermentation in a total of 3.5 L of *Fusarium* defined medium (FDM). Each flask was inoculated with 1% spore suspension of O.D. 0.08–0.1, as described earlier in Section 2.5, and incubated at 25 °C with 150 rpm for one week. Extraction of beauvericin was carried out as described in Section 2.6. The final dry residue was dissolved in 36 mL of acetonitrile. This extract was used for further study.

2.9. Determination of the Antibacterial Activity

The *Fusarium tardicrescens* NFCCI 5201 extract containing BEA dissolved in acetonitrile (500 µg/mL) was used to check antimicrobial activity. The antibacterial activity was tested against a panel of test organisms, including *Escherichia coli* MTCC 739, *Bacillus subtilis* MTCC 121, *Micrococcus luteus* MTCC 2470, *Raoultella planticola* MTCC 530, *Pseudomonas aeruginosa* MTCC 2453, *Staphylococcus aureus* MLS16 MTCC 2940 and *Staphylococcus aureus* MTCC 96. Cultures were grown on nutrient agar plates and incubated at 37 °C for 48 h. After incubation, one to two pure colonies of each culture were transferred to 10 mL Muller hinton broth (MHB) and incubated for 4–5 h at 37 °C. Further, optical density was adjusted to 1 using broth. The cultures were diluted to a cell density of 1×10^6 cells/mL. The test was carried out using a resazurin-based turbidometric assay in a 96-well microtiter plate. All wells of rows (A–G) were filled with 100 µL Muller hinton broth. The first well of each row was filled with 100 µL of *Fusarium tardicrescens* NFCCI 5201 extract containing BEA (500 µg/mL), which was mixed well, and then 100 µL of the mixture from the first well was transferred to the second well of the vertical row. This serial dilution was continued until the tenth well; lastly, 100 µL of the mixture was discarded from the tenth well. The final concentration of the extract was half of the original concentration in each well. Ten µL of each bacterial suspension was added in vertical rows individually, except for the twelfth well (sterility control). One hundred µL of antibiotic ampicillin (10 µg/50 µL) was used, except *Pseudomonas aeruginosa* MTCC 2453, in which neomycin (30 µg/50 µL) was added in the eleventh well of each row (positive control). After 48 hrs of incubation at 37 °C, 25 µL of resazurin (0.01%) was added to all wells and incubated at 37 °C for 75 min in the dark. Color changes were observed and recorded. The lowest concentration before the color change was considered as the minimum inhibitory concentration for that particular test organism [47,48].

2.10. Determination of the Individual and Combined Effect of Fusarium tardicrescens NFCCI 5201 Extract Containing Beauvericin and Amphotericin B on Pathogenic Fungi of Agricultural Importance

The antifungal potential of the individual and combined effect of *F. tardicrescens* NFCCI 5201 extract containing beauvericin and amphotericin B on agriculturally important pathogenic fungi was carried out using the poisoned food technique as described earlier [49]. Fungal pathogens *Pythium* sp. NFCCI 3482, *Geotrichum candidum* NFCCI 3744, *Rhizoctonia solani* NFCCI 4327, *Rhizopus* sp. NFCCI 2108, *Sclerotium rolfsii* NFCCI 4263, *Bipolaris sorokiniana* NFCCI 4690 and *Absidia* sp. NFCCI 2716 were selected for the antifungal study. These pathogenic fungi were grown on PDA and incubated at 25 ± 2 °C for 5–7 days.

Agar discs with mycelia (5 mm in diameter) were cut out from the actively growing regions of the 5–7 days old pure cultures using a sterile cork borer and aseptically inoculated at the center of four different Petri plates, as follows:

1. Control plate containing PDA;
2. PDA supplemented with amphotericin B (40 µg/mL);
3. PDA supplemented with *F. tardicrescens* NFCCI 5201 extract containing beauvericin (40 µg/mL);
4. PDA supplemented with amphotericin B (20 µg/mL) and *F. tardicrescens* NFCCI 5201 extract containing beauvericin (20 µg/mL).

The growth of fungal colonies on different media was measured between 3 and 7 days depending upon the growth of fungi. Each treatment was repeated thrice, and the plates were incubated at 25 °C. The diameter of the fungal colony was measured after 3 and 7 days of incubation. The percentage inhibition of the mycelial growth of the test fungi was calculated using the formula described earlier [50]. Inhibition of mycelial growth (%) = $(dc - dt)/dc \times 100$, where dc is the mean diameter (in mm) of the colony in the control sample and dt is the diameter (in mm) of the colony in the treatment.

2.11. Antioxidant Activity of Extract of Fusarium tardicrescens NFCCI 5201 Containing Beauvericin

The extract was tested for in vitro antioxidant activity using the 2,2-diphenyl-1-picrylhydrazyl (DPPH) radical scavenging method.

The concentration of the extract as well as standard solutions (control) glucose (Himedia) (negative) and ascorbic acid (Sigma) (positive) used were 1, 0.9, 0.8, 0.7, 0.6, 0.5, 0.4, 0.3, 0.2 and 0.1 mg/mL in acetonitrile. The extract or standard solution (10 µL) was added to DPPH in acetonitrile solution (200 µL, 100 µM) in a 96-well microtiter plate (Tarsons Products (P) Ltd., Kolkata, India). After incubation at 37 °C for 30 min, the absorbance of each solution was determined at 517 nm using a Synergy HT Multi-detection microplate reader (BioTek, Winooski, VT, USA).

The radical scavenging activity was calculated by the following formula:

Radical scavenging activity (%) = (OD control − OD sample)/OD control × 100 [51,52]. Three replicates were maintained for each treatment.

3. Results

3.1. Phylogenetic Analysis

The nrDNA (ITS), translation elongation factor 1-alpha (*Tef1-α*), 28S large subunit of the nrDNA (LSU), RNA polymerase second largest subunit (*Rpb2*), beta-tubulin (*β-tub*) and calmodulin (*CaM*) loci sequence alignments were together used to confirm the resolution of the isolates used in this study. The concatenated file contained sequence data of 186 taxa. Alignment contained 6944 columns, 3815 distinct patterns, 2049 parsimony-informative, 950 singleton sites and 3944 constant sites. TIM2e + R4 was found to be the best-fit model of 286 models tested and was chosen based on the Bayesian Information Criterion (BIC). The phylogeny was inferred using the Maximum Likelihood Method based on the model mentioned above. The log-likelihood of the consensus tree was –59368.419. Rate parameters: A-C: 1.27988, A-G: 2.69656, A-T: 1.27988, C-G: 1.00000, C-T: 5.18110, G-T: 1.00000; Base frequencies: A: 0.250, C: 0.250, G: 0.250, T: 0.250; Site proportion and rates: (0.587, 0.115) (0.171, 1.101) (0.200, 2.369) (0.042, 6.470) (Figure 3).

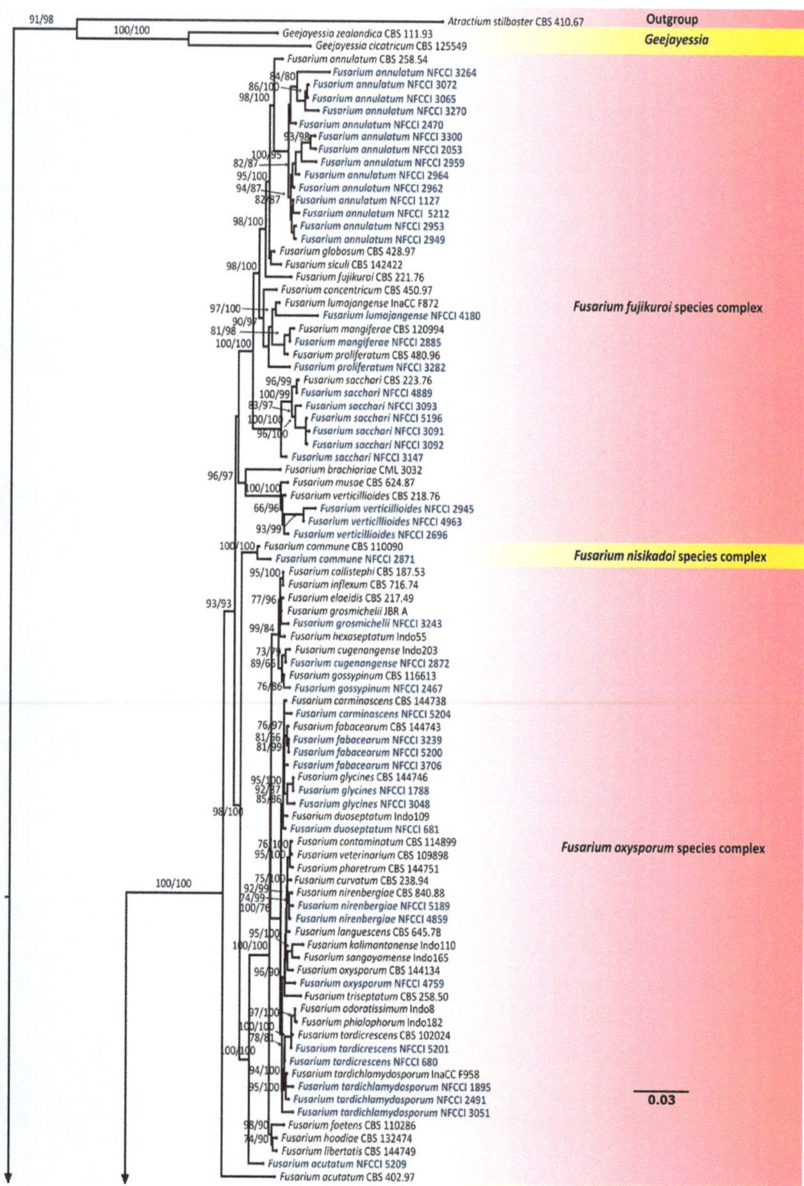

Figure 3. *Cont.*

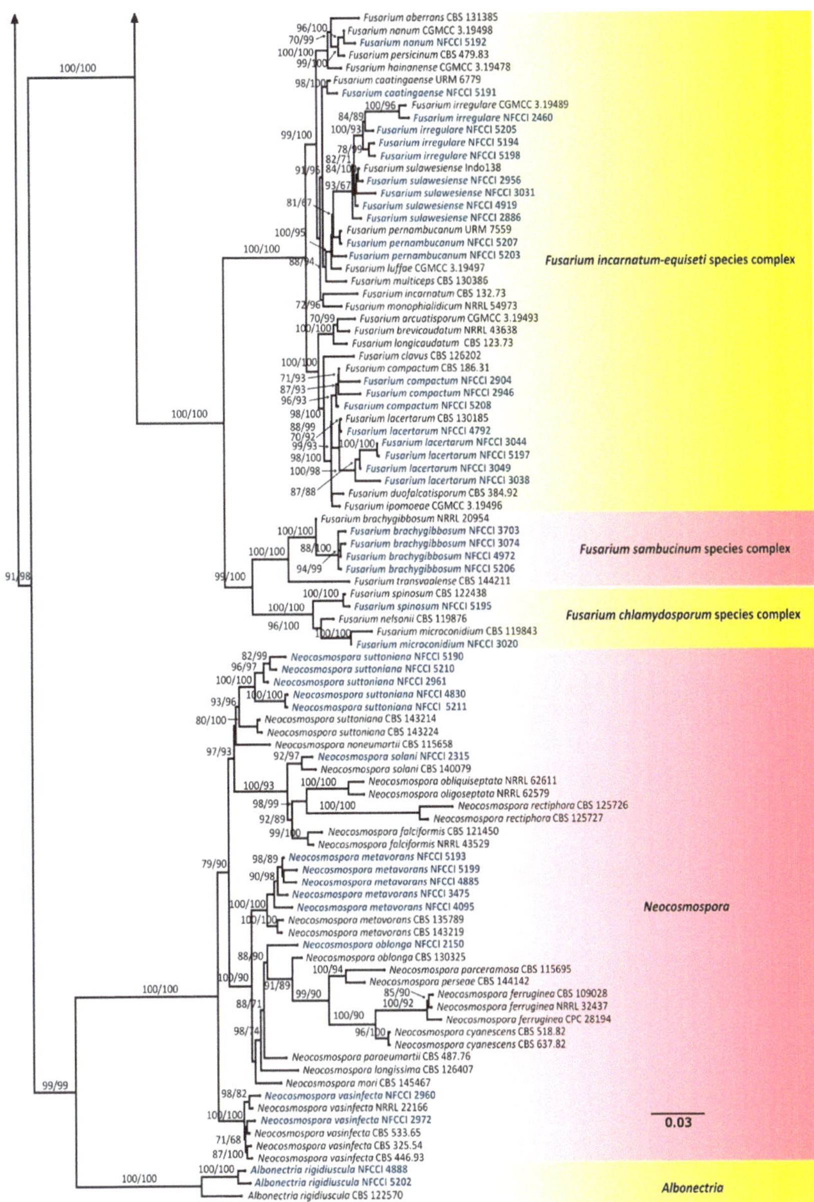

Figure 3. Maximum-likelihood (IQ-TREE-ML) consensus tree inferred from the combined ITS, LSU, *Rpb2*, *β-tub*, *CaM* and *Tef-1α* multiple sequence alignment of genus *Fusarium* and related genera. Numbers at the branches indicate statistical support values (UFBS and SH-aLRT). The scale bar indicates expected changes per site. The tree is rooted to *Atractium stilbaster* (CBS 410.67). Isolates used in this study are shown in blue bold.

Of the isolates, 88 were found to belong to *Fusarium* and related genera, i.e., *Albonectria* and *Neocosmospora*. Strains used in this study were found to represent 35 species, including *Albonectria rigidiuscula*, *Fusarium acutatum*, *F. annulatum*, *F. brachygibbosum*, *F. caatingaense*,

F. carminascens, *F. commune*, *F. compactum*, *F. cugenangense*, *F. duoseptatum*, *F. fabacearum*, *F. glycines*, *F. gossypinum*, *F. grosmichelii*, *F. irregulare*, *F. lacertarum*, *F. lumajangense*, *F. mangiferae*, *F. microconidium*, *F. nanum*, *F. nirenbergiae*, *F. oxysporum*, *F. pernambucanum*, *F. proliferatum*, *F. sacchari*, *F. spinosum*, *F. sulawesiense*, *F. tardichlamydosporum*, *F. tardicrescens*, *F. verticillioides*, *Neocosmospora metavorans*, *N. oblonga*, *N. solani*, *N. suttoniana* and *N. vasinfecta*.

Fusarium isolates were found to represent six species complexes. The species falling into a particular species complex are included in brackets, viz. *Fusarium sambucinum* species complex [*F. brachygibbosum*], *Fusarium chlamydosporum* species complex [*F. microconidium*, *F. spinosum*], *Fusarium incarnatum-equiseti* species complex [*F. caatingaense*, *F. compactum*, *F. irregulare*, *F. lacertarum*, *F. nanum*, *F. pernambucanum*, *F. sulawesiense*], *Fusarium oxysporum* species complex [*F. carminascens*, *F. cugenangense*, *F. duoseptatum*, *F. fabacearum*, *F. glycines*, *F. gossypinum*, *F. grosmichelii*, *F. nirenbergiae*, *F. oxysporum*, *F. tardichlamydosporum*, *F. tardicrescens*], *Fusarium nisikadoi* species complex [*F. commune*] and *Fusarium fujikuroi* species complex [*F. acutatum*, *F. annulatum*, *F. lumajangense*, *F. mangiferae*, *F. proliferatum*, *F. sacchari*, *F. verticillioides*].

Interestingly, out of 35 species reported in this study, 17 species, *Fusarium caatingaense*, *F. carminascens*, *F. compactum*, *F. cugenangense*, *F. duoseptatum*, *F. fabacearum*, *F. glycines*, *F. gossypinum*, *F. grosmichelii*, *F. lumajangense*, *F. microconidium*, *F. nanum*, *F. nirenbergiae*, *F. sulawesiense*, *F. tardichlamydosporum*, *F. tardicrescens*, *Neocosmospora oblonga* and *N. suttoniana*, were found to be new records from India [53,54], most of them being reported for the first time from a new host in this study (Table 4).

Table 4. New records of *Fusarium* from India along with their host detail.

Identity	This Study		Earlier Reports	
	NFCCI No.	**Host**	**Host**	**Reference**
Fusarium caatingaense	NFCCI 5191	Pomegranate	*Dactylopius opuntiae*	[55]
Fusarium carminascens	NFCCI 5204	Potato	*Zea mays*	[56]
Fusarium compactum	NFCCI 2946	*Aegle marmelos*	*Poa annua*	[https://www.ncbi.nlm.nih.gov/nuccore accessed on 29 April 2022, [57–62]]
	NFCCI 2904		Soil	
	NFCCI 5208	*Zingiber officinale*	Spinach Wild rocket Cultivated rocket Lettuce *Quercus suber* Wheat Apple fruit cultivar Idared and Pink lady Banana corm and root rot Safflower (*Carthamus tinctorius L.*) Wheat soil Grasses	
Fusarium cugenangense	NFCCI 2872	Wilted tomato plant (Root)	*Crocus* sp. *Gossypium barbadense* Human toe nail *Vicia faba* *Musa* sp. var. Pisang Kepok *Gossypium* sp.	[56,63]
Fusarium duoseptatum	NFCCI 681	*Azadirachta indica* (Endophyte)	*Musa sapientum* cv. Pisang Ambon *Musa* sp. var. Pisang Rastali *M. acuminata* var. Dwarf Cavendish *M. acuminata* var. Pisang Ambon *Musa* sp. var. Pisang Raja *Musa* sp. var. Pisang Hawa *Musa* sp. var. Pisang Awak *Musa* sp. var. Pisang Susu *Musa* sp. var. Pisang Keling	[56,63]
Fusarium fabacearum	NFCCI 3239	Marigold seeds	*Zea mays* *Glycine max*	[56]
	NFCCI 3706	Castor root		
	NFCCI 5200	Pea		

Table 4. *Cont.*

Identity	This Study		Earlier Reports	
	NFCCI No.	**Host**	**Host**	**Reference**
Fusarium glycines	NFCCI 3048	Wilted cumin plant	*Linum usitatissium* *Ocimum basilicum* *Glycine max*	[56]
	NFCCI 1788	Soil		
Fusarium gossypinum	NFCCI 2467	Bottle gourd	*Gossypium hirsutum*	[56]
Fusarium grosmichelii	NFCCI 3243	Cow pea	Banana corm *M. acuminata* var. Pisang Ambon *Musa* sp. var. Pisang Awak *M. acuminata* var. Pisang Ambon Lumut *M. acuminata* var. Cavendish *Musa* sp. var. Pisang Siem Jumbo *M. acuminata* var. Pisang Ambon Kuning *Musa* sp. var. Pisang Kepok	[63,64]
Fusarium lumajangense	NFCCI 4180	Soil	*Musa* sp. var. Pisang Raja Nangka *Musa acuminata* var. Pisang Mas Kirana	[65]
Fusarium microconidium	NFCCI 3020	Soil	Unknown	[11]
Fusarium nanum	NFCCI 5192	*Zingiber officinale*	*Musa nana* (Leaves) *Solanum lycopersicum* *Musa acuminata* Oat Soil *Triticum* sp. *Sorghum* sp.	[[66], https://www.ncbi. nlm.nih.gov/nuccore accessed on 29 April 2022]
Fusarium nirenbergiae	NFCCI 5189	Soil	*Secale cereale* *Musa* sp.	[56]
	NFCCI 4859	*Dillenia indica*	*S. tuberosum* *Solanum lycopersicum* *Passiflora edulis* *Chrysanthemum* sp. *Bouvardia longiflora* *Dianthus caryophyllus* *Agathosma betulina* Tulip roots Amputated human toe Human leg ulcer	
Fusarium sulawesiense	NFCCI 2956	Cotton field	*Musa acuminata* var. Pisang Cere (AAA) *Oryza* sp.	[[11,67–70], https://www.ncbi. nlm.nih.gov/nuccore accessed on 29 April 2022]
	NFCCI 2886	*Azadirachta indica* (Endophyte)	*Smilax corbularia* *Acalypha insulana*	
	NFCCI 3031	*Piberbettle leaf* (endophyte)	*Alocasia odora* *Ipomoea batatas* *Musa nana* *Musa paradisiacal* Plum leaf *Triticum aestivum* Soil *Oryza sativa* *Aleurocanthus woglumi* *Phaseolus lunatus* Eucalyptus *Carica papaya* *Prosopis* sp. *Cucumis melo* *Galia melon* *Bixa orellana* *Gossypium hirsutum* *Sorghum vulgare* *Musa sampientum* var. Robusta Mango leaf Rhizosphere soil (*Bromus tectorum*) Soybean	
	NFCCI 4919	Soil		

Table 4. *Cont.*

Identity	This Study		Earlier Reports	
	NFCCI No.	Host	Host	Reference
Fusarium tardichlamydosporum	NFCCI 3051	Wilted cumin plant	*M. sapientum* cv. Pisang Awak Legor	[56,63]
	NFCCI 1895		*Musa* sp. var. Monthan *M. acuminata* var. Pisang Barangan	
	NFCCI 2491	Soil	*Musa* sp. var. Bluggoe *M. acuminata* var. Lady finger *Musa* sp. var. Ney Poovan *Musa* sp. var. Pisang Awak Legor	
Fusarium tardicrescens	NFCCI 680	*Azadirachta indica* (Endophyte)	*Musa* sp. var. Harare *Cicer* sp.	[63]
	NFCCI 5201	Dead bark	*Raphanus* sp.	
Neocosmospora oblonga	NFCCI 2150	Soil	Human eye Carbonatite	[[11], https://www.ncbi. nlm.nih.gov/nuccore accessed on 29 April 2022]
Neocosmospora suttoniana	NFCCI 5190	Ginger	Human wound	[[11], https://www.ncbi. nlm.nih.gov/nuccore accessed on 29 April 2022]
	NFCCI 2961	Pigeon pea root (Wilted)	Equine eye Soil	
	NFCCI 5210	Chilli fruit	*Homo sapiens*	
	NFCCI 4830	Cotton rot	*Gossypium hirsutum* *Podocnemis unifilis*	
	NFCCI 5211	Onion plants (Damping-off)	Human cornea Human skin leukemic Human blood leukemia Human blood	

3.2. Detection of Beauvericin Produced

3.2.1. Thin Layer Chromatography

Thin layer chromatography analysis of fungal extracts along with the standard beauvericin showed spots at the same retention factor as the standard beauvericin, depicting the possible presence of beauvericin in the fungal extracts (Figure 4).

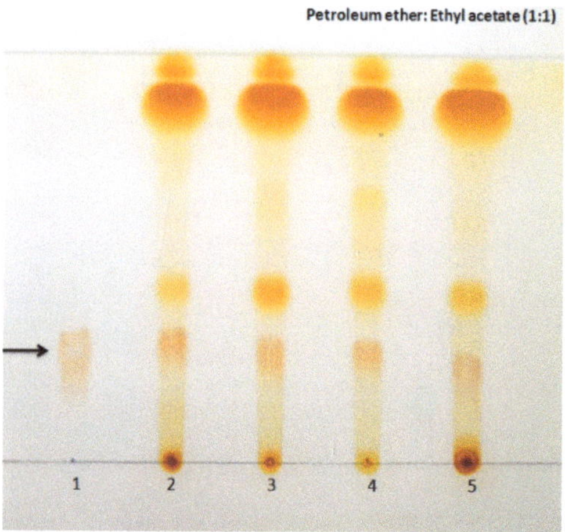

Figure 4. Thin layer chromatography developed in petroleum ether: Ethyl acetate (1:1) showing spots of beauvericin detected using iodine vapours. Black arrow shows spots of beauvericin (1) Standard beauvericin (500 µg/mL) (2) *Fusarium tardicrescens* NFCCI 5201 (3) *F. cugenangense* NFCCI 2872 (4) *Neocosmospora vasinfecta* NFCCI 2960 (5) *Fusarium annulatum* NFCCI 2962.

3.2.2. High-Performance Liquid Chromatography

There was a linear correlation between the concentration of the standard beauvericin and the areas of the peak in HPLC chromatogram. The retention time of standard beauvericin was found to be 9.1 min (Figure 5). Biomass produced by various isolates varied from 4.41 to 14.17 g/L of FDM. Among 88, 50 isolates (56%) were found to be capable of beauvericin production which varied from 0.01 to 15.82 mg/g of biomass. Table 5 shows the mycelial biomass, BEA content of mycelial biomass and final medium pH after a week's fermentation of different *Fusarium* isolates in *Fusarium* defined medium. *F. tardicrescens* NFCCI 5201 showed maximum beauvericin production of 15.82 mg/g of biomass. *F. carminascens* NFCCI 5204 and *F. fabacearum* NFCCI 5200 also produced a significant amount of beauvericin: 13.54 and 14.25 mg/g of biomass, respectively (Figure 6). The final pH of the fermented broth varied from 6.38 to 8.96 for different isolates. The correlation coefficient between biomass produced and the beauvericin produced was 0.2692; that between the pH of the fermented broth and the beauvericin produced was 0.16, indicating very weak and/or no association in both cases.

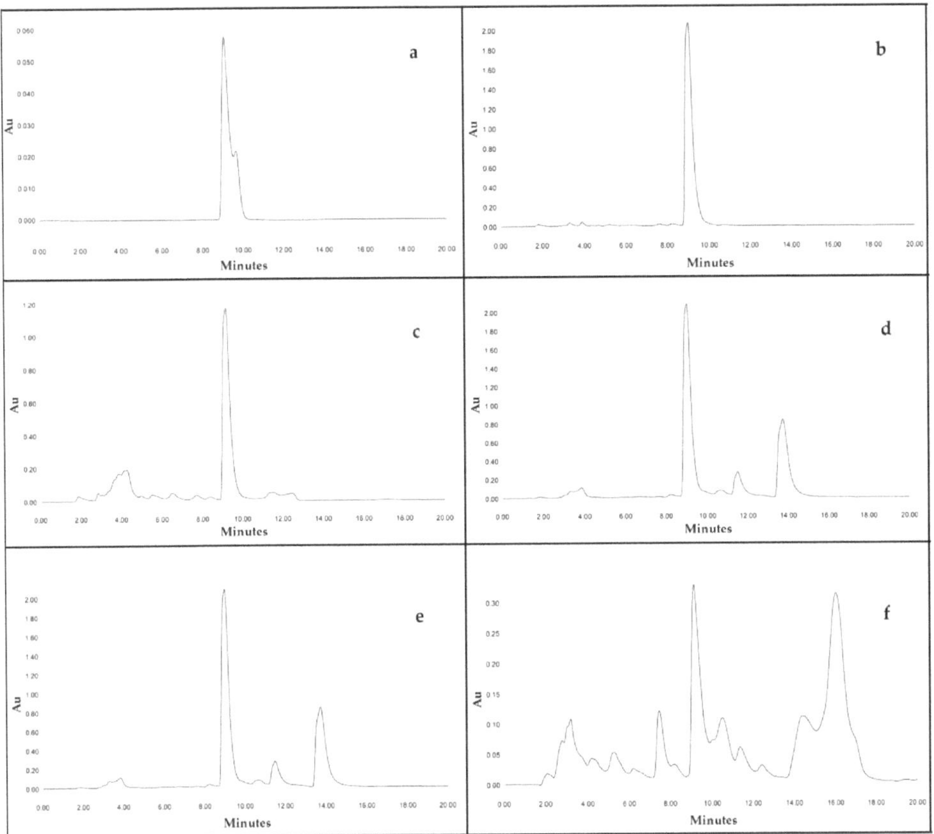

Figure 5. HPLC chromatograms of (**a**) standard beauvericin (500 μg/mL) (**b**) *Fusarium fabacearum* NFCCI 5200 extract (**c**) *F. sacchari* NFCCI 4889 extract (**d**) *F. carminascens* NFCCI 5204 extract (**e**) *F. tardicrescens* NFCCI 5201 extract (**f**) *F. cugenangense* NFCCI 2872 extract.

Table 5. Mycelial biomass and BEA content of *Fusarium* isolates and their pH in the *Fusarium* defined medium (on day 7).

Identity	NFCCI No.	Biomass (g/L)	BEA Content (mg/g)	Final pH of Medium
Fusarium nirenbergiae	NFCCI 5189	6.61 ± 0.26	2.93 ± 0.12	7.73 ± 0.32
Fusarium annulatum	NFCCI 3264	6.72 ± 0.37	0.97 ± 0.14	7.48 ± 0.18
Fusarium lacertarum	NFCCI 3044	8.07 ± 0.52	0.66 ± 0.14	7.65 ± 0.24
Neocosmospora suttoniana	NFCCI 5190	7.14 ± 0.08	-	7.63 ± 0.22
Fusarium fabacearum	NFCCI 3239	6.22 ± 0.31	-	7.77 ± 0.4
Fusarium sacchari	NFCCI 3147	6.34 ± 0.10	-	7.6 ± 0.06
Fusarium caatingaense	NFCCI 5191	9.26 ± 0.26	0.33 ± 0.03	7.63 ± 0.03
Neocosmospora solani	NFCCI 2315	8.06 ± 0.65	0.34 ± 0.07	7.61 ± 0.12
Fusarium annulatum	NFCCI 3072	6.83 ± 0.09	-	7.75 ± 0.15
Fusarium glycines	NFCCI 3048	5.64 ± 0.12	2.00 ± 0.44	7.39 ± 0.24
Fusarium annulatum	NFCCI 3300	7.10 ± 0.34	0.66 ± 0.05	7.62 ± 0.54
Fusarium annulatum	NFCCI 2964	7.07 ± 0.19	1.11 ± 0.18	7.5 ± 0.13
Fusarium mangiferae	NFCCI 2885	7.97 ± 0.28	-	7.8 ± 0.15
Fusarium grosmichelii	NFCCI 3243	6.37 ± 0.59	1.44 ± 0.083	7.57 ± 0.21
Fusarium annulatum	NFCCI 2962	7.42 ± 0.06	3.26 ± 0.23	7.49 ± 0.17
Fusarium nanum	NFCCI 5192	6.40 ± 0.18	-	7 ± 0.14
Fusarium sacchari	NFCCI 3093	8.61 ± 0.83	2.31 ± 0.58	7.32 ± 0.22
Fusarium sulawesiense	NFCCI 2956	6.44 ± 0.37	-	7.44 ± 0.29
Neocosmospora metavorans	NFCCI 5193	6.25 ± 0.46	-	7.34 ± 0.17
Fusarium sulawesiense	NFCCI 2886	5.89 ± 0.27	-	7.34 ± 0.12
Fusarium annulatum	NFCCI 2959	6.13 ± 0.49	-	7.29 ± 0.18
Fusarium sulawesiense	NFCCI 3031	8.99 ± 0.67	-	7.31 ± 0.48
Fusarium compactum	NFCCI 2946	8.89 ± 0.11	0.08 ± 0.04	7.44 ± 0.42
Neocosmospora vasinfecta	NFCCI 2960	6.64 ± 0.41	4.11 ± 0.17	7.44 ± 0.13
Fusarium brachygibbosum	NFCCI 3703	5.74 ± 0.32	0.15 ± 0.06	8.79 ± 0.04
Fusarium irregulare	NFCCI 5194	6.58 ± 0.48	1.88 ± 0.06	8.5 ± 0.20
Neocosmospora oblonga	NFCCI 2150	5.06 ± 0.19	-	8.67 ± 0.37
Neocosmospora metavorans	NFCCI 3475	6.51 ± 0.22	-	8.64 ± 0.06
Fusarium commune	NFCCI 2871	8.15 ± 0.18	-	8.67 ± 0.18
Neocosmospora metavorans	NFCCI 4095	6.71 ± 0.37	-	8.96 ± 0.28
Fusarium spinosum	NFCCI 5195	9.15 ± 0.64	-	8.9 ± 0.07
Fusarium microconidium	NFCCI 3020	8.52 ± 0.13	-	8.82 ± 0.08
Fusarium cugenangense	NFCCI 2872	7.01 ± 0.07	4.47 ± 0.51	8.69 ± 0.15
Fusarium irregulare	NFCCI 2460	9.04 ± 0.13	-	8.84 ± 0.03
Fusarium annulatum	NFCCI 3065	6.22 ± 0.42	0.20 ± 0.06	8.93 ± 0.17
Fusarium tardicrescens	NFCCI 680	7.12 ± 0.34	0.47 ± 0.07	8.87 ± 0.05
Fusarium sacchari	NFCCI 5196	9.18 ± 0.18	0.60 ± 0.06	7.87 ± 0.05
Fusarium lacertarum	NFCCI 5197	7.79 ± 2.1	0.49 ± 0.17	8.03 ± 0.02
Fusarium irregulare	NFCCI 5198	11.30 ± 4.8	-	8.36 ± 0.15
Fusarium annulatum	NFCCI 2470	10.71 ± 0.29	-	7.87 ± 0.13
Fusarium sacchari	NFCCI 3091	5.68 ± 0.17	-	7.28 ± 0.05
Neocosmospora metavorans	NFCCI 5199	5.35 ± 0.57	-	7.42 ± 0.17
Fusarium brachygibbosum	NFCCI 3074	7.80 ± 0.69	0.01 ± 0.05	7.72 ± 0.21
Neocosmospora suttoniana	NFCCI 2961	10.24 ± 0.7	0.67 ± 0.07	8.17 ± 0.45
Fusarium lacertarum	NFCCI 3049	7.22 ± 0.51	2.39 ± 0.39	7.19 ± 0.16
Fusarium sacchari	NFCCI 3092	5.79 ± 0.14	-	7.26 ± 0.15
Fusarium annulatum	NFCCI 1127	5.56 ± 0.61	-	6.92 ± 0.09
Fusarium fabacearum	NFCCI 3706	7.30 ± 0.47	1.00 ± 0.20	8.6 ± 0.05
Fusarium compactum	NFCCI 2904	6.45 ± 0.18	3.33 ± 0.51	7.02 ± 0.04
Fusarium lacertarum	NFCCI 3038	7.59 ± 0.09	0.16 ± 0.06	8.07 ± 0.15
Fusarium annulatum	NFCCI 2953	7.38 ± 0.15	2.21 ± 0.12	7.85 ± 0.17

Table 5. *Cont.*

Identity	NFCCI No.	Biomass (g/L)	BEA Content (mg/g)	Final pH of Medium
Fusarium tardichlamydosporum	NFCCI 3051	7.99 ± 0.12	2.19 ± 0.18	7.92 ± 0.13
Fusarium duoseptatum	NFCCI 681	6.37 ± 0.28	-	6.86 ± 0.12
Fusarium annulatum	NFCCI 3270	5.14 ± 0.19	4.84 ± 0.36	7.24 ± 0.17
Fusarium verticillioides	NFCCI 2945	5.43 ± 1.2	-	7.18 ±0.3
Neocosmospora vasinfecta	NFCCI 2972	7.61 ± 2.1	-	7.32 ± 0.02
Fusarium annulatum	NFCCI 2053	4.64 ± 0.09	3.78 ± 0.55	7.73 ± 0.5
Fusarium proliferatum	NFCCI 3282	6.51 ± 3.2	0.16 ± 0.19	7.55 ± 0.12
Fusarium fabacearum	NFCCI 5200	10.07 ± 0.17	14.25 ± 0.28	8.11 ± 0.16
Fusarium tardicrescens	NFCCI 5201	11.23 ± 0.38	15.82 ± 0.54	8.2 ± 0.11
Fusarium sacchari	NFCCI 4889	14.17 ± 2.7	5.95 ± 0.61	8.23 ± 0.04
Albonectria rigidiuscula	NFCCI 5202	14.03 ± 0.07	0.16 ± 0.04	7.39 ± 0.13
Fusarium tardichlamydosporum	NFCCI 1895	7.80 ± 1.8	0.12 ± 0.05	7.27 ± 0.24
Fusarium pernambucanum	NFCCI 5203	9.90 ± 0.15	1.04 ± 0.11	8.09 ± 0.15
Fusarium carminascens	NFCCI 5204	11.15 ± 2.4	13.54 ± 0.62	8.44 ± 0.19
Fusarium nirenbergiae	NFCCI 4859	6.52 ± 0.08	-	7.5 ± 0.32
Fusarium irregulare	NFCCI 5205	9.32 ± 1.6	-	7.59 ± 0.27
Fusarium brachygibbosum	NFCCI 4972	11.03 ± 0.15	-	7.66 ± 0.21
Fusarium brachygibbosum	NFCCI 5206	8.52 ± 2.9	-	7.66 ± 0.51
Fusarium pernambucanum	NFCCI 5207	9.80 ± 2.6	1.43 ± 0.16	7.22 ± 0.18
Fusarium oxysporum	NFCCI 4759	7.69 ± 0.14	2.22 ± 0.43	7.12 ± 0.16
Fusarium verticillioides	NFCCI 4963	10.55 ± 3.7	-	7.55 ± 0.14
Fusarium verticillioides	NFCCI 2696	11.05 ± 2.7	-	7.85 ± 0.05
Fusarium compactum	NFCCI 5208	4.41 ± 0.13	1.146 ± 0.13	6.67 ± 0.07
Fusarium gossypinum	NFCCI 2467	4.59 ± 0.17	0.27 ± 0.08	8.29 ± 1.5
Fusarium acutatum	NFCCI 5209	4.69 ± 0.15	0.29 ± 0.06	6.38 ± 0.7
Neocosmospora suttoniana	NFCCI 5210	7.50 ± 3.7	0.90 ± 0.08	7.86 ± 0.36
Neocosmospora metavorans	NFCCI 4885	7.12 ± 2.5	0.86 ± 0.07	7.94 ± 0.59
Fusarium glycines	NFCCI 1788	11.20 ± 0.13	1.05 ± 0.10	8.04 ± 0.16
Fusarium tardichlamydosporum	NFCCI 2491	13.59 ± 0.19	1.38 ± 0.11	8 ± 0.08
Albonectria rigidiuscula	NFCCI 4888	13.77 ± 0.27	1.09 ± 0.08	8.13 ± 0.8
Neocosmospora suttoniana	NFCCI 4830	4.67 ± 2.6	1.01 ± 0.05	7.65 ±0.02
Neocosmospora suttoniana	NFCCI 5211	8.56 ± 3.1	-	8 ± 0.07
Fusarium annulatum	NFCCI 5212	10.94 ± 1.2	-	8.14 ± 0.4
Fusarium lumajangense	NFCCI 4180	8.26 ± 2.7	1.07 ± 0.39	8.18 ±0.05
Fusarium lacertarum	NFCCI 4792	9.66 ± 1.8	-	8.23 ± 0.1
Fusarium annulatum	NFCCI 2949	6.70 ± 0.27	-	8.16 ± 0.02
Fusarium sulawesiense	NFCCI 4919	8.72 ± 0.44	-	8.25 ± 0.14

- Not Detected; All values represent means ± S.E.M.

3.2.3. High-Resolution Mass Spectrometry (HRMS)

The HRMS results showed molecular ion peaks at 806.3999 (m/z) [M + Na]$^+$, indicating the presence of beauvericin ($C_{45}H_{57}N_3O_9$) in the fungal extracts (Figure 7). Similar (m/z) [M + Na]$^+$ 806.3956 have been reported for beauvericin ($C_{45}H_{57}N_3O_9$) [71]. It was observed that the fungal extract contains, majorly, beauvericin, as maximum ion intensity was observed with m/z 806.3999.

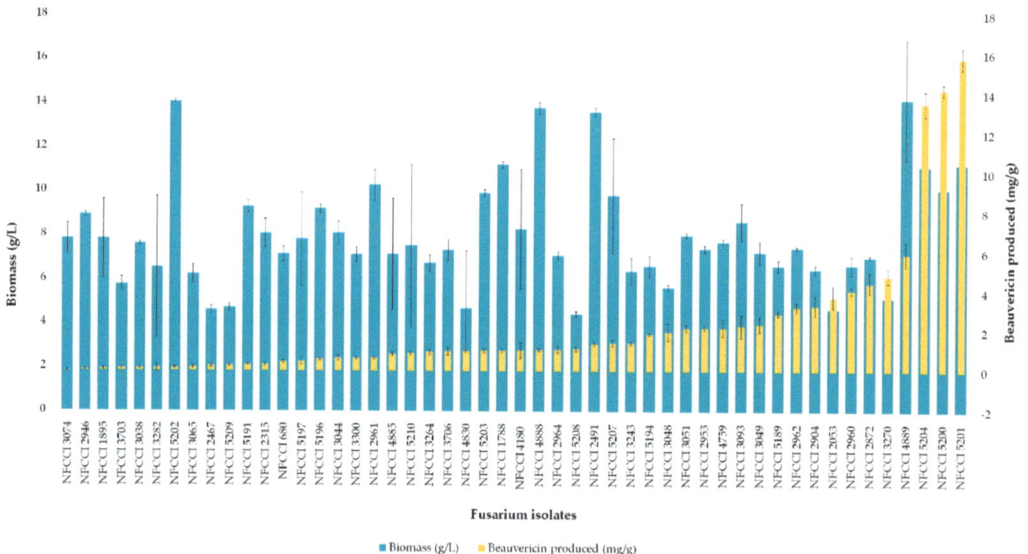

Figure 6. Mycelial biomass and BEA content produced by different *Fusarium* isolates in the *Fusarium* defined medium (on day 7).

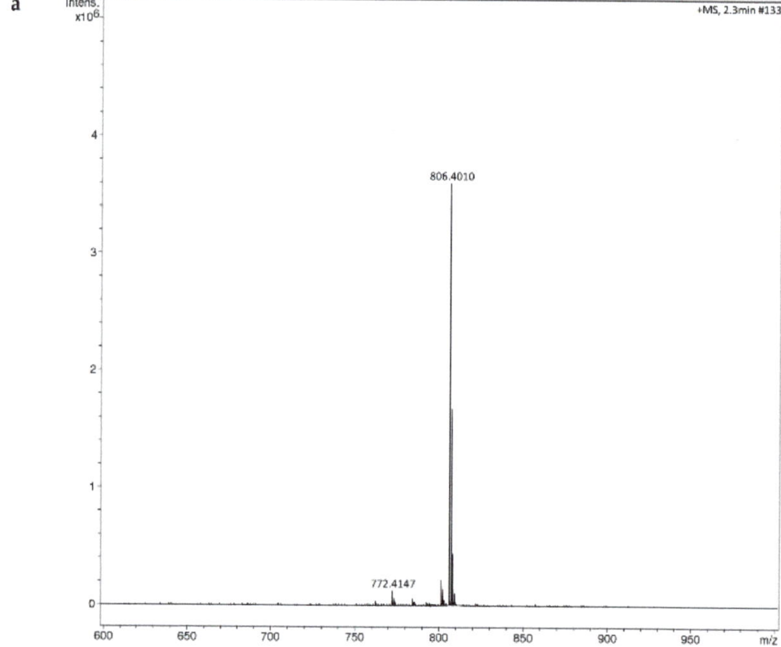

Figure 7. *Cont.*

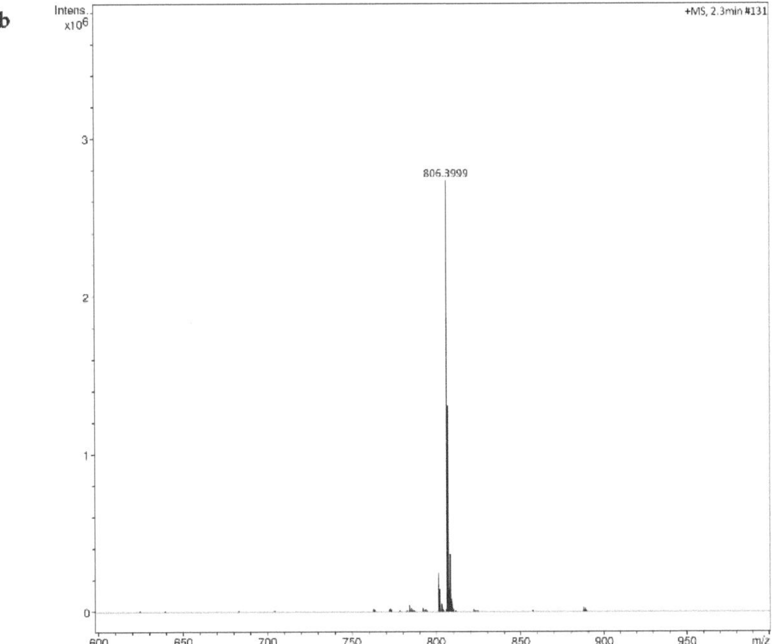

Figure 7. HRMS spectra from higher-collision dissociation of the [M + Na]⁺ ions of fungal extract, (**a**) *Fusarium sacchari* NFCCI 4889 (**b**) *F. tardicrescens* NFCCI 5201.

3.3. Large Scale Fermentation

Fermentation of 3.5 liters of *Fusarium* defined medium resulted in 36.27 gms of biomass which, on extraction, resulted in 144 mg of crude which was dissolved in 36 mL of acetonitrile, making the final concentration 4 mg/mL. This extract was subjected to HPLC; it was found that beauvericin is the major compound in the extract, as the area covered by it was found to be nearly 64–68%. This stock was diluted as per requirement for further studies.

3.4. Antimicrobial and MIC of Crude Extract of *Fusarium tardicrescens* NFCCI 5201

The extract showed promising results against *Staphylococcus aureus* MLS16 MTCC 2940 and *Micrococcus luteus* MTCC 2470. The MIC of the crude extract of *F. tardicrescens* NFCCI 5201 was found to be 15.63 μg/mL against *Micrococcus luteus* MTCC 2470 and 62.5 μg/mL against *Staphylococcus aureus* MLS16 MTCC 2940 (Figure 8). The extract did not show any antimicrobial activity against *Escherichia coli* MTCC 739, *Bacillus subtilis* MTCC 121, *Raoultella planticola* MTCC 530, *Pseudomonas aeruginosa* MTCC 2453 and *Staphylococcus aureus* MTCC 96.

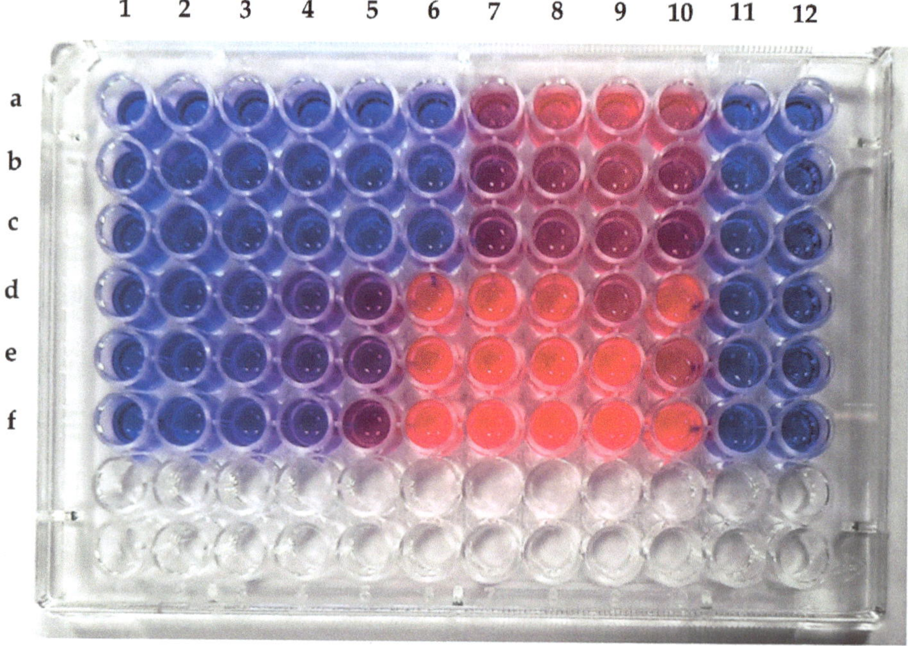

Figure 8. Microtiter plate showing MIC of crude extract of *F. tardicrescens* NFCCI 5201 against *Staphylococcus aureus* MLS16 MTCC 2940 (Rows a–c) and *Micrococcus luteus* MTCC 2470 (Rows d–f).

3.5. Individual and Combined Effect of Extract of Fusarium tardicrescens NFCCI 5201 Containing Beauvericin and Amphotericin B on Pathogenic Fungi

The antifungal activity was observed as a reduction in the mycelial growth of pathogenic fungi in poisoned plates when compared to the control plates (Figure 9). The extract of *F. tardicrescens* NFCCI 5201 containing beauvericin (40 µg/mL) showed good antifungal activity against plant pathogenic fungi, *Rhizoctonia solani* NFCCI 4327, *Sclerotium rolfsii* NFCCI 4263, *Geotrichum candidum* NFCCI 3744 and *Pythium* sp. NFCCI 3482 showed a % inhibition of 84.31, 49.76, 38.22 and 35.13. Amphotericin B (40 µg/mL) inhibited *Geotrichum candidum* NFCCI 3744, *Rhizoctonia solani* NFCCI 4327, *Sclerotium rolfsii* NFCCI 4263 and *Absidia* sp. NFCCI 2716, showing % inhibition of 71.76, 55.72, 56.82 and 48.68. Interestingly, the results were remarkable when the extracts of *F. tardicrescens* NFCCI 5201 containing beauvericin (20 µg/mL) and amphotericin B (20 µg/mL) were used in combination against *Rhizopus* sp. NFCCI 2108, *Sclerotium rolfsii* NFCCI 4263, *Bipolaris sorokiniana* NFCCI 4690 and *Absidia* sp. NFCCI 2716, showing % inhibition of 50.35, 79.37, 48.07 and 76.72%, respectively. Individually, extract of *F. tardicrescens* NFCCI 5201 containing beauvericin (40 µg/mL) showed % inhibition of 1.68, 49.76, 2.56 and 7.9 against *Rhizopus* sp. NFCCI 2108, *Sclerotium rolfsii* NFCCI 4263, *Bipolaris sorokiniana* NFCCI 4690 and *Absidia* sp. NFCCI 2716. Individually, amphotericin B (40 µg/mL) showed % inhibition of 11.51, 56.82, 1.28 and 48.67 against *Rhizopus* sp. NFCCI 2108, *Sclerotium rolfsii* NFCCI 4263, *Bipolaris sorokiniana* NFCCI 4690 and *Absidia* sp. NFCCI 2716. The extract of *F. tardicrescens* NFCCI 5201 containing beauvericin (40 µg/mL) showed better results than amphotericin B in the case of *Pythium* sp. NFCCI 3482 and *Rhizoctonia solani* NFCCI 4327 (Figure 10).

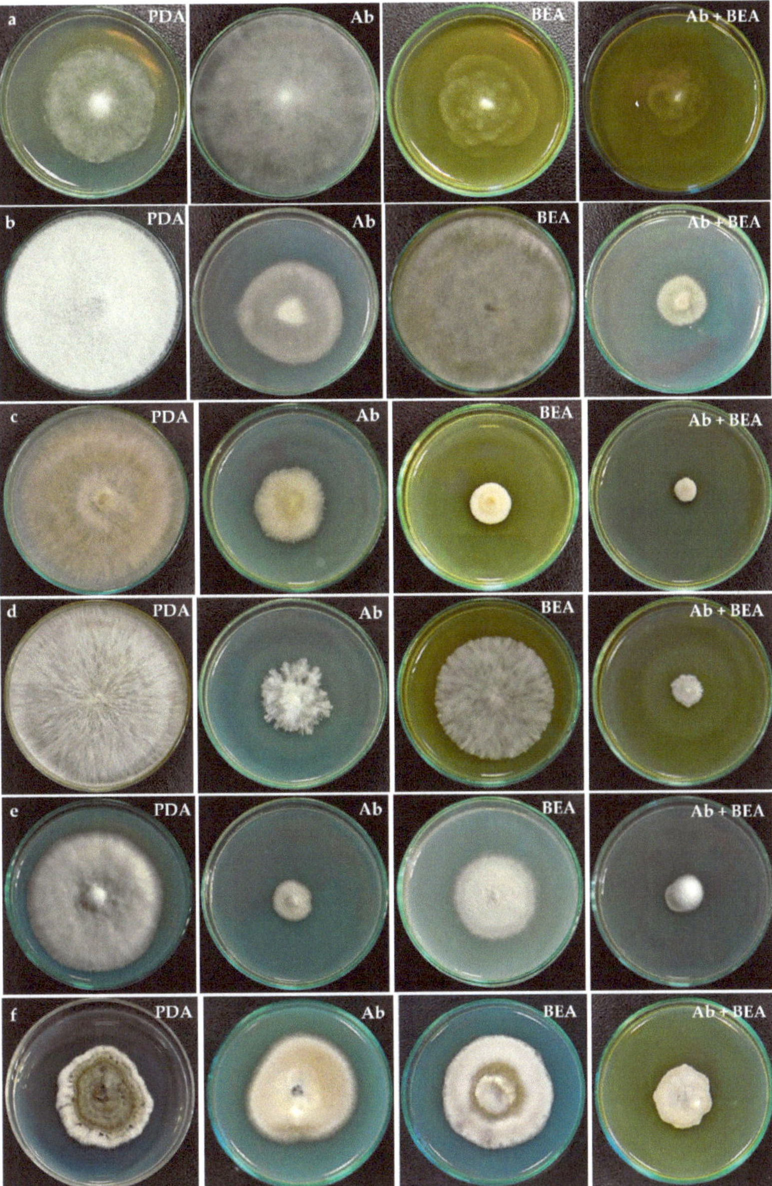

Figure 9. Antifungal activity of individual and combined effect of extract of *F. tardicrescens* NFCCI 5201 containing beauvericin and amphotericin B on various pathogenic fungi (**a**) *Pythium* sp. NFCCI 3482 (**b**) *Rhizopus* sp. NFCCI 2108 (**c**) *Rhizoctonia solani* NFCCI 4327 (**d**) *Sclerotium rolfsii* NFCCI 4263 (**e**) *Geotrichum candidum* NFCCI 3744 (**f**) *Bipolaris sorokiniana* NFCCI 4690. PDA: Potato dextrose agar (Control); Ab: Potato dextrose agar containing amphotericin B (40 μg/mL); BEA: Potato dextrose agar having extract of *F. tardicrescens* NFCCI 5201 containing beauvericin (40 μg/mL); Ab + BEA: Potato dextrose agar having amphotericin B (20 μg/mL) and extract of *F. tardicrescens* NFCCI 5201 containing beauvericin (20 μg/mL).

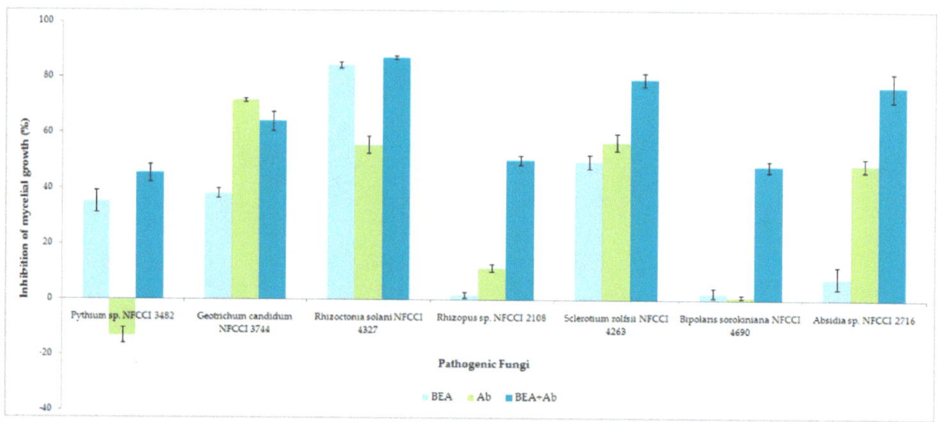

Figure 10. Inhibition of mycelial growth (%) of individual and combined effect of extract of *F. tardicrescens* NFCCI 5201 containing beauvericin and amphotericin B on various pathogenic fungi using food poison technique.

3.6. Antioxidant Effect of Extract of Fusarium tardicrescens NFCCI 5201 Containing Beauvericin

Figure 11 shows the free radical scavenging activity of the extract of *F. tardicrescens* NFCCI 5201 containing beauvericin and the standard ascorbic acid when tested for in vitro antioxidant activity. The extract of *F. tardicrescens* NFCCI 5201 containing beauvericin showed good satisfactory dose-dependent DPPH radical scavenging activity with an IC_{50} value of 0.675 mg/mL when tested with standard ascorbic acid, which showed IC_{50} value of 0.146 mg/mL.

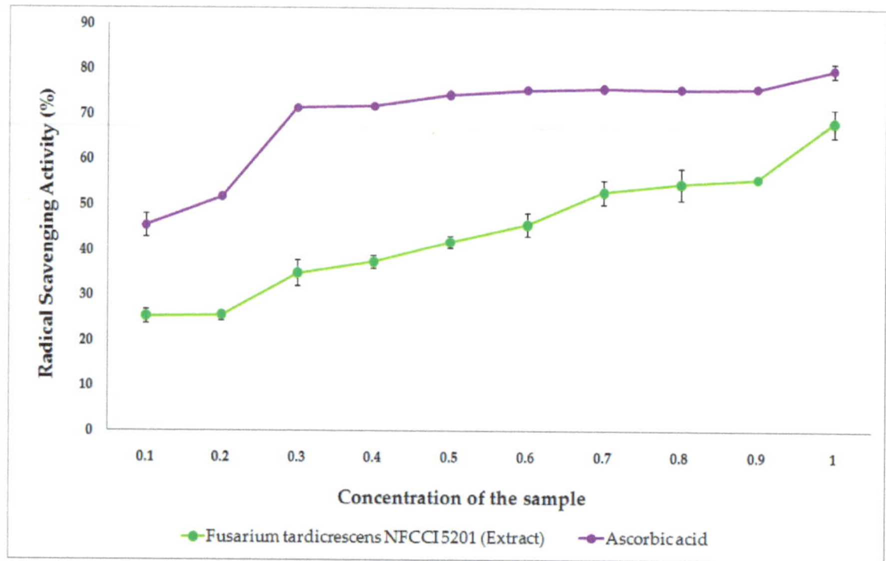

Figure 11. The free radical scavenging activity of the extract of *F. tardicrescens* NFCCI 5201 containing beauvericin (mg/mL) and the standard ascorbic acid (mg/mL) when tested for in vitro antioxidant activity.

4. Discussion

One of the greatest hindrances in studying *Fusarium* has been the ambiguous nomenclature or incorrect species names of the isolates because of the limitations in recognition of the species based on morphology [72]. Being the world's most important pathogens, its prophylaxis and management needs correct and rapid identification [73]. As mentioned earlier, ITS and LSU frequently fail to distinguish at the species level and, preferably, *Tef1-α* and *Rpb2* can be used to distinguish isolates at the species level. Literature reveals that Indian *Fusarium* isolates have been mostly identified on the basis of morphology. Very few studies were found where the *Fusarium* isolates had been identified using molecular studies. Wherever molecular studies were performed, they were based on the ITS gene region; very limited studies used either *Tef1-α* or *Rpb2* or both of these gene regions [74]. To our knowledge, this is the first study from India where *Fusarium* isolates have been identified on the basis of six gene regions, nrDNA (ITS), *translation elongation factor 1-alpha* (*Tef1-α*), 28S large subunit of the nrDNA (LSU), *RNA polymerase second largest subunit* (*Rpb2*), *beta-tubulin* (*β-tub*) and *calmodulin* (*CaM*). Interestingly, in this study, 17 species, *Fusarium caatingaense*, *F. carminascens*, *F. compactum*, *F. cugenangense*, *F. duoseptatum*, *F. fabacearum*, *F. glycines*, *F. gossypinum*, *F. grosmichelii*, *F. lumajangense*, *F. microconidium*, *F. nanum*, *F. nirenbergiae*, *F. sulawesiense*, *F. tardichlamydosporum*, *F. tardicrescens*, *Neocosmospora oblonga* and *N. suttoniana* were found to be new records for India, most of them being reported from new hosts [53,54].

BEA is a potent bioactive compound possessing antimicrobial, anti-insecticidal, antitumor and antiplatelet activities at extremely low concentration and with unique uncharacterized active mechanisms. A review of the literature indicates that no study was undertaken in India on BEA production from Indian species of *Fusarium,* except [43] who had reported BEA production from two isolates of *Fusarium* viz., *F. anthophilum* (Host: Sugarcane; 1300 μg/g) and *F. nygamai* (Host: *Cajanus indicus*; 3 μg/g) [43]. This work was undertaken in Bari, Italy. This suggests that Indian *Fusarium* isolates have remained unexplored for their capability of BEA production and its applicability.

Albonectria rigidiuscula, *F. acutatum*, *F. annulatum*, *F. brachygibbosum*, *F. caatingaense*, *F. carminascens*, *F. compactum*, *F. cugenangense*, *F. fabacearum*, *F. glycines*, *F. gossypinum*, *F. grosmichelii*, *F. irregulare*, *F. lacertarum*, *F. lumajangense*, *F. nirenbergiae*, *F. oxysporum*, *F. pernambucanum*, *F. proliferatum*, *F. sacchari*, *F. tardichlamydosporum*, *F. tardicrescens*, *N. metavorans*, *N. solani*, *N. suttoniana* and *N. vasinfecta* were found to be the positive producers of BEA in this study, many of them being reported as BEA producers for the first time in this study. *Fusarium tardicrescens* NFCCI 5201 showed maximum beauvericin production of 15.82 mg/g of biomass. *Fusarium carminascens* NFCCI 5204 and *F. fabacearum* NFCCI 5200 also produced a significantly ample amount of beauvericin, 13.54 and 14.25 mg/g of biomass, respectively, all three of them being reported for the first time as significant BEA producers in this study.

In earlier reports, BEA was detected in cultures of *Fusarium moniliforme*, *F. semitectum* [75], *F. subglutinans*, *F. thapsinum* [76], *F. sambucinum*, *F. acuminatum*, *F. equiseti*, *F. longipes*, *F. anthophilum*, *F. oxysporum*, *F. poae*, *F. avenaceum*, *F. beomiforme*, *F. dlamini*, *F. bulbicola*, *F. nygamai* [43], *F. chlamydosporum*, *F. solani*, *F. proliferatum* and *F. sacchari* [77].

The crude extract of *F. tardicrescens* NFCCI 5201 showed promising results against *Staphylococcus aureus* MLS16 MTCC 2940 and *Micrococcus luteus* MTCC 2470, with MIC values of 15.63 μg/mL against *Micrococcus luteus* MTCC 2470 and 62.5 μg/mL against *Staphylococcus aureus* MLS16 MTCC 2940. BEA from *F. oxysporum* had a potent inhibitory effect on the growth of pathogenic *Staphylococcus aureus* with a MIC of 3.91 μM [23]. Earlier reports reveal that inhibitory concentration (IC_{50}) values of BEA against *Bacillus subtilis*, *Staphylococcus haemolyticus*, *Pseudomonas lachrymans*, *Agrobacterium tumefaciens*, *Escherichia coli* and *Xanthomo vesicatoria* by a 96-well microplate broth dilution–MTT assay ranged between 18.45 and 70.41 μg/mL [78]. BEA has been known to inhibit bacteria, including *Bacillus pumilus* (0.1 μg of BEA per disk), several other species of *Bacillus* and *Paenibacillus* (1 μg of BEA per disk), *P. validus*, *Bifidobacterium adolescentis*, *Clostridium perfringens*, *Eubacterium biforme*, *Peptostreptococcus anaerobius* and *P. productus* (25 μg of

BEA per disk) [18]. Strong antimicrobial activity has been reported against *Staphylococcus aureus, Salmonella typhimurium* and *Bacillus cereus* with MIC values of 3.91 µM, 6.25 and 3.12 µg/mL, respectively [19,23].

With BEA being a high score ABC inhibitor, it has been reported to show strong synergy with some azole compounds (miconazole, ketoconazole) against *Candida albicans* and *Candida parapsilosis* (in vitro as well as in vivo). It has been reported that BEA individually fails to show antifungal activity, but it was found to synergize the effect of ketoconazole [25].

In this study, the antifungal activity was observed as a reduction in the mycelial growth of pathogenic fungi of agricultural importance in poisoned plates when compared to the control plates. The extract of *F. tardicrescens* NFCCI 5201 containing beauvericin (40 µg/mL) showed good antifungal activity against pathogenic fungi *Rhizoctonia solani* NFCCI 4327, *Sclerotium rolfsii* NFCCI 4263, *Geotrichum candidum* NFCCI 3744 and *Pythium* sp. NFCCI 3482 showing % inhibition of 84.31, 49.76, 38.22 and 35.13. Interestingly, when the extract of *F. tardicrescens* NFCCI 5201 containing beauvericin (20 µg/mL) and amphotericin B (20 µg/mL) were used in combination, they produced a synergized antifungal effect against *Rhizopus* sp. NFCCI 2108, *Sclerotium rolfsii* NFCCI 4263, *Bipolaris sorokiniana* NFCCI 4690 and *Absidia* sp. NFCCI 2716, showing % inhibition of 50.35, 79.37, 48.07 and 76.72%, respectively. Individually, the extract of *F. tardicrescens* NFCCI 5201 containing beauvericin (40 µg/mL) showed a % inhibition of 1.68, 49.76, 2.56 and 7.9 against *Rhizopus* sp. NFCCI 2108, *Sclerotium rolfsii* NFCCI 4263, *Bipolaris sorokiniana* NFCCI 4690 and *Absidia* sp. NFCCI 2716, and amphotericin B (40 µg/mL) showed a % inhibition of 11.51, 56.82, 1.28 and 48.67 against *Rhizopus* sp. NFCCI 2108, *Sclerotium rolfsii* NFCCI 4263, *Bipolaris sorokiniana* NFCCI 4690 and *Absidia* sp. NFCCI 2716. The extract of *F. tardicrescens* NFCCI 5201 containing beauvericin (40 µg/mL) showed better results than amphotericin B in the case of *Pythium* sp. NFCCI 3482, *Rhizoctonia solani* NFCCI 4327. This is the first report on the testing of BEA against filamentous, agriculturally important fungal pathogens. The synergistic effect of BEA along with amphotericin B against filamentous agriculturally important pathogens has also been studied for the first time. Earlier studies focused on the synergistic effect of BEA and azoles (miconazole, ketoconazole) on yeast (*Candida albicans* and *Candida parapsilosis*). This study focuses on the synergistic effect of fungal extracts and polyene (amphotericin B) on agriculturally important filamentous fungal pathogens. Moreover, the extract of *F. tardicrescens* NFCCI 5201 containing beauvericin showed good dose-dependent DPPH radical scavenging activity with an IC_{50} value of 0.675 mg/mL. Antioxidant activity has been reported from metabolites produced by *Fusarium oxysporum* [79]. The DPPH-radicals' scavenging activity has been reported from *F. solani* with IC_{50} value of 24 µg/mL [80].

5. Conclusions

Fusarium spp. have been known globally for centuries for their pathogenic behavior; however, they can prove to be a treasure trove because of their capability to produce other interesting metabolites of great applicability in various sectors such as food, agriculture and medicine. As shown in this study, the antifungal activity became synergized when a half dosage of the fungal extract and amphotericin B was used. Therefore, this study indicates that concerted efforts are required, in which high-throughput screening needs to be carried out with dosage of the BEA and the other antifungal compound, which can possibly be used at a large scale for calculating the optimum dosage required. Similar studies can be carried out in the field of medicine where the dosage is optimized and the delivery of the drug combinations is also targeted.

Supplementary Materials: The following supporting information can be downloaded at https://www.mdpi.com/article/10.3390/jof8070662/s1, Figure S1: Fermentation of different isolates in *Fusarium* defined media after a week's fermentation (a) Control (Uninnoculated FDM medium) (b) *Fusarium fabacearum* NFCCI 5200 (c) *F. annulatum* NFCCI 3300 (d) *F. sulawesiense* NFCCI 4919 (e) *F. sacchari* NFCCI 4889 (f) *F. annulatum* NFCCI 5212 (g) *F. nirenbergiae* NFCCI 5189 (h) *F. tardicrescens* NFCCI 5201 (i) *F. lacertarum* NFCCI 4792 (j) *F. lumajangense* NFCCI 4180 (k) *F. pernambucanum* NFCCI 5203 (l) *F. duoseptatum* NFCCI 681 (m) *F. tardichlamydosporum* NFCCI 1895 (n) *Neocosmospora solani*

NFCCI 2315 (o) *F. mangiferae* NFCCI 2885 (p) *F. annulatum* NFCCI 2964 (q) *F. sacchari* NFCCI 3147 (r) *F. grosmichelii* NFCCI 3243; Table S1: GenBank Accession No. and other details of the isolates used in the phylogenetic study.

Author Contributions: Conceptualization, S.R. and S.K.S.; methodology, S.R. and S.K.S.; formal analysis, S.R. and S.K.S.; investigation, S.R.; resources, S.K.S.; writing—original draft preparation, S.R.; writing—review and editing, S.R., S.K.S. and L.D.; supervision, S.K.S. All authors have read and agreed to the published version of the manuscript.

Funding: This research received no external funding.

Institutional Review Board Statement: Not applicable.

Informed Consent Statement: Not applicable.

Data Availability Statement: Not applicable.

Acknowledgments: We thank Prashant Dhakephalkar, Director MACS-Agharkar Research Institute, Pune for providing the necessary facilities and encouragement to carry out the research work. Shiwali Rana acknowledges the University Grants Commission (U.G.C.), New Delhi for granting a Senior Research Fellowship (SRF), and S.P. Pune University, Pune for registering for the Ph.D. degree. We acknowledge technical support and help by Deepak Kumar Maurya and Subhash Gaikwad of Agharkar Research Institute, Pune. Laurent Dufossé deeply acknowledges the strong financial support from Conseil Régional de Bretagne and Conseil Régional de La Réunion for research activities dedicated to microbial biotechnology.

Conflicts of Interest: The authors declare no conflict of interest.

References

1. Gupta, A.K.; Baran, R.; Summerbell, R.C. *Fusarium* infections of the skin. *Curr. Opin. Infect. Dis.* **2000**, *13*, 121–128. [CrossRef]
2. Summerell, B.A. Resolving *Fusarium*: Current status of the genus. *Annu. Rev. Phytopathol.* **2019**, *57*, 323–339. [CrossRef]
3. Taylor, J.W.; Jacobson, D.J.; Kroken, S.; Kasuga, T.; Geiser, D.M.; Hibbett, D.S.; Fisher, M.C. Phylogenetic species recognition and species concepts in fungi. *Fungal Genet. Biol.* **2000**, *31*, 21–32. [CrossRef]
4. Schoch, C.L.; Seifert, K.A.; Huhndorf, S.; Robert, V.; Spouge, J.L.; Levesque, C.A.; Chen, W.; Bolchacova, E.; Voigt, K.; Crous, P.W.; et al. Nuclear ribosomal internal transcribed spacer (ITS) region as a universal DNA barcode marker for Fungi. *Proc. Natl. Acad. Sci. USA* **2012**, *109*, 6241–6246. [CrossRef]
5. Odonnell, K.; Cigelnik, E. Two divergent intragenomic rDNA ITS2 types within a monophyletic lineage of the fungus *Fusarium* are nonorthologous. *Mol. Biol. Evol.* **1997**, *7*, 103–116. [CrossRef]
6. O'Donnell, K.; Cigelnik, E.; Nirenberg, H.I. Molecular systematics and phylogeography of the *Gibberella fujikuroi* species complex. *Mycologia* **1998**, *90*, 465–493. [CrossRef]
7. O'Donnell, K.; Sutton, D.A.; Rinaldi, M.G.; Gueidan, C.; Crous, P.W.; Geiser, D.M. Novel multilocus sequence typing scheme reveals high genetic diversity of human pathogenic members of the *Fusarium incarnatum-F. equiseti* and *F. chlamydosporum* species complexes within the United States. *J. Clin. Microbiol.* **2009**, *47*, 3851–3861. [CrossRef]
8. O'Donnell, K. Molecular phylogeny of the *Nectria haematococca-Fusarium solani* species complex. *Mycologia* **2000**, *92*, 919–938. [CrossRef]
9. Nalim, F.A.; Samuels, G.J.; Wijesundera, R.L.; Geiser, D.M. New species from the *Fusarium solani* species complex derived from perithecia and soil in the Old-World tropics. *Mycologia* **2011**, *103*, 1302–1330. [CrossRef]
10. Herron, D.A.; Wingfield, M.J.; Wingfield, B.D.; Rodas, C.A.; Marincowitz, S.; Steenkamp, E.T. Novel taxa in the *Fusarium fujikuroi* species complex from *Pinus* spp. *Stud. Mycol.* **2015**, *80*, 131–150. [CrossRef]
11. Crous, P.W.; Lombard, L.; Sandoval-Denis, M.; Seifert, K.A.; Schroers, H.J.; Chaverri, P.; Gene, J.; Guarro, J.; Hirooka, Y.; Bensch, K.; et al. *Fusarium*: More than a node or a foot-shaped basal cell. *Stud. Mycol.* **2021**, *98*, 100116. [CrossRef]
12. Zhang, T.; Jia, X.P.; Zhuo, Y.; Liu, M.; Gao, H.; Liu, J.T.; Zhang, L.X. Cloning and characterization of a novel 2-ketoisovalerate reductase from the beauvericin producer *Fusarium proliferatum* LF061. *BMC Biotechnol.* **2012**, *12*, 55. [CrossRef]
13. Sood, S.; Sandhu, S.S.; Mukherjee, T.K. Pharmacological and therapeutic potential of beauvericin: A Short Review. *J. Proteom. Bioinform.* **2017**, *10*, 18–23. [CrossRef]
14. Hilgenfeld, R.; Saenger, W. Structural chemistry of natural and synthetic ionophores and their complexes with cations. *Top. Curr. Chem.* **1982**, *101*, 1–82.
15. Ovchinnikov, Y.A.; Ivanov, V.T.; Evstratov, A.E.; Mikhaleva, I.I.; Bystrov, V.F.; Portnova, S.L.; Balashova, T.A.; Meshcheryakova, E.N.; Tulchinsky, V.M. The enniatins ionophores, conformation and ion binding properties. *Int. J. Pept. Protein Res.* **1974**, *6*, 465–498. [CrossRef]
16. Steinrauf, L.K. Beauvericin and the other enniatins. *Met. Ions Biol. Syst.* **1985**, *19*, 139–171.
17. Lin, H.I.; Lee, Y.J.; Chen, B.F.; Tsai, M.C.; Lu, J.L.; Chou, C.J.; Jow, G.M. Involvement of Bcl-2 family, cytochrome c and caspase 3 in induction of apoptosis by beauvericin in human non-small cell lung cancer cells. *Cancer Lett.* **2005**, *230*, 248–259. [CrossRef]

18. Castlebury, L.A.; Sutherland, J.B.; Tanner, L.A.; Henderson, A.L.; Cerniglia, C.E. Short Communication: Use of a bioassay to evaluate the toxicity of beauvericin to bacteria. *World J. Microbiol. Biotechnol.* **1999**, *15*, 131–133. [CrossRef]

19. Dzoyem, J.P.; Melong, R.; Tsamo, A.T.; Maffo, T.; Kapche, D.; Ngadjui, B.T.; McGaw, L.J.; Eloff, J.N. Cytotoxicity, antioxidant and antibacterial activity of four compounds produced by an endophytic fungus *Epicoccum nigrum* associated with *Entada abyssinica*. *Rev. Bras. Farmacogn.* **2017**, *27*, 251–253. [CrossRef]

20. Fotso, J.; Smith, J.S. Evaluation of beauvericin toxicity with the bacterial bioluminescence assay and the Ames mutagenicity bioassay. *J. Food Sci.* **2003**, *68*, 1938–1941. [CrossRef]

21. Meca, G.; Sospedra, I.; Soriano, J.M.; Ritieni, A.; Moretti, A.; Manes, J. Antibacterial effect of the bioactive compound beauvericin produced by *Fusarium proliferatum* on solid medium of wheat. *Toxicon* **2010**, *56*, 349–354. [CrossRef]

22. Nilanonta, C.; Isaka, M.; Kittakoop, P.; Trakulnaleamsai, S.; Tanticharoen, M.; Thebtaranonth, Y. Precursor-directed biosynthesis of beauvericin analogs by the insect pathogenic fungus *Paecilomyces tenuipes* BCC 1614. *Tetrahedron* **2002**, *58*, 3355–3360. [CrossRef]

23. Zhang, H.W.; Ruan, C.F.; Bai, X.L.; Zhang, M.; Zhu, S.S.; Jiang, Y.Y. Isolation and identification of the antimicrobial agent beauvericin from the endophytic *Fusarium oxysporum* 5-19 with NMR and ESI-MS/MS. *Biomed Res. Int.* **2016**, *2016*, 1084670. [CrossRef]

24. Wu, X.F.; Xu, R.; Ouyang, Z.J.; Qian, C.; Shen, Y.; Wu, X.D.; Gu, Y.H.; Xu, Q.; Sun, Y. Beauvericin ameliorates experimental colitis by inhibiting activated T Cells via downregulation of the PI3K/Akt signaling pathway. *PLoS ONE* **2013**, *8*, e83013. [CrossRef]

25. Fukuda, T.; Arai, M.; Tomoda, H.; Omura, S. New beauvericins, potentiators of antifungal miconazole activity, produced by Beauveria sp FKI-1366-II. Structure elucidation. *J. Antibiot.* **2004**, *57*, 117–124. [CrossRef]

26. Shekhar-Guturja, T.; Tebung, W.A.; Mount, H.; Liu, N.N.; Kohler, J.R.; Whiteway, M.; Cowen, L.E. Beauvericin potentiates azole activity via inhibition of multidrug efflux, blocks *Candida albicans* morphogenesis, and is effluxed via Yor1 and circuitry controlled by Zcf29. *Antimicrob. Agents Chemother.* **2016**, *60*, 7468–7480. [CrossRef]

27. Zhang, L.X.; Yan, K.Z.; Zhang, Y.; Huang, R.; Bian, J.; Zheng, C.S.; Sun, H.X.; Chen, Z.H.; Sun, N.; An, R.; et al. High-throughput synergy screening identifies microbial metabolites as combination agents for the treatment of fungal infections. *Proc. Natl. Acad. Sci. USA* **2007**, *104*, 4606–4611. [CrossRef]

28. Campos, F.F.; Sales, P.A.; Romanha, A.J.; Araujo, M.S.S.; Siqueira, E.P.; Resende, J.M.; Alves, T.M.A.; Martins, O.A.; dos Santos, V.L.; Rosa, C.A.; et al. Bioactive endophytic fungi isolated from *Caesalpinia echinata* Lam. (Brazilwood) and identification of beauvericin as a trypanocidal metabolite from *Fusarium* sp. *Mem. Inst. Oswaldo Cruz* **2015**, *110*, 65–74. [CrossRef]

29. Aamir, S.; Sutar, S.; Singh, S.K.; Baghela, A. A rapid and efficient method of fungal genomic DNA extraction, suitable for PCR based molecular methods. *Plant Pathol. Quar.* **2015**, *5*, 74–81. [CrossRef]

30. White, T.J.; Bruns, T.; Lee, S.; Taylor, J. Amplification and direct sequencing of fungal ribosomal RNA genes for phylogenetics. In *PCR Protocols: A Guide to Methods and Applications*; Innes, M.A., Gelfand, D.H., Sninsky, J.J., White, T.J., Eds.; Academic Press: Cambridge, MA, USA, 1990; pp. 315–322.

31. O'Donnell, K.; Kistler, H.C.; Cigelnik, E.; Ploetz, R.C. Multiple evolutionary origins of the fungus causing Panama disease of banana: Concordant evidence from nuclear and mitochondrial gene genealogies. *Proc. Natl. Acad. Sci. USA* **1998**, *95*, 2044–2049. [CrossRef]

32. Vilgalys, R.; Sun, B.L. Ancient and recent patterns of geographic speciation in the oyster mushroom *Pleurotus* revealed by phylogenetic analysis of ribosomal DNA-sequences. *Proc. Natl. Acad. Sci. USA* **1994**, *91*, 7832. [CrossRef] [PubMed]

33. Vilgalys, R.; Hester, M. Rapid genetic identification and mapping of enzymatically amplified ribosomal DNA from several *Cryptococcus* species. *J. Bacteriol.* **1990**, *172*, 4238–4246. [CrossRef] [PubMed]

34. Liu, Y.J.J.; Whelen, S.; Benjamin, D.H. Phylogenetic relationships among ascomycetes: Evidence from an RNA polymerase II subunit. *Mol. Biol. Evol.* **1999**, *16*, 1799–1808. [CrossRef] [PubMed]

35. Watanabe, M.; Yonezawa, T.; Lee, K.; Kumagai, S.; Sugita-Konishi, Y.; Goto, K.; Hara-Kudo, Y. Molecular phylogeny of the higher and lower taxonomy of the *Fusarium* genus and differences in the evolutionary histories of multiple genes. *BMC Evol. Biol.* **2011**, *11*, 322. [CrossRef] [PubMed]

36. Peterson, S.W. Multilocus DNA sequence analysis shows that *Penicillium biourgeianum* is a distinct species closely related to *P. brevicompactum* and *P. olsonii*. *Mycol. Res.* **2004**, *108*, 434–440. [CrossRef]

37. Katoh, K.; Standley, D.M. MAFFT Multiple sequence alignment software version 7: Improvements in performance and usability. *Mol. Biol. Evol.* **2013**, *30*, 772–780. [CrossRef]

38. Larsson, A. AliView: A fast and lightweight alignment viewer and editor for large data sets. *Bioinformatics* **2014**, *30*, 3276–3278. [CrossRef]

39. Kalyaanamoorthy, S.; Minh, B.Q.; Wong, T.K.F.; von Haeseler, A.; Jermiin, L.S. ModelFinder: Fast model selection for accurate phylogenetic estimates. *Nat. Methods* **2017**, *14*, 587. [CrossRef]

40. Nguyen, L.T.; Schmidt, H.A.; von Haeseler, A.; Minh, B.Q. IQ-TREE: A fast and effective stochastic algorithm for estimating maximum-likelihood phylogenies. *Mol. Biol. Evol.* **2015**, *32*, 268–274. [CrossRef]

41. Aberkane, A.; Cuenca-Estrella, M.; Gomez-Lopez, A.; Petrikkou, E.; Mellado, E.; Monzón, A.; Rodriguez-Tudela, J.L. Comparative evaluation of two different methods of inoculum preparation for antifungal susceptibility testing of filamentous fungi. *J. Antimicrob. Chemother.* **2002**, *50*, 719–722. [CrossRef]

42. Lee, H.; Kang, J.; Kim, B.; Park, S.; Lee, C. Statistical optimization of culture conditions for the production of enniatins H, I, and MK1688 by *Fusarium oxysporum* KFCC 11363P. *J. Biosci. Bioeng.* **2011**, *111*, 279–285. [CrossRef] [PubMed]

43. Logrieco, A.; Moretti, A.; Castella, G.; Kostecki, M.; Golinski, P.; Ritieni, A.; Chelkowski, J. Beauvericin production by *Fusarium* species. *Appl. Environ. Microbiol.* **1998**, *64*, 3084–3088. [CrossRef] [PubMed]

44. Touchstone, J.C. *Practice of Thin Layer Chromatography*, 3rd ed.; John Wiley and Sons, Inc.: Hoboken, NJ, USA, 1992; pp. 1–14.

45. Fessenden, R.J.; Fessenden, J.S.; Feist, P. *Organic Laboratory Techniques*, 3rd ed.; Cengage Learning: Boston, MA, USA; Brooks/Cole Thomson Leaning, Inc.: Belmont, CA, USA, 2001; pp. 133–138.

46. Xu, L.J.; Liu, Y.S.; Zhou, L.G.; Wu, J.Y. Optimization of a liquid medium for beauvericin production in *Fusarium redolens* Dzf2 mycelial culture. *Biotechnol. Bioprocess Eng.* **2010**, *15*, 460–466. [CrossRef]

47. Teh, C.H.; Nazni, W.A.; Nurulhusna, A.; Norazah, A.; Lee, H.L. Determination of antibacterial activity and minimum inhibitory concentration of larval extract of fly via resazurin-based turbidometric assay. *BMC Microbiol.* **2017**, *17*, 36. [CrossRef]

48. Foerster, S.; Desilvestro, V.; Hathaway, L.J.; Althaus, C.L.; Unemo, M. A new rapid resazurin-based microdilution assay for antimicrobial susceptibility testing of *Neisseria gonorrhoeae*. *J. Antimicrob. Chemother.* **2017**, *72*, 1961–1968. [CrossRef]

49. Adjou, E.S.; Kouton, S.; Dahouenon-Ahoussi, E.; Sohounhloue, C.K.; Soumanou, M.M. Antifungal activity of *Ocimum canum* essential oil against toxinogenic fungi isolated from peanut seeds in post-harvest in Benin. *Int. Res. J. Biol.* **2012**, *1*, 20–26.

50. Philippe, S.; Soua¨ıbou, F.; Guy, A.; Sébastien, D.T.; Boniface, Y.; Paulin, A.; Issaka, Y.; Dominique, S. Chemical composition and antifungal activity of essential oil of fresh leaves of *Ocimum gratissimum* from Benin against six mycotoxigenic fungi isolated from traditional cheese *wagashi*. *Res. J. Biol. Sci.* **2012**, *1*, 22–27.

51. Jinesh, V.K.; Jaishree, V.; Badami, S.; Shyam, W. Comparative evaluation of antioxidant properties of edible and non-edible leaves of *Anethum graveolens* Linn. *Indian J. Nat. Prod. Reso.* **2010**, *1*, 168–173.

52. Elizabeth, K.; Rao, M.W.A. Oxygen radical scavenging activity of *Curcumin*. *Int. J. Pharmaceu.* **1990**, *58*, 237–240.

53. Jamaluddin; Goswami, M.S.; Ojha, B.M. *Fungi of India 1989–2001*; Scientific Publishers: Jodhpur, India, 2004; pp. 1–326.

54. Manoharachary, C.; Atri, N.S.; Devi, T.P.; Kamil, D.; Singh, S.K.; Singh, A.P. *Bilgrami's Fungi of India List and References (1988–2020)*; Today & Tomorrow's Printers and Publishers: New Delhi, India, 2022; pp. 1–475.

55. Santos, A.C.D.; Trindade, J.V.C.; Lima, C.S.; Barbosa, R.D.; da Costa, A.F.; Tiago, P.V.; Oliveira, N. Morphology, phylogeny, and sexual stage of *Fusarium caatingaense* and *Fusarium pernambucanum*, new species of the *Fusarium incarnatum-equiseti* species complex associated with insects in Brazil. *Mycologia* **2019**, *111*, 244–259. [CrossRef]

56. Lombard, L.; Sandoval-Denis, M.; Lamprecht, S.C.; Crous, P.W. Epitypification of *Fusarium oxysporum*—Clearing the taxonomic chaos. *Persoonia* **2019**, *43*, 1–47. [CrossRef]

57. Vanwyk, P.S.; LOS, O.; Palterl, G.D.C.; Marasas, W.F.O. Geographic distribution and pathogenicity of *Fusarium* species associated with crown rot of wheat in the orange free state, South Africa. *Phytophylactica* **1987**, *19*, 271–274.

58. Matic, S.; Tabone, G.; Gullino, M.L.; Garibaldi, A.; Guarnaccia, V. Emerging leafy vegetable crop diseases caused by the *Fusarium incarnatum-equiseti* species complex. *Phytopathol. Mediterr.* **2020**, *59*, 303–317. [CrossRef]

59. Sever, Z.; Ivić, D.; Kos, T.; Miličević, T. *Fusarium* rot in stored apples in Croatia *Arh. Hig. Rada. Toksikol.* **2012**, *63*, 463–470. [CrossRef]

60. Frisullo, S.; Logrieco, A.; Moretti, A.; Grammatikaki, G.; Bottalico, A. Banana Corm and Root Rot by *Fusarium compactum*, in Crete. *Phytopathol. Mediterr.* **1994**, *33*, 78–82.

61. Mazhar, M.A.; Jahangir, A.Z.; Banihashemi, Z.; Izadpanah, K. First report of isolation and pathogenicity of *Fusarium compactum* on safflower. *Iran. J. Plant Pathol.* **2013**, *49*, 277.

62. Brizuela, A.M.; De la Lastra, E.; Marin-Guirao, J.I.; Galvez, L.; de Cara-Garcia, M.; Capote, N.; Palmero, D. *Fusarium* consortium populations associated with *Asparagus* crop in Spain and their role on field decline syndrome. *J. Fungi* **2020**, *6*, 336. [CrossRef]

63. Maryani, N.; Lombard, L.; Poerba, Y.S.; Subandiyah, S.; Crous, P.W.; Kema, G.H.J. Phylogeny and genetic diversity of the banana *Fusarium* wilt pathogen *Fusarium oxysporum* f. sp. *cubense* in the Indonesian centre of origin. *Stud. Mycol.* **2019**, *92*, 155–194. [CrossRef]

64. Ujat, A.; Vadamalai, G.; Hattori, Y.; Nakashima, C.; Wong, C.; Zulperi, D. Current classification and diversity of *Fusarium* species complex, the causal pathogen of *Fusarium* wilt disease of banana in Malaysia. *Agronomy* **2021**, *11*, 1955. [CrossRef]

65. Maryani, N.; Sandoval-Denis, M.; Lombard, L.; Crous, P.W.; Kema, G.H.J. New endemic *Fusarium* species hitch-hiking with pathogenic *Fusarium* strains causing Panama disease in small-holder banana plots in Indonesia. *Persoonia* **2019**, *43*, 48–69. [CrossRef]

66. Wang, M.M.; Chen, Q.; Diao, Y.Z.; Duan, W.J.; Cai, L. *Fusarium incarnatum-equiseti* complex from China. *Persoonia* **2019**, *43*, 70–89. [CrossRef]

67. Lu, Y.A.; Qiu, J.B.; Wang, S.F.; Xu, J.H.; Ma, G.Z.; Shi, J.R.; Bao, Z.H. Species diversity and toxigenic potential of *Fusarium incarnatum-equiseti* species complex isolates from rice and soybean in China. *Plant Dis.* **2021**, *105*, 2628–2636. [CrossRef]

68. Yi, R.H.; Lian, T.; Su, J.J.; Chen, J. First report of internal black rot on *Carica papaya* fruit caused by *Fusarium sulawesiense* in China. *Plant Dis.* **2022**, *106*, 319. [CrossRef]

69. Lu, M.; Zhang, Y.; Li, Q.; Huang, S.; Tang, L.; Chen, X.; Guo, T.; Mo, J.; Ma, L. First report of leaf blight caused by *Fusarium pernambucanum* and *Fusarium sulawesiense* on Plum in Sichuan, China. *Plant Dis.* **2022**. [CrossRef]

70. Araújo, M.B.M.; Moreira, G.M.; Nascimento, L.V.; Nogueira, G.D.; Nascimento, S.R.D.; Pfenning, L.H.; Ambrosio, M.M.D. *Fusarium* rot of melon is caused by several *Fusarium* species. *Plant Pathol.* **2021**, *70*, 712–721. [CrossRef]

71. Urbaniak, M.; Stepien, L.; Uhlig, S. Evidence for naturally produced beauvericins containing N-Methyl-Tyrosine in Hypocreales fungi. *Toxins* **2019**, *11*, 182. [CrossRef]

72. Geiser, D.M.; Jimenez-Gasco, M.D.; Kang, S.C.; Makalowska, I.; Veeraraghavan, N.; Ward, T.J.; Zhang, N.; Kuldau, G.A.; O'Donnell, K. *FUSARIUM*-ID v. 1.0: A DNA sequence database for identifying *Fusarium*. *Eur. J. Plant Pathol.* **2004**, *110*, 473–479. [CrossRef]
73. Wingfield, M.J.; De Beer, Z.W.; Slippers, B.; Wingfield, B.D.; Groenewald, J.Z.; Lombard, L.; Crous, P.W. One fungus, one name promotes progressive plant pathology. *Mol. Plant Pathol.* **2012**, *13*, 604–613. [CrossRef]
74. Aiyaz, M.; Thimmappa Divakara, S.; Mudili, V.; George Moore, G.; Kumar Gupta, V.; i Yli-Mattila, T.; Chandra Nayaka, S.; Ramachandrappa Niranjana, S. Molecular diversity of seed-borne *Fusarium* species associated with maize in India. *Curr. Genom.* **2016**, *17*, 132–144. [CrossRef]
75. Gupta, S.; Krasnoff, S.B.; Underwood, N.L.; Renwick, J.A.A.; Roberts, D.W. Isolation of beauvericin as an insect toxin from *Fusarium semitectum* and *Fusarium moniliforme* var. *subglutinans*. *Mycopathologia* **1991**, *155*, 185–189. [CrossRef]
76. Moretti, A.; Logrieco, A.; Bottalico, A.; Ritieni, A.; Fogliano, V.; Randazzo, G. Diversity in beauvericin and fusaproliferin production by different populations of *Gibberella fujikuroi* (*Fusarium* section Liseola). *Sydowia* **1997**, *48*, 44–56.
77. Azliza, I.N.; Hafizi, R.; Nurhazrati, M.; Salleh, B. Production of major mycotoxins by *Fusarium* species isolated from wild grasses in peninsular Malaysia. *Sains Malays.* **2014**, *43*, 89–94.
78. Xu, L.; Wang, J.; Zhao, J.; Li, P.; Shan, T.; Wang, J.; Li, X.; Zhou, L. Beauvericin from the endophytic fungus, *Fusarium redolens*, isolated from *Dioscorea zingiberensis* and its antibacterial activity. *Nat. Prod. Commun.* **2010**, *5*, 811–814. [CrossRef]
79. Caicedo, N.H.; Davalos, A.F.; Puente, P.A.; Rodriguez, A.Y.; Caicedo, P.A. Antioxidant activity of exo-metabolites produced by *Fusarium oxysporum*: An endophytic fungus isolated from leaves of *Otoba gracilipes*. *Open Microbiol. J.* **2019**, *8*, e903. [CrossRef]
80. Menezes, B.S.; Solidade, L.S.; Conceição, A.A.; Santos Junior, M.N.; Leal, P.L.; de Brito, E.S.; Canuto, K.M.; Mendonça, S.; de Siqueira, F.G.; Marques, L.M. Pigment production by *Fusarium solani* BRM054066 and determination of antioxidant and anti-inflammatory properties. *AMB Express* **2020**, *10*, 117. [CrossRef]

Journal of **Fungi**

MDPI

Article

Isolation and Characterization of a Novel Hydrophobin, Sa-HFB1, with Antifungal Activity from an Alkaliphilic Fungus, *Sodiomyces alkalinus*

Anastasia E. Kuvarina [1], Eugene A. Rogozhin [1,2], Maxim A. Sykonnikov [1], Alla V. Timofeeva [3], Marina V. Serebryakova [3], Natalia V. Fedorova [3], Lyudmila Y. Kokaeva [4], Tatiana A. Efimenko [1], Marina L. Georgieva [1,4] and Vera S. Sadykova [1,*]

[1] Laboratory of Taxonomic Study and Collection of Cultures of Microorganisms, Gause Institute of New Antibiotics, St. Bolshaya Pirogovskaya, 11, 119021 Moscow, Russia; nastena.lysenko@mail.ru (A.E.K.); rea21@list.ru (E.A.R.); sukonnikoff.maxim@yandex.ru (M.A.S.); efimen@inbox.ru (T.A.E.); i-marina@yandex.ru (M.L.G.)

[2] Shemyakin and Ovchinnikov Institute of Bioorganic Chemistry, RAS, St. Miklukho-Maklaya, 16/10, 117997 Moscow, Russia

[3] Belozersky Institute of Physico-Chemical Biology, Lomonosov Moscow State University, 119991 Moscow, Russia; 2.04.cochon@gmail.com (A.V.T.); mserebr@mail.ru (M.V.S.); fedorova@belozesky.msu.ru (N.V.F.)

[4] Faculty of Biology, Lomonosov Moscow State University, 1-12 Leninskie Gory, 119234 Moscow, Russia; kokaeval@gmail.com

* Correspondence: sadykova_09@mail.ru

Citation: Kuvarina, A.E.; Rogozhin, E.A.; Sykonnikov, M.A.; Timofeeva, A.V.; Serebryakova, M.V.; Fedorova, N.V.; Kokaeva, L.Y.; Efimenko, T.A.; Georgieva, M.L.; Sadykova, V.S. Isolation and Characterization of a Novel Hydrophobin, Sa-HFB1, with Antifungal Activity from an Alkaliphilic Fungus, *Sodiomyces alkalinus. J. Fungi* 2022, 8, 659. https://doi.org/10.3390/jof8070659

Academic Editor: Laurent Dufossé

Received: 3 May 2022
Accepted: 20 June 2022
Published: 23 June 2022

Publisher's Note: MDPI stays neutral with regard to jurisdictional claims in published maps and institutional affiliations.

Abstract: The adaptations that alkaliphilic microorganisms have developed due to their extreme habitats promote the production of active natural compounds with the potential to control microorganisms, causing infections associated with healthcare. The primary purpose of this study was to isolate and identify a hydrophobin, Sa-HFB1, from an alkaliphilic fungus, *Sodiomyces alkalinus*. A potential antifungal effect against pathogenic and opportunistic fungi strains was determined. The MICs of Sa-HFB1 against opportunistic and clinical fungi ranged from 1 to 8 µg/mL and confirmed its higher activity against both non- and clinical isolates. The highest level of antifungal activity (MIC 1 µg/mL) was demonstrated for the clinical isolate *Cryptococcus neoformans* 297 m. The hydrophobin Sa-HFB1 may be partly responsible for the reported antifungal activity of *S. alkalinus*, and may serve as a potential source of lead compounds, meaning that it can be developed as an antifungal drug candidate.

Keywords: alkaliphilic fungi; antifungal peptide; *Sodiomyces alkalinus*; hydrophobin

1. Introduction

Extremophiles are a unique group of microorganisms capable of producing numerous bioactive compounds [1–3]. Alkaliphiles are extremophiles whose optimum rate of growth is at a high pH (9 to 11) [4,5]. Past decades of research have shown that the diversity of alkaliphilic fungi is less when compared to that of alkaliphilic bacteria. By now, the list of known alkaliphilic fungi (it is noteworthy that all are within the Ascomycota division) includes about 30 species [5–7]. Members of the genus *Sodiomyces* were isolated from habitats with alkaline conditions (pH of $\geq$10), such as soda salterns and the edges of the soda lakes of Russia, Mongolia, Kenya, and Tanzania. All species of this genus (*S. tronii, S. magadii*, and *S. alkalinus*) are obligate alkaliphiles [5,8,9].

The demand for novel and improved medicine from biological sources to cater to the biopharmaceutical sector has increased significantly in recent years. Investigating unexplored ecological units on the globe synergizes with the concept of investigating the least or not explored species of microbes [10]. Alkaliphilic fungi show a great potency to produce enzymes and antibiotics due to their ability to survive and persist in extreme environmental conditions [11–13]. From previous studies on alkaliphilic and alkali-tolerant fungi, the *Emericellopsis* genus was found to produce novel peptides with antimicrobial and antitumor activity [14,15].

The screening of metabolites of eight obligate alkaliphilic strains of *Sodiomyces alkalinus* obtained from soda soils has revealed antifungal activity against different fungal taxa, including human pathogenic isolates. As a result of the determination of the spectrum and yield of antibiotic compounds, a promising producer of *Sodiomyces alkalinus* was selected from the most active strains: 8KS17-10. This producer exhibited antifungal activity against opportunistic fungi, as well as pathogenic clinical isolates of molds and yeasts—pathogens of systemic mycoses. We have previously identified compounds as antimicrobial peptides based on the totality of the identified structural features (molecular weight, absorption ratio at certain wavelengths) [16].

In this study, we focus on the purification, identification, and antifungal activity of a novel biomolecule that has been identified as a hydrophobin and denoted as Sa-HFB1. Hydrophobins are a large family of small, cysteine-rich proteins produced by filamentous fungi. HFBs are characterized by their small size and their amphipathic nature at hydrophilic/hydrophobic interfaces, and are currently attracting great interest from the biotechnology industry [17]. This is the first time that the antifungal activity of a hydrophobin has been described.

2. Materials and Methods

2.1. Strains of Sodiomyces alkalinus

The objects of study were two alkaliphilic strains of *S. alkalinus* Grum-Grzhimaylo, Debets, and Bilanenko (Plectosphaerellaceae, Glomerellales, Hypocreomycetidae, Sordariomycetes, Pezizomycotina, Ascomycota): the F11 ex-type strain (=CBS 110278) [9] and the 8KS17-10 strain (=SLF 0117.0810). The 8KS17-10 strain was identified as *S. alkalinus* based on morphological features (Supplementary Figure S1) and on a comparison of the sequences within the ITS region of strain 8KS17-10 with those of an ex-type strain (NCBI ID ON146325 and JX158405, respectively, with a 99.51% similarity score) and sequences in the GenBank database (https://www.ncbi.nlm.nih.gov/genbank accessed on 10 April 2022) [9]. The cultures were obtained from the Fungi of Extreme Conditions collection of the Department of Mycology and Algology, Faculty of Biology, Lomonosov Moscow State University (Moscow, Russia).

2.2. Cultivation of the Sodiomyces alkalinus and the Extraction of the Hydrophobin

Fungi were cultivated according to the previous protocol on a special alkaline medium, containing, per liter, the following: (1) Na_2CO_3—24 g, $NaHCO_3$—6 g, $NaCl$—5 g, KNO_3—1 g, and K_2HPO_4—1 g; and (2) malt extract—17 g, yeast extract—1 g. Components 1 and 2 were autoclaved separately for 20 min at 120 °C and mixed together after cooling, which resulted in a final pH of 10 [9]. Fungi were grown in 500 mL flasks for 14 days at 26 °C in Erlenmeyer flasks [15]. The culture liquid (CL) was separated by filtration through membrane filters MF-Millipore, Merk (Darmstadt, Germany) on a Seitz funnel under vacuum. To isolate the antibiotic substances, the CL of the producers was extracted three times with ethyl acetate in an organic solvent/CL ratio of 1:5. The extracts were evaporated under vacuum on a Rotavapor rotary evaporator (Buchi, Flawil, Switzerland) to dryness at 42 °C, the residue was dissolved in aqueous 70% ethanol, and the alcohol concentrates were obtained.

2.3. Purification and Identification of the Hydrophobin

2.3.1. HPLC Analysis

Further separation of the active fractions (after extraction) was carried out via analytical reversed-phase high-performance liquid chromatography (RP-HPLC) with an XBridge 5 μm 130 A column with a size of 250 × 4.6 mm (Waters, Maynooth, Ireland) in a growing linear gradient of the acetonitrile concentration as a mobile phase (eluent A, 0.1% trifluoroacetic acid (TFA) (HPLC, Sigma-Aldrich, Darmstadt, Germany)) (in water MQ; eluent B, 80% acetonitrile in 0.1% aqueous TFA) at a flow rate of 950 L/min. Ultragradient acetonitrile (Panreac, Barcelona, Spain) and TFA (HPLC, Sigma-Aldrich, Darmstadt, Germany) were used for the RP-HPLC. The substances to be separated were determined at a wavelength of 214 nm in the concentration gradient of eluent B: 16–28% in 12 min; 28–55% in 27 min; 55–75% in 20 min; and 75–85% in 10 min. This was followed by isocratic elution for 25 min. The rechromatography of an active compound was performed by using a Jupiter C5 4.6 × 250 mm analytical HPLC column (Phenomenex, Torrance, CA, USA) at the same gradient conditions and at a flow rate of 1 mL/min. The absorbance (D) was determined at a wavelength of 214 nm and a mobile phase flow rate of 4 mL/min. The fractions obtained during the RP-HPLC, which correspond to individual peaks, were collected manually, and the excess of the organic solvent (acetonitrile) was then removed via evaporation in a Speed-Vac vacuum centrifuge (Thermo Fisher Scientific, Waltham, MA, USA) and lyophilized (Labconco, Kansas, MO, USA) to remove residual amounts of TFA [16]. The spectrum of the antimicrobial action of the substances contained in the fractions was determined via disk diffusion, as described below.

2.3.2. RP-HPLC Analysis of the Hydrophobin Sa-HFB1 Tryptic Hydrolysate

The hydrophobin obtained after the trypsinolysis reaction mixture was analyzed with the RP-HPLC method described in [18].

2.3.3. Matrix-Assisted Laser Desorption Ionization Mass Spectrography (MALDI-MS) and Tandem Mass Spectography (MS I MS)

Analysis was performed on a MALDI time-of-flight (ToF) mass spectrometer (Ultraflex II Bruker, Bremen, Germany) equipped with a neodymium-doped (Nd) laser. The $[M+H]^+$ molecular ions were measured in the reflector mode; the accuracy of the mass peak measurement was within 0.005%. The identification of the proteins was carried out by a peptide fingerprint search using Mascot software (Bremen, Germany) [19] through the NCBL mammalian protein database with the indicated accuracy. The search allowed for the possible oxidation of methionine by environmental oxygen and the modification of cysteine with acrylamide; where a score was >71, protein matches were considered significant (pb0.05).

2.3.4. Hydrophobin Sa-HFB1 Peptides Preparation for MALDI-TOF MS Analysis

An aliquot of the peptide solution (1 μL) obtained by trypsinolysis was applied on the steal target, where it was mixed up with 0.3 μL of 2,5-dexidroxobenzoic acid (Sigma-Aldrich, Darmstadt, Germany) in a solution of 20 mg/mL 20% acetonitrile (Sigma-Aldrich, Darmstadt, Germany) in 0.5% TFA (Sigma-Aldrich, Darmstadt, Germany). Drops on the target were dried at the atmospheric temperature and pressure.

2.3.5. Hydrophobin (Sa-HFB1) Derivate Preparation

A sample of the hydrophobin Sa-HFB1 in ethanol (20 μL) with a concentration of 1 μg/μL was entirely dried in a rotor evaporator (Labconco, Kansas, MO, USA). Then, 32 μL of a 100 mM NH_4HCO_3 water solution was added, and the mixture was heated up for 3 h. After this was performed, 4 μL of 2 M DTT (Serva, Heidelberg, Germany) was added, and it was stored at room temperature for 30 min. Finally, a portion of 0.6 M IAA (Sigma-Aldrich, St. Louis, MO, USA) in a volume of 4 μL was added, and the total solution was stored for 30 min at room temperature in full darkness.

2.3.6. Hydrophobin (Sa-HFB1) Proteolysis

To the solution obtained in the derivatization reaction process described in the previous paragraph, 40 μL of modified trypsin (Promega, Madison, WI, USA) in a 100 mM NH_4HCO_3 solution was added at a ratio of enzyme to substrate of 1:50. Thus, the obtained mixture was stored at 37 °C for 1 h, after which some trypsin was additionally added to the reaction mixture, and then the solution was stored for 4 h at 37 °C.

2.3.7. Amino Acid Analysis of a Hydrophobin Sa-HFB1 Sample

Before analysis, a probe was used to perform protein hydrolysis with the method presented in [20], and then was analyzed in agreement with the method exhibited in [21].

2.3.8. Edman Sequencing

Automated N-terminal sequencing was performed on a PSSQ-33A sequencer (Shimadzu Corp., Kyoto, Japan) according to the manufacturer's protocol. The identification of the PTH derivatives of amino acid residues was conducted was carried out using the LabSolutions software version 1.10(ROM 3.0) (Shimadzu Corp., Kyoto, Japan).

2.4. Sequences Analysis and Primer Design

Based on the hydrophobin Sa-HFB1 DNA sequence obtained from the *S. alkalinus* F11 whole-genome sequence (NCBI ID XM_028612330.1 and NW_021167086.1), new primers were designed to amplify the corresponding genomic regions in various strains of *Sodiomyces* species. Sequences were aligned using the Geneious program (Geneious version 7.1.5; Biomatters Ltd., Auckland, New Zealand) [22] with the default settings for multiple alignments. All of the primers used were designed by using the PRIMER3 online tool [23] and Geneious software (Geneious version 7.1.5; Biomatters Ltd., Auckland, New Zealand) Primers were designed from the intron boundaries of the exons to amplify the coding regions. The criteria for the selection of the best primer combination was a clear sequence chromatogram. The final primer pair used in the study was Hyd1F (5′-AATCCCTTATCTTTTCCAGCTTAACC-3′) and Hyd580r (5′-CAGACACGACGAGTTCGACA-3′).

2.5. Hydrophobin Gene Sequencing

The genomic DNA was isolated using a set of DNeasy PowerSoil Kit reagents (Qiagen Inc., Carlsbad, CA, USA), following the manufacturer's instructions. The final volume of the 50 μL PCR mix included the following: 25 μL of 2X PCR Master Mix (Thermo Fisher Scientific, Waltham, MA, USA), 0.5 μM of each primer, and 1–100 ng of isolated DNA and water (nuclease-free). PCR was performed according to the following scheme: (1) 94 °C for 5 min, (2) 33 cycles with alternating temperature intervals: 94 °C for 1 min, 60 °C for 1 min, and 72 °C for 1 min, and (3) 72 °C for 7 min. The DNA fragments were sequenced by the Sanger method on an Applied Biosystems 3500 Series Genetic Analyzer. Sequence data were submitted to GenBank (NCBI ID ON149453).

2.6. Biological Assays

Antifungal Activity

The antifungal activity was measured by the disk diffusion method. Disks of 6 mm diameter, containing 40 μL of sample, were deposited on PDA agar plates (Sigma-Aldrich, St. Louis, MO, USA). The spectrum of the antifungal activity of the CL, extracts, and individual compounds was determined on test cultures of mycelial and yeast fungi from the collection of cultures of the Gause Institute of New Antibiotics (Moscow, Russia). Opportunistic mold and yeast test cultures of the fungal species *Aspergillus fumigatus* KPB F-37, *Penicillium brevicompactum* VKM F-4481, *A. niger* INA 00760, and *Candida albicans* ATCC 2091 were used. The diameters of the inhibition zones were measured after 24 h at 28 °C. The sensitivity of the test organism was controlled with standard disks containing amphotericin B (AmpB), itraconazole (IZ), fluconazole (FZ), and voriconazole (VOR) (40 μg/disk) as positive controls.

The spectrum of antifungal action was also evaluated on clinical isolates of molds and yeasts pathogens of opportunistic pneumomycosis of the bronchi and lungs—in tuberculosis patients with multi-resistance to the antibiotics–azoles used in clinical practice, from the collection of the mycological laboratory of the Moscow City Scientific and Practical

Center for Tuberculosis Control (Russia): *Candida albicans* 1582 m, *Pichia kudriavzevii* (former *C. glabrata*) 1402 m, *Nakaseomyces glabrataa* 1447 m, *C. parapsilopsis* 571 m, *C. tropicalis* 156 m, *Cryptococcus neoformans* 297 m, *Aspergillus fumigatus* 390 m, and *A. niger* 219 m.

The minimal inhibitory concentration (MIC) value of each individual compound was determined using the broth twofold microdilution method according to the CLSI/NCCLS documents M27-A3 and M38-A2. The MIC was determined for the yeast fungi *Candida albicans* ATCC 2091, *C. albicans* 1582 m, and *Cryptococcus neoformans* 297 m, as well as for 48 h for the molds *Aspergillus niger* INA 00760, *A. fumigatus* KPB F-37, and *A. fumigatus* 390 m. The minimum inhibitory concentrations (MIC) were determined by a two-fold serial microdilution method in a liquid culture medium RPMI 1640 with Lglutamine without sodium bicarbonate for fungi in accordance with the requirements of the Institute of Clinical and Laboratory Standards [24,25].

The experiments were carried out in three to five replicates. Statistical processing of the results and the assessment of the reliability of the differences in mean values were carried out according to the Student's *t*-test for a probability level of at least 95% with Microsoft Excel 2010 and Statistica 10.0 software (Palo Alto, CA, USA).

3. Results

3.1. Isolation of the Active Polypeptide from Culture Broth Extract

According to previous data, a high level of active compounds from culture liquids was obtained for the *S. alkalinus* 8KS17-10 strain. Analytical separation by reversed-phase HPLC for the ethyl acetate extract was carried out according to the data described earlier [16]. The maximum amount of an individual peak was determined for the 8KS17-10 strain, and it was selected for further investigations. It was shown that an active compound can be attributed to the group of peptides based on the totality of the identified structural features (molecular weight, absorption ratio at certain wavelengths). Previously, a component with a retention time of 31.171 min with strong antifungal activity against a number of clinical fungal and yeast isolates (*Candida* spp., *Aspergillus* spp., and *Cryptococcus neoformans*) was shown [16].

The peptide production yield from the culture liquid and mycelium extracts of *S. alkalinus* on the alkaline medium was achieved in a period of fermentation in 14 days of about 25.54 ± 1.4 mg/L. To obtain the compound for further structural analyses we made three single-type chromatographic separations, and the target peak was collected manually (Figure 1). In the following steps, this compound was rechromatographed to confirm its purity, and an individual molecule was prepared for investigation by analytical methods.

3.2. Structural Analysis

Previously, it has been suggested that the molecule studied is typical of a polypeptide; therefore, its initial structure analysis was conducted by automated Edman sequencing. After ten cycles, a predominant N-terminal amino acid sequence for the analyzed peptide was composed, [1]TYIAXPISLY[10] (Supplementary Figure S3). Searching for potential homologies among NCBI databases using the BLASTP algorithm led to complete matching with the fungal hydrophobin F11 (NCBI ID ROT36721.1/XP_028464527.1). These data were obtained using previously annotated whole-genome shotgun sequences for the ex-type strain of *S. alkalinus*, F11 [26].

The complete matchup for the N-terminal sequence of the isolated peptide with the *S. alkalinus* fungal hydrophobin F11 allowed us to specify that the fifth amino acid residue is cysteine, which can be visualized as a PTH-s-beta-4-pyridylethylated derivative [27,28]. The isolated polypeptide is found and denominated as Sa-HFB1. Hence, the complete amino acid sequence for the precursor of *S. alkalinus*, the class II hydrophobin F11 (NCBI ID XP_028464527.1), is [1]MKFIAVVAALTASLAMAAPTESSTDTTYIA**CPISLYGNAQCCATDI LGLANLDCESPTDVPRDAGHFQRTCADVGKRARCCAIPVLGQALLCIQP**AGAN[99] [26]. Based on NCBI annotation data for ID XP_028464527.1, the current protein precursor has a calculated molecular mass of 10181.6 Da; residues 1–30 are represented a signal peptide, whereas residues 31–95 form a mature hydrophobin sequence (marked in bold), in which the cleaved primary peptide bond is located between Tre26 and Tre27. This allowed for the identification of one new hydrophobin of class II, denoted as Sa-HFB1. Thus, Sa-HFB1 typically has up to five additional amino acid residues at the N-terminus. For the identification of any possible homologies for the complete hydrophobin sequence in genome NCBI ID PRJNA196044, we used a local blast. As a result of the search, only one gene was found to correspond to the hydrophobin (located in the genomic scaffold SODALscaffold8). A further search in the transcriptional data confirmed this fact (according to scaffold SODALDRAFT_334916, NCBI ID XM_028612330.1).

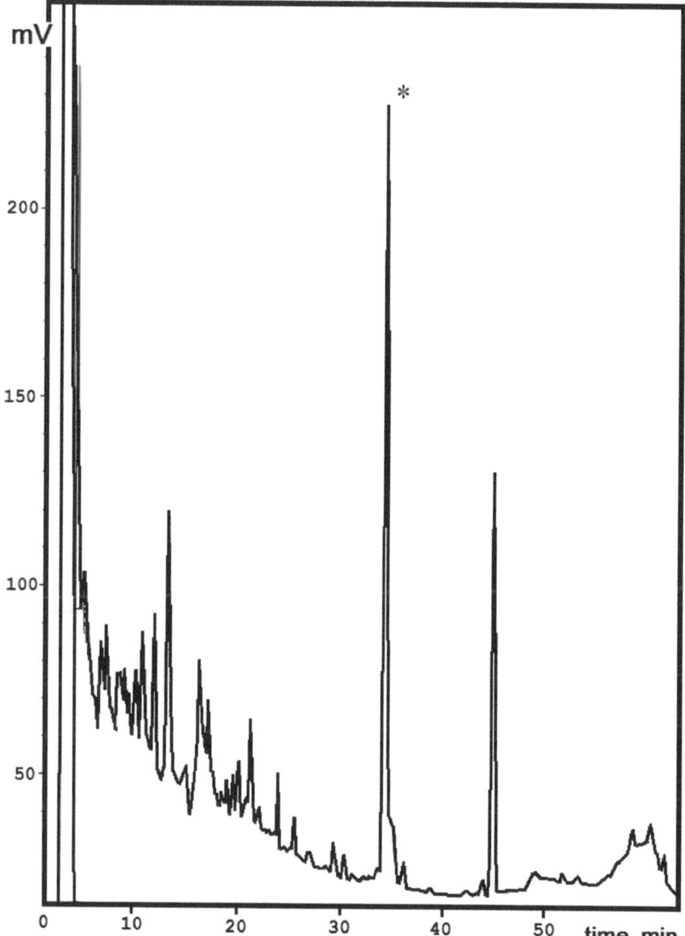

Figure 1. Isolation of Sa-HFB1 from *S. alkalinus* ethyl acetate concentrate. The target peak is indicated by in asterisk.

As a result of the full-genome sequencing of the *S. alkalinus* F11 strain, the nucleotide sequence that encodes the hydrophobin was recently discovered and translated into an amino acid sequence in silico, since a bioinformatics assay was carried out.

Some general information calculated on the template of the predicted amino acid sequence of the hydrophobin, its molecular weight and amino acid composition, is presented in short on Figure 2.

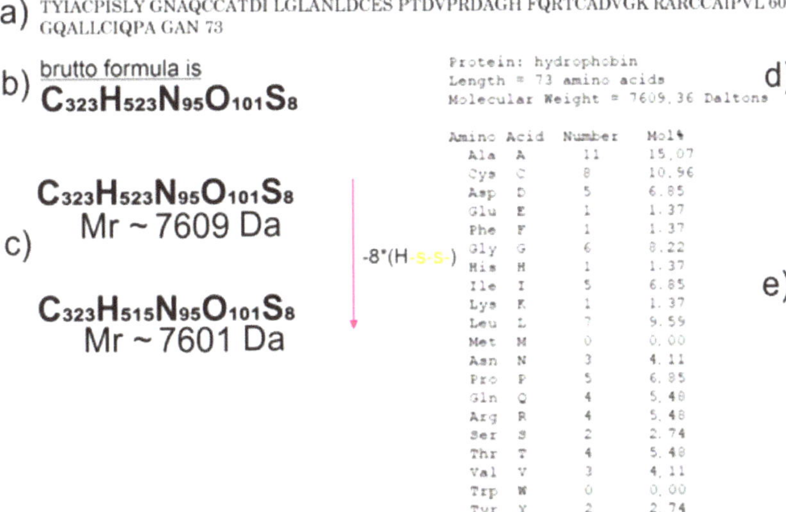

Figure 2. Physicochemical properties of the hydrophobin. (**a**) The sequence of the hydrophobin, annotated under the result of the full-genome sequencing of the ex-type strain of *Sodiomyces alkalinus*, F11. (**b**) Atom composition of the goal protein, impressed in numbers of corresponding atoms. (**c**) Chemical transition in mass and atom composition due to all of the disulfide bonds forming. (**d**) Information about the exact mass of the protein. (**e**) The amino acid composition of the hydrophobin.

To confirm the sequence of the hydrophobin and to examine the conservation of this protein, a sample of the hydrophobin isolate obtained by EtOAc, produced by the *S. alkalinus* 8KS17-10 strain and subsequent chromatographic purification, was analyzed.

3.3. MALDI-TOF MS

Individual fractions were analyzed with the MALDI-TOF MS method to obtain information about masses, consisting of a matrix. The result of the mass spectrometry is depicted in Figure 3a. Several peaks on the specter were received, but only the major one with a mass of 7588 Da ($[M+H]^+$) was of interest. One of the rest peaks with a mass of 3794 Da is a binary ionized form of the major peak ($[M+H]^{++}$), and the others were considered to be any homologues of the hydrophobin.

One real problem after the analyses concerned the mass difference between the annotated hydrophobin with a mass of 7609 Da (with the formation of disulfide bonds—7601 Da) and 7588 Da, presented at the specter. To dissolve the problem, an idea about one amino acid change in the protein was proposed, taking into account 14 Da as the difference between the major peak mass and the annotated sequence mass with a full set of disulfide bonds.

For an analysis, a table with the molecular mass differences between different natural amino acids was created, and the goal mass of 14 Da was discovered and colored in red in Figure 3b. It provided us with an opportunity to propose a set of appropriate amino acid changes, presented in Figure 3c.

3.4. Amino Acid Analysis

As a method for amino acid quantification in an analytical solution, an amino acid analysis was performed. Chromatograms of a standard amino acid solution and of the hydrophobin, obtained via acid hydrolysis, are presented in Figure 4a,b, respectively. In a table in Figure 4c, the amount of amino acids and the numbers of residues in the announced sequence of the hydrophobin (real values), as well as the number of residues as evaluated from the result of the amino acid analysis, assuming that only one protein exists (theoretical values), are shown. In Figure 4d, they are shown graphically as well.

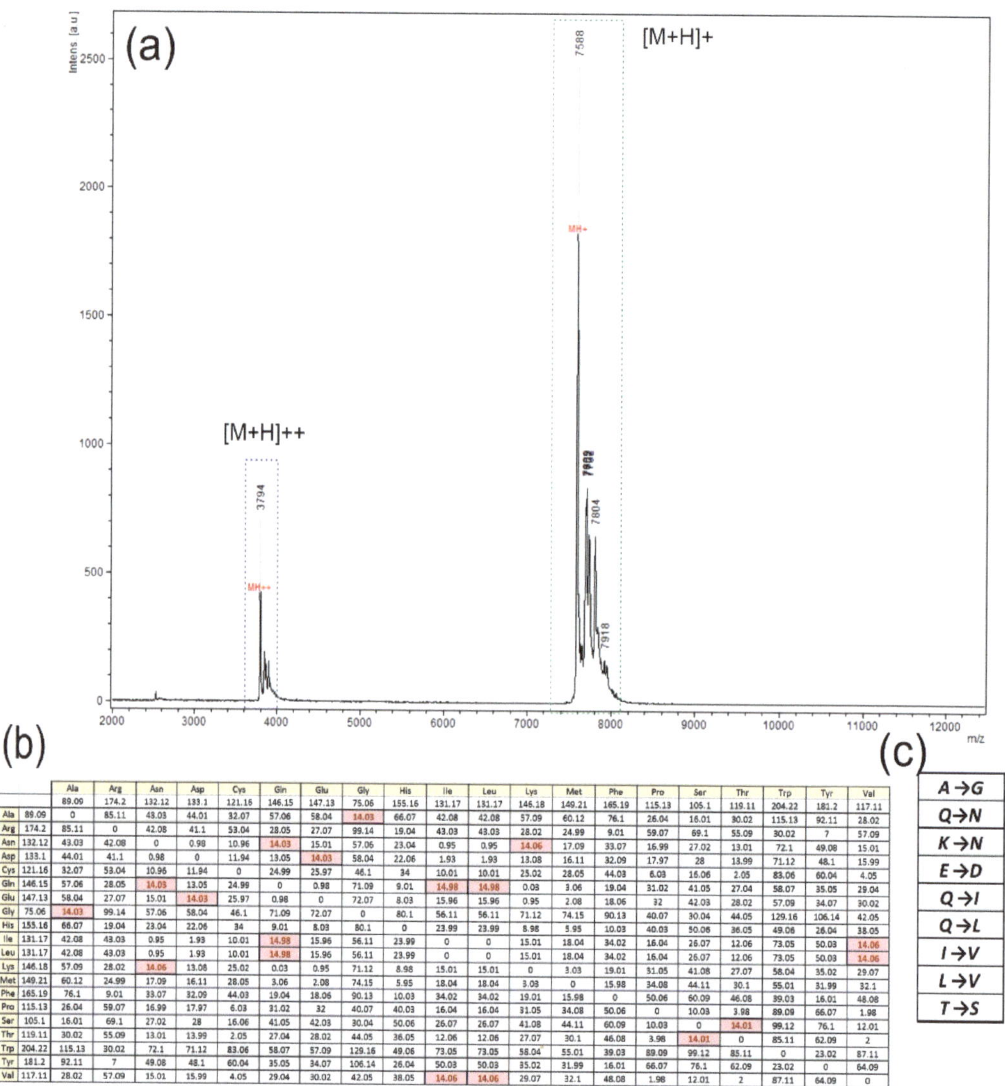

		Ala	Arg	Asn	Asp	Cys	Gln	Glu	Gly	His	Ile	Leu	Lys	Met	Phe	Pro	Ser	Thr	Trp	Tyr	Val
		89.09	174.2	132.12	133.1	121.16	146.15	147.13	75.06	155.16	131.17	131.17	146.18	149.21	165.19	115.13	105.1	119.11	204.22	181.2	117.11
Ala	89.09	0	85.11	43.03	44.01	32.07	57.06	58.04	14.03	66.07	42.08	42.08	57.09	60.12	76.1	26.04	16.01	30.02	115.13	92.11	28.02
Arg	174.2	85.11	0	42.08	41.1	53.04	28.05	27.07	99.14	19.04	43.03	43.03	28.02	24.99	9.01	59.07	69.1	55.09	30.02	7	57.09
Asn	132.12	43.03	42.08	0	0.98	10.96	14.03	15.01	57.06	23.04	0.95	0.95	14.06	17.09	33.07	16.99	27.02	13.01	72.1	49.08	15.01
Asp	133.1	44.01	41.1	0.98	0	11.94	13.05	14.03	58.04	22.06	1.93	1.93	13.08	16.11	32.09	17.97	28	13.99	71.12	48.1	15.99
Cys	121.16	32.07	53.04	10.96	11.94	0	24.99	25.97	46.1	34	10.01	10.01	25.02	28.05	44.03	6.03	16.06	2.05	83.06	60.04	4.05
Gln	146.15	57.06	28.05	14.03	13.05	24.99	0	0.98	71.09	9.01	14.98	14.98	0.03	3.06	19.04	31.02	41.05	27.04	58.07	35.05	29.04
Glu	147.13	58.04	27.07	15.01	14.03	25.97	0.98	0	72.07	8.03	15.96	15.96	0.95	2.08	18.06	32	42.03	28.02	57.09	34.07	30.02
Gly	75.06	14.03	99.14	57.06	58.04	46.1	71.09	72.07	0	80.1	56.11	56.11	71.12	74.15	90.13	40.07	30.04	44.05	129.16	106.14	42.05
His	155.16	66.07	19.04	23.04	22.06	34	9.01	8.03	80.1	0	23.99	23.99	8.98	5.95	10.03	40.03	50.06	36.05	49.06	26.04	38.05
Ile	131.17	42.08	43.03	0.95	1.93	10.01	14.98	15.96	56.11	23.99	0	0	15.01	18.04	34.02	16.04	26.07	12.06	73.05	50.03	14.06
Leu	131.17	42.08	43.03	0.95	1.93	10.01	14.98	15.96	56.11	23.99	0	0	15.01	18.04	34.02	16.04	26.07	12.06	73.05	50.03	14.06
Lys	146.18	57.09	28.02	14.06	13.08	25.02	0.03	0.95	71.12	8.98	15.01	15.01	0	3.03	19.01	31.05	41.08	27.07	58.04	35.02	29.07
Met	149.21	60.12	24.99	17.09	16.11	28.05	3.06	2.08	74.15	5.95	18.04	18.04	3.03	0	15.98	34.08	44.11	30.1	55.01	31.99	32.1
Phe	165.19	76.1	9.01	33.07	32.09	44.03	19.04	18.06	90.13	10.03	34.02	34.02	19.01	15.98	0	50.06	60.09	46.08	39.03	16.01	48.08
Pro	115.13	26.04	59.07	16.99	17.97	6.03	31.02	32	40.07	40.03	16.04	16.04	31.05	34.08	50.06	0	10.03	3.98	89.09	66.07	1.98
Ser	105.1	16.01	69.1	27.02	28	16.06	41.05	42.03	30.04	50.06	26.07	26.07	41.08	44.11	60.09	10.03	0	14.01	99.12	76.1	12.01
Thr	119.11	30.02	55.09	13.01	13.99	2.05	27.04	28.02	44.05	36.05	12.06	12.06	27.07	30.1	46.08	3.98	14.01	0	85.11	62.09	2
Trp	204.22	115.13	30.02	72.1	71.12	83.06	58.07	57.09	129.16	49.06	73.05	73.05	58.04	55.01	39.03	89.09	99.12	85.11	0	23.02	87.11
Tyr	181.2	92.11	7	49.08	48.1	60.04	35.05	34.07	106.14	26.04	50.03	50.03	35.02	31.99	16.01	66.07	76.1	62.09	23.02	0	64.09
Val	117.11	28.02	57.09	15.01	15.99	4.05	29.04	30.02	42.05	38.05	14.06	14.06	29.07	32.1	48.08	1.98	12.01	2	87.11	64.09	0

(c)

A → G
Q → N
K → N
E → D
Q → I
Q → L
I → V
L → V
T → S

Figure 3. Analysis of peaks with the MALDI-TOF MS method. (**a**) The spectrum with two fields of peaks is presented. The main peak, with a mass of 7588 Da, is highlighted with others in one set in the dotted rectangle colored in green, and another set of peaks with the main peak mass of 3794 Da is highlighted with the blue dotted rectangle. Symbols [M+H]$^+$ and [M+H]$^{++}$ mean the masses of the molecular ions plus proton, the single-charged and double-charged ions, respectively. (**b**) The table, including information about the molecular mass differences between each of the natural amino acids; results are presented in Da units, and the goal of 14 Da, meaning between corresponding amino acids, is colored in red. (**c**) The set of possible amino acid single changes in the hydrophobin, calculated over the molecular masses, is presented.

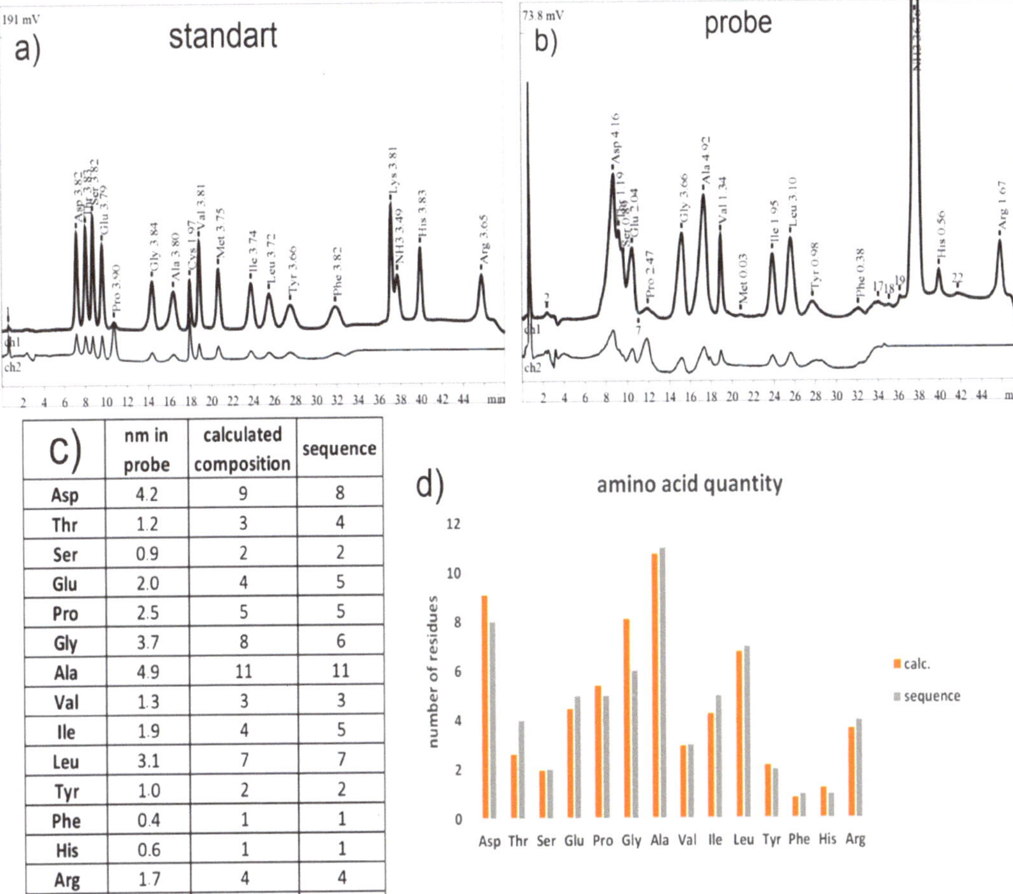

c)	nm in probe	calculated composition	sequence
Asp	4.2	9	8
Thr	1.2	3	4
Ser	0.9	2	2
Glu	2.0	4	5
Pro	2.5	5	5
Gly	3.7	8	6
Ala	4.9	11	11
Val	1.3	3	3
Ile	1.9	4	5
Leu	3.1	7	7
Tyr	1.0	2	2
Phe	0.4	1	1
His	0.6	1	1
Arg	1.7	4	4
total	29.3	64	64

Figure 4. The results of the amino acids' analysis. (**a**) Chromatogram of the standard mixture of amino acids examined at two wavelengths: 440 nm (ch1) and 590 nm (ch2). (**b**) Chromatogram of the real amino acid mixture, obtained via the hydrolysis of the hydrophobin with mineral acids (ch1 and ch2 are the same as those in (**a**)). (**c**) A table with the amino acids' quantification and transformation into numbers of amino acids in a hypothetical protein in a sample (calculated composition), and the amino acid numbers in a query annotated sequence (sequence). Some of the amino acids are not presented in the table (Lys, Met, Trp, and Cys) due to their degradation; still, the mineral acid hydrolysis take place, and the others (Gln and Asn), at the same conditions, transform into Glu and Asp, correspondingly. (**d**) Information about the amino acid distribution in the protein examined via amino acid analysis (orange bars) and the residue numbers in the annotated sequence (grey) is graphically presented.

The assumption that there was only one protein in a probe was generated to convert the quantity characteristics of amino acids into a number form for the next correlation analysis between the theoretical and real numbers of the corresponding amino acids. In addition, it must be reported that some amino acids degrade; still, the hydrolyses take place, and they are not presented in the results of the amino acids' analysis. This study shows the same pattern of amino acid distribution in the theoretical and real sequences as well as the low level of dispersion in numbers of residues (Figure 4d). Taking all of this information into account, it appears that all of the proteins comprised in a probe are of the same nature and are consistent with the annotated sequences.

In addition, every mismatch in the numbers of the theoretical and real amino acid residues in addition to all of the possible amino acid changes were pointed out (Figure 5a). To provide any quantity characteristic controls, the balance conservation of the full amino acid number in a protein, the score function *d*, was proposed:

$$d = \sqrt{\left(N_1^{theoretical} - N_1^{real}\right)^2 + \left(N_2^{theoretical} - N_2^{real}\right)^2}$$

a)

1 → 2	N1 theory	N1 practice	N2 theory	N2 practice
Glu → Asp	5	4	8	9
Lys → Asn	1	0	8	9
Ala → Gly	11	11	6	8
Glu → Asn	5	4	8	9
Gln → Asn	5	4	8	9
Gln → Ile	5	4	5	4
Gln → Leu	5	4	7	7
Ile → Val	5	4	3	3
Leu → Val	7	7	3	3
Thr → Ser	4	3	2	2

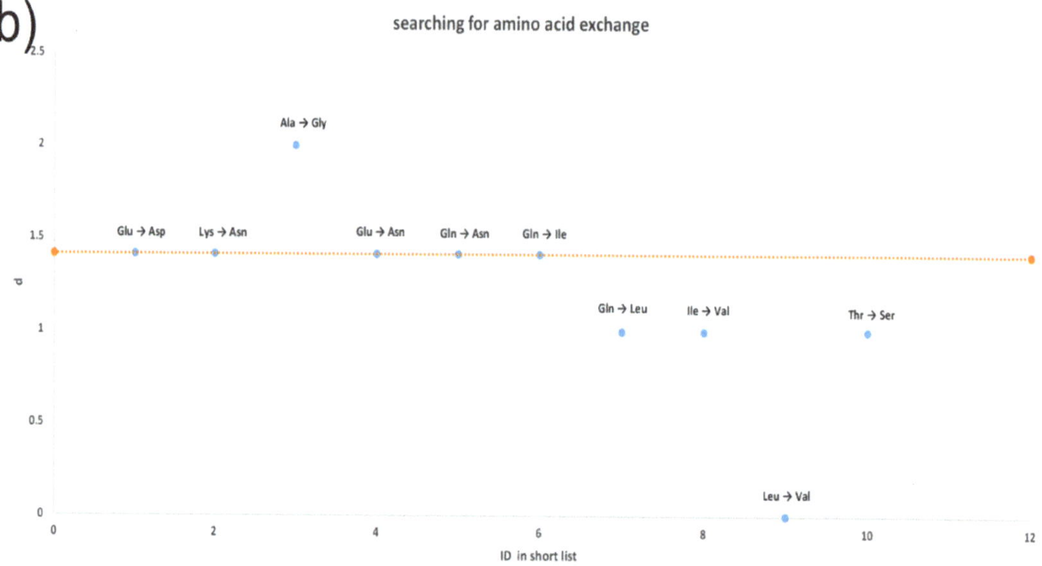

Figure 5. Amino acid change research over the amino acid analysis results. (**a**) A table including all of the possible amino acid changes, where the initial amino acid, "1" (presented in the annotated sequence), changed to another one, "2" (presented in the results of research). N_1 and N_2 are the numbers of corresponding amino acids. (**b**) The *d* function plot is depicted. Each amino acid change is displayed on the plot with the calculated meaning of the *d* function. The *d* function value, related to the normal proportion of a one-letter change in the protein, is marked on the plot as a yellow dotted straight line.

N_1 is the number of any amino acid, which was changed to another one with the number N_2. The *"theoretical"* and *"real"* indexes characterize the ways of calculating amino acid numbers. The first represents calculations based on amino acid analysis results, while the second is based on annotated sequences of hydrophobins.

This procedure was performed to reduce the number of proposed amino acid changes.

For all of the changes, presented in a table (Figure 5a), the meaning of the d function was calculated, and these changes were marked as dots on the plot (Figure 5b). The meaning of the d function, related to the normal balance of one amino acid change in the protein, was marked as a yellow dotted straight line on the plot (Figure 5b). Thus, all of the dots of amino acid changes, which are lying on the normal line (Figure 5b), are likely to be related to the examined protein.

3.5. Proteolysis and Amino Acid Sequencing

To study the hydrophobin in detail and to confirm the amino acid sequence of this protein, a procedure of sequencing with MALDI-TOF MS I MS was performed.

Trypsin was chosen as the enzyme for proteolysis due to the existence of several trypsinolisys sites in the protein sequence. Solving the problem of sites' access to protease, cysteine residues were reduced with DTT and were modified with IAA (Figure 6a). Furthermore, IAA was able to modify the side chain amino group of lysine and the N-terminus of the polypeptide chain, as well the cysteine residues, and not only in a 1:1 proportion (Figure 6b). Therefore, this information was taken in account.

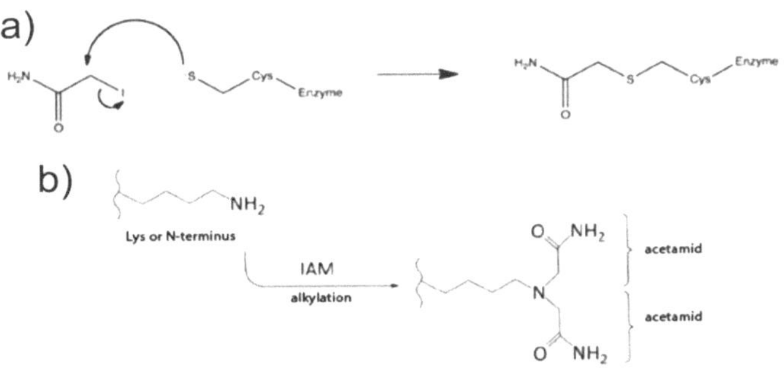

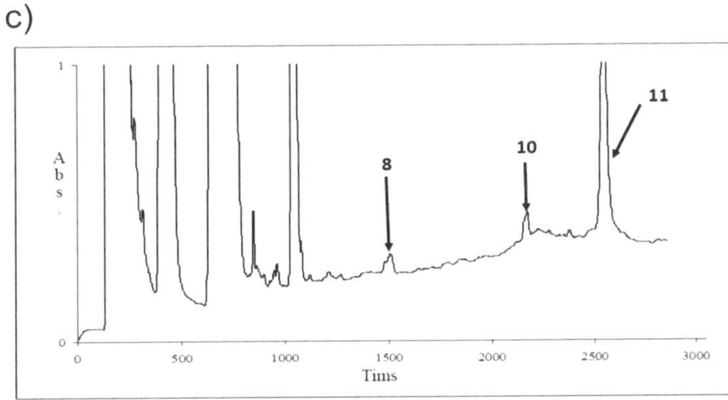

Figure 6. Protein modification and trypsinolisys results. (**a**) The chemical reaction of cysteine modification with IAA is presented. (**b**) The adverse reactions of the lysine side chain amino group and the N-terminus of the polypeptide modification with IAA are depicted. (**c**) An RP-HPLC chromatogram of the sample, performed after the trypsinolisys took place. The collected peaks are marked with corresponding numbers.

The probe, after modification, was analyzed with the RP-HPLC method, and all of the marked peaks (Figure 6c) were collected to perform MALDI-TOF MS and MS I MS analyses. Modifying the radical of IAA is presented in Figure 7a, with some physicochemical information and the character of subsequent fragmentation in the MS I MS method (Figure 7b).

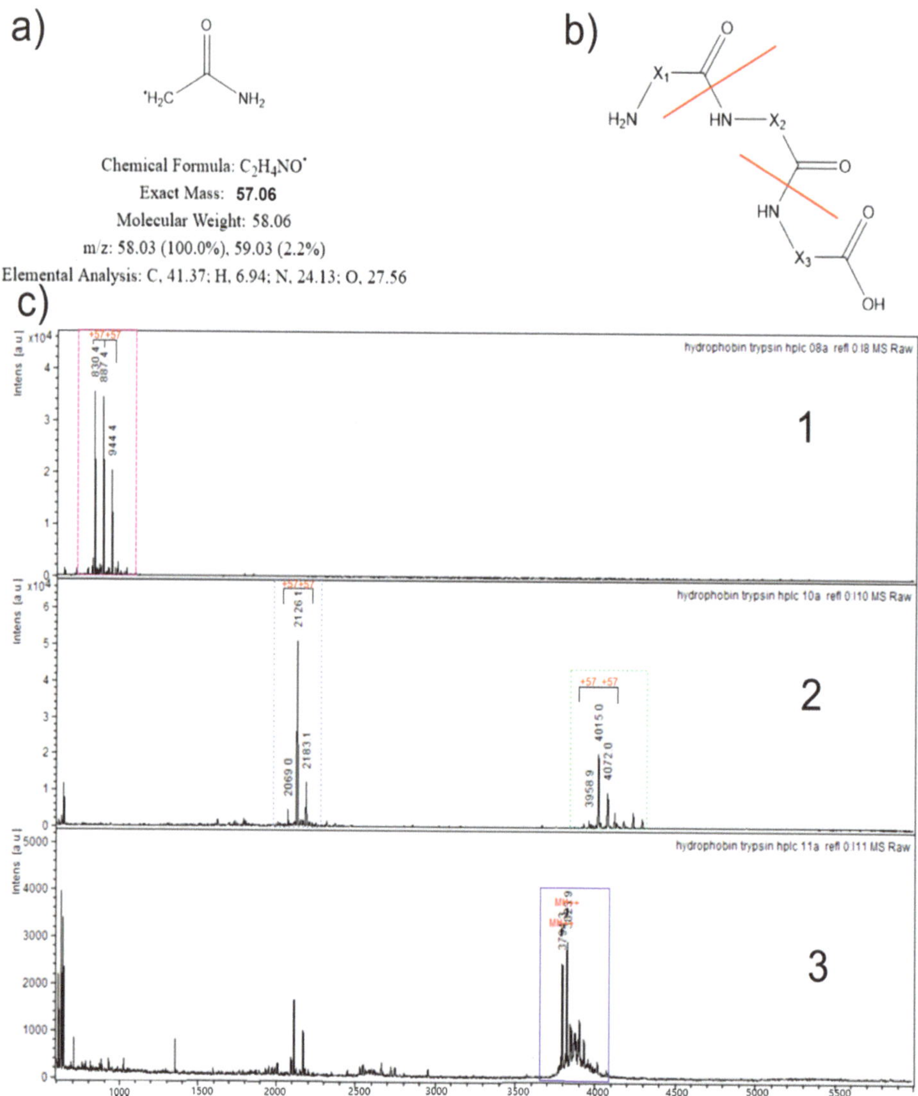

Figure 7. MALDI-TOF MS analysis of the collected fractions. (a) An acetamide radical as a modifying unit is presented with some physicochemical information. (b) The character of probable fragmentation in the MS I MS specters and the broken bonds are marked with red lines. (c) The mass specters of chromatographically collected fractions are presented. Peaks within every set that differ between each other by 57 Da (modifying unit) are marked with black lines and indexes "+57" above the peaks. For specters 1 and 2, the peak sets were highlighted in dotted rectangles colored in corresponding colors. The set in specter 3 is highlighted in a continuous rectangle as the mass of the unmodified hydrophobin.

All of the collected fractions were analyzed with the MALDI-TOF MS method, and the corresponding specters are represented in Figure 7c (1, 2, and 3). The first specter includes one set of peaks, and the second one includes two sets. The third specter includes one set, but the mass (m/z) values indicate an unmodified hydrophobin. Inside every set, the difference in the mass between the closest peaks seems to be modifying the difference in IAA (57 Da).

All of the main peaks from the sets (1, 2, and 3) were subsequently analyzed by performing the MS I MS method, and the corresponding specters are presented in Figure 8a–c. The main peaks were chosen as the results of an annotated sequence analysis with Mascot service free web software for MALDI-TOF ion prediction [19]. They were peaks with masses of 4013 Da, 829 Da, and 2125 Da (including the masses of modifiers), and the result of the Mascot analysis is presented in Figure 8d. All of the theoretical ions (Figure 8d) were found in the specters (Figure 8a–c).

Every specter of the MS I MS analysis of the corresponding masses, 829, 2125, and 4013 Da, was sequenced, using tools of Bruker Daltonics flexAnalysis 3.4 software (Bruker, Billerica, MA, USA). The first two masses were aligned to the corresponding fragment (marked on the MS I MS specters) of the annotated sequence, but the mass of 4013 also seems to be the fragment with one amino acid change in the 29 position (Figure 8c).

Thus, an amino acid change was observed in the entire sequence fragment for the *S. alkalinus* 8KS17-10 strain, different from the one for the full-genome sequencing of the ex-type strain of *S. alkalinus*, F11. The registered change is E29D, and this result correlates with the results of the MALDI-TOF MS analysis of the probe in addition to the amino acid analysis.

3.6. Sequence Analysis of the S. alkalinus Hydrophobin

The genome of *S. alkalinus* (NCBI ID GCA_003711515) contains one gene encoding for a hydrophobin. No already-characterized hydrophobins of different Ascomycetes species show significant similarities in their protein and nucleotide structures. The gene has a distinct structure, containing three introns and two exons. The corresponding DNA sequence information was translated into polypeptide sequences and showed a predicted protein size of 99 residues.

A nucleotide analysis of the hydrophobin gene, using designed Hyd1F/Hyd580r primers of the F11 strain, show 100% similarity with the known hydrophobin sequence from the annotated genome of the *S. alkalinus* F11 ex-type strain (Figure 9).

The *S. alkalinus* 8KS17-10 strain sequences of the hydrophobin gene (NCBI ID ON149453) contained two substitutions: The substitution of C to A at the 369 position was in an intron, although it did not alter the amino acid sequence. The other, a G to C substitution at codon 165, was in an exon, resulting in a glutamic acid (Glu or E) to aspartic acid (Asp or D) replacement (Figure 10).

3.7. Antifungal Activity

Previously, we have tested the antimicrobial activity of eight *S. alkalinus* strains against 10 strains of Gram-positive and Gram-negative bacteria, in addition to fungi, including either sensitive or drug-resistant isolates from ATCC, as well as resistant clinical isolates. All of the *S. alkalinus* strains demonstrated high activity against yeast and mold fungi, including collection cultures and multidrug-resistant clinical isolates [16]. The highest amount of Sa-HFB1 and antifungal activity was detected for the *S. alkalinus* 8KS17-10 strain (Supplementary Figure S3).

The antifungal activity of Sa-HFB1 against opportunistic fungi and multidrug-resistant clinical isolates is shown in Table 1. The clinical pathogenic isolates *Cryptococcus neoformans*, *Nakaseomyces glabrataa*, and *C. tropicalis* were selected based on their resistance phenotypes, including their resistance to caspofungin, micafungin, fluconazole, and flucytosine. The antifungals AmpB, FZ, and VOR were used as reference drugs. Sa-HFB1 inhibits the whole panel of opportunistic and clinical isolates with the broth twofold microdilution method.

The activity of Sa-HFB1 against *Aspergilllus* spp. was comparable to the reference polyene cyclic drug AmB. The inhibition zones for all of the clinical *Candida* isolates were found to be 12–14 mm for Sa-HFB1, while *Nakaseomyces glabrataa* 1447 m and *C. tropicalis* 156 m were inactive to AmpB, FZ, and VOR, respectively.

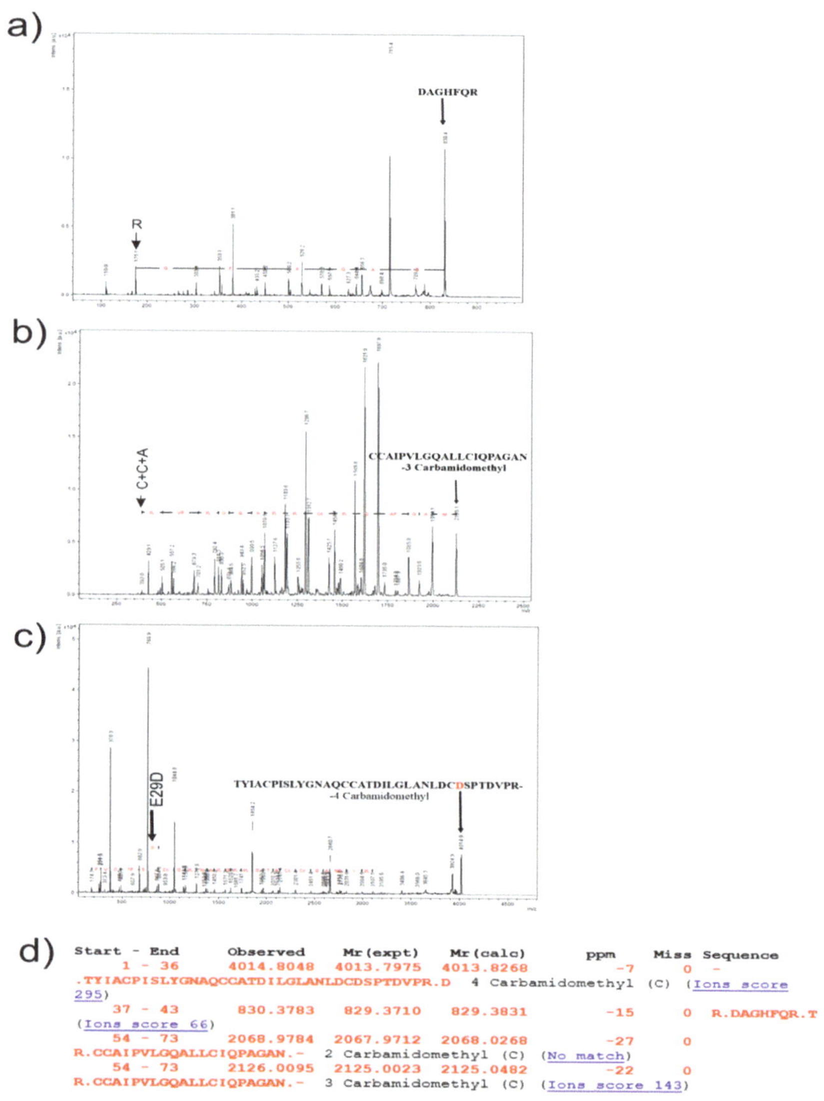

d)

Start - End	Observed	Mr(expt)	Mr(calc)	ppm	Miss	Sequence
1 - 36	4014.8048	4013.7975	4013.8268	-7	0	-

.TYIACPISLYGNAQCCATDILGLANLDCDSPTDVPR.D   4 Carbamidomethyl (C) (Ions score 295)

| 37 - 43 | 830.3783 | 829.3710 | 829.3831 | -15 | 0 | R.DAGHFQR.T |

(Ions score 66)

| 54 - 73 | 2068.9784 | 2067.9712 | 2068.0268 | -27 | 0 | |

R.CCAIPVLGQALLCIQPAGAN.-   2 Carbamidomethyl (C) (No match)

| 54 - 73 | 2126.0095 | 2125.0023 | 2125.0482 | -22 | 0 | |

R.CCAIPVLGQALLCIQPAGAN.-   3 Carbamidomethyl (C) (Ions score 143)

Figure 8. Sequencing of the hydrophobin. (**a–c**) MALDI-TOF MS | MS specters of different fragments of the hydrophobin, obtained with the utilization of trypsinolisys and separated with the RP-HPLC method, are presented. The characters of fragments are pointed out in black above the rightmost peak in every specter. In addition, the annotation to other peaks is exhibited with black labels above the corresponding peaks, and the amino acid change is indicated with a label and arrow. The process of sequencing is exhibited with a black line joining together the corresponding peaks with a red annotation in amino acid (acids) symbol(s). (**d**) The probable fragments of trypsinolisys with different types of modification, which were predicted with Mascot service free web software for MALDI-TOF ion prediction correlated with the masses exhibited on the specters above, are presented.

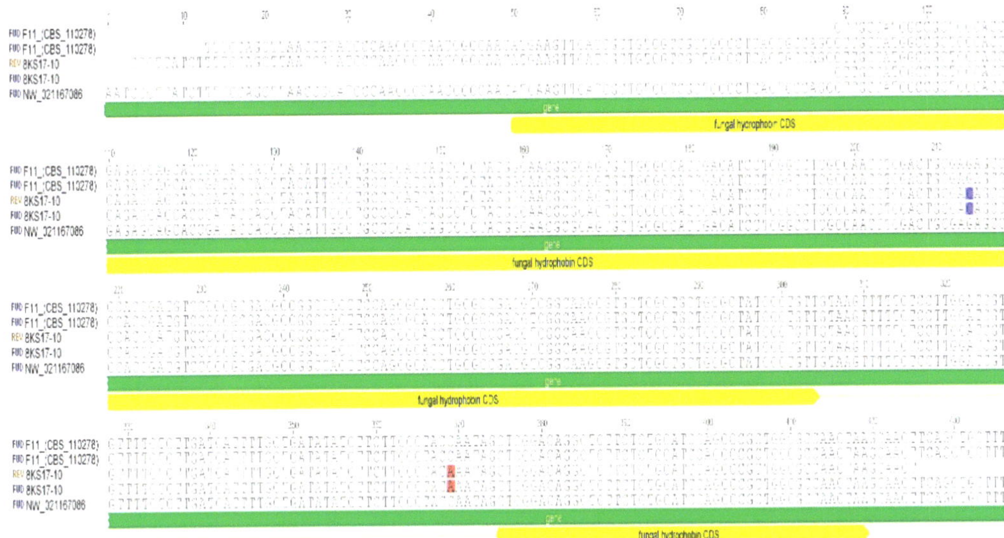

Figure 9. Alignment of hydrophobin DNA sequences. The alignment shows single-nucleotide polymorphisms (SNPs) in the *S. alkalinus* 8KS17-10 strain. An exon substitution is marked by blue highlighting, and an intron SNP by red highlighting.

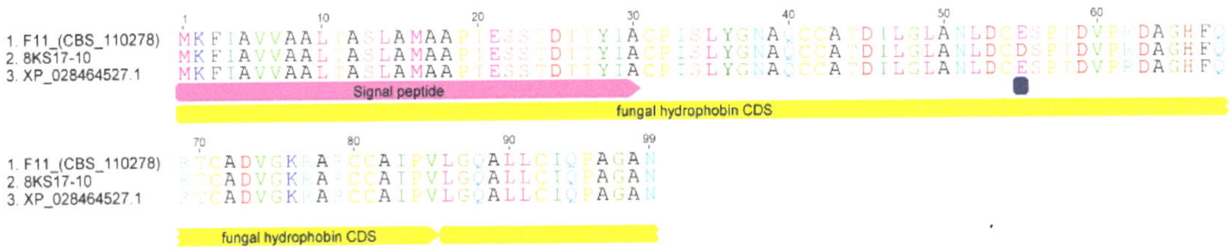

Figure 10. Alignment of the hydrophobin's amino acid sequences. The alignment shows an E to D amino acid replacement at position 29 in the mature hydrophobin sequence (amino acids 1–30 are signal peptides).

Table 1. The activity of purified Sa-HFB1 on the growth of opportunistic and clinical pathogenic fungi isolates measured by a disk diffusion assay.

	Zone, mm			
	Compound, 40 µg/disk			
Strain	**Sa-HFB1**	**AmpB**	**FZ**	**VOR**
Candida albicans 1582 m	15 ± 0.1	10 ± 0.6	0	10 ± 0.6
Pichia kudriavzevii 1402 m	16 ± 0.3	15 ± 0.1	0	10 ± 0.6
Nakaseomyces glabrataa 1447 m	12.5 ± 0.2	0	0	0
C. tropicalis 156 m	14 ± 0.1	0	0	0
C. parapsilosis 571 m	14 ± 0.2	18 ± 0.3	0	0
Cryptococcus neoformans 297 m	30 ± 0.1	18 ± 0.6	0	0
Aspergillus fumigatus 390 m	12 ± 0.5	9 ± 0.6	0	0
A. niger 219	14 ± 0.1	15 ± 0.8	0	11 ± 0.6
Penicillium brevicompactum VKM F-4481	15 ± 0.5	17 ± 0.4	0	10 ± 0.9

The MICs of Sa-HFB1 against opportunistic and clinical fungi ranged from 1 to 8 µg/mL and confirmed higher activity against both opportunistic and clinical isolates than has previously been reported (Table 2). The highest level of antifungal activity (MIC 1 µg/mL) was demonstrated for *Cryptococcus neoformans* 297 m.

Table 2. Minimal inhibitory concentration of Sa-HFB1 compared with target antifungal drugs.

Minimal Inhibitory Concentration (MIC, µg/mL)				
Compound				
Strain	Sa-HFB1	AmpB	FZ	IZ
Aspergillus niger 219	8	1	R	4
A. fumigatus 390 m	8	1	R	0.5
Candida albicans ATCC 2091	4	1	>64	4
C. albicans 1582 m	4	2	>64	4
Cryptococcus neoformans 297 m	1	0.5	16	0.5

R—resistant; AmpB—amphotericin B; FZ—fluconazole; and IZ—itraconazole.

4. Discussion

The population around the globe is affected every year by mycoses caused by pathogenic fungi, which can be classified as subcutaneous, cutaneous, and systemic. Clinical pathogens, such as *Candida*, *Cryptococcus*, and *Aspergillus*, lead to severe fungal infections and are a common cause of nosocomial infections, reaching mortality rates of up to 40% [29,30]. Due to the notorious increase in the incidence of IFIs globally, which result in up to 1.7 million deaths annually, numerous strategies have been employed to optimize the treatment of these deadly infections. The main reason for limited mycosis treatment options is the development of antifungal resistance or multidrug resistance (MDR), which can abolish treatment options. There are different factors leading to drug resistance and therapeutic failure: Firstly, patients with immunosuppression are more likely to fail at responding to antifungal therapy due to the lack of assistance of a robust immune response against an infection. Secondly, long-term or repeated drug exposure can also lead to the emergence of resistance. Furthermore, being exposed to agricultural fungicides with identical molecular targets to those of systemic antifungal drugs can increase the development of resistant organisms. All drug classes can develop antifungal resistance, and some fungal species can even show resistant activity against all antifungal classes. At present, researchers direct their attention to the examination of plant and microbial isolates, secondary metabolites, and newly synthesized molecules.

Hydrophobins (HFBs) are small proteins found only in filamentous fungi (Dikarya) [31–37]. HFBs are characterized by their small size and amphipathic nature at hydrophilic/hydrophobic interfaces, and are currently attracting great interest from the biotechnology industry [38–41]. A central attraction is the ability of a hydrophobin monolayer to reverse the nature of a surface from hydrophobic to hydrophilic, and vice versa [31,37,42]. These layers cover fungal bodies and spores in water-repelling coats [37] and influence spore dispersal, stress resistance, development, and biotic interactions [43–45]. In pathogenic fungi, such as *Aspergillus fumigatus* [46,47], *Metarhizium brunneum*, and *M. acridum* [43,44], HFBs are considered to be virulence factors because they reduce the exposure of pathogen-associated molecular patterns (PAMPs) and antigens to receptors of the immune system [43–45,48]. HFBs are also involved in symbiotic interactions, such as those between lichens and mycorrhizae [17,49–51]. HFBs play a role in development and morphogenesis in the majority of filamentous fungi and influence spore properties [33,34,52]. To reveal HFBs that are associated with the sporulation of *T. harzianum* CBS 226.95 and *T. guizhouense* NJAU 4742, the expression of respective genes during the three stages of fungal development has been tested. The results showed that two genes were highly expressed during the formation of aerial mycelium and remained highly active during conidiation [53–56].

Among biocontrol fungi, some studies have revealed that the participation of HFBs in biocontrol processes could contribute to plant growth and elicit plant defense reactions.

T. asperellum mutants that lack the TasHyd1 hydrophobin gene are severely impaired in terms of root attachment and colonization, and these phenotypes are recovered by the complementation of TasHyd1, indicating that this protein contributes to the interactions between *Trichoderma* and its host plant [55,57–59]. In *Clonostachys rosea*, the Hyd3 class II hydrophobin can influence root colonization ability [45]. Similarly, the HYTLO1 class II hydrophobin in *T. longibrachiatum* can directly inhibit both the spore germination and hyphal elongation of *Botrytis cinerea* and *Alternaria alternata* in vitro, as well as enhance tomato plantlet development. HYTLO1, which has multiple roles in and effects on treated plants, was able to trigger a nicotinic acid adenine dinucleotide phosphate-mediated Ca^{2+} signaling pathway in *Lotus japonicus*, highlighting a possible mechanism underlying its action. Zhang et al. demonstrated the biocontrol functions of HFB2-6, a class II hydrophobin of *T. asperellum* ACCC30536 [53,56]. The HFB2-6 hydrophobin gene was expressed in *Escherichia coli*, and the rHFB2-6 recombinant protein was found to affect the transcription of poplar defense-related genes. HFBII-4 from *T. asperellum* can enhance the resistance of *Populus davidiana* × *P. alba* var. *pyramidalis* to *A. alternata* phytopathogenic fungi. In summary, the class II hydrophobin gene HFBII-4 upregulated the expression of growth-related, disease resistance, and defense response genes in PdPap poplars [44,56].

Fungi growing at extreme pH values are of scientific interest for the general study of fungal adaptive evolution, as well as for the evaluation of their potential in producing commercially valuable substances. Obviously, fungi adapting to alkalinity must have metabolic pathways that have become modified with respect to those seen in related neutrophilic fungi [1–3,5,9,60]. Alkaliphilic *Sodiomyces* species are fundamentally different from alkali-tolerant ones in terms of their mechanisms of adaptation. They accumulate trehalose in the cytosol and phosphatidic acids in the membrane lipids, whereas alkali-tolerant fungi contain these compounds in low amounts. In addition, adaptations to alkaline environments are required for structures involved in exporting metabolites such as antibiotics, for domains of membrane transporters exposed to ambient environments, and for the regulation of gene expression by an ambient pH. The life cycle of *S. alkalinus* and some possible mechanisms of its adaptation to a high pH combined with salinization have been investigated recently at the cytomorphological level [61], while the biochemical mechanisms have not been studied sufficiently yet [62–64]. Our experiments to evaluate the antifungal activity and identification of an active hydrophobin might offer a clue as to the possible ecological role and biological activity of alkaliphilic *S. alkalinus* in alkaline soils.

The structural, MALDI MS | MS, and NCBI annotation data analyses helped to identify Sa-HFB1 as a class II hydrophobin. Hydrophobin genes may have SNPs, resulting in different amino acid sequences in synthesized HFBs of different strains. The Sa-HFB1 production yield of about 25.54 ± 1.4 mg/L from the culture liquid of *S. alkalinus* on an alkaline medium was achieved in a period of fermentation of 14 days. This value has been found to be typical of class II HFBs secreted by wild-type fungi into an extracellular medium [40,41,51,58,65]. Previous research has reported the amount of HFBII of wild-type *T. reesei* to be around 30 mg/L, and around 200 mg/L for genetically engineered *T. reesei* [38]. This is the first time that *S. alkalinus* was explored for the isolation of HFBs.

In our previous studies, we demonstrated the antifungal activity of different species of conditionally pathogenic molds of *Aspergillus* spp. for the *S. alkalinus* 8KS17-10 strain. We have previously identified the compound as an antimicrobial peptide based on the totality of the identified structural features (molecular weight, absorption ratio at certain wavelengths). In the present study, we also showed that the active compound is an antimicrobial peptide and highly effective towards pathogenic molds. Due to the strong antifungal property exhibited by Sa-HFB1, it has been evaluated to further antifungal activity against pathogenic clinical fungi and yeasts. To the best of our knowledge, this is the first report of a direct antifungal effect of class II HFBs from alkaliphilic fungi against non- and clinical pathogen's isolates.

The hydrophobin, produced by the *S. alkalinus* 8KS17-10 strain, was isolated and puri-fied chromatographically, analyzed, and sequenced using a trypsinolisys assay in addition to the MALDI-TOF MS and MS ǀ MS methods. As a result, we report the convergence of amino acids between the experimental and annotated sequences, obtained by the genome sequencing of the *S. alkalinus* F11 strain, to within one amino acid in the 29 position. Thus, we declare more concretely the change of glutamic acid to aspartic acid in the 29 position of the protein (E29D), as well as the fact that this region has a low level of residue conservation, so amino acids can differ there from strain to strain.

5. Conclusions

There are *Sodiomyces* fungi that can secrete active molecules, such as HFBs. Therefore, discovering novel FBs from unstudied alkaliphilic fungi and understanding the possible role of these molecules is essential to realize the full biotechnological potential of these proteins. In this study, we isolated and identified, for the first time, a class II hydrophobin from *S. alkalinus*, denominated as Sa-HFB1. Moreover, this study is the first to report its antifungal activity towards opportunistic and pathogenic fungi.

Supplementary Materials: The following supporting information can be downloaded at: https://www.mdpi.com/article/10.3390/jof8070659/s1, Figure S1: *Sodiomyces alkalinus* Grum-Grzhimaylo, Debets & Bilanenko: A—growth on the alkaline agar plate; B–D scanning electron microscopy: B—conidiophores with 3-5 branches and conidia; C—cliestothecium ascomata, creeping vegetative hyphae and conidiophores; D—ascospores and multilayered peridium of ascomata; Figure S2: Activity of Sa-HFB1 against pathogenic clinical isolates of *Cryptococcus neoformans* 297 m (A), *Candida albicans* 1582 m (B) and *Aspergillus fumigatus*; Figure S3: Chromatograms of PTH derivatives of amino acids in N-terminal automated sequencing.

Author Contributions: Conceptualization, V.S.S. and M.L.G.; methodology, E.A.R., M.L.G. and V.S.S.; software, A.E.K.; validation, A.E.K., M.A.S.; formal analysis, M.L.G. and E.A.R.; investigation, A.E.K., T.A.E., L.Y.K., N.V.F., M.V.S. and V.S.S.; data curation, A.E.K.; writing—original draft preparation, A.E.K., T.A.E., L.Y.K., M.V.S., A.V.T., M.L.G. and V.S.S.; writing—review and editing, A.E.K., M.L.G., V.S.S.; visualization, V.S.S.; supervision, V.S.S. and M.L.G.; project administration, V.S.S.; funding acquisition, A.E.K. and V.S.S. All authors have read and agreed to the published version of the manuscript.

Funding: This study was supported by the Russian Science Foundation (grant no. 22-25-00353).

Institutional Review Board Statement: Not applicable.

Informed Consent Statement: Not applicable.

Data Availability Statement: All sequence data are available in NCBI GenBank following the accession numbers in the manuscript.

Conflicts of Interest: The authors declare no conflict of interest.

References

1. Zhang, X.; Li, S.J.; Li, J.J.; Liang, Z.Z.; Zhao, C.Q. Novel natural products from extremophilic fungi. *Mar. Drugs* **2018**, *16*, 194. [CrossRef] [PubMed]
2. Ibrar, M.; Ullah, M.W.; Manan, S.; Farooq, U.; Rafiq, M.; Hasan, F. Fungi from the extremes of life: An untapped treasure for bioactive compounds. *Appl. Microbiol. Biotechnol.* **2020**, *104*, 2777–2801. [CrossRef] [PubMed]
3. Mamo, G.; Mattiasson, B. Alkaliphiles: The versatile tools in biotechnology. In *Alkaliphiles in Biotechnology. Advances in Biochemical Engineering/Biotechnology*; Mamo, G., Mattiasson, B., Eds.; Springer Nature: Cham, Switzerland, 2020; Volume 172, p. 51. [CrossRef]
4. Horikoshi, K. Alkaliphiles. In *Extremophiles*; Springer: Tokyo, Japan, 2016; Volume 4, pp. 53–78. [CrossRef]
5. Grum-Grzhimaylo, A.A.; Georgieva, M.L.; Bondarenko, S.A.; Debets, A.J.M.; Bilanenko, E.N. On the diversity of fungi from soda soils. *Fungal Divers.* **2016**, *76*, 27–74. [CrossRef]
6. Kevbrin, V.V. Isolation and cultivation of alkaliphiles. In *Alkaliphiles in Biotechnology. Advances in Biochemical Engineering/Biotechnology*; Mamo, G., Mattiasson, B., Eds.; Springer Nature: Cham, Switzerland, 2020; Volume 172, pp. 53–84. [CrossRef]
7. Wei, Y.; Zhang, S.H. Abiostress resistance and cellulose degradation abilities of haloalkaliphilic fungi: Applications for saline–alkaline remediation. *Extremophiles* **2018**, *22*, 155–164. [CrossRef] [PubMed]

8. Bilanenko, E.N.; Sorokin, D.; Georgieva, M.L.; Kozlova, M.V. *Heleococcum alkalinum* a new alkali-tolerant ascomycete from saline soda soils. *Mycotaxon* **2005**, *91*, 497–507.

9. Grum-Grzhimaylo, A.A.; Debets, A.J.M.; van_Diepeningen, A.D.; Georgieva, M.L.; Bilanenko, E.N. *Sodiomyces alkalinus*, a new holomorphic alkaliphilic ascomycete within the Plectosphaerellaceae. *Persoonia* **2013**, *31*, 147–158. [CrossRef]

10. Deshmukh, S.K.; Dufossé, L.; Chhipa, H.; Saxena, S.; Mahajan, G.B.; Gupta, M.K. Fungal endophytes: A potential source of antibacterial compounds. *J. Fungi* **2022**, *8*, 164. [CrossRef]

11. Kladwang, W.; Bhumirattana, A.; Hywel-Jones, N. Alkaline-tolerant fungi from Thailand. *Fungal Divers.* **2003**, *13*, 69–83.

12. Tiquia-arashiro, S.M.; Grube, M. *Fungi in Extreme Environments: Ecological Role and Biotechnological Significance*; Grube, M., Ed.; Springer Nature: Cham, Switzerland, 2019; p. 626. [CrossRef]

13. Alkin, N.; Dunaevsky, Y.; Elpidina, E.; Beljakova, G.; Tereshchenkova, V.; Filippova, I.; Belozersky, M. Proline-specific fungal peptidases: Genomic analysis and identification of secreted DPP4 in alkaliphilic and alkalitolerant fungi. *J. Fungi* **2021**, *7*, 744. [CrossRef]

14. Kuvarina, A.E.; Gavryushina, I.A.; Kulko, A.B.; Ivanov, I.A.; Rogozhin, E.A.; Georgieva, M.L.; Sadykova, V.S. The Emericellipsins A-E from an alkalophilic fungus *Emericellopsis alkalina* show potent activity against multi-drug-resistant pathogenic fungi. *J. Fungi* **2021**, *7*, 153. [CrossRef]

15. Rogozhin, E.A.; Sadykova, V.S.; Baranova, A.A.; Vasilchenko, A.S.; Lushpa, V.A.; Mineev, K.S.; Georgieva, M.L.; Kul'Ko, A.B.; Krasheninnikov, M.E.; Lyundup, A.V.; et al. A Novel lipopeptaibol Emericellipsin A with antimicrobial and antitumor activity produced by the extremophilic fungus *Emericellopsis alkaline*. *Molecules* **2018**, *23*, 2785. [CrossRef] [PubMed]

16. Kuvarina, A.E.; Georgieva, M.L.; Rogozhin, E.A.; Kulko, A.B.; Gavryushina, I.A.; Sadykova, V.S. Antimicrobial potential of the alkalophilic fungus *Sodiomyces alkalinus* and selection of strains–producers of new antimicotic compound. *Appl. Biochem. Microbiol.* **2021**, *57*, 86–93. [CrossRef]

17. Bayry, J.; Aimanianda, V.; Guijarro, J.I.; Sunde, M.; Latgé, J.P. Hydrophobins-unique fungal proteins. *PLoS Pathog.* **2012**, *8*, e1002700. [CrossRef] [PubMed]

18. Galkina, S.I.; Fedorova, N.V.; Serebryakova, M.V.; Arifulin, E.A.; Stadnichuk, V.I.; Gaponov, T.V.; Baratova, L.A.; Sudina, G.F. Inhibition of the GTPase dynamin or actin depolymerisation initiates outward plasma membrane tabulation/vesiculation (cytoneme formation) in neutrophils. *Biol. Cell* **2015**, *107*, 144–158. [CrossRef]

19. Mascot Search Engine. Protein Identification Software for Mass Spectrometry. Available online: https://www.matrixscience.com (accessed on 10 April 2022).

20. Tsugita, A.; Scheffler, J.J. A rapid method for acid hydrolysis of protein with a mixture of trifluoroacetic acid and hydrochloric acid. *Eur. J. Biochem.* **1982**, *124*, 585–588. [CrossRef]

21. Trofimova, L.; Ksenofontov, A.; Mkrtchyan, G.; Graf, A.; Baratova, L.; Bunik, V. Quantification of rat brain. Amino acids: Analysis of the data consistency. *Curr. Anal. Chem.* **2016**, *12*, 349–356. [CrossRef]

22. Geneious® 7.1.15. Bioinformatics Software for Sequence Data Analysis. Available online: https://www.geneious.com (accessed on 10 April 2022).

23. Primer3 Online Tool (Version 4.1.0) Software. Available online: http://primer3.ut.ee (accessed on 10 April 2022).

24. *CLSI M27-A3*; Reference Method for Broth Dilution Antifungal Susceptibility Testing of Yeasts. Clinical and Laboratory Standards Institute: Pittsburgh, PA, USA, 2013.

25. *CLSI M38-A2*; Reference Method for Broth Dilution Antifungal Susceptibility Testing of Filamentous Fungi; Approved standard. 2nd ed.; Clinical and Laboratory Standards Institute: Pittsburgh, PA, USA, 2008.

26. Grum-Grzhimaylo, A.A.; Falkoski, D.L.; van den Heuvel, J.; Valero-Jiménez, C.A.; Min, B.; Choi, I.G.; Lipzen, A.; Daum, C.G.; Aanen, D.K.; Tsang, A.; et al. The obligate alkalophilic soda-lake fungus *Sodiomyces alkalinus* has shifted to a protein diet. *Mol. Ecol.* **2018**, *27*, 4808–4819. [CrossRef]

27. Fullmer, C.S. Identification of cysteine-containing peptides in protein digests by high-performance liquid chromatography. *Anal. Biochem.* **1984**, *142*, 336–339. [CrossRef]

28. Aitken, A. Analysis of cysteine residues and disulfide bonds. *Methods Mol. Biol.* **1994**, *32*, 351–360. [CrossRef]

29. Shadrivova, O.; Gusev, D.; Vashukova, M.; Lobzin, D.; Gusarov, V.; Zamyatin, M.; Zavrazhnov, A.; Mitichkin, M.; Borzova, Y.; Kozlova, O.; et al. COVID-19-associated pulmonary Aspergillosis in Russia. *J. Fungi* **2021**, *7*, 59. [CrossRef]

30. Revie, N.M.; Iyer, K.R.; Robbins, N.; Cowen, L.E. Antifungal drug resistance: Evolution, mechanisms and impact. *Curr. Opin. Microbiol.* **2018**, *45*, 70–76. [CrossRef] [PubMed]

31. Li, X.; Wang, F.; Liu, M.; Dong, C. Hydrophobin CmHYD1 is involved in conidiation, infection and primordium formation, and regulated by GATA transcription factor CmAreA in edible fungus, *Cordyceps militaris*. *J. Fungi* **2021**, *7*, 674. [CrossRef] [PubMed]

32. Wösten, H.A. Hydrophobins: Multipurpose proteins. *Annu. Rev. Microbiol.* **2001**, *55*, 625–646. [CrossRef] [PubMed]

33. Tao, Y.X.; Chen, R.L.; Yan, J.J.; Long, Y.; Tong, Z.J.; Song, H.B.; Xie, B.G. A hydrophobin gene, Hyd9, plays an important role in the formation of aerial hyphae and primordia in *Flammulina filiformis*. *Gene* **2019**, *706*, 84–90. [CrossRef]

34. Sammer, D.; Krause, K.; Gube, M.; Wagner, K.; Kothe, E. Hydrophobins in the life cycle of the ectomycorrhizal basidiomycete *Tricholoma vaccinum*. *PLoS ONE* **2016**, *11*, e0167773. [CrossRef]

35. Linder, M.B.; Szilvay, G.R.; Nakari-Setälä, T.; Penttilä, M.E. Hydrophobins: The protein-amphiphiles of filamentous fungi. *FEMS Microbiol. Rev.* **2005**, *29*, 877–896. [CrossRef]

36. Wang, L.; Lu, C.; Fan, M.; Liao, B. *Coriolopsis trogii* hydrophobin genes favor a clustering distribution and are widely involved in mycelial growth and primordia formation. *Gene* **2021**, *802*, 145863. [CrossRef]

37. Xu, D.; Wang, Y.; Keerio, A.A.; Ma, A. Identification of hydrophobin genes and their physiological functions related to growth and development in *Pleurotus ostreatus*. *Microbiol. Res.* **2021**, *247*, 126723. [CrossRef]

38. Khalesi, M.; Zune, Q.; Telek, S.; Riveros-Galan, D.; Verachtert, H.; Toye, D.; Gebruers, K.; Derdelinckx, G.; Delvigne, F. Fungal biofilm reactor improves the productivity of hydrophobin HFBII. *Biochem. Eng. J.* **2014**, *88*, 171–178. [CrossRef]

39. Kulkarni, S.S.; Nene, S.N.; Joshi, K.S. Exploring malted barley waste for fungi producing surface active proteins like hydrophobins. *SN Appl. Sci.* **2020**, *2*, 1884. [CrossRef]

40. Siddiquee, R.; Choi, S.S.C.; Lam, S.S.; Wang, P.; Qi, R.; Otting, G.; Sunde, M.; Kwan, A.H. Cell-free expression of natively folded hydrophobins. *Protein Expr. Purif.* **2020**, *170*, 105591. [CrossRef] [PubMed]

41. Landeta-Salgado, C.; Cicatiello, P.; Lienqueo, M.E. Mycoprotein and hydrophobin like protein produced from marine fungi *Paradendryphiella salina* in submerged fermentation with green seaweed *Ulva* spp. *Algal Res.* **2021**, *56*, 102314. [CrossRef]

42. Mosbach, A.; Leroch, M.; Mendgen, K.W.; Hahn, M. Lack of evidence for a role of hydrophobins in conferring surface hydrophobicity to conidia and hyphae of *Botrytis cinerea*. *BMC Microbiol.* **2011**, *11*, 10. [CrossRef]

43. Sevim, A.; Donzelli, B.G.; Wu, D.; Demirbag, Z.; Gibson, D.M.; Turgeon, B.G. Hydrophobin genes of the entomopathogenic fungus, *Metarhizium brunneum*, are differentially expressed and corresponding mutants are decreased in virulence. *Curr. Genet.* **2012**, *58*, 79–92. [CrossRef] [PubMed]

44. Jiang, Z.Y.; Ligoxygakis, P.; Xia, Y.X. HYD3, a conidial hydrophobin of the fungal entomopathogen *Metarhizium acridum* induces the immunity of its specialist host locust. *Int. J. Biol. Macromol.* **2020**, *165*, 1303–1311. [CrossRef] [PubMed]

45. Dubey, M.K.; Jensen, D.F.; Karlsson, M. Hydrophobins are required for conidial hydrophobicity and plant root colonization in the fungal biocontrol agent *Clonostachys rosea*. *BMC Microbiol.* **2014**, *14*, 18. [CrossRef] [PubMed]

46. Brown, N.A.; Ries, L.N.A.; Reis, T.F.; Rajendran, R.; dos Santos, R.A.C.; Ramage, G.; Riano-Pachon, D.M.; Goldman, G.H. RNAseq reveals hydrophobins that are involved in the adaptation of *Aspergillus nidulans* to lignocellulose. *Biotechnol. Biofuels* **2016**, *9*, 145. [CrossRef] [PubMed]

47. Valsecchi, I.; Dupres, V.; Stephen, V.E.; Guijarro, J.I.; Gibbons, J.; Beau, R.; Bayry, J.; Coppee, J.Y.; Lafont, F.; Latge, J.P.; et al. Role of hydrophobins in *Aspergillus fumigatus*. *J. Fungi* **2018**, *4*, 2. [CrossRef]

48. Linder, M.B. Hydrophobins: Proteins that self-assemble at interfaces. *Curr. Opin. Colloid Interface Sci.* **2009**, *14*, 356–363. [CrossRef]

49. Casarrubia, S.; Daghino, S.; Kohler, A.; Morin, E.; Khouja, H.R.; Daguerre, Y.; Veneault-Fourrey, C.; Martin, F.M.; Perotto, S.; Martino, E. The hydrophobin-like OmSSP1 may be an effector in the ericoid mycorrhizal symbiosis. *Front. Plant Sci.* **2018**, *9*, 546. [CrossRef]

50. Ball, S.R.; Kwan, A.H.; Sunde, M. Hydrophobin rodlets on the fungal cell wall. *Curr. Top. Microbiol. Immunol.* **2020**, *425*, 29–51. [CrossRef] [PubMed]

51. Salgado, C.L.; Cicatiello, P.; Stanzione, I.; Medino, D.; Mora, I.B.; Gomez, C.; Lienqueo, M.E. The growth of marine fungi on seaweed polysaccharides produces cerato-platanin and hydrophobin self-assembling proteins. *Microbiol. Res.* **2021**, *251*, 126835. [CrossRef] [PubMed]

52. Li, X.; Wang, F.; Xu, Y.Y.; Liu, G.J.; Dong, C.H. Cysteine-rich hydrophobin gene family: Genome wide analysis, phylogeny and transcript profiling in *Cordyceps militaris*. *Int. J. Mol. Sci.* **2021**, *22*, 643. [CrossRef] [PubMed]

53. He, R.; Li, C.; Feng, J.; Zhang, D. A class II hydrophobin gene, Trhfb3, participates in fungal asexual development of *Trichoderma reesei*. *FEMS Microbiol. Lett.* **2017**, *364*, fnw297. [CrossRef]

54. Khalesi, M.; Jahanbani, R.; Riveros-Galan, D.; Sheikh-Hassani, V.; Sheikh-Zeinoddin, M.; Sahihi, M.; Winterburn, J.; Derdelinckx, G.; Moosavi-Movahedi, A.A. Antioxidant activity and ACE-inhibitory of class II hydrophobin from wild strain *Trichoderma reesei*. *Int. J. Biol. Macromol.* **2016**, *91*, 174–179. [CrossRef]

55. Ruocco, M.; Lanzuise, S.; Lombardi, N.; Woo, S.L.; Vinale, F.; Marra, R.; Varlese, R.; Manganiello, G.; Pascale, A.; Scala, V.; et al. Multiple roles and effects of a novel *Trichoderma hydrophobin*. *Mol. Plant-Microbe Interact.* **2015**, *28*, 167–179. [CrossRef]

56. Zhang, H.; Ji, S.; Guo, R.; Zhou, C.; Wang, Y.; Fan, H.; Liu, Z. Hydrophobin HFBII-4 from *Trichoderma asperellum* induces antifungal resistance in poplar. *Braz. J. Microbiol.* **2019**, *50*, 603–612. [CrossRef]

57. Cai, F.; Gao, R.; Zhao, Z.; Ding, M.; Jiang, S.; Yagtu, C.; Zhu, H.; Zhang, J.; Ebner, T.; Mayrhofer-Reinhartshuber, M.; et al. Evolutionary compromises in fungal fitness: Hydrophobins can hinder the adverse dispersal of conidiospores and challenge their survival. *ISME J.* **2020**, *14*, 2610–2624. [CrossRef]

58. Vereman, J.; Thysens, T.; Derdelinckx, G.; Impe, J.V.; de Voorde, I.V. Extraction and spray drying of class II hydrophobin HFBI produced by *Trichoderma reesei*. *Process Biochem.* **2019**, *77*, 159–163. [CrossRef]

59. Winandy, L.; Hilpert, F.; Schlebusch, O.; Fischer, R. Comparative analysis of surface coating properties of five hydrophobins from *Aspergillus nidulans* and *Trichoderma reseei*. *Sci. Rep.* **2018**, *8*, 12033. [CrossRef]

60. Kuvarina, A.E.; Gavryushina, I.A.; Sykonnikov, M.A.; Efimenko, T.A.; Markelova, N.N.; Bilanenko, E.N.; Bondarenko, S.A.; Kokaeva, L.Y.; Timofeeva, A.V.; Serebryakova, M.V.; et al. Exploring peptaibol's profile, antifungal, and antitumor activity of Emericellipsin A of *Emericellopsis* species from soda and saline soils. *Molecules* **2022**, *27*, 1736. [CrossRef] [PubMed]

61. Kozlova, M.V.; Ianutsevich, E.A.; Danilova, O.A.; Kamzolkina, O.V.; Tereshina, V.M. Lipids and soluble carbohydrates in the mycelium and ascomata of alkaliphilic fungus *Sodiomyces alkalinus*. *Extremophiles* **2019**, *23*, 487–493. [CrossRef] [PubMed]

62. Bondarenko, S.A.; Ianutsevich, E.A.; Danilova, O.A.; Grum-Grzhimaylo, A.A.; Kotlova, E.R.; Kamzolkina, O.V.; Bilanenko, E.N.; Tereshina, V.M. Membrane lipids and soluble sugars dynamics of the alkaliphilic fungus *Sodiomyces tronii* in response to ambient pH. *Extremophiles* **2017**, *21*, 743–754. [CrossRef] [PubMed]

63. Bondarenko, S.A.; Yanutsevich, E.A.; Sinitsyna, N.A.; Georgieva, M.L.; Bilanenko, E.N.; Tereshina, V.M. Dynamics of the cytosol soluble carbohydrates and membrane lipids in response to ambient pH in alkaliphilic and alkalitolerant fungi. *Microbiology* **2018**, *87*, 21–32. [CrossRef]

64. Bondarenko, S.A.; Georgieva, M.L.; Bilanenko, E.N. Fungi inhabiting the coastal zone of lake Magadi. *Contemp. Probl. Ecol.* **2018**, *11*, 439–448. [CrossRef]

65. Sallada, N.D.; Dunn, K.J.; Berger, B.W. A structural and functional role for disulfide bonds in a class II hydrophobin. *Biochemistry* **2017**, *57*, 645–653. [CrossRef] [PubMed]

Journal of **Fungi**

Article

Insecticidal Efficacy of *Metarhizium anisopliae* Derived Chemical Constituents against Disease-Vector Mosquitoes

Perumal Vivekanandhan [1,2,*], Kannan Swathy [1], Amarchand Chordia Murugan [1] and Patcharin Krutmuang [2,3,4,*]

1 Society for Research and Initiatives for Sustainable Technologies and Institutions, Grambharti, Amarapur, Gujarat-382735, India; swathykannan.23@gmail.com (K.S.); amarchand.chordia@gmail.com (A.C.M.)
2 Department of Entomology and Plant Pathology, Faculty of Agriculture, Chiang Mai University, Chiang Mai 50200, Thailand
3 Innovative Agriculture Research Center, Faculty of Agriculture, Chiang Mai University, Chiang Mai 50200, Thailand
4 Research Center of Microbial Diversity and Sustainable Utilization, Faculty of Science, Chiang Mai University, Chiang Mai 50200, Thailand
* Correspondence: mosqvk@gmail.com (P.V.); patcharink26@gmail.com (P.K.)

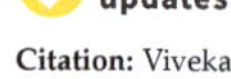

Citation: Vivekanandhan, P.; Swathy, K.; Murugan, A.C.; Krutmuang, P. Insecticidal Efficacy of *Metarhizium anisopliae* Derived Chemical Constituents against Disease-Vector Mosquitoes. *J. Fungi* **2022**, *8*, 300. https://doi.org/10.3390/jof8030300

Academic Editor: Laurent Dufossé

Received: 27 January 2022
Accepted: 7 March 2022
Published: 15 March 2022

Publisher's Note: MDPI stays neutral with regard to jurisdictional claims in published maps and institutional affiliations.

Abstract: Insecticides can cause significant harm to both terrestrial and aquatic environments. The new insecticides derived from microbial sources are a good option with no environmental consequences. *Metarhizium anisopliae* (mycelia) ethyl acetate extracts were tested on larvae, pupae, and adult of *Anopheles stephensi* (Liston, 1901), *Aedes aegypti* (Meigen, 1818), and *Culex quinquefasciatus* (Say, 1823), as well as non-target species *Eudrilus eugeniae* (Kinberg, 1867) and *Artemia nauplii* (Linnaeus, 1758) at 24 h post treatment under laboratory condition. In bioassays, *Metarhizium anisopliae* extracts had remarkable toxicity on all mosquito species with LC_{50} values, 29.631 in *Ae. aegypti*, 32.578 in *An. stephensi* and 48.003 in Cx. *quinquefasciatus* disease-causing mosquitoes, in *A. nauplii* shows (5.33–18.33 %) mortality were produced by the *M. anisopliae* derived crude extract. The LC_{50} and LC_{90} values were, 620.481; 6893.990 µg/mL. No behavioral changes were observed. A low lethal effect was observed in *E. eugeniae* treated with the fungi metabolites shows a 14.0 % mortality. The earthworm *E. eugeniae* mid-gut histology revealed that *M. anisopliae* extracts had no more harmful effects on the epidermis, circular muscle, setae, mitochondrion, and intestinal lumen tissues than chemical pesticides. By Liquid chromatography mass spectrometry (LC-MS) analysis, camphor (25.4 %), caprolactam (20.68 %), and monobutyl phthalate (19.0 %) were identified as significant components of *M. anisopliae* metabolites. Fourier transform infrared (FT-IR) spectral investigations revealed the presence of carboxylic acid, amides, and phenol groups, all of which could be involved in mosquito toxicity. The *M. anisopliae* derived chemical constituents are effective on targeted pests, pollution-free, target-specific, and are an alternative chemical insecticide.

Keywords: *Metarhizium anisopliae*; *Artemia nauplii*; *Eudrilus eugeniae*; mosquitoes; target specific; green pesticides

1. Introduction

Mosquitoes are a major health problem because they transmit diseases such as malaria, dengue fever, yellow fever, and the Zika virus, which impact 700 million people each year and kill over one million people [1]. Adult mosquito control has been the primary approach for avoiding disease transmission, and it usually requires the use of synthetic pesticides and repellents, primarily organophosphates and pyrethroids [2]. Mosquitoes have evolved resistance to organophosphate and synthetic pyrethroids [3–6]. The chemical pesticides have been accumulated in our green ecosystem soil and waterbodies as well as food chains [5]. For insect pest management, entomopathogenic bacteria, fungi, and nematodes are considered effective microbial control methods for insect pests [7–9].

Entomopathogenic fungi produce secondary metabolites that could be used as a source for biopesticide development [10–14]. The entomopathogenic fungi *Metarhizium anisopliae* secondary metabolites, in particular, are known to be effective biopesticides for the control of *Aedes aegypti (Meigen, 1818)* mosquitoes [7,15], and other fungi such as *Tolypocladium* [13], *Beauveria* [9], *Fusarium* [10,11], and Lagenidium *giganteum* [14] have also been bioprospected as insect pests. Several biopesticides have been hampered by the fact that they are slow acting, taking anywhere from a few days to a week to exhibit action, which has hampered their commercialization [16]. Furthermore, their effects on non-target organisms are understudied [7].

The brine shrimp, *Artemia* sp. (Anostraca: Artemiidae), is a branchiopod crustacean that can tolerate salinities of up to 250 gL^{-1}/L [17]. *Artemia* are commonly employed for the evaluation of marine contamination by synthetic chemicals because of their high sensitivity to chemicals or other toxicants [17], and *Artemia nauplii* (Linnaeus, 1758), an important component of the aquatic ecosystem, are regarded as indicators for environmental toxicity [17,18]. Earthworms, therefore, are considered to be bio-indicators of terrestrial ecosystems and are frequently used as biomarkers for assessing the environmental toxicity of chemical contaminants [19,20]. In the present study, we investigated the toxicity of secondary metabolites extracted from *Metarhizium anisopliae* (Metschn, 1879) strains and their toxicity effect was evaluated against disease-vector mosquitoes *Aedes Aegypti* (Meigen, 1818), *Anopheles Stephensi* (Liston, 1901), and *Culex quinquefasciatus (Say, 1823)*, as well as their toxicity against non-target organisms, such as earthworm *Eudrilus eugeniae* (Kinberg, 1867) *and brine shrimp Artemia nauplii* (Linnaeus, 1758).

2. Materials and Methods

2.1. Fungal Cultures

M. anisopliae, was isolated and collected from a soil sample from the Eastern Ghats of Tamil Nadu, India (Latitudes 11°30′ and 22° N, and longitudes 76°50′ and 86°30′ E). Morphological and 18s rDNA sequencing was used to identify fungi cultures. The gene sequences were submitted to the National Center for Biotechnology Information (NCBI, Data Base Accession No is: MH165400.1).

2.2. Mass Culturing of M. anisopliae

Metarhizium anisopliae was cultured on Potato Dextrose Broth (PDB), as a medium for fungal growth. Sixteen 500 mL conical flasks, each containing 250 mL of PDB (dextrose 8 g, peptone 2 g and distilled water 250 mL), were autoclaved at 15 psi for 25 min. Chloramphenicol antibiotics (150 mg/mL) (Sigma-Aldrich Chemicals Private Limited, Bangalore) was added to the culture medium to prevent bacterial contamination. The cultures were allowed to grow for 20 days, and spore concentration was counted using a hemocytometer. A concentration of 1×10^7 spores/mL of *M. anisopliae* conidia were transferred to the culturing medium using an inoculation needle. The culture medium was maintained at the optimized culture conditions (pH 7.0, temperature 28 ± 5 °C) for 30 days.

2.3. Extraction of Secondary Metabolites

Fungus mycelial biomass was washed with distilled water after 20 days to eliminate culture medium components. *Metarhizium anisopliae* biomass was cold extracted with ethyl acetate to extract the biologically active chemical constituents under laboratory conditions. The ethyl acetate solvent was fully pooled with fungal biomass and left for 25 days. Then, the organic phase (light-yellow color) was separated after 25 days using a separating funnel, and the solvent was evaporated using a rotary evaporator at 45 °C.

2.4. Larval Collection and Maintenance

The Institute of Vector Control and Zoonoses at Hosur, Tamil Nadu, India. The mosquito egg masses per species were separately placed in a plastic tray (22 cm × 27 cm × 12 cm) (wonder, India) in dechlorinated tap water. The containers were transferred to room temperature with $28 \pm 2\ °C$, 70–80% RH relative humidity and 12:12 (L:D) photoperiod and kept in it for 10–15 days. Each stage (larvae, pupae, and adult) of the mosquitos were taken for bioassay. During this process, mosquito larvae were fed with 0.5 g Tetra Bit (Pellet Fish Food) in each container, and adults were given a 10% sugar solution as a feeding source.

2.5. Non-Target Organisms

The *Eudrilus eugeniae* stock were maintained under laboratory condition at a room temperature of $27 \pm 2\ °C$. *Artemia nauplii* larvae were kept in 1000 mL of saltwater with a salinity of 30 ppt in a culture medium with a pH range of (7–8). An aspirator was used to provide oxygen.

2.6. Mosquitocidal Bioassays

The fungal metabolites larvicidal and pupicidal efficacy was assessed using the World Health Organization protocol [21]. Stock solutions of fungal extract were dissolved in Dimethylsulfoxide (DMSO) (Sigma-Aldrich, India) at a concentration of 10% w/v (10 µg of extracts in 100 mL of DMSO) and diluted to five different concentrations: 10, 15, 30, 50, and 75 µg/mL. Twenty-five 4th instar larvae and pupae were each transferred to 249 mL of tap water with 1 mL of different concentrations of fungal extract and replicated three times. Dead insects were counted 24 h. As a negative control, DMSO at a concentration of 10% w/v was used.

The adulticidal activity was evaluated following methods described by the Centers for Disease Control and Prevention [22]. Twenty-five newly emerged adults of *A. aegypti*, *A. stephensi* and *C. quinquefasciatus* were exposed to different concentrations (25, 50, 100, 150 and 200 µg/mL) of *M. anisopliae* secondary metabolites. Metabolites solutions were dispensed to the screw cap bottle of 80 mL, and for solvent evaporation, it was air dried over-night. In the control treatment, adult mosquitoes were exposed to DMSO (0.1%). Mortality was recorded post 24 h of treatment. A cotton ball soaked with a 10% glucose solution was used as a food source for mosquitoes. Three replicates for each concentration were performed (n = 450).

2.7. Non-Target Bioassays

The effects of fungal metabolites on earthworms *E. eugeniae* was tested in an artificial soil composed of 15% sphagnum peat, 25% kaolinite clay, and 77 % fine sand. To keep the pH at 5.9, a few drops of $CaCO_3$ were added. The water content was reduced to 30% of the dry weight. Fungi metabolites from *M. anisopliae* were put into the artificial soil at concentrations of 50 g/mL and 75 µg/mL. The 15 *E. eugeniae* larvae were then moved to a plastic container (375 mm $\times$ 300 mm $\times$ 75 mm) containing 1 kg of sterile artificial soil, which was then sealed with a plastic lid to keep the worms from escaping. Dead worms were counted 24–h after exposure. Monocrotophos was used as a positive control, while the negative control was free of fungal metabolites. Each treatment was replicated three times.

On brine shrimp *A. nauplii*, the toxicity of fungal secondary metabolites was determined as follows: Mature *A. nauplii* were collected with a hand pipette and utilized in toxicity tests on *A. nauplii* with different concentrations of *M. anisopliae* secondary metabolites (10, 15, 30, 50, and 75 µg/mL). As a negative control, the DMSO solution was employed. After 24 h of treatment, the *A. nauplii* dead mortality was calculated. Each concentration was tested three times, with each replicate containing 25 mature *A. nauplii*.

2.8. Fourier Transformed Infrared Spectroscopy Analysis

FT-IR analysis was conducted for the identification of the functional groups presents in the crude fungal metabolites. Two mg of fungi metabolites were properly mixed in 75 mg KBr; KBr acts as a binding agent on cleaned micro mortar and pestle. The mixed component was made into KBr pellets formed at low pressure. The KBr pellets were taken for FT-IR analysis using a BRUKER FT-IR spectrometer. FT-IR spectra scanning range was from 500 to 4000 cm^{-1}.

2.9. Liquid Chromatography-Mass Spectrophotometer Analysis

The chemical components profiling of crude fungal extracts was completed through the use of a Bruker Daltonik Impact II ESI-Q-TOF system (Bremen, Germany), ready with a Bruker Daltonik Elute, Ultra High Performance Liquid Chromatography (UHPLC) system (Bremen, Germany), in each positive (M + H) and negative (M − H) electrospray ionisation modes. Chromatographic separation was carried out on a Bruker Daltonik (Bremen, Germany) C18 reversed segment column (2.1 mm, 1.8 m, 120) at 30 °C, with an autosampler temperature of 8 °C and a total run time of 20 min, using water/methanol (90:10%) as eluent with five mM ammonium formate and 0.1% formic acid. The crude extract was dissolved in 2.0 mL of DMSO, and the quantity was multiplied to 50 mL with acetonitrile prior to centrifugation at 4000 rpm for two min and injection. The composition of the samples became mounted with the aid of figuring out the m/z ratio when it comes to the retention length of the utilised standards.

2.10. Statistical Analysis

The mortality rate was corrected using Abbot formula [23]. The dead *A. nauplii*, mosquito larvae, pupae, and adults were counted separately 24 h after treatment, and LC_{50} and LC_{90} were estimated using probit analysis. The SPSS-16.00 programme [18] was used to conduct all of the analyses.

3. Results

3.1. M. anisopliae metabolites against Ae. aegypti, An. stephensi, Cx. quinquefasciatus Mosquitoes

M. anisopliae crude metabolites treatments, at the tested concentrations (10, 15, 30, 50 and 75 µg/mL), caused significant mortality against *Ae. aegypti* larvae (ranging from 17.33 to 95.33%), pupae (ranging from 13.66 to 76.00%), and adults (ranging from 7.00 to 65.00%) (Figure 1; Table 1). The probit model indicated that, *Ae. aegypti* larvae are more susceptible to the *M. anisopliae* crude metabolites than the pupae and adults with an LC_{50}-29.631, 45.530, and 62.589 µg/mL for larvae, pupae, and adults, respectively (Table 1).

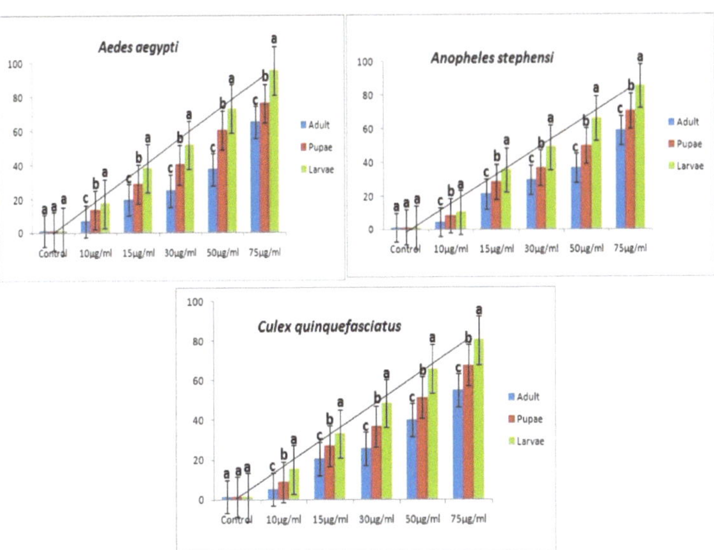

Figure 1. Larvicidal, pupicidal and adulticidal activities of *M. anisopliae* derived extract against larvae, pupae, and adult of *Ae. aegypti*, *An. stephensi* and *Cx. quinquefasciatus* vectors. Bars with the identical lower case letters do not differ significantly ($p > 0.05$).

Table 1. Mosquitocidal activities of *M. anisopliae* ethyl acetate crude extract against larvae, pupae, and adults of three mosquito species at 24 h after treatments.

Mosquito	Stage	N = Insect Number	LC_{50} (LCL-UCL)	LC_{90} (LCL-UCL)	χ^2 (df = 12)
Ae. aegypti	Larvae	450	29.631 (25.440–36.833)	80.560 (74.910–87.001)	5.673
	Pupae	450	45.530 (39.920–51.532)	103.430 (98.571–109.642)	4.041
	Adult	450	62.589 (57.439–67.991)	123.775 (115.679–129.002)	6.090
An. stephensi	Larvae	450	32.578 (27.871–35.900)	88.003 (82.717–93.966)	5.214
	Pupae	450	52.491 (46.913–56.331)	98.110 (95.332–105.88)	1.287
	Adult	450	70.235 (66.057–75.339)	150.921 (141.883–157.991)	3.002
Cx. quinquefasciatus	Larvae	450	48.003 (41.771–53.994)	96.883 (93.880–103.439)	6.454
	Pupae	450	69.017 (64.771–74.000)	158.881 (151.875–164.640)	0.989
	Adult	450	73.937 (66.383–78.382)	180.440 (176.003–189.337)	7.046

na is total number of larvae, pupae and adult used per each species, 25 per replicate, three replicates were carried out, five concentrations were tested; LC_{50} = lethal concentration killing 50% of exposed organisms; LC_{90} = lethal concentration killing 90% of exposed organisms; LCL = 95% lower confidence limits; UCL = 95% upper confidence limits; χ^2 = chi square; df = degrees of freedom; SD = Standard deviation.

Mortality of *An. stephensi* larvae varied from 10.33 to 85.33%, for pupae, 8.33 to 70.33%, and adult (from 4.33 to 58.66%) (Figure 1; Table 1). As for *An. stephensi* the susceptibility of the larvae to the *M. anisopliae* crude metabolites was higher than the pupae and adults (LC_{50} = 32.578, 52.491, and 70.235 μg/mL for larvae, pupae, and adults, respectively) (Table 1). Similarly, larvae mortality of *Cx. quinquefasciatus* varied from 8.66 to 80.33%; pupal from 6.00 to 61.00%, for adult 21.00 to 54.66% (Figure 1; Table 1). The toxicity of the *M. anisopliae* crude metabolites was higher for the *Cx. quinquefasciatus* larvae, (LC_{50} = 48.003 μg/mL) than it was for the pupae (LC_{50} = 69.017 μg/mL) or for the adults (LC_{50}, 73.937 μg/mL) (Table 1).

3.2. Non-Target Organisms

Entomopathogenic fungi *M. anisopliae* constituents showed a minimal effect on non-targeted *A. nauplii*. This study clearly shows (5.33–18.33 %) mortality were produced by the *M. anisopliae* derived crude extract (Table 2; Figure 2). The LC_{50} and LC_{90} values were 620.481; 6893.990 μg/mL (Table 2). No behavioral changes were observed during the treatment with fungal extracts.

A low lethal effect was observed in *E. eugeniae* treated with the fungi metabolites; 14.0% mortality were observed in those treated with *M. anisopliae* secondary metabolites at 30 days after treatments. The highest earthworm mortality was observed in Monocrotophos pesticide treatment that shows 87.33 % mortality. Furthermore, the chemical treatment epidermis, intestinal and body wall thickness was reduced by the chemical (Figure 3; Tables 3 and 4).

Table 2. Toxicity of *M. anisopliae* secondary metabolites on *A. nauplii* at 24 h after treatments.

Mosquito (na = 450)	Concentration (µg/mL)	% Mortality ± SD	LC_{50} (LCL-UCL)	LC_{90} (LCL-UCL)	χ^2 (df = 12)
M. anisopliae	Control	1.33 ± 0.5			
	10	5.33 ± 0.5			
	15	12.66 ± 1.0	620.481 (612.550–635.779)	6893.990 (6587.612–7432.900)	1.599
	30	15.0 ± 0.5			
	50	13.33 ± 1.0			
	75	18.33 ± 0.5			

na = total number of *A. nauplii* used per each species, 25 per replicate, three replicates were carried out, five concentrations were tested; LC_{50} = lethal concentration killing 50% of exposed organisms; LC_{90} = lethal concentration killing 90 % of exposed organisms; LCL = 95 % lower confidence limits; UCL = 95 % upper confidence limits; χ^2 = chi square; df = degrees of freedom; SD = Standard deviation.

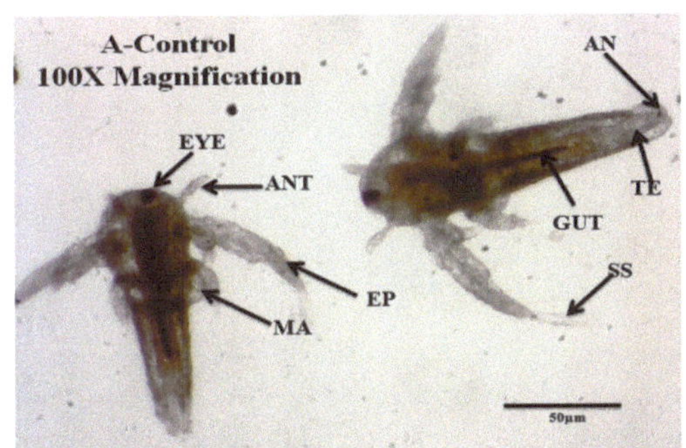

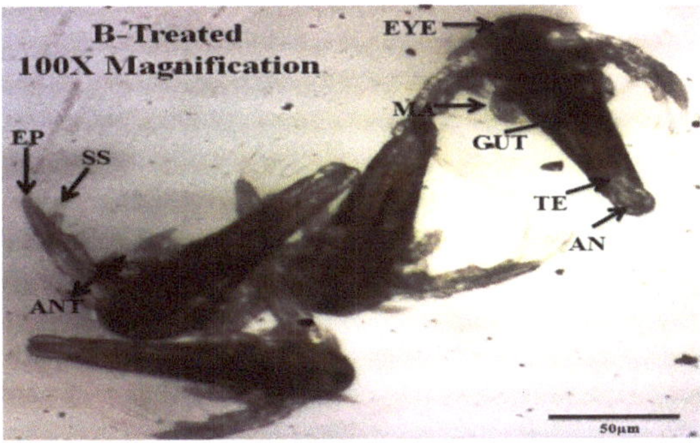

(**A**)   (**B**)

Figure 2. Morphological changes of *A. nauplii* exposed of *M. anisopliae* secondary metabolites at post 24 h of treatment. (**A**). Control (not treated fungal extract), (**B**). *M. anisopliae* secondary metabolites treated *A. nauplii* have no morphological changes were observed. (AN-1: Antennae 1, AN-2: Antennae 2, EYE: eye, EP: exopod, MA: mandible, GUT: gut, TE: telson, AN: anus, SS: swimming setae, ANT: antenna).

Table 3. Mortality of *E. eugeniae* after the treatment of *M. anisopliae* crude extract and Monocrotophos at post 24 h treatments. The identical lower case letters do not differ significantly ($p > 0.05$).

Treatment	Concentration (µg/mL)	% Mortality ± SD
M. anisopliae	Control	1.33 ± 0.5 [a]
	50	4.66 ± 1.0 [b]
	75	14.00 ± 1.1 [c]
Monocrotophos	Control	1.33 ± 0.5 [a]
	50	50.00 ± 0.5 [b]
	75	87.33 ± 0.5 [c]

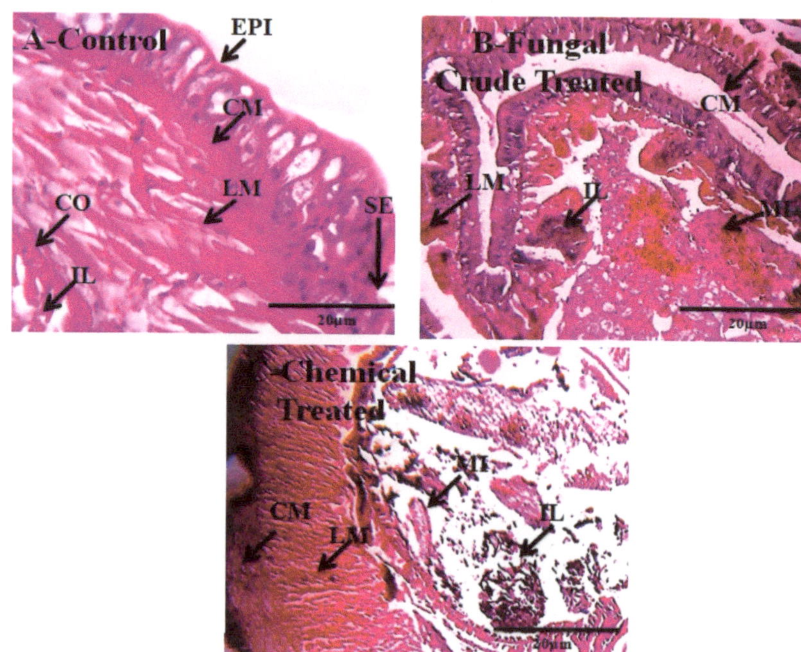

Figure 3. The *M. anisopliae* secondary metabolites (200 μg/mL) were exposed *E. eugeniae* and after 30 days of treatment, the earthworm gut tissues were sectioned for histopathological evaluation and magnified at 40× under a light microscope. (**A**) is control (without fungal crude extract treatment); (**B**) is fungal secondary metabolites treated; and (**C**) is Monocrotophos 200 ppm/kg treated. In the control and entomopathogenic fungi crude extract treatments, no changes were observed, but chemical pesticide treatment of several gut tissues morphology and shapes changed in the lumen tissues was entirely spoiled compared with control (EPI-epidermis, SE-setae, IL-intestinal lumen, LM-longitudinal muscle, CO-coelom, CM-circular muscle, MI-mitochondrion).

Table 4. Thickness of the epidermis, intestinal epithelium, and body wall of earthworms after the 30 days treatment of *M. anisopliae* crude extract. The identical lower case letters do not differ significantly ($p > 0.05$).

Treatments	E. eugeniae		
	Epidermis (μm) ± SD	Intestinal Epithelium (μm) ± SD	Body Wall (μm) ± SD
Control	37.13 ± 0.0 [a]	71.14 ± 0.5 [a]	280.12 ± 0.0 [a]
M. anisopliae	36.51 ± 0.5 [b]	70.55 ± 0.5 [b]	279.10 ± 0.0 [b]
Monocrotophos	23.32 ± 0.5 [c]	55.15 ± 1.1 [c]	210.12 ± 0.5 [c]

3.3. LC-MS and FT-IR Analysis

LC-MS analysis results of *M. anisopliae* extract showed the presence of two major chemical constituents, and retension time namely Camphor (21.08), Caprolactam (21.66) and Monobutyl phthalate (23.90) (Figure 4; Table 5).

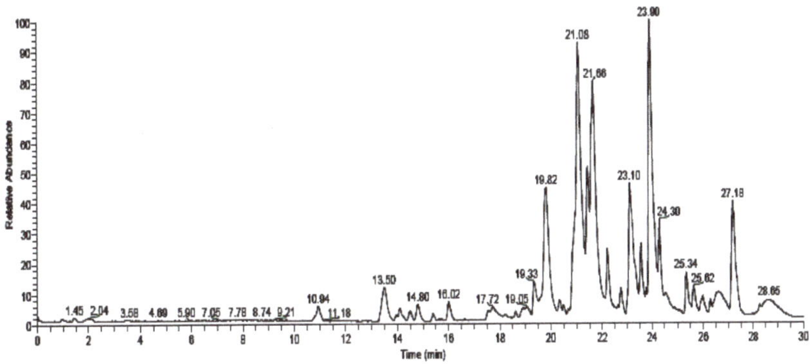

Figure 4. Chemical constituents were identified from *M. anisopliae* secondary metabolites using LC-MS analysis.

Table 5. The *M. anisopliae* ethyl acetate crude extract chemical constituents were identified using LC-MS analysis.

S. No	Retention Time	Molecular Formula	Molecular Weight	Compound Name	Compound Structure
1	19.82	$C_{37}H_{67}NO_{13}$	733.46124	(-)-Erythromycin	
2	21.08	$C_{10}H_{16}O$	152.12012	(-)-Camphor	
3	21.66	$C_6H_{11}NO$	113.08406	Caprolactam	
4	23.10	$C_{16}H_{30}O_4$	286.21441	2,2,4-Trimethyl-1,3-pentadienol diisobutyrate	
5	23.90	$C_{12}H_{14}O_4$	222.08921	Monobutyl phthalate	
6	24.30	$C_{20}H_{38}O_2$	310.28718	Ethyl oleate	
7	27.18	$C_{16}H_{22}O_4$	278.15181	Dibutyl phthalate	

FT-IR showed the presence of functional groups such as, O–H stretching (3457.62 cm^{-1}), O–H stretching (2854.91 cm^{-1}) and the medium peak C=O stretching (1679.00 cm^{-1}) (Figure 5; Table 6).

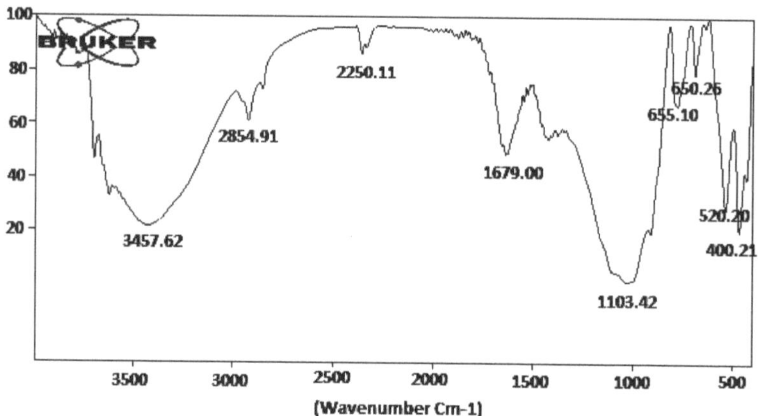

Figure 5. The major functional group was identified from *M. anisopliae* secondary metabolites using FT-IR analysis.

Table 6. The major functional group was identified from *M. anisopliae* ethyl acetate crude extract using FT-IR analysis.

S. No	Observed Wavenumber (cm^{-1})	Functional Group	Bonding Pattern
1	3457.62	O–H stretch	Phenols
2	2854.91	O–H stretch	Carboxylic acids
3	2250.11	-C C- stretch	Alkynes
4	1679.00	C=O stretch	Aldehydes
5	1103.42	C-H wag	Alkyl halides
6	655.10	C-H bends	Aromatics
7	650.25	C-H bends	Aromatics
8	520.20	C-Br stretch	Alkyl halides
9	400.21	C-Br stretch	Alkyl halides

4. Discussion

Recently, there has been a great interest in the use of biologically derived pesticides as an alternative to synthetic chemicals [9,10]. Entomopathogenic fungi-derived toxins have several advantages over synthetic pesticides in that they kill mosquitos at different stages in both laboratory and environmental conditions, have lower toxic effects on non-target organisms, and remain stable for several months in extreme cold and hot conditions [9,10,14]. In this study, we evaluated the toxic effects of secondary metabolites isolated from *M. anisopliae* strains against larvae, pupae, and adults of the disease-vector mosquitoes *Ae. aegypti*, *An. stephensi* and *Cx. quinquefasciatus*, and we assessed their target specificity and environmental safety by testing the extracts against the aquatic and terrestrial non-target species *A. naupli* L. and *E. eugeniae*.

Fungal secondary metabolites showed clear toxicity against all the tested instars of the mosquitoes and much lower toxicity against the non-target organisms. In the present study, *M. anisopliae* crude metabolites showed high toxicity towards the larvae, pupae, and adults of *A. aegypti*, *A. stephensi* and *C. quinquefasciatus* mosquitoes at 24 h post treatment under laboratory conditions (Figure 1; Table 1). In line with our results, previous studies on entomopathogenic fungal derived pesticides from several species of *Metarhizium*, *Fusarium*, *Aspergillus*, *Trichoderma* and *Lecanicillium* showed that they are effective against medical and agricultural insect pests [24]. Soni and Prakash [25] reported that *Chrysosporium keratinophilum* derived secondary metabolites have strong larvicidal activity against *C. quinquefasciatus* and *A. stephensi* mosquito larvae, while [26] reported that different fungal metabolites cause strong larvicidal activity against larvae of *A. stephensi* and *C. quinquefasciatus*. Similarly, *Metarhiziumanisopliae*, *Aspergillus flavus*, *Fusarium oxysporum*, *Verticillium lecanii*, *Paecilomyces fumosoroseus*, *Beauveria bassiana*, and *Fusarium moniliforme* and their toxins have been shown to produce remarkable mosquitocidal potential on larvae, pupae, and adult mosquitoes [9–11,27]. *C. tropicum*, *C. clavisporus* and *F. oxysporum* culture filtrates showed strong larvicidal activity against *A. stephensi*, *A. aegypti* and *C. quinquefasciatus* [10,11,22,28], and secondary metabolites of *A. fumigatus* showed strong larvicidal activity against larvae of *A. aegypti* [29].

On the contrary, in our study, we observed low toxicity of the fungal metabolites against non-target species such as *A. nauplii* and *E. eugeniae* (Tables 2–4; Figures 2 and 3). Similarly, [30] reported few swimming speed alterations in *Artemia* adults after their treatment by different toxins. A similar study about the effects of the fungi secondary metabolites from *Penicillium daleae* on *Artemia*, observed morphological changes in eye shape, eye color, and eye fading [31]. These results suggest that secondary metabolites from different fungi may produce lower levels of toxicity to non-target organisms. For this reason, the assessment of the lower effects of fungal secondary metabolites in aquatic and terrestrial ecosystems on non-target species is of prime importance. The chemical composition of the secondary metabolites extracted from the *M. anisopliae* entomopathogenic fungi analysed in this study is in accord with previous research by [9,10] and by [32], who observed similar kinds of chemical constituents (Figure 4; Table 5). Previously, [9,10] reported that *B. bassiana* and *F. oxysporum* derived crude metabolites had the same chemical constituents showing a strong larvicidal activity on *A. aegypti*, *A. stephensi* and *C. quinquefasciatus* larvae. In this study, FT-IR analyses showed the presence of phenols, biogenic amines, and carboxylic acids, which may be involved in the toxic effects on mosquitoes (Figure 5; Table 6).

Similarly, previous studies showed that the metabolites of entomopathogenic fungi are constituted by components belonging to several chemical classes (phenols, alcohols, carboxylic acids, misc, aromatics, phosphoramide, and disulfides), which may be involved in the mosquitocidal effects [9–11,26,33–36]. The strong mosquitocidal activity and the low toxic effect on non-target organisms exhibited by *M. anisopliae* indicate that, besides entomopathogenic fungal conidia, their metabolites may also have a significant role in efficient microbial-derived mosquito control tools that can be used in mosquito control programmes as effective, cheaper, biodegradable, target-specific alternatives to chemical insecticides. Further research into the single crude metabolite chemical constituents under laboratory and semi-field conditions may result in the development of effective *M. anisopliae* derived bio-pesticides.

5. Conclusions

The strong mosquitocidal activity and the low toxic effect on non-target organisms exhibited by *M. anisopliae* indicate that, besides entomopathogenic fungal conidia, their metabolites may also have a significant role in efficient microbial-derived mosquito control tools that can be used in mosquito control programmes as effective, cheaper, biodegradable, target-specific alternatives to chemical insecticides. Further research into the single crude metabolite chemical constituents under laboratory and semi-field conditions may result in the development of effective *M. anisopliae* derived biopesticides.

Author Contributions: Conceptualization: P.V., K.S. and P.K.; Data curation: P.V. and K.S.; Formal analysis: P.V., K.S., P.K. and A.C.M.; Funding acquisition: P.K.; Investigation: P.V.; Methodology: P.V.; Project administration: P.V., K.S. and P.K.; Resources: P.V. and K.S.; Software: K.S. and P.K.; Supervision: P.V. and K.S.; Validation: P.V., K.S. and P.K.; Visualization: K.S.; Writing—original draft: P.V., K.S. and P.K.; Writing—review & editing: P.V., K.S., P.K. and A.C.M. All authors have read and agreed to the published version of the manuscript.

Funding: This research received no external funding.

Institutional Review Board Statement: Not applicable for this present studies and does not involving humans or animals.

Informed Consent Statement: Not applicable for this present studies and does not involving humans or animals.

Data Availability Statement: The data that support the findings of this present study are available from the corresponding author upon reasonable request.

Acknowledgments: Authors thank to Ananthanarayanan Yuvaraj, Department of Zoology for his valuable suggestions for histopathological studies and also thank to Periyar University, Tamil, Nadu, India for making available the infrastructure and resources for this study. This research was partially supported by Chiang Mai University, Thailand.

Conflicts of Interest: All the authors state that they do not have any conflict of interest.

Ethical Statement: This article does not contain any studies with human participants performed by any of the authors. All applicable international, national, and institutional guidelines for the care and use of animals were followed.

References

1. Huang, W.; Wang, S.; Jacobs-Lorena, M. Use of microbiota to fight mosquito-borne disease. *Front. Genet.* **2020**, *11*, 196. [CrossRef] [PubMed]
2. Chareonviriyaphap, T.; Bangs, M.J.; Suwonkerd, W.; Kongmee, M.; Corbel, V.; Ngoen-Klan, R. Review of insecticide resistance and behavioral avoidance of vectors of human diseases in Thailand. *Parasites Vectors* **2013**, *6*, 280. [CrossRef] [PubMed]
3. Ghosh, A.; Chowdhury, N.; Chandra, G. Plant extracts as potential mosquito larvicides. *Indian J. Med. Res.* **2012**, *135*, 581–598. [PubMed]
4. Busvine, J.R. *Recommended Methods for Measurement of Pest Resistance to Pesticides*; FAO: Rome, Italy, 1980.
5. Vivekanandhan, P.; Thendralmanikandan, A.; Kweka, E.J.; Mahande, A.M. Resistance to temephos in *Anopheles stephensi* larvae is associated with increased cytochrome P450 and α-esterase genes overexpression. *Int. J. Trop. Insect Sci.* **2021**, *41*, 2543–2548. [CrossRef]
6. Vatandoost, H.; Hanafi-Bojd, A.A. Indication of pyrethroid resistance in the main malaria vector, *Anopheles stephensi* from Iran. *Asian Pac. J. Trop. Med.* **2012**, *5*, 722–726. [CrossRef]
7. Li, Q.Q.; Loganath, A.; Chong, Y.S.; Tan, J.; Obbard, J.P. Persistent organic pollutants and adverse health effects in humans. *J. Toxicol. Environ. Health Part A* **2006**, *69*, 1987–2005.
8. Sarwar, M. Biopesticides: An effective and environmental friendly insect-pests inhibitor line of action. *Int. J. Eng. Adv. Res. Technol.* **2015**, *1*, 10–15.
9. Morales-Rodriguez, A.; Peck, D.C. Synergies between biological and neonicotinoid insecticides for the curative control of the white grubs *Amphimallon majale* and *Popillia japonica*. *Biol. Control* **2009**, *51*, 169–180. [CrossRef]
10. Ruiu, L.; Satta, A.; Floris, I. Emerging entomopathogenic bacteria for insect pest management. *Bull. Insectol.* **2013**, *66*, 181–186.
11. Bojke, A.; Tkaczuk, C.; Stepnowski, P.; Gołębiowski, M. Comparison of volatile compounds released by entomopathogenic fungi. *Microbiol. Res.* **2018**, *214*, 129–136. [CrossRef]
12. Rai, D.; Updhyay, V.; Mehra, P.; Rana, M.; Pandey, A.K. Potential of entomopathogenic fungi as biopesticides. *Indian J. Sci. Res. Technol.* **2014**, *2*, 7–13.
13. Zhang, L.; Fasoyin, O.E.; Molnár, I.; Xu, Y. Secondary metabolites from hypocrealean entomopathogenic fungi: Novel bioactive compounds. *Nat. Prod. Rep.* **2020**, *37*, 1181–1206. [CrossRef]
14. Darbro, J.M.; Thomas, M.B. Spore persistence and likelihood of aeroallergenicity of entomopathogenic fungi used for mosquito control. *Am. J. Trop. Med. Hyg.* **2009**, *80*, 992–997. [CrossRef]
15. Islam, W.; Adnan, M.; Shabbir, A.; Naveed, H.; Abubakar, Y.S.; Qasim, M.; Tayyab, M.; Noman, A.; Nisar, M.S.; Khan, K.A.; et al. Insect-fungal-interactions: A detailed review on entomopathogenic fungi pathogenicity to combat insect pests. *Microb. Pathog.* **2021**, *159*, 105122. [CrossRef]
16. Vyas, N.; Dua, K.K.; Prakash, S. Efficacy of *Lagenidium giganteum* metabolites on mosquito larvae with reference to nontarget organisms. *Parasitol. Res.* **2007**, *101*, 385–390. [CrossRef]

17. Vivekanandhan, P.; Swathy, K.; Thomas, A.; Kweka, E.J.; Rahman, A.; Pittarate, S.; Krutmuang, P. Insecticidal Efficacy of Microbial-Mediated Synthesized Copper Nano-Pesticide against Insect Pests and Non-Target Organisms. *Int. J. Environ. Res. Public Health* **2021**, *18*, 10536. [CrossRef]

18. Vivekanandhan, P.; Arunthirumeni, M.; Vengateswari, G.; Shivakumar, M.S. 5 Bioprospecting of Novel Fungal Secondary Metabolites for Mosquito Control. In *Microbial Control of Vector-Borne Diseases*; Taylor & Francis: Raton, FL, USA, 2018; pp. 61–89.

19. Lu, Y.; Yu, J. A Well-Established Method for the Rapid Assessment of Toxicity Using *Artemia* spp. Model. In *Assessment and Management of Radioactive and Electronic Wastes*; IntechOpen: London, UK, 2019; pp. 1–15. [CrossRef]

20. World Health Organization. *Guidelines for Laboratory and Field Testing of Mosquito Larvicides. Communicable Disease Control, Prevention and Eradication*; WHO, Pesticide Evaluation Scheme; WHO: Geneva, Switzerland, 2005; pp. 1–219.

21. World Health Organization. Global programme to eliminate lymphatic filariasis-progress report on mass drug administration in 2016. *Wkly. Epidemiol. Rec.* **2016**, *85*, 365–372.

22. Norris, E.J.; Bloomquist, J.R. Nutritional status significantly affects toxicological endpoints in the CDC bottle bioassay. *Pest Manag. Sci.* **2022**, *78*, 743–748. [CrossRef]

23. Abbott, W.S. A method of computing the effectiveness of an insecticide. *J. Econ. Entomol.* **1925**, *18*, 265–267. [CrossRef]

24. Thakur, N.; Kaur, S.; Tomar, P.; Thakur, S.; Yadav, A.N. Microbial biopesticides: Current status and advancement for sustainable agriculture and environment. In *New and Future Developments in Microbial Biotechnology and Bioengineering*; Elsevier: Amsterdam, The Netherlands, 2020; pp. 243–282.

25. Soni, N.; Prakash, S. Effect of *Chrysosporium keratinophilum* metabolites against *Culex quinquefasciatus* after chromatographic purification. *Parasitol. Res.* **2010**, *107*, 1329–1336. [CrossRef]

26. Soni, N.; Prakash, S. Entomopathogenic fungus generated Nanoparticles for enhancement of efficacy in *Culex quinquefasciatus* and *Anopheles stephensi*. *Asian Pac. J. Trop. Dis.* **2012**, *2*, S356–S361. [CrossRef]

27. Vivekanandhan, P.; Karthi, S.; Shivakumar, M.S.; Benelli, G. Synergistic effect of entomopathogenic fungus *Fusarium oxysporum* extract in combination with temephos against three major mosquito vectors. *Pathog. Glob. Health* **2018**, *112*, 37–46. [CrossRef]

28. Vivekanandhan, P.; Deepa, S.; Kweka, E.J.; Shivakumar, M.S. Toxicity of *Fusarium oxysporum*-VKFO-01 Derived Silver Nanoparticles as Potential Inseciticide Against Three Mosquito Vector Species (Diptera: Culicidae). *J. Clust. Sci.* **2018**, *29*, 1139–1149. [CrossRef]

29. Balumahendhiran, K.; Vivekanandhan, P.; Shivakumar, M.S. Mosquito control potential of secondary metabolites isolated from *Aspergillus flavus* and *Aspergillus fumigatus*. *Biocatal. Agric. Biotechnol.* **2019**, *21*, 101334. [CrossRef]

30. Manfra, L.; Canepa, S.; Piazza, V.; Faimali, M. Lethal and sublethal endpoints observed for Artemia exposed to two reference toxicants and an ecotoxicological concern organic compound. *Ecotoxicol. Environ. Saf.* **2016**, *123*, 60–64. [CrossRef]

31. Uwizeyimana, H.; Wang, M.; Chen, W.; Khan, K. The eco-toxic effects of pesticide and heavy metal mixtures towards earthworms in soil. *Environ. Toxicol. Pharmacol.* **2017**, *55*, 20–29. [CrossRef]

32. Dubey, N.K.; Shukla, R.; Kumar, A.; Singh, P.; Prakash, B. Global scenario on the application of natural products in integrated pest management programmes. *Nat. Prod. Plant Pest. Manag.* **2011**, *1*, 1–20.

33. Libralato, G.; Prato, E.; Migliore, L.; Cicero, A.M.; Manfra, L. A review of toxicity testing protocols and endpoints with *Artemia* spp. *Ecol. Indic.* **2016**, *69*, 35–49. [CrossRef]

34. Libralato, G. The case of *Artemia* spp. in nanoecotoxicology. *Mar. Environ. Res.* **2014**, *101*, 38–43. [CrossRef]

35. Gao, Q.; Jin, K.; Ying, S.H.; Zhang, Y.; Xiao, G.; Shang, Y.; Duan, Z.; Hu, X.; Xie, X.Q.; Zhou, G.; et al. Genome sequencing and comparative transcriptomics of the model entomopathogenic fungi *Metarhizium anisopliae* and *M. acridum*. *PLoS Genet.* **2011**, *7*, e1001264. [CrossRef]

36. Vasantha-Srinivasan, P.; Senthil-Nathan, S.; Thanigaivel, A.; Edwin, E.; Ponsankar, A.; Selin-Rani, S.; Pradeepa, V.; Sakthi-Bhagavathy, M.; Kalaivani, K.; Hunter, W.B.; et al. Developmental response of *Spodoptera litura* Fab. to treatments of crude volatile oil from *Piper betle* L. and evaluation of toxicity to earthworm, *Eudrilus eugeniae* Kinb. *Chemosphere* **2016**, *155*, 336–347. [CrossRef] [PubMed]

Journal of
Fungi

Article

Genome Sequencing and Analysis of *Trichoderma* (Hypocreaceae) Isolates Exhibiting Antagonistic Activity against the Papaya Dieback Pathogen, *Erwinia mallotivora*

Amin-Asyraf Tamizi [1], Noriha Mat-Amin [1], Jack A. Weaver [2], Richard T. Olumakaiye [2], Muhamad Afiq Akbar [3], Sophie Jin [2], Hamidun Bunawan [3,*] and Fabrizio Alberti [2,*]

1 Agri-Omics and Bioinformatics Programme, Biotechnology and Nanotechnology Research Centre, Malaysian Agricultural Research and Development Institute Headquarters (MARDI), Serdang 43400, Selangor, Malaysia; aminasyraf@mardi.gov.my (A.-A.T.); noriha@mardi.gov.my (N.M.-A.)

2 School of Life Sciences, University of Warwick, Gibbet Hill Road, Coventry CV4 7AL, UK; jack.weaver@warwick.ac.uk (J.A.W.); richard.olumakaiye@warwick.ac.uk (R.T.O.); sophie.jin@warwick.ac.uk (S.J.)

3 Institute of Systems Biology, Universiti Kebangsaan Malaysia (UKM), Bangi 43600, Selangor, Malaysia; muhdafiq.akbar@gmail.com

* Correspondence: hamidun.bunawan@ukm.edu.my (H.B.); f.alberti@warwick.ac.uk (F.A.)

Abstract: *Erwinia mallotivora*, the causal agent of papaya dieback disease, is a devastating pathogen that has caused a tremendous decrease in Malaysian papaya export and affected papaya crops in neighbouring countries. A few studies on bacterial species capable of suppressing *E. mallotivora* have been reported, but the availability of antagonistic fungi remains unknown. In this study, mycelial suspensions from five rhizospheric *Trichoderma* isolates of Malaysian origin were found to exhibit notable antagonisms against *E. mallotivora* during co-cultivation. We further characterised three isolates, *Trichoderma koningiopsis* UKM-M-UW RA5, UKM-M-UW RA6, and UKM-M-UW RA3a, that showed significant growth inhibition zones on plate-based inhibition assays. A study of the genomes of the three strains through a combination of Oxford nanopore and Illumina sequencing technologies highlighted potential secondary metabolite pathways that might underpin their antimicrobial properties. Based on these findings, the fungal isolates are proven to be useful as potential biological control agents against *E. mallotivora,* and the genomic data opens possibilities to further explore the underlying molecular mechanisms behind their antimicrobial activity, with potential synthetic biology applications.

Keywords: biocontrol; fungus; *Trichoderma*; Gram-negative; *Erwinia*

Citation: Tamizi, A.-A.; Mat-Amin, N.; Weaver, J.A.; Olumakaiye, R.T.; Akbar, M.A.; Jin, S.; Bunawan, H.; Alberti, F. Genome Sequencing and Analysis of *Trichoderma* (Hypocreaceae) Isolates Exhibiting Antagonistic Activity against the Papaya Dieback Pathogen, *Erwinia mallotivora. J. Fungi* **2022**, *8*, 246. https://doi.org/10.3390/jof8030246

Academic Editor: Laurent Dufossé

Received: 25 January 2022
Accepted: 26 February 2022
Published: 28 February 2022

Publisher's Note: MDPI stays neutral with regard to jurisdictional claims in published maps and institutional affiliations.

1. Introduction

Papaya dieback disease (PDD), or bacterial crown rot (BCR), is a bacterial-caused disease that has negatively affected papaya crops (*Carica papaya* L.) in Malaysia, reducing the national papaya export up to 70% [1]. Initially, *Erwinia papayae* was reported to be the causal agent of PDD [2], though later Mat Amin et al. [3] molecularly validated the pathogen to have the closest match to *E. mallotivora* Goto strain DSM 4565 and this was further supported through genome sequencing [4]. Most of the Malaysian papaya cultivated varieties (cultivars) such as 'Eksotika', 'Sekaki', and 'Setiawan' are highly susceptible to this Gram-negative pathogen [5]. Once infected, water soak symptoms become prevalent on papaya petioles, leaves, and the apical stem, and the tree can succumb to the disease within one to two weeks [5–7]. In addition to physical diagnosis, Mohd Said et al. [8] developed a DNA-based detection method for *E. mallotivora* to further assist in early detection of the disease and Ramachandran et al. [9] conducted a repetitive element PCR fingerprinting (rep-PCR) study to provide an accurate and rapid identification method on various local *E. mallotivora* isolates. Currently, there is no effective cure for PDD from the moment the symptoms appear on the trees. Hence, farmers rely on preventive measures and good management practices to contain and minimise the spread of the disease to neighbouring trees and plots [5,10,11]. Recently, *E. mallotivora* was reported to be affecting papaya in the Philippines and Indonesia, making this species a pathogen of concern that could affect the papaya industry in the entire Southeast Asia region [7,12]. A similar disease termed as papaya black rot has been recently detected in Japan and the pathogen was characterised to be from the genus *Erwinia* [13].

Numerous studies have been conducted in recent years to understand the pathogenesis of *E. mallotivora* and find solutions to curb the disease. About four years after the identification of the pathogen, the draft genome sequence of *E. mallotivora* was published and this has initiated more research [4]. Wee et al. [14], Hamid et al. [15], Juri et al. [16,17], Abu-Bakar et al. [18], Abu Bakar et al. [19], and Tamizi et al. [20] have conducted in-depth studies to understand the host–pathogen interactions and pathogenicity of *E. mallotivora* at the molecular level, while Noor Shahida et al. [6] described the infection from a physiological perspective. The use of conventional breeding and genetic engineering to develop new papaya lines that would be resistant to *E. mallotivora* is still in the research pipelines [21–25]. Nevertheless, the effort to combat the disease should be augmented with other approaches, including the use of effective microorganisms and native microbial biocontrol.

Mat Amin et al. [26,27] discovered several *Bacillus* spp. from soil to be antagonistic towards *E. mallotivora*, and this effort has pioneered the search for potential antagonistic microbes against *E. mallotivora* that have been ravaging papaya trees. Later, a biocontrol inoculant containing an undisclosed species of *Bacillus* sp. known as 'Dieback Buster 95' was launched and claimed to be highly efficient in the prevention of PDD [28,29]. In addition to these, native endophytic bacteria (NEB) isolated from papaya were reported to be capable of suppressing the growth of *E. mallotivora* [30,31].

While these potential biocontrols have been isolated from the kingdom of bacteria, fungal species possessing antagonistic effects against *E. mallotivora* remain unexplored. Fungi of the genus *Trichoderma* have been widely studied for their biological control properties against pathogenic fungi and other bacterial species [32–35]. They are known plant growth-promoting fungi (PGPF), abundant in soil and easily culturable [36,37]. According to Guo et al. [37] and Khan et al. [38], *Trichoderma* species produced diverse compounds and active metabolites during their interaction with competitors. This makes an effort to single out a potential antimicrobial compound from a particular *Trichoderma* species relatively challenging. Rush et al. [39] produced a systemic review on *Trichoderma* bioprospecting which discussed the application of genome technology in screening for natural products and biocontrol compounds. In the present study, we have discovered five *Trichoderma* isolates that showed antimicrobial activity against the *E. mallotivora* strain BT-MARDI, of which three were further characterised through genome sequencing to investigate the possible mechanisms behind their antimicrobial properties.

2. Materials and Methods

2.1. Preparation of Pathogen

The papaya dieback pathogen, *E. mallotivora* strain BT-MARDI, was obtained from the Biotechnology and Nanotechnology Research Centre, at the Malaysian Agricultural Research and Development Institute (MARDI) in Serdang, Selangor, Malaysia. A stock culture of *E. mallotivora* was maintained in 15% (w/v) glycerol at $-80\,^\circ$C. Working cultures were established by streaking the glycerol stock onto Luria Bertani (LB) agar plates (10 g/L tryptone, 5 g/L yeast extract, 10 g/L sodium chloride, 15 g/L agar) and incubating at 30 $^\circ$C for 48 h. Fresh single colonies were then picked and cultured in 50 mL LB medium (10 g/L tryptone, 5 g/L yeast extract, 10 g/L sodium chloride) in a 250 mL flask, shaking at 180 rpm for 24 h.

2.2. Isolation of Antagonist Fungi

The rhizosphere soil samples were randomly collected from healthy papaya plants (wild and cultivated) at five different locations (Table 1) in Peninsular Malaysia. Then, soil dilution was performed and used to grow culturable fungi, particularly *Trichoderma*, on potato dextrose agar (PDA) supplemented with selective agents. Briefly, 50 g of soil sample were diluted in 50 mL of sterile distilled water. Serial dilution was performed at 1/10, 1/100, 1/1000, and 1/10,000. Four hundred microliters of the diluted soil suspension were spread on modified PDA (PDA; 20 g/L glucose, 4 g/L potato extract, 20 g/L agar) medium containing 0.3 g/L chloramphenicol, 0.1 g/L streptomycin, and 0.02 g/L Rose Bengal dye [40]. The plates were incubated at 25 °C for 3 to 14 days to allow fast and medium–fast growing fungi to appear. The emerging fungal colonies were then purified by single spore isolation [41] and maintained on PDA slants at 4 °C. Long-term storage of fungi was achieved by cutting a 5 mm^2 of mycelial agar plug from freshly growing cultures, which was stored in 30% (*w/v*) glycerol at −80 °C.

Table 1. Origin of the soil samples used for the isolation of the fungi identified in this study. The soil samples were collected from wild and cultivated healthy papaya populations in different states (Selangor, Negeri Sembilan, Pahang, and Kedah) of Peninsular Malaysia.

Soil Sample ID	Location Site, State	Coordinates
Mardi-2018-TK	Tanjong Karang, Selangor	3°26′27.8″ N 101°08′50.1″ E
Mardi-2018-R	Rembau, Negeri Sembilan	2°34′26.0″ N 102°05′48.5″ E
Mardi-2018-P	Raub, Pahang	3°49′19.3″ N 101°50′38.4″ E
Mardi-2018-C	Mardi Serdang, Selangor	2°59′24.9″ N 101°41′49.5″ E
Mardi-2018-BK	Baling, Kedah	5°40′05.6″ N 100°49′47.5″ E

2.3. Antagonistic Activity of Fungal Isolates against E. mallotivora Strain BT-MARDI

The antagonistic activity of fungal isolates against *E. mallotivora* was screened in vitro using the agar well diffusion method on PDA plates [31,42,43]. An agar plug (5 mm^2) from each five-day old fungal culture was inoculated into 100 mL potato dextrose broth (PDB; 20 g/L glucose, 4 g/L potato extract) in 500 mL flasks and grown for seven days at 25 °C with agitation at 180 rpm. A lawn of *E. mallotivora* was prepared by spreading 100 μL of the *E. mallotivora* inoculum over the entire agar surface. A hole with a diameter of 6 mm was punched aseptically with a sterile cork borer and 100 μL of the fungal suspension were dispensed into the well. A kanamycin solution (100 μg/mL) and PDB, 100 μL of each, were used as the positive and negative controls, respectively. After overnight incubation at 28 °C, the plates were observed for zone of inhibition formation on the *E. mallotivora* lawn. The experiment was repeated in triplicate for each isolate to record the average diameter of the inhibition zone. The data for fungal isolates exhibiting significant antagonism compared to the positive control was analysed using IBM SPSS Statistics version 28.0.1.1 (14).

2.4. Genomic DNA Extraction

Cultures of the three selected fungal isolates, UKM-M-UW RA3a, UKM-M-UW RA5, and UKM-M-UW RA6 (hereafter referred to as RA3a, RA5, and RA6), were set up in 25 mL of PDB from PDA plates and placed in 250 mL conical flasks. The fungal cultures were incubated at 25 °C for 72 h, shaking at 200 rpm. The mycelia from the broth were pelleted by centrifugation at 8000× *g* for 5 min, and subsequently washed twice with sterile distilled water to remove any residual medium. High molecular weight (HMW) genomic DNA extraction was carried out via cryogenic grinding using a sterile mortar and pestle. Fungal mycelia were ground into a fine powder in liquid nitrogen; thereafter, 200 mg of the powder were transferred aseptically into a microcentrifuge tube for immediate use. A GenElute Plant Genomic DNA miniprep kit (Sigma-Aldrich, Burlington, MA, USA) was used for DNA extraction, following the manufacturer's instructions. HMW genomic DNA was concentrated using ethanol precipitation. Briefly, 1/10 volume of 3 M sodium acetate, pH 5.2, was added, followed by three volumes of 100% ethanol. Samples were inverted, incubated at $-20\ °C$ overnight, then spun at $13,000\times g$ for 30 min at $4\ °C$. Supernatant was decanted, the DNA pellet was air dried, and then resuspended in nuclease-free water (Thermo Fisher, Waltham, MA, USA). Gel electrophoresis (1% *w/v* agarose gel with 1X GelRed), a Nanodrop ND1000 (Thermo Fisher, USA), and a Qubit (dsDNA Broad range, Invitrogen, Waltham, MA, USA) were used to assess the concentration (Table S1) and quality of the HMW genomic DNA extracted from the fungal strains.

2.5. Identification of Fungi Based on Mycelia Morphology, PCR Amplification, and Sanger Sequencing

Mycelial morphology was originally used to identify the class/genus of the fungal isolates based on the colour of the mycelia and spore characteristics. Subsequently, PCR amplification of target sequences was carried out using purified genomic DNA as a template to identify the fungal isolates of interest, RA3a, RA5, and RA6. PCR was carried out using Q5 High-Fidelity 2X Master Mix (NEB, Ipswich, MA, USA) according to the manufacturer's instructions, using the primers listed in Table S2. The amplicons were purified upon gel electrophoresis (1% *w/v* agarose gel with 1X GelRed) using the Gene-JET PCR purification kit (Thermo Fisher, USA) following the manufacturer's instructions. One hundred nanograms of the purified PCR product was used for Sanger sequencing through Eurofins-GATC, which enabled for the phylogenetic confirmation of the strains. The sequences were trimmed to obtain diagnostic fragments for molecular identification as detailed in Kopchinskiy et al. [44] prior to submission to BLASTN [45].

2.6. DNA Library Preparation and Sequencing through Oxford Nanopore Technology (ONT)

Utilising the extracted HMW genomic DNA, preparation of the library was performed using the native barcoding expansion kit (EXP-NBD104 and EXP-NBD114) and the ligation sequencing kit (SQK-LSK109) for multiplex genomic DNA sequencing of the fungal isolates RA3a, RA5, and RA6. NBD01 was used to tag RA3a, NBD02 for RA6, and NBD03 for RA5. The manufacturer's protocol was followed with minor modifications introduced for both DNA repair and end-prep stages, by increasing the incubation time and temperature after washing with ethanol to 15 min at $37\ °C$. Furthermore, at the native barcode ligation step, the incubation time and temperature were increased to 30 min at $37\ °C$. To enrich DNA fragments of 3kb or longer, the Long Fragment Buffer was used for the DNA clean-up. An amount of $1.2\ \mu g$ of each genomic DNA were used for the library preparation, resulting in a final amount of 350 ng of DNA at the end of the adapter ligation and clean-up step. The resulting DNA library containing the sequencing buffer and loading beads were loaded on the primed SpotON flow cell. Nanopore sequencing was performed on MinION (ONT) with a FLO-MIN-106 R9.4 flow-cell (ONT). The MinION (ONT) multiplex sequencing reaction was run for 18 h.

2.7. Genome Assembly and Error Corrections

A *de novo* strategy was employed for sequencing each of the three genomes. The raw data produced by nanopore was base-called and simultaneously demultiplexed using Guppy version 4.4.1 available to ONT users via https://community.nanoporetech.com (accessed on 23 January 2022). The config file used in Guppy was dna_r9.4.1_450bps_hac.cfg, the remaining parameters used were the default settings. Guppy was also used to trim the barcodes from the reads during the base-calling, depositing each genome's reads in a directory as .fastq files. The .fastq files for each read were merged into one .fastq file and used as the input for Flye version 2.8.2 with the nano-raw mode selected, which was used to assemble the draft genomes [46]. The draft assembly was then polished multiple times, by aligning the base-called raw reads in the merged file against the draft genome using Minimap2 version 2.11 [47] before correcting errors using Racon version 1.4.20 [48], where matching bases were assigned a score of 8 and mismatches a score of -6, with a gap penalty of -8 and a window size of 500. The output from Racon was aligned to the raw reads again and the process was repeated three times for each assembly. Further corrections were made using Medaka version 1.2.1 utilising model r941_min_high_g360.

Polishing the genome continued by using Illumina short-read data generated through Apical Scientific Sdn. Bhd. (Seri Kembangan, Selangor, Malaysia). This was done by aligning the Illumina data to the output from Medaka using Bowtie2 version 2.4.2 [49]. Bowtie2 was run with the following settings, -D (number of attempts at extension before skipping to the next task) 20, -R (number of seeds looked at before moving to the next task) 3, -N (maximum number of mismatches in alignments) 1, -L (length of seed substrings) 20, and -i (interval between seed substrings) S,1,0.50, all other settings were left at the default parameters. Polishing was performed with Pilon version 1.23 [50], the output from Pilon was then re-aligned to the Illumina reads and the draft genome polished again; after Pilon had been run four times, the final output was analysed using BUSCO version 5.0.0 [51], comparing the three genomes against the Hypocreales_odb10 database. Unless specified above, parameters within programs were left at their default setting.

2.8. Genome Annotations

Functional annotation was performed using the Funannotate version 1.8.3 pipeline [52]. Funannotate was also able to clean (remove small repetitive contigs by using minimap2 or mummer) and perform mask repeats (softmasking of low complexity and short period tandem repeats using tantan). The prediction of functional elements was done using a seed species of *Verticillium longisporum* and the BUSCO *Sordariomycetes* database; the minimum number of training models required was dropped to 100, and the Optimize Augustus setting was turned on. Funannotate predict uses AUGUSTUS, snap, GlimmerHMM, and tRNAscan-SE in order to predict genes for proteins and tRNAs, as well as other programs to generate the inputs for those mentioned above and the final outputs. The predictions were then run through Phobius [53] and AntiSMASH [54], contained within Funannotate. Finally, the annotation function was used to create the final genomes. BUSCO was run again to look at the predicted proteins generated, once again using the Hypocreales_odb10 database. All parameters not mentioned above were left on their default setting within the software package.

CMscan [55] was used with StructRNAfinder [56] for screening the presence of potential non-coding RNA. The Rfam database [57] was used as the input database for StructRNAfinder with the default parameters.

2.9. Comparative Genomics and Phylogenomic Analysis

Pairwise genomic similarities between our isolates with addition of nine additional genomes from other *Trichoderma* spp. (obtained from the NCBI database) were calculated using FastANI, with assembled genomic sequences as input [58]. The pairwise genomic similarities were then visualised using the Intervene Shiny package [59]. Next, to reconstruct the phylogenomic association between several strains of *Trichoderma* spp., single-copy orthologous proteins were first identified from the predicted proteome sequence via OrthoFinder v2.2.7 with default settings [60]. Then, amino acid sequences from single-copy orthologous proteins identified were aligned using MAFFT [61]. This alignment was then used for maximum likelihood phylogenomic tree reconstruction via the IQ-TREE program, using the JTT+F+G4 model and 1000 ultrafast bootstrap replications [62]. iTOL was used for phylogenomic tree visualisation [63]. Finally, OrthoVenn2 web server was used for comparison of orthologous proteins from our isolates with the predicted proteome as input [64].

3. Results

3.1. Antagonism of Fungal Isolates against E. mallotivora Strain BT-MARDI

In this study, a total of 128 filamentous fungal and yeast isolates were cultured from five soil samples collected in various locations in Peninsular Malaysia (Table 1). Out of these, only 17 fungal isolates that grew mycelia in less than 14 days on PDA plates—

species putatively belonging mainly to the genera *Trichoderma, Fusarium, Aspergillus,* and *Penicillium* based on morphological identification—were observed to have antagonistic activity against *E. mallotivora* strain BT-MARDI. According to plate-based inhibition assays, five fast-growing fungal isolates that required less than seven days to form a full mycelial lawn on PDA plates were confirmed to exhibit visible antagonism (Figure 1) and selected for further analysis. The taxonomic affiliation of these isolates was originally deduced based on mycelium morphology, with all five strains, or isolates, putatively assigned to the genus *Trichoderma* (Figure 2). All isolates exhibited inhibition of the growth of *E. mallotivora* (Table 2). Based on the Tukey's test, RA5, RA6, and RA3a had a significant antagonistic activity when compared to the positive control (kanamycin) (Figure 3).

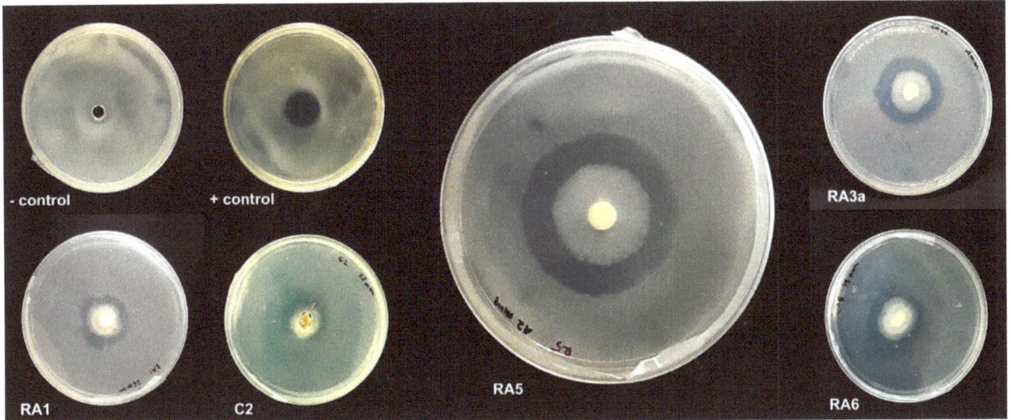

Figure 1. Plate-based growth inhibition assays of *E. mallotivora* through well diffusion method. All fungal isolates displayed notable growth inhibition zones. Potato dextrose broth (PDB) and kanamycin were used as the negative and positive controls, respectively.

Table 2. Antagonistic potential of the *Trichoderma* isolates against *E. mallotivora*. The negative (−) control consists of 100 µL of PDB, the positive (+) control of 100 µL of kanamycin (100 µg/mL).

| Treatment | Diameter of Inhibition Zone (mm) | | | | | Fungal Genus (Based on Morphology) | Origin of the Fungal Isolate |
| | Replicate | | | Mean | Standard Deviation | | |
	1	2	3				
− control	0	0	0	0	0	-	-
+ control	17	17	17	17.0	0	-	-
Isolate UKM-M-UW RA5	42	39	42	41.0	1.73	*Trichoderma*	Rembau, Negeri Sembilan
Isolate UKM-M-UW RA6	33	26	34	31.0	4.36	*Trichoderma*	Rembau, Negeri Sembilan
Isolate UKM-M-UW RA3a	34	28	27	29.7	3.79	*Trichoderma*	Rembau, Negeri Sembilan
Isolate UKM-M-UW RA1	20	23	26	23.0	3.0	*Trichoderma*	Rembau, Negeri Sembilan
Isolate UKM-M-UW C2	15	22	23	20.0	4.36	*Trichoderma*	Serdang, Selangor

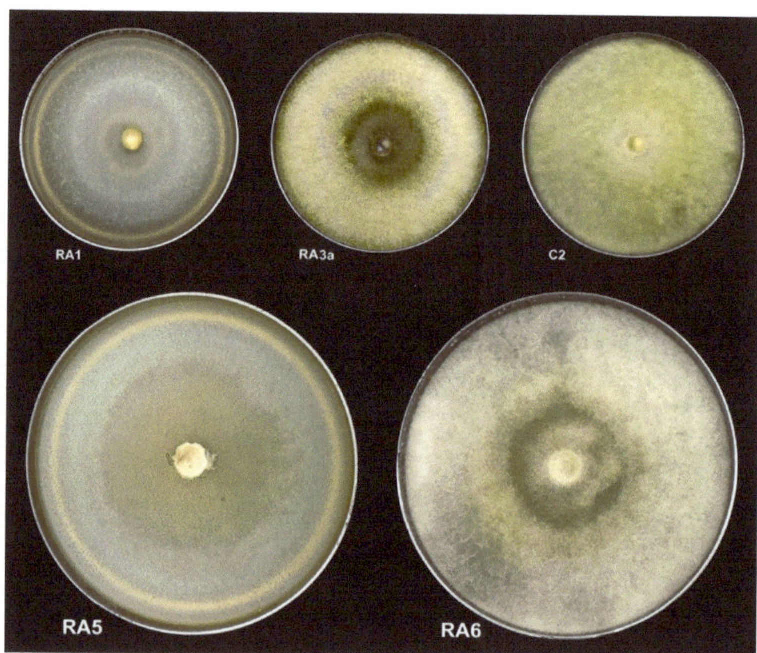

Figure 2. Physical morphology of the five fungal isolates RA1, RA3a, C2, RA5, and RA6 cultured on PDA plates growing filamentous mycelia across the agar after five days of incubation. All isolates show the morphological characteristics of *Trichoderma*—white filaments and, as the isolates mature, faint olive-green spores.

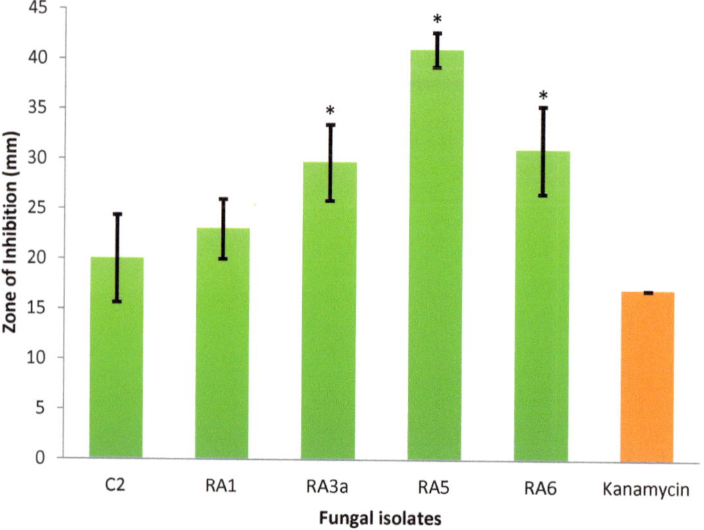

Figure 3. In vitro screening for antagonism of five fast-growing fungal isolates (*Trichoderma* spp.) against *E. mallotivora* 'BT-MARDI' using the agar well diffusion method. Values are means of triplicates and the standard deviations are indicated by the error bars. Isolates with significant antagonism (compared to kanamycin) were determined using the Tukey's test (N = 18, $p < 0.05$) and are marked with an asterisk (*).

3.2. Genomic DNA Extraction

High molecular weight (HMW) genomic DNA extraction was performed on the three fungal isolates, RA3a, RA5, and RA6, that showed significant inhibition on *E. mallotivora*. To improve the genomic DNA yield and quality, the extracted fungal DNA was subjected to ethanol precipitation and resuspended in 35 μL of nuclease-free water. The purity and concentration of the DNA was assessed with a Nanodrop ND1000 (Thermo Fisher, Boston, MA, USA) and Qubit (dsDNA Broad range, Invitrogen, Boston, MA, USA), with all preparations yielding more than 100 ng/μL HMW DNA according to Qubit readings (Table S1). The HMW DNA was further visualised via agarose gel electrophoresis (Figure S1). Overall, these analyses proved that the HMW DNA extracted from the three isolates was suitable to be sequenced through Oxford nanopore and Illumina whole-genome sequencing.

3.3. Molecular Identification of Strains

The three fungal isolates RA3a, RA5, and RA6 were identified through PCR amplification and Sanger sequencing of three DNA barcodes [65]: internal transcribed spacer (ITS) [66], translation elongation factor 1 alpha (*tef1*) gene [67] and RNA polymerase B subunit II (*rpb2*) gene [68]. Amplification of each sample yielded a single DNA fragment, the sequence of which was analysed through BLASTN [45]. The sequencing of the ITS marker allowed us to confidently assign all three fungal isolates to the *Trichoderma* genus. Further sequencing of the *tef1* and *rpb2* genes enabled us to identify all three fungal strains to be from a single species, *T. koningiopsis* (Table 3).

Table 3. Molecular identification of the three fungal isolates RA3a, RA5 and RA6 based on Sanger sequencing of the three DNA barcodes ITS, *Tef1* and *Rpb2*.

Isolate	Locus	Closest Match Organism	NCBI Accession Number	Coverage (%)	Identity (%)
UKM-M-UW RA3a	ITS	*Trichoderma sp.* strain ZMQRS9	MT446202.1	100	99.83
	Tef1	*Trichoderma koningiopsis* strain LESF360	KT278986.1	100	99.65
	Rpb2	*Trichoderma koningiopsis* isolate Tkois1	MT081443.1	100	99.77
UKM-M-UW RA5	ITS	*Trichoderma koningiopsis* strain 18ASMA001	MT520621.1	100	100
	Tef1	*Trichoderma koningiopsis* strain VSL155	MT058870.1	100	99.33
	Rpb2	*Trichoderma koningiopsis* isolate Tkois1	MT081443.1	100	99.60
UKM-M-UW RA6	ITS	*Trichoderma koningiopsis* strain 18ASMA001	MT520621.1	100	100
	Tef1	*Trichoderma koningiopsis* strain LESF360	KT278986.1	100	100
	Rpb2	*Trichoderma koningiopsis* isolate Tkois1	MT081443.1	100	100

3.4. Genome Sequencing

The genomic DNA of the three fungal isolates, RA3a, RA5, and RA6, were sequenced using a combination of Oxford nanopore and Illumina whole-genome sequencing. This generated a significant quantity of both long-read and short-read data which gave a high degree of coverage for each of the three genomes. From the nanopore data, after barcodes had been separated and trimmed, we had generated an estimated 3,176,623,257 bases for strain RA3a, 3,973,025,475 bases for RA5, and 2,406,065,838 bases for RA6. This was complemented by the Illumina sequencing which produced 3,352,912,241 bases for RA3a, and 3,365,209,722 and 2,697,750,185 for RA5 and RA6, respectively. The final draft genome sizes for each organism were as follows: RA3a 36.53 Mb, RA5 36.48 Mb, and RA6 36.47 Mb. This is consistent with the size of other *Trichoderma* genome assemblies which tend to be 31–40 Mb in size [69]. As shown in Table 4, the genomes contain 14 (RA3a), 11 (RA5), and 13 (RA6) contigs, the majority of these contigs may be putatively assembled at the chromosome level. Closely related species of *Trichoderma* have seven chromosomes [70], and our assemblies also included the mitochondrial DNA.

Table 4. Details of genome assembly statistics of the three *Trichoderma* isolates RA3a, RA5, and RA6.

Parameter	RA3a	RA5	RA6
Number of contigs	14	11	13
Total contigs length	36,531,570	36,477,170	36,470,223
Mean contig size	2,609,397.86	3,316,106.36	2,805,401.77
Contig size first quartile	1,043,387	3,650,583	981,951
Median contig size	2,049,512	3,895,316	3,855,011
Contig size third quartile	5,555,030	6,876,866	5,268,312
Longest contig	6,903,293	6,995,056	6,877,006
Shortest contig	6075	6406	5219
Contigs > 500 nt	14 (100%)	11 (100%)	13 (100%)
Contigs > 1K nt	14 (100%)	11 (100%)	13 (100%)
Contigs > 10K nt	13 (92.86%)	10 (90.91%)	12 (92.31%)
Contigs > 100K nt	11 (78.57%)	8 (72.73%)	10 (76.92%)
Contigs > 1M nt	10 (71.43%)	7 (63.64%)	8 (61.54)
N50	5,555,030	5,554,967	3,979,290
L50	3	3	4
N80	2,447,863	3,862,469	3,855,011
L80	6	6	6

Using BUSCO [71] to assess the genome assemblies for core conserved genes across the order Hypocreales, we were able to generate the results presented in Tables 5 and 6. Table 5 shows the conserved genes identified within the scaffold of our assemblies and Table 6 shows the conserved proteins identified in the predicted proteome of the annotated genomes of each species. An outline of the predicted non-coding RNA is shown in Table S3, the full set of results from StructRNAfinder are publicly available at https://osf.io/vsbc2/ (Accessed on 17 February 2022).

Table 5. Scaffold BUSCO: dataset Hypocreales odb10 for the genomes of the fungal isolates RA3a, RA5, and RA6.

BUSCO Scaffold Stat	RA3a	RA5	RA6
Percentage BUSCO	97.7%	97.7%	97.8%
Complete BUSCO's	4392	4391	4394
Complete and single copy BUSCO's	4378	4379	4381
Complete and duplicate BUSCO's	14	12	13
Fragmented BUSCO's	20	20	20
Missing BUSCO's	82	83	80
Total BUSCO groups searched		4494	

Table 6. Proteins BUSCO: dataset Hypocreales_odb10 for the predicted proteomes of the fungal isolates RA3a, RA5, and RA6.

BUSCO Scaffold Stat	RA3a	RA5	RA6
Percentage BUSCO	92.3%	88.0%	87.4%
Complete BUSCO's	4146	3955	3927
Complete and single copy BUSCO's	4137	3946	3922
Complete and duplicate BUSCO's	9	9	5
Fragmented BUSCO's	108	219	245
Missing BUSCO's	240	320	322
Total BUSCO groups searched		4494	

3.5. Genome Analysis

While the gene function of the predicted genes in each of the three genomes was assigned using Funannotate, the predicted proteome of each strain was run again through Eggnog-mapper v2.1.4-2. This yielded the following results, where for strain RA3a of the 8,951 predicted proteins, 86.7% could have a COG (Clusters of Orthologous Groups) category assigned; similarly, of the 8964 proteins predicted in strain RA5, 86.7% could have COG categories assigned, and 86.2% of the 9124 predicted proteins from RA6 could be too.

The distribution of these predicted proteins across the COG categories is shown in Figure 4. Distribution of proteins across the categories did not differ significantly across any of the three genomes, and while the most frequently mapped category was "Function Unknown", of those that could be placed into a category of known function, the five most common in order of decreasing predicted protein count within the categories were "Intracellular trafficking, secretion, and vesicular transport", "Amino acid transport and metabolism", "Secondary metabolite biosynthesis, transport and catabolism", "Post-translational modification, protein turnover, chaperones", and "Carbohydrate transport and metabolism". The presence of over 475 proteins associated with secondary metabolites in each genome is promising and these deserve further examinations, particularly in the context of the antagonistic activity of the fungal isolates against *E. mallotivora* and, potentially, other microorganisms. It is also interesting that no mobilome elements associated with transposons and prophages were detected, and less than 100 predicted proteins were identified in each genome that were associated with any of the following categories: "Extracellular structures", "Cell motility", "Nuclear structure", and "Defence mechanisms".

Figure 5 shows that the eggnog mapper was also able to assign gene ontology terms; often multiple terms are assigned to individual proteins, as well as enzyme commission numbers, a variety of matches to multiple KEGG databases, and also to BRITE hierarchies. A limited number of predicted proteins (no more than 2% in any genomes) were also associated with matches to the CAZy database of carbohydrate active enzymes, as well as matches to BiGG IDs. More promising was that around 85% of genes in each genome were matched to proteins in the pfam database. There was considerable overlap between these categories, however; most genes are close to identical across each of the three genomes. All data generated using eggnog mapper are included in the Dataset S1.

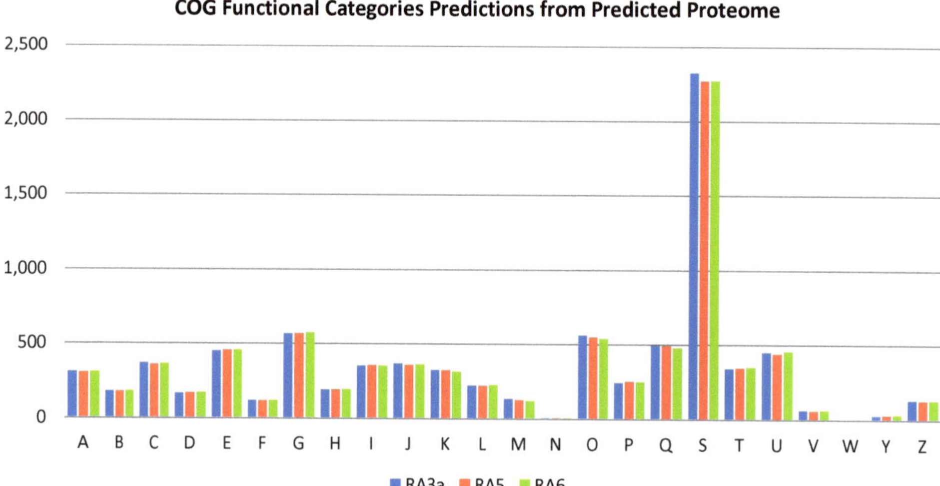

Figure 4. Distribution of predicted proteins in the three fungal isolates RA3a, RA5, and RA6 across the different COG categories. A: RNA processing and modification. B: Chromatin structure and dynamics. C: Energy production and conversion. D: Cell cycle control, cell division, chromosome partitioning. E: Amino acid transport and metabolism. F: Nucleotide transport and metabolism. G: Carbohydrate transport and metabolism. H: Coenzyme transport and metabolism. I: Lipid transport and metabolism. J: Translation, ribosomal structure. and biogenesis. K: Transcription. L: Replication, recombination. and repair. M: Cell wall/membrane/envelope biogenesis. N: Cell motility. O: Posttranslational modification, protein turnover, chaperones. P: Inorganic ion transport and metabolism. Q: Secondary metabolites biosynthesis, transport. and catabolism. S: Function unknown. T: Signal transduction mechanisms. U: Intracellular trafficking, secretion, and vesicular transport. V: Defense mechanisms. W: Extracellular structures. Y: Nuclear structure. Z: Cytoskeleton. The following returned no hits: R: General function prediction only; X: Mobilome- prophages, transposons.

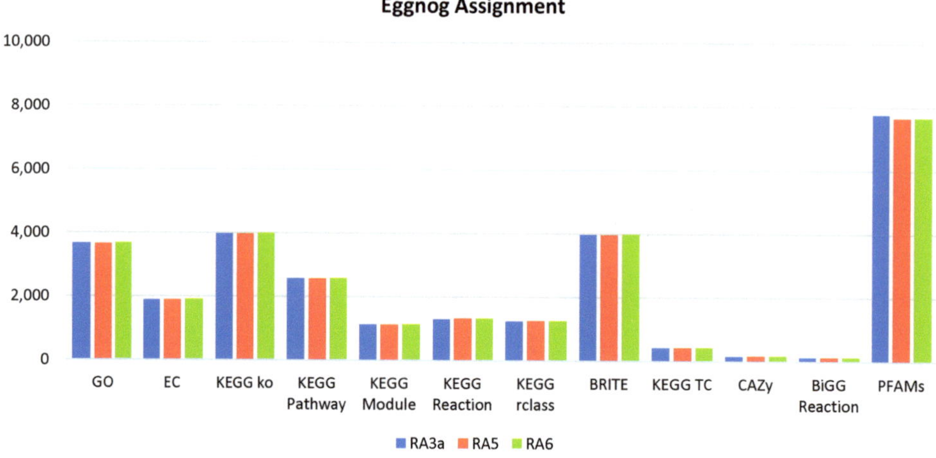

Figure 5. Assignment of predicted proteins in the three fungal isolates RA3a, RA5, and RA6 using eggnog mapper to multiple databases.

3.6. Comparative Genomics and Phylogenomic Analysis

Pairwise comparison of genomic similarities between our isolates was performed by calculating the average nucleotide identity (ANI) values. The three isolates, RA3a, RA5, and RA6, showed high genomic similarities between each other (ANI value of 99%) in addition to the *T. koningiopsis* POS7 isolate, which shared an ANI value of 96% (Figure 6). Conversely, the three isolates showed lower genomic similarities with other *Trichoderma* spp. (79–89% of ANI value). These results further confirm the identity of our isolates as belonging to the *T. koningiopsis* species.

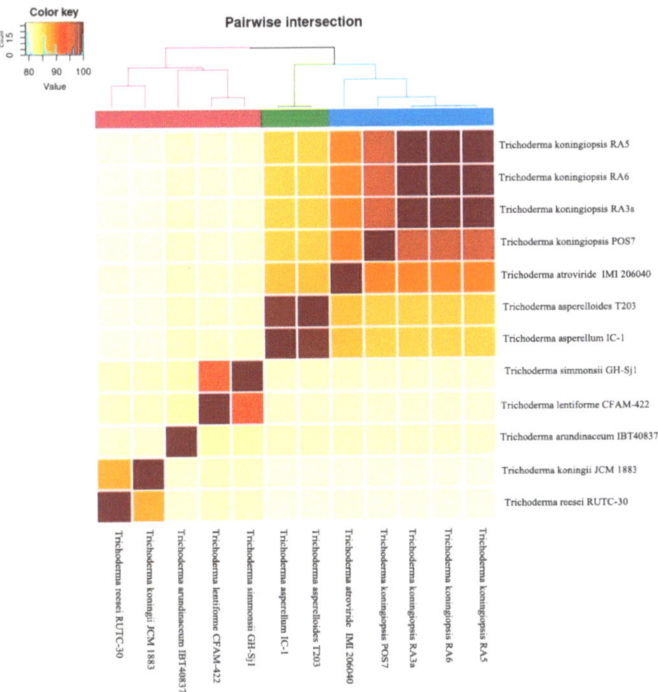

Figure 6. The ANI values based on the FastANI algorithm for *Trichoderma* spp. related genomes. The clustering was done based on a Euclidean distance matrix.

Further analysis using OrthoFinder suggested that there is a total of 12,270 orthogroups present among the predicted proteomes used in our analysis. Among that, a total of 3392 of the orthologous proteins exist as single-copy and alignment of their amino acids sequence was used for phylogenomic tree construction. The phylogenomic tree generated shows that the three isolates formed a distinct monophyletic group and shared a recent common ancestor within *T. atroviride* IMI206040 (Figure S2). The robustness of the phylogenomic tree generated was confirmed as all the branches showed 100% bootstrap values.

Next, the predicted protein sequences from our isolates were compared with each other using the OrthoVenn2 web server. All the predicted protein sequences extracted from our genomes were further grouped into 9225 orthologous proteins. Among these, a total of 8245 protein groups were shared by all isolates. These proteins accounted for 86.64 (RA3a), 89.09 (RA5), and 89.00% (RA6) of the total orthologous protein group encoded in each genome (Figure S3). Therefore, OrthoVenn2 analysis confirmed that most of the proteins encoded for by the three isolates are shared.

3.7. Secondary Metabolite Clusters

The whole genome sequences of the *Trichoderma* isolates RA3a, RA5, and RA6 were mined for putative biosynthetic gene clusters (BGCs) using AntiSMASH [54]. A total number of 124 clusters were identified across the three genomes; 43 BGCs were identified in isolate RA3a, 40 in isolate RA5, and 41 BGCs in isolate RA6. The strains varied in the putative classes they encode (Table 7 and Dataset S2). Non-ribosomal peptide synthetase (NRPS) clusters are the most dominant across the isolates followed by polyketide synthase (PKS) clusters, and Terpene and hybrid NRPS-PKS clusters. The number of BGCs from the sequenced fungal strains (RA3a, RA5, and RA6) was compared to the number of BGCs from other members of the *Trichoderma* genus (Figure 7), as predicted by AntiSMASH. This revealed that the fungal strains isolated in this study have the potential to be talented producers of polyketides and non-ribosomal peptide synthases, in line with other members of the *Trichoderma* genus.

Table 7. Biosynthetic gene clusters predicted through AntiSMASH analysis for the genomes of the *Trichoderma* isolates RA3a, RA5, and RA6.

Fungal Strain	Total Clusters	NRPS-Like	PKS	Terpene	Hybrid NRPS/PKS	Hybrid PKS/Terpene
UKM-M-UW RA3a	43	17	14	7	4	1
UKM-M-UW RA5	40	16	11	8	4	1
UKM-M-UW RA6	41	17	12	7	4	1

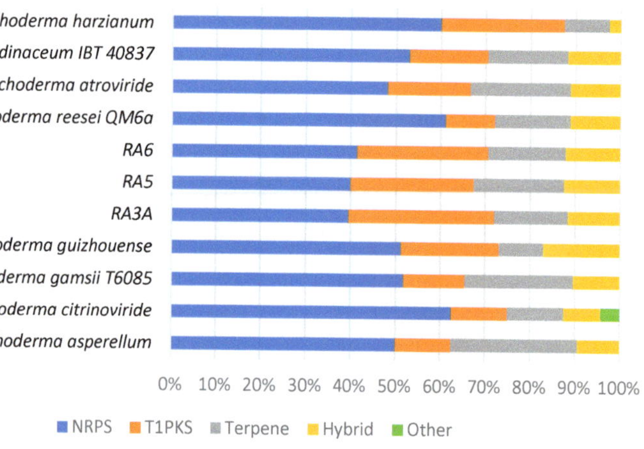

Figure 7. Comparison of the BGCs found in the genomes of the fungal strains RA3a, RA5, and RA6 with other members of the *Trichoderma* genus. BGC prediction was performed through AntiSMASH.

We then screened for clusters that were consistently present across the three strains, which could be indicative of their shared bioactivity against *E. mallotivora*. As predicted by AntiSMASH, we could find BGCs with low-to-high similarity to those for known bioactive metabolites, namely fusaric acid, naphthopyrone, neurosporin A, ascochlorin, and clavaric acid (Table S4 and Dataset S3). Further manual BlastP analysis of the putative neurosporin A BGC, revealed that this cluster may in fact code for the biosynthesis of a salicylaldehyde-related compound. This observation is based on the identification within the BGC of a hr-PKS megasynthase and tailoring enzymes, all showing high homology with genes of the *vir* gene cluster, which is involved in the biosynthesis of the antimicrobial compounds trichoxide and virensols from the biocontrol fungus *T. virens* [72] (Table S5 and Figure S4).

4. Discussion

Trichoderma spp. are easily culturable and generally considered to be harmless to humans and animals—except for *T. longibrachiatum* [73,74]. Members of this genus have been used extensively in agriculture as eco-friendly biological control agents as they are capable to control (or suppress) other microorganisms directly or indirectly [37,38,75,76]. It is reported that various *Trichoderma* spp. are effective biocontrol means against many fungal pathogens and some bacterial pathogens [77]. Previous works explored and applied the idea of using microbial species to control *E. mallotivora* [26–31]; nevertheless, the use of *Trichoderma* as a fungal antagonist against the PDD pathogen has not been reported. In this study, we demonstrated the potential of five *Trichoderma* strains to cause growth inhibition of *E. mallotivora* strain BT-MARDI. *Trichoderma* offers an advantage over many other microbes (especially bacteria) in terms of fast rhizosphere colonisation, involvement in the soil nutrient cycle, and excellent viability after an extended storage period (>12 months), thus making the fungus more efficient and appealing to the farmers [78,79]. In addition to their antagonistic features against pathogens, some species of *Trichoderma* are even capable of inducing plant defence mechanisms, which could be another important advantage [80].

To further gain insights on the antagonism of our isolates, we performed whole-genome sequencing on three fast-growing *Trichoderma* strains that showed significant inhibition of *E. mallotivora* to aid in the elucidation of the metabolites that might play a role in this interaction. From the sequencing results, we believe that the annotated draft genomes we produced can be treated as full open reading frames. Mutations and evolutionary divergence can explain some of the fragmented BUSCO scores, and even well-conserved genes can still be lost in some lineages [81]. The large coverage of the long read nanopore sequencing allowed for a robust first draft genome to be created, with few fragments. However, nanopore read base calling remains less accurate than Illumina, around 95% [82], and while modern base calling software harnessing neural networks are improving this, it is still necessary to polish, with higher accuracy, short read Illumina sequencing data to create a robust genome assembly, in order to remove the indels and miscalled bases present in nanopore reads that otherwise lead to frameshifts and fragmentation of genes.

The genus *Trichoderma* is a well-recognised group of filamentous fungi known for their production of secondary metabolites, especially as talented producers of bioactive peptides, polyketides, plant growth regulators, enzymes, siderophores, and other antibiotics [38,81,83], and has been credited for its biocontrol activity as antifungal and antibacterial. For example, peptaibols are a well-studied class of peptide natural products from *Trichoderma*, synthesised by NRPS modules, producing a linear peptide consisting of dialkylated amino acids, isovaline, amino isobutyric acid (Aib), an acetylated N-terminus, and a C-terminal amino alcohol [38]. They are credited for their antimicrobial properties, as well as their ability to induce systemic resistance in plants against microbial invasion. Another major class of *Trichoderma* natural products with biocontrol activity are koninginins. This family of compounds were first isolated from *T. koningii* and exhibited antifungal activity [38,84]. Interestingly, koninginins have also been isolated from *T. koningiopsis* in other studies and are reported to exhibit antifungal properties against *Fusarium* spp., *Plectosphaerella cucumerina*, and *Alternaria panax* [85]. Prominent examples of polyketides isolated from *Trichoderma* spp. include pyrones and pyridines [38,86]. Considering the diversity of bioactive molecules isolated from the genus—for a report about *Trichoderma* natural products we refer the reader to the review by Shenouda and Cox [86]—and given the vast biosynthetic potential emerged from AntiSMASH analysis conducted in our study, the three *Trichoderma* strains (RA3a, RA5, and RA6) have high potential to produce bioactive molecules that will warrant their use as biocontrol agents against plant pathogens. Noteworthy, we have identified BGCs conserved across the three fungal isolates, including one for the biosynthesis of putative salicylaldehyde-related compounds, which are known for their antimicrobial properties, such as those that have been isolated from *T. virens* [72] and *T. citrinum* [87]. Heterologous expression and/or knockout of the salicylaldehyde BGC homologues from RA3a, RA5, and RA6 will be needed to reveal the structure of the corresponding metabolites and their potential role in the bioactivity of the fungal isolates.

The dogma of biosynthetic studies remains that many fungal biosynthetic clusters are silent under standard laboratory conditions, which makes their full exploitation challenging [88,89]. Advances in synthetic biology via heterologous expression or genome editing may help us uncover the biosynthetic potential of the gene clusters [90,91].

5. Conclusions

In this study, we have isolated fungal strains from soil samples collected in the rhizosphere of healthy papaya trees from different locations in Peninsular Malaysia, with the aim to identify a potential biological control agent capable of suppressing *E. mallotivora*, the pathogen that is responsible for the ongoing PDD outbreak in Malaysia and surrounding countries. The three *Trichoderma* isolates, UKM-M-UW RA3a, UKM-M-UW RA5, and UKM-M-UW RA6, have shown significant inhibition of the growth of *E. mallotivora* from plate-based bioassays, and molecular identification allowed us to assign them to the *T. koningiopsis* species. Whole-genome sequencing was performed, thereby providing a platform for their biosynthetic exploitation, with the goal of linking secondary metabolites to biosynthetic gene clusters. Biosynthetic gene clusters homologous to those for known bioactive metabolites were identified and found to be conserved across the three isolates, opening the way for future exploration of the biosynthetic potential of these fungi. With the growing need for greener alternatives to chemical pesticides, the biosynthetic studies on natural products from *Trichoderma* spp. is expected to grow, which may give rise to a new generation of biocontrol agents with an enormous impact in the agrochemical sector.

Supplementary Materials: The following are available online at https://www.mdpi.com/article/10.3390/jof8030246/s1, Table S1: Concentration of the extracted gDNA measured through Qubit. Figure S1: HMW genomic DNA of the three Trichoderma isolates. Table S2: Primers and PCR amplicon sizes for molecular identification of isolates. Table S3: Details on the number of predicted non-coding RNAs identified in each genome. Figure S2: Phylogenomic tree from alignment of 3392 single-copy orthologous proteins from selected Trichoderma spp. Figure S3: Shared and unique orthologous protein group among isolates sequenced in this study as inferred from OrthoVenn2 web server analysis. Table S4: Distribution of predicted BGCs found across the fungal genomes. Table S5: Putative salicylaldehyde cluster found across the three sequenced fungal genomes. Figure S4: Protein sequence comparison of the putative salicylaldehyde cluster found in the genome of strains RA6, RA5, and RA3a. Dataset S1: Genome analysis through eggnog mapper. Dataset S2: AntiSMASH analysis of genomes. Dataset S3: Details of secondary metabolite clusters conserved across the genomes of strains RA3a, RA5, and RA6.

Author Contributions: Conceptualization, H.B., F.A. and N.M.-A.; methodology, H.B., M.A.A., N.M.-A., A.-A.T., J.A.W., R.T.O., S.J. and F.A.; software, F.A., J.A.W., S.J., M.A.A. and R.T.O.; validation, H.B., F.A. and N.M.-A.; formal analysis, N.M.-A., J.A.W., H.B., S.J. and F.A.; investigation, N.M.-A., A.-A.T., H.B., J.A.W. and R.T.O.; resources, H.B., F.A.; data curation, J.A.W.; writing—original draft preparation, A.-A.T., N.M.-A. and R.T.O.; writing—review and editing, H.B., F.A., J.A.W., A.-A.T. and M.A.A.; visualization, J.A.W., R.T.O., S.J. and A.-A.T.; supervision, H.B. and F.A.; project administration, H.B. and F.A.; funding acquisition, F.A and H.B. All authors have read and agreed to the published version of the manuscript.

Funding: This research was jointly funded by Universiti Kebangsaan Malaysia (UKM) through a Geran Galakan Penyelidikan-Industri grant [GGPI-2016-002], and the University of Warwick through an International Partnership Award. F.A. was supported by a Leverhulme Trust Early Career Fellowship [ECF-2018-691] and by a UKRI Future Leaders Fellowship [MR/V022334/1]. J.A.W. was supported by scholarship from the the Engineering and Physical Sciences Research Council and the Biotechnology and Biological Sciences Research Council [EP/L016494/1] through the Centre for Doctoral Training in Synthetic Biology (SynBioCDT).

Institutional Review Board Statement: Not applicable.

Informed Consent Statement: Not applicable.

Data Availability Statement: The genome sequences generated in this study were submitted to GenBank. BioSample metadata is publicly available in the NCBI BioSample database (http://www.ncbi.nlm.nih.gov/biosample/, last accessed on 25 January 2022) under accession numbers SAMN23731967, SAMN23731968, and SAMN23731969 for strains RA3a, RA5, and RA6 respectively. The annotated genomes are publicly available in the NCBI Genbank database (https://www.ncbi.nlm.nih.gov/genbank/, last accessed on 25 January 2022) under accession numbers JAJPEM000000000, JAJPEL000000000, and JAJPEK000000000 for strains RA3a, RA5, and RA6 respectively. The results of the StructRNAfinder analysis are publicly available at https://osf.io/vsbc2/ (last accessed on 25 January 2022).

Acknowledgments: The authors would like to acknowledge Maathavi Kannan and Iqbal M. Noor for their assistance in the laboratory and fieldworks.

Conflicts of Interest: The authors declare no conflict of interest.

References

1. FAOSTAT (Food and Agriculture Organisation of the United Nations Statistics Division). Available online: https://www.fao.org/3/ca5688en/CA5688EN.pdf (accessed on 20 July 2021).
2. Maktar, N.H.; Kamis, S.; Mohd Yusof, F.Z.; Hussain, N.H. *Erwinia papayae* causing papaya dieback in Malaysia. *Plant Pathol.* **2008**, *57*, 774. [CrossRef]
3. Mat Amin, N.; Bunawan, H.; Ahmad Redzuan, R.; Jaganath, I.B.S. *Erwinia Mallotivora* sp., a new pathogen of papaya (*Carica papaya*) in Peninsular Malaysia. *Int. J. Mol. Sci.* **2010**, *12*, 39–45. [CrossRef] [PubMed]
4. Ahmad Redzuan, R.; Abu Bakar, N.; Rozano, L.; Badrun, R.; Mat Amin, N.; Mohd Raih, M.F. Draft genome sequence of *Erwinia mallotivora* BT-MARDI, causative agent of papaya dieback disease. *Genome Announc.* **2014**, *2*, e00375-14. [CrossRef]
5. Supian, S.; Saidi, N.B.; Wee, C.-Y.; Abdullah, M.P. Antioxidant-mediated response of a susceptible papaya cultivar to a compatible strain of *Erwinia mallotivora*. *Physiol. Mol. Plant Pathol.* **2017**, *98*, 37–45. [CrossRef]
6. Noor Shahida, Y.; Awang, Y.; Sijam, K.; Noriha, M.A.; Satar, M.G.M. Biochemical changes and leaf photosynthesis of *Erwinia mallotivora* infected papaya (*Carica papaya*) aeedlings. *Am. J. Plant Physiol.* **2016**, *11*, 12–22. [CrossRef]
7. Pathania, N.; Justo, V.; Magdalita, P.; de la Cueva, F.; Herradura, L.; Waje, A.; Lobres, A.; Cueto, A.; Dillon, N.; Vawdrey, L.; et al. Integrated Disease Management Strategies for the Productive, Profitable and Sustainable Production of High Quality Papaya Fruit in the Southern Philippines and Australia—Final Report. Available online: https://www.aciar.gov.au/publication/technical-publications/integrated-disease-management-strategies-productive-profitable-and-sustainable (accessed on 20 July 2021).
8. Mohd Said, N.A.; Abu Bakar, N.; Lau, H.Y. Label-free detection of *Erwinia mallotivora* DNA for papaya dieback disease using electrochemical impedance spectroscopy approach. *ASM Sci. J.* **2020**, *4*, 1–8.
9. Ramachandran, K.; Manaf, U.A.; Zakaria, L. Molecular characterization and pathogenicity of *Erwinia* spp. associated with pineapple [*Ananas comosus* (L.) Merr.] and papaya (*Carica papaya* L.). *J. Plant Prot. Res.* **2015**, *55*, 396–404. [CrossRef]
10. Bunawan, H.; Baharum, S.N. Papaya dieback in Malaysia: A step towards a new insight of disease resistance. *Iran J. Biotechnol.* **2015**, *13*, 1–2. [CrossRef]
11. Mohd Khairil, J.; Muhammad Munzir, M. Experiences in managing bacterial dieback disease of papaya in Malaysia. *Acta Hortic.* **2014**, *1022*, 125–132. [CrossRef]
12. Suharjo, R.; Oktaviana, H.A.; Aeny, T.N.; Ginting, C.; Wardhana, R.A.; Nugroho, A.; Ratdiana, R. *Erwinia mallotivora* is the causal agent of papaya bacterial crown rot disease in Lampung Timur, Indonesia. *Plant Prot. Sci.* **2021**, *57*, 122–133. [CrossRef]
13. Hanagasaki, T.; Yamashiro, M.; Gima, K.; Takushi, T.; Kawano, S. Characterization of *Erwinia* sp. causing black rot of papaya (*Carica papaya*) first recorded in Okinawa Main Island, Japan. *Plant Pathol.* **2021**, *70*, 932–942. [CrossRef]
14. Wee, C.Y.; Muhammad Hanam, H.; Mohd Waznul Adly, M.Z.; Khairun, H.N. Expression of defense-related genes in papaya seedling infected with *Erwinia mallotivora* using real-time PCR. *J. Trop. Agric. Fd. Sci.* **2014**, *42*, 73–82.
15. Hamid, M.H.; Rozano, L.; Chien Yeong, W.; Abdullah, J.O.; Saidi, N.B. Analysis of MAP kinase MPK4/MEKK1/MKK genes of *Carica papaya* L. comparative to other plant homologues. *Bioinformation* **2017**, *13*, 31–41. [CrossRef] [PubMed]
16. Juri, N.M.; Samsuddin, A.F.; Abdul-Murad, A.M.; Tamizi, A.A.; Shaharuddin, N.A.; Abu-Bakar, N. Discovery of pathogenesis related and effector genes of *Erwinia mallotivora* in *Carica papaya* (Eksotika I) seedlings via transcriptomic analysis. *Int. J. Agric. Biol.* **2020**, *23*, 1021–1032. [CrossRef]
17. Juri, N.M.; Samsuddin, A.F.; Abdul-Murad, A.M.; Tamizi, A.A.; Hassan, M.A.; Abu-Bakar, N. In silico analysis and functional characterization of FHUB, a component of *Erwinia mallotivora* ferric hydroxamate uptake system. *J. Teknol.* **2020**, *82*, 83–90. [CrossRef]
18. Abu-Bakar, N.; Juri, N.M.; Abu-Bakar, R.A.H.; Sohaime, M.Z.; Badrun, R.; Sarip, J.; Hassan, M.A.; Ahmad, K. Recombinant protein foliar application activates systemic acquired resistance and increases tolerance against papaya dieback disease. *Asian J. Agric. Rural Dev.* **2021**, *11*, 1–9. [CrossRef]

19. Abu Bakar, N.S.; Saidi, N.B.; Rozano, L.; Abdullah, M.P.; Supian, S. In-silico characterization and expression analysis of NB-ARC genes in response to *Erwinia mallotivora* in *Carica papaya*. *Sains Malays.* **2021**, *50*, 2591–2602. [CrossRef]
20. Tamizi, A.-A.; Abu-Bakar, N.; Samsuddin, A.-F.; Rozano, L.; Ahmad-Redzuan, R.; Abdul-Murad, A.-M. Characterisation and mutagenesis study of an alternative sigma factor gene (*hrpL*) from *Erwinia mallotivora* reveal its central role in papaya dieback disease. *Biology* **2020**, *9*, 323. [CrossRef] [PubMed]
21. Sekeli, R.; Hamid, M.H.; Razak, R.A.; Wee, C.Y.; Ong-Abdullah, J. Malaysian *Carica papaya* L. var. Eksotika: Current research strategies fronting challenges. *Front. Plant Sci.* **2018**, *9*, 1380. [CrossRef]
22. Sekeli, R.; Nazaruddin, N.H.; Tamizi, A.A.; Mat Amin, N.; Wee, C.Y.; Sarip, J.; Abdullah, N.; Saidi, N.I.; Abdul Razak, R.; Zulkifli, Z. Enhancing Eksotika papaya resistance to dieback disease through quorum quenching. *J. Trop. Plant Physiol.* **2019**, *11*, 1–9.
23. Mohd-Azhar, H.; Sarip, J.; Ghazali, N.F.; Mohd Razikin, M.Z.; Mariatulqabtiah, A.R. Tolerance level of grafted papaya plants against papaya dieback disease. *Malays. Appl. Biol.* **2021**, *50*, 95–103.
24. Mohd Azhar, H.; Johari, S.; Nur Sulastri, J.; Razali, M.; Muhammad Zulfa, M.R.; Noor Faimah, G.; Mariatulqabtiah, A.R. Field performance of selected papaya hybrids for tolerance to dieback disease. *J. Trop. Agric. Fd. Sci.* **2020**, *48*, 25–33.
25. Sarip, J.; Radzuan, S.M.; Ghazali, M.F.; Norasiah, R. Viorica: A Papaya Variety Highly Tolerant to Dieback Disease. In Proceedings of the International Congress of the Malaysian Society for Microbiology, Penang, Malaysia, 7–10 December 2015.
26. Mat Amin, N.; Nor Rahim, M.Y.; Ahmad Redzuan, R.; Tamizi, A.A.; Bunawan, H. Acyl homoserine lactonase genes from *Bacillus* species isolated from tomato rhizosphere soil in Malaysia. *Res. J. Appl. Sci.* **2016**, *11*, 656–659.
27. Mat Amin, N.; Nor Rahim, M.Y.; Ahmad Redzuan, R.; Tamizi, A.A.; Bunawan, H. Quorum quenching bacteria isolated from rice and tomato rhizosphere soil in Malaysia. *J. Appl. Biol. Sci.* **2016**, *10*, 61–63.
28. Blog Rasmi MARDI. Available online: https://blogmardi.wordpress.com/2017/08/02/ (accessed on 6 December 2021).
29. All Cosmos. Available online: https://allcosmos.com/services-view/dieback-buster-95-2dbottle/ (accessed on 6 December 2021).
30. Rivarez, M.P.S.; Parac, E.P.; Dimasingkil, S.F.M.; Magdalita, P.M. Influence of native endophytic bacteria on the growth and bacterial crown rot tolerance of papaya (*Carica papaya*). *Eur. J. Plant Pathol.* **2021**, *161*, 593–606. [CrossRef]
31. Mohd Taha, M.D.; Mohd Jaini, M.F.; Saidi, N.B.; Abdul Rahim, R.; Md Shah, U.K.; Mohd Hashim, A. Biological control of *Erwinia mallotivora*, the causal agent of papaya dieback disease by indigenous seed-borne endophytic lactic acid bacteria consortium. *PLoS ONE* **2019**, *14*, e0224431. [CrossRef]
32. Mukhopadhyay, R.; Kumar, D. *Trichoderma*: A beneficial antifungal agent and insights into its mechanism of biocontrol potential. *Egypt J. Biol. Pest Control* **2020**, *30*, 133. [CrossRef]
33. Harman, G.E. Overview of mechanisms and uses of *Trichoderma* spp. *Phytopathology* **2006**, *96*, 190–194. [CrossRef]
34. Keswani, C.; Mishra, S.; Sarma, B.K.; Singh, S.P.; Singh, H.B. Unraveling the efficient applications of secondary metabolites of various *Trichoderma* spp. *Appl. Microbiol. Biotechnol.* **2014**, *98*, 533–544. [CrossRef] [PubMed]
35. Contreras-Cornejo, H.A.; Macías-Rodríguez, L.; Del-Val, E.; Larsen, J. Ecological functions of *Trichoderma* spp. and their secondary metabolites in the rhizosphere: Interactions with plants. *FEMS Microbiol. Ecol.* **2016**, *92*, fiw036. [CrossRef] [PubMed]
36. Bitas, V.; Kim, H.S.; Bennett, J.W.; Kang, S. Sniffing on microbes: Diverse roles of microbial volatile organic compounds in plant health. *Mol. Plant Microbe Interact.* **2013**, *26*, 835–843. [CrossRef]
37. Guo, Y.; Ghirardo, A.; Weber, B.; Schnitzler, J.P.; Benz, J.P.; Rosenkranz, M. *Trichoderma* species differ in their volatile profiles and in antagonism toward ectomycorrhiza *Laccaria bicolor*. *Front. Microbiol.* **2019**, *10*, 891. [CrossRef]
38. Khan, R.A.A.; Najeeb, S.; Hussain, S.; Xie, B.; Li, Y. Bioactive secondary metabolites from *Trichoderma* spp. against phytopathogenic fungi. *Microorganisms* **2020**, *8*, 817. [CrossRef] [PubMed]
39. Rush, T.A.; Shrestha, H.K.; Gopalakrishnan Meena, M.; Spangler, M.K.; Ellis, J.C.; Labbé, J.L.; Abraham, P.E. Bioprospecting *Trichoderma*: A systematic roadmap to screen genomes and natural products for biocontrol applications. *Front. Fungal Biol.* **2021**, *2*, 716511. [CrossRef]
40. Vargas Gil, S.; Pastor, S.; Marcha, G.J. Quantitative isolation of biocontrol agents *Trichoderma* spp., *Gliocladium* spp. and actinomycetes from soil with culture media. *Microbiol. Res.* **2009**, *164*, 196–205. [CrossRef] [PubMed]
41. Zhang, K.; Yuan-Ying, S.; Lei, C. An optimized protocol of single spore isolation for fungi. *Cryptogam. Mycol.* **2013**, *34*, 349–356. [CrossRef]
42. Magaldi, S.; Mata-Essayag, S.; Hartung de Capriles, C.; Perez, C.; Colella, M.T.; Olaizola, C.; Ontiveros, Y. Well diffusion for antifungal susceptibility testing. *Int. J. Infect. Dis.* **2004**, *8*, 39–45. [CrossRef]
43. Valgas, C.; de Souza, S.M.; Smânia, E.F.A.; Smânia, A., Jr. Screening methods to determine antibacterial activity of natural products. *Braz. J. Microbiol.* **2007**, *38*, 369–380. [CrossRef]
44. Kopchinskiy, A.; Komoń, M.; Kubicek, C.P.; Druzhinina, I.S. TrichoBLAST: A multilocus database for *Trichoderma* and *Hypocrea* identifications. *Mycol. Res.* **2005**, *109*, 658–660. [CrossRef] [PubMed]
45. Altschul, S.F.; Gish, W.; Miller, W.; Myers, E.W.; Lipman, D.J. Basic local alignment search tool. *J. Mol. Biol.* **1990**, *215*, 403–410. [CrossRef]
46. Kolmogorov, M.; Yuan, J.; Lin, Y.; Pevzner, P.A. Assembly of long, error-prone reads using repeat graphs. *Nat. Biotechnol.* **2019**, *37*, 540–546. [CrossRef] [PubMed]
47. Li, H. Minimap2: Pairwise alignment for nucleotide sequences. *Bioinformatics* **2018**, *34*, 3094–3100. [CrossRef]
48. Vaser, R.; Sović, I.; Nagarajan, N.; Šikić, M. Fast and accurate de novo genome assembly from long uncorrected reads. *Genome Res.* **2017**, *27*, 737–746. [CrossRef]

49. Langmead, B.; Salzberg, S.L. Fast gapped-read alignment with Bowtie 2. *Nat. Methods* **2021**, *9*, 357–359. [CrossRef]
50. Walker, B.J.; Abeel, T.; Shea, T.; Priest, M.; Abouelliel, A.; Sakthikumar, S.; Cuomo, C.A.; Zeng, Q.; Wortman, J.; Young, S.K.; et al. Pilon: An integrated tool for comprehensive microbial variant detection and genome assembly improvement. *PLoS ONE* **2014**, *9*, e112963. [CrossRef]
51. Seppey, M.; Manni, M.; Zdobnov, E.M. BUSCO: Assessing genome assembly and annotation completeness. *Methods Mol. Biol.* **2019**, *1962*, 227–245. [CrossRef] [PubMed]
52. Palmer, J.M.; Stajich, J. Funannotate v1.8.1: Eukaryotic Genome Annotation. Available online: https://zenodo.org/record/4054 262#.YhxHjOpBxPY (accessed on 20 August 2021).
53. Käll, L.; Krogh, A.; Sonnhammer, E.L.L. A combined transmembrane topology and signal peptide prediction method. *J. Mol. Biol.* **2004**, *338*, 1027–1036. [CrossRef]
54. Blin, K.; Shaw, S.; Steinke, K.; Villebro, R.; Ziemert, N.; Lee, S.Y.; Medema, M.H.; Weber, T. antiSMASH 5.0: Updates to the secondary metabolite genome mining pipeline. *Nucleic Acids Res.* **2019**, *47*, W81–W87. [CrossRef]
55. Burge, S.W.; Daub, J.; Eberhardt, R.; Tate, J.; Barquist, L.; Nawrocki, E.P.; Eddy, S.R.; Gardner, P.P.; Bateman, A. Rfam 11.0: 10 years of RNA families. *Nucleic Acids Res.* **2013**, *41*, D226–D232. [CrossRef]
56. Arias-Carrasco, R.; Vásquez-Morán, Y.; Nakaya, H.I.; Maracaja-Coutinho, V. StructRNAfinder: An automated pipeline and web server for RNA families prediction. *BMC Bioinform.* **2018**, *19*, 55. [CrossRef]
57. Griffiths-Jones, S.; Moxon, S.; Marshall, M.; Khanna, A.; Eddy, S.R.; Bateman, A. Rfam: Annotating non-coding RNAs in complete genomes. *Nucleic Acids Res.* **2005**, *33*, D121–D124. [CrossRef] [PubMed]
58. Jain, C.; Rodriguez-R, L.M.; Phillippy, A.M.; Konstantinidis, K.T.; Aluru, S. High throughput ANI analysis of 90K prokaryotic genomes reveals clear species boundaries. *Nat. Commun.* **2018**, *9*, 5114. [CrossRef] [PubMed]
59. Khan, A.; Mathelier, A. Intervene: A tool for intersection and visualization of multiple gene or genomic region sets. *BMC Bioinform.* **2017**, *18*, 287. [CrossRef] [PubMed]
60. Emms, D.M.; Kelly, S. OrthoFinder: Phylogenetic orthology inference for comparative genomics. *Genome Biol.* **2019**, *20*, 238. [CrossRef]
61. Katoh, K.; Standley, D.M. MAFFT multiple sequence alignment software version 7: Improvements in performance and usability. *Mol. Biol. Evol.* **2013**, *30*, 772–780. [CrossRef]
62. Nguyen, L.T.; Schmidt, H.A.; Von Haeseler, A.; Minh, B.Q. IQ-TREE: A fast and effective stochastic algorithm for estimating maximum-likelihood phylogenies. *Mol. Biol. Evol.* **2015**, *32*, 268–274. [CrossRef]
63. Letunic, I.; Bork, P. Interactive Tree of Life (iTOL) v4: Recent updates and new developments. *Nucleic Acids Res.* **2019**, *47*, W256–W259. [CrossRef] [PubMed]
64. Xu, L.; Dong, Z.; Fang, L.; Luo, Y.; Wei, Z.; Guo, H.; Zhang, G.; Gu, Y.Q.; Coleman-Derr, D.; Xia, Q.; et al. OrthoVenn2: A web server for whole-genome comparison and annotation of orthologous clusters across multiple species. *Nucleic Acids Res.* **2019**, *47*, W52–W58. [CrossRef]
65. Cai, F.; Druzhinina, I.S. In honor of John Bissett: Authoritative guidelines on molecular identification of *Trichoderma*. *Fungal Divers.* **2021**, *107*, 1–69. [CrossRef]
66. Schoch, C.L.; Seifert, K.A.; Huhndorf, S.; Robert, V.; Spouge, J.L.; Levesque, C.A.; Chen, W.; Bolchacova, E.; Voigt, K.; Crous, P.W.; et al. Nuclear ribosomal internal transcribed spacer (ITS) region as a universal DNA barcode marker for Fungi. *Proc. Natl. Acad. Sci. USA* **2012**, *109*, 6241–6246. [CrossRef]
67. Druzhinina, I.; Kubicek, C.P. Species concepts and biodiversity in *Trichoderma* and *Hypocrea*: From aggregate species to species clusters? *J. Zhejiang Univ. Sci. B* **2005**, *6*, 100–112. [CrossRef]
68. Liu, Y.J.; Whelen, S.; Hall, B.D. Phylogenetic relationships among ascomycetes: Evidence from an RNA polymerse II subunit. *Mol. Biol. Evol.* **1999**, *16*, 1799–1808. [CrossRef] [PubMed]
69. Kubicek, C.P.; Steindorff, A.S.; Chenthamara, K.; Manganiello, G.; Henrissat, B.; Zhang, J.; Cai, F.; Kopchinskiy, A.G.; Kubicek, E.M.; Kuo, A.; et al. Evolution and comparative genomics of the most common *Trichoderma* species. *BMC Genomics* **2019**, *20*, 485. [CrossRef] [PubMed]
70. Druzhinina, I.S.; Kopchinskiy, A.G.; Kubicek, E.M.; Kubicek, C.P. A complete annotation of the chromosomes of the cellulase producer *Trichoderma reesei* provides insights in gene clusters, their expression and reveals genes required for fitness. *Biotechnol. Biofuels* **2016**, *9*, 75. [CrossRef] [PubMed]
71. El Komy, M.H.; Saleh, A.A.; Eranthodi, A.; Molan, Y.Y. Characterization of novel *Trichoderma asperellum* isolates to select effective biocontrol agents against tomato Fusarium wilt. *Plant Pathol. J.* **2015**, *31*, 50–60. [CrossRef]
72. Liu, L.; Tang, M.-C.; Tang, Y. Fungal Highly Reducing Polyketide Synthases Biosynthesize Salicylaldehydes That Are Precursors to Epoxycyclohexenol Natural Products. *J. Am. Chem. Soc.* **2019**, *141*, 19538–19541. [CrossRef]
73. Munoz, F.M.; Demmler, G.J.; Travis, W.R.; Ogden, A.K.; Rossmann, S.N.; Rinaldi, M.G. *Trichoderma longibrachiatum* infection in a pediatric patient with aplastic anemia. *J. Clin. Microbiol.* **1997**, *35*, 499–503. [CrossRef] [PubMed]
74. Sautour, M.; Chrétien, M.L.; Valot, S.; Lafon, I.; Basmaciyana, L.; Legouge, C.; Verrier, T.; Gonssaud, B.; Abou-Hanna, H.; Dalle, F.; et al. First case of proven invasive pulmonary infection due to *Trichoderma longibrachiatum* in a neutropenic patient with acute leukemia. *J. Mycol. Med.* **2018**, *28*, 659–662. [CrossRef] [PubMed]
75. Waghunde, R.R.; Shelake, R.M.; Sabalpara, A.N. *Trichoderma*: A significant fungus for agriculture and environment. *Afr. J. Agric. Res.* **2016**, *11*, 1952–1965. [CrossRef]

76. Zin, N.A.; Badaluddin, N.A. Biological functions of *Trichoderma* spp. for agriculture applications. *Ann. Agric. Sci.* **2020**, *65*, 168–178. [CrossRef]
77. Thapa, S.; Rai, N.; Limbu, A.K.; Joshi, A. Impact of *Trichoderma* sp. in agriculture: A mini-review. *J. Biol. Today's World* **2020**, *9*, 227.
78. Singh, A.; Sarma, B.K.; Singh, H.B.; Upadhyay, R.S. Trichoderma: A silent worker of plant rhizosphere. In *Biotechnology and Biology of Trichoderma*, 1st ed.; Gupta, V.K., Schmoll, M., Herrera-Estrella, A., Upadhyay, R.S., Druzhinina, I., Tuohy, M.G., Eds.; Elsevier: Amsterdam, The Netherlands, 2014; pp. 533–542.
79. Stocco, M.; Mónaco, C.; Cordo, C. A comparison of preservation methods for *Trichoderma harzianum* cultures. *Rev. Iberoam. Micol.* **2010**, *27*, 213. [CrossRef]
80. Sood, M.; Kapoor, D.; Kumar, V.; Sheteiwy, M.S.; Ramakrishnan, M.; Landi, M.; Araniti, F.; Sharma, A. *Trichoderma*: The "secrets" of a multitalented biocontrol agent. *Plants* **2020**, *9*, 762. [CrossRef]
81. Simão, F.A.; Waterhouse, R.M.; Ioannidis, P.; Kriventseva, E.V.; Zdobnov. E.M. BUSCO: Assessing genome assembly and annotation completeness with single-copy orthologs. *Bioinformatics* **2015**, *31*, 3210–3212. [CrossRef] [PubMed]
82. Zhang, Y.Z.; Akdemir, A.; Tremmel, G.; Imoto, S.; Miyano, S.; Shibuya, T.; Yamaguchi, R. Nanopore basecalling from a perspective of instance segmentation. *BMC Bioinform.* **2020**, *21*, 136. [CrossRef]
83. Li, M.F.; Li, G.H.; Zhang, K.Q. Non-volatile metabolites from *Trichoderma* spp. *Metabolites* **2019**, *9*, 58. [CrossRef]
84. Parker, S.R.; Cutler, H.G.; Schreiner, P.R. Koninginin C: A biologically active natural product from *Trichoderma koningii*. *Biosci. Biotechnol. Biochem.* **1995**, *59*, 1126–1127. [CrossRef]
85. Liu, K.; Yang, Y.B.; Chen, J.L.; Miao, C.P.; Wang, Q.; Zhou, H.; Chen, Y.W.; Li, Y.Q.; Ding, Z.T.; Zhao, L.X. Koninginins N-Q, polyketides from the endophytic fungus *Trichoderma koningiopsis* harbored in *Panax notoginseng*. *Nat. Prod. Bioprospect.* **2016**, *6*, 49–55. [CrossRef] [PubMed]
86. Shenouda, M.L.; Cox, R.J. Molecular methods unravel the biosynthetic potential of *Trichoderma* species. *RSC Adv.* **2021**, *11*, 3622–3635. [CrossRef]
87. Berkaew, P.; Soonthornchareonnon, N.; Salasawadee, K.; Chanthaket, R.; Isaka, M. Aurocitrin and Related Polyketide Metabolites from the Wood-Decay Fungus *Hypocrea* sp. BCC 14122. *J. Nat. Prod.* **2008**, *71*, 902–904. [CrossRef] [PubMed]
88. Hyde, K.D.; Xu, J.; Rapior, S.; Jeewon, R.; Lumyong, S.; Niego, A.G.T.; Abeywickrama, P.D.; Aluthmuhandiram, J.V.S.; Brahamanage, R.S.; Brooks, S.; et al. The amazing potential of fungi: 50 ways we can exploit fungi industrially. *Fungal Divers.* **2019**, *97*, 1–136. [CrossRef]
89. Kjærbølling, I.; Mortensen, U.H.; Vesth, T.; Andersen, M.R. Strategies to establish the link between biosynthetic gene clusters and secondary metabolites. *Fungal Genet. Biol.* **2019**, *130*, 107–121. [CrossRef]
90. Alberti, F.; Foster, G.D.; Bailey, A.M. Natural products from filamentous fungi and production by heterologous expression. *Appl. Microbiol. Biotechnol.* **2017**, *101*, 493–500. [CrossRef] [PubMed]
91. Harvey, C.J.B.; Tang, M.; Schlecht, U.; Horecka, J.; Fischer, C.R.; Lin, H.C.; Li, J.; Naughton, B.; Cherry, J.; Miranda, M.; et al. HEx: A heterologous expression platform for the discovery of fungal natural products. *Sci. Adv.* **2018**, *4*, eaar5459. [CrossRef] [PubMed]

Journal of **Fungi**

MDPI

Article

Absolute Configuration Determination of Two Diastereomeric Neovasifuranones A and B from *Fusarium oxysporum* R1 by a Combination of Mosher's Method and Chiroptical Approach

Zhiyang Fu [1], Yuanyuan Liu [1], Meijie Xu [1], Xiaojun Yao [2], Hong Wang [1] and Huawei Zhang [1,3,*]

[1] School of Pharmaceutical Sciences, Zhejiang University of Technology, Hangzhou 310014, China; fuzhiyoung@163.com (Z.F.); liuyuan_0507@163.com (Y.L.); xumeijie1230@163.com (M.X.); hongw@zjut.edu.cn (H.W.)
[2] Department of Chemistry and Chemical Engineering, Lanzhou University, Lanzhou 730000, China; xjyao@lzu.edu.cn
[3] Key Laboratory of Marine Fishery Resources Exploitment & Utilization of Zhejiang Province, Hangzhou 310014, China
* Correspondence: hwzhang@zjut.edu.cn; Tel.: +86-571-88320913

Abstract: Endophytic fungi are one of prolific sources of bioactive natural products with potential application in biomedicine and agriculture. In our continuous search for antimicrobial secondary metabolites from *Fusarium oxysporum* R1 associated with traditional Chinese medicinal plant *Rumex madaio* Makino using one strain many compounds (OSMAC) strategy, two diastereomeric polyketides neovasifuranones A (**3**) and B (**4**) were obtained from its solid rice medium together with *N*-(2-phenylethyl)acetamide (**1**), 1-(3-hydroxy-2-methoxyphenyl)-ethanone (**2**) and 1,2-*seco*-trypacidin (**5**). Their planar structures were unambiguously determined using 1D NMR and MS spectroscopy techniques as well as comparison with the literature data. By a combination of the modified Mosher's reactions and chiroptical methods using time-dependent density functional theory-electronic circular dichroism (TDDFT-ECD) and optical rotatory dispersion (ORD), the absolute configurations of compounds **3** and **4** are firstly confirmed and, respectively, characterized as (4*S*,7*S*,8*R*), (4*S*,7*S*,8*S*). Bioassay results indicate that these metabolites **1–5** exhibit weak inhibitory effect on *Helicobacter pylori* 159 with MIC values of $\geq$16 µg/mL. An in-depth discussion for enhancement of fungal metabolite diversity is also proposed in this work.

Keywords: endophytic fungus; *Fusarium oxysporum*; secondary metabolite; absolute configuration; chiroptical method; Mosher's reaction

check for **updates**

Citation: Fu, Z.; Liu, Y.; Xu, M.; Yao, X.; Wang, H.; Zhang, H. Absolute Configuration Determination of Two Diastereomeric Neovasifuranones A and B from *Fusarium oxysporum* R1 by a Combination of Mosher's Method and Chiroptical Approach. *J. Fungi* **2022**, *8*, 40. https://doi.org/10.3390/jof8010040

Academic Editor: Laurent Dufossé

Received: 8 December 2021
Accepted: 30 December 2021
Published: 31 December 2021

Publisher's Note: MDPI stays neutral with regard to jurisdictional claims in published maps and institutional affiliations.

1. Introduction

The assignment of absolute configuration (AC) is one of the most challenging tasks in the structure elucidation of chiral natural products. Application of Mosher's method or quantum mechanical calculation of chiroptical properties had proved to be practical and reliable, including time-dependent density functional theory-electronic circular dichroism (TDDFT-ECD) and optical rotatory dispersion (ORD) [1–4]. However, some difficulties and uncertainties still exist in determining ACs of molecules with high conformational flexibility using single method. Therefore, a combined application of these approaches is necessary and has been shown to be valid in some cases, such as chenopodolans B and D [5,6], sapinofuranones B and C [7] and ent-thailandolide B [8]. A growing number of evidence indicates that the genus *Fusarium* is one rich source of secondary metabolites with a wide variety of chemical structures and biological properties [9]. In our continuous search for antimicrobial secondary metabolites from the endophytic strain *F. oxysporum* R1 associated with traditional Chinese medicinal plant *Rumex madaio* Makino using one strain many compounds (OSMAC) strategy [10,11], chemical study of the ethyl acetate

extract of its rice medium resulted in the isolation of five known compounds including
N-(2-phenylethyl)acetamide (**1**), 1-(3-hydroxy-2-methoxyphenyl)-ethanone (**2**), neovasifuranones A (**3**) and B (**4**) and 1,2-*seco*-trypacidin (**5**) (Figure 1). Compounds **3** and **4** are
diastereomeric polyketides originally isolated from the phytopathogenic strain *Neocosmospora vasinfecta* NHL2298 [12,13] and later found to be produced by the soil-derived strain
Penicillium sp. SYPF7381 [14] and the endophytic fungus *Aspergillus japonicus* CAM231
from *Garcinia preussii* [15]. However, their ACs are still unassigned. Herein the present
work highlights on assignment of ACs in **3** and **4** by a combined application of Mosher's
method and quantum mechanical calculation of chiroptical (ECD and ORD) properties.

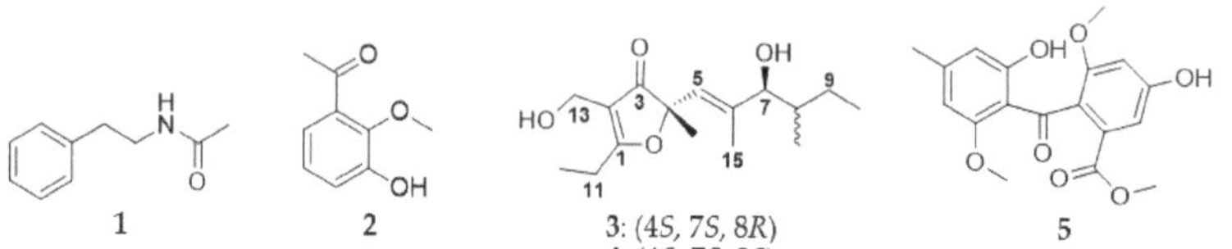

Figure 1. Chemical structures of compounds **1–5** from *Fusarium oxysporum* R1.

2. Materials and Methods

2.1. General

The NMR spectra were determined on Bruker Avance DRX600 instruments (600 MHz
for ^{1}H and 150.92 MHz for ^{13}C NMR) (Bruker, Fällande, Switzerland). ESIMS were obtained
with an Agilent 6210 LC/TOF-MS spectrometer (Agilent Technologies, Santa Clara, CA,
USA). Optical rotation and CD spectra were performed on JASCO P-2000 polarimeter and
JASCO J-1500 spectrometer (JASCO, Fukuoka, Japan). UV and IR spectra were measured
through a Hitachi-UV-3000 spectrometer (Hitachi, Tokyo, Japan) and a Nexus 870 spectrometer (Thermo-Nicolet, Madison, WI, USA), respectively. Reverse phase HPLC was
carried out on an Essentia LC-16P apparatus (Essentia, San Diego, CA, USA) fitted with a
preparative HPLC column (Phenomenex Gemini-NX C18, 50 mm £ 21.2 mm, 5 mm) or a
semi-preparative column (Phenomenex Synergi Hydro-RP, 250 × 10 mm, 4 μm). Acetonitrile and H_2O used in HPLC system were chromatographic grade, and all other chemicals
were analytical.

2.2. Biological Material

The endophytic fungal strain R1 was isolated from the healthy plant *R. madaio* Makino
collected off the coastal region of Putuo Island, China [16], and molecularly identified as *F.
oxysporum* according to its 18S rDNA gene sequence (GenBank accession No. MF376147)
and deposited at China General Microbiological Culture Collection Centre (CGMCC no.
17763) [11].

2.3. Fermentation, Extraction and Isolation

The strain R1 grown on potato dextrose agar (PDA) media was inoculated into 500 mL
Erlenmeyer flasks containing 200 mL potato dextrose broth (PDB) medium, and shaken
for 3 days at 200 rpm and 30 °C. The fermentation was performed in Erlenmeyer flasks
(50 × 1 L) with sterilized rice (160 g) and tap water (320 mL). After autoclaving at 121 °C
for 20 min, each flask was inoculated with 5% seed cultures and then incubated at room
temperature under static conditions for 30 days. The fermented rice of each flask was
extracted with 500 mL EtOAc by an ultrasonic instrument for 20 min, 3 times followed by
filtration using gauze. All filtrate was combined and evaporated under vacuum to dryness,
affording the crude extract (approximate 19 g). Then the extract was quickly separated
using HPLC on a preparative column to afford six fractions A–F, and further purified using
a semi-preparative column for subdivision [17]. Compound **1** (1.8 mg, t_R = 7.2 min) and
compound **2** (1.8 mg, t_R = 8.2 min) were obtained from fraction A with 30% CH_3CN/H_2O
with a flow rate of 3.0 mL/min at 210 nm. Compound **3** (23.4 mg, t_R = 9.0 min) and

compound **4** (8.2 mg, t_R = 10.5 min) were isolated from fraction C with 40% CH_3CN/H_2O. Compound **5** (4.6 mg, t_R = 10.7 min) was purified from fraction D with 50% CH_3CN/H_2O.

2.4. Preparation of (R)- and (S)-MTPA Esters of Compounds 3 and 4

Compound **3** (1.2 mg, 4.26 µmol) was transferred into a NMR tube and dried under vacuum. Pyridine-d_5 (0.5 mL) and (S)-(-)-α-methoxy-α-(trifluoromethyl)phenylacetyl chloride (5 µL, 26.5 µmol) were added under a N_2 gas stream, and the NMR tube was shaken carefully to mix the sample and the MTPA chloride. The acylation was achieved at 20 °C for 36 h [18], suggesting that compound **3** was entirely transformed into the desired product (R)-MTPA ester derivative **3a**: ^{1}H NMR (600 MHz, pyridine-d_5) δ_H 5.902 (1H, d, H-5), 5.404 (1H, d, H-7), 1.322 (1H, m, H-8), 1.177 (1H, m, H-9a), 1.291 (1H, m, H-9b), 1.541 (3H, s, H-14), 1.847 (3H, s, H-15), 1.167 (3H, d, H-16). In the same way, compound **3** was treated with (R)–MTPA chloride in pyridine-d_5 to give the expected (S)-MTPA ester derivative **3b**: ^{1}H NMR (600 MHz, pyridine-d_5) δ_H 5.971 (1H, d, H-5), 5.849 (1H, d, H-7), 1.292 (1H, m, H-8), 1.145 (1H, m, H-9a), 1.123 (1H, m, H-9b), 1.571 (3H, s, H-14), 1.875 (3H, s, H-15), 1.092 (3H, d, H-16), shown as Figures S12 and S13 and Table S1.

By the same method described above, compound **4** (0.5 mg, 1.77 µmol) was, respectively, acylated using (S)- and (R)-(-)-α-methoxy-α-(trifluoromethyl)phenylacetyl chloride (3 µL, 15.9 µmol), which resulted in products of (R)-MTPA ester **4a** and (S)-MTPA ester **4b**, respectively. **4a**: ^{1}H NMR (600 MHz, pyridine-d_5) δ_H 6.008 (1H, d, H-5), 4.057 (1H, d, H-7), 1.712 (1H, m, H-8), 1.272 (1H, m, H-9a), 1.269 (1H, m, H-9b), 1.988 (3H, s, H-14), 1.590 (3H, s, H-15), 1.206 (3H, d, H-16). **4b**: ^{1}H NMR (600 MHz, pyridine-d_5) δ_H 6.009 (1H, d, H-5), 4.058 (1H, d, H-7), 1.171 (1H, m, H-8), 1.271 (1H, m, H-9a), 1.268 (1H, m, H-9b), 1.989 (3H, s, H-14), 1.590 (3H, s, H-15), 1.206 (3H, d, H-16), shown as Figures S14 and S15 and Table S2.

2.5. Computational Section for Compound 3

To determine the absolute configurations of C-4 and C-8 in **3**, time-dependent density functional theory (TDDFT) method as a useful tool was applied for theoretical calculations of ECD spectra [19,20]. The conformational searches were carried out using Spartan software with the preliminary Merck Molecular Force Field (MMFF) in a 10.0 kcal mol^{-1} energy window [21]. All the obtained conformers were reoptimized at the B3LYP/6-31+G (d, p) level with the IEFPCM solvent model for methanol, and eight, twenty, nineteen and twenty conformers for **3**-(4R, 8R), **3**-(4R, 8S), **3**-(4S, 8R) and **3**-(4S, 8S) with a Boltzmann population above 1% were obtained, respectively, shown as Figures S16–S19. The vibrational frequencies of these conformers were also calculated in the M06-2X/6-311++g (d, p) level, demonstrating all conformers are true minima. Then, these conformers were subjected to calculate the ECD spectra using the TDDFT method with the PBE0 functional and the def-TZVP basis set in the same solvent model, and the rotatory strength for a total of 60 exited states were considered. The Boltzmann-weighted ECD spectra from the ZPVE-corrected M06-2X/6-311++g (d, p) energies were generated in GaussView 6.0.16 software, and the results were represented in Figures S16–S19. All calculations were implemented in the Gaussian 16 package [22].

2.6. Antimicrobial Assay

Antimicrobial activity was investigated according to the agar dilution method described by Unemo and coworkers [23], ampicillin was used as a positive standard. Clinical strain *H. pylori* 159 was obtained from biopsy sample of gastritis patient. Isolation and identification of *H. pylori* 159 were used standard protocols on basis of colony appearance, Gram staining, and positive reactions in the rapid urease test [24]. Additionally, 10% fetal calf serum (FCS) brain heart infusion (BHI, Becton Dickinson, Sparks, NV, USA) broth or 5% FCS Columbia blood agar (Oxoid, Basingstoke, UK), supplemented with Dent selective supplement (Oxoid), were used for routinely culture of H. pylori strains. Incubation of strains were under microaerophilic conditions (10% CO_2, 85% N_2, and 5% O_2 and 90% relative humidity) using a double-gas CO_2

incubator (Binder, model CB160, Tuttlingen, Germany) at 37 °C for 48 to 72 h. Three replicates were performed for every antimicrobial assay.

Anti-*H. pylori* activities were carried out according to broth microdilution assay [25]. *H. pylori* cultures in the exponential phase of growth were diluted ten times in BHI broth and inoculated into each well containing 100 μL test compounds. The final concentration of *H. pylori* was 5×10^5 to 1×10^6 CFU/mL. After incubated in a microaerophilic atmos-phere at 37 °C for 3 days, the plates were examined visually. Antimicrobial activity testing of pure compound followed Antimicrobial Susceptibility Testing Standards outlined by the Clinical and Laboratory Standards Institute (CLSI) document M07-A7 (Clinical and Laboratory Standards Institute 2008) against strain *H. pylori* 159. MIC value indicated the minimum inhibitory concentration for each compound.

3. Results

3.1. Structure Elucidation

By careful comparison of the ^{1}H and ^{13}C NMR and ESI-MS spectral data with literature (Table 1, Figures S1 and S2), the chemical structures of compounds **3** and **4** were, respectively, identified as neovasifuranone A and neovasifuranone B [12–15], while compounds **1**, **2** and **5** were, respectively, characterized as *N*-(2-phenylethyl)acetamide [26], 1-(3-hydroxy-2-methoxyphenyl)-ethanone [27] and 1,2-seco-trypacidin [28].

Table 1. NMR spectral data for compounds **3** and **4** (^{1}H, 600 MHz and ^{13}C 150 MHz).

Position	Compound 3 (in DMSO-d_6)		Compound 4 (in CDCl$_3$)	
	δ_C	δ_H (in ppm, *J* in Hz)	δ_C	δ_H (in ppm, *J* in Hz)
1	189.6		190. 9	
2	112.1		112.4	
3	203.9		206.6	
4	87.7		89.1	
5	121.7	5.37 (1H, *m*)	123.2	5.42 (1H, *s*)
6	143.5		144.1	
7	78.0	3.60 (1H, *t*, *J* = 4.2)	81.4	3.70 (1H, *d*, *J* = 6.9)
7-OH		8.31 (1H, *s*)		
8	36.9	1.39 (1H, *m*)	37.5	1.50 (1H, *m*)
9	25.9	1.06 (1H, *m*)	26.4	1.05 (1H, *m*)
		1.31 (1H, *m*)		1.31 (1H, *m*)
10	11.6	0.84 (3H, *t*, *J* = 7.2)	11.9	0.87 (3H, *t*, *J* = 7.5)
11	21.9	2.62 (1H, *q*, *J* = 7.2)	22.9	2.65 (2H, *m*)
		2.67 (1H, *q*, *J* = 7.8)		
12	10.5	1.16 (3H, *t*, *J* = 7.2)	10.9	1.24 (3H, *t*, *J* = 7.5)
13	50.5	4.01 (2H, *s*)	53.1	4.24 (2H, *m*)
14	24.0	1.37 (3H, *s*)	24.5	1.47 (3H, *s*)
15	13.6	1.60 (3H, *d*, *J* = 0.6)	13.5	1.64 (3H, *d*, *J* = 1.3)
16	13.7	0.71 (3H, *d*, *J* = 6.6)	14.3	0.84 (3H, *d*, *J* = 6.7)

The absolute configurations of compounds **3** and **4** were further confirmed by a combination of modified Mosher's reactions and calculated electronic circular dichroism (ECD) and optical rotatory dispersion (ORD) analysis. Since the more abundant compound **3** possessed two readily acylable hydroxyl groups at C-7 and C-13, its (*S*)- and (*R*)-MTPA esters (**3a** and **3b**) were prepared as detailed elsewhere [29–31]. The chemical shift deviations ($\Delta\delta_{S-R}$, Figure 2) calculated from the ^{1}H NMR spectral data of **3a** and **3b** indicated the presence of a 7*S*-configuration. Obviously, the calculated ECD spectra for 3-(4*S*, 8*R*) and 3-(4*S*, 8*S*) are similar to their experimental ECD spectra (Figure 3). Furthermore, computed ORD values for 3-(4*S*, 8*R*) and 3-(4*S*, 8*S*) under the 589.3 nm are, respectively, −163° and −39°, whereas the experimental ORD value for **3** is −140°, which closely agrees with the calculated ORD value of 3-(4*S*, 8*R*) [32,33]. Therefore, the absolute configuration of **3** is unambiguously established as (4*S*, 7*S*, 8*R*).

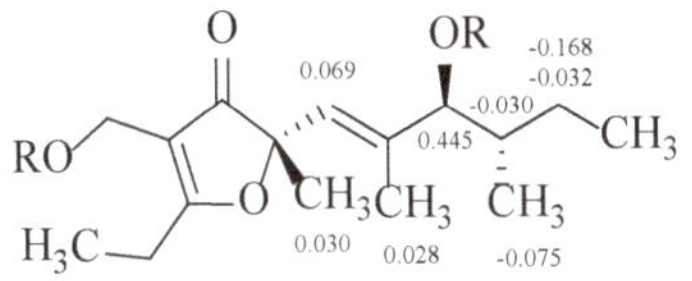

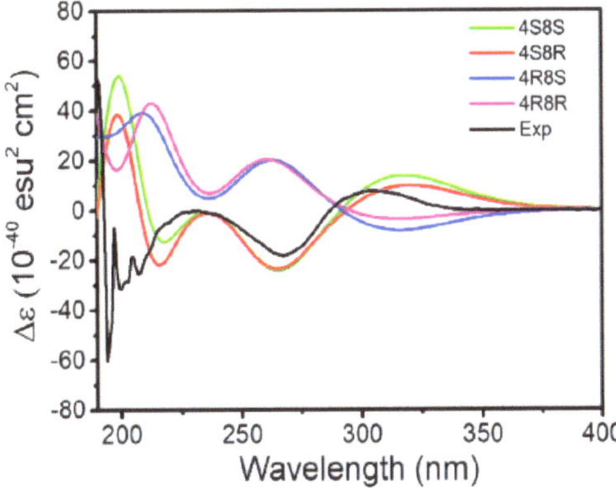

3a: R= (*R*)-MTPA
3b: R= (*S*)-MTPA

Figure 2. $\Delta\delta_{S-R}$ values for MTPA esters of compounds **3a** and **3b**.

Figure 3. Calculated and experimental ECD spectra of compound **3**.

As far as compound **4** concerned, the Mosher's reaction result has a similar trend with compound **3**. The chemical shift deviations ($\Delta\delta_{S-R}$, Figure 4) calculated from the ^{1}H NMR spectral data of **4a** and **4b** indicated the presence of a 7*S*-configuration in **4**. By comparison of ECD spectrum of compounds **3** with that of **4** (Figure 5), they had very similar cotton effects, which one valley at 202 nm and a peak at 306 nm were respectively shown in the first negative and the positive cotton effect regions, and the other elliptical valley was apparent at 267 nm in the negative cotton effect region [34]. Furthermore, the experimental ORD value for **4** is −92°, which is similar to that of its isomer 3-(4*S*, 8*S*), suggesting that three chiral centers at C-4, C-7 and C-8 in **4** are *S* configurations. Accordingly, the absolute configuration of **4** is undoubtedly characterized as (4*S*, 7*S*, 8*S*).

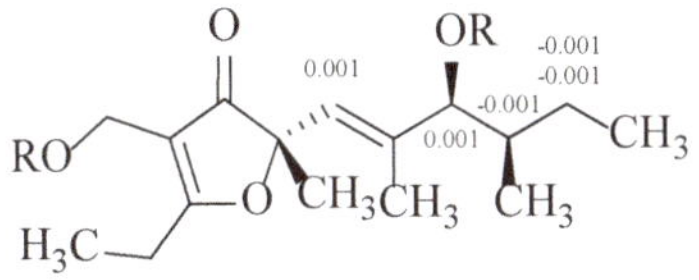

4a: R= (*R*)-MTPA
4b: R= (*S*)-MTPA

Figure 4. $\Delta\delta_{S-R}$ values for MTPA esters of compounds **4a** and **4b**.

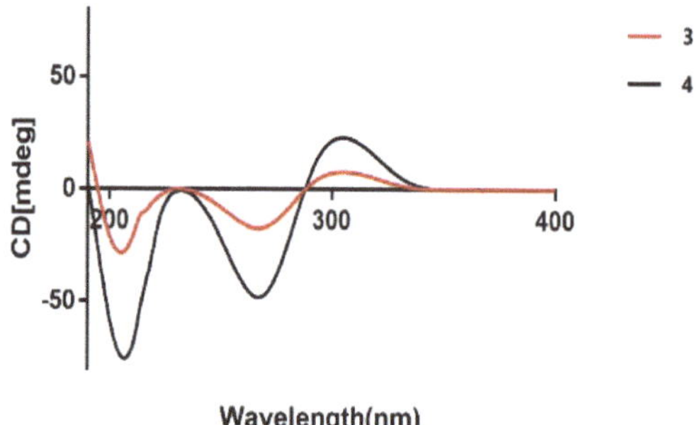

Figure 5. Experimental ECD spectra of compounds **3** and **4**.

3.2. Antimicrobial Activity

Antimicrobial tests were carried out on one of the most serious pathogenic bacteria *Helicobacter pylori* 159. The results indicated that none of these compounds **1–5** had remarkable inhibitory effect on *H. pylori* 159, which MIC values are no less than 16 μg/mL (Table 2).

Table 2. In vitro anti-*Helicobacter pylori* effects of compounds **1-5**.

Compound	MIC Value (μg/mL)
	Helicobacter pylori **159**
1	>16
2	>16
3	>16
4	>16
5	16
Ampicillin sodium	4

4. Discussion

To the best of our knowledge, more than 50% of the currently used drugs are chiral compounds. The enantiomers of the same drug have the same physical and chemical properties, but they exhibit differences in pharmacokinetics, pharmacodynamics and toxicity [35]. Owing to inherent structural and stereochemical complexity, fungal secondary metabolites play a significant role in drug discovery and development processes. In this study, the absolute configurations of two flexible molecules neovasifuranones A (**3**) and B (**4**) from *F. oxysporum* R1 were firstly determined by a combination of Mosher's reactions and quantum mechanical calculation of chiroptical (ECD and ORD) properties. These findings will assist in further analysis of structure-activity relationship of compounds **3** and **4**.

Endophytic fungi are one of important sources of bioactive secondary metabolites with potential application in biomedicine and agriculture [36,37]. *Fusarium* microorganisms are ubiquitous in nature including terrestrial and marine environments and plants. Genome sequencing and analysis indicate that these microbes possess a great number of secondary metabolites biosynthetic gene clusters (BGCs), including polyketide synthetase, non-ribosomal peptide synthetase and terpene synthetase [9]. However, most of these cryptic BGCs are not expressed under conventional culture conditions, which result in an unfavorable trend that the number of novel natural products from the genus *Fusarium* has been decreasing in the past decade. Therefore, more efforts should be made to awaken their silent BGCs to produce novel functional biomolecules using OSMAC strategy and advanced interdisciplinary technology, such as genome mining, metabonomics, gene heteroexpression and functional characterization [38,39].

Supplementary Materials: The following are available online at https://www.mdpi.com/article/10.3390/jof8010040/s1, Figure S1: (+) ESI-MS spectrum of compound **3**; Figure S2: (+) ESI-MS spectrum of **4**; Figure S3: IR (KBr) spectrum of **3**; Figure S4: IR (KBr) spectrum of **4**; Figure S5: ^{1}H NMR (600 MHz, (CD$_3$)$_2$SO) spectrum of **3**; Figure S6: ^{13}C NMR (151 MHz, (CD$_3$)$_2$SO) spectrum of **3**; Figure S7: ^{1}H NMR (600 MHz, CDCl$_3$) spectrum of **4**; Figure S8: ^{13}C NMR (151 MHz, CDCl$_3$) spectrum of **4**; Figure S9: DEPT-135 (CDCl$_3$) spectrum of **4**; Figure S10: UV spectrum of **3** in MeOH; Figure S11: UV spectrum of **4** in MeOH; Figure S12: ^{1}H NMR (600 MHz, pyridine-d_5) spectrum of (*R*)-MTPA of **3**; Figure S13: ^{1}H NMR (600 MHz, pyridine-d_5) spectrum of (*S*)-MTPA of **3**; Figure S14: ^{1}H NMR (600 MHz, pyridine-d_5) spectrum of (*R*)-MTPA of **4**; Figure S15: ^{1}H NMR (600 MHz, pyridine-d_5) spectrum of (*S*)-MTPA of **4**; Table S1: ^{1}H NMR spectral data for **3** and two MTPA esters **3a** and **3b** (^{1}H, 600 MHz); Table S2: NMR spectral data for **4** and two MTPA esters **4a** and **4b** (^{1}H, 600 MHz); Figure S16: Optimized conformers ($\geq$1%) of **3**-(4*R*, 8*R*); Figure S17: Optimized conformers ($\geq$1%) of **3**-(4*R*, 8*S*); Figure S18: Optimized conformers ($\geq$1%) of **3**-(4*S*, 8*S*); Figure S19: Optimized conformers ($\geq$1%) of **3**-(4*S*, 8*R*).

Author Contributions: Conceptualization, project administration and funding acquisition, H.Z.; methodology, Z.F. and Y.L.; software, X.Y.; formal analysis, Z.F. and H.Z.; investigation, Z.F., Y.L. and M.X.; resources, H.Z.; data curation, H.W.; writing—original draft preparation, Z.F.; writing—review and editing, X.Y. and H.Z. All authors have read and agreed to the published version of the manuscript.

Funding: This work was co-financially supported by the National Key Research and Development Program of China (2018YFC0311004), the National Natural Science Foundation of China (41776139) and the Fundamental Research Fund for the Provincial Universities of Zhejiang of China (RF-C2019002).

Institutional Review Board Statement: Not applicable.

Informed Consent Statement: Not applicable.

Data Availability Statement: Not applicable.

Conflicts of Interest: The authors declare no conflict of interest.

References

1. Zhou, Z.F.; Kurtán, T.; Yang, X.H. Penibruguieramine A, a novel pyrrolizidine alkaloid from the endophytic fungus *Penicillium* sp. GD6 associated with chinese mangrove *Bruguiera gymnorrhiza*. *Org. Lett.* **2014**, *45*, 1390–1393. [CrossRef]
2. Sun, L.L.; Li, W.S.; Li, J. Uncommon diterpenoids from the south china sea soft coral *Sinularia humilis* and their stereochemistry. *J. Org. Chem.* **2021**, *86*, 3367–3376. [CrossRef]
3. Ye, F.; Zhu, Z.D.; Chen, J.S. Xishacorenes A–C, diterpenes with bicyclo[3.3.1]nonane nucleus from the Xisha soft coral *Sinularia polydactyla*. *Org. Lett.* **2017**, *19*, 4183–4186. [CrossRef]
4. Seco, J.M.; QuinOá, E.; Riguera, R. The assignment of absolute configuration by NMR. *Chem. Rev.* **2004**, *104*, 17–118. [CrossRef]
5. Evidente, M.; Cimmino, A.; Zonno, M.C.; Masi, M.; Berestetskyi, A.; Santoro, E.; Superchi, S.; Vurro, M.; Evidente, A. Phytotoxins produced by *Phoma chenopodiicola*, a fungal pathogen of *Chenopodium album*. *Phytochemistry* **2015**, *117*, 482–488. [CrossRef] [PubMed]
6. Evidente, M.; Cimmino, A.; Zonno, M.C.; Masi, M.; Santoro, E.; Vergura, S.; Berestetskiy, A.; Superchi, S.; Vurro, M.; Evidente, A. Chenopodolans E and F, two new furopyrans produced by *Phoma chenopodiicola* and absolute configuration determination of chenopodolan B. *Tetrahedron* **2016**, *72*, 8502–8507. [CrossRef]
7. Mazzeo, G.; Cimmino, A.; Masi, M.; Longhi, G.; Maddau, L.; Memo, M.; Evidente, A.; Abbate, S. Importance and difficulties in the use of chiroptical methods to assign the absolute configuration of natural products: The case of phytotoxic pyrones and furanones produced by *Diplodia corticola*. *J. Nat. Prod.* **2017**, *80*, 2406–2415. [CrossRef]
8. El-Elimat, T.; Figueroa, M.; Raja, H.A.; Alnabulsi, S.; Oberlies, N.H. Coumarins, dihydroisocoumarins, a dibenzo-alpha-pyrone, a meroterpenoid, and a merodrimane from *Talaromyces amestolkiae*. *Tetrahedron Lett.* **2021**, *72*, 153067. [CrossRef]
9. Li, M.Z.; Yu, R.L.; Bai, X.L.; Wang, H.; Zhang, H.W. Fusarium: A treasure trove of bioactive secondary metabolites. *Nat. Prod. Rep.* **2020**, *37*, 1568–1588. [PubMed]
10. Yu, R.; Li, M.; Wang, Y. Chemical investigation of a co-culture of *Aspergillus fumigatus* D and *Fusarium oxysporum* R1. *Rec. Nat. Prod.* **2020**, *15*, 130–135. [CrossRef]
11. Chen, J.; Bai, X.; Hua, Y.; Zhang, H.; Wang, H. Fusariumins C and D, two novel antimicrobial agents from *Fusarium oxysporum* ZZP-R1 symbiotic on *Rumex madaio* Makino. *Fitoterapia* **2019**, *134*, 1–4. [CrossRef]
12. Furumoto, T.; Fukuyama, K.; Hamasaki, T. Neovasipyrones and neovasifuranones: Four new metabolites related to neovasinin, a phytotoxin of the fungus *Neocosmospora vasinfecta*. *Phytochemistry* **1995**, *40*, 745–751. [CrossRef]

13. Furumoto, T.; Hamasaki, T.; Nakajima, H. Biosynthesis of phytotoxin neovasinin and its related metabolites, neovasipyrones A and B and neovasifuranones A and B, in the phytopathogenic fungus *Neocosmospora vasinfecta*. *Cheminform* **1999**, *30*, 131–136.
14. Feng, Q.M.; Li, X.Y.; Li, B.X. Isolation and identification of two new compounds from the *Penicillium* sp. SYPF7381. *Nat. Prod. Res.* **2019**, *34*, 1–7. [CrossRef] [PubMed]
15. Jouda, J.B.; Fopossi, J.; Kengne, F.M. Secondary metabolites from *Aspergillus japonicus* CAM231, an endophytic fungus associated with *Garcinia preussii*. *Nat. Prod. Res.* **2017**, *31*, 861–869. [CrossRef] [PubMed]
16. Wei, Y.L.; Wu, J.; Deng, Y.P. Optimization of extraction process of total flavonoids from *Rumex madaio* Makino. *Food Res. Dev.* **2015**, *14*, 20–24. (In Chinese)
17. Zhang, H.; Loveridge, S.T.; Tenney, K.; Crews, P. A new 3-alkylpyridine alkaloid from the marine sponge *Haliclona* sp. and its cytotoxic activity. *Nat. Prod. Res.* **2015**, *30*, 1262–1265. [CrossRef]
18. Zhang, H.W.; Zhang, J.; Hu, S.; Zhang, Z.J.; Zhu, C.J.; Ng, S.W.; Tan, R.X. Ardeemins and cytochalasins from *Aspergillus terreus* residing in *Artemisia annua*. *Planta Med.* **2010**, *76*, 1616–1621. [CrossRef]
19. Lin, S.; Shi, T.; Chen, K.Y.; Zhang, Z.X.; Shan, L.; Shen, Y.H.; Zhang, W.D. Cyclopenicillone, a unique cyclopentenone from the cultures of *Penicillium decumbens*. *Chem. Commun.* **2011**, *47*, 10413–10415. [CrossRef]
20. Suárez-Ortiz, G.A.; Cerda-García-Rojas, C.M.; Fragoso-Serrano, M.; Pereda-Miranda, R. Complementarity of DFT calculations, NMR anisotropy, and ECD for the configurational analysis of brevipolides K–O from *Hyptis brevipes*. *J. Nat. Prod.* **2017**, *80*, 181–189. [CrossRef]
21. Agarwal, M.; Frank, M.I. SPARTAN: A software tool for parallelization bottleneck analysis. In Proceedings of the 2009 ICSE Workshop on Multicore Software Engineering, Vancouver, BC, Canada, 18 May 2009; pp. 56–63.
22. Frisch, M.J.; Trucks, G.W.; Schlegel, H.B.; Scuseria, G.E.; Robb, M.A.; Cheeseman, J.R.; Scalmani, G.; Barone, V.; Mennucci, B.; Petersson, G.A.; et al. *Gaussian 2016, Revision, A. 03*; Gaussian, Inc.: Wallingford, CT, USA, 2016.
23. Unemo, M.; Fasth, O.; Fredlund, H.; Limnios, A.; Tapsall, J. Phenotypic and genetic characterization of the 2008 WHO *Neisseria gonorrhoeae* reference strain panel intended for global quality assurance and quality control of gonococcal antimicrobial resistance surveillance for public health purposes. *J. Antimicro. Chemother.* **2009**, *63*, 1142–1151. [CrossRef] [PubMed]
24. Castellote, J.; Guardiola, J.; Porta, F.; Falco, A. Rapid urease test: Effect of preimmersion of biopsy forceps in formalin. *Gastrointest. Endosc.* **2001**, *53*, 744–746. [CrossRef]
25. Covacci, A.; Censini, S.; Bugnoli, M.; Petracca, R.; Burroni, D.; Macchia, G.; Massone, A.; Papini, E.; Xiang, Z.; Figura, N.; et al. Molecular characterization of the 128-kDa immunodominant antigen of *Helicobacter pylori* associated with cytotoxicity and duodenal ulcer. *Proc. Natl. Acad. Sci. USA* **1993**, *90*, 5791–5795. [CrossRef]
26. Daoud, N.N.; Foster, H.A. Antifungal activity of *Myxococcus* species 1 production, physiochemical and biological properties of antibiotics from *Myxococcus flavus* S110 (Myxobacterales). *Microbios* **1993**, *73*, 173–184.
27. Giri, A.; Zelinkova, Z.; Wenzl, T. Experimental design-based isotope-dilution SPME-GC/MS method development for the analysis of smoke flavouring products. *Food Addit. Contam.* **2017**, *34*, 2069–2084. [CrossRef] [PubMed]
28. Liu, R.; Zhu, W.M.; Zhang, Y.P.; Zhu, T.J.; Liu, H.B.; Fang, Y.C.; Gu, Q.Q. A new diphenyl ether from marine-derived fungus *Aspergillus* sp. B-F-2. *J. Antibiot.* **2006**, *59*, 362–365. [CrossRef]
29. Ohtani, I.; Kusumi, T.; Kashman, Y. High-field FT NMR application of Mosher's method. The absolute configurations of marine terpenoids. *J. Am. Chem. Soc.* **1991**, *113*, 4092–4096. [CrossRef]
30. Kusumi, T.; Fujita, Y.; Ohtani, I. Anomaly in the modified Mosher's method: Absolute configurations of some marine cembranolides. *Tetrahedron Lett.* **1991**, *32*, 2923–2926. [CrossRef]
31. Konno, K.; Fujishima, T.; Liu, Z. Determination of absolute configuration of 1,3-diols by the modified Mosher's method using their di-MTPA esters. *Chirality* **2010**, *14*, 72–80. [CrossRef] [PubMed]
32. Ma, Z.; Hano, Y.; Feng, Q. Determination of the absolute stereochemistry of lupane triterpenoids by fucofuranoside method and ORD spectrum. *Tetrahedron Lett.* **2004**, *45*, 3261–3263. [CrossRef]
33. Dreyer, D.L. Citrus bitter principles-III: Application of ORD and CD to stereochemical problems. *Tetrahedron* **1968**, *24*, 3273–3283. [CrossRef]
34. Lee, S.; Hoshino, M.; Fujita, M. Cycloelatanene A and B: Absolute configuration determination and structural revision by the crystalline sponge method. *Chem. Sci.* **2017**, *8*, 1547–1550. [CrossRef]
35. Sanganyado, E.; Lu, Z.; Fu, Q. Chiral pharmaceuticals: A review on their environmental occurrence and fate processes. *Water Res.* **2017**, *124*, 527–542. [CrossRef]
36. Venugopalan, A.; Srivastava, S. Endophytes as in vitro production platforms of high value plant secondary metabolites. *Biotechnol. Adv.* **2015**, *33 Pt 1*, 873–887. [CrossRef]
37. Zhang, H.W.; Bai, X.L.; Zhang, M.; Chen, J.W.; Wang, H. Bioactive natural products from endophytic microbes. *Nat. Prod. J.* **2018**, *8*, 86–108. [CrossRef]
38. Pan, R.; Bai, X.L.; Chen, J.W.; Zhang, H.W.; Wang, H. Exploring structural diversity of microbe secondary metabolites using OSMAC strategy: A literature review. *Front. Microbiol.* **2019**, *10*, 294–313. [CrossRef] [PubMed]
39. Atanasov, A.G.; Zotchev, S.B.; Dirsch, V.M.; The International Natural Product Sciences Taskforce; Supuran, C.T. Natural products in drug discovery: Advances and opportunities. *Nat. Rev. Drug Discov.* **2021**, *20*, 200–216. [CrossRef] [PubMed]

Article

Volatiles Produced by Yeasts Related to *Prunus avium* and *P. cerasus* Fruits and Their Potentials to Modulate the Behaviour of the Pest *Rhagoletis cerasi* Fruit Flies

Raimondas Mozūraitis [1,*], Violeta Apšegaitė [1], Sandra Radžiutė [1], Dominykas Aleknavičius [1], Jurga Būdienė [1], Ramunė Stanevičienė [2], Laima Blažytė-Čereškienė [1], Elena Servienė [2] and Vincas Būda [1]

[1] Laboratory of Chemical and Behavioural Ecology, Institute of Ecology, Nature Research Centre, Akademijos Str. 2, LT-08412 Vilnius, Lithuania; violeta.apsegaite@gamtc.lt (V.A.); sandra.radziute@gamtc.lt (S.R.); dominykas.aleknavicius@gamtc.lt (D.A.); jurga.budiene@gamtc.lt (J.B.); laima.blazyte@gamtc.lt (L.B.-Č.); vincas.buda@gamtc.lt (V.B.)

[2] Laboratory of Genetics, Institute of Botany, Nature Research Centre, Akademijos Str. 2, LT-08412 Vilnius, Lithuania; ramune.staneviciene@gamtc.lt (R.S.); elena.serviene@gamtc.lt (E.S.)

* Correspondence: raimondas.mozuraitis@gamtc.lt; Tel.: +370-5-272-92-42

Abstract: Yeast produced semiochemicals are increasingly used in pest management programs, however, little is known on which yeasts populate cherry fruits and no information is available on the volatiles that modify the behaviour of cherry pests including *Rhagoletis cerasi* flies. Eighty-two compounds were extracted from the headspaces of eleven yeast species associated with sweet and sour cherry fruits by solid phase micro extraction. Esters and alcohols were the most abundant volatiles released by yeasts. The multidimensional scaling analysis revealed that the odour blends emitted by yeasts were species-specific. *Pichia kudriavzevii* and *Hanseniaspora uvarum* yeasts released the most similar volatile blends while *P. kluyveri* and *Cryptococcus wieringae* yeasts produced the most different blends. Combined gas chromatographic and electroantennographic detection methods showed that 3-methybutyl acetate, 3-methylbutyl propionate, 2-methyl-1-butanol, and 3-methyl-1-butanol elicited antennal responses of both *R. cerasi* fruit fly sexes. The two-choice olfactometric tests revealed that *R. cerasi* flies preferred 3-methylbutyl propionate and 3-methyl-1-butanol but avoided 3-methybutyl acetate. Yeast-produced behaviourally active compounds indicated a potential for use in pest monitoring and control of *R. cerasi* fruit flies, an economically important pest of cherry fruits.

Keywords: yeasts; microorganisms; Diptera; Tephritidae; volatiles; attractant; repellent; electroantennography; pest management; behaviour modification

Citation: Mozūraitis, R.; Apšegaitė, V.; Radžiutė, S.; Aleknavičius, D.; Būdienė, J.; Stanevičienė, R.; Blažytė-Čereškienė, L.; Servienė, E.; Būda, V. Volatiles Produced by Yeasts Related to *Prunus avium* and *P. cerasus* Fruits and Their Potentials to Modulate the Behaviour of the Pest *Rhagoletis cerasi* Fruit Flies. *J. Fungi* **2022**, *8*, 95. https://doi.org/10.3390/jof8020095

Academic Editor: Laurent Dufossé

Received: 9 December 2021
Accepted: 6 January 2022
Published: 19 January 2022

1. Introduction

Carposphere is a specific habitat populated by bacterial and fungal microorganisms including yeasts [1,2]. Berries and fruits are rich in carbohydrates and often bear the most diverse microbiome in a phyllosphere [2]. Carposphere microbiota are determined by a variety of factors such as environmental conditions, host genotype, berry developmental stage, and interactions with other organisms sharing a habitat [3–7]. Microorganisms associated with fruits and berries interact with insects that use these habitats for feeding and oviposition. Insects and yeasts could come into diverse relationships ranging from amensal to commensal and mutualistic [8]. Yeasts provide essential nutrients missing in sugar-rich berries that insects cannot produce while insects transfer yeasts from one substrate to another [1,9,10]. In addition to berry and fruit-related odours, yeast-produced volatiles are used by insects to acquire information about habitat quality and host choice [11–14]. The behaviour modifying effect of yeast volatiles has the potential for use in integrated pest management programs increasing the efficiency of attractive lures and serving as repellents in the push–pull pest control strategy [15].

The purpose of the study was to characterize volatile blends emitted by cultivable yeasts populating the fruit surface of sweet and sour cherries and to determine semiochemicals that modify the behaviour of *R. cerasi* fruit flies, the most important pest of cherry fruits.

2. Materials and Methods

2.1. Yeast Sampling, Culturing, and Identification

Yeast species have been isolated from sweet cherry (*Prunus avium* L.) and sour cherry (*Prunus cerasus* L.) (Rosales: Rosaceae) fruits collected during June–July of 2018–2020 from private plantations located in the Vilnius region (GPS coordinates: 54°45′08.2″ N, 25°17′10.0″ E; 54°46′14.4″ N 25°21′04.1″ E; 54°41′19.9″ N 25°26′20.6″ E), Klaipėda region (GPS coordinates: 55°36′1.66″ N 21°36′3.8″ E; 55°34′50.6″ N 21°14′58.1″ E) and Alytus region (GPS coordinates: 54°23′43.8″ N, 23°56′18.7″ E) of Lithuania. Isolation and culturing methods have been described in detail by Stanevičienė et al. [16]. Generally, cultivable yeasts are isolated by direct rinsing of fruits with MD medium (2% dextrose, 1% (NH4)2SO4, 0.09% KH2PO4, 0.05% MgSO4, 0.023% K2HPO4, 0.01% NaCl, 0.01% $CaCl_2$) or by applying fermentation-based enrichment. After cultivation on YPD-agar plates (1% yeast extract, 1% peptone, 2% dextrose, 2% agar), morphologically distinct colonies proceeded molecular analysis. For taxonomic identification of the isolates, the ITS1-5.8S rDNA-ITS2 region or D1/D2 region of 26S rDNA were PCR-amplified as described in Stanevičienė et al. [16]. To assess their taxonomic position, the resolved sequences were compared with those available in the current version of the GenBank database at the National Centre for Biotechnology Information (NCBI) (Table A1).

2.2. Insects

European cherry fruit flies, *Rhagoletis cerasi* (L.) (Diptera: Tephritidae), were collected as pupae from soil under sweet and sour cherry trees in April of 2019–2021 at private orchards in Vilnius (GPS coordinates: 54°45′24.0″ N 25°03′02.4″ E) and Kaunas (GPS coordinates: 54°54′13.8″ N 23°48′07.9″ E) districts, Lithuania. Reactivation of the pupae took place in a climate chamber "Fitotron" under 20–24 °C, 16L:8D (light:dark) photoperiod, and 65–75% relative humidity. Each pupa was placed in an individual 14 mL glass vial bearing wet 3 cm^2 filter paper inside and closed by foam stoppers. The filter paper was humidified periodically to keep the humidity up inside the vial. After emergence, the adults were kept within the same vials in the room under 18–20 °C, a natural daylight photoperiod, 50–60% relative humidity, and fed on 10% sugar solution in water. The flies possessing an ovipositor were attributed to females. After sexing, each individual was kept in a separate vial under the identical conditions as described above.

2.3. Sampling and Analysis of Volatiles Produced by Yeasts

For sampling of volatile organic compounds, yeasts were selected based on isolation frequency over the different ripening stages of cherries (Figure A1). The methods have been described in detail by Lukša et al. [7]. In summary, overnight grown yeast cells (50 μL) at the concentration of about 3–5 × 10^7 cells/mL were placed on the surface of YPD-agar medium and cultivated for two days at 25 °C. The solid-phase micro-extraction (SPME) technique was used to sample the headspace volatiles produced by yeasts. For sampling background volatiles, YPD-agar plates without yeast were used as control samples. The SPME needle was placed above the yeast culture through a small hole drilled in a Petri dish; the purified fibre coated with a polydimethylsiloxane-divinylbenzene absorbent (65 mm coating layer thickness) was exposed to the headspace for 60 min at room temperature. The volatiles collected on the fibre were desorbed for 2 min in the injection liner of a gas chromatograph (GC).

GC and mass spectrometer (MS) were used to analyse the collected volatiles. The compounds were separated by a DB-Wax column under the subsequent temperature program: isothermal at 40 °C for 1 min and afterwards gradually increased to 200 °C at a rate of 5 °C/min, then to 240 °C at a rate of 10 °C/min, and maintained isothermally for 11 min. The GC injector was run isothermally at 240 °C. Helium served as a carrier gas. The relative amount of each of the compounds was determined based on the area of the chromatographic peak. The volatile compounds were identified by comparing their mass spectra and retention indexes with those presented in a NIST version 2.0 mass spectral library and those of the available synthetic standards. C_8–C_{28} n-alkanes were used to calculate the retention indexes of the volatiles.

2.4. Gas Chromatography-Electroantennogram Detection

Gas chromatographic and electroantennogram detection (EAD) techniques were applied to determine yeasts produced olfactory active volatiles to *R. cerasi* flies. A detailed description of the GC-EAD setup as well as the procedure has been published by Būda et al. [17]. Briefly, the GC was set up with a polar DB-Wax column. The injector and the detector were run at 240 °C. The oven temperature was maintained at 40 °C for 1 min; afterwards, it was raised to 240 °C at a rate of 10 °C/min, then maintained isothermally for 13 min. Hydrogen, at a flow rate of 1.5 mL/min, was used as a carrier gas. At the end of the GC column, a splitter divided an eluent into two equal parts, allowing simultaneous flame ionisation (FID) and EAD detection of the separated volatiles. A nitrogen make-up gas at 5 mL/min flow rate was added to increase FID sensitivity. The part of an eluent allocated to EAD was mixed with charcoal filtered and humidified air flowing at 0.5 m/s through a glass tube over antenna preparation. Glass capillary electrodes were used. The EAD and the FID signals were registered simultaneously, saved, and analysed. Before EAD recording, the antenna was stimulated with 1 μg of 3-methyl-1-butanol to check sensitivity. Four to seven days old flies were used in the tests. Each antenna tested was from a different fly. In total, 21 antennae of males and 18 antennae of females were used.

2.5. Electroantennogram Dose-Response

The same electrophysiological recording setup and the antennal preparation technique were used to record electroantennogram (EAG) dose–responses of male and female flies to the synthetic EAD active compounds: 3-methylbutyl acetate; 3-methylbutyl propionate; 2-methyl-1-butanol; and 3-methyl-1-butanol.

The compounds were tested at the doses of 10^{-5}, 10^{-4}, 10^{-3}, 10^{-2}, and 10^{-1} mg applied in 10 μL hexane on filter paper (5 × 45 mm). The compounds were selected randomly, and five doses of each compound were tested in ascending order. A solvent blank (10 μL of hexane after evaporation) was tested as a control stimulus both at the beginning and the end of stimulation with each compound. Each EAD test was replicated 13 times, and each antenna used was from a different fly. The EAG response (R) to the EAD-active compound dose was calculated according to the formula $R = RA - (RC_1 + RC_2)/2$, where RA is the EAG response to the EAD active compound, and RC1 and RC2 are EAG responses to the first and the second control stimuli, respectively.

2.6. Behavioural Assay

To test the behavioural choice of the flies to the synthetic EAD active compounds versus the control, a Y-tube olfactometer [18] (25 cm main tube, 17 cm arms, 110° branching angle, the inner diameter of each arm and main tube 5 cm) was used. The olfactometer was placed in a fume cupboard. Four T8/840, Colourlux plus, 18 W tube type lamps (NARVA Lichtquellen GmbH + Co. KG, Brand-Erbisdorf, Germany) covered with a white, mat, plastic shield (65 cm length, 42 cm width) at a distance of 23 cm were placed in front of the Y tube of the olfactometer. For the fruit flies, positive phototaxis is characteristic, and the light slightly stimulated the insects to move towards the light source. Each arm of the olfactometer was connected to a glass tube that contained either the stimulus or control. A purified air delivery system CADS-4CPP (Sigma Scientific LLC, Micanopy, FL, USA) was used to push air at a rate of 0.5 L/min through each arm.

The synthetic 3-methylbutyl propionate, 2-methyl-1-butanol, and 3-methyl-1-butanol were dissolved in hexane while paraffin oil was used to dissolve 3-methylbutyl acetate and the four component mixture. Behaviour modifying effect was assessed at the few doses by dispensing the 10 μL of the solution on a filter paper strip (5 × 40 mm). The proportion of EAD active components in the mixture used in the bioassay was based on the proportion of EAD active compounds determined in the sample of *H. uvarum* yeasts. The synthetic mixture consisted of 3-methylbutyl acetate 0.55 mg, 3-methylbutyl propionate 0.05 mg, 2-methyl-1-butanol 0.012 mg, and 3-methyl-1-butanol 0.28 mg per 10 μL of paraffin oil. After 0.5 min of solvent evaporation (only applicable for samples that were dissolved in hexane and no evaporation was carried out when paraffin oil was used), the filter paper strip was placed in the glass tube connected to one arm of the olfactometer. The same size filter paper was treated either with 10 μL of hexane or with paraffin oil and was placed in the other arm serving as the control. After each test, the olfactometer was taken apart and the glassware was cleaned with hexane, soaked overnight in distilled water, and dried for 2 h in an oven, raising the temperature to 200 °C. Silicone parts of the Y-tube olfactometer were cleaned with hexane, soaked overnight in distilled water, and air-dried or replaced between the tests.

A single fly was released into the Y olfactometer at the end of the main tube. The duration within which a fly must have reached the branch point was set to 15 min. A fly was considered to have made a choice when it reached the distal end of the glass tube containing either a stimulus or a control (solvent after evaporation), irrespectively of whether the fly switched arms or not before reaching the odour source. The fly was considered as not making a choice if none of the arms was chosen within 15 min. After every five tests, the positions of the two Y-tube arms were reversed. All insects were observed individually and used in a bioassay only once. The tests were carried out at 23 ± 2 °C, 60% RH, between 10 h AM and 5 h PM local time.

2.7. Statistical Analysis

A nonparametric Mann–Whitney U test was applied to evaluate differences in the volatile amounts between the yeast and control samples. To assess and visualise the associations between odour blends of eleven yeast species and volatile compounds, a multidimensional scaling (MDS) analysis with a Bray–Curtis index was performed on absolute amounts expressed as areas under chromatographic peaks using R (version 4.0.2) and Rstudio (version 1.3.959), with the metaMDS function in the vegan package (version 2.5–6), and the results were visualised using ggplot2 (version 3.3.2). Prior to analysis, the data were log-transformed. Dendrogram of the odour blends was obtained by cluster analysis based on Euclidean distance using the same way transformed data as in the MDS analysis. The clustering was carried out based on the average of quantified volatile compounds from three different isolates per species. The different clusters were identified by visually evaluating the clustering. Paired *t* test was applied to compare EAG amplitudes of *R. cerasi* antennae of males versus females at each dose tested. To evaluate the choices of flies to the synthetic EAD active compounds versus the control, the total number of flies that made a choice was analysed with a χ2 test (observed vs. expected). All the analysis except MDS was performed using Statistica 6.0 software (StatSoft, Inc., Tulsa, OK, USA).

3. Results

3.1. Composition of Yeast Produced Volatile Blends

Analysis of yeast produced volatiles revealed 82 compounds that were exclusively present in the headspace of eleven yeast species (Table A1) or occurred at significantly larger amounts compared to those of the control samples. The esters represented by 41 compounds were accounted as the most abounded group of volatiles released by yeasts followed by 18 alcohols, nine compounds bearing aromatic moiety, eight ketones, six fatty acids, four terpenoids, three lactones, and one each of isothiocyanate, furane, and sulphide functional group was detected once (Table 1). Compounds bearing ester, alcohol, aromatic, ketone, and fatty acid moiety were detected in the volatile blends of all yeast species while terpenoid, lactone, isothiocyanoate, furane, and sulphide type volatiles were emitted by the yeast of single or few species (Figure 1).

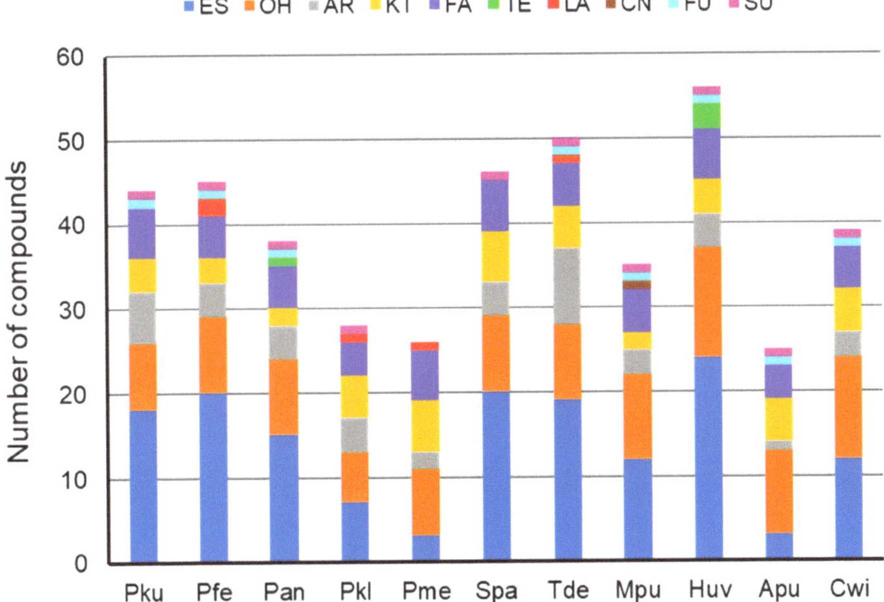

Figure 1. Chemical diversity of the volatile blends produced by yeasts. Pku–*Pichia kudriavzevii*, Pfe–*P. fermentans*, Pan-*P. anomala*, Pkl-*P. kluyveri*, Pme-*P. membranifaciens*, Spa-*Saccharomyces paradoxus*, Tde-*Torulaspora delbrueckii*, Mpu-*Metschnikowia pulcherrima*, Huv-*Hanseniaspora uvarum*, Apu-*Aureobasidium pullulans*, Cwi-*Cryptococcus wieringae*. Functional group of volatiles: ES–ester; OH–alcohol; AR–aromatic; KT–ketone; FA–fatty acid; TE–terpenoid; LA–lactone; CN–isothiocyanoate; FU—furane; SU–sulphide.

Nonmetric multidimensional scaling analysis showed that volatile blends of all eleven species grouped in a species-specific manner (Figure 2).

P. kudriavzevii and *H. uvarum* yeasts released the most similar volatile blends and together with *P. anomala* as well as *M. pulcherrima* yeasts formed a distinct cluster. Yeasts of *P. fermentans*, *P. membranifaciens*, *A. pullulans*, *T. delbrueckii*, and *S. paradoxus* clustered in another group. The most different blends were produced by *P. kluyveri* and *C. wieringae* yeasts (Figure 3).

3.2. EAD Active Compounds

GC–EAD analyses of the headspace collections from five yeast species representing three fruit ripening stages showed that antennae of *R. cerasi* flies, a pest of cherry fruits, responded to 3-methybutyl acetate, 3-methylbutyl propionate, 2-methyl-1-butanol, and 3-methyl-1-butanol (Figure 4, Table 2). Antennae of both *R. cerasi* sexes responded to all four EAD active compounds (Table 2).

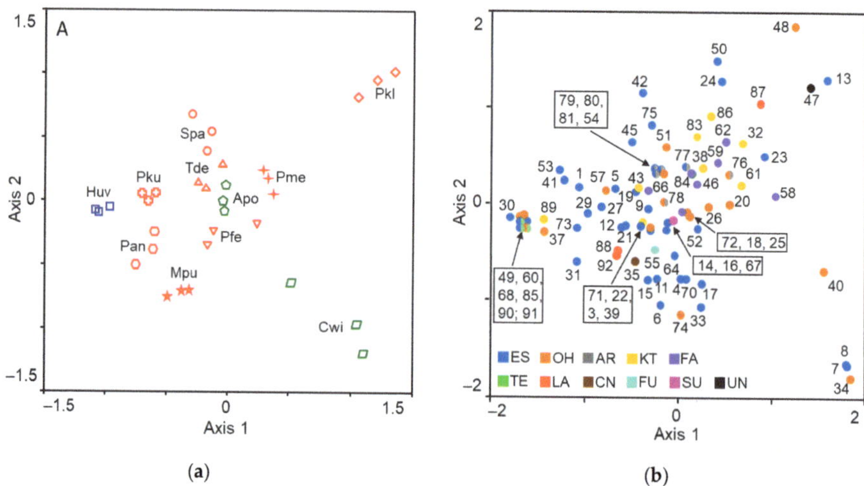

(a)   (b)

Figure 2. Multidimensional scaling plots: (**a**) eleven yeast species (each represented by three different isolates); (**b**) volatile compounds produced by yeasts. Volatiles were sampled from the headspace by SPME. Pku—*Pichia kudriavzevii*, Pfe—*P. fermentans*, Pan—*P. anomala*, Pkl—*P. kluyveri*, Pme—*P. membranifaciens*, Spa—*Saccharomyces paradoxus*, Tde—*Torulaspora delbrueckii*, Mpu—*Metschnikowia pulcherrima*, Huv—*Hanseniaspora uvarum*, Apu—*Aureobasidium pullulans*, Cwi—*Cryptococcus wieringae*. Apu and Cwi yeasts are more common on unripe fruits and their symbols are coloured green, Huv yeasts are the most common on medium-ripe and ripe fruits and are indicated by the blue colour, and the red colour represents yeasts, the most common on ripe fruits. The name of volatiles indicated by numbers are listed in Table 1. Functional group of volatiles: ES—ester; OH—alcohol; AR—aromatic; KT—ketone; FA—fatty acid; TE—terpenoid; LA—lactone; CN—isothiocyanoate; FU—furane; SU—sulphide; UN—unidentified.

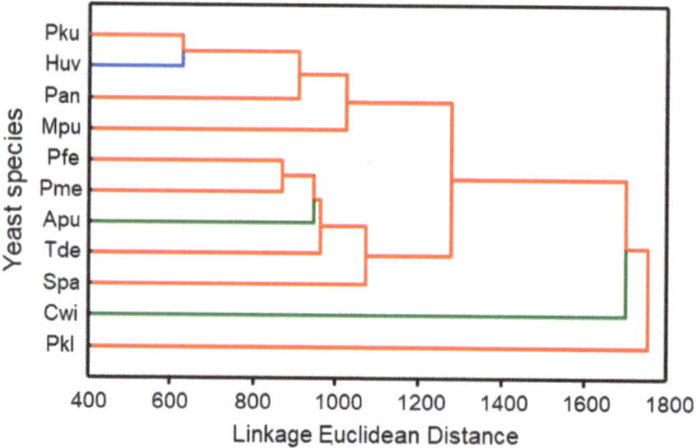

Figure 3. Dendrogram of odour blends sampled by SPME from the headspace of eleven yeast species. The dendrogram was obtained by cluster analysis based on Euclidean distance. The clustering was carried out based on the average of quantified volatile compounds from three different isolates per species. The different clusters were identified by visually evaluating the clustering. Pku—*Pichia kudriavzevii*, Pfe—*P. fermentans*, Pan—*P. anomala*, Pkl—*P. kluyveri*, Pme—*P. membranifaciens*, Spa—*Saccharomyces paradoxus*, Tde—*Torulaspora delbrueckii*, Mpu—*Metschnikowia pulcherrima*, Huv—*Hanseniaspora uvarum*, Apu—*Aureobasidium pullulans*, Cwi—*Cryptococcus wieringae*. Apu and Cwi yeasts are more common on unripe fruits and are represented by the green colour, Huv yeasts are the most common on medium-ripe and ripe fruits and are indicated by the blue colour, and the red colour represents yeasts, the most common on ripe fruits.

Table 1. Odour blends of eleven yeast species and controls sampled from a headspace by the SPME technique.

No	Compound	CAS No[3]	RI[4]	GR[5]	Control	*A. pullulans*	*C. wieringae*	*H. uvarum*	*P. kudriavzevii*	*P. fermentans*	*P. anomala*	*P. kluyveri*	*P. membranifac*	*S. paradoxus*	*T. delbrueckii*	*M. pulcherrima*
1	Ethyl acetate	141-78-6	>900	ES[6]	11[19] ± 6	0	0	2256 ± 657	2092 ± 150	1.6 ± 1.2	2816 ± 116	0	0	118 ± 76	202 ± 55	1586 ± 765
2	3-Methylbutanal *[1]	590-86-3	927	AL[7]	38 ± 8	0	0	0	0	0	0	0	0	15 ± 8	0	0
3	Ethanol	64-17-5	944	OH[8]	0	707 ± 106	48 ± 11	386 ± 34	425 ± 5	152 ± 23	39 ± 23	0	0	2594 ± 1683	468 ± 19	677 ± 44
4	Ethyl propionate	1105-37-3	953	ES	0	0	0.9 ± 0.3	139 ± 38	149 ± 9	0	147 ± 14	0	0	1.00 ± 0.05	63 ± 8	65 ± 25
5	Ethyl 2-methylpropionate	97-62-1	960	ES	0	0	0	0	0	0.9 ± 0.7	0	0	0	0	0	6.3 ± 3.6
6	Propyl acetate	109-60-4	969	ES	0	0	0	8.0 ± 0.7	0	0	99 ± 11	0	0	0	0	0
7	Methyl 2-methylbutanoate	868-57-5	989	ES	0	0	0.8 ± 0.2	0	0	0	0	0	0	0	0	0
8	Methyl 3-methylbutanoate	556-24-1	996	ES	0	0	0.7 ± 0.2	0	0	0	0	0	0	5.4 ± 2.0	12 ± 3	8.9 ± 2.1
9	2-Methylprop-1-yl acetate	110-19-0	994	ES	0	0	0	15 ± 5	91 ± 6	0	59 ± 19	0.4 ± 0.4	0	5.4 ± 2.0	12 ± 3	8.9 ± 2.1
10	Toluene *	108-88-3	1013	AR[9]	6.3 ± 1.0	1.9 ± 0.1	1.7 ± 0.2	1.3 ± 0.6	0.8 ± 0.3	2.2 ± 0.3	0.5 ± 0.1	1.8 ± 0.8	3.8 ± 0.2	3.8 ± 1.7 ns	2.2 ± 0.7	1.6 ± 0.1
11	Ethyl butanoate	105-54-4	1015	ES	0	0	0	0.8 ± 0.5	1.1 ± 0.1	0.5 ± 0.4	0.6 ± 0.1	0	0	8.8 ± 0.8	1.2 ± 0.3	0.5 ± 0.4
12	Ethyl 2-methylbutanoate	7452-79-1	1035	ES	0	3.6 ± 0.5	1.6 ± 0.3	1.3 ± 0.5	1.2 ± 0.2	4.3 ± 3.7	3.8 ± 0.9	0	0	6.6 ± 0.2	2.6 ± 0.2	3.5 ± 0.5
13	3-Methylbutyl formate	110-45-2	1053	ES	0	0	0	0	0	0	0	1.0 ± 0.4	0	0	0	0
14	Ethyl 3-methylbutanoate	108-64-5	1053	ES	0	0.4 ± 0.1	0.10 ± 0.02	0	0	3.9 ± 2.8	0	0	0	1.7 ± 0.4	0.7 ± 0.1	0.6 ± 0.2
15	Butyl acetate	123-86-4	1055	ES	0	0	0	12 ± 2	1.8 ± 0.2	0	46 ± 4	0	0	0	0.7 ± 0.4	0.15 ± 0.03
16	2-Methylpropyl propionate	540-42-1	1064	ES	0	0	0.4 ± 0.1	1.1 ± 0.2	3.7 ± 0.5	1.7 ± 0.9	3.6 ± 0.8	496 ± 496	0	0.8 ± 0.1	3.5 ± 0.4	0.4 ± 0.2
17	2-Methylpropyl 2-methylpropionate	97-85-8	1074	ES	0	0	0.2 ± 0.1	0	0	0.3 ± 0.1	0	0	0	0	0	0
18	2-Methylpropanol	78-83-1	1093	OH	0	114 ± 6	143 ± 12	53 ± 9	61 ± 7	106 ± 17	25 ± 1	56 ± 14	26 ± 03	97 ± 6	56 ± 06	102 ± 20
19	3-Methylbutyl acetate	123-92-2	1109	ES	5.4 ± 0.2	0	4.3 ± 1.1 ns	658 ± 88	1202 ± 167	117 ± 84	1934 ± 948	70 ± 50	0	188 ± 55	623 ± 252	49 ± 19
20	1-Butanol *	71-36-3	1144	OH	0.8 ± 0.03	2.9 ± 0.6	18 ± 5	4.4 ± 0.6	0.3 ± 0.2	0.4 ± 0.4 ns	1.6 ± 0.6	0	0	2.2 ± 0.3	0.4 ± 0.2 ns	4.5 ± 1.0
21	2-Methylpropyl 2-methylbutanoate	2445-67-2	1166	ES	0	0	0	0	0	0.7 ± 0.1	0	0	0	0	0	0
22	2-Heptanone	110-43-0	1171	KT[10]	0	10 ± 8	8.0 ± 1.7	0.5 ± 0.1	0	2.6 ± 1.1	0.2 ± 0.04	1.9 ± 0.9	3.3 ± 0.6	2.6 ± 0.6	4.8 ± 0.6	0.4 ± 0.4
23	3-Methylbutyl propionate	105-68-0	1178	ES	0.4 ± 0.02	0	0.3 ± 0.2 ns	84 ± 24	23 ± 4	20 ± 18	71 ± 8	2937 ± 1962	1.6 ± 0.7	11 ± 1	98 ± 11	1.4 ± 0.5
24	3-Methylbutyl 2-methylpropionate	2050-01-3	1184	ES	0	0	0	0	0	4.5 ± 1.8	0	0	0.5 ± 0.2	0.3 ± 0.1	12 ± 3	0
25	2-Methyl-1-butanol	137-32-6	1205	OH	0	80 ± 11	262 ± 64	129 ± 25	201 ± 61	209 ± 19	96 ± 11	154 ± 15	29 ± 4	172 ± 53	199 ± 37	193 ± 4
26	3-Methyl-1-butanol	123-51-3	1212	OH	0.8 ± 0.1	640 ± 86	889 ± 297	461 ± 84	874 ± 74	1876 ± 117	416 ± 21	568 ± 118	687 ± 169	1418 ± 93	1787 ± 61	644 ± 226
27	Ethyl hexanoate	123-66-0	1225	ES	0	0.8 ± 0.4	0	0.6 ± 0.1	0.7 ± 0.2	3.6 ± 3.6	0	0	0	16 ± 9	0.1 ± 0.03	0
28	Styrene *	100-42-5	1237	AR	11 ± 4	2.4 ± 2.0	19 ± 5 ns	0.5 ± 0.2	3.5 ± 2.9 ns[20]	5.2 ± 2.6 ns	19 ± 24 ns	14 ± 0.7 ns	7.2 ± 2.1 ns	9 ± 2 ns	10 ± 3 ns	11 ± 6 ns
29	3-Methylbutyl butanoate	106-27-4	1256	ES	0	0	0	0.4 ± 0.1	0	0.5 ± 0.02	0	0	0	0	0.4 ± 0.1	0
30	Hexyl acetate	142-92-7	1265	ES	0	0	0	0.45 ± 0.03	0	0	0	0	0	0	0	0
31	3-Methylbutyl 2-methylbutanoate	27625-35-0	1272	ES	0	0	0	0	0	3.0 ± 0.8	1.5 ± 0.3	0	0	0	0	0
32	2-hydroxy-3-butanone	513-86-0	1274	KT	0	2.2 ± 0.9	3.6 ± 0.4	10 ± 1	0	0	0	0.7 ± 0.1	0	6.7 ± 0.7	2.2 ± 0.2	17 ± 2
33	3-Methylbutyl 3-methylbutanoate	659-70-1	1288	ES	0	0	0.4 ± 0.2	0	0	8.9 ± 3.9	0.7 ± 0.1	0	0	0	0	0
34	2-Methylpentanol	105-30-6	1299	OH	0	0	13 ± 3	0	0	0	0	0	0	0	0	0
35	2-Methylpropyl isothiocyanate	591-82-2	1304	CN[11]	0	0	0	0	0	0	0	0	0	0	0	0.4 ± 0.1
36	2,5- Dimethyl pyrazine *	123-32-0	1314	PY[12]	9.1 ± 0.5	1.2 ± 0.2	22 ± 13 ns	3.8 ± 0.6	3.4 ± 0.4	4.8 ± 0.2	2.0 ± 0.6	6.0 ± 3.1 ns	7.6 ± 0.4 ns	6.4 ± 0.5 ns	3.9 ± 0.5	4.6 ± 0.6
37	2-Heptanol	543-49-7	1321	OH	0	2.0 ± 1.3	0	0.50 ± 0.01	0	0	0	0	0	0	0	0.9 ± 0.7
38	6-Methyl 5-hepten-2-one	110-93-0	1327	KT	0.4 ± 0.1	4.4 ± 3.8	103 ± 23	0.29 ± 0.03 ns	0	0.7 ± 0.1	0	0.8 ± 0.5 ns	2.8 ± 1.0	1.1 ± 0.2	0.5 ± 0.4 ns	0
39	Ethyl heptanoate	106-30-9	1331	ES	0	0	0	0	0.4 ± 0.1	0	0	0	0	0	0	0
40	1-Hexanol	111-27-3	1346	OH	0	0	0.5 ± 0.2	0	0	0	0	0	4.3 ± 1.9	0.3 ± 0.2	0	0
41	Heptyl acetate	112-06-1	1369	ES	0	0	0	0.7 ± 0.1	0	0	0	0	0	1.0 ± 0.4	0	0
42	2-Ethylhexyl acetate	103-09-3	1382	ES	0	0	0	0	0	0	0	0	0	5.2 ± 4.2	2.8 ± 0.6	0
43	Nonan-2-one	821-55-6	1383	KT	0	49 ± 28	13 ± 4	0.5 ± 0.2	0.3 ± 0.1	2.3 ± 1.0	1.7 ± 0.4	1.3 ± 0.2	2.9 ± 1.0	5.2 ± 4.2	2.8 ± 0.6	0
44	2,3,5-Trimethyl pyrazine *	14667-55-1	1397	PY	0.6 ± 0.1	0.6 ± 0.1 ns	0.8 ± 0.1 ns	0.58 ± 0.01 ns	0.6 ± 0.04 ns	0.6 ± 0.1 ns	0.6 ± 0.1 ns	0.4 ± 0.2 ns	0.7 ± 0.2 ns	0.8 ± 0.1 ns	0.5 ± 0.4 ns	0.60 ± 0.02 ns
45	Ethyl octanoate	106-32-1	1427	ES	0	0	0	4.1 ± 0.1	3.7 ± 0.6	0.2 ± 0.2	0	0	5.6 ± 5.2	0	0	0
46	Acetic acid	64-19-7	1439	FA[13]	0.3 ± 0.01	0.7 ± 0.01	4.5 ± 3.5 ns	69 ± 11	5.3 ± 3.9 ns	0.6 ± 0.1	7.9 ± 3.0	0	0.8 ± 0.1	0.79 ± 0.04	0.77 ± 0.04	1.1 ± 0.4
47	Unknown		1440		0	0	0	0	0	0	0	10 ± 4	0	0	0	0
48	1-Octen-3-ol	3391-86-4	1450	OH	0	0	0	0	0	0	0	0	0.12 ± 0.01	0	0	0
49	1-Heptanol	111-70-6	1454	OH	0	0	0	0.8 ± 0.2	0	0	0	0	0	0	0	0
50	3-Methylbutyl hexanoate	2198-61-0	1457	ES	0	0	0	0	0	0	0	0	0.25 ± 0.01	0.8 ± 0.2	0	0
51	6-Methyl-5-hepten-2-ol	1569-60-4	1460	OH	0	1.1 ± 0.7	12 ± 3	0	0	0	0	0	0	0	1.2 ± 0.3	0
52	MHMP [2]	40348-72-9	1462	ES	0	0	0	0	0	0	0	0	0	1.3 ± 0.9	0	0
53	Octyl acetate	112-14-1	1472	ES	0	0	0	0.3 ± 0.1	0	0	0	0	0	1.3 ± 0.9	0	0
54	2-Ethylhexanol	104-76-7	1483	OH	0.3 ± 0.03	1.5 ± 0.1	1.8 ± 0.2	0.8 ± 0.1	0.6 ± 0.1	0.6 ± 0.1	1.3 ± 0.4	0.8 ± 0.4 ns	1.7 ± 0.8	0.9 ± 0.1	1.1 ± 0.2	0.71 ± 0.03
55	Acetylfuran	1192-62-7	1496	FU[14]	0	0.3 ± 0.1	0.3 ± 0.2	0.6 ± 0.1	0.2 ± 0.1	0.6 ± 0.3	0.7 ± 0.3	0	0	0.5 ± 0.1	0.5 ± 0.1	0.19 ± 0.02
56	Benzaldehyde *	100-52-7	1505	AR, AL	9.9 ± 2.7	2.0 ± 1.6	0.4 ± 0.1	0.4 ± 0.1	0.28 ± 0.04	0.29 ± 0.04	0.18 ± 0.02	1.0 ± 0.2	0.20 ± 0.02	0.23 ± 0.01	0.20 ± 0.03	0.3 ± 0.1
57	2-Nonanol	628-99-9	1524	OH	0	0.05 ± 0.04	0	1.1 ± 0.2	0	0.2 ± 0.1	1.0 ± 0.4	0.9 ± 0.2	1.6 ± 0.3	2.3 ± 0.4	2.4 ± 0.5	0.3 ± 0.1 ns
58	Propionic acid	79-09-4	1526	FA	0.3 ± 0.04	0	2.9 ± 2.0 ns	6.7 ± 1.7	2.2 ± 0.1	0.7 ± 0.5 ns	9.9 ± 2.2	8.3 ± 1.6	20 ± 3	1.8 ± 1.2 ns	1.9 ± 0.8 ns	1.9 ± 1.0 ns
59	2-Methylpropionic acid	79-31-2	1557	FA	4.9 ± 0.5	0	12 ± 6 ns	34 ± 5	26 ± 2	13 ± 7	9.9 ± 2.2	8.3 ± 1.6	20 ± 3	1.8 ± 1.2 ns	1.9 ± 0.8 ns	1.9 ± 1.0 ns
60	2-Decanol	1120-06-5	1574	OH	0	0	0	0.7 ± 0.2	0	0	0	0.2 ± 0.1	1.1 ± 0.2	0.20 ± 0.02	0.8 ± 0.2	0
61	2-Undecanone	112-12-9	1594	KT	0	1.1 ± 0.6	0.6 ± 0.1	0	0	0	0	0.3 ± 0.2	1.1 ± 0.2	0.2 ± 0.1	0.21 ± 0.04	0.3 ± 0.2
62	Butanoic acid	107-92-6	1614	FA	1.0 ± 0.1	0.2 ± 0.04	2.6 ± 1.8 ns	8.5 ± 1.7	2.7 ± 0.1	0.7 ± 0.5 ns	9.0 ± 2.1	0.3 ± 0.2	1.1 ± 0.1 ns	0.2 ± 0.1	0.21 ± 0.04	0.3 ± 0.2
63	Acetophenone *	98-86-2	1625	AR, KT	1.8 ± 0.2	0.7 ± 0.1	0.7 ± 0.1	1.1 ± 0.1	0.8 ± 0.04	0.4 ± 0.4	0.9 ± 0.1	1.1 ± 0.8 ns	1.3 ± 0.1 ns	1.6 ± 1.0 ns	1.5 ± 0.1 ns	0.6 ± 0.1
64	Ethyl decanoate	110-38-3	1632	ES	0	0	0	4.0 ± 1.2	0.9 ± 0.1	0.5 ± 0.5	0.2 ± 0.1	0.2 ± 0.1	0	2.3 ± 2.2	1.7 ± 0.4	0
65	2-Furanmethanol *	98-00-0	1649	OH	1.8 ± 0.1	1.7 ± 1.5 ns	1.2 ± 0.4 ns	0	0	0	0	0	0	0	1.7 ± 0.4	0

Table 1. *Cont.*

No	Compound	CAS No[3]	RI[4]	GR[5]	Control	A. pullulans	C. wieringae	H. uvarum	P. kudriavzevii	P. fermentans	P. anomala	P. kluyveri	P. membranifac	S. paradoxus	T. delbrueckii	M. pulcherrima
66	3-Methylbutanoic acid	503-74-2	1658	FA	20 ± 2	0.26 ± 0.02	58 ± 39	261 ± 66	136 ± 9	37 ± 25 ns	44 ± 12	21 ± 6 ns	71 ± 14	6.8 ± 5.6	4.5 ± 2.9	4.6 ± 4.1
67	3-Hydroxypropyl methylsulphide	505-10-2	1700	SU [15]	0	0.18 ± 0.03	0.4 ± 0.1	0.33 ± 0.04	2.0 ± 1.0	1.3 ± 0.7	6.6 ± 1.1	0.9 ± 0.4	0	0.8 ± 0.2	7.5 ± 2.2	0.5 ± 0.2
68	Geranyl acetate	16409-44-2	1764	TE [16], ES	0	0	0	0.7 ± 0.2	0	0	0	0	0	0	0	0
69	Methoxy-phenyl-oxime *	67160-14-9	1768	IM [17]	3.1 ± 0.7	1.2 ± 0.2	2.9 ± 0.4 ns	0.6 ± 0.3	1.3 ± 0.1	0.7 ± 0.2	0.43 ± 0.02	1.7 ± 0.4 ns	1.0 ± 0.3	0.6 ± 0.2	0.6 ± 0.2	1.4 ± 0.4
70	Ethyl 2-phenylacetate	119-43-7	1767	AR, ES	0	0	0	0	0.6 ± 0.2	0	0.3 ± 0.1	0	0	0	0	0
71	2-Phenylethyl acetate	103-45-7	1794	AR, ES	0	0	0	148 ± 13	353 ± 48	2.1 ± 1.6	88 ± 8	7.8 ± 6.3	0	8.0 ± 3.4	65 ± 17	11 ± 8
72	Hexanoic acid	122-70-3	1816	AR, ES	0	0	0.8 ± 0.1	0	0.4 ± 0.1	0.3 ± 0.2	0	0	0	0.6 ± 0.1	0	0
73	Ethyl dodecanoate	106-33-2	1823	ES	0	0	0	1.7 ± 1.0	0.1 ± 0.03	0.3 ± 0.1	0	0	0	0	0	0
74	Geraniol	106-24-1	1842	TEOH	0	0	0.5 ± 0.1	0	0	0	0.5 ± 0.1	0	0	0	0	0
75	3-Methylbutyl decanoate	2306-91-4	1843	ES	0	0	0	0	0	0	0	0	0	0.2 ± 0.1	0	0
76	Phenyl methanol	100-51-6	1854	AR, OH	0.7 ± 0.04	0	0.7 ± 0.2 ns	5.1 ± 0.9	2.9 ± 0.5	1.8 ± 1.0 ns	1.2 ± 0.3 ns	1.2 ± 0.2	0.6 ± 0.2 ns	0.7 ± 0.2 ns	1.1 ± 0.2 ns	2.2 ± 0.3
77	2-Phenylethyl propionate	122-70-3	1862	AR, ES	0	0	0	18 ± 4	6.3 ± 0.7	0	1.1	1.1 ± 0.9	0	0	172 ± 38	0
78	2-Phenyl ethanol	60-12-8	1890	AR, OH	0.2 ± 0.04	362 ± 102	20 ± 2	336 ± 46	567 ± 19	865 ± 206	235 ± 17	432 ± 57	814 ± 17	271 ± 83	448 ± 87	490 ± 320
79	2-Phenylethyl butanoate	103-52-6	1949	AR, ES	0	0	0	0	0	0	0	0	0	0	0.2 ± 0.1	0
80	2-Phenylethyl 2-methylbutanoate	24817-51-4	1955	AR, ES	0	0	0	0	0	0	0	0	0	0	4.3 ± 1.8	0
81	2-Phenylethyl 3-methylbutanoate	140-26-1	1973	AR, ES	0	0	0	0	0	0	0	0	0	0	0.11 ± 0.02	0
82	Phenol *	87-66-1	1979	AR, OH	0.4 ± 0.04	0.38 ± 0.01 ns	0.5 ± 0.1 ns	0.15 ± 0.01	0.07 ± 0.03	0.4 ± 0.1 ns	0.5 ± 0.1 ns	0.2 ± 0.1 ns	0.6 ± 0.2 ns	0.49 ± 0.02	0.3 ± 0.1 ns	0.30 ± 0.04 ns
83	Pentadecan-2-one	2345-28-0	2010	KT	0	0	0	0	0.2 ± 0.1	0	0	0	1.2 ± 0.3	0.08 ± 0.06	0	0
84	Octanoic acid	124-07-2	2065	FA	0.3 ± 0.02	1.0 ± 0.2	0	0.3 ± 0.1 ns	0.6 ± 0.1	0	0	0	0.27 ± 0.03 ns	0.20 ± 0.01	0	0
85	3-Methylbutyl dodecanoate	6309-51-9	2089	ES	0	0	0	2.3 ± 0.4	0	0	0	0	0	0	0	0
86	Hexadecan-2-one	18787-63-8	2118	KT	0	0	0	0	0.4 ± 0.1	0	0	0	0.6 ± 0.2	0	0	0
87	gamma-Decalactone	706-14-9	2125	LA [18]	0	0	0	0	0	0	0	0	0	0	0	0
88	6-Pentyl-5,6 dihydro-2H-pyran-2-one	54814-64-1	2222	LA	0	0	0	0	0	2.3 ± 1.1	0	1.2 ± 0.2	5.2 ± 1.2	0	1.3 ± 0.8	0
89	Heptadecan-2-one	2922-51-2	2230	KT	0	0	0	0	2.1 ± 0.6	0	0	0	0	0	0	0
90	Farnesyl acetate	4128-17-0	2247	TE, ES	0	0	0	1.1 ± 0.2	0	0	0	0	0	0	0	0
91	Farnesol	4602-84-0	2287	TE, OH	0	0	0	1.7 ± 0.3	0	0	0	0	0	0	0	0
92	4-Hydroxy-6-pentyloxan-2-one	36555-25-6	2456	LA	0	0	0	0	0	1.2 ± 0.4	0	0	0	0	0	0

[1] Compounds indicated by star mark were excluded from multivariate analysis, [2] MHMP-methyl 2-hydroxy-4-methylpentanoate, [3] CAS No–chemical abstract service number, [4] RI–retention index (DB-Wax fused silica capillary column 30 m × 0.25 mm i.d., 0.25 μm film thickness), [5] GR—group of a chemical compound, [6] ES—ester, [7] AL—aldehyde, [8] OH—alcohol, [9] KT—ketone, [10] AR—aromatic, [11] CN—isothiocyanoate, [12] PY—pyrazine, [13] FA—fatty acid, [14] FU—furane, [15] sulphide, [16] TE—terpenoid, [17] IM—imine, [18] LA—lactone, [19] mean ± standard error of the mean (means are the absolute amounts expressed as areas under the chromatographic peaks and have to be read as numbers times 1000), [20] ns—not significantly different compare to control (Nonparametric Mann–Whitney U test, *p* < 0.05), three different isolates of each yeast species have been used; control samples were obtained by collecting background volatiles from YPD-agar plates without yeast.

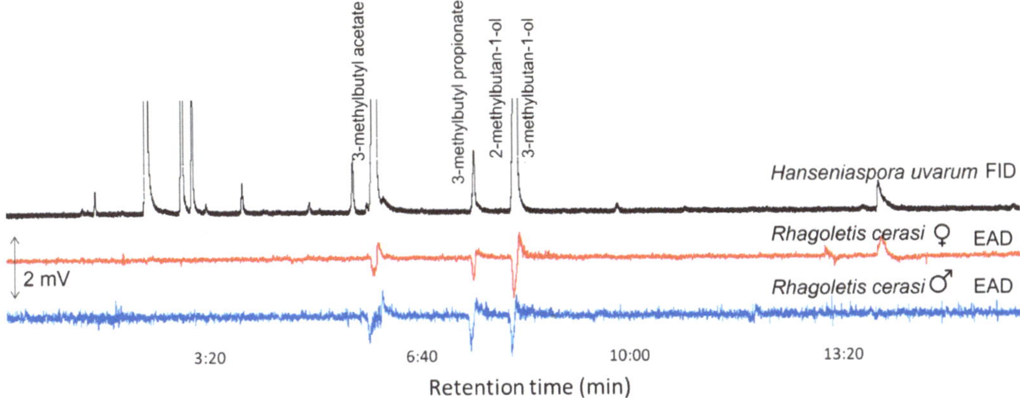

Figure 4. GC-EAD response of male and female *Rhagoletis cerasi* to headspace volatiles of *Hanseniaspora uvarum* yeast. FID, flame ionisation detector; EAD, electroantennographic detector; DB-Wax capillary column (30 m × 0.25 mm × 0.25 μm; Agilent Technologies, Santa Clara, CA, USA.)

Table 2. Electroantennographic responses of *R. cerasi* flies to volatiles present in the headspace of five yeast species.

Yeast Species	3-MBA [1]		3-MBP [2]		2-MBOH [3]		3-MBOH [4]	
	Male	Female	Male	Female	Male	Female	Male	Female
Aureobasidium pullulans	- [5]				3 (3)	3 (3)	3 (3)	3 (3)
Cryptococcus wieringae					4 (4)	3 (3)	4 (4)	3 (3)
Hanseniaspora uvarum	4 [6] (4) [7]	4 (4)	4 (4)	3 (4)	4 (4)	4 (4)	4 (4)	4 (4)
Pichia kudriavzevii	5 (7)	4 (4)	5 (7)	4 (4)	7 (7)	4 (4)	7 (7)	4 (4)
Metschnikowia pulcherrima	1 (3)	3 (4)			3 (3)	4 (4)	3 (3)	4 (4)
Total	10 (14)	11 (12)	9 (11)	7 (8)	21 (21)	18 (18)	21 (21)	18 (18)

[1] 3-Methylbutyl acetate, [2] 3-methylbutyl propionate, [3] 2-methyl-1-butanol, [4] 3-methyl-1-butanol, [5] compound was not detected in the headspace, [6] Number of antennae responded, [7] Number of antennae tested.

Unripe fruits associated with yeast-like fungi, *A. pullulans*, and *C. wieringae* yeasts produced two alcohols, 2-methyl-1-butanol and 3-methyl-1-butanol, which elicited antennae responses of *R. cerasi* flies. *H. uvarum*, the most common yeast species inhabiting medium-ripe fruits as well *P. kudriavzevii* attributed to microbiota of ripe fruits in addition to two EAD active alcohols produced two esters, 3-methybutyl acetate and 3-methylbutyl propionate, which evoked antennae responses. Ripe fruits associated yeasts, *M. pulcherrima* released three EAD active volatiles 3-methybutyl acetate, 2-methyl-1-butanol, and 3-methyl-1-butanol (Table 2).

3.3. EAG Dose–Response

The antennographic responses of females to 3-methybutyl acetate and 3-methylbutyl propionate at the moderate and the higher doses were of significantly higher amplitudes compared to the responses of males (Figure 5a,b). Females showed a higher sensitivity to 2-methyl-1-butanol than males at all doses tested, and significant differences were revealed at the doses 10^{-5}, 10^{-3}, and 10^{-2} mg (Figure 5c). Significant stronger responses of females than males were recorded to 3-methyl-1-butanol at the dose 10^{-4} mg, while antennae stimulation with the higher doses revealed stronger responses of males compared to the responses of females (Figure 5d).

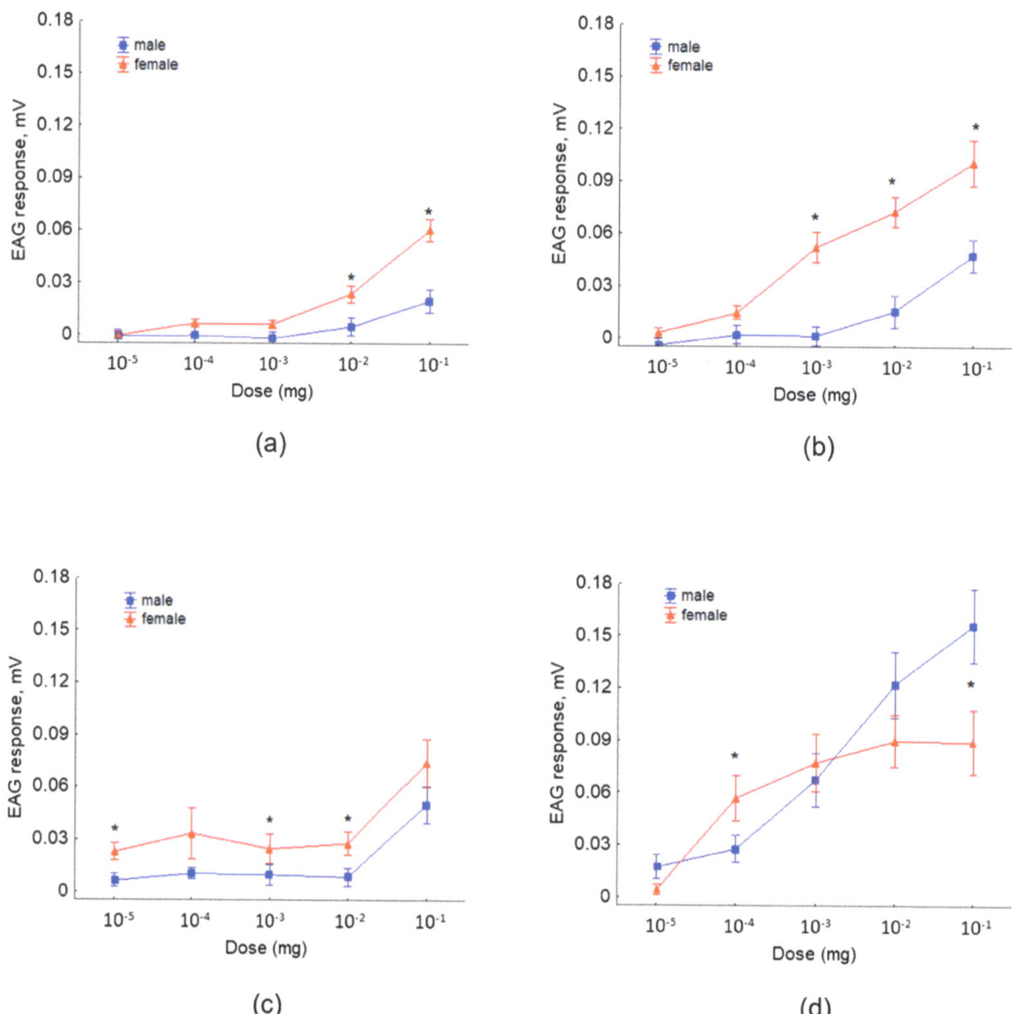

Figure 5. EAG responses (mean amplitude $\pm$ standard error (SE), mV) of *R. cerasi* male and female antennae to different doses (10^{-5} to 10^{-1} mg) of synthetic: (**a**) 3-methylbutyl acetate, (**b**) 3-methylbutyl propionate, (**c**) 2-methyl-1-butanol, and (**d**) 3-methyl-1-butanol. The asterisk denotes significant differences in EAG responses between sexes (paired *t* test, $p < 0.05$); each EAD test was replicated 13 times and each antenna used was from a different fly.

3.4. Behavioural Tests in Olfactometer

In the two-choice tests, *R. cerasi* males showed no preference to 3-methylbutyl acetate at the dose 10^{-2} mg, while females significantly avoided the olfactometer arm bearing 3-methylbutyl acetate at the dose 10^{-3} mg. Fruit flies of both sexes significantly preferred 3-methylbutyl propionate provided at the dose of 10^{-2} mg for males and at the dose 10^{-3} mg for females (Figure 6). Flies of both sexes do not discriminate between olfactory arms bearing 2-methyl-1-butanol. Males chose 3-methyl-1-butanol at the dose 10^{-2} mg and showed no preference to a 10 times lower dose. Females preferred the alcohol at the dose of 10^{-4} mg, while a 10 times higher dose obliterated the preference.

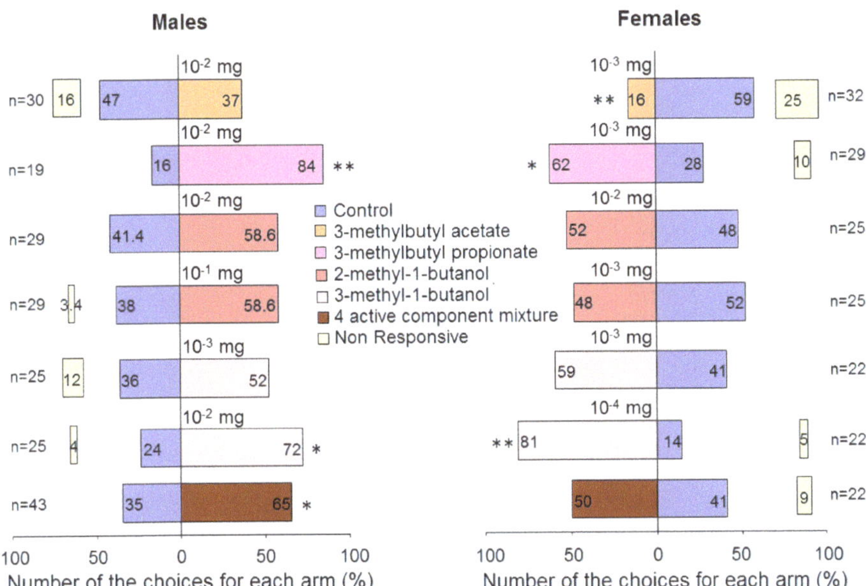

Figure 6. Behavioural responses of *Rhagoletis cerasi* flies in Y-tube olfactometer to EAD active volatiles. The mixture consisted of 3-methylbutyl acetate 0.55 mg, 3-methylbutyl propionate 0.05 mg, 2-methyl-1-butanol 0.012 mg, and 3-methyl-1-butanol 0.28 mg. n–number of flies tested. Stars denote significant differences in behavioural responses of flies between stimulus and control: * $p < 0.05$ and ** $p < 0.01$ by χ^2 test.

4. Discussion

Yeasts produce a wide range of volatile compounds [19]. Common volatiles at large amounts emitted by different yeast species originate from primary metabolism pathways and these compounds are considered as side products or waste compounds [20]. Ethanol and ethyl acetate are the best-known examples of such volatile metabolites. Our analysis showed that yeasts of *S. paradoxus* species released high amounts of ethanol while *P. anomala*, *P. kudriavzevii*, *H. uvarum*, and *M. pulcherrima* yeasts produced the largest amounts of ethyl acetate, an ester derived from ethanol. Ethanol or ethyl acetate were the major components in the volatile bouquets of these yeast species. Interestingly, we did not detect ethanol and ethyl acetate in the headspaces of *P. kluyveri* and *P. membranifaciens* yeasts. More structurally diverse volatiles released by yeasts are derived from amino or fatty acid synthesis and degradation as well as from terpene biosynthetic pathways [19,20]. Analysis of volatile profiles of eleven yeast species isolated from sweet and sour cherry fruits showed that esters and alcohols were the most numerous groups of volatiles.

The multidimensional scaling analysis revealed that volatile blends of all yeast species were clearly separated from each other, therefore, the amount and composition of volatiles characterised yeast species. Our results are consistent with previous findings that volatile profiles of microorganisms are species-specific, reflecting specific metabolic activities of the particular microorganisms [19].

The ecological role of microbial volatiles falls in two major groups: they are important as semiochemicals mediating information flow at intra- and inter-specific levels; and function as promoters or inhibitors of microbial growth [21–23]. Yeast volatiles play an essential role in distribution of yeast from one habitat to another by attracting vector insects. Our data showed that 3-methybutyl acetate, 3-methylbutyl propionate, 2-methyl-1-butanol, and 3-methyl-1-butanol emitted by yeasts populating *P. avium* and *P. cerasus* fruits elicited electroantennographic responses in both sexes of *R. cerasi* fruit flies, a common pest of cherry fruits. Recently, it was reported that ten volatiles emitted by *P. kudriavzevii* yeast species including 3-methybutyl acetate, 3-methylbutyl propionate, and 3-methyl-1-butanolelicited antennal responses in males and females of closely related *R. batava* flies while 2-methyl-1-butanol was not EAD active [24].

Females showed the higher sensitivity to 3-methybutyl acetate, 3-methylbutyl propionate, and 2-methyl-1-butanol than males and the more pronounced difference in antennal sensitivity was recorded at higher doses of both esters. As far as we know, higher antennae sensitivity of females to microbial volatiles has not been reported in other tephritid fruit flies.

Two EAD active compounds 3-methylbutyl propionate and 3-methyl-1-butanol stimulated both sexes of *R. cerasi* flies to choose the olfactometer arm bearing these volatiles versus control without stimulus, while 2-methyl-1-butanol did not significantly affect the choice of flies. 3-Methyl-1-butanol is the end product of degradation of the amino acid leucine [20] and is found in volatile bouquets of many yeast species [20,25]. In our samplings, 3-methyl-1-butanol at various amounts was isolated from the emissions of all eleven yeast species. As an attractant, 3-methyl-1-butanol functions in five dipteran species [26] including one tephritid species [25]. Attractiveness of 3-methylbutyl propionate has not been reported for any dipteran species.

The two-choice test showed that *R. cerasi* females avoided the olfactometer hand bearing 3-methybutyl acetate at the dose 10^{-2} mg, while males did not significantly discriminate stimulus versus the control at even higher 10^{-1} mg dose. *R. cerasi* females prefer to lay eggs in cherries at the stage of colour change from green to yellow [27] (i.e., just at the beginning of ripening). At the early maturation stage of fruits, yeast-like fungi from *Aureobasidium* genus, yeasts from *Cryptococcus*, *Taphrina*, *Cladosporium*, and some other genus are more prevalent compared to yeast from *Pichia*, *Metschnikowia*, *Saccharomyces*, and *Torulaspora* genus inhabiting ripe fruits [3,7,28]. Analysis of chromatographic profiles of yeast emitted volatiles revealed that unripe fruits associated with yeast-like fungi, *Aureobasidium pullulans*, and *Cryptococcus wieringae* yeasts produced none or very low amounts of 3-methylbutyl acetate, acting as a repellent to *R. cerasi* females. 3-Methybutyl acetate is a ubiquitous volatile in the odour bouquets of ripe and fermenting fruit-yeast complexes [29], which is in agreement with our data showing that fermenting yeasts emitted large amounts of this acetate. Contrary to *R. cerasi*, many *Drosophila* species oviposit in ripe fruits and berries and 3-methylbutyl acetate released by ripe fruits associated yeasts functions as an attractant [29,30].

Ammonium acetate is reported as the most efficient food attractant for *R. cerasi* flies [31]. The number of yeast-based commercially available attractive lures such as brewer's yeast waste, baker's yeast (*Saccharomyces cerevisiae*), and Torula yeast (*Candida utilis*) have been used in tephritid pest control programs [32], however, none of the formulations tested showed a potential to control *R. cerasi* fruit flies. A possible explanation is that all lures released high amounts of 3-methybutyl acetate, the repellent to *R. cerasi* fruit flies. Our data suggest that yeast species selected for an efficient lure to target *R. cerasi* pests should release high amounts of 3-methylbutyl propionate and/or 3-methyl-1-butanol and do not emit 3-methylbutyl acetate. Moreover, our data provide a background for the application of behaviour modifying semiochemicals in push-pull and other integrated pest management techniques to control *R. cerasi* fruit flies.

5. Conclusions

The odour blends emitted by yeasts were species-specific. 3-Methybutyl acetate, 3-methylbutyl propionate, 3-methylbutanol, and 2-methyl-1-butanol released by yeasts populating *P. avium* and *P. cerasus* fruits elicited electroantennographic responses and modulated behaviour of *R. cerasi* fruit flies, a common pest of cherry fruits. Therefore, these olfactory and behaviourally active compounds show potential for use in integrated pest management techniques to control *R. cerasi* fruit flies.

Author Contributions: Conceptualisation, R.M., E.S. and V.B.; Methodology, D.A., S.R., V.A. and R.S.; Software, R.M., S.R. and L.B.-Č.; Validation, V.A., S.R., L.B.-Č., E.S. and R.M.; Formal analysis, V.A., S.R., J.B., L.B.-Č., E.S. and R.M.; Investigation, D.A., S.R., V.A., R.S. and L.B.-Č.; Resources, R.M., V.B. and E.S.; Data curation, V.A., S.R., L.B.-Č. and E.S.; writing—original draft preparation, R.M.; writing—review and editing, R.M., S.R., L.B.-Č., V.B. and E.S.; visualisation, R.M., S.R. and L.B.-Č.; Supervision, R.M. and V.B.; Project administration, R.M. and L.B.-Č.; Funding acquisition, R.M. All authors have read and agreed to the published version of the manuscript.

Funding: This research was funded by EUROPEAN SOCIAL FUND, grant number 09.3.3-LMT-K-712-01-0099 under grant agreement with the Research Council of Lithuania (LMTLT).

Institutional Review Board Statement: Not applicable.

Informed Consent Statement: Not applicable.

Data Availability Statement: Not applicable.

Acknowledgments: We thank A. Amšiejus and L. Levulytė for the kind permission to collect puparia in their cherry orchards. Thanks are also given to I. Vepštaitė-Monstavičė and J. Lukša for providing cherry fruits.

Conflicts of Interest: The authors declare no conflict of interest. The funders had no role in the design of the study; in the collection, analyses, or interpretation of data; in the writing of the manuscript, or in the decision to publish the results.

Appendix A

Table A1. List of yeast isolates analysed in this study.

Yeast Species	Strain	GenBank Reference	Identity (%)
Aureobasidium pullulans	PA-4-13	MN400109	100
	PC-5-28	HQ909088	99.6
	PC-4-8	MW361317	99.8
Cryptococcus wieringae	PC-4-6	KF981864	100
	PC-4-9	KF981864	100
	PA-5-41	KF981864	100
Metschnikowia pulcherrima	PC-4-5	MF574308	100
	PA-5-13	KT029787	99.7
	PA-5-47	MK352050	100
Hanseniaspora uvarum	PA-5-27	MF062209	100
	PC-3-8	MK352020	99.9
	PC-5-12	KY103573	100
Pichia kudriavzevii	PA-1-15	MF685423	100
	PC-4-25	MF685411	100
	PC-4-19.3	MG015972	99.6
Pichia fermentans	PC-5-47	FJ713081	100
	PC-4-19.1	MF462777	100
	PA-4-39	KY104537	100
Pichia kluyveri	PA-4-30	JX103190	100
	PA-4-34	KY108823	99.8
	PC-4-35	KC510043	100
Pichia membranifaciens	PC-3-36	JX188207	100
	PC-2-71	FJ231461	100
	PA-2-55	JX188207	99.6
Saccharomyces paradoxus	PA-5-14	FJ713072	100
	PC-3-33	FJ713072	100
	PC-3-59	KY105204	99.9

Table A1. *Cont.*

Yeast Species	Strain	GenBank Reference	Identity (%)
Torulaspora delbrueckii	PA-4-16	KY105641	100
	PA-5-17	MN371902	99.8
	PC-2-42	MK352012	100
Pichia anomala	PC-5-5	KJ527050	100
	PA-5-18	MH248067	100
	PC-5-24	MK343437	100

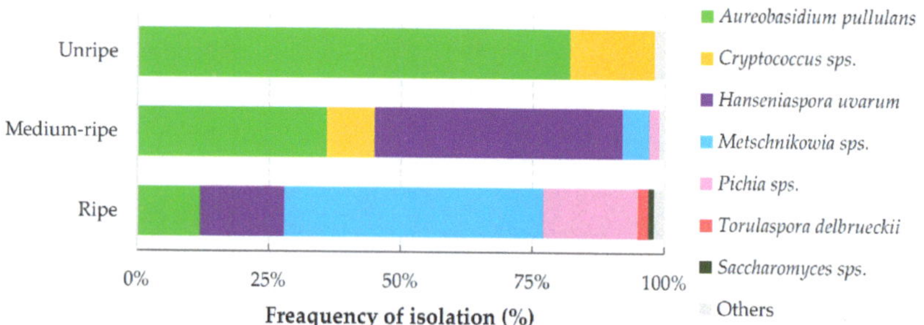

Figure A1. Frequency of yeast isolation among cherry samples at different ripening stages. *C. wieringae* dominated among *Cryptococcus* genus species; *M. pulcherrima* frequently occurred among *Metschnikowia* yeasts; *Pichia* genus were represented by *P. anomala*, *P. kudriavzevii*, *P. kluyveri*, *P. fermentans*, and *P. membranifaciens* yeasts; *S. paradoxus* occurred among *Saccharomyces* yeasts; others represented yeasts occurred in low frequency.

References

1. Starmer, W.T.; Lachance, M.-A. Chapter 6-Yeast Ecology. In *The Yeasts*, 5th ed.; Kurtzman, C.P., Fell, J.W., Boekhout, T., Eds.; Elsevier: London, UK, 2011; pp. 65–83. [CrossRef]
2. Allard, S.M.; Ottesen, A.R.; Brown, E.W.; Micallef, S.A. Insect exclusion limits variation in bacterial microbiomes of tomato flowers and fruit. *J. Appl. Microbiol.* **2018**, *125*, 1749–1760. [CrossRef] [PubMed]
3. Barata, A.; Malfeito-Ferreira, M.; Loureiro, V. The microbial ecology of wine grape berries. *Int. J. Food Microbiol.* **2012**, *153*, 243–259. [CrossRef] [PubMed]
4. Singh, P.; Gobbi, A.; Santoni, S.; Hansen, L.H.; This, P.; Peros, J.P. Assessing the impact of plant genetic diversity in shaping the microbial community structure of *Vitis vinifera* phyllosphere in the Mediterranean. *Front. Life Sci.* **2018**, *11*, 35–46. [CrossRef]
5. Vepstaite-Monstavic, I.; Luksa, J.; Staneviclene, R.; Strazdaite-Zieliene, Z.; Yurchenko, V.; Serva, S.; Serviene, E. Distribution of apple and blackcurrant microbiota in Lithuania and the Czech Republic. *Microbiol. Res.* **2018**, *206*, 1–8. [CrossRef] [PubMed]
6. Kioroglou, D.; Kraeva-Deloire, E.; Schmidtke, L.M.; Mas, A.; Portillo, M.C. Geographical origin has a greater impact on grape berry fungal community than grape variety and maturation state. *Microorganisms* **2019**, *7*, 669. [CrossRef]
7. Luksa, J.; Vepstaite-Monstavice, I.; Apsegaite, V.; Blazyte-Cereskiene, L.; Staneviciene, R.; Strazdaite-Zieliene, Z.; Ravoityte, B.; Aleknavicius, D.; Buda, V.; Mozuraitis, R.; et al. Fungal microbiota of sea buckthorn berries at two ripening stages and volatile profiling of potential biocontrol yeasts. *Microorganisms* **2020**, *8*, 456. [CrossRef]
8. Stefanini, I. Yeast-insect associations: It takes guts. *Yeast* **2018**, *35*, 315–330. [CrossRef] [PubMed]
9. Blackwell, M. Made for each other: Ascomycete yeasts and insects. *Microbiol. Spectr.* **2017**, *5*, 18. [CrossRef] [PubMed]
10. Janson, E.M.; Stireman, J.O.; Singer, M.S.; Abbot, P. Phytophagous insect-microbe mutualisms and adaptive evolutionary diversification. *Evolution* **2008**, *62*, 997–1012. [CrossRef]
11. Dobzhansky, T.; Cooper, D.M.; Phaff, H.J.; Knapp, E.P.; Carson, H.L. Studies on the ecology of *Drosophila* in the Yosemite region of California Differential attraction of species of *Drosophila* to different species of yeasts. *Ecology* **1956**, *37*, 544–550. [CrossRef]
12. Date, P.; Dweck, H.K.M.; Stensmyr, M.C.; Shann, J.; Hansson, B.S.; Rollmann, S.M. Divergence in olfactory host plant preference in *D-mojavensis* in response to cactus host use. *PLoS ONE* **2013**, *8*, 10. [CrossRef] [PubMed]
13. Keesey, I.W.; Knaden, M.; Hansson, B.S. Olfactory specialization in *Drosophila suzukii* supports an ecological shift in host preference from rotten to fresh fruit. *J. Chem. Ecol.* **2015**, *41*, 121–128. [CrossRef]
14. Scheidler, N.H.; Liu, C.; Hamby, K.A.; Zalom, F.G.; Syed, Z. Volatile codes: Correlation of olfactory signals and reception in *Drosophila*-yeast chemical communication. *Sci. Rep.* **2015**, *5*, 14059. [CrossRef]
15. Holighaus, G.; Rohlfs, M. Fungal allelochemicals in insect pest management. *Appl. Microbiol. Biotechnol.* **2016**, *100*, 5681–5689. [CrossRef]

16. Staneviciene, R.; Luksa, J.; Strazdaite-Zieliene, Z.; Ravoityte, B.; Losinska-Siciuniene, R.; Mozuraitis, R.; Serviene, E. Mycobiota in the carposphere of sour and sweet cherries and antagonistic features of potential biocontrol yeasts. *Microorganisms* **2021**, *9*, 1423. [CrossRef] [PubMed]

17. Buda, V.; Blazyte-Cereskiene, L.; Radziute, S.; Apsegaite, V.; Stamm, P.; Schulz, S.; Aleknavicius, D.; Mozuraitis, R. Male-produced (-)-delta-heptalactone, pheromone of fruit fly *Rhagoletis batava* (Diptera: Tephritidae), a sea buckthorn berries pest. *Insects* **2020**, *11*, 138. [CrossRef] [PubMed]

18. Hare, D.J. Bioassay methods with terrestrial invertebrates. In *Methods in Chemical Ecology: Bioassay Methods*; Millar, J.G., Haynes, K.F., Eds.; Kluwer Academic Publishers: Norwell, MA, USA, 2000; pp. 212–270.

19. Lemfack, M.C.; Nickel, J.; Dunkel, M.; Preissner, R.; Piechulla, B. mVOC: A database of microbial volatiles. *Nucleic Acids Res.* **2014**, *42*, D744–D748. [CrossRef] [PubMed]

20. Ebert, B.E.; Halbfeld, C.; Blank, L.M. Exploration and Exploitation of the Yeast Volatilome. *Curr. Metab.* **2017**, *5*, 102–118. [CrossRef]

21. Schmidt, R.; Cordovez, V.; de Boer, W.; Raaijmakers, J.; Garbeva, P. Volatile affairs in microbial interactions. *ISME J.* **2015**, *9*, 2329–2335. [CrossRef]

22. Becher, P.G.; Hagman, A.; Verschut, V.; Chakraborty, A.; Rozpedowska, E.; Lebreton, S.; Bengtsson, M.; Flick, G.; Witzgall, P.; Piskur, J. Chemical signaling and insect attraction is a conserved trait in yeasts. *Ecol. Evol.* **2018**, *8*, 2962–2974. [CrossRef]

23. Kai, M.; Haustein, M.; Molina, F.; Petri, A.; Scholz, B.; Piechulla, B. Bacterial volatiles and their action potential. *Appl. Microbiol. Biotechnol.* **2009**, *81*, 1001–1012. [CrossRef]

24. Mozuraitis, R.; Aleknavicius, D.; Vepstaite-Monstavice, I.; Staneviciene, R.; Emami, S.N.; Apsegaite, V.; Radziute, S.; Blazyte-Cereskiene, L.; Serviene, E.; Buda, V. Hippophae rhamnoides berry related *Pichia kudriavzevii* yeast volatiles modify behaviour of *Rhagoletis batava* flies. *J. Adv. Res.* **2020**, *21*, 71–77. [CrossRef] [PubMed]

25. Vitanovic, E.; Aldrich, J.R.; Boundy-Mills, K.; Cagalj, M.; Ebeler, S.E.; Burrack, H.; Zalom, F.G. Olive Fruit Fly, *Bactrocera oleae* (Diptera: Tephritidae), attraction to volatile compounds produced by host and insect-associated yeast strains. *J. Econ. Entomol.* **2020**, *113*, 752–759. [CrossRef] [PubMed]

26. Davis, T.S.; Landolt, P.J. A survey of insect assemblages responding to volatiles from a ubiquitous fungus in an agricultural landscape. *J. Chem. Ecol.* **2013**, *39*, 860–868. [CrossRef]

27. Daniel, C.; Grunder, J. Integrated management of European cherry fruit fly *Rhagoletis cerasi* (L.): Situation in Switzerland and Europe. *Insects* **2012**, *3*, 956–988. [CrossRef]

28. Ghanbarzadeh, B.; Sampiao, J.P.; Arzanlou, M. Grape maturity significantly influences yeast community on grape berries: Basidiomycetous yeasts are dominant colonizers of immature grape berries in northwestern Iran. *Nova Hedwig.* **2021**, *113*, 191–206. [CrossRef]

29. Revadi, S.; Vitagliano, S.; Stacconi, M.V.R.; Ramasamy, S.; Mansourian, S.; Carlin, S.; Vrhovsek, U.; Becher, P.G.; Mazzoni, V.; Rota-Stabelli, O.; et al. Olfactory responses of *Drosophila suzukii* females to host plant volatiles. *Physiol. Entomol.* **2015**, *40*, 54–64. [CrossRef]

30. Cha, D.H.; Gill, M.A.; Epsky, N.D.; Werle, C.T.; Adamczyk, J.J.; Landolt, P.J. From a non-target to a target: Identification of a fermentation volatile blend attractive to *Zaprionus indianus*. *J. Appl. Entomol.* **2015**, *139*, 114–122. [CrossRef]

31. Katsoyannos, B.I.; Papadopoulos, N.T.; Stavridis, D. Evaluation of trap types and food attractants for *Rhagoletis cerasi* (Diptera: Tephritidae). *J. Econ. Entomol.* **2000**, *93*, 1005–1010. [CrossRef]

32. Biasazin, T.D.; Chernet, H.T.; Herrera, S.L.; Bengtsson, M.; Karlsson, M.F.; Lemmen-Lechelt, J.K.; Dekker, T. Detection of volatile constituents from food lures by Tephritid fruit flies. *Insects* **2018**, *9*, 119. [CrossRef]

Journal of *Fungi*

Article

Effect of *Issatchenkia terricola* WJL-G4 on Deacidification Characteristics and Antioxidant Activities of Red Raspberry Wine Processing

Hongying He [1], Yuchen Yan [1], Dan Dong [1], Yihong Bao [1,2], Ting Luo [3], Qihe Chen [4,*] and Jinling Wang [1,2,*]

1   School of Forestry, Northeast Forestry University, No. 26, Hexing St., Harbin 150040, China; Nuyoah0820@163.com (H.H.); yanyuchen69619@nefu.edu.cn (Y.Y.); 957331735@nefu.edu.cn (D.D.); baoyihong@163.com (Y.B.)
2   Key Laboratory of Forest Food Resources Utilization of Heilongjiang Province, No. 26, Hexing St., Harbin 150040, China
3   State Key Laboratory of Food Science and Technology, Nanchang University, No. 999, Xuefu St., Nanchang 330047, China; ting.luo@ncu.edu.cn
4   Department of Food Science and Nutrition, Zhejiang University, Hangzhou 310058, China
*   Correspondence: chenqh@zju.edu.cn (Q.C.); wangjinling@nefu.edu.cn (J.W.)

Citation: He, H.; Yan, Y.; Dong, D.; Bao, Y.; Luo, T.; Chen, Q.; Wang, J. Effect of *Issatchenkia terricola* WJL-G4 on Deacidification Characteristics and Antioxidant Activities of Red Raspberry Wine Processing. *J. Fungi* **2022**, *8*, 17. https://doi.org/10.3390/jof8010017

Academic Editor: Laurent Dufossé

Received: 18 November 2021
Accepted: 24 December 2021
Published: 27 December 2021

Publisher's Note: MDPI stays neutral with regard to jurisdictional claims in published maps and institutional affiliations.

Abstract: Our previous study isolated a novel *Issatchenkia terricola* WJL-G4, which exhibited a potent capability of reducing citric acid. In the current study, *I. terricola* WJL-G4 was applied to decrease the content of citric acid in red raspberry juice, followed by the red raspberry wine preparation by *Saccharomyces cerevisiae* fermentation, aiming to investigate the influence of *I. terricola* WJL-G4 on the physicochemical properties, organic acids, phenolic compounds and antioxidant activities during red raspberry wine processing. The results showed that after being treated with *I. terricola* WJL-G4, the citric acid contents in red raspberry juice decreased from 19.14 ± 0.09 to 6.62 ± 0.14 g/L, which was further declined to 5.59 ± 0.22 g/L after *S. cerevisiae* fermentation. Parameters related to CIELab color space, including L*, a*, b*, h°, and ΔE* exhibited the highest levels in samples after *I. terricola* WJL-G4 fermentation. Compared to the red raspberry wine pretreated without deacidification (RJO-SC), wine pretreated by *I. terricola* WJL-G4 (RJIT-SC) exhibited significantly decreased contents of gallic acid, cryptochlorogenic acid, and arbutin, while significantly increased contents of caffeic acid, sinapic acid, raspberry ketone, quercitrin, quercetin, baicalein, and rutin. Furthermore, the antioxidant activities including DPPH· and ABTS⁺· radical scavenging were enhanced in RJIT-SC group as compared to RJO-SC. This work revealed that *I. terricola* WJL-G4 had a great potential in red raspberry wine fermentation.

Keywords: red raspberry wine; *Issatchenkia terricola* WJL-G4; deacidification; phenolic compounds; antioxidant activity

1. Introduction

Red raspberry (*Rubus idaeus* L.) is known as an important commercial fruit with high contents of health-beneficial constituents and unique flavor [1]. It was reported that numerous bioactive components such as phenolics, flavones, anthocyanins, and vitamins are present in mature raspberry fruits, and these components contributed to the antioxidant capacities of these fruits [2]. Studies showed that red raspberry has positive effects on biological activities linked to reducing the risk for cancer, cardiovascular disease, diabetes mellitus and Alzheimer's disease [3–5]. Polyphenol extracts from red raspberry whole fruit or pulp attenuated high-fat diet induced obesity in mice [6,7].

The red raspberry has been an emerging fruit crop in China over the last decades, although, for thousands of years, dried red raspberries produced in eastern China have been used for medicinal purposes. In the recent years, red raspberry has been widely cultivated in China, because of its unique flavor, taste, texture, and health benefits [8]. The production of red raspberries is steadily increasing [9]. However, industrial development of red raspberry products got restricted due to the limitation in preservation and processing techniques [10]. In the international market, the processed products of red raspberry include food drinks [11], baked products [12], leisure ready-to-eat products [13], frozen foods [14], while the research on red raspberry products in China is limited.

Red raspberry wine fermentation has been an alternative method to preserve nutrition and extend shelf life of red raspberry, not only effectively retaining the original flavor and active substances with antioxidant, hypoglycemia, prevention of cardiovascular diseases functions, but also enriching the market of processed red raspberry products. Red raspberry wine has the characteristics of pleasing sensory qualities, high nutritional value, and strong antioxidant function, which is in line with the green and healthy standards of modern people. However, high content of organic acid in red raspberry, mainly citric acid (22.87 g/L, accounting for 90.9% of the total acid) [15], caused excessive sour taste of red raspberry wine, and led to growth difficultly of microorganisms during fermentation. Therefore, to reduce the content of citric acid in the red raspberry wine has become an urgent problem.

The most common biological deacidification methods, including malo-lactic fermentation and malo-alcohol fermentation, can not apply to degrade citric acid [16,17]. Simultaneously, physical and chemical deacidification lead to poor quality, such as unstable wine body, bitter taste, loss of flavor substances, beneficial ingredients and color adsorption [18]. Therefore, biodegradation of citric acid has gradually become a new breakthrough in the process of winemaking. In our previous study, we isolated a novel characterized *I. terricol*a WJL-G4 from red raspberry fruits, which exhibited a potent capability of reducing the total acid contents from 25.16 g/L to 9.96 g/L within 8 d static fermentation of red raspberry juice [18,19]. However, the influence of *I. terricol*a WJL-G4 on red raspberry wine processing is not clarified. Other researches reported citric acid degradation strain JT-1-3, which was identified to be *Pichia fermentans* and showed strong ability to degrade citric acid [20] and improve kiwifruit wine quality. In our previous research, the fermentation of red raspberry juice with *I. terricola* WJL-G4 potently reduced citric acid concentration and increased the content of total flavonoid [15].

Based on the above research, we hypothesized that fermentation with *I. terricola* WJL-G4 could reduce citric acid content and improve the quality of red raspberry wine. In this work, the effects of deacidification using *I. terricola* WJL-G4 followed by *S. cerevisiae* wine fermentation on the physicochemical properties, organic acids, phenolic compounds and antioxidant activities of red raspberry wine were investigated in order to evaluate the use potential of *I. terricola* WJL-G4 on the characteristics of red raspberry wine.

2. Materials and Methods

2.1. Materials

Red raspberries of the cultivar "Autumn Bless", one abundant raspberry species in Northeast China, were obtained from a local farm (Harbin, China) in August 2020. The fruits with similar maturity were hand-picked from different trees and then snap-frozen. Frozen red raspberries (40 kg) were mixed, transported to the laboratory in ice box and stored at −80 °C until use. All the standards for high performance liquid chromatography (HPLC) analysis, including rutin, quercetin, chlorogenic acid, ellagic acid, raspberry ketone, epicatechin, catechinic acid, syringic acid, gallic acid, caffeic acid, luteolin, citric acid, L-malic acid, succinic acid, α-ketoglutarate, and so on (HPLC grade), were obtained from Shanghai YuanYe Biotechnology (Shanghai, China).

2.2. Yeast Strains and Culture Media

I. terricola WJL-G4 was screened and isolated from fresh red raspberry (Chinese patent No. 2019113316700) after morphological and molecular biological identification and stored at China General Microbiological Culture Collection Center (CGMCC), No. 18712 [19]. *S.*

cerevisiae BV818 (*S. cerevisiae* BV818) was purchased from Angel Yeast Co. Ltd. (Yichang, China). Wallerstein Laboratory (WL) citric acid medium (WLC) (1% citric acid, 0.5% yeast extract, 0.05% Mg_2SO_4) was used to cultivate *I. terricola* WJL-G4. Yeast extract peptone dextrose medium (YPD) (1% yeast extract, 2% yeast peptone, 2% glucose, *w/v*) was utilized to cultivate *S. cerevisiae* BV818.

2.3. Samples Preparation

To keep the consistency of sampling, at least 6 kg red raspberry fruits were milled after natural thawing. After an enzymatic treatment (pectinase, 40,000 U/g, enzyme:substrate = 0.02:100, *w/w*) at a temperature of 50 °C for 2 h, the obtained red raspberry pulp was incubated at 90 °C for 5 min for enzyme deactivation, followed by filtration using an 8-layer gauze and pasteurization at 80 °C for 15 min. Fermentation was performed in 250 mL glass bottles after yeasts were inoculated according to the designed protocols (next paragraph). Wine fermentation temperature was controlled between 24 °C and 26 °C (sample size was 3).

Subsequently, the red raspberry juice was equally divided into five different treatment samples (Figure 1): (i) RJ: red raspberry juice; (ii) RJO: red raspberry juice with oscillation (28 °C, 36 h, 240 r/min) and without yeasts; (iii) RJIT: acid-reducing fermentation of RJ with *I. terricola* WJL-G4 (28 °C, 36 h, 240 r/min, 0.1% inoculation); (iv) RJO-SC: soluble solid content of RJO was adjusted to 24 °Brix, then followed by wine fermentation with *S.cerevisiae* BV818 (24–26 °C, 15 d, 0.1% inoculation) and (v) RJIT-SC: soluble solid content of RJIT was adjusted to 24 °Brix, then followed by wine fermentation with *S.cerevisiae* BV818 (24–26 °C, 15 d, 0.1% inoculation). RJ, RJO, RJIT and RJO-SC were set as the control groups compared to RJIT-SC. After treatment, the supernatant was harvested with centrifugation at 4000 r/min for 15 min for further analysis.

2.4. Physicochemical Properties

Total sugar, reducing sugar, soluble solid content (SSC), alcoholic degree, pH value, and titratable acid were assayed applying the methods recommended by the International Organization of Vine and Wine [21]. Color measurements of samples were determined using the CIELab systems, as measured by the CM-5 Spectrophotometer color meter (Konica Minoita, Japan). L^* indicates lightness, where white = 100 and black = 0. L* (lightness), a* (from green to red), b* (from blue to yellow), C* (chroma or saturation-a vector from grey to saturated color) as square root of $(a^{*2} + b^{*2})$, and h° (color shade) was calculated as arctan (b*/a*), $\Delta E^* = [(\Delta L^*)^2 + (\Delta a^*)^2 + (\Delta b^*)^2]^{1/2}$ in CIELab unit [22].

2.5. Determination of Organic Acids

The organic acids in samples were analyzed by using Agilent 1260 Infinity II HPLC equipped with a DAD detector (Agilent Technologies Inc., Santa Clara, CA, USA). Separation and quantification of organic acids were performed at 210 nm with an Agilent Poroshell 120 EC-C18 column (4.6 × 150 mm, 4 μm). Two eluents, filtered through a 0.45 μm durapore membrane pore filter were used as mobile phases: eluent A: 0.5% (*w/v*) KH_2PO_4 (pH 2.3), eluent B: methanol, the elution condition was isostatic elution with A:B of 97:3 (*v/v*). Injection volume was 10 μL, flow rate was 0.7 mL/min, and column temperature was 35 °C.

2.6. Determination of the Total Phenol Contents

Total phenol contents of samples were determined using Folin-Ciocalteu method [23] with minor modifications. Briefly, samples were firstly tenfold diluted in distilled water (1:9, *v/v*). Diluted samples (1 mL) were mixed with 4 mL of sodium carbonate solution (7.5%, *w/v*) and 5 mL of 10% (*v/v*) Folin-Ciocalteu reagent using distilled water and was allowed to stand at room temperature for 1 h. The absorbance of mixtures was measured at 765 nm. A calibration curve was obtained using 0–100 μg gallic acid (GAE) /mL and was used to calculate the total phenol contents of samples. Total phenol contents in samples were calculated by regression equation and expressed as mg GAE/L.

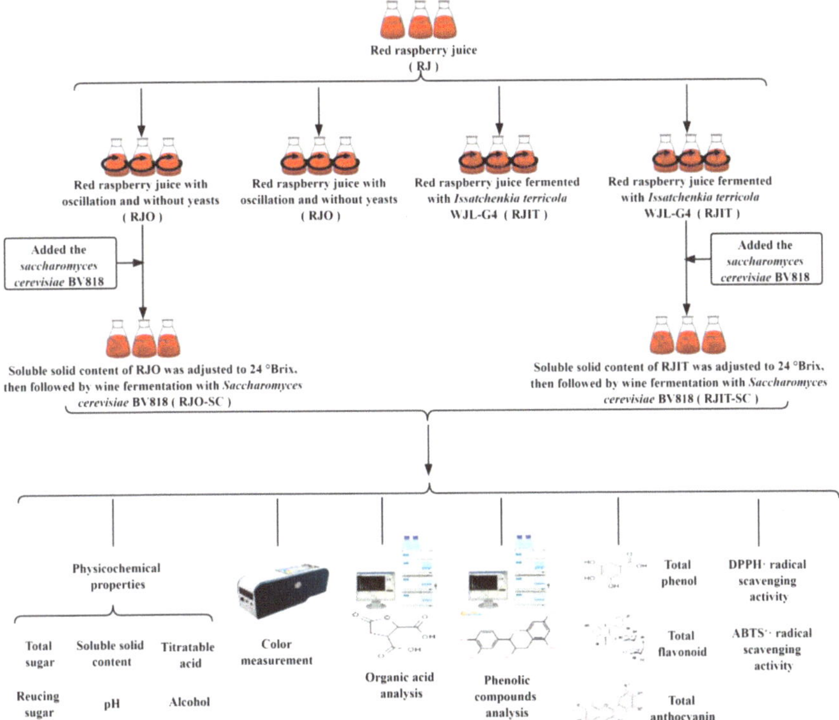

Figure 1. Experimental design of this work. RJ: red raspberry juice; RJO: red raspberry juice with oscillation (28 °C, 36 h, 240 r/min) and without yeasts; RJO-SC: soluble solid content of RJO was adjusted to 24 °Brix, then followed by wine fermentation with *S.cerevisiae* BV818 (24–26 °C, 15 d, 0.1% inoculation); RJIT: acid-reducing fermentation of RJ with *I. terricola* WJL-G4 (28 °C, 36 h, 240 r/min, 0.1% inoculation); and RJIT-SC: soluble solid content of RJIT was adjusted to 24 °Brix, then followed by wine fermentation with *S.cerevisiae* BV818 (24–26 °C, 15 d, 0.1% inoculation).

2.7. Determination of the Total Flavonoid Contents

Total flavonoid contents of samples were evaluated using aluminium nitrate non-ahydrate [24]. Rutin standard (28 mg, dried to a constant weight at 120 °C) was diluted with 70% (v/v) ethanol solution in a 50 mL volumetric flask. Then, 0.1, 0.2, 0.4, 0.6, 0.8 and 1.0 mL of rutin solutions were respectively mixed with 0.4 mL of 5% (w/v) sodium nitrite solution in 25 mL calibration test tubes, followed by the addition of 0.4 mL of 10% (w/v) aluminium nitrate solutions. After standing for 6 min, 4 mL aliquot of 1 M sodium hydroxide solution (NaOH) was further added. The mixtures were diluted to 25 mL with 70% (v/v) ethanol solution. Ten minutes later, the absorbance of rutin standard solutions was measured at 510 nm with reagent blank as reference and regression equation was calculated. To determine total flavonoid contents in samples, 1 mL of sample solution was put into a 25 mL calibration test tube and treated according to the above-mentioned operation. Total flavonoid contents in samples were calculated by regression equation and expressed as mg Rutin/L.

2.8. Determination of the Total Anthocyanin Contents

Total anthocyanin contents of samples were directly quantified by pH differential method [25]. Briefly, 1 mL samples were kept in constant volume to 10 mL with potassium chloride solution or sodium acetate solution (pH 1.0: 1.49 g KCl dissolved in 100 mL distilled water and adjusted to pH 1.0 with 1 M HCl, or sodium acetate buffer pH 4.5: 1.64 g $CH_3COONa \cdot 3H_2O$ dissolved in 10 mL distilled water and adjusted to pH 4.5 with 1 M HCl). Then the absorbance was measured both at 520 and 700 nm after incubation for 20 min.

$$C(mg/L) = \frac{A \times M}{\varepsilon \times L} \times DF \times 1000$$

A = absorption value, M = molecular weight (449 g/mol for cyd-3-glc), ε = molar extinction coefficient (26,900 L/mol cm for cyd-3-glc at pH 1.0), DF = dilution factor, and, L = path length of the cuvette (1 cm).

2.9. Identification of Phenolic Compounds by HPLC

The phenolic compounds confirmation analysis by HPLC (1260 Infinity ll, Agilent Technologies, Santa Clara, CA, USA) was accomplished by Agilent ZORBAX Eclipse Plus C18 column (250 × 4.6 mm, 5 μm). The phenolic compounds were separated under gradient conditions with a flow rate of 0.8 mL/min. Column temperature was 35 °C and volume injection was 10 μL. The wavelength was 280 nm. The solvents used in the elution process were an aqueous solution of 100% methanol (phase A) and 0.02% (*v/v*) formic acid (phase B). The samples were eluted according to the linear gradient, the phase-time program was as follows: 0–10% A (0–5 min), 10–20% A (5–10 min), 10–35% A (10–20 min), 35–40% (20–35 min), 40–75% A (35–40 min), 75–10% A (40–45 min). The peak areas of the samples were compared with those of the internal standards and quantified [26].

2.10. Determination of Antioxidant Activities

2.10.1. DPPH Free Radical Scavenging Ability

DPPH· free radical scavenging abilities were performed using the method described by Ekumah [27] with slight modifications. DPPH· solution (1 mL, 0.025 g/L) was added to 0.5 mL sample, allowed to stand for 30 min in the dark at room temperature and the absorbance was recorded as sample group. Methyl alcohol (1 mL) was added to 0.5 mL sample and reacted for 30 min in the dark at room temperature and the absorbance was recorded as control group. Methyl alcohol (0.5 mL) and 1 mL DPPH· solution were mixed as the blank group. All absorbance was measured at a wavelength of 517 nm. The radical scavenging activity was calculated as follows:

$$\text{Radical scavenging activity (\%)} = \left(1 - \frac{A_{sample} - A_{control}}{A_{blank}}\right) \times 100\% \tag{1}$$

A_{sample}: The absorbance of the sample group at 517 nm;
$A_{control}$: The absorbance of the control group at 517 nm;
A_{blank}: The absorbance of the blank group at 517 nm.

2.10.2. ABTS$^+$ Free Radical Scavenging Ability

ABTS$^+$·free radical scavenging abilities of the samples were measured as previously described with slight modifications [28]. Briefly, 7 mM ABTS$^+$·solution was mixed with 140 mM potassium persulfate aqueous solution to generate ABTS$^+$·radical cation. The mixture was kept in the dark at room temperature for 12 h. Before the measurements, the resultant ABTS$^+$·solution was diluted with ethanol to an absorbance of 0.70 ± 0.02 at 734 nm. Sample (0.5 mL) was mixed with 1 mL of the diluted ABTS$^+$ solution and the absorbance was recorded as sample group. Sample (0.5 mL) was mixed with 1 mL of the ethanol and the absorbance was recorded as control group. Ethanol (0.5 mL) mixed with 1 mL of the diluted ABTS$^+$· solution was used as blank group. Afterwards, the absorbance was measured at 734 nm. The radical scavenging activity represented the percentage of ABTS$^+$·free radical inhibition and was calculated with Formula (1).

A_{sample}: The absorbance of the sample group at 734 nm;
$A_{control}$: The absorbance of the control group at 734 nm;
A_{blank}: The absorbance of the blank group at 734 nm.

2.11. Statistical Analyses

Each experiment was repeated three times, and the results were expressed as the means ± standard deviation. SPSS 20.0 (V.26.0, SPSS Inc., Chicago, IL, USA) was used to analyze variance, and Duncan's multiple comparison test was used to analyze the differences between the experimental groups. $p < 0.05$ was considered statistically significant.

3. Results and Discussion

3.1. Effect of I. terricola WJL-G4 on Physicochemical Properties of Red Raspberry Wines

The physicochemical properties of samples with different treatments (RJ, RJO, RJIT, RJO-SC, RJIT-SC) showed significant differences (Table 1). The contents of total sugar and reducing sugar declined in RJIT, RJO-SC and RJIT-SC, because energy from carbohydrate metabolism is crucially important for yeasts growth and reproduction during fermentation [29]. The decline in SSC and total sugar coincided with the great consumption of sugar by *I. terricola* WJL-G4 and *S. cerevisiae* BV818. The alcoholic degree of red raspberry wines, including RJO-SC and RJIT-SC, reached 9.93 ± 0.47 and 10.23 ± 0.87 (%, v/v), respectively. The results indicated that *I. terricola* WJL-G4 produced a limited amount of ethanol, which was only 3.90 ± 0.12 (%, v/v) in RJIT. Because of the inherent properties of non-*Saccharomyces cerevisiae*, its alcohol-producing capacity was generally lower than that of *S. cerevisiae* [30]. The titratable acid content of RJO-SC was 19.40 ± 0.15 g/L, slightly lower than that of RJO (22.86 ± 0.21 g/L), while significantly higher than that of RJIT and RJIT-SC with mean values of 11.83 ± 0.97 and 9.22 ± 0.17 g/L, respectively. On the other hand, RJIT and RJIT-SC showed significantly higher pH values in comparison with RJO and RJO-SC. These results indicated that *I. terricola* WJL-G4 fermentation reduced titratable acid and increased pH values of red raspberry wines.

Table 1. Effect of *I. terricola* WJL-G4 on physicochemical properties of red raspberry wines.

	RJ	RJO	RJIT	RJO-SC	RJIT-SC
Total sugar/(g·L−1)	61.79 ± 5.33 [a]	55.56 ± 1.29 [b]	6.19 ± 0.41 [c]	6.89 ± 0.74 [c]	6.43 ± 0.43 [c]
Reducing sugar/(g·L−1)	42.67 ± 3.06 [a]	39.64 ± 1.76 [b]	3.68 ± 0.24 [c]	4.28 ± 0.23 [c]	4.00 ± 0.46 [c]
Soluble solid content/(°Brix)	9.15 ± 0.60 [a]	8.00 ± 0.00 [b]	3.00 ± 0.00 [e]	5.63 ± 0.75 [c]	4.75 ± 0.42 [d]
Alcohol/(%, v/v)	n.d.	n.d.	3.90 ± 0.12 [c]	9.93 ± 0.47 [b]	10.23 ± 0.87 [a]
Titratable acid/(g·L^{-1})	23.56 ± 1.70 [a]	22.86 ± 0.21 [ab]	11.83 ± 0.97 [c]	19.40 ± 0.15 [b]	9.22 ± 0.17 [d]
pH	2.86 ± 0.01 [e]	2.88 ± 0.01 [d]	3.54 ± 0.12 [b]	3.02 ± 0.03 [c]	3.86 ± 0.10 [a]
L*	16.99 ± 0.16 [c]	19.7 ± 0.02 [b]	24.76 ± 0.93 [a]	20.04 ± 1.17 [b]	24.95 ± 0.99 [a]
a*	47.50 ± 0.01 [d]	50.95 ± 0.01 [c]	56.55 ± 0.87 [a]	50.74 ± 1.56 [c]	53.66 ± 0.81 [b]
b*	28.97 ± 0.01 [c]	33.62 ± 0.16 [b]	42.17 ± 1.74 [a]	34.26 ± 1.89 [b]	41.32 ± 1.53 [a]
C*	55.63 ± 0.01 [c]	68.86 ± 0.02 [a]	70.61 ± 1.75 [a]	61.23 ± 2.35 [b]	67.73 ± 1.57 [a]
h°	31.37 ± 0.01 [c]	33.41 ± 0.12 [b]	36.76 ± 0.71 [a]	34.01 ± 0.64 [b]	37.39 ± 0.96 [a]
ΔE*	n.a.	6.39 ± 0.13 [b]	17.86 ± 0.15 [a]	6.91 ± 2.69 [b]	15.93 ± 1.99 [a]

Values were given as the means ± standard deviation ($n = 3$), and the different letters within each row were significantly different ($p < 0.05$). CIE Lab coordinates: L* (lightness), a* (from green to red), b* (from blue to yellow), C* (chroma or saturation), and h° (hue angle). $\Delta E^* = [(\Delta L^*)^2 + (\Delta a^*)^2 + (\Delta b^*)^2]^{1/2}$ in CIELab unit. n.d.: Not detected. n.a.: Not applicable.

Color is also a basic quality of fruit drinks. ΔE* reflects the relation of total color difference among L*, a* and b* [31]. The values of CIELab in different samples were reported in Table 1. Among samples, RJIT and RJIT-SC had significantly higher values of L*, a*, b*, h°, and ΔE*, which was 24.76 ± 0.93, 56.55 ± 0.87, 42.17 ± 1.74, 36.76 ± 0.71, and 17.86 ± 0.15 in RJIT and 24.95 ± 0.99, 53.66 ± 0.81, 41.32 ± 1.53, 37.39 ± 0.96, and 15.93 ± 1.99 in RJIT-SC, respectively, compared to other samples, indicating that *I. terricola* WJL-G4 fermentation significantly promoted the color of fermented samples. Non-*Saccharomyces cerevisiae* has been reported to play a positive role in wine color. Morata and Benito reported that fermentation with non-*Saccharomyces cerevisiae* strains with high activity of ethyl p-hydroxycinnamate decar-decarboxylase (HCDC) could increase the formation of styrene-pyranoanthocyanins in wine fermentation [32]. Sequential fermentation with *Pichia guilliermondii* and *S. cerevisiae* was used to enhance the production of styrene-pyranoanthocyanins, thus making the color of wine more stable [33].

3.2. Effect of I. terricola WJL-G4 on Organic Acids Contents

Organic acids have been reported to be important factors influencing aroma, taste and microbiological stability in wine [34]. The contents of 7 organic acids in samples were presented in Table 2. Except for succinic acid, other organic acids in the wines (RJO-SC and RJIT-SC) showed a downward trend compared to RJ. Compared to RJO-SC, the concentrations of oxalic acid, tartaric acid, α-ketoglutaric acid in RJIT-SC were decre- ased from 0.57 ± 0.03 to 0.45 ± 0.01 g/L, 0.49 ± 0.00 to 0.16 ± 0.00 g/L, and 0.28 ± 0.13 to 0.08 ± 0.00 g/L, respectively; especially, the contents of citric acid were reduced from 17.14 ± 0.16 to 5.59 ± 0.22 g/L. In addition, compared with RJ (0.69 ± 0.05 g/L), succinic acid concentrations in RJIT and RJIT-SC were increased to 1.74 ± 0.17 and 1.42 ± 0.22 g/L, respectively. During aging, succinic acid will combine with other substances to form ethyl succinic acid, which is beneficial to the aroma of wine, accentuates wine's flavor and vinous character [35]. The contents of organic acids in RJO and RJIT showed a similar trend compared to that in RJO-SC and RJIT-SC, indicating that *I. terricola* WJL-G4 had the ability to metabolize organic acids, especially citric acid, when there was glucose or other absorbable carbon sources, therefore, it could be concluded that *I. terricola* WJL-G4 reduced organic acids significantly in red raspberry wine, yet *S. cerevisiae* BV818 had limited capacity to transform organic acids. Compared to our previous study, which showed that static fermentation with *I. terricola* WJL-G4 could degrade citric acid from 22.87 to 9.25 g/L in red raspberry juice after 8 days [15], *I. terricola* WJL-G4 showed more efficient potential to reduce citric acid during wine fermentation with a shorter time period and better acid-degradation ability with oscillation. The non-*Saccharomyce* yeast *Pichia fermentans* JT-1-3 was reported to have the ability to degrade citric acid from 12.3 to 11.0 g/L in kiwifruit wine, and to degrade citric acid by 43.35% in blueberry wine [20,36]. The reduction of organic acids may reduce the formation of "fermentation bouquet", leading to a better wine taste, quality and purchase desire [37].

Table 2. Effect of *I. terricola* WJL-G4 on organic acids contents of red raspberry wines.

	RJ	RJO	RJIT	RJO-SC	RJIT-SC
Citric acid/$(g \cdot L^{-1})$	19.14 ± 0.09 [a]	18.80 ± 0.40 [a]	6.62 ± 0.14 [c]	17.14 ± 0.16 [b]	5.59 ± 0.22 [d]
Malic acid/$(g \cdot L^{-1})$	1.25 ± 0.03 [a]	1.10 ± 0.22 [ab]	0.39 ± 0.04 [c]	0.57 ± 0.18 [c]	0.94 ± 0.08 [b]
Oxalic acid/$(g \cdot L^{-1})$	0.92 ± 0.01 [a]	0.92 ± 0.07 [a]	0.47 ± 0.01 [c]	0.57 ± 0.03 [b]	0.45 ± 0.01 [c]
Tartaric acid/$(g \cdot L^{-1})$	0.56 ± 0.01 [b]	0.79 ± 0.06 [a]	0.39 ± 0.03 [c]	0.49 ± 0.00 [b]	0.16 ± 0.00 [d]
Succinic acid/$(g \cdot L^{-1})$	0.69 ± 0.05 [c]	0.81 ± 0.02 [c]	1.74 ± 0.17 [a]	0.09 ± 0.05 [d]	1.42 ± 0.22 [b]
α-Ketoglutaric acid/$(g \cdot L^{-1})$	0.31 ± 0.12 [a]	0.22 ± 0.15 [ab]	0.16 ± 0.00 [bc]	0.28 ± 0.13 [ab]	0.08 ± 0.00 [c]
Fumaric acid/$(g \cdot L^{-1})$	0.02 ± 0.00 [b]	0.03 ± 0.00 [a]	n.d.	n.d.	n.d.

Values were given as the means $\pm$ standard deviation ($n = 3$), and the different letters within each row were significantly different ($p < 0.05$). n.d.: Not detected.

3.3. Effect of I. terricola WJL-G4 on Total Phenol, Flavonoid and Total Anthocyanin Contents of Red Raspberry Wines

Polyphenols is a critical parameter of wine quality, which are not only related to the color, astringency, bitterness and other flavor quality parameters of wine, but also have good biological and antioxidant properties [38]. As shown in Figure 2, the initial content of total phenol in the RJ sample was the highest of 1555.10 ± 29.47 mg GAE/L. By the end of different processing, the total phenol contents in RJO, RJIT, RJO-SC and RJIT-SC were 1476.40 ± 8.58, 1330.55 ± 37.06, 1342.72 ± 19.11 and 1184.65 ± 26.12 mg GAE/L, respectively. The results indicated that both *I. terricola* WJL-G4 and *S. cerevisiae* could reduce the total phenol contents in red raspberry. Shu et al. [39] found that a detectable drop of almost 20% in phenolic contents after wine fermentation was probably due to the adsorption of phenols onto yeast cell walls and the reaction with cell wall proteins [40]. The decrease of phenolic acids in raspberry juice during fermentation might be due to microbial decomposition of macromolecular phenolic acids into small molecules or adherence of phenolic acids to yeast cell walls [41,42]. The highest amounts of total flavonoid were found in RJ and RJIT (Figure 2), consistent with our previous results [19]. Interestingly, RJIT-SC retained more total flavonoid content than RJO-SC. It could be concluded that the total flavonoid content in RJIT-SC was increased by *I. terricola* WJL-G4 fermentation.

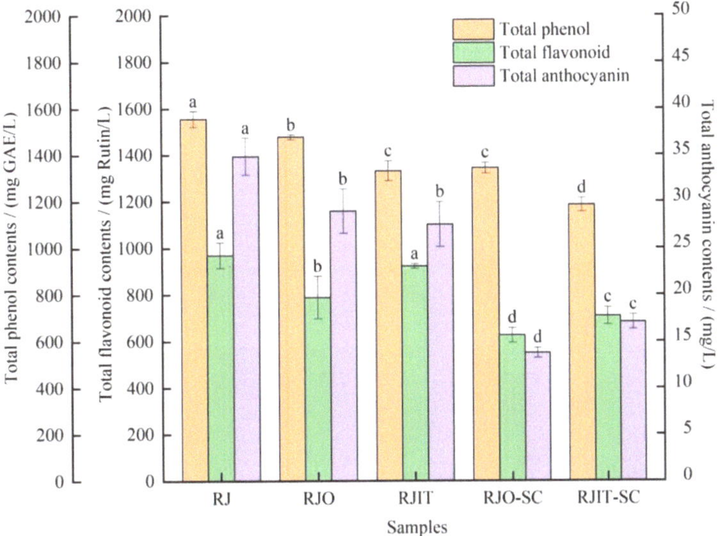

Figure 2. Effect of *I. terricola* WJL-G4 on total phenol, total flavonoid and total anthocyanin contents of red raspberry wines. Values were given as the means ± standard deviation (*n* = 3), and the different letters within the same color were significantly different (*p* < 0.05).

Total anthocyanin contents in RJO and RJIT decreased from 34.86 ± 1.72 to 28.98 ± 1.94 and 27.54 ± 2.74 mg/L, respectively, compared to RJ, and showed no significant difference, as shown in Figure 2. Interestingly, total anthocyanin content in RJIT-SC was significantly higher than that in RIO-SC, although both were significantly decreased after wine fermentation. Anthocyanins have a variety of biological activities, including antioxidant, anti-inflammatory, anti-cardiovascular disease, anti-skin damage and protection of the reproductive system [43]. However, anthocyanins are sensitive to light, pH, temperature and other factors during fermentation and aging, resulting in poor preservation of sensory properties of wine and its antioxidative effects [25]. It was found that after 15 d and 30 d of storage, anthocyanin contents decreased by 66.4% and 90.0%, respectively [44]. The degradation of anthocyanin and polymerization between anthocyanin and other constituents (e.g., protein) might lead to the loss of total anthocyanin contents in raspberry juice during storage [45,46]. Yang et al. [47] found that fermentation significantly reduced the total contents of anthocyanin from 422.1 mg/L in strawberry juice to 235.5 mg/L in fermentation beverage. The loss of total anthocyanin during fermentation can be influenced by multiple factors. Anthocyanin may interact with other flavonoids, forming more stable pyranoanthocyanins, and the decrease in polarity of these compounds was accompanied by a decrease in solubility [48]. Therefore, a conclusion might be drawn that *I. terricola* WJL-G4 exhibited protective effects on total anthocyanin contents during wine fermentation.

3.4. Effect of I. terricola WJL-G4 on Phenolic Compounds of Red Raspberry Wines

Eighteen phenolic compounds, including 10 phenolic acids and 8 flavonoids, were identified and quantified by HPLC, as shown in Table 3. Of the compounds tested, only quercitrin was not identified in RJ and RJO. The phenolic profile varied between samples. Most phenolic compound contents were found decreased after fermentation in RJIT, RJO-SC and RJIR-SC, especially in RJO-SC and RJIT-SC. However, contents of caffeic acid, quercitrin, quercetin and raspberry ketone increased significantly after *I. terricola* WJL-G4 fermentation in both RJIT and RJIT-SC, compared to RJ, RJO and RJO-SC. Ellagic acid contents showed a slight decrease after wine fermentation. Compared to RJO-SC, contents of gallic acid, cryptochlorogenic acid and arbutin decreased significantly, yet caffeic acid, sinapic acid, rutin, quercitrin, quercetin, baicalein and raspberry ketone increased significantly in RJIT-SC.

Table 3. Effect of *I. terricola* WJL-G4 on phenolic compounds of red raspberry wines.

Category	Contents/(mg·L^{-1})				
	RJ	**RJO**	**RJIT**	**RJO-SC**	**RJIT-SC**
Gallic acid	9.99 ± 0.30 [a]	9.41 ± 0.12 [a]	7.53 ± 0.23 [b]	7.47 ± 0.51 [b]	6.40 ± 1.12 [c]
Cryptochlorogenic acid	90.92 ± 9.71 [a]	95.17 ± 3.98 [a]	54.45 ± 7.86 [b]	88.56 ± 7.34 [a]	46.56 ± 6.77 [b]
p-Hydroxybenzonic acid	73.73 ± 8.12 [a]	59.90 ± 5.44 [b]	45.40 ± 4.03 [c]	42.89 ± 1.57 [c]	38.89 ± 3.02 [c]
Chlorogenic acid	107.58 ± 4.10 [a]	92.06 ± 2.25 [b]	75.51 ± 9.99 [c]	45.86 ± 1.06 [d]	49.06 ± 5.11 [d]
Neochlorogenic acid	311.23 ± 44.71 [a]	260.18 ± 20.93 [b]	164.34 ± 7.95 [c]	57.80 ± 7.42 [d]	32.69 ± 5.70 [d]
Caffeic acid	14.18 ± 0.13 [cd]	14.79 ± 0.74 [c]	21.21 ± 0.61 [a]	13.38 ± 0.11 [d]	16.43 ± 0.90 [b]
Syringate	1.90 ± 0.05 [ab]	1.98 ± 0.06 [a]	1.74 ± 0.02 [c]	1.88 ± 0.04 [b]	1.83 ± 0.09 [bc]
p-Coumaric acid	2.33 ± 0.11 [b]	4.27 ± 0.35 [a]	4.48 ± 0.31 [a]	1.46 ± 0.13 [c]	1.48 ± 0.12 [c]
Ellagic acid	30.45 ± 0.12 [a]	30.49 ± 0.11 [a]	30.36 ± 0.08 [a]	30.04 ± 0.18 [b]	29.99 ± 0.14 [b]
Sinapic acid	88.73 ± 4.90 [a]	83.46 ± 1.47 [a]	72.74 ± 5.49 [b]	29.71 ± 5.28 [d]	54.11 ± 9.49 [c]
Arbutin	99.93 ± 7.70 [a]	94.28 ± 5.03 [a]	72.45 ± 2.14 [c]	85.50 ± 5.37 [b]	66.93 ± 5.18 [c]
Rutin	11.29 ± 0.53 [a]	11.83 ± 0.57 [a]	12.13 ± 0.78 [a]	2.48 ± 0.43 [c]	3.45 ± 0.38 [b]
Quercitrin	n.d.	n.d.	4.69 ± 0.22 [a]	3.05 ± 0.30 [c]	3.67 ± 0.20 [b]
Quercetin	3.66 ± 0.11 [bc]	3.98 ± 0.21 [b]	4.75 ± 0.87 [a]	2.99 ± 0.44 [c]	4.52 ± 0.41 [ab]
Luteolin	0.88 ± 0.06 [b]	0.90 ± 0.03 [b]	0.76 ± 0.05 [b]	1.49 ± 0.32 [a]	1.53 ± 0.26 [a]
Kaempferol	-	-	-	-	-
Baicalein	3.95 ± 0.18 [c]	4.13 ± 0.23 [a]	3.64 ± 0.06 [b]	3.13 ± 0.11 [c]	3.71 ± 0.35 [b]
Raspberry ketone	5.48 ± 0.08 [c]	5.45 ± 0.03 [c]	20.19 ± 0.67 [a]	3.76 ± 0.60 [d]	12.71 ± 1.27 [b]

Values were given as the means ± standard deviation (*n* = 3), and the different letters within each row were significantly different (*p* < 0.05). n.d.: Not detected. -: the content' was too low to calculate.

Raspberry ketone, as the secondary metabolite of raspberry, was significantly increased in RJIT and RJIT-SC. Raspberry ketone contents in RJ, RJO, RJIT, RJO-SC and RJIT-SC were 5.48 ± 0.08, 5.45 ± 0.03, 20.19 ± 0.67, 3.76 ± 0.60 and 12.71 ± 1.27 mg/L, respectively. It might be concluded that fermentation with *I. terricola* WJL-G4 significantly increased the contents of raspberry ketone. Raspberry ketone is an important characteristic aroma component in ripe raspberries [49]. In addition, raspberry ketone has been used in the pharmaceutical makeup industry because of its skin-whiting property [50].

3.5. Effect of I. terricola WJL-G4 on DPPH and ABTS$^+$·Radical Scavenging Activities of RED Raspberry Wines

Phenolic compounds exhibit different biological activities, especially antioxidant activities. They protect biomolecules from oxidative damage through free radical mediated reactions and inhibit oxidative chain reactions in a variety of ways, including direct quenching of reactive oxygen species, inhibition of enzymes and chelating metal ions [51]. Herein, different in vitro antioxidant activities were tested and shown in Table 4. A concentration effect relationship was found in all samples between concentrations and antioxidant activities. In particular, RJ showed the highest DPPH· and ABTS$^+$· radical scavenging activities with IC50 values of 1.51 ± 0.04 and 13.25 ± 0.14 mg/mL, respectively, followed by RJIT with IC50 value of 1.84 ± 0.07 and 14.49 ± 0.24 mg/mL, respectively. The DPPH· and ABTS$^+$·radical scavenging IC50 values in RJIT-SC were 2.10 ± 0.06 and 16.38 ± 0.13 mg/mL, compared to that of RJO-SC with 2.33 ± 0.06 and 17.26 ± 0.17 mg/mL. We might draw a conclusion that the fermentation treatment by *I. terricola* WJL-G4 could enhance the antioxidant activities of fermented red raspberry wine.

Table 4. Effect of *I. terricola* WJL-G4 on DPPH· and ABTS$^+$· radical scavenging activities of red raspberry wines.

	RJ	RJO	RJIT	RJO-SC	RJIT-SC	VC
IC50 for DPPH/(mg/mL)	1.51 ± 0.04 [d]	1.90 ± 0.24 [bc]	1.84 ± 0.07 [c]	2.33 ± 0.06 [a]	2.10 ± 0.06 [bc]	0.016 ± 0.01 [f]
IC50 for ABTS$^+$/(mg/mL)	13.25 ± 0.14 [e]	15.16 ± 0.16 [c]	14.49 ± 0.24 [d]	17.26 ± 0.17 [a]	16.38 ± 0.13 [b]	0.028 ± 0.01 [f]

Values were given as the means $\pm$ standard deviation ($n = 3$), and the different letters within each row were significantly different ($p < 0.05$). VC was Vitamin C (ascorbic acid).

Su and Wang [20] found that during juice storage, the changes of antioxidant activity was likely to be associated with the degradation pathways of anthocyanin. Reports [52] showed that the total phenol and total flavonoid contents of wines exhibited the strongest correlations with antioxidant properties but that total anthocyanin contents exhibited weaker correlations. He et al. [53] found that hawthorn wine showed positive correlations between antioxidant activities and total anthocyanin, total phenol and total flavonoid contents, and the strongest correlation was observed between antioxidant activity and total phenol content.

The present data were basically consistent with some of the results mentioned above, as shown in Figure 3. There was a good correlation between the antioxidant activities and the bioactive components of samples. The IC50 of DPPH· and ABTS$^+$ free radical scavenging activities had significantly negative correlation with the contents of total flavonoid and total anthocyanin, with the correlation coefficients of -0.95, -0.96 and -0.98, -0.97 ($p < 0.05$), respectively; and they were negatively correlated with total phenol content with the correlation coefficients of -0.68 ($p > 0.05$). Total phenol had positive correlation with total flavonoid and total anthocyanin with the correlation coefficients of 0.57 and 0.79 ($p > 0.05$), respectively, yet total flavonoid had significantly positive correlation with total anthocyanin (correlation coefficient of 0.92, $p < 0.05$). This result confirmed with the previous conclusion that the higher the contents of total phenol, total flavonoid and total anthocyanin, the higher the antioxidant activity will be observed. The results showed that the contents of these active substances were important factors affecting the antioxidant capacity of red raspberry wine.

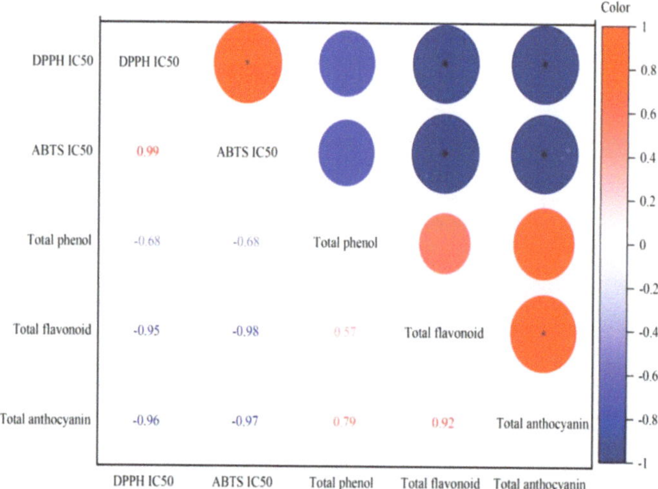

Figure 3. Correlation between total phenol, total flavonoid, total anthocyanin contents and antioxidant activities. The * indicated significantly different ($p < 0.05$). The number represented the Pearson correlation coefficient.

4. Conclusions

In this work, the fermentation with *I. terricola* WJL-G4 could significantly degrade citric acid, improve the color of the wine, enhance contents of total flavonoid and total anthocyanin, thus enhanced the antioxidant activities of red raspberry wine. Red raspberry wine produced with *I. terricola* WJL-G4 might have an improved taste due to citric acid reduction. In conclusion, *I. terricola* WJL-G4 showed great potential to be applied in red raspberry or other fruit wine production with high levels of citric acid. Further work may focus on the effect of *I. terricola* WJL-G4 fermentation on the aroma compounds of red raspberry wine during the wine processing.

Author Contributions: H.H.: Conceptualization, Investigation, Methodology, Validation, Formal analysis, Writing-Original Draft; Y.Y.: Investigation, Methodology, Writing-Original Draft; D.D.: Investigation, Methodology; Y.B.: Writing-Review and Editing; T.L.: Writing-Review and Editing; Q.C. and J.W.: Conceptualization, Writing-Review and Editing, Funding acquisition, Project administration. All authors have read and agreed to the published version of the manuscript.

Funding: We are grateful to the financial support by Heilongjiang Tongsheng Food Technology Co., Ltd., (Xiangfang City, China) (2019TSYF01) and Major Science and Technology Research Plan of Yunnan Province (202002AA1000056-3).

Institutional Review Board Statement: Not applicable.

Informed Consent Statement: Not applicable.

Data Availability Statement: Not applicable.

Conflicts of Interest: The authors declare no conflict of interest.

References

1. González-Orozco, B.D.; Mercado-Silva, E.M.; Castaño-Tostado, E.; Vázquez-Barrios, M.E.; Rivera-Pastrana, D.M. Effect of short-term controlled atmospheres on the postharvest quality and sensory shelf life of red raspberry (*Rubus idaeus* L.). *CyTA-J. Food* **2020**, *18*, 352–358. [CrossRef]
2. Yang, F.; Zhang, Z.; Tong, L. The complete chloroplast genome sequence of *Rubus amabilis* Focke. *Mitochondrial DNA Part B* **2020**, *5*, 1975–1976. [CrossRef]
3. Rao, A.V.; Snyder, D.M. Raspberries and human health: A review. *J. Agric. Food Chem.* **2010**, *58*, 3871–3883. [CrossRef] [PubMed]

4. Li, M.; Liu, Y.; Yang, G.; Sun, L.; Song, X.; Chen, Q.; Bao, Y.; Luo, T.; Wang, J. Microstructure, physicochemical properties, and adsorption capacity of deoiled red raspberry pomace and its total dietary fiber. *LWT* **2022**, *153*, 112478. [CrossRef]
5. Burton-Freeman, B.M.; Sandhu, A.K.; Edirisinghe, I. Red raspberries and their bioactive polyphenols: Cardiometabolic and neuronal health links. *Adv. Nutr.* **2016**, *7*, 44–65. [CrossRef] [PubMed]
6. Xian, Y.; Fan, R.; Shao, J.; Toney, A.M.; Chung, S.; Ramer-Tait, A.E. Polyphenolic fractions isolated from red raspberry whole fruit, pulp, and seed differentially alter the gut microbiota of mice with diet-induced obesity. *J. Funct. Foods* **2021**, *76*, 104288. [CrossRef]
7. Luo, T.; Miranda-Garcia, O.; Adamson, A.; Sasaki, G.; Shay, N. Development of obesity is reduced in high-fat fed mice fed whole raspberries, raspberry juice concentrate, and a combination of the raspberry phytochemicals ellagic acid and raspberry ketone. *J. Berry Res.* **2016**, *6*, 213–223. [CrossRef]
8. Zhang, W.; Lao, F.; Bi, S.; Pan, X.; Pang, X.; Hu, X.; Liao, X.; Wu, J. Insights into the major aroma-active compounds in clear red raspberry juice (*Rubus idaeus* L. cv. Heritage) by molecular sensory science approaches. *Food Chem.* **2021**, *336*, 127721. [CrossRef]
9. Aaby, K.; Skaret, J.; Røen, D.; Sønsteby, A. Sensory and instrumental analysis of eight genotypes of red raspberry (*Rubus idaeus* L.) fruits. *J. Berry Res.* **2019**, *9*, 483–498. [CrossRef]
10. Teng, H.; Fang, T.; Lin, Q.; Song, H.; Liu, B.; Chen, L. Red raspberry and its anthocyanins: Bioactivity beyond antioxidant capacity. *Trends Food Sci. Technol.* **2017**, *66*, 153–165. [CrossRef]
11. Gagneten, M.; Corfield, R.; Mattson, M.G.; Sozzi, A.; Leiva, G.; Salvatori, D.; Schebor, C. Spray-dried powders from berries extracts obtained upon several processing steps to improve the bioactive components content. *Powder Technol.* **2019**, *342*, 1008–1015. [CrossRef]
12. Šarić, B.; Dapčević-Hadnađev, T.; Hadnađev, M.; Sakač, M.; Mandić, A.; Misan, A.; Škrobot, D. Fiber concentrates from raspberry and blueberry pomace in gluten-free cookie formulation: Effect on dough rheology and cookie baking properties. *J. Texture Stud.* **2018**, *50*, 124–130. [CrossRef] [PubMed]
13. Ozcelik, M.; Ambros, S.; Heigl, A.; Dachmann, E.; Kulozik, U. Impact of hydrocolloid addition and microwave processing condition on drying behavior of foamed raspberry puree. *J. Food Eng.* **2019**, *240*, 83–91. [CrossRef]
14. Pichler, A.; Pozderović, A.; Moslavac, T.; Popović, K. Influence of sugars, modified starches and hydrocolloids addition on colour and thermal properties of raspberry cream fillings. *Pol. J. Food Nutr. Sci.* **2017**, *67*, 49–58. [CrossRef]
15. Chen, S.; Tang, L.; Feng, J.; Sun, L.; Wang, J. Isolation and identification of yeast capable of efficiently degrading citric acid and its acid degradation characteristics in red raspberry juice. *Food Sci.* **2020**, *41*, 133–139. [CrossRef]
16. Malherbe, S.; Tredoux, A.G.J.; Nieuwoudt, H.H.; du Toit, M. Comparative metabolic profiling to investigate the contribution of *O. oeni* MLF starter cultures to red wine composition. *J. Ind. Microbiol. Biotechnol.* **2012**, *39*, 477–494. [CrossRef]
17. Romano, P.; Brandolini, V.; Ansaloni, C.; Menziani, E. The production of 2,3-butanediol as a differentiating character in wine yeasts. *World J. Microbiol. Biotechnol.* **1998**, *14*, 649–653. [CrossRef]
18. Vera, E.; Ruales, J.; Dornier, M.; Sandeaux, J.; Persin, F.; Pourcelly, G.; Vaillant, F.; Reynes, M. Comparison of different methods for deacidification of clarified passion fruit juice. *J. Food Eng.* **2003**, *59*, 361–367. [CrossRef]
19. Chen, S.; Tang, Y.; Dong, D.; Jiang, Y.; He, H.; Wang, J. Variation of bioactive components in red raspberry juice during fermentation for deacidification. *Food Sci.* **2021**, *42*, 241–248. [CrossRef]
20. Zhong, W.; Chen, T.; Yang, H.; Li, E. Isolation and selection of non-*Saccharomyces* yeasts being capable of degrading citric acid and evaluation its effect on kiwifruit wine fermentation. *Fermentation* **2020**, *6*, 25. [CrossRef]
21. International Organisation of Vine and Wine Home Page. Available online: https://www.oiv.int (accessed on 24 September 2020).
22. Wrolstad, R.E.; Durst, R.W.; Lee, J. Tracking color and pigment changes in anthocyanin products. *Trends Food Sci. Technol.* **2005**, *16*, 423–428. [CrossRef]
23. Su, D.; Wang, Z.; Dong, L.; Huang, F.; Zhang, R.; Jia, X.; Wu, G.; Zhang, M. Impact of thermal processing and storage temperature on the phenolic profile and antioxidant activity of different varieties of lychee juice. *LWT* **2019**, *116*, 108578. [CrossRef]
24. Liu, G.; Sun, J.; He, X.; Tang, Y.; Li, J.; Ling, D.; Li, C.-B.; Li, L.; Zheng, F.; Sheng, J.; et al. Fermentation process optimization and chemical constituent analysis on longan (*Dimocarpus longan* Lour.) wine. *Food Chem.* **2018**, *256*, 268–279. [CrossRef] [PubMed]
25. Li, X.; Zhang, L.; Peng, Z.; Zhao, Y.; Wu, K.; Zhou, N.; Yan, Y.; Ramaswamy, H.S.; Sun, J.; Bai, W. The impact of ultrasonic treatment on blueberry wine anthocyanin color and its In-vitro anti-oxidant capacity. *Food Chem.* **2020**, *333*, 127455. [CrossRef]
26. Liu, B.; Yuan, D.; Li, Q.; Zhou, X.; Wu, H.; Bao, Y.; Lu, H.; Luo, T.; Wang, J. Changes in organic acids, phenolic compounds, and antioxidant activities of lemon juice fermented by *Issatchenkia terricola*. *Molecules* **2021**, *26*, 6712. [CrossRef] [PubMed]
27. Ekumah, J.-N.; Ma, Y.; Akpabli-Tsigbe, N.D.K.; Kwaw, E.; Jie, H.; Quaisie, J.; Manqing, X.; Nkuma, N.A.J. Effect of selenium supplementation on yeast growth, fermentation efficiency, phytochemical and antioxidant activities of mulberry wine. *LWT* **2021**, *146*, 111425. [CrossRef]
28. Tandee, K.; Kittiwachana, S.; Mahatheeranont, S. Antioxidant activities and volatile compounds in longan (*Dimocarpus longan* Lour.) wine produced by incorporating longan seeds. *Food Chem.* **2021**, *348*, 128921. [CrossRef] [PubMed]
29. Markowski, J.; Baron, A.; Le Quéré, J.-M.; Płocharski, W. Composition of clear and cloudy juices from French and Polish apples in relation to processing technology. *LWT* **2015**, *62*, 813–820. [CrossRef]
30. Liu, J.; Liu, M.; Ye, P.; Lin, F.; Huang, J.; Wang, H.; Zhou, R.; Zhang, S.; Zhou, J.; Cai, L. Characterization of major properties and aroma profile of kiwi wine co-cultured by *Saccharomyces* yeast (*S. cerevisiae*, *S. bayanus*, *S. uvarum*) and *T. delbrueckii*. *Eur. Food Res. Technol.* **2020**, *246*, 807–820. [CrossRef]

31. Wibowo, S.; Essel, E.A.; de Man, S.; Bernaert, N.; van Droogenbroeck, B.; Grauwet, T.; van Loey, A.; Hendrickx, M. Comparing the impact of high pressure, pulsed electric field and thermal pasteurization on quality attributes of cloudy apple juice using targeted and untargeted analyses. *Innov. Food Sci. Emerg. Technol.* **2019**, *54*, 64–77. [CrossRef]

32. Morata, A.; Benito, S.; Loira, I.; Palomero, F.; González, M.; Suárez-Lepe, J. Formation of pyranoanthocyanins by *Schizosaccharomyces pombe* during the fermentation of red must. *Int. J. Food Microbiol.* **2012**, *159*, 47–53. [CrossRef] [PubMed]

33. Swiegers, J.; Bartowsky, E.; Henschke, P.; Pretorius, I. Yeast and bacterial modulation of wine aroma and flavour. *Aust. J. Grape Wine Res.* **2005**, *11*, 139–173. [CrossRef]

34. Wei, J.; Zhang, Y.; Yuan, Y.; Dai, L.; Yue, T. Characteristic fruit wine production via reciprocal selection of juice and non-*Saccharomyces* species. *Food Microbiol.* **2019**, *79*, 66–74. [CrossRef]

35. Tredoux, A.; de Villiers, A.; Májek, P.; Lynen, F.; Crouch, A.; Sandra, P. Stir bar sorptive extraction combined with GC-MS analysis and chemometric Methods methods for the classification of South African wines according to the volatile composition. *J. Agric. Food Chem.* **2008**, *56*, 4286–4296. [CrossRef]

36. Zhong, W.; Liu, S.; Yang, H.; Li, E. Effect of selected yeast on physicochemical and oenological properties of blueberry wine fermented with citrate-degrading *Pichia fermentans*. *LWT* **2021**, *145*, 111261. [CrossRef]

37. Ohira, S.-I.; Kuhara, K.; Shigetomi, A.; Yamasaki, T.; Kodama, Y.; Dasgupta, P.; Toda, K. On-line electrodialytic matrix isolation for chromatographic determination of organic acids in wine. *J. Chromatogr. A* **2014**, *1372*, 18–24. [CrossRef] [PubMed]

38. Luo, L.; Cui, Y.; Zhang, S.; Li, L.; Suo, H.; Sun, B. Detailed phenolic composition of Vidal grape pomace by ultrahigh-performance liquid chromatography–tandem mass spectrometry. *J. Chromatogr. B* **2017**, *1068–1069*, 201–209. [CrossRef] [PubMed]

39. Sun, S.Y.; Jiang, W.G.; Zhao, Y.P. Evaluation of different *Saccharomyces cerevisiae* strains on the profile of volatile compounds and polyphenols in cherry wines. *Food Chem.* **2011**, *127*, 547–555. [CrossRef]

40. Li, H.; Huang, J.; Wang, Y.; Wang, X.; Ren, Y.; Yue, T.; Wang, Z.; Gao, Z. Study on the nutritional characteristics and antioxidant activity of dealcoholized sequentially fermented apple juice with *Saccharomyces cerevisiae* and *Lactobacillus plantarum* fermentation. *Food Chem.* **2021**, *363*, 130351. [CrossRef] [PubMed]

41. McCue, P.P.; Shetty, K. Phenolic antioxidant mobilization during yogurt production from soymilk using Kefir cultures. *Process. Biochem.* **2005**, *40*, 1791–1797. [CrossRef]

42. Morata, A.; Loira, I.; Heras, J.M.; Callejo, M.J.; Tesfaye, W.; González, C.; Suárez-Lepe, J.A. Yeast influence on the formation of stable pigments in red winemaking. *Food Chem.* **2016**, *197*, 686–691. [CrossRef]

43. Li, X.; Yao, Z.; Yang, D.; Jiang, X.; Sun, J.; Tian, L.; Hu, J.; Wu, B.; Bai, W. Cyanidin-3-O-glucoside restores spermatogenic dysfunction in cadmium-exposed pubertal mice via histone ubiquitination and mitigating oxidative damage. *J. Hazard. Mater.* **2020**, *387*, 121706. [CrossRef] [PubMed]

44. Chen, J.-Y.; Du, J.; Li, M.-L.; Li, C.-M. Degradation kinetics and pathways of red raspberry anthocyanins in model and juice systems and their correlation with color and antioxidant changes during storage. *LWT* **2020**, *128*, 109448. [CrossRef]

45. Millet, M.; Poupard, P.; Guilois-Dubois, S.; Poiraud, A.; Fanuel, M.; Rogniaux, H.; Guyot, S. Heat-unstable apple pathogenesis-related proteins alone or interacting with polyphenols contribute to haze formation in clear apple juice. *Food Chem.* **2020**, *309*, 125636. [CrossRef] [PubMed]

46. Tong, Y.; Deng, H.; Kong, Y.; Tan, C.; Chen, J.; Wan, M.; Wang, M.; Yan, T.; Meng, X.; Li, L. Stability and structural characteristics of amylopectin nanoparticle-binding anthocyanins in Aronia melanocarpa. *Food Chem.* **2020**, *311*, 125687. [CrossRef]

47. Yang, W.; Liu, S.; Marsol-Vall, A.; Tähti, R.; Laaksonen, O.; Karhu, S.; Yang, B.; Ma, X. Chemical composition, sensory profile and antioxidant capacity of low-alcohol strawberry beverages fermented with *Saccharomyces cerevisiae* and *Torulaspora delbrueckii*. *LWT* **2021**, *149*, 111910. [CrossRef]

48. Benito, S.; Morata, A.; Palomero, F.; González, M.; Suárez-Lepe, J. Formation of vinylphenolic pyranoanthocyanins by *Saccharomyces cerevisiae* and *Pichia guillermondii* in red wines produced following different fermentation strategies. *Food Chem.* **2011**, *124*, 15–23. [CrossRef]

49. Aprea, E.; Biasioli, F.; Gasperi, F. Volatile compounds of raspberry fruit: From analytical methods to biological role and sensory impact. *Molecules* **2015**, *20*, 2445–2474. [CrossRef]

50. Lin, V.C.-H.; Ding, H.-Y.; Kuo, S.-Y.; Chin, L.-W.; Wu, J.-Y.; Chang, T.-S.; Lin, C.-H.V. Evaluation of in vitro and in vivo depigmenting activity of raspberry ketone from *Rheum officinale*. *Int. J. Mol. Sci.* **2011**, *12*, 4819–4835. [CrossRef] [PubMed]

51. Restuccia, D.; Sicari, V.; Pellicanò, T.; Spizzirri, U.; Loizzo, M. The impact of cultivar on polyphenol and biogenic amine profiles in Calabrian red grapes during winemaking. *Food Res. Int.* **2017**, *102*, 303–312. [CrossRef]

52. Li, H.; Wang, X.; Li, Y.; Li, P.; Wang, H. Polyphenolic compounds and antioxidant properties of selected China wines. *Food Chem.* **2009**, *112*, 454–460. [CrossRef]

53. He, G.; Sui, J.; Du, J.; Lin, J. Characteristics and antioxidant capacities of five hawthorn wines fermented by different wine yeasts. *J. Inst. Brew.* **2013**, *119*, 321–327. [CrossRef]

Journal of
Fungi

A Comprehensive Insight into Fungal Enzymes: Structure, Classification, and Their Role in Mankind's Challenges

Hamada El-Gendi [1], Ahmed K. Saleh [2], Raied Badierah [3,4], Elrashdy M. Redwan [3,5], Yousra A. El-Maradny [5] and Esmail M. El-Fakharany [5,*]

1 Bioprocess Development Department, Genetic Engineering and Biotechnology Research Institute, City of Scientific Research and Technological Applications (SRTA-City), Universities and Research Institutes Zone, New Borg El-Arab, Alexandria 21934, Egypt; elgendi1981@yahoo.com
2 Cellulose and Paper Department, National Research Centre, El-Tahrir St., Dokki, Giza 12622, Egypt; asrk_saleh@yahoo.com
3 Biological Science Department, Faculty of Science, King Abdulaziz University, P.O. Box 80203, Jeddah 21589, Saudi Arabia; rbadierah@kau.edu.sa (R.B.); rredwan@gmail.com (E.M.R.)
4 Medical Laboratory, King Abdulaziz University Hospital, King Abdulaziz University, P.O. Box 80203, Jeddah 21589, Saudi Arabia
5 Protein Research Department, Genetic Engineering and Biotechnology Research Institute, City of Scientific Research and Technological Applications (SRTA-City), New Borg EL-Arab, Alexandria 21934, Egypt; dr.yousraadel@yahoo.com
* Correspondence: esmailelfakharany@yahoo.co.uk

Citation: El-Gendi, H.; Saleh, A.K.;
Badierah, R.; Redwan, E.M.;
El-Maradny, Y.A.; El-Fakharany, E.M.
A Comprehensive Insight into Fungal
Enzymes: Structure, Classification,
and Their Role in Mankind's
Challenges. *J. Fungi* 2022, 8, 23.
https://doi.org/10.3390/jof8010023

Academic Editor: Laurent Dufossé

Received: 22 November 2021
Accepted: 25 December 2021
Published: 28 December 2021

Publisher's Note: MDPI stays neutral
with regard to jurisdictional claims in
published maps and institutional affiliations.

Abstract: Enzymes have played a crucial role in mankind's challenges to use different types of biological systems for a diversity of applications. They are proteins that break down and convert complicated compounds to produce simple products. Fungal enzymes are compatible, efficient, and proper products for many uses in medicinal requests, industrial processing, bioremediation purposes, and agricultural applications. Fungal enzymes have appropriate stability to give manufactured products suitable shelf life, affordable cost, and approved demands. Fungal enzymes have been used from ancient times to today in many industries, including baking, brewing, cheese making, antibiotics production, and commodities manufacturing, such as linen and leather. Furthermore, they also are used in other fields such as paper production, detergent, the textile industry, and in drinks and food technology in products manufacturing ranging from tea and coffee to fruit juice and wine. Recently, fungi have been used for the production of more than 50% of the needed enzymes. Fungi can produce different types of enzymes extracellularly, which gives a great chance for producing in large amounts with low cost and easy viability in purified forms using simple purification methods. In the present review, a comprehensive trial has been advanced to elaborate on the different types and structures of fungal enzymes as well as the current status of the uses of fungal enzymes in various applications.

Keywords: fungi; enzymes; structure and classification; function; applications

1. Introduction

Enzymes play an important role in lowering the energy of activation and accelerating numerous biological reactions that are crucial in nourishing life without causing any permanent modifications. Enzymes are widely distributed in all forms of living organisms, including animal, plant, and microbial sources. The inability of plant and animal sources to satisfy industrial demands of enzymes directed attention to microbial sources [1]. Microbial enzymes production is faster, cost-effective, scalable, and amenable to genetic manipulations [2]. Among microbial sources, fungi represent an interesting source for industrial enzymes, sharing with more than one has of enzymes in the market [3]. Fungal enzymes are characterized by high production potency, easier purification and separation requirements, especially for filamentous fungi, and efficient catalysis with desired stability against harsh conditions [3]. Furthermore, the uses of fungi in a great number of traditional preparations, such as brewing and baking from ancient periods, offer a safe and firm context for their recent applications [3]. Enzymes have an important role in mankind's challenges to use biological systems for a wide range of dedications. Currently, fungal enzymes are accounted for more than 50% of the total enzymes market [3]. This huge market share is largely attributed to a few species of *Aspergillus, Trichoderma, Rhizopus,* and *Penicillium* genera that fulfill the commercial-scale requirements for enzymes production (Table 1). The fast-growing enzymes market forced the continuous search for novel enzymes producers with desirable industrial characteristics. Recently, mushroom cultivation has represented a promising competitor in enzymes production in terms of higher productivity and lower invested cost [4]. Attributed to the nutritional value and health benefits of the edible mushroom, its global production pace is very fast, recording 8.99 million tons in 2018 [4]. Mushrooms are mostly saprophytic fungi and derive their nutrient from surrounding cellulosic material; hence, lignocellulolytic enzymes are highly characterized in most mushroom cells, including cellulases, β-glucosidases lignin peroxidases, and laccases [5,6]. This diverse enzymatic machine enhanced the ability of mushrooms to grow in diverse low-cost agriculture wastes inspiring their industrial cultivation for nutrition and enzyme production [4].

Table 1. General application of fungal enzymes in different fields.

Application	Field	Fungal Name	Enzyme Name	Enzyme Use	Ref.
Industries	Food and beverage	*Aspergillus oryzae, Aspergillus oryzae* CCT 3940, and *Fusariumculmorum* ASP-87	L-asparaginases	Reduce the acrylamide formation in potato chips or French fries, bakery products, and coffee by degradation of L-asparagine	[7–9]
		Myceliophthora thermophilia	Laccases	Dough conditioner	[10]
		Aspergillus niger DFR-5	Xylanases	Improve yield and clarity of pineapple juice	[11]
		Aspergillus niger	Pectinases	Improve the quantity of the extracted Orange juice	[12]
		Talaromyces leycettanus	Pectinases	Efficiency in pectin degradation from grape juice	[13]
	Pulp and paper	*Pycnoporus cinnabarinus*	Laccases	Improve the brightness and strength properties of the pulp	[14]
		Trametes villosa	Laccases	Internal sizing of paper by use of laccase and hydrophobic compounds	[15]
		Trichoderma reesei QM9414	Xylanases	Eco-friendly of biobleaching of Kraft pulp of sugarcane straw	[16]
		Trichoderma viride VKF-3, *Fusariumequiseti* MF-3, and *Aspergillus japonicus* MF-1	Cellulases, xylanases, laccases, and lipases	Treatment enhances the brightness, deinking and reduces the heavy metals in the newspaper pulp	[17]
		Rhizopus oryzae MUCL 28168 and *Fusarium solani*	Tannases	Detoxification of coffee pulp by reduction of caffeine and tannins	[18]
		Aspergillus niger, Phanerochaete chrysosporium, and *Pycnoporus cinnabarinus*	Feruloyl esterases, Mn^{2+}-oxidizing peroxidases, and laccases	Decrease the final lignin content of flax pulp and improvement of pulp brightness	[19]

Table 1. *Cont.*

Application	Field	Fungal Name	Enzyme Name	Enzyme Use	Ref.
	Textile	*Aspergillus niger* CKB and *Trichoderma reesei* ATCC 24449	Cellulases	Textile waste hydrolysis for recovery of glucose and polyester	[20,21]
		Trichoderma longibrachiatum KT693225	Xylanases	Desizing, bioscouring, and biofinishing of cellulosic fabrics (textile) without adding any additives	[22]
		Aspergillus	Amylases	Desizing of cotton fibers by removal of starch from the surface of textile fibers	[23]
		Candida orthopsilosis	Pectinases	Bioscouring of cotton fibers	[24]
		Aspergillus niger and *Penicillium*	Glucose oxidases and catalases	Removal of hydrogen peroxide from cotton bioprocessing	[25]
		Chaetomium globosum IMA1	Lignin peroxidases laccases and manganese peroxidases	Decolorization of the industrial textile effluent	[26]
Environment	Biodegradation	*Irpex lacteus* and *Pleurotusostreatus*	Manganese-peroxidases and laccases	Biodegradation of chlorhexidine and octenidine as antimicrobial compounds used in oral careproducts	[27]
		Marasmius sp.	Laccases	Degrade lignin by oxidizing the phenolic and non-phenolic compounds to produce dimers, oligomers, and polymers	[28]
		Mucor circinelloides	Lipases, laccases, and peroxidases	Biodegradation of diesel oil hydrocarbons	[29]
	Bioremediation	*Penicillium* sp.	Enzymatic reduction by the *mer* operon	The fungal enzyme could detoxify mercury (II) by extracellular sequestration via adsorption and precipitation	[30]
		Coriolopsis gallica	Laccases	Bioremediation of pollutants such as bisphenol, diclofenac, and 17-a-ethinylestradiol in real samples from the AQUIRIS wastewater	[31]
		Aspergillus flavus FS4 and *Aspergillus fumigates* FS6	Extracellular enzymes	Fungal consortium used for removal of chromium and cadmium	[32]
		Pycnoporus sanguineus	Laccases	Fugal laccase was immobilized on calcium and copper alginate/chitosan beads and used for the removal of 17 a-ethinylestradiol	[33]
		Thermomyces lanuginosus	Chitinases	Biocontrol agent against larvae of *Eldana saccharina* and fungi of *Aspergillus* sp., *Mucor* sp., and *Fusarium verticillioides*	[34]
	Decolorization	*Phanerochaete chrysosporium* CDBB 686	Ligninolytic enzymes	Decolorization of Congo red, Poly R-478, and Methyl green	[35]
		Geotrichum candidum	Peroxidases and laccases	Decolorization of methyl orange, Congo red, trypan blue, and Eriochrome black T	[36]
		Coprinopsis cinerea	Laccases	High indigo dye decolorization	[37]
		Trametes sp. SYBC-L4	Laccases	Decolorization of Congo red, aniline blue, and indigo carmine	[38]
		Phanerochaete chrysosporium	Manganese peroxidase	Decolorization of AO7 or CV pigment	[39]

Table 1. *Cont.*

Application	Field	Fungal Name	Enzyme Name	Enzyme Use	Ref.
Biomedical	Antimicrobial	32 different isolated fungi identified by morphological characteristics and internal transcribed spacer sequence analysis	Amylases, proteases, pectinases, xylanases, cellulases, chitinases, and lipases	Antimicrobial activity against pathogenic organisms by agar diffusion assays	[40]
		Aspergillus oryzae and *Aspergillus flavipes*	Proteases	Production of bioactive peptides from bovine and goat milk and the generated peptides tested against bacteria and fungi	[41]
		Trichoderma harzianum	Chitinases	Degradation of chitosan to form chitosan-oligosaccharides and used as antimicrobial against pathogenic organisms	[42]
	Anticancer	*Trichoderma viride* AUMC 13021	Chitinases	Antitumor efficiency of chitinase against different types of cancer cell line	[43]
		Trichoderma harzianum	Chitinases	Chitosan-oligosaccharides used as anticancer compounds, which inhibited the growth of cervical cancer cells at concentration of 4 mg/mL and significantly reduced the survival rate of the cells	[42]
		Aspergillus terreus	L-asparaginases	The synthesized zinc oxide conjugated L-asparaginase nanobiocomposite on MCF-7 cell line using MTT assay	[44]
	Antioxidant	*Aspergillus flavus*	Catalases	Antioxidant system plays a crucial role in fungal development, aflatoxins biosynthesis, and virulence	[45]
		Pleurotus columbinus, P. foridanus, Aspergillus fumigatus, and *Paecilomyces variotii*	Peroxidases and catalases	Production of enzymatic antioxidant from peels of banana, pomegranate, and orange	[46]
		Chytridiomycetes sp.	Ligninases	During biodegradation of lignin, the fungi synthesize bioactive compounds such as mycophenolic acid, dicerandrol C, phenyl acetates, anthrax quinones, benzo furans, and alkenyl phenols that have antioxidant activities	[47,48]

In general, enzymes are classified under seven classes, including transferases, oxidoreductases, lyases, hydrolases, ligases, isomerases, and translocases [49]. Fungal cells revealed both intracellular and extracellular enzymatic activity. Fungi are considered natural decomposers, and consequently, they are allowed to produce a great number of extracellular enzymes needed for bioconversion of a wide range of substrates and complexes [50]. Other extracellular enzymes may participate in fungal protection against hazardous compounds that exist naturally or result from the substrate hydrolysis process. Commercially important fungal enzymes belong to hydrolases and oxidoreductases. Among extracellular fungal enzymes, laccases and peroxidases, such as manganese peroxidase and lignin peroxidase, represent the two main subclasses of fungal enzymes that have been investigated for the degradation of xenobiotics and removal of harmful phenolic components from industrial environment and wastewater [51]. The current review discusses the different types of enzymes produced by fungi involved in many biotechnological applications. The structure of the fungal enzymes is also discussed. Numerous biotechnological uses of fungal enzymes, including industrial, medical, bioremediation, and environmental applications, are critically reviewed.

2. Fungal Enzymes Structure and Function

Enzymes are proteins in nature, formed originally from a long chain of amino acids linked together through peptide linkage. Once the primary amino acid chain is formed, subsequent multistep processes take place, collectively known as post-translation modification, where this long chain is modified into its three-dimensional structure (Figure 1). Sequence and nature of the amino acid in various enzymes are varied and have an obvious relationship to enzyme activity and stability. Not all enzyme molecules get involved in the catalytic reaction, but only very few amino acids of (2–4) are called the active site [52], where the remaining part (called Apoenzyme) does not participate directly. Most enzymes are synthesized inactive and activated to holoenzyme (the functional form of an enzyme) with different organic/or inorganic factors. The organic enzyme activators are usually non-proteinaceous molecules known as coenzymes, e.g., vitamins, whereas the inorganic molecules are called cofactors, e.g., zinc, iron, calcium, and copper [52]. The accumulation of enzymes end products may terminate the enzyme activity in a process called feedback inhibition. The feedback enzyme inhibition is a regulatory mechanism initiated by cells to control the amount needed from the enzyme end products [53]. Generally, enzymes are very specific catalysts; however, some enzymes have broad substrate specificity and could work upon multiple substrates. Different theories were proposed to elucidate the enzyme–substrate interaction and recognition, but the most popular one is the lock and key theory proposed by Emil Fischer [54]. This theory elucidates the enzyme and substrate recognition as unique specific complementation between enzyme active site and substrate resembling lock and key. Unfortunately, this theory could not explain the transition state stabilization, where enzyme configuration was altered to fit the proceeding catalytic process [52,55]. To explain these configuration changes in the enzyme active site, a flexible enzyme active site concept was proposed by Koshland, establishing the enzymes Induced Fit Model. In this model, the active site is flexible, where the substrate can modify its orientation to achieve the perfect fit [56]. In Michaelis–Menten's theory, two phases of catalytic reaction were proposed, where in the first phase, an enzyme-substrate complex was established, and then, the enzyme and products were liberated in the second phase [57].

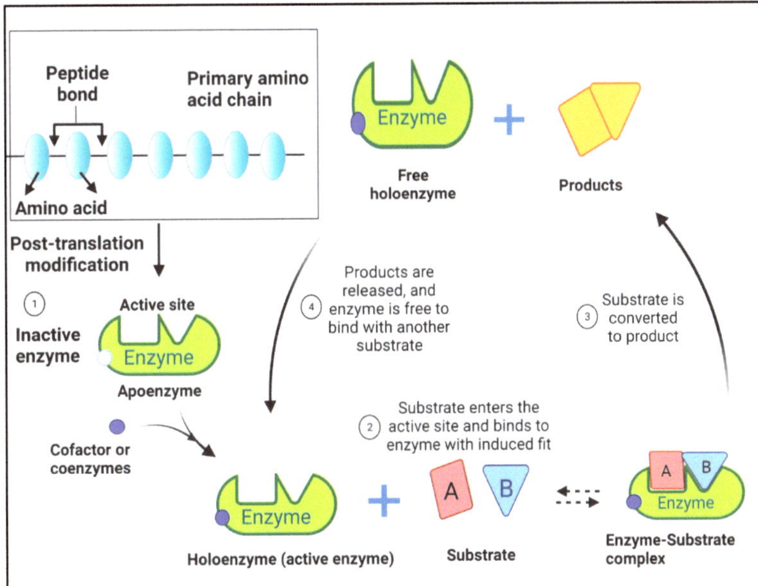

Figure 1. Schematic illustration for enzyme structure, activation, and steps of enzyme and substrate interaction.

3. Fungal Enzyme Nomenclature and Classifications

Nowadays, fungal enzymes represent more than one-half of the market enzymes sharing in the different industrial sectors [3]. Due to the vast number of enzymes discovered to date, many criteria were proposed to organize and classify enzymes, including their origin, structure, or substrate. The first rational enzymes nomenclature and classification system were firstly proposed in the 1950s, based upon the enzyme catalytic reaction [58]. The International Union of Pure and Applied Chemistry (IUPAC) adopted the new system, where enzymes are classified into six groups: oxidoreductases, transferases, hydrolases, lyases, isomerases, ligases, and translocases [59]. This classification is a numerically based system where each catalytic reaction (enzymes group) is given a number called commission number (EC number) to specify its enzyme members [52].

3.1. Oxidoreductases (EC 1)

Oxidoreductases are enzymes catalyzing the oxidation/reduction (redox) reaction through the transfer of an electron from donor to an electron acceptor. Generally, redox reactions are numerous in the living cells and included in many vital biological processes such as tricarboxylic acid cycle, glycolysis, amino acid metabolism, and oxidative phosphorylation reactions; hence, oxidoreductases are among the housekeeping enzymes essential for cell's life and activity. Recently, fungal oxidoreductases were characterized to play a major role in pathogenicity and protection against host defense mechanisms [60]. Yu et al. reported the importance of oxidoreductase-like protein Olp1 in *Cryptococcus neoformans* for sexual reproduction through enhancing the meiotic division and its essential role in the pathogenicity by protecting the fungal cells from lithium–ion toxicity [60]. Benzoquinone oxidoreductase is another reported fungal virulence factor produced by *Beauveria bassiana* (Entomopathogenic fungus) to overcome the exposure to benzoquinone from the host cell [61]. To date, more than twenty classes of oxidoreductases have been recognized; however, the most studied and reported oxidoreductases include dehydrogenases (transfer of hydrogen to an electron acceptor), oxygenase (the final electron acceptor is oxygen), and peroxidases (the final electron acceptor is peroxides) [2]. Commercially applied oxygenases include monooxygenases, dioxygenases, and laccases. Fungal laccases are copper-containing oxygenases and play an important role in fungal physiological growth, including sporulation, pigmentation, and aid in plant pathogenesis [62]. Laccases have a high potential to oxidize a wide range of phenolic and non-phenolic compounds, independently of any cofactors [2,62]. The high efficiency of fungal laccases, in addition to cofactor-independence nature, nominated them to several industrial applications such as stain removal [63], paper industry [64], medical applications [65], and biosensor manufacturing [66]. Though microbial laccases were reported from bacteria [67,68], white-rot fungi are the main source of laccases with commercial importance [69]. Peroxidases are oxidoreductases where peroxide compound (H_2O_2) acts as the final electron acceptor [70]. The majority of peroxidases recognized to date are metal-dependent enzymes, especially iron; however, metal-independent peroxidases were also reported [70]. Two of the most extensively reported peroxidases are lignin peroxidases (LiP) and manganese-peroxidases (MnP). Basidiomycetes, especially white-rot fungi, are characterized by their high ability to degrade lignin attributed to the extracellular production of lignin peroxidases and manganese-peroxidases [71]. Lignin peroxidases (LiP) catalyze the oxidation of a wide range of phenolic compounds; hence, they have several industrial applications [72], in contrast to manganese-peroxidase, which has narrow substrate specificity and low oxidation potential [70].

3.2. Transferases (EC 2)

Transferases catalyze the transfer or exchange of certain groups (amino group) among many compounds [73]. This process has a vital role in creating essential amino acids for protein synthesis in all cells [74]. Some transferases such as glutathione transferase may assist in environmental and oxidative stress tolerance in pathogenic fungi [75]. Glutathione transferases catalyze the reversible transfer (incorporation/elimination) of glutathione group to/from different compounds [76]. Incorporation or elimination of glutathione group resulted in more soluble fewer toxic compounds that can be handled by the fungal cells. Fungal glutathione transferases secreted by fungal cells during host pathogenesis to tolerate oxidative stress created by host and/or toxic wood-derived molecules [77,78]. Fungal glutathione transferases have wide substrate specificity; therefore, they have the potential to detoxify many toxic xenobiotic compounds [76,78,79]. The production of glutathione transferases was reported from many fungi, including *Alternaria brassicicola* [78], *Phanerochaete chrysosporium* [80], and *Saccharomyces cerevisiae* [81]. Another commercially important transferase is fructosyltransferase, which catalyzes the transformation of sucrose into fructooligosaccharides [82]. Fructooligosaccharides are commercially important due to their numerous health benefits as lowering blood pressure and flourishing the growth of beneficial microflora while inhibiting the other pathogenic ones [83]. The sweetness effect and viscous texture of fructooligosaccharides encourage their application in the food industry [82]. Fructosyltransferase production was widely reported from the Aspergillus genus, including *Aspergillus oryzae* [84], *Aspergillus niger* [85], *Aspergillus aculeatus* [86]. However, [82] reported applying *Aureobasidium pullulans* for fructosyltransferase commercial production.

3.3. Hydrolases (EC 3)

Hydrolases are the most extensively studied groups of enzymes; they catalyze the hydrolysis of their substrate through the addition of water. To date, hydrolases represent the most commercially marketed enzymes due to their wide application in different industrial sectors [3,52,87]. Fungal proteases, amylases, lipases, and cellulases represent the most commercially demanded enzymes [88]; hence, continuous improvement in production efficiency with lowered cost are mandatory [89].

Proteases play an important role in fungal physiology to digest extracellular large peptides and also in defense mechanisms against attaching pathogens [90]. Based upon the amino acid in the enzyme active site, proteases could be categorized into different types, including serine, asparagine, cysteine, aspartic, and metalloproteases [90]. Serine and metalloprotease are the most studied types among all proteases and are usually produced from microbial origins [90]. Fungal proteases are privileged over that of bacteria for easier purification process and less hazardous application [91,92]. Filamentous fungi, especially that of *Aspergillus* sp., are characterized by their high capacity for protease production [89,93,94]. Other fungal genera also reported for their potency regarding proteases production, including *Penicillium* sp. [95], *Fusarium* sp. [96,97], and *Pichia farinosa* [98].

Amylase is the world premiere in enzyme production for commercial application and was firstly applied medicinally in treating digestive disorders [99]. Amylases could be classified into α, β, and γ-Amylases depending on the attaching site in the starch molecules and the nature of the resulting products. α-Amylases are calcium-dependent metalloenzymes that act randomly on the starchy substrates yielding maltose and maltotriose from amylose or glucose and dextrin from amylopectin [87]. β-Amylases hydrolyze 1,4-glycosidic bonds in the carbohydrate chain, yielding one maltose unit at a time. They are extensively important in plants, especially in the seed ripping process, but they are also reported from the microbial origin [100,101]. γ-Amylases resemble the other two types of amylases in hydrolysis activity toward 1,4-glycosidic linkages, unlike the two forms characterized with 1,6-glycosidic linkages hydrolysis activity and preferring acidic environment pH 3 [87]. *Aspergillus niger* is considered the potent commercial α-Amylase producer among all filamentous fungi [87]. Many other fungi were reported for their capacity to produce different types of amylases, including *Aspergillus oryzae* [102], *Aspergillus terreus* [103], *Fusarium solani* [104], and *Penicillium citrinum* [105].

Lipases are a group of hydrolytic enzymes that act by hydrolysis of triacylglycerol yielding fatty acid and glycerol [106]. Lipases also catalyze the reverse reaction by esterification of glycerol and fatty acid. Microbial lipase is produced by many bacteria *Bacillus licheniformis* [107], *Geobacillus thermodenitrificans* [108], *Pseudomonas aeuriginosa* [109], and fungal cells including *Aspergillus niger* [110,111], *Penicillium verrucosum* [112], *Fusarium solani* [113], *Arthrographis curvata*, and *Rhodosporidium babjevae* [114]. The remarkable specificity and thermal stability, in addition to the easier recovery procedure of the extracellular fungal lipase over other microbial sources, attracted great attention from the application point of view [115–117]. Lipases are implemented in vast commercial applications, including detergents and cosmetics additives, fine chemical production, medical application, paper pitching, leather de-fating, wastewater treatment, and biodiesel production [118–120]. The application of lipase in biodiesel production, as an ecofriendly alternative for traditional fuel, intensifies the research in diminishing the production cost and enhancing the enzyme efficiency. In this regard, solid-state fermentation of agricultural wastes represents a step forward cost reduction [110], while protein engineering has been recently applied for enhancing lipase efficiency [121].

The continual necessity for renewable ecofriendly fuel sources intensified the research in cellulose-degrading enzymes. The worldwide abundance of cellulosic material represents a promising ideal source for energy, hence the extensive studies and application of cellulolytic enzymes [48], ranking them among the most worldwide marketing enzyme [122]. Cellulose, hemicellulose, and lignin are the main components of most agricultural wastes representing 40–50%, 25–30%, and 15–20% for cellulose, hemicellulose, and lignin, respectively [48]. Most fungi have the complete enzymatic system (Endoglucanases, Cellobiohydrolases, B-glucosidases, and Xylanases) to degrade this complex cellulosic material for nutrition [123]. *Trichoderma reesei* is widely applied for the commercial production of cellulases [48], but other fungi also represent potent cellulase producers, including *Aspergillus niger* [124], *Saccharomyces cerevisiae* [125], and *Aspergillus brasiliensis* [126]. Xylan, a complex polysaccharide, is also a major component of hemicellulose; hence, xylanases play an important role in the efficient hydrolysis of plant cellulolytic material [127]. Regarding the diverse and complex structure of Xylan, its hydrolysis required a group of synergistically working enzymes (xylanolytic system) for complete degradation [128]. Filamentous fungi are characterized by the required xylanolytic system for complete xylan degradation, especially that of *Trichoderma reesei* [128,129], *Aspergillus oryzae* [130], and *Aspergillus flavus* [131]. Vast numbers of other hydrolytic enzymes are produced by fungi with wide commercial applications, including fungal phytase and pectinase. Phytase act upon phytic acid (plant inorganic phosphorus), dissolving its content of insoluble inorganic phosphorus makes it available for different physiological processes. Fungal phytase is available for different commercial applications from *Aspergillus niger* [132]. Pectinase also has important industrial applications, especially in food and paper industries, produced commercially through fungi [133].

The worldwide availability of chitin, representing the second abundant natural polymer next to cellulose, renders chitinases among the most important enzymes [134]. Fungal chitinases belong to hydrolytic enzymes with an extracellular role in chitin decomposition, where intracellularly involved in cell wall lysis and reconstruction in addition to protein deglycosylation [135]. Chitin is one of the main structures in pests' cell walls and some plant-pathogenic fungi [136]. Hence, chitinases are widely applied in agriculture for plant infection control to overcome the conventional chemical fungi/insecticidal hazardous [137]. In addition, the chitin degradation products (chitosan and chitooligosaccharides) are implemented for various medical applications [134]. Trichoderma viride chitinases were reported with acidoplic, thermostable, and wide chitinolytic activities for commercial applications under the name of Usukizyme [134,138]. Chitinases were reported within various fungal species, including thermostable chitinases (maximum activity between 60 and 55 °C) through *Myceliophthora thermophila* [139], and *Humicola grisea* [140] with thermotolerant/mesophilic chitinases (maximum activity between 40 and 37 °C) through *Penicillium oxalicum* [137] and *Trichoderma koningiopsis* [141].

3.4. Lyases (EC 4)

Lyases are a group of enzymes that catalyze the addition or elimination reaction. The result of this type of addition or elimination is a new compound with either cyclic structure or new double bonds [73]. Though lyases usually require only one substrate to act in one direction, two substrates are necessary to reverse this reaction. Normally, lyases have very limited substrate specificity; however, the independence of expensive cofactors encourages their commercial applications in many industries. One of the most reported lyases is pectin lyase catalyzes random degradation of pectin through the elimination of water molecules yielding one unite of galacturonan with unsaturated double bonds [142]. Pectin lyase plays a major role in fungal pathogenicity, though degrading the pectic materials holding the plant cell wall together [142]. The role of pectin lyase in fungal pathogenicity was reported in many plant pathogenic fungi, such as *Clonostachys rosea* [143], *Cylindrocarpon destructans* [142], and *Penicillium canescens* [144]. Alginate lyase represents another important fungal lyase; it catalyzes the degradation of alginate to defined oligosaccharides. Alginate lyase was reported in fungi, including terrestrial fungus *Aspergillus oryzae* [145] and marine fungus *Paradendryphiella salina* [146].

3.5. Isomerases (EC 5)

Isomerases are a group of enzymes that catalyze the rearrangement of its substrate structure through the interchange of a specific group within the same compound [52]. One of the most widely studied and applied fungal isomerases is glucose/xylose isomerase, representing the third marketing enzyme with protease and amylase [147]. Glucose/xylose isomerase reversely converts D-glucose and D-Xylose into D-fructose and D-Xylulose, respectively [148,149]. Though glucose/xylose isomerase was extensively reported from many bacteria as *Bacillus licheniformis* [147], *Serratia marcescens* [150], *Caldicellulosiruptor bescii* [151], and *Streptomyces lividans* [152], the fungal enzyme was reported from a few species of *Aspergillus* genus [153]. Marshall and Kooi reported for the first time the ability of xylose isomerase from *Pseudomonas hydrophila* to convert D-glucose into D-fructose [154]. Production of high fructose corn syrup through glucose isomerase has wide application in the food industry [148]. With emerging of white biotechnology and biofuel concepts, great attention was directed toward xylose isomerase as a tool to overcome the xylose accumulation issue. The hydrolysis of lignocelluloses is a prerequisite for the growth of *Saccharomyces cerevisiae* and biofuel production. The monosaccharide in hydrolysate liquor is a mixture of glucose (60–70%) and xylose (30–40%) [155], while *Saccharomyces cerevisiae* does not have the necessary system to assimilate the xylose [84]. Taking into account that assimilation of xylose in bacteria is one-step process and independent in any cofactor, contrary to that of fungi which is multi-steps and dependent on expensive cofactors [155], many trials were conducted to insert the gene for bacterial xylose isomerase in the genome of *Saccharomyces cerevisiae* and were first achieved with the xylose isomerase gene from *Termus thermophilus* [156]. Recently, the expression of xylose isomerase from *Clostridium phytofermentans* in *Saccharomyces cerevisiae* greatly enhanced xylose fermentation [157].

3.6. Ligases (EC 6)

Ligases, in general, are those classes of enzymes that catalyze the joining of two compounds through the formation of new bonds. Different types of ligases may be classified according to the nature of bonds established in the new compounds, such as carbon–carbon bonds, carbon–nitrogen bonds, carbon–sulfur bonds, carbon–oxygen bonds, nitrogen–metal bonds, and phosphoric–ester bonds [73]. To date, most of the reported ligases are intracellular, where their main function is modifying the cellular nucleic acid content. The unique structure and mechanism of some fungal ligases represent a promising target for developing new antifungal drugs. The tRNA ligase (Trl1) represents a good candidate for such drugs. The tRNA ligase (Trl1) plays an important role in repairing the RNA breaks in fungal cells [158]. Fungal tRNA ligases (Trl1) were reported to have a similar structure in different pathogenic fungi [159,160]. Most bacteria and mammalian cells have different biochemical mechanism and structure of tRNA ligase from that reported in fungi [160]; therefore, an extensive research directed toward fungal tRNA ligases (Trl1) as a target for new antifungal drugs [160]. Apart from the molecular level, ligases were also reported to have other roles in different fungi. Ubiquitin ligases (E3s) catalyze the ligation of ubiquitin (low molecular weight protein) to the target protein (a process known as ubiquitination). Conjugation of protein to ubiquitin determines its fate inside the cell, whether to be degraded, activated, or relocated [161,162]. Unicellular fungi, including *Saccharomyces cerevisiae*, represent an ideal model for fully understanding the evolutionary background for the ubiquitination system [163].

4. Application of Fungal Enzymes

The present review discusses the advancement in fungal enzyme technology for different applications. A comprehensive list of fields, fungal source of enzymes, enzymes names, and the wide range of their application is conceded. Table 1 and Figure 2 give an overview of applications of fungal enzymes in different application fields.

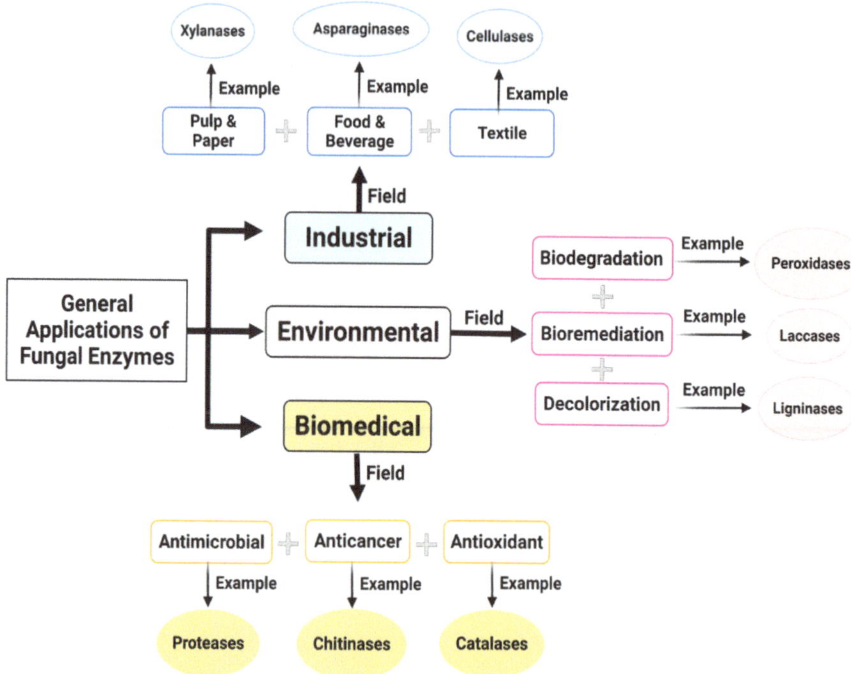

Figure 2. Schematic diagram of general application of fungal enzymes.

4.1. Industrial Applications

4.1.1. Food and Beverage

Food enzymes are also provided by microbial fermentation (bacteria, fungi, yeasts, actinomycetes, and algae), for which fungal strains are widely used. The application of fungal enzymes in food production offers a promising approach to enhance the preservation and shelf life of foods without affecting the characteristics of the organoleptic and nutritional content of foods [164]. The common fungal enzyme used in food applications is L-asparaginase which is considered GRAS (generally recognized as safe), which is used as a food additive to prevent the formation of acrylamide generated from the reaction between the food components as amino acids and reducing sugars at high temperatures and low humidity. This response often creates desirable color, flavor, and aroma compounds [165,166]. To minimize acrylamide levels in food, L-asparaginase assists with mitigation strategies from two aspects, intervention with the Maillard reaction or by converting L-asparagine into l-aspartic acid (non-toxic) to prevent precursors formation without altering the nutritional value, quality, or taste of the final product [167–169]. Single or consortium fungal enzymes were applied to improve the quality, clarity, and yield of fruit juice that ensure consumer appeal [11].

4.1.2. Pulp and Paper

The government's policies impose sustained pressure on the paper industry to protect the environment from hazards and diseases by replacing the chemical bleaching processes with greener ecofriendly alternatives. Enzymes such as xylanases and laccases obtained from fungi have the ability for bio-bleaching of agriculture wastes-based pulps by cleaving the β-1, 4 backbone of the complex plant cell wall for papermaking [170,171]. Ecofriendly papermaking depends on the two processing, Bio-pulping followed by bio-bleaching. Bio-pulping is the pre-treatment of agriculture-waste-based pulp by lignin-degrading enzymes before the routine pulping process. The process of bio-pulping is technologically feasible and cost-effective, reduces energy costs, and improves paper strength properties [172]. Filamentous fungi, especially members of white-rot fungi are the main source for lignin-degrading enzymes, whoever those with cellulases-free activity are important for the paper industry [173]. *Ceriporiopsis subvermispora, Ganoderma australe Phellinus pini,* and *Phlebia tremellosa* are three white-rot fungi with potent cellulases-free lignin-degrading activities suitable for the bio-pulping process where *Phanerochaete chrysosporium* is the model white-rot fungi with a complete enzymes system for lignin mineralization [174]. Bio-bleaching refers to the complete removal of lignin from bio-pulping wastes by the fungal enzyme to achieve white and bright pulp appropriate for papermaking. Bio-bleaching refers to the complete removal of lignin from bio-pulping wastes by the fungal enzyme to achieve white and bright pulp appropriate for papermaking. In addition to improving paper quality, bio-bleaching reduces the applied chemicals that diminish effluent toxicity and pollution [170]. Xylanases from *Fusarium equiseti* MF-3 and *Talaromyces thermophilus* were successfully applied for the bio-pulping process with a significant reduction in the organo-chloro compounds applied in the conventional bleaching process [17,175].

4.1.3. Textile

Fungal enzymes play an important role in enhancing textile processing as well as valorizing the waste generated from streams of fibers, textile, and clothing manufacturing processes to value products [176]. Of the 7000 known enzymes, only about 75 are commonly used in textile manufacturing [177]. Xylanases are among the most common enzymes used for their effectiveness in textile processing at several stages, including desizing, bio-scouring (removing waxy material from fibers), suitable for coloration with low conductive fabrics, bio-bleaching, and bio-finishing of cellulosic fabrics to improve the textile quality [22]. The main mechanism for using fungal enzymes in textile processing, removal or degradation of undesirable components such as lignin, pectin, starch, excess H_2O_2 after bleaching, waxes by xylanases, or polygalacturonases or amylases form a fibers-based textile. Cellulases for cotton remove projecting fibers selectively, and this action is known as bio-polishing. In contrast, the textile wastes contain 35–40% of cellulosic fibers, which is utilized as a carbon sole for the production of bioethanol, biocellulose, etc., after being degraded by cellulases enzyme originated from *Trichoderma reesei* ATCC 24,449 [20]. Another application of fungal enzymes, which degrade the textile dyes by using *Pleurotus ostreatus* and *Pleurotus eryngii*, had the activity of laccase, lignin peroxidase, manganese peroxidase, xylanase, cellulase, and lipase for removal of days after 24 h [178]. The fungal enzymes applications textile is illustrated in Figure 3.

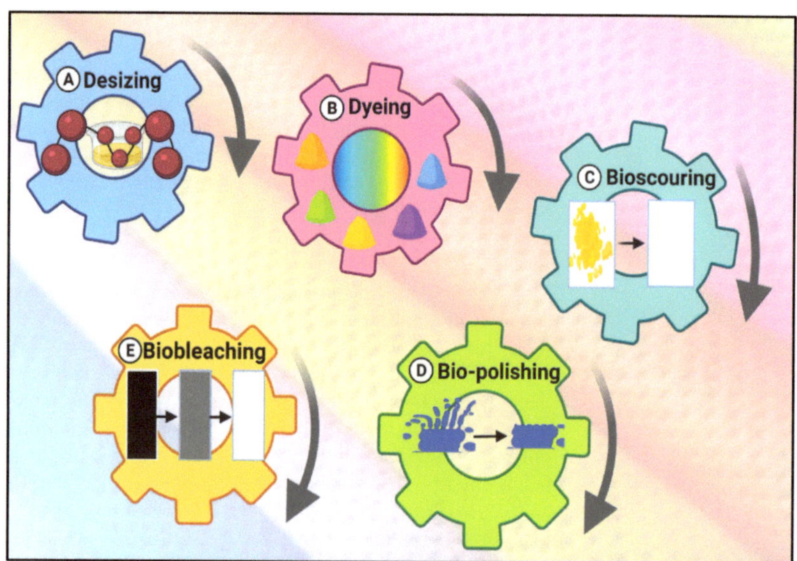

Figure 3. Fungal enzyme applications for textiles. (**A**) Desizing agents such as amylases; (**B**) dyeing and dye diffusers such as proteases; (**C**) textile processing and bioscouring of cotton fibers such as pectinases; (**D**) biofinishing and biopolishing such as cellulases; and (**E**) biobleaching and enzymatic rinse process after reactive dyeing such as peroxidases.

4.2. Environmental Applications

Bioremediation

With the continuous increase in the human being population, production, and consumption, industrial and agricultural wastes emerged as cumulative pressing environmental challenges that threaten human health and welfare. Waste management through fungal enzymes arises as promising solutions toward green and sustainable environments with the potential for generating new valuable products in a less consuming energy approach [179,180]. Currently, polyethylene-based materials (plastic packaging) are heavily consumed commercially attributed to their relatively low cost, and hence environmentally accumulated at a very high rate. Polyethylene-derived material revealed a high resistance for biological bioremediation; therefore, developing an effective approach for polyethylene removal is a pressing necessity from an environmental perspective. In this regard, a mixture of fungal oxidoreductases including laccase, manganese-peroxidase, lignin peroxidase, and unspecified peroxygenase was firstly studied (in silico) for eco-friendly polyethylene degradation [181]. The molecular docking results indicate the importance and essential role of laccase and unspecified peroxygenase in the polyethylene degradation process [181].

In the same regard, the widespread contamination with polyaromatic hydrocarbons from industrial effluents represents a huge challenge for human and animal health. In nature, polyaromatic hydrocarbons usually react to surrounding compounds to form exceptional multi-complex structures, such as aliphatic alkanes/alkenes, or halogenated hydrocarbons, potentially more toxic than the original structures [182]. Long-term exposure to polyaromatic hydrocarbons associated with many health complications included hepatic and renal toxicity and lung function damage, with a high rate of cancer development [183]. To overcome these problems, *Aspergillus oryzae* and *Mucor irregularis* were isolated from crude oil and reported as hydrocarbon degradation by secretion of several enzymes such as laccase, manganese peroxidase, and lignin peroxidase [184].

Recently, fungal laccases were widely applied for the bioremediation of synthetic textile dyes [185] by altering the dye molecules into non-colored, safer, and ecofriendly structures. Dyes-based industries (textile, paper, paint, and tannery) usually discharge their wastewater into water systems without any prior treatment and are heavily loaded with many unused synthetic dyes. These dyes imply many serious impacts in the aquatic environment, including increasing water turbidity and inhibiting the growth of some organisms like algae. In addition, most of these dyes are carcinogenic with high hepatocellular and renal toxicity upon long-term exposure [186]. There are two major mechanisms for decolorization of dyes using fungi, which are biosorption and biodegradation. The biosorption is essentially committed for dead cells. In this mechanism, the association of physiochemical interactions such as adsorption, deposition, and ion exchange is capable of decolorization. The biodegradation is particularly committed for living cells because they can produce the lignin-modifying enzymes, laccase, manganese peroxidase, and lignin peroxidase [187]. Fungal species (including *Alternaria*, *Aspergillus*, and *Cladosporium*) are capable of producing extracellular enzymes (lignin peroxidases, manganese peroxidases, and laccases) that can biodegrade complex synthetic dyes (Congo red, Poly R-478, Methyl green, indigo carmine, etc.) which widen the industrial applications of fungal enzymes toward the sustainable environment [35,37,188].

With vast developments in agriculture techniques, the worldwide accumulation of lignocellulosic agricultural wastes also represents a considerable environmental challenge. As discussed above, fungi naturally genetically inherited diverse enzymes (hydrolases and oxidoreductases) for the ecofriendly valorization of cellulosic material into valuable industrial products like bioethanol. Collectively, the application of fungal enzymes revealed dual sustainable benefits from an industrial quality and environmental perspective.

4.3. Biomedical Applications

4.3.1. Antimicrobial

Fungi are well known for the production of antimicrobial agents, industrial enzymes, and microbial biomass. They have evolved the ability to synthesize novel metabolites, including natural products and enzymes, which could be used for novel antimicrobial. The antimicrobial activity of the fungal enzyme may be classified into two categories: direct and indirect antimicrobial agents. First, fungal enzymes such as cellulases, amylases, and lipases were used directly for antimicrobial activity [40], while indirect antimicrobial activity depended on generating intermediate compounds with antimicrobial activity through fungal enzymes [42]. Direct antimicrobial activity includes rupturing the cell membrane, resulting in the loss of intracellular elements necessary for life. Furthermore, fungal enzymes stop DNA synthesis, inhibit essential bacterial enzymes, block bacterial receptors, and inhibit the electron transport chain [189,190], as illustrated in Figure 4. In addition, many fungal enzymes have the ability to inhibit both DNA and RNA viruses. The purified oyster mushroom laccase extracted from *Pleurotus ostreatus* showed potent antiviral activity against hepatitis C Virus (HCV) [191]. Other fungal enzymes, such as ribonucleases, can exert an alternative antiviral activity against hepatitis B virus (HBV) [192] and human immunodeficiency virus type 1 (HIV-1) [193]. Regarding many fungal crude extracts and their enzymes, the antiviral effect of fungal lectin against HIV was the most investigated virus. El-Maradny et al. [194] indicated that the purified Trichogin protein from *Tricholoma giganteum* was found to inhibit the HIV reverse transcriptase (RT) enzyme with the lowest IC_{50} (half maximal inhibitory concentration) value determined to be 83 nM [195], followed by the purified lectin from *Russula delica* with IC_{50} value of 0.26 [75] and the purified laccase from *Tricholoma mongolicum* with IC_{50} value of 0.62 μM [196]. Lectins bind with viral envelope glycoproteins and inhibit the entry of many viruses, such as HIV, thus inhibiting virus attachment and blocking its entry via inducing the virus conformational rearrangements [197]. Ma et al. [198] revealed that the purified lectin from *Agrocybe aegerita* was found to be a nontoxic protein and served as an effective adjuvant with influenza vaccine against H9N2 viruses in infected mice model through binding with viral envelope glycoprotein.

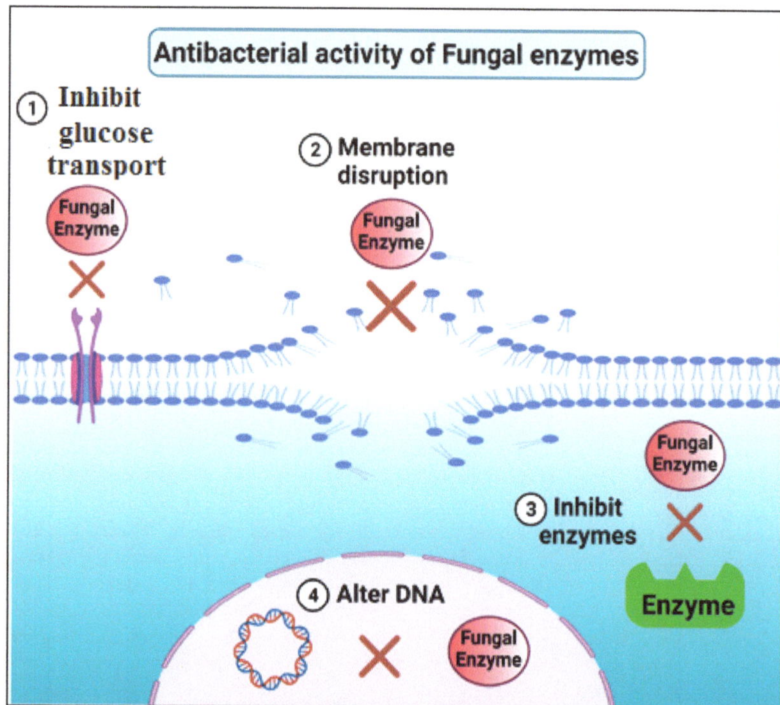

Figure 4. Schematic illustration represents the molecular mechanisms of the main antibacterial effects of fungal enzymes, including 1. alter membrane permeability and inhibition of glucose transport, 2. bacterial membrane damage and disruption, causing cytoplasm leakage, 3. inhibit bacterial enzymes, and 4. inhibit or alter cell division and the DNA replication process.

4.3.2. Anticancer

Fungi produce a diverse set of natural products as secondary metabolites, including polyketides, nonribosomal peptides, terpenes, chitosan-oligosaccharides, polyketide–terpene hybrids, and polyketide–nonribosomal peptide hybrids by the action of fungal enzymes. Those secondary metabolites have been a rich source for therapeutics used as antibiotics (penicillin), cholesterol-lowering agents (lovastatin), and immune suppressants (cyclosporine) [199]. Chitosan-oligosaccharides are low-molecular-weight chitosan derivatives with interesting anticancer applications, which are produced from the *Trichoderma harzianum* chitinolytic system by the action of extracellular enzymes. As anticancer compounds, chitosan-oligosaccharides inhibited the growth of cervical cancer Hela cells at mild concentration and reduced the survival cells rate significantly [42]. The specific mechanism of chitosan preventing tumor growth is unclear, although it might be linked to chitosan electrostatic charges, alterations in tumor cell permeability, and modulation of tumor factor expression, including metalloproteinase-9 or/and vascular endothelial growth factors [200]. Endophytic fungi anticancer toxal (Paclitaxel) production has also been reported from *Penicillium aurantiogriseum* (NRRL 62431) [201], *Cladosporium oxysporum* [202], *Phoma medicaginis* [203] by the action of fungal enzymes. Paclitaxel is the world's first billion-dollar anticancer drug and is used to treat breast, ovarian, and lung cancer [204]. Taxol was the first microtubule targeting drug to be disclosed in the literature, and its major mode of action was to disrupt microtubule dynamics, resulting in mitotic arrest and cell death [205], as presented in Figure 5. Additionally, various fungal enzymes act as prodrug activators for anticancer drugs [206]. The enzyme α-L-rhamnosidase, derived from

Acremonium species, is used in the one-step rhamnosylation of anticancer drugs such as 2′-deoxy-5-fluorouridine, cytosine arabinoside, and hydroxyurea [207]. Surprisingly, these prodrugs only had a lethal impact on breast cancer MDA-MB-231 cells after being exposed to -L-rhamnosidase, showing that they are delivered to tumor locations specifically. The yeast cytosine deaminase enzyme is another model of enzyme-activated prodrug treatment. This enzyme has been studied in the conversion of the non-toxic prodrug 5-fluorocytosine (5-FC) to the cytotoxic 5-fluorouracil (5-FU) [208].

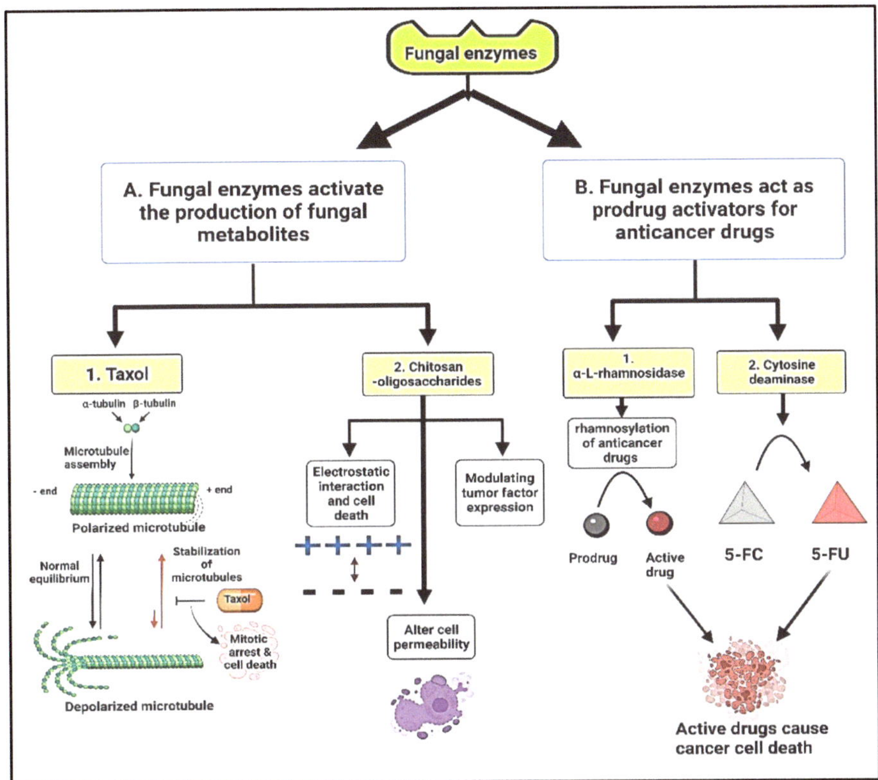

Figure 5. Schematic illustration represents the mechanism of anticancer effect of the fungal enzymes, 5-fluorocytosine (5-FC), and 5-fluorouracil (5-FU).

4.3.3. Antioxidant

Antioxidants are important compounds that enhance the living organisms to avoid oxidative stress originated by reactive oxygen species, which are harmful compounds, causing damage and altering the structure of the carbohydrates, nucleic acids, lipids, and proteins [209]. Fungi can profoundly participate in the production of antioxidants metabolites, including phenolics, flavonoids, β-glucans, and steroids [210]. The main mechanism of fungal enzyme antioxidants such as catalases, peroxidases, superoxide dismutases, and transferases becomes activated to defend the drastic condition by converting reactive oxygen species into oxygen and water [211]. Phenolic compounds are natural antioxidants and bind to the cellulose matrixes of the plant cell walls [212]. Fungi are the widely promising organisms that produce lignocellulolytic enzymes responsible for plant cell walls degradation [213].

5. Toward Safe, Sustainable Production and Application of Fungal Enzymes

Industrial implementation of the fungal enzymes represents a sustainable, eco-friendly, and energy-saving solution for many environmental and quality aspects compared to the currently applied conventional chemicals/physical approaches [214]. However, some biosafety aspects should be considered in the fungal enzyme applications and productions. Recently, a sensitization toward the fungal enzymes was estimated to be 1–15% among populations, while in the enzyme production workers, the risk increased [215,216]. Bakery and detergent-packaging employees are clear examples of the prevalence of fungal enzymes allergy [216]. From the commercial point of view, the application of genetically modified fungal cells for enzyme production is more reliable and applicable in terms of energy-saving and nutrient consumption [217]; however, the biosafety and regulations for genetically modified enzymes are still challenging. Many criteria govern the application of genetically modified organisms for enzyme production, including genetic stability, absence of any potential virulence, and antibiotic resistance factors [218].

The huge cell mass and byproducts generated during enzyme production through fungi also represent a considerable environmental risk and imply an additional cost for management and control [219]. Several attempts were applied to implement such byproducts as bio-fertilizers [220] or in synthetic dye and heavy metal removal [221]. Nikkilä and his colleagues evaluated the nutritional quality of the aqueous and alkaline treated fungal-biomass resulting as a byproduct from the enzyme industry. The extracted fungal-biomass samples revealed high contents of both polysaccharides and glycoproteins with carbohydrate and amino acid contents of 37.5 and 27.6%, respectively, for alkaline-treated fungal biomass, where the aqueous ex-traction yielded 6.6 and 61.3% of carbohydrate and amino acid contents, respectively [219]. In the same regard, mushroom cultivation at the industrial level produces huge amounts of a byproduct called spent mushroom substrate comprises a mixture of remaining substrate and mushroom mycelia [222]. This byproduct represents a potential challenge for the mushroom growing industry and to the environment. As per literature, there is growing interest in the implementation of the spent mushroom substrate in valuable bioproducts and enzymes synthesis [223]. The enzymes production capacity of six *Trichoderma* spp. were evaluated using 36% cellulases-hydrolyzed spent mushroom substrate. As a sole substrate, spent mushroom substrate supported several enzymes production through *Trichoderma atroviride* isolate T42, compared to *Aspergillus niger*, including β-glucosidases (five isoforms) and twelve isoforms of both endocellulases and xylanases [224].

6. Conclusions and Future Perspectives

In recent times, it has been recognized that the production of microbial enzymes is retaining an advanced topic of innovation. The use of enzymes is still better than the use of chemical processes as they can carry out several reaction types of importance with more considerable efficiency under mild circumstances. Among microbial enzymes, fungal enzymes are at the best locations, owing to their ability to be carried out in large quantities and much faster as well as at affordable costs. Production of the fungal enzyme has had an incredible influence not only on industrial sectors but also on biochemical synthesis processes and healthcare services. The progress of biotechnology in these fields offers different approaches to reduce the cost of enzyme production economically, enhance methods of gene discovery, and decrease the time of enzymes production through accelerating the growth of production hosts. Furthermore, the increased access toward production systems of the fungal enzyme also raises the availability and sustainability of manufacturing dealing with low-value and bulk products. Usage of microbial enzymes, especially fungal enzymes, has significantly established their importance in processing and quality improvement. Rapid progress has happened in enzyme marketing over recent periods, owing to the development of industrial and biotechnological uses. Present developments in protein modeling and engineering, besides advancements in recombination technology, have developed enzyme commercialization and production. Three main applications led to an increase in the marketing of fungal enzymes, including technical, food industry, and animal feed sectors. The technical sector includes enzymes that are used in search and analysis, such as cellulases, proteases, and amylases. Moreover, these enzymes are comprehensively utilized in the textile, leather, starch, detergent, and personal care, as well as paper and pulp industries. The second big market sector is for the use of fungal enzymes in the food industry, such as lipases and pectinases. The third essential one is the feed sector which includes several enzymes such as xylanases, β-glucanases, and phytase. There is no doubt that in the near future, the utilizing of fungal enzymes will be expanded rapidly in several fields such as food processing, bio-pulping, chemical conversions, carbohydrate conversions, cleaning, animal feed, and food additives, as well as used in detoxification and the removal of pollutants from the environment. In addition, in the case of the use of other processes other than enzymatic uses, various industries produce byproducts, including wastes and contaminants, which cause environmental problems and can be harmful to nature. However, the use of fungal enzymes, which break down some particular substrates into amino acids that are useful, biodegradable, and naturally can be easily recycled in the surrounding environment. Additionally, the use of enzymatic processes leads to increasing overall yields and the formation of well-made products with avoiding undesirable byproducts. All these reasons give fungal enzymes advanced importance for enabling many industrial areas to be increased efficiency. Moreover, the usage of fungal enzymes offers the development and production of more friendly main products in the environment and acting via using less raw materials, energy, and water. In addition, the usage of fungal enzymes in many processes leads to the production of less waste and hazardous products. Although there has been a great revolution in exploring novel fungal enzymes, it is also an important demand for assuming a new screening program of novel strains to optimize the enzyme production with new characteristics, as the current identified fungi not more than 5% of the fungal flora. The combination of the modem knowledge of biotechnology and molecular biology can offer a better-quality yield and an efficient approach with novel features.

Author Contributions: H.E.-G.: Conceptualization, formal analysis, data curation, collected literature data, wrote and edited the manuscript. A.K.S.: Conceptualization, formal analysis, data curation, collected literature data, wrote and edited the manuscript. R.B.: Conceptualization, formal analysis, data curation, collected literature data, wrote and edited the manuscript. E.M.R.: Conceptualization, formal analysis, data curation, collected literature data, wrote and edited the manuscript. Y.A.E.-M.: Conceptualization, formal analysis, data curation, collected literature data, wrote and edited the manuscript. E.M.E.-F.: Conceptualization, formal analysis, data curation, collected literature data, wrote and edited the manuscript. All authors have read and agreed to the published version of the manuscript.

Funding: This research received no external funding.

Conflicts of Interest: The authors declared no potential conflict of interest with respect to the research, authorship, and/or publication of this article.

References

1. Guerrand, D. Economics of food and feed enzymes: Status and prospectives. In *Enzymes in Human and Animal Nutrition: Principles and Perspectives*; Nunes, C.S., Kumar, V., Eds.; Academic Press: London, UK, 2018; pp. 487–514.
2. Singh, R.; Singh, T.; Pandey, A. Microbial enzymes—An overview. In *Advances in Enzyme Technology*; Singh, R.S., Singhania, R.R., Pandey, A., Larroche, C., Eds.; Elsevier: Amsterdam, The Netherlands, 2019; pp. 1–40.
3. Kango, N.; Jana, U.K.; Choukade, R. Fungal Enzymes: Sources and Biotechnological Applications. In *Advancing Frontiers in Mycology & Mycotechnology: Basic and Applied Aspects of Fungi*; Satyanarayana, T., Deshmukh, S.K., Deshpande, M.V., Eds.; Springer: Singapore, 2019; pp. 515–538; ISBN 9789811393488.
4. Kumla, J.; Suwannarach, N.; Sujarit, K.; Penkhrue, W.; Kakumyan, P.; Jatuwong, K.; Vadthanarat, S.; Lumyong, S. Cultivation of Mushrooms and Their Lignocellulolytic Enzyme Production Through the Utilization of Agro-Industrial Waste. *Molecules* **2020**, *25*, 2811. [CrossRef]
5. Da Luz, J.M.R.; Nunes, M.D.; Paes, S.A.; Torres, D.P.; da Silva, M.D.C.S.; Kasuya, M.C.M. Lignocellulolytic enzyme production of Pleurotus ostreatus growth in agroindustrial wastes. *Braz. J. Microbiol.* **2012**, *43*, 1508–1515. [CrossRef]

6. Fen, L.; Xuwei, Z.; Nanyi, L.; Puyu, Z.; Shuang, Z.; Xue, Z.; Pengju, L.; Qichao, Z.; Haiping, L. Screening of Lignocellulose-Degrading Superior Mushroom Strains and Determination of Their CMCase and Laccase Activity. *Sci. World J.* **2014**, *2014*, 763108. [CrossRef] [PubMed]

7. Dias, F.F.G.; Junior, S.B.; Hantao, L.W.; Augusto, F.; Sato, H.H. Acrylamide mitigation in French fries using native L-asparaginase from Aspergillus oryzae CCT 3940. *LWT—Food Sci. Technol.* **2017**, *76*, 222–229. [CrossRef]

8. Meghavarnam, A.K.; Janakiraman, S. Evaluation of acrylamide reduction potential of L-asparaginase from Fusarium culmorum (ASP-87) in starchy products. *LWT* **2018**, *89*, 32–37. [CrossRef]

9. Pedreschi, F.; Mariotti, S.; Granby, K.; Risum, J. Acrylamide reduction in potato chips by using commercial asparaginase in combination with conventional blanching. *LWT—Food Sci. Technol.* **2011**, *44*, 1473–1476. [CrossRef]

10. Renzetti, S.; Courtin, C.; Delcour, J.; Arendt, E. Oxidative and proteolytic enzyme preparations as promising improvers for oat bread formulations: Rheological, biochemical and microstructural background. *Food Chem.* **2010**, *119*, 1465–1473. [CrossRef]

11. Pal, A.; Khanum, F. Efficacy of xylanase purified from Aspergillus niger DFR-5 alone and in combination with pectinase and cellulase to improve yield and clarity of pineapple juice. *J. Food Sci. Technol.* **2011**, *48*, 560–568. [CrossRef] [PubMed]

12. Salim, D.; Anwar, Z.; Zafar, M.; Anjum, A.; Bhatti, K.H.; Irshad, M. Pectinolytic cocktail: Induced yield and its exploitation for lignocellulosic materials saccharification and fruit juice clarification. *Food Biosci.* **2018**, *22*, 154–164. [CrossRef]

13. Li, Y.; Wang, Y.; Tu, T.; Zhang, D.; Ma, R.; You, S.; Wang, X.; Yao, B.; Luo, H.; Xu, B. Two acidic, thermophilic GH28 polygalacturonases from Talaromyces leycettanus JCM 12802 with application potentials for grape juice clarification. *Food Chem.* **2017**, *237*, 997–1003. [CrossRef] [PubMed]

14. Ibarra, D.; Camarero, S.; Romero, J.; Martínez, M.J.; Martínez, A.T. Integrating laccase–mediator treatment into an industrial-type sequence for totally chlorine-free bleaching of eucalypt kraft pulp. *J. Chem. Technol. Biotechnol.* **2006**, *81*, 1159–1165. [CrossRef]

15. Garcia-Ubasart, J.; Esteban, A.; Vila, C.; Roncero, M.B.; Colom, J.F.; Vidal, T. Enzymatic treatments of pulp using laccase and hydrophobic compounds. *Bioresour. Technol.* **2011**, *102*, 2799–2803. [CrossRef] [PubMed]

16. Campioni, T.S.; de Jesus Moreira, L.; Moretto, E.; Nunes, N.S.S.; de Oliva Neto, P. Biobleaching of Kraft pulp using fungal xylanases produced from sugarcane straw and the subsequent decrease of chlorine consumption. *Biomass Bioenergy* **2019**, *121*, 22–27. [CrossRef]

17. Nathan, V.K.; Rani, M.E.; Gunaseeli, R.; Kannan, N.D. Enhanced biobleaching efficacy and heavy metal remediation through enzyme mediated lab-scale paper pulp deinking process. *J. Clean. Prod.* **2018**, *203*, 926–932. [CrossRef]

18. Londoño-Hernandez, L.; Ruiz, H.A.; Ramírez, T.C.; Ascacio, J.A.; Rodríguez-Herrera, R.; Aguilar, C.N. Biocatalysis and Agricultural Biotechnology Fungal detoxification of coffee pulp by solid-state fermentation. *Biocatal. Agric. Biotechnol.* **2020**, *23*, 101467 [CrossRef]

19. Sigoillot, C.; Camarero, S.; Vidal, T.; Record, E.; Asther, M.; Pérez-Boada, M.; Martínez, M.J.; Sigoillot, J.-C.; Asther, M.; Colom, J.F.; et al. Comparison of different fungal enzymes for bleaching high-quality paper pulps. *J. Biotechnol.* **2005**, *115*, 333–343. [CrossRef]

20. Wang, H.; Kaur, G.; Pensupa, N.; Uisan, K.; Du, C.; Yang, X.; Lin, C.S.K. Textile waste valorization using submerged filamentous fungal fermentation. *Process. Saf. Environ. Prot.* **2018**, *118*, 143–151. [CrossRef]

21. Hu, Y.; Du, C.; Leu, S.-Y.; Jing, H.; Li, X.; Lin, C.S.K. Valorisation of textile waste by fungal solid state fermentation: An example of circular waste-based biorefinery. *Resour. Conserv. Recycl.* **2018**, *129*, 27–35. [CrossRef]

22. Abd El Aty, A.A.; Saleh, S.A.; Eid, B.M.; Ibrahim, N.A.; Mostafa, F.A. Thermodynamics characterization and potential textile applications of Trichoderma longibrachiatum KT693225 xylanase. *Biocatal. Agric. Biotechnol.* **2018**, *14*, 129–137. [CrossRef]

23. Aggarwal, R.; Dutta, T.; Sheikh, J. Extraction of amylase from the microorganism isolated from textile mill effluent vis a vis desizing of cotton. *Sustain. Chem. Pharm.* **2019**, *14*, 100178. [CrossRef]

24. Aggarwal, R.; Dutta, T.; Sheikh, J. Extraction of pectinase from Candida isolated from textile mill effluent and its application in bio-scouring of cotton. *Sustain. Chem. Pharm.* **2020**, *17*, 100291. [CrossRef]

25. Farooq, A.; Ali, S.; Abbas, N.; Fatima, G.A.; Ashraf, M.A. Comparative performance evaluation of conventional bleaching and enzymatic bleaching with glucose oxidase on knitted cotton fabric. *J. Clean. Prod.* **2013**, *42*, 167–171. [CrossRef]

26. Manai, I.; Miladi, B.; El Mselmi, A.; Smaali, I.; Ben Hassen, A.; Hamdi, M.; Bouallagui, H. Industrial textile effluent decolourization in stirred and static batch cultures of a new fungal strain Chaetomium globosum IMA1 KJ472923. *J. Environ. Manag.* **2016**, *170*, 8–14. [CrossRef]

27. Linhartová, L.; Michalíková, K.; Šrédlová, K.; Cajthaml, T. Biodegradability of Dental Care Antimicrobial Agents. *Molecules* **2020**, *25*, 400. [CrossRef] [PubMed]

28. Vantamuri, A.B.; Kaliwal, B.B. Purification and characterization of laccase from Marasmius species BBKAV79 and effective decolorization of selected textile dyes. *3 Biotech* **2016**, *6*, 189. [CrossRef] [PubMed]

29. Marchut-Mikolajczyk, O.; Kwapisz, E.; Wieczorek, D.; Antczak, T.Z. Biodegradation of diesel oil hydrocarbons enhanced with Mucor circinelloides enzyme preparation. *Int. Biodeterior. Biodegrad.* **2015**, *104*, 142–148. [CrossRef]

30. Chang, J.; Shi, Y.; Si, G.; Yang, Q.; Dong, J.; Chen, J. The bioremediation potentials and mercury(II)-resistant mechanisms of a novel fungus Penicillium spp. DC-F11 isolated from contaminated soil. *J. Hazard. Mater.* **2020**, *396*, 122638. [CrossRef] [PubMed]

31. Nair, R.R.; Demarche, P.; Agathos, S.N. Formulation and characterization of an immobilized laccase biocatalyst and its application to eliminate organic micropollutants in wastewater. *New Biotechnol.* **2013**, *30*, 814–823. [CrossRef]

32. Talukdar, D.; Jasrotia, T.; Sharma, R.; Jaglan, S.; Kumar, R.; Vats, R.; Kumar, R.; Mahnashi, M.H.; Umar, A. Evaluation of novel indigenous fungal consortium for enhanced bioremediation of heavy metals from contaminated sites. *Environ. Technol. Innov.* **2020**, *20*, 101050. [CrossRef]
33. Garcia, L.; Lacerda, M.; Thomaz, D.; de Souza Golveia, J.; Pereira, M.; de Souza Gil, E.; Schimidt, F.; Santiago, M. laccase-alginate-chitosan-based matrix toward 17 α-ethinylestradiol removal. *Prep. Biochem. Biotechnol.* **2019**, *49*, 375–383. [CrossRef]
34. Zhang, M.; Puri, A.K.; Govender, A.; Wang, Z.; Singh, S.; Permaul, K. The multi-chitinolytic enzyme system of the compost-dwelling thermophilic fungus Thermomyces lanuginosus. *Process. Biochem.* **2015**, *50*, 237–244. [CrossRef]
35. Sosa-Martínez, J.D.; Balagurusamy, N.; Montañez, J.; Peralta, R.A.; Moreira, R.D.F.P.M.; Bracht, A.; Peralta, R.M.; Morales-Oyervides, L. Synthetic dyes biodegradation by fungal ligninolytic enzymes: Process optimization, metabolites evaluation and toxicity assessment. *J. Hazard. Mater.* **2020**, *400*, 123254. [CrossRef] [PubMed]
36. Rajhans, G.; Sen, S.K.; Barik, A.; Raut, S. Elucidation of fungal dye-decolourizing peroxidase (DyP) and ligninolytic enzyme activities in decolourization and mineralization of azo dyes. *J. Appl. Microbiol.* **2020**, *129*, 1633–1643. [CrossRef] [PubMed]
37. Yin, Q.; Zhou, G.; Peng, C.; Zhang, Y.; Kües, U.; Liu, J.; Xiao, Y.; Fang, Z. The first fungal laccase with an alkaline pH optimum obtained by directed evolution and its application in indigo dye decolorization. *AMB Express* **2019**, *9*, 151. [CrossRef]
38. Li, H.-X.; Zhang, R.-J.; Tang, L.; Zhang, J.-H.; Mao, Z.-G. In vivo and in vitro decolorization of synthetic dyes by laccase from solid state fermentation with Trametes sp. SYBC-L4. *Bioprocess Biosyst. Eng.* **2014**, *37*, 2597–2605. [CrossRef]
39. Emami, E.; Zolfaghari, P.; Golizadeh, M.; Karimi, A.; Lau, A.; Ghiasi, B.; Ansari, Z. Effects of stabilizers on sustainability, activity and decolorization performance of Manganese Peroxidase enzyme produced by Phanerochaete chrysosporium. *J. Environ. Chem. Eng.* **2020**, *8*, 104459. [CrossRef]
40. Liu, S.; Ahmed, S.; Zhang, C.; Liu, T.; Shao, C.; Fang, Y. Diversity and antimicrobial activity of culturable fungi associated with sea anemone Anthopleura xanthogrammica. *Electron. J. Biotechnol.* **2020**, *44*, 41–46. [CrossRef]
41. Zanutto-Elgui, M.R.; Vieira, J.C.S.; do Prado, D.Z.; Buzalaf, M.A.R.; de Magalhães Padilha, P.; de Oliveira, D.E.; Fleuri, L.F. Production of milk peptides with antimicrobial and antioxidant properties through fungal proteases. *Food Chem.* **2018**, *278*, 823–831. [CrossRef]
42. Rafael, O.-H.D.; Fernándo, Z.-G.L.; Abraham, P.-T.; Alberto, V.-L.P.; Guadalupe, G.-S.; Pablo, P.J. Production of chitosan-oligosaccharides by the chitin-hydrolytic system of Trichoderma harzianum and their antimicrobial and anticancer effects. *Carbohydr. Res.* **2019**, *486*, 107836. [CrossRef] [PubMed]
43. Abu-Tahon, M.A.; Isaac, G.S. Anticancer and antifungal efficiencies of purified chitinase produced from Trichoderma viride under submerged fermentation. *J. Gen. Appl. Microbiol.* **2020**, *66*, 32–40. [CrossRef]
44. Baskar, G.; Chandhuru, J.; Fahad, K.S.; Praveen, A.S.; Chamundeeswari, M.; Muthukumar, T. Anticancer activity of fungal L-asparaginase conjugated with zinc oxide nanoparticles. *J. Mater. Sci. Mater. Med.* **2015**, *26*, 43. [CrossRef]
45. Zhu, Z.; Yang, M.; Bai, Y.; Ge, F.; Wang, S. Antioxidant-related catalaseCTA1regulates development, aflatoxin biosynthesis, and virulence in pathogenic fungusAspergillus flavus. *Environ. Microbiol.* **2020**, *22*, 2792–2810. [CrossRef]
46. El-Katony, T.M.; El-Dein, M.M.N.; El-Fallal, A.A.; Ibrahim, N.G.; Mousa, M. Substrate–fungus interaction on the enzymatic and non-enzymatic antioxidant activities of solid state fermentation system. *Bioresour. Bioprocess.* **2020**, *7*, 28. [CrossRef]
47. Maitan-Alfenas, G.P.; Visser, E.M.; Guimarães, V.M. Enzymatic hydrolysis of lignocellulosic biomass: Converting food waste in valuable products. *Curr. Opin. Food Sci.* **2015**, *1*, 44–49. [CrossRef]
48. Gupta, V.K.; Kubicek, C.P.; Berrin, J.-G.; Wilson, D.W.; Couturier, M.; Berlin, A.; Filho, E.X.; Ezeji, T. Fungal Enzymes for Bio-Products from Sustainable and Waste Biomass. *Trends Biochem. Sci.* **2016**, *41*, 633–645. [CrossRef]
49. Jeske, L.; Placzek, S.; Schomburg, I.; Chang, A.; Schomburg, D. BRENDA in 2019: A European ELIXIR core data resource. *Nucleic Acids Res.* **2019**, *47*, D542–D549. [CrossRef]
50. Berbee, M.L.; James, T.Y.; Strullu-Derrien, C. Early Diverging Fungi: Diversity and Impact at the Dawn of Terrestrial Life. *Annu. Rev. Microbiol.* **2017**, *71*, 41–60. [CrossRef]
51. Verma, M.; Thakur, M.; Randhawa, J.; Sharma, D.; Thakur, A.; Meehnian, H.; Jana, A. Biotechnological applications of fungal enzymes with special reference to bioremediation. In *Environmental Biotechnology*; Gothandam, K.M., Ranjan, S., Lichtfouse, E., Dasgupta, N., Eds.; Springer: Cham, Switzerland, 2020; pp. 221–247; ISBN 9783030381967.
52. Gurung, N.; Ray, S.; Bose, S.; Rai, V. A Broader View: Microbial Enzymes and Their Relevance in Industries, Medicine, and Beyond. *BioMed Res. Int.* **2013**, *2013*, 329121. [CrossRef]
53. Datta, P.; Gest, H. Control of enzyme activity by concerted feedback inhibition. *Proc. Natl. Acad. Sci. USA* **1964**, *52*, 1004–1009. [CrossRef]
54. Lemieux, R.; Spohr, U. How Emil Fischer was led to the lock and key concept for enzyme specificity. *Adv. Carbohydr. Chem. Biochem.* **1994**, *50*, 1–20.
55. Bhatia, S. Introduction to enzymes and their applications. In *Introduction to Pharmaceutical Biotechnology: Enzymes, Proteins and Bioinformatics*; Bhatia, S., Ed.; IOP Publishing: Bristol, UK, 2018; pp. 1–29; ISBN 9780750316026.
56. Koshland, D.E. Das Schüssel-Schloß-Prinzip und die Induced-fit-Theorie. *Angew. Chem.* **1994**, *106*, 2468–2472. [CrossRef]
57. Johnson, K.A.; Goody, R.S. The Original Michaelis Constant: Translation of the 1913 Michaelis–Menten Paper. *Biochemistry* **2011**, *50*, 8264–8269. [CrossRef] [PubMed]
58. Boyce, S.; Tipton, K.F. Enzyme Classification and Nomenclature. *Encycl. Life Sci.* **2013**, *1*, 1–11.
59. arrett, A. Enzyme Nomenclature. Recommendations 1992. *Eur. J. Biochem.* **1995**, *232*, 1–6. [CrossRef]

60. Yu, Q.-K.; Han, L.-T.; Wu, Y.-J.; Liu, T.-B. The Role of Oxidoreductase-Like Protein Olp1 in Sexual Reproduction and Virulence of *Cryptococcus neoformans*. *Microorganisms* **2020**, *8*, 1730. [CrossRef]

61. Pedrini, N.; Ortiz-Urquiza, A.; Huarte-Bonnet, C.; Fan, Y.; Juárez, M.P.; Keyhani, N.O. Tenebrionid secretions and a fungal benzoquinone oxidoreductase form competing components of an arms race between a host and pathogen. *Proc. Natl. Acad. Sci. USA* **2015**, *112*, E3651–E3660. [CrossRef] [PubMed]

62. Góralczyk-Bińkowska, A.; Jasińska, A.; Długoński, A.; Płociński, P.; Długoński, J. Laccase activity of the ascomycete fungus Nectriella pironii and innovative strategies for its production on leaf litter of an urban park. *PLoS ONE* **2020**, *15*, e0231453. [CrossRef]

63. Sayyed, R.Z.; Bhamare, H.M.; Sapna; Marraiki, N.; Elgorban, A.M.; Syed, A.; El-Enshasy, H.A.; Dailin, D.J. Tree bark scrape fungus: A potential source of laccase for application in bioremediation of non-textile dyes. *PLOS ONE* **2020**, *15*, e0229968. [CrossRef]

64. Sharma, D.; Chaudhary, R.; Kaur, J.; Arya, S.K. Greener approach for pulp and paper industry by Xylanase and Laccase. *Biocatal. Agric. Biotechnol.* **2020**, *25*, 101604. [CrossRef]

65. Shin, S.K.; Hyeon, J.E.; Joo, Y.-C.; Jeong, D.W.; You, S.K.; Han, S.O. Effective melanin degradation by a synergistic laccase-peroxidase enzyme complex for skin whitening and other practical applications. *Int. J. Biol. Macromol.* **2019**, *129*, 181–186. [CrossRef]

66. Sorrentino, I.; Giardina, P.; Piscitelli, A. Development of a biosensing platform based on a laccase-hydrophobin chimera. *Appl. Microbiol. Biotechnol.* **2019**, *103*, 3061–3071. [CrossRef]

67. Mehandia, S.; Sharma, S.C.; Arya, S.K. Isolation and characterization of an alkali and thermostable laccase from a novel Alcaligenes faecalis and its application in decolorization of synthetic dyes. *Biotechnol. Rep.* **2020**, *25*, e00413. [CrossRef] [PubMed]

68. Fernández-Fernández, M.; Sanromán, M. Ángeles; Moldes, D. Recent developments and applications of immobilized laccase. *Biotechnol. Adv.* **2013**, *31*, 1808–1825. [CrossRef] [PubMed]

69. Wang, J.; Chang, F.; Tang, X.; Li, W.; Yin, Q.; Yang, Y.; Hu, Y. Bacterial laccase of Anoxybacillus ayderensis SK3-4 from hot springs showing potential for industrial dye decolorization. *Ann. Microbiol.* **2020**, *70*, 51. [CrossRef]

70. Periasamy, D.; Mani, S.; Ambikapathi, R. White Rot Fungi and Their Enzymes for the Treatment of Industrial Dye Effluents. In *Recent Advancement in White Biotechnology through Fungi*; Yadav, A., Singh, S., Mishra, S., Gupta, A., Eds.; Springer: Berlin/Heidelberg, Germany, 2019; pp. 73–100; ISBN 9783030255060.

71. Manavalan, T.; Manavalan, A.; Heese, K. Characterization of Lignocellulolytic Enzymes from White-Rot Fungi. *Curr. Microbiol.* **2015**, *70*, 485–498. [CrossRef] [PubMed]

72. Erden, E.; Ucar, M.C.; Gezer, T.; Pazarlioglu, N.K. Screening for ligninolytic enzymes from autochthonous fungi and applications for decolorization of Remazole Marine Blue. *Braz. J. Microbiol.* **2009**, *40*, 346–353. [CrossRef] [PubMed]

73. Manubolu, M.; Goodla, L.; Pathakoti, K.; Malmlöf, K. Enzymes as direct decontaminating agents-mycotoxins. *Enzym. Hum. Anim. Nutr. Princ. Perspect.* **2018**, 313–330. [CrossRef]

74. Mehta, P.; Hale, T.I.; Christen, P. Aminotransferases: Demonstration of homology and division into evolutionary subgroups. *JBIC J. Biol. Inorg. Chem.* **1993**, *214*, 549–561. [CrossRef] [PubMed]

75. Mathieu, Y.; Prosper, P.; Favier, F.; Harvengt, L.; Didierjean, C.; Jacquot, J.-P.; Morel-Rouhier, M.; Gelhaye, E. Diversification of Fungal Specific Class A Glutathione Transferases in Saprotrophic Fungi. *PLOS ONE* **2013**, *8*, e80298. [CrossRef]

76. Hayes, J.D.; Flanagan, J.U.; Jowsey, I.R. Glutathione transferases. *Annu. Rev. Pharmacol. Toxicol.* **2005**, *45*, 51–88. [CrossRef] [PubMed]

77. Rahmanpour, S.; Backhouse, D.; Nonhebel, H. Induced tolerance of *Sclerotinia sclerotiorum* to isothiocyanates and toxic volatiles from *Brassica* species. *Plant Pathol.* **2009**, *58*, 479–486. [CrossRef]

78. Calmes, B.; Morel-Rouhier, M.; Bataillé-Simoneau, N.; Gelhaye, E.; Guillemette, T.; Simoneau, P. Characterization of glutathione transferases involved in the pathogenicity of Alternaria brassicicola. *BMC Microbiol.* **2015**, *15*, 123. [CrossRef]

79. Al-Madboly, L.; Ali, S.M.; El Fakharany, E.M.; Ragab, A.E.; Khedr, E.G.; Elokely, K.M. Stress-Based Production, and Characterization of Glutathione Peroxidase and Glutathione S-Transferase Enzymes from Lactobacillus plantarum. *Front. Bioeng. Biotechnol.* **2020**, *8*, 8. [CrossRef] [PubMed]

80. Morel, M.; Ngadin, A.A.; Droux, M.; Jacquot, J.-P.; Gelhaye, E. The fungal glutathione S-transferase system. Evidence of new classes in the wood-degrading basidiomycete Phanerochaete chrysosporium. *Cell. Mol. Life Sci.* **2009**, *66*, 3711–3725. [CrossRef] [PubMed]

81. Garcerá, A.; Barreto, L.; Piedrafita, L.; Tamarit, J.; Herrero, E. Saccharomyces cerevisiae cells have three Omega class glutathione S-transferases acting as 1-Cys thiol transferases. *Biochem. J.* **2006**, *398*, 187–196. [CrossRef] [PubMed]

82. Lateef, A.; Oloke, J.; Gueguim-Kana, E.; Raimi, O. Production of fructosyltransferase by a local isolate of Aspergillus nigerin both submerged and solid substrate media. *Acta Aliment.* **2012**, *41*, 100–117. [CrossRef]

83. Nakakuki, T. Development of Functional Oligosaccharides in Japan. *Trends Glycosci. Glycotechnol.* **2003**, *15*, 57–64. [CrossRef]

84. Cunha, J.T.; Soares, P.O.; Romaní, A.; Thevelein, J.M.; Domingues, L. Biotechnology for Biofuels Xylose fermentation efficiency of industrial Saccharomyces cerevisiae yeast with separate or combined xylose reductase/xylitol dehydrogenase and xylose isomerase pathways. *Biotechnol. Biofuels* **2019**, *12*, 20. [CrossRef] [PubMed]

85. Mao, S.; Liu, Y.; Yang, J.; Ma, X.; Zeng, F.; Zhang, Z.; Wang, S.; Han, H.; Qin, H.; Lu, F. Cloning, expression and characterization of a novel fructosyltransferase from Aspergillus niger and its application in the synthesis of fructooligosaccharides. *RSC Adv.* **2019**, *9*, 23856–23863. [CrossRef]

86. Ghazi, I.; Fernandez-Arrojo, L.; Garcia-Arellano, H.; Ferrer, M.; Ballesteros, A.O.; Plou, F.J. Purification and kinetic characterization of a fructosyltransferase from Aspergillus aculeatus. *J. Biotechnol.* **2007**, *128*, 204–211. [CrossRef] [PubMed]

87. Saranraj, P.; Stella, D. Fungal amylase—A review. *Int. J. Microbiol. Res.* **2013**, *4*, 203–211. [CrossRef]

88. Singh, S. Aspergillus Enzymes for Textile Industry. In *New and Future Developments in Microbial Biotechnology and Bioengineering*; Elsevier BV: Amsterdam, The Netherlands, 2016; pp. 191–198.

89. Chimbekujwo, K.I.; Ja'Afaru, M.I.; Adeyemo, O.M. Purification, characterization and optimization conditions of protease produced by Aspergillus brasiliensis strain BCW. *Sci. Afr.* **2020**, *8*, e00398. [CrossRef]

90. Yike, I. Fungal Proteases and Their Pathophysiological Effects. *Mycopathologia* **2011**, *171*, 299–323. [CrossRef] [PubMed]

91. Germano, S.; Pandey, A.; Osaku, C.A.; Rocha, S.N.; Soccol, C.R. Characterization and stability of proteases from Penicillium sp. produced by solid-state fermentation. *Enzym. Microb. Technol.* **2003**, *32*, 246–251. [CrossRef]

92. Vishwanatha, K.S.; Rao, A.G.A.; Singh, S.A. Acid protease production by solid-state fermentation using Aspergillus oryzae MTCC 5341: Optimization of process parameters. *J. Ind. Microbiol. Biotechnol.* **2010**, *37*, 129–138. [CrossRef] [PubMed]

93. El-Shora, H.M.; Metwally, M.A.-A. Production, purification and characterisation of proteases from whey by some fungi. *Ann. Microbiol.* **2008**, *58*, 495–502. [CrossRef]

94. De Souza, P.M.; Bittencourt, M.L.D.A.; Caprara, C.C.; de Freitas, M.; de Almeida, R.P.C.; Silveira, D.; Fonseca, Y.M.; Ferreira Filho, E.X.; Pessoa Junior, A.; Magalhães, P.O. A biotechnology perspective of fungal proteases. *Braz. J. Microbiol.* **2015**, *46*, 337–346. [CrossRef]

95. Benluvankar, V.; Jebapriya, G.R.; Gnanadoss, J.J. Protease production by Penicillium sp. LCJ228 under solid state fermentation using groundnut oilcake as substrate. *Int. J. Life Sci. Pharma Res.* **2015**, *5*, 12–19.

96. Pekkarinen, A.; Mannonen, L.; Jones, B.; Niku-Paavola, M.-L. Production of Proteases by Fusarium Species Grown on Barley Grains and in Media Containing Cereal Proteins. *J. Cereal Sci.* **2000**, *31*, 253–261. [CrossRef]

97. Al-Askar, A.; Abdulkhair, W.; Rashad, Y. Production, Purification and Production, Purification and Optimization of Protease by *Fusarium solani* under Solid State Fermentation and Isolation of Protease Inhibitor Protein from *Rumex vesicarius* L. *J. Pure Appl. Microbiol.* **2014**, *8*, 239–250.

98. Kim, J.Y. Isolation of Protease-producing Yeast, Pichia farinosa CO-2 and Characterization of Its Extracellular Enzyme. *J. Korean Soc. Appl. Biol. Chem.* **2010**, *53*, 133–141. [CrossRef]

99. Zakowski, J.J.; Bruns, D.E. Biochemistry of Human Alpha Amylase Isoenzymes. *CRC Crit. Rev. Clin. Lab. Sci.* **1985**, *21*, 283–322. [CrossRef] [PubMed]

100. Amin, M.; Bhatti, H.N.; Pervin, F. Production, partial purification and thermal characterization of β-Amylase from Fusarium solani in solid state fermentation. *J. Chem. Soc. Pakistan* **2008**, *30*, 480–485.

101. Oseni, O.A. Activity of β-Amylase in Some Fungi Strains Isolated from Forest Soil in South-Western Nigeria. *Br. Biotechnol. J.* **2014**, *4*, 1–9. [CrossRef]

102. Silano, V.; Barat Baviera, J.M.; Bolognesi, C.; Cocconcelli, P.S.; Crebelli, R.; Gott, D.M.; Grob, K.; Lampi, E.; Mortensen, A.; Rivière, G.; et al. Safety evaluation of the food enzyme α-amylase from Aspergillus oryzae (strain DP-Bzb41). *EFSA J.* **2019**, *17*, e05899. [CrossRef]

103. Ahmed, N.E.; El Shamy, A.R.; Awad, H.M. Optimization and immobilization of amylase produced by Aspergillus terreus using pomegranate peel waste. *Bull. Natl. Res. Cent.* **2020**, *44*, 109. [CrossRef]

104. Bakri, Y.; Jawhar, M.; Arabi, M.I.E. Enhanced amylase production by Fusarium solani in solid state fermentation. *Pak. J. Sci. Ind. Res. Ser. B Biol. Sci.* **2014**, *57*, 123–128. [CrossRef]

105. Shruthi, B.R.; Achur, R.N.H.; Nayaka Boramuthi, T. Optimized Solid-State Fermentation Medium Enhances the Multienzymes Production from Penicillium citrinum and Aspergillus clavatus. *Curr. Microbiol.* **2020**, *77*, 2192–2206. [CrossRef] [PubMed]

106. Kanmani, P.; Aravind, J.; Kumaresan, K. An insight into microbial lipases and their environmental facet. *Int. J. Environ. Sci. Technol.* **2014**, *12*, 1147–1162. [CrossRef]

107. Bora, L. Purification and Characterization of Highly Alkaline Lipase from Bacillus licheniformis MTCC 2465: And Study of its Detergent Compatibility and Applicability. *J. Surfactants Deterg.* **2014**, *17*, 889–898. [CrossRef]

108. Abdel-Fattah, Y.; Soliman, N.A.; Yousef, S.M.; El-Helow, E.R. Application of experimental designs to optimize medium composition for production of thermostable lipase/esterase by Geobacillus thermodenitrificans AZ1. *J. Genet. Eng. Biotechnol.* **2012**, *10*, 193–200. [CrossRef]

109. Ilesanmi, O.I.; Adekunle, A.E.; Omolaiye, J.A.; Olorode, E.M.; Ogunkanmi, A.L. Isolation, optimization and molecular characterization of lipase producing bacteria from contaminated soil. *Sci. Afr.* **2020**, *8*, e00279. [CrossRef]

110. Nema, A.; Patnala, S.H.; Mandari, V.; Kota, S.; Devarai, S.K. Production and optimization of lipase using Aspergillus niger MTCC 872 by solid-state fermentation. *Bull. Natl. Res. Cent.* **2019**, *43*, 82. [CrossRef]

111. Putri, D.N.; Khootama, A.; Perdani, M.S.; Utami, T.S.; Hermansyah, H. Optimization of Aspergillus niger lipase production by solid state fermentation of agro-industrial waste. *Energy Rep.* **2020**, *6*, 331–335. [CrossRef]

112. Kempka, A.P.; Lipke, N.L.; Da Luz Fontoura Pinheiro, T.; Menoncin, S.; Treichel, H.; Freire, D.M.G.; Di Luccio, M.; De Oliveira, D. Response surface method to optimize the production and characterization of lipase from Penicillium verrucosum in solid-state fermentation. *Bioprocess Biosyst. Eng.* **2008**, *31*, 119–125. [CrossRef]

113. Geoffry, K.; Achur, R.N. Optimization of novel halophilic lipase production by Fusarium solani strain NFCCL 4084 using palm oil mill effluent. *J. Genet. Eng. Biotechnol.* **2018**, *16*, 327–334. [CrossRef]

114. El Aamri, L.; Hafidi, M.; Scordino, F.; Krasowska, A.; Lebrihi, A.; Orlando, M.G.; Barresi, C.; Criseo, G.; Barreca, D.; Romeo, O. Arthrographis curvata and Rhodosporidium babjevae as New Potential Fungal Lipase Producers for Biotechnological Applications. *Braz. Arch. Biol. Technol.* **2020**, *63*, 1–9. [CrossRef]

115. Salihu, A.; Bala, M.; Bala, S.M. Application of Plackett-Burman Experimental Design for Lipase Production by Aspergillus niger Using Shea Butter Cake. *ISRN Biotechnol.* **2013**, *2013*, 718352. [CrossRef] [PubMed]

116. Fleuri, L.F.; de Oliveira, M.C.; de Lara Campos Arcuri, M.; Capoville, B.L.; Pereira, M.S.; Delgado, C.H.O.; Novelli, P.K. Production of fungal lipases using wheat bran and soybean bran and incorporation of sugarcane bagasse as a co-substrate in solid-state fermentation. *Food Sci. Biotechnol.* **2014**, *23*, 1199–1205. [CrossRef]

117. Mehta, A.; Bodh, U.; Gupta, R. Fungal lipases: A review. *J. Biotech Res.* **2017**, *8*, 58–77.

118. Selvakumar, P.; Sivashanmugam, P. Optimization of lipase production from organic solid waste by anaerobic digestion and its application in biodiesel production. *Fuel Process. Technol.* **2017**, *165*, 1–8. [CrossRef]

119. Gopinath, S.C.B.; Anbu, P.; Lakshmipriya, T.; Hilda, A. Strategies to Characterize Fungal Lipases for Applications in Medicine and Dairy Industry. *BioMed Res. Int.* **2013**, *2013*, 154549. [CrossRef]

120. Chandra, P.; Enespa; Singh, R.; Arora, P.K. Microbial lipases and their industrial applications: A comprehensive review. *Microb. Cell Fact.* **2020**, *19*, 169. [CrossRef]

121. Hwang, H.T.; Qi, F.; Yuan, C.; Zhao, X.; Ramkrishna, D.; Liu, D.; Varma, A. Lipase-catalyzed process for biodiesel production: Protein engineering and lipase production. *Biotechnol. Bioeng.* **2014**, *111*, 639–653. [CrossRef]

122. Meyer, V.; Andersen, M.R.; Brakhage, A.A.; Braus, G.H.; Caddick, M.X.; Cairns, T.C.; De Vries, R.P.; Haarmann, T.; Hansen, K.; Hertz-Fowler, C.; et al. Current challenges of research on filamentous fungi in relation to human welfare and a sustainable bio-economy: A white paper. *Fungal Biol. Biotechnol.* **2016**, *3*, 6. [CrossRef] [PubMed]

123. Várnai, A.; Mäkelä, M.R.; Djajadi, D.T.; Rahikainen, J.; Hatakka, A.; Viikari, L. Carbohydrate-binding modules of fungal cellulases: Occurrence in nature, function, and relevance in industrial biomass conversion. *Adv. Appl. Microbiol.* **2014**, *88*, 103–165.

124. Abdullah, A.; Hamid, H.; Christwardana, M.; Hadiyanto, H. Optimization of Cellulase Production by Aspergillus niger ITBCC L74 with Bagasse as Substrate using Response Surface Methodology. *HAYATI J. Biosci.* **2018**, *25*, 115. [CrossRef]

125. Amadi, O.C.; Egong, E.J.; Nwagu, T.N.; Okpala, G.; Onwosi, C.O.; Chukwu, G.C.; Okolo, B.N.; Agu, R.C.; Moneke, A.N. Process optimization for simultaneous production of cellulase, xylanase and ligninase by Saccharomyces cerevisiae SCPW 17 under solid state fermentation using Box-Behnken experimental design. *Heliyon* **2020**, *6*, e04566. [CrossRef] [PubMed]

126. Cekmecelioglu, D.; Demirci, A. Production of Cellulase and Xylanase Enzymes Using Distillers Dried Grains with Solubles (DDGS) by Trichoderma reesei at Shake-Flask Scale and the Validation in the Benchtop Scale Bioreactor. *Waste Biomass-Valorization* **2020**, *11*, 6575–6584. [CrossRef]

127. Polizeli, M.L.T.M.; Rizzatti, A.C.S.; Monti, R.; Terenzi, H.F.; Jorge, J.A.; Amorim, D.S. Xylanases from fungi: Properties and industrial applications. *Appl. Microbiol. Biotechnol.* **2005**, *67*, 577–591. [CrossRef]

128. Silva, L.; Terrasan, C.R.F.; Carmona, E.C. Purification and characterization of xylanases from Trichoderma inhamatum. *Electron. J. Biotechnol.* **2015**, *18*, 307–313. [CrossRef]

129. Khalaji, A.; Sedighi, M.; Vahabzadeh, F. Optimization and Kinetic Evaluation of Acetylxylan Esterase and Xylanase Production by Trichoderma reesei Using Corn Cob Xylan. *Environ. Process.* **2020**, *7*, 885–909. [CrossRef]

130. Atalla, S.M.M.; Ahmed, N.E.; Awad, H.M.; El Gamal, N.G.; El Shamy, A.R. Statistical optimization of xylanase production, using different agricultural wastes by Aspergillus oryzae MN894021, as a biological control of faba bean root diseases. *Egypt. J. Biol. Pest Control.* **2020**, *30*, 125. [CrossRef]

131. Nikhil, B.; Dharmesh, A.; Priti, T. International Journal of Research in Chemistry and Environment Production of Xylanase by Aspergillus flavus FPDN1 on Pearl millet bran: Optimization of Culture Conditions and Application in Bioethanol Production. *Int. J. Res. Chem. Environ.* **2012**, *2*, 204–210.

132. Corrêa, T.L.R.; de Araújo, E.F. Fungal phytases: From genes to applications. *Braz. J. Microbiol.* **2020**, *51*, 1009–1020. [CrossRef]

133. Kc, S.; Upadhyaya, J.; Joshi, D.R.; Lekhak, B.; Chaudhary, D.K.; Pant, B.R.; Bajgai, T.R.; Dhital, R.; Khanal, S.; Koirala, N.; et al. Production, Characterization, and Industrial Application of Pectinase Enzyme Isolated from Fungal Strains. *Fermentation* **2020**, *6*, 59. [CrossRef]

134. Mathew, G.M.; Madhavan, A.; Arun, K.B.; Sindhu, R.; Binod, P.; Singhania, R.R.; Sukumaran, R.K.; Pandey, A. Thermophilic Chitinases: Structural, Functional and Engineering Attributes for Industrial Applications. *Appl. Biochem. Biotechnol.* **2021**, *193*, 142–164. [CrossRef] [PubMed]

135. Hartl, L.; Zach, S.; Seidl-Seiboth, V. Fungal chitinases: Diversity, mechanistic properties and biotechnological potential. *Appl. Microbiol. Biotechnol.* **2011**, *93*, 533–543. [CrossRef]

136. Elsoud, M.M.A.; El Kady, E.M. Current trends in fungal biosynthesis of chitin and chitosan. *Bull. Natl. Res. Cent.* **2019**, *43*, 59. [CrossRef]

137. Xie, X.-H.; Fu, X.; Yan, X.-Y.; Peng, W.-F.; Kang, L.-X. A Broad-Specificity Chitinase from *Penicillium oxalicum* k10 Exhibits Antifungal Activity and Biodegradation Properties of Chitin. *Mar. Drugs* **2021**, *19*, 356. [CrossRef] [PubMed]
138. Omumasaba, C.A.; Yoshida, N.; Ogawa, K. Purification and characterization of a chitinase from Trichoderma viride. *J. Gen. Appl. Microbiol.* **2001**, *47*, 53–61. [CrossRef] [PubMed]
139. Krolicka, M.; Hinz, S.W.A.; Koetsier, M.J.; Joosten, R.; Eggink, G.; Van Den Broek, L.A.M.; Boeriu, C.G. Chitinase Chi1 from Myceliophthora thermophila C1, a Thermostable Enzyme for Chitin and Chitosan Depolymerization. *J. Agric. Food Chem.* **2018**, *66*, 1658–1669. [CrossRef] [PubMed]
140. Kumar, M.; Brar, A.; Vivekanand, V.; Pareek, N. Production of chitinase from thermophilic Humicola grisea and its application in production of bioactive chitooligosaccharides. *Int. J. Biol. Macromol.* **2017**, *104*, 1641–1647. [CrossRef]
141. Baldoni, D.B.; Antoniolli, Z.I.; Mazutti, M.A.; Jacques, R.J.S.; Dotto, A.C.; de Oliveira Silveira, A.; Ferraz, R.C.; Soares, V.B.; de Souza, A.R.C. Chitinase production by Trichoderma koningiopsis UFSMQ40 using solid state fermentation. *Braz. J. Microbiol.* **2020**, *51*, 1897–1908. [CrossRef] [PubMed]
142. Sathiyaraj, G.; Srinivasan, S.; Kim, H.-B.; Subramaniyam, S.; Lee, O.R.; Kim, Y.-J.; Yang, D.C. Screening and optimization of pectin lyase and polygalacturonase activity from ginseng pathogen Cylindrocarpon Destructans. *Braz. J. Microbiol.* **2011**, *42*, 794–806. [CrossRef] [PubMed]
143. Atanasova, L.; Dubey, M.; Grujić, M.; Gudmundsson, M.; Lorenz, C.; Sandgren, M.; Kubicek, C.P.; Jensen, D.F.; Karlsson, M. Evolution and functional characterization of pectate lyase PEL12, a member of a highly expanded Clonostachys rosea polysaccharide lyase 1 family. *BMC Microbiol.* **2018**, *18*, 178. [CrossRef]
144. Semenova, M.V.; Sinitsyna, O.A.; Morozova, V.V.; Fedorova, E.A.; Gusakov, A.V.; Okunev, O.N.; Sokolova, L.M.; Koshelev, A.V.; Bubnova, T.V.; Vinetskii, Y.P.; et al. Use of a preparation from fungal pectin lyase in the food industry. *Appl. Biochem. Microbiol.* **2006**, *42*, 598–602. [CrossRef]
145. Singh, R.P.; Gupta, V.; Kumari, P.; Kumar, M.; Reddy, C.R.K.; Prasad, K.; Jha, B. Purification and partial characterization of an extracellular alginate lyase from Aspergillus oryzae isolated from brown seaweed. *Environ. Boil. Fishes* **2010**, *23*, 755–762. [CrossRef]
146. Pilgaard, B.; Wilkens, C.; Herbst, F.-A.; Vuillemin, M.; Rhein-Knudsen, N.; Meyer, A.S.; Lange, L. Proteomic enzyme analysis of the marine fungus Paradendryphiella salina reveals alginate lyase as a minimal adaptation strategy for brown algae degradation. *Sci. Rep.* **2019**, *9*, 12338. [CrossRef] [PubMed]
147. Nwokoro, O. Studies on the production of glucose isomerase by Bacillus licheniformis. *Pol. J. Chem. Technol.* **2015**, *17*, 84–88. [CrossRef]
148. Crueger, A.; Crueger, W. Glucose transforming enzymes. In *Microbial Enzymes and Biotechnology*; Fogarty, W.M., Kelly, C.T., Eds.; Springer: Dordrecht, The Netherlands, 1990; pp. 177–226; ISBN 978-94-009-0765-2.
149. Bhosale, S.H.; Rao, M.B.; Deshpande, V.V. Molecular and industrial aspects of glucose isomerase. *Microbiol. Rev.* **1996**, *60*, 280–300. [CrossRef] [PubMed]
150. Sharma, H.K.; Xu, C.; Qin, W. Isolation of Bacterial Strain with Xylanase and Xylose/Glucose Isomerase (GI) Activity and Whole Cell Immobilization for Improved Enzyme Production. *Waste Biomass-Valorization* **2021**, *12*, 833–845. [CrossRef]
151. Dai, C.; Miao, T.; Hai, J.; Xiao, Y.; Li, Y.; Zhao, J.; Qiu, H.; Xu, B. A Novel Glucose Isomerase fromCaldicellulosiruptor besciiwith Great Potentials in the Production of High-Fructose Corn Syrup. *BioMed Res. Int.* **2020**, *2020*, 1871934. [CrossRef]
152. Rengasamy, S.; Subramanian, M.R.; Perumal, V.; Ganeshan, S.; Al Khulaifi, M.M.; Al-Shwaiman, H.A.; Elgorban, A.M.; Syed, A.; Thangaprakasam, U. Purification and kinetic behavior of glucose isomerase from Streptomyces lividans RSU. *Saudi J. Biol. Sci.* **2020**, *27*, 1117–1123. [CrossRef] [PubMed]
153. Sayyed, R.Z.; Shimpi, G.B.; Chincholkar, S.B. Constitutive production of extracellular glucose isomerase by an osmophillic Aspergillus sp. under submerged conditions. *J. Food Sci. Technol.* **2010**, *47*, 496–500. [CrossRef] [PubMed]
154. Marshall, R.O.; Kooi, E.R. Enzymatic conversion of D-glucose to D-fructose. *Science* **1957**, *125*, 648–649. [CrossRef] [PubMed]
155. Kwak, S.; Jin, Y.-S. Production of fuels and chemicals from xylose by engineered Saccharomyces cerevisiae: A review and perspective. *Microb. Cell Factories* **2017**, *16*, 82. [CrossRef] [PubMed]
156. Lilius, S.T.A. Ethanolic Fermentation of Xylose with Saccharomyces cerevisiae Harboring the Thermus thermophilus xylA Gene, Which Expresses an Active Xylose (Glucose) Isomerase. *Microb. Cell Factories* **1996**, *62*, 4648–4651.
157. Brat, D.; Boles, E.; Wiedemann, B. Functional Expression of a Bacterial Xylose Isomerase in Saccharomyces cerevisiae. *Appl. Environ. Microbiol.* **2009**, *75*, 2304–2311. [CrossRef] [PubMed]
158. Sidrauski, C.; Cox, J.S.; Walter, P. tRNA Ligase Is Required for Regulated mRNA Splicing in the Unfolded Protein Response. *Cell* **1996**, *87*, 405–413. [CrossRef]
159. Banerjee, A.; Ghosh, S.; Goldgur, Y.; Shuman, S. Structure and two-metal mechanism of fungal tRNA ligase. *Nucleic Acids Res.* **2018**, *47*, 1428–1439. [CrossRef] [PubMed]
160. Remus, B.S.; Schwer, B.; Shuman, S. Characterization of the tRNA ligases of pathogenic fungi Aspergillus fumigatus and Coccidioides immitis. *RNA* **2016**, *22*, 1500–1509. [CrossRef] [PubMed]
161. Kerscher, O.; Felberbaum, R.; Hochstrasser, M. Modification of Proteins by Ubiquitin and Ubiquitin-Like Proteins. *Annu. Rev. Cell Dev. Biol.* **2006**, *22*, 159–180. [CrossRef] [PubMed]
162. Kulathu, Y.; Komander, D. Atypical ubiquitylation — the unexplored world of polyubiquitin beyond Lys48 and Lys63 linkages. *Nat. Rev. Mol. Cell Biol.* **2012**, *13*, 508–523. [CrossRef] [PubMed]

163. Marín, I. Origin and evolution of fungal HECT ubiquitin ligases. *Sci. Rep.* **2018**, *8*, 6419. [CrossRef] [PubMed]
164. Da Silva, R.R. Enzyme technology in food preservation: A promising and sustainable strategy for biocontrol of post-harvest fungal pathogens. *Food Chem.* **2019**, *277*, 531–532. [CrossRef]
165. Thomas, A.; Thomas, A. Acrylamide—A Potent Carcinogen in Food. *Int. J. Sci. Res.* **2014**, *3*, 177–188.
166. Cladière, M.; Camel, V. The Maillard reaction and food safety: Focus on acrylamide. *Environ. Risques St.* **2017**, *16*, 31–43. [CrossRef]
167. Swanston, J. *Acrylamide Mitigation in Foods—An Update Acrylamide Mitigation in Foods—An Update*; Leatherhead Food Research: Epsom, UK, 2019.
168. Batool, T.; Makky, E.A.; Jalal, M.; Yusoff, M. A Comprehensive Review on l-Asparaginase and Its Applications. *Appl. Biochem. Biotechnol.* **2015**, *178*, 900–923. [CrossRef]
169. Orabi, H.M.; El Fakharany, E.; Abdelkhalek, E.S.; Sidkey, N.M. L-Asparaginase And L-Glutaminase: Sources, Production, And Applications in Medicine and Industry. *J. Microbiol. Biotechnol. Food Sci.* **2019**, *9*, 179–190. [CrossRef]
170. Walia, A.; Guleria, S.; Mehta, P.; Chauhan, A.; Parkash, J. Microbial xylanases and their industrial application in pulp and paper biobleaching: A review. *3 Biotech* **2017**, *7*, 1–12. [CrossRef] [PubMed]
171. Singh, R.S.; Singh, T.; Pandey, A. Microbial enzymes–An overview. In *Biomass, Biofuels, Biochemicals: Advances in Enzyme Technology*; Elsevier: Amsterdam, The Netherlands, 2019; ISBN 9780444641144.
172. Khonzue, P.; Laothanachareon, T.; Rattanaphan, N.; Tinnasulanon, P.; Apawasin, S.; Paemanee, A.; Ruanglek, V.; Tanapongpipat, S.; Champreda, V.; Eurwilaichitr, L. Optimization of Xylanase Production fromAspergillus nigerfor Biobleaching of Eucalyptus Pulp. *Biosci. Biotechnol. Biochem.* **2011**, *75*, 1129–1134. [CrossRef]
173. Kumar, A.; Gautam, A.; Dutt, D. Bio-pulping: An energy saving and environment-friendly approach. *Phys. Sci. Rev.* **2020**, *5*. [CrossRef]
174. Weng, C.; Peng, X.; Han, Y. Depolymerization and conversion of lignin to value-added bioproducts by microbial and enzymatic catalysis. *Biotechnol. Biofuels* **2021**, *14*, 84. [CrossRef] [PubMed]
175. Maalej-Achouri, I.; Guerfali, M.; Romdhane, I.B.-B.; Gargouri, A.; Belghith, H. The effect of Talaromyces thermophilus cellulase-free xylanase and commercial laccase on lignocellulosic components during the bleaching of kraft pulp. *Int. Biodeterior. Biodegrad.* **2012**, *75*, 43–48. [CrossRef]
176. Pensupa, N.; Leu, S.; Hu, Y.; Du, C.; Liu, H.; Jing, H.; Wang, H.; Lin, C. Recent Trends in Sustainable Textile Waste Recycling Methods: Current Situation and Future Prospects. In *Chemistry and Chemical Technologies in Waste Valorization*; Lin, C.S.K., Ed.; Springer: Cham, Switzerland, 2017; pp. 189–228.
177. Araújo, R.; Casal, M.; Cavaco-Paulo, A. Application of enzymes for textile fibres processing. *Biocatal. Biotransform.* **2008**, *26*, 332–349. [CrossRef]
178. Lopes, L.S.; Vieira, N.; da Luz, J.M.R.; Marliane de Cássia, S.S.; Cardoso, W.S.; Kasuya, M.C.M. Production of fungal enzymes in Macaúba coconut and enzymatic degradation of textile dye. *Biocatal. Agric. Biotechnol.* **2020**, *26*, 101651. [CrossRef]
179. Velázquez-Fernández, J.B.; Muñiz-Hernández, S. *Bioremediation: Processes, Challenges and Future Prospects*; Nova Science Publishers, Inc.: Hauppauge, NY, USA, 2014; ISBN 9781629485157.
180. Zhang, S.; Gedalanga, P.B.; Mahendra, S. Advances in bioremediation of 1,4-dioxane-contaminated waters. *J. Environ. Manag.* **2017**, *204*, 765–774. [CrossRef] [PubMed]
181. Santacruz-Juárez, E.; Buendia-Corona, R.E.; Ramírez, R.E.; Sánchez, C. Fungal enzymes for the degradation of polyethylene: Molecular docking simulation and biodegradation pathway proposal. *J. Hazard. Mater.* **2021**, *411*, 125118. [CrossRef]
182. Steliga, T. Role of Fungi in Biodegradation of Petroleum Hydrocarbons. *Pol. J. Environ. Stud.* **2012**, *21*, 471–479.
183. Sun, K.; Song, Y.; He, F.; Jing, M.; Tang, J.; Liu, R. A review of human and animals exposure to polycyclic aromatic hydrocarbons: Health risk and adverse effects, photo-induced toxicity and regulating effect of microplastics. *Sci. Total Environ.* **2021**, *773*, 145403. [CrossRef]
184. Asemoloye, M.D.; Tosi, S.; Daccò, C.; Wang, X.; Xu, S.; Marchisio, M.A.; Gao, W.; Jonathan, S.G.; Pecoraro, L. Hydrocarbon Degradation and Enzyme Activities of *Aspergillus oryzae* and *Mucor irregularis* Isolated from Nigerian Crude Oil-Polluted Sites. *Microorganisms* **2020**, *8*, 1912. [CrossRef] [PubMed]
185. Sanghi, R.; Dixit, A.; Verma, P.; Puri, S. Design of reaction conditions for the enhancement of microbial degradation of dyes in sequential cycles. *J. Environ. Sci.* **2009**, *21*, 1646–1651. [CrossRef]
186. Rápó, E.; Tonk, S. Factors Affecting Synthetic Dye Adsorption; Desorption Studies: A Review of Results from the Last Five Years (2017–2021). *Molecules* **2021**, *26*, 5419. [CrossRef]
187. Karnwal, A.; Singh, S.; Kumar, V.; Sidhu, G.; Dhanjal, D.; Datta, S.; Amin, D.; Saini, M.; Singh, J. Fungal enzymes for the textile industry. In *Recent Advancement in White Biotechnology through Fungi*; Yadav, A., Mishra, S., Singh, S., Gupta, A., Eds.; Springer: Cham, Switzerland, 2019; pp. 459–482; ISBN 978-3-030-10479-5.
188. El Fakharany, E.; Hassan, M.A.; Taha, T.H. Production and Application of Extracellular Laccase Produced by Fusarium oxysporum EMT. *Int. J. Agric. Biol.* **2016**, *18*, 939–947. [CrossRef]
189. Li, K.; Guan, G.; Zhu, J.; Wu, H.; Sun, Q. Antibacterial activity and mechanism of a laccase-catalyzed chitosan–gallic acid derivative against Escherichia coli and Staphylococcus aureus. *Food Control* **2019**, *96*, 234–243. [CrossRef]
190. Fuglsang, C.C.; Johansen, C.; Christgau, S.; Adler-Nissen, J. Antimicrobial enzymes: Applications and future potential in the food industry. *Trends Food Sci. Technol.* **1995**, *6*, 390–396. [CrossRef]

191. El Fakharany, E.; Haroun, B.M.; Ng, T.; Redwan, E. Oyster Mushroom Laccase Inhibits Hepatitis C Virus Entry into Peripheral Blood Cells and Hepatoma Cells. *Protein Pept. Lett.* **2010**, *17*, 1031–1039. [CrossRef]
192. Li, H.-C.; Huang, E.-Y.; Su, P.-Y.; Wu, S.-Y.; Yang, C.-C.; Lin, Y.-S.; Chang, W.-C.; Shih, C. Nuclear Export and Import of Human Hepatitis B Virus Capsid Protein and Particles. *PLOS Pathog.* **2010**, *6*, e1001162. [CrossRef] [PubMed]
193. Zhang, R.; Zhao, L.; Wang, H.; Ng, T.B. A novel ribonuclease with antiproliferative activity toward leukemia and lymphoma cells and HIV-1 reverse transcriptase inhibitory activity from the mushroom, Hohenbuehelia serotina. *Int. J. Mol. Med.* **2014**, *33*, 209–214. [CrossRef]
194. El-Maradny, Y.A.; El-Fakharany, E.M.; Abu-Serie, M.M.; Hashish, M.H.; Selim, H.S. Lectins purified from medicinal and edible mushrooms: Insights into their antiviral activity against pathogenic viruses. *Int. J. Biol. Macromol.* **2021**, *179*, 239–258. [CrossRef] [PubMed]
195. He, M.; Su, D.; Liu, Q.; Gao, W.; Kang, Y. Mushroom lectin overcomes hepatitis B virus tolerance via TLR6 signaling. *Sci. Rep.* **2017**, *7*, 5814. [CrossRef]
196. Wang, H. Purification and characterization of a laccase from the edible wild mushroom Tricholoma mongolicum. *J. Microbiol. Biotechnol.* **2010**, *20*, 1069–1076. [CrossRef] [PubMed]
197. Barrientos, L.G.; Matei, E.; LaSala, F.; Delgado, R.; Gronenborn, A.M. Dissecting carbohydrate–Cyanovirin-N binding by structure-guided mutagenesis: Functional implications for viral entry inhibition. *Protein Eng. Des. Sel.* **2006**, *19*, 525–535. [CrossRef] [PubMed]
198. Ma, L.-B.; Xu, B.-Y.; Huang, M.; Sun, L.-H.; Yang, Q.; Chen, Y.-J.; Yin, Y.-L.; He, Q.-G.; Sun, H. Adjuvant effects mediated by the carbohydrate recognition domain of Agrocybe aegerita lectin interacting with avian influenza H9N2 viral surface glycosylated proteins. *J. Zhejiang Univ. Sci. B* **2017**, *18*, 653–661. [CrossRef]
199. Li, J.W.-H.; Vederas, J.C. Drug Discovery and Natural Products: End of an Era or an Endless Frontier? *Science* **2009**, *325*, 161–165. [CrossRef]
200. Liaqat, F.; Eltem, R. Chitooligosaccharides and their biological activities: A comprehensive review. *Carbohydr. Polym.* **2018**, *184*, 243–259. [CrossRef]
201. Yang, Y.; Zhao, H.; Barrero, R.A.; Zhang, B.; Sun, G.; Wilson, I.W.; Xie, F.; Walker, K.D.; Parks, J.W.; Bruce, R.; et al. Genome sequencing and analysis of the paclitaxel-producing endophytic fungus Penicillium aurantiogriseum NRRL. *BMC Genom.* **2014**, *15*, 69. [CrossRef] [PubMed]
202. Raj, K.G.; Manikandan, R.; Arulvasu, C.; Pandi, M. Anti-proliferative effect of fungal taxol extracted from Cladosporium oxysporum against human pathogenic bacteria and human colon cancer cell line HCT 15. *Spectrochim. Acta Part A Mol. Biomol. Spectrosc.* **2015**, *138*, 667–674. [CrossRef]
203. Zaiyou, J.; Li, M.; Xiqiao, H. An endophytic fungus efficiently producing paclitaxel isolated from *Taxus wallichiana* var. *mairei*. *Medicine* **2017**, *96*, e7406. [CrossRef]
204. Ranjan, A.; Singh, R.K.; Singh, M. Metabolic versatility of fungi as a source for anticancer compounds. In *Evolutionary Diversity as a Source for Anticancer Molecules*; Srivastava, A.K., Kannaujiya, V.K., Singh, R.K., Singh, D., Eds.; Elsevier: London, UK, 2020; pp. 191–207; ISBN 9780128217108.
205. Gallego-Jara, J.; Lozano-Terol, G.; Sola-Martínez, R.A.; Cánovas-Díaz, M.; de Diego Puente, T. A Compressive Review about Taxol®: History and Future Challenges. *Molecules* **2020**, *25*, 5986. [CrossRef]
206. How, C.W.; Ong, Y.S.; Low, S.S.; Pandey, A.; Show, P.L.; Foo, J.B. How far have we explored fungi to fight cancer? *Semin. Cancer Biol.* **2021**, in press. [CrossRef] [PubMed]
207. Xu, L.; Liu, X.; Li, Y.; Yin, Z.; Jin, L.; Lu, L.; Qu, J.; Xiao, M. Enzymatic rhamnosylation of anticancer drugs by an α-l-rhamnosidase from Alternaria sp. L1 for cancer-targeting and enzyme-activated prodrug therapy. *Appl. Microbiol. Biotechnol.* **2019**, *103*, 7997–8008. [CrossRef] [PubMed]
208. Stolworthy, T.S.; Korkegian, A.M.; Willmon, C.; Ardiani, A.; Cundiff, J.; Stoddard, B.L.; Black, M.E. Yeast Cytosine Deaminase Mutants with Increased Thermostability Impart Sensitivity to 5-Fluorocytosine. *J. Mol. Biol.* **2008**, *377*, 854–869. [CrossRef]
209. Aklakur, M. Natural antioxidants from sea: A potential industrial perspective in aquafeed formulation. *Rev. Aquac.* **2016**, *10*, 385–399. [CrossRef]
210. Butkhup, L.; Samappito, W.; Jorjong, S. Evaluation of bioactivities and phenolic contents of wild edible mushrooms from northeastern Thailand. *Food Sci. Biotechnol.* **2017**, *27*, 193–202. [CrossRef] [PubMed]
211. Rafi, S.; Shoaib, A.; Awan, Z.A.; Rizvi, N.B.; Nafisa; Shafiq, M. Chromium tolerance, oxidative stress response, morphological characteristics, and FTIR studies of phytopathogenic fungus Sclerotium rolfsii. *Folia Microbiol.* **2016**, *62*, 207–219. [CrossRef]
212. Abdel-Azeem, M.; El-Maradny, Y.; Othman, A.; Abdel-Azeem, A. Endophytic Fungi as a Source of New Pharmaceutical Biomolecules. In *Industrially Important Fungi for Sustainable Development. Fungal Biology*; Abdel-Azeem, A.M., Yadav, A.N., Yadav, N., Sharma, M., Eds.; Springer: Cham, Switzerland, 2021; pp. 115–151; ISBN 9783030856038.
213. Filipe, D.; Fernandes, H.; Castro, C.; Peres, H.; Oliva-Teles, A.; Belo, I.; Salgado, J.M. Improved lignocellulolytic enzyme production and antioxidant extraction using solid-state fermentation of olive pomace mixed with winery waste. *Improv. Lignocellulolytic* **2020**, *14*, 78–91. [CrossRef]
214. Arnau, J.; Yaver, D.; Hjort, C.M. Strategies and Challenges for the Development of Industrial Enzymes Using Fungal Cell Factories. In *Grand Challenges in Biology and Biotechnology*; Nevalainen, H., Ed.; Springer: Cham, Switzerland, 2020; pp. 179–210; ISBN 9783030295417.

215. Biagini, R.; MacKenzie, B.; Sammons, D.; Smith, J.; Striley, C.; Robertson, S.; Snawder, J. Evaluation of the prevalence of anti-wheat-, anti—flour dust-, and anti-alpha amylase-specific IgE antibodies in blood donors. *J. Allergy Clin. Immunol.* **2003**, *111*, S95. [CrossRef]
216. Green, B.J.; Beezhold, D.H. Industrial Fungal Enzymes: An Occupational Allergen Perspective. *J. Allergy* **2011**, *2011*, 682574. [CrossRef]
217. Nielsen, P.H.; Oxenbøll, K.M.; Wenzel, H. Cradle-to-Gate Environmental Assessment of Enzyme Products Produced Industrially in Denmark by Novozymes A/S. *Int. J. Life Cycle Assess.* **2007**, *12*, 432–438. [CrossRef]
218. Sutay Kocabaş, D.; Grumet, R. Evolving regulatory policies regarding food enzymes produced by recombinant microorganisms. *GM Crop. Food* **2019**, *10*, 191–207. [CrossRef] [PubMed]
219. Nikkilä, I.; Waldén, M.; Maina, N.H.; Tenkanen, M.; Mikkonen, K.S. Fungal Cell Biomass from Enzyme Industry as a Sustainable Source of Hydrocolloids. *Front. Chem. Eng.* **2020**, *2*, 1–12. [CrossRef]
220. Owaid, M.N.; Abed, I.A.; Al-Saeedi, S.S.S. Applicable properties of the bio-fertilizer spent mushroom substrate in organic systems as a byproduct from the cultivation of *Pleurotus* spp. *Inf. Process. Agric.* **2017**, *4*, 78–82. [CrossRef]
221. Ayele, A.; Haile, S.; Alemu, D.; Kamaraj, M. Comparative Utilization of Dead and Live Fungal Biomass for the Removal of Heavy Metal: A Concise Review. *Sci. World J.* **2021**, *2021*, 5588111. [CrossRef]
222. Phan, C.-W.; Sabaratnam, V. Potential uses of spent mushroom substrate and its associated lignocellulosic enzymes. *Appl. Microbiol. Biotechnol.* **2012**, *96*, 863–873. [CrossRef]
223. Rajavat, A.S.; Rai, S.; Pandiyan, K.; Kushwaha, P.; Choudhary, P.; Kumar, M.; Chakdar, H.; Singh, A.; Karthikeyan, N.; Bagul, S.Y.; et al. Sustainable use of the spent mushroom substrate ofPleurotus floridafor production of lignocellulolytic enzymes. *J. Basic Microbiol.* **2020**, *60*, 173–184. [CrossRef]
224. Grujić, M.; Dojnov, B.; Potočnik, I.; Duduk, B.; Vujčić, Z. Spent mushroom compost as substrate for the production of industrially important hydrolytic enzymes by fungi Trichoderma spp. and Aspergillus niger in solid state fermentation. *Int. Biodeterior. Biodegrad.* **2015**, *104*, 290–298. [CrossRef]

Review

Fungal-Derived Mycoprotein and Health across the Lifespan: A Narrative Review

Emma Derbyshire 

Nutritional Insight, Epsom KT17 2AA, Surrey, UK; emma@nutritional-insight.co.uk

Abstract: Mycoprotein is a filamentous fungal protein that was first identified in the 1960s. A growing number of publications have investigated inter-relationships between mycoprotein intakes and aspects of human health. A narrative review was undertaken focusing on evidence from randomized controlled trials, clinical trials, intervention, and observational studies. Fifteen key publications were identified and undertaken in early/young adulthood, adulthood (mid-life) or older/advanced age. Main findings showed that fungal mycoprotein could contribute to an array of health benefits across the lifespan including improved lipid profiles, glycaemic markers, dietary fibre intakes, satiety effects and muscle/myofibrillar protein synthesis. Continued research is needed which would be worthwhile at both ends of the lifespan spectrum and specific population sub-groups.

Keywords: ageing populations; fungi; glycaemic markers; human health; lifespan; lipids; mycoprotein

Citation: Derbyshire, E. Fungal-Derived Mycoprotein and Health across the Lifespan: A Narrative Review. *J. Fungi* **2022**, *8*, 653. https://doi.org/10.3390/jof8070653

Academic Editor: Laurent Dufossé

Received: 26 May 2022
Accepted: 15 June 2022
Published: 22 June 2022

1. Introduction

Globally there has been a marked increase in longevity, however, double burdens of energy excess and undernutrition mean that physical deterioration, reduced life quality and increased medical costs are increasingly reported [1]. Projections of global human populations and the number of 'peak people' suggest that the population may plateau within the next 10 years [2]. Despite this, in the 1930s the global population was just a mere 2 billion which is anticipated to reach around 9.1 billion by 2050, with 2.1 billion expected to be aged 60 years or older [3–5]. The cumulative changes in population growth and transition towards ageing populations pose challenges to food systems, which need to meet the nutritional demands of expanding and ageing populations [6].

Shifts towards sustainable living mean that consumers are seeking to obtain food proteins that are non-animal derived yet aesthetically appealing with suitable nutritive value [7]. This includes ageing populations where sustainable protein sources could help to facilitate muscle mass and strength [6]. Fungal food proteins have been available for several decades, yet their health role(s) across the lifespan are only beginning to be fully appreciated [8]. Fungi are a predominant and diverse component of the Earths ecosystems [9]. Increasingly, fungi are being seen as a 'Third Kingdom' of organisms and valuable food protein outside the classical categories of animal and plant-derived foods [10–12].

There is some evidence that certain dietary patterns, such as veganism could have lifespan enhancing potential [13], however, for fungal-derived food proteins less is known about potential benefits across the lifespan. The present narrative review examines evidence looking at the potential roles of fungal mycoprotein consumption across the lifespan.

Fungal-Derived Mycoprotein

Mycoprotein is a fungal-derived whole-food protein, filamentous fungal biomass and well-established meat analogue [14,15]. This fungal protein was first identified in the 1960s when it was derived from the soil-dwelling saprotrophic non-pathogenic micro fungus *Fusarium venenatum A3/5* [16,17]. Today mycoprotein is produced vertically in air lift pressure cycle fermenters 50 metres high, placing minimal demands on arable land compared with animal and plant-sourced proteins [14,18,19]. The use of microbial strains and specific substrates under various conditions (temperature, pH, relative humidity, moisture, inoculum age and size) has been found to yield mycoprotein biomasses produced using submerged fermentation with a high nutritional value [20]. Modelling has shown that, per unit of mass, cell-cultured foods such as mycoprotein have a lower environmental footprint that animal-derived proteins [21]. A recent paper using model projections found that substituting 20% of per-capita ruminant meat consumption with fermentation-derived microbial protein would (by 2050) reduce annual deforestation and related carbon dioxide emissions by about half, reduce methane emissions and offset expansions in global pasture areas [22].

From a nutritional stance mycoprotein provides a range of nutrients of value across the lifespan. It provides the nine main essential amino acids and has a protein digestibility-corrected amino acid score of 0.996 indicating that it is a high-quality protein [11,23]. According to European Commission standards mycoprotein can be classified as being 'high in fibre 'due to it providing at least 6g of fibre per 100 g [24,25]. It is also low in both total and saturated fat and contains negligible amounts of cholesterol [26]. Mycoprotein provides a range of micronutrients including vitamin B12, riboflavin, folate, phosphorous, choline, zinc and manganese [27].

2. Materials and Methods

A narrative review was undertaken using the National Library of Medicine, National Center for Biotechnology Information (PubMed.gov (accessed on 12 May 2022)) database with the predefined keyword 'mycoprotein'.

The literature search was finalized on 9th June 2022. The selection criteria were limited to randomized controlled trials, clinical trials, intervention studies and observational studies. The search was limited to humans, and those published in English language. Fifteen key publications were retrieved and are discussed in the present narrative review (Table 1). An additional search for publications was conducted by reviewing reference lists. Abstract or poster presentations were excluded.

Table 1. Mycoprotein & Health Across the Lifespan: Key Studies.

Life Stage (Author, Year, Location)	Population (Sample Size, Age, Health)	Study Design	Methods	Health Outcome(s)	Main Findings
Early and young adulthood					
Coelho et al. (2021) UK [28]	n = 20, 24 years, recreationally active.	Randomised, parallel-group trial.	A 7-d fully controlled diet where lunch & dinner contained either meat/fish or mycoprotein as the source of dietary protein.	Plasma lipidome.	Substituting meat/fish for mycoprotein twice daily for 1 week resulted in a reduction in cholesterol-containing lipoproteins.
Coelho et al. (2020) UK [29]	n = 10, 25 years	Randomized, controlled, double-blind, crossover trial.	Consumed a mixed-meal containing nucleotide-depleted mycoprotein or high-nucleotide mycoprotein on two separate visits.	Postprandial glucose, Serum insulin, Serum uric acid	The nucleotide-rich mixed-meal increased serum uric acid concentrations for ~12 h, but had no effects on postprandial blood glucose or serum insulin levels.
Dunlop et al. (2017) UK [30]	n = 12, 28 years, healthy young males.	Randomised, single-blind, cross-over study.	Volunteers consumed a test drink containing either 20 g milk protein or a bolus of mycoprotein (20 g, 40 g, 60 g or 80 g).	Postprandial hyperaminoacidaemia, hyperinsulinaemia	Mycoprotein ingestion resulted in slower but more sustained hyperinsulinaemia & hyperaminoacidaemia compared with milk when protein matched.
Monteyne et al. (2020a) UK [31]	n = 10, 22 years, healthy young males.	Randomized, double-blind, parallel-group study.	Participants received infusions of L-phenylalanine and ingested either 31 g milk protein or 70 g mycoprotein following a bout of unilateral resistance-type exercise	Protein synthesis rates	Mycoprotein ingestion stimulated resting & postexercise MPS rates to a greater extent than a leucine-matched bolus of milk protein.
Monteyne et al. (2020b) UK [32]	n = 10, 22 years, young males	Randomized, double-blind, parallel-group study.	Participants received infusions of L-phenylalanine ingested with either 70 g mycoprotein or 35 g BCAA-enriched mycoprotein following a bout of unilateral resistance exercise.	Protein synthesis rates	The lower-dose BCAA-enriched mycoprotein stimulated resting and postexercise MPS rates, but to a lesser extent compared with the ingestion of a BCAA-matched 70-g mycoprotein bolus.
Turnbull & Ward (1995) UK [33]	n = 19, 22.8 ± 3.55 years	Single meal study periods x2, crossover design	Milkshakes provided containing mycoprotein or a control.	Glycemia, insulinemia	Glycemia was significantly reduced 60 min after mycoprotein ingestion. Insulinemia was significantly reduced 30 min (19% reduction) and 60 min (36% reduction) after mycoprotein ingestion.
Turnbull et al. (1993) UK [34]	n = 13 24.8 ± 7.8 years, female.	3-day study periods x2	Subjects ingested an isoenergetic meal providing mycoprotein or chicken.	Energy intake, appetite	Food consumption and desire to eat decreased after mycoprotein compared with chicken consumption.
Udall et al. (1984) USA [35]	n = 100, 19.9–25.6 years	30-day double-blind cross-over study	Cookies with or without 20 g of *Fusarium graminearium* were ingested.	Tolerance, cholesterol levels	There was a decrease in serum cholesterol during the *F graminearium* study.

Table 1. *Cont.*

Life Stage (Author, Year, Location)	Population (Sample Size, Age, Health)	Study Design	Methods	Health Outcome(s)	Main Findings
Adulthood (mid-life)					
Bottin et al. (2016) UK [36]	$n = 55$, 18–65 years, overweight & obese adults.	Two randomised-controlled trials.	Consumed a test meal containing low (44 g), medium (88 g) or high (132 g) mycoprotein or isoenergetic chicken.	Postprandial insulin release	Mycoprotein ingestion reduced energy intake & insulin release in overweight volunteers.
Cherta-Murillo et al. (2021) UK [37]	$n = 5507$ free-living adults	Observational data used from the UK NDNS years 2008/9 to 2016/17.	Cross-sectional secondary analysis of the UK NDNS years 2008/9 to 2016/17.	Fibre intake, energy density intake, BMI, fasting glucose, glycated HbA1c	Mycoprotein consumers had higher dietary fibre intakes, lower glycaemic markers, energy density intake & BMI than non-consumers.
Ruxton & McMillan (2010) UK [38]	$n = 21$, 17–58 years	6-week non-blinded, controlled intervention	Asked to eat mycoprotein (88g wet weight mycoprotein), daily for six weeks.	Total cholesterol levels, glucose levels	Good compliance with the mycoprotein-rich diet appeared to significantly lower total blood & LDL cholesterol.
Turnbull et al. (1992) UK [39]	$n = 21$, aged 25–61 years staff & students.	8-week study.	The experimental group was fed cookies containing mycoprotein and the control group a cookie without mycoprotein	Blood lipids	Total cholesterol was reduced by 0.95 mmol/L in the mycoprotein versus 0.46 mmol/L in the control group. LDL was reduced by 0.84 mmol/L in the mycoprotein group versus 0.34 mmol/L in the control group.
Turnbull et al. (1990) UK [40]	$n = 17$, 19–48 years, staff & students.	3-week study	The experimental group was fed mycoprotein instead of meat and the control diet included meat.	Blood lipids	LDL declined in the mycoprotein group by 9%.
Williamson et al. (2006) USA [41]	$n = 42$, 18–50 years, pre-menopausal overweight females.	3-test day interventions.	At lunch, isocaloric pasta preloads, containing mycoprotein, tofu, or chicken provided.	Eating behaviour, hunger, satiety	Mycoprotein & tofu versus the chicken preload were associated with lower food intake after the preload at lunch indicating satiating properties.
Older/Advanced age					
Monteyne et al. (20021) UK [42]	$n = 10$, 68 ± 2 years, healthy older adults.	Randomized, parallel-group, controlled trial.	3-day isocaloric high-protein (1.8 g·kg body mass^{-1}·d^{-1}) diet, where the protein was from mycoprotein providing 57% of daily protein intake.	Myofibrillar protein synthesis rates	Mycoprotein & omnivorous protein sources supported rested & exercised daily myofibrillar protein synthesis rates in healthy older adults ingesting a high-protein diet.

Key: BCAA, branched chain amino acids; BMI, Body Mass Index; HbA1c, glycated hemoglobin; LDL, low-density lipoprotein; MPS, muscle protein synthesis; NDNS, National Diet & Nutrition Survey; UK, United Kingdom.

3. Fungal Mycoprotein across the Lifespan

3.1. Early and Young Adulthood

Early adulthood has been defined as ages 20 to 24 years and young adulthood ages 26 to 31 years [43,44]. Certain lifestyle factors, such as preventing weight gain in early adulthood may be important in preventing premature death later in life [45]. Additionally, other work has shown that higher intakes of dietary fibre during early adulthood in females was associated with reduced breast cancer risk [46]. Like adolescence, early adulthood is a life stage where opportunities for health are vast [47].

Eight studies have evaluated the effects of mycoprotein consumption in relation to markers of health in early and young adulthood [28–35]. A range of outcomes were studied including effects of ingestion on the plasma lipidome [28], cholesterol levels [35] protein synthesis rates [31,32], postprandial amino acid, glucose and insulin levels [29,30,33] and energy intake and appetite [34].

Coelho et al. (2021) undertook a 7-day fully controlled diet where lunch and dinner contained either mycoprotein or meat/fish as the control group [28]. After 1-week mycoprotein ingestion by adults aged 24 ± 1 year significantly reduced total plasma cholesterol, free cholesterol, low-density lipoprotein cholesterol and high-density lipoprotein-2 cholesterol [28]. Another randomized trial by the same team conducted with ten healthy young adults found that nucleotide-rich mycoprotein ingestion (as a mixed-meal) did not influence postprandial blood glucose nor serum insulin concentrations [29].

Monteyne et al. (2020) allocated participants (22 ± 1 year) to receive ingest either 70 g mycoprotein or 31 g milk protein following a bout of resistance exercise [31]. Results showed that the myoprotein consumption enhanced postexercise and resting muscle protein synthesis rates to a greater extent than the control [31]. Similarly, work by the same research group conducted with young males (mean age 22 years) found that a 35 g dose of branched chain amino acid (BCAA) enriched mycoprotein stimulated muscle protein synthesis rates at rest and postexercise but to a lesser extent than the 70 g mycoprotein bolus (BCAA matched) in healthy young men [32].

Earlier research by Dunlop and colleagues (2017) provided healthy young males (mean age 28 years) a test drink containing either a bolus of mycoprotein (20, 40, 60 or 80 g) or 20 g of milk protein which were mass matched [30]. Mycoprotein resulted in slower and more sustained hyperinsulinaemia and hyperaminoacidaemia compared with milk, indicating that this dietary protein could have the potential to facilitate rates of muscle protein synthesis [30]. Turnbull & Ward (1995) recruited young adults (mean age 22.8 years) providing them with milkshakes with or without mycoprotein and finding that glycemia and insulinemia were significantly reduced at 60 and 30 min respectively after ingestion [33]. In other earlier work Turnbull et al. (1993) recruited thirteen females aged 24.8 ± 7.9 years finding that desire to eat and levels of food consumption reduced after myoprotein ingestion, compared with chicken as a control [34].

3.2. Adulthood (Mid-Life)

Mid-life is often referred to as 'the halfway mark' [48]. Health outcomes later in life are often determined and underpinned by activities that occurred in midlife [49]. A growing body of evidence has linked certain dietary factors, including lower protein intakes during adulthood to increased risk of muscle loss (sarcopenia), functional limitations and increased the risk of falls [50,51]. Interestingly, other research has suggested that changes in gut microbiota i.e., *Bifidobacterium* reduction begin to change as early as midlife [52]. At least six studies have investigated the effects of mycoprotein ingestion of markers of health during adulthood [36–41].

Cross-sectional research analysing data from free-living adults taking part in the United Kingdom National Diet and Nutrition Survey showed that mycoprotein consumers had significantly higher intakes of dietary fibre, improved diet quality and lower markers of glycaemia [37].

Bottin et al. (2016) undertook a randomized controlled trial with overweight and obese adults (18 to 65 years) finding that mycoprotein ingestion (44, 88 or 132 g) reduced insulin levels and energy intake compared with a chicken control [36]. Williamson and researchers (2006) found that a mycoprotein and tofu preload had satiating effects amongst 41 pre-menopausal overweight females, compared with a chicken preload [41]. A 6-week intervention trial by Ruxton & McMillan (2010) found that mycoprotein consumption (88g daily, wet weight) led to significant reductions in total and low-density lipoprotein cholesterol, particularly amongst those with high levels at baseline [38]. Larger and longer blinded trials in community settings would build on these results.

Earlier research undertaken by Turnbull et al. (1992) recruited adults aged 25 to 61 years finding that mycoprotein ingestion (26.9 g dry weight/day over 8-weeks reduced total and low-density lipoprotein cholesterol compared with control biscuits [39]. In a shorter 3-week study adults 19–48 years received 191 g mycoprotein daily distributed over lunch and dinner, with a 13% reduction in plasma cholesterol observed compared with the meat control [40].

3.3. Older/Advanced Age

A range of definitions have been used to encompass older age. The term 'old-old' has been used to define those aged 80 years and over [53]. Other publications use the term 'advanced age' to study those typically over 65 years of age [54]. Age, however, is highly subjective, particularly at this life stage. For example, in one study perceived mean age of 'old age' was around 69 years, with some participants reporting feeling younger than their chronological age [55].

The World Health Organization forecasts that the proportion of people aged 85 years and over in the European region is predicted to increase from 14 million to 19 million by 2020, and double further to 40 million by 2050 [56]. On a global scale an estimated 2 billion people are expected to be older than 65 years by 2050, potential having heavy impacts on health and social care sectors if frailty is not addressed [57]. It is well established that good nutrition can support healthy ageing, helping to get ahead of age-related declines in physiology [58].

Research has now investigated the effects of mycoprotein ingestion in older age. A randomised controlled trial recruited older adults (n = 19; 66 $\pm$ 1 year) allocating these to consume a 3-day isocaloric high-protein diet deriving protein from mycoprotein or animal protein [42]. The team found that the fungal (vegan-derived) mycoprotein could facilitate rested and exercised daily myofibrillar protein synthesis rates similarly to that amongst those on the omnivorous diet [42].

4. Discussion

Genetic, environmental and lifestyle factors which include nutrition form an integral part of our health [59]. Multimorbidity defined as the coexistence of two or more health conditions is increasingly being linked to considerable economic burdens [60]. This is emerging to be one of the greatest challenges to health services, partially driven by ageing populations [61]. It has been estimated that by 2035 17% of the population in the United Kingdom alone is expected to have four or more chronic conditions, with functional decline often tending to form part of this [61,62].

From a dietary stance lifespan health can be influenced by certain nutrients [63]. For example, it has been proposed that shifts away from animal-derived food proteins providing high levels of methionine and movements towards alternative plant sources may benefit longevity and metabolic health [63]. There is also a plethora of evidence linking dietary fibre intake to metabolic health, colonic health and reduced cardiovascular and mortality risk [64]. Mycoprotein has been positioned as a fungal protein that is high-fibre, protein rich and produced sustainably [65]. Given this, it has been questioned whether fungal-derived food protein such as mycoprotein should be included more prominently within food-based dietary guidelines [11,12].

A growing body of evidence from systemic reviews shows that acute mycoprotein ingestion reduces energy intake and insulinaemia [33,37,66] and has the ability to lower circulating cholesterol levels, improve postprandial glycaemic control and facilitate satiety [33,35,38,65,67]. Evidence from the present narrative review further demonstrates that fungal mycoprotein could benefit health across the lifespan.

In early and young adulthood benefits have been observed on the plasma lipidome [28], in sustaining insulin and amino acids levels [30], attenuating the desire to eat (inducing satiety) [34] and stimulating resting and postexercise muscle protein synthesis rates [32]. In adulthood most research has been conducted on healthy adult populations. Observational evidence shows that mycoprotein consumers have higher dietary quality scores, fibre intakes and lower glycaemic markers and a lower body mass index [37]. Trials conducted on overweight adults show that mycoprotein ingestion was associated with reduced energy intake and insulin release [36] and satiety effects [41]. Several other publications found that mycoprotein ingestion improved blood lipid levels [35,38–40]. Research with older adults (mean age 68 years) showed that induced myofibrillar protein synthesis rates aligned with those followed an omnivorous diet [42]. Continued research is now needed across additional population groups. Larger studies would be worthwhile as trial sample sized ranged from ten to 100. This could have contributed to discrepancies between study findings.

Overall, these are interesting findings implying that fungal-derived mycoprotein has an important role to play in health across the lifespan. Regarding, mechanistic effects the ingredient structure of mycoprotein is what appears to lower lipolysis and bind bile salts which, in turn, is thought to lower blood lipid levels [68]. Regarding the high bioacessibility of protein from fungal mycoprotein, this is thought to be attributed to porous cell walls which facilitate the diffusion of proteases [69].

A publication modelling fibre intake in childhood showed one daily portion of mycoprotein shapes for children aged 4-to-10 years and 11-to-18 years could contribute to approximately a quarter of the daily fibre recommendation [70]. It would be useful to conduct additional randomized controlled trials in childhood and ageing populations. There is also scope to investigate inter-relationships between fungal mycoprotein ingestion and markers of health at specific lifespan phases such as gestation. Great opportunity lies in accruing evidence for non-traditional dietary proteins such a mycoprotein given rising demands for health sustainably-produced protein foods [71].

5. Conclusions

This narrative review has described the roles of fungal mycoprotein across the lifespan. Given expanding and ageing populations coupled with growing multimorbidity's the integration of fungal mycoprotein with daily diets could help to diversify protein intakes and have potential beneficial implications for health across the lifespan. Further research at both ends of the lifespan spectrum would be worthwhile.

Funding: This research was funded by Marlow Foods Ltd., Stokesley, UK.

Institutional Review Board Statement: Not applicable.

Informed Consent Statement: Not applicable.

Data Availability Statement: Not applicable.

Acknowledgments: Thank you to Hannah Theobald and Tim Finnigan at Quorn Foods for reviewing the article prior to submission.

Conflicts of Interest: Marlow Foods reviewed the paper but played no role in writing the publication.

References

1. Wickramasinghe, K.; Mathers, J.C.; Wopereis, S.; Marsman, D.S.; Griffiths, J.C. From lifespan to healthspan: The role of nutrition in healthy ageing. *J. Nutr. Sci.* **2020**, *9*, e33. [CrossRef] [PubMed]
2. Bystroff, C. Footprints to singularity: A global population model explains late 20th century slow-down and predicts peak within ten years. *PLoS ONE* **2021**, *16*, e0247214. [CrossRef] [PubMed]
3. WHO. Ageing and Health. 2018. Available online: https://www.who.int/health-topics/ageing#tab=tab_1 (accessed on 12 May 2022).
4. Short, R.V. Population growth and global warming. *Facts Views Vis. Obgyn* **2009**, *1*, 27–28. [PubMed]
5. UN. 9.7 Billion on Earth by 2050, but Growth Rate Slowing, Says New UN Population Report. Available online: https://www.un.org/en/academic-impact/97-billion-earth-2050-growth-rate-slowing-says-new-un-population-report (accessed on 12 May 2022).
6. Lonnie, M.; Hooker, E.; Brunstrom, J.M.; Corfe, B.M.; Green, M.A.; Watson, A.W.; Williams, E.A.; Stevenson, E.J.; Penson, S.; Johnstone, A.M. Protein for Life: Review of Optimal Protein Intake, Sustainable Dietary Sources and the Effect on Appetite in Ageing Adults. *Nutrients* **2018**, *10*, 360. [CrossRef]
7. Thavamani, A.; Sferra, T.J.; Sankararaman, S. Meet the Meat Alternatives: The Value of Alternative Protein Sources. *Curr. Nutr. Rep.* **2020**, *9*, 346–355. [CrossRef]
8. Schweiggert-Weisz, U.; Eisner, P.; Bader-Mittermaier, S.; Osen, R. Food proteins from plants and fungi. *Curr. Opin. Food Sci.* 2020; Epub ahead of print. [CrossRef]
9. Bahram, M.; Netherway, T. Fungi as mediators linking organisms and ecosystems. *FEMS Microbiol. Rev.* **2022**, *46*, fuab058. [CrossRef]
10. Naranjo-Ortiz, M.A.; Gabaldon, T. Fungal evolution: Diversity, taxonomy and phylogeny of the Fungi. *Biol. Rev. Camb Philos. Soc.* **2019**, *94*, 2101–2137. [CrossRef]
11. Derbyshire, E. Food-Based Dietary Guidelines and Protein Quality Definitions-Time to Move Forward and Encompass Mycoprotein? *Foods* **2022**, *11*, 647. [CrossRef]
12. Derbyshire, E.J. Is There Scope for a Novel Mycelium Category of Proteins alongside Animals and Plants? *Foods* **2020**, *9*, 1151. [CrossRef]
13. Norman, K.; Klaus, S. Veganism, aging and longevity: New insight into old concepts. *Curr. Opin. Clin. Nutr. Metab. Care* **2020**, *23*, 145–150. [CrossRef] [PubMed]
14. Finnigan, T.J.A.; Wall, B.T.; Wilde, P.J.; Stephens, F.B.; Taylor, S.L.; Freedman, M.R. Mycoprotein: The Future of Nutritious Nonmeat Protein, a Symposium Review. *Curr. Dev. Nutr.* **2019**, *3*, nzz021. [CrossRef] [PubMed]
15. Souza Filho, P.F.; Andersson, D.; Ferreira, J.A.; Taherzadeh, M.J. Mycoprotein: Environmental impact and health aspects. *World J. Microbiol. Biotechnol.* **2019**, *35*, 147. [CrossRef] [PubMed]
16. King, R.; Brown, N.A.; Urban, M.; Hammond-Kosack, K.E. Inter-genome comparison of the Quorn fungus Fusarium venenatum and the closely related plant infecting pathogen Fusarium graminearum. *BMC Genomics* **2018**, *19*, 269. [CrossRef]
17. Finnigan, T.J.A. Chapter 13 – Mycoprotein: Origins, production and properties. In *Handbook of Food Proteins*; Woodhead Publishing Series in Food Science, Technology and Nutrition; Woodhead Publishing ltd.: Sawston, UK, 2011.
18. Finnigan, T.; Needham, L.; Abbott, C. Mycoprotein: A healthy new proteinwith a low environmental impact. In *Sustainable Protein Sources*; Nadathur, S.R., Wanasundara, J.P.D., Scanlin, L., Eds.; Academic Press: London, UK, 2017; pp. 305–325.
19. Upcraft, T.; Wei-Chien, T.; Johnson, R.; Finnigan, T.J.A.; Hung, N.; Hallett, J.; Guo, M. Protein from renewable resources: Mycoprotein production from agricultural residues. *Green Chem* **2021**, *23*, 5150–5165. [CrossRef]
20. Ahmad, M.; Farooq, S.; Alhamooud, Y.; Li, C.; Zhang, H. A review on mycoprotein: History, nutritional composition, production methods, and health benefits. *Trends Food Sci. Technol.* **2022**, *121*, 14–29. [CrossRef]
21. Tuomisto, H.L. Mycoprotein produced in cell culture has environmental benefits over beef. *Nature* **2022**, *605*, 34–35. [CrossRef]
22. Humpenoder, F.; Bodirsky, B.L.; Weindl, I.; Lotze-Campen, H.; Linder, T.; Popp, A. Projected environmental benefits of replacing beef with microbial protein. *Nature* **2022**, *605*, 90–96. [CrossRef]
23. Edwards, G.D.; Cummings, J. The protein quality of mycoprotein. *Proc. Nutr. Soc.* **2010**, *69*, OCE4. [CrossRef]
24. EC. COMMISSION DIRECTIVE 2008/100/EC of 28 October 2008 amending Council Directive 90/496/EEC on nutrition labelling for foodstuffs as regards recommended daily allowances, energy conversion factors and definitions. *Off. J. Eur. Union L* **2008**, 285/9.
25. EFSA. Regulation (EC) No 1924/2006 of the european parliament and of the council of 20 December 2006 on nutrition and health claims made on foods. *Off. J. Eur. Union* **2006**, L 404/409.
26. Derbyshire, E. Mycoprotein: Nutritional and Health Properties. *Nutr. Today* **2019**, *54*, 1–9. [CrossRef]
27. Derbyshire, E.; Finnigan, T.J.A. Chapter 16—Mycoprotein: A futuristic portrayal. In *Future Foods Global Trends, Opportunities, and Sustainability Challenges*; Elsevier Academic Press: Cambridge, MA, USA, 2022; pp. 287–303.
28. Coelho, M.O.C.; Monteyne, A.J.; Dirks, M.L.; Finnigan, T.J.A.; Stephens, F.B.; Wall, B.T. Daily mycoprotein consumption for 1 week does not affect insulin sensitivity or glycaemic control but modulates the plasma lipidome in healthy adults: A randomised controlled trial. *Br J Nutr* **2021**, *125*, 147–160. [CrossRef]
29. Coelho, M.O.C.; Monteyne, A.J.; Kamalanathan, I.D.; Najdanovic-Visak, V.; Finnigan, T.J.A.; Stephens, F.B.; Wall, B.T. Short-Communication: Ingestion of a Nucleotide-Rich Mixed Meal Increases Serum Uric Acid Concentrations but Does Not Affect Postprandial Blood Glucose or Serum Insulin Responses in Young Adults. *Nutrients* **2020**, *12*, 1115. [CrossRef]

30. Dunlop, M.V.; Kilroe, S.P.; Bowtell, J.L.; Finnigan, T.J.A.; Salmon, D.L.; Wall, B.T. Mycoprotein represents a bioavailable and insulinotropic non-animal-derived dietary protein source: A dose-response study. *Br. J. Nutr.* **2017**, *118*, 673–685. [CrossRef]

31. Monteyne, A.J.; Coelho, M.O.C.; Porter, C.; Abdelrahman, D.R.; Jameson, T.S.O.; Jackman, S.R.; Blackwell, J.R.; Finnigan, T.J.A.; Stephens, F.B.; Dirks, M.L.; et al. Mycoprotein ingestion stimulates protein synthesis rates to a greater extent than milk protein in rested and exercised skeletal muscle of healthy young men: A randomized controlled trial. *Am. J. Clin. Nutr.* **2020**, *112*, 318–333. [CrossRef] [PubMed]

32. Monteyne, A.J.; Coelho, M.O.C.; Porter, C.; Abdelrahman, D.R.; Jameson, T.S.O.; Finnigan, T.J.A.; Stephens, F.B.; Dirks, M.L.; Wall, B.T. Branched-Chain Amino Acid Fortification Does Not Restore Muscle Protein Synthesis Rates following Ingestion of Lower- Compared with Higher-Dose Mycoprotein. *J. Nutr.* **2020**, *150*, 2931–2941. [CrossRef] [PubMed]

33. Turnbull, W.H.; Ward, T. Mycoprotein reduces glycemia and insulinemia when taken with an oral-glucose-tolerance test. *Am. J. Clin. Nutr.* **1995**, *61*, 135–140. [CrossRef] [PubMed]

34. Turnbull, W.H.; Walton, J.; Leeds, A.R. Acute effects of mycoprotein on subsequent energy intake and appetite variables. *Am. J. Clin. Nutr.* **1993**, *58*, 507–512. [CrossRef]

35. Udall, J.N.; Lo, C.W.; Young, V.R.; Scrimshaw, N.S. The tolerance and nutritional value of two microfungal foods in human subjects. *Am. J. Clin. Nutr.* **1984**, *40*, 285–292. [CrossRef]

36. Bottin, J.H.; Swann, J.R.; Cropp, E.; Chambers, E.S.; Ford, H.E.; Ghatei, M.A.; Frost, G.S. Mycoprotein reduces energy intake and postprandial insulin release without altering glucagon-like peptide-1 and peptide tyrosine-tyrosine concentrations in healthy overweight and obese adults: A randomised-controlled trial. *Br. J. Nutr.* **2016**, *116*, 360–374. [CrossRef] [PubMed]

37. Cherta-Murillo, A.; Frost, G.S. The association of mycoprotein-based food consumption with diet quality, energy intake and non-communicable diseases' risk in the UK adult population using the National Diet and Nutrition Survey (NDNS) years 2008/2009-2016/2017: A cross-sectional study. *Br. J. Nutr.* **2021**. [CrossRef]

38. Ruxton, C. The impact of mycoprotein on blood cholesterol levels: A pilot study. *Br. Food J.* **2010**, *112*, 109. [CrossRef]

39. Turnbull, W.H.; Leeds, A.R.; Edwards, D.G. Mycoprotein reduces blood lipids in free-living subjects. *Am. J. Clin. Nutr.* **1992**, *55*, 415–419. [CrossRef]

40. Turnbull, W.H.; Leeds, A.R.; Edwards, G.D. Effect of mycoprotein on blood lipids. *Am. J. Clin. Nutr.* **1990**, *52*, 646–650. [CrossRef]

41. Williamson, D.A.; Geiselman, P.J.; Lovejoy, J.; Greenway, F.; Volaufova, J.; Martin, C.K.; Arnett, C.; Ortego, L. Effects of consuming mycoprotein, tofu or chicken upon subsequent eating behaviour, hunger and safety. *Appetite* **2006**, *46*, 41–48. [CrossRef] [PubMed]

42. Monteyne, A.J.; Dunlop, M.V.; Machin, D.J.; Coelho, M.O.C.; Pavis, G.F.; Porter, C.; Murton, A.J.; Abdelrahman, D.R.; Dirks, M.L.; Stephens, F.B.; et al. A mycoprotein-based high-protein vegan diet supports equivalent daily myofibrillar protein synthesis rates compared with an isonitrogenous omnivorous diet in older adults: A randomised controlled trial. *Br. J. Nutr.* **2021**, *126*, 674–684. [CrossRef]

43. Das, J.K.; Salam, R.A.; Thornburg, K.L.; Prentice, A.M.; Campisi, S.; Lassi, Z.S.; Koletzko, B.; Bhutta, Z.A. Nutrition in adolescents: Physiology, metabolism, and nutritional needs. *Ann. N. Y. Acad. Sci.* **2017**, *1393*, 21–33. [CrossRef] [PubMed]

44. Lawrence, E.M.; Mollborn, S.; Hummer, R.A. Health lifestyles across the transition to adulthood: Implications for health. *Soc. Sci. Med.* **2017**, *193*, 23–32. [CrossRef] [PubMed]

45. Chen, C.; Ye, Y.; Zhang, Y.; Pan, X.F.; Pan, A. Weight change across adulthood in relation to all cause and cause specific mortality: Prospective cohort study. *BMJ* **2019**, *367*, l5584. [CrossRef] [PubMed]

46. Farvid, M.S.; Eliassen, A.H.; Cho, E.; Liao, X.; Chen, W.Y.; Willett, W.C. Dietary Fiber Intake in Young Adults and Breast Cancer Risk. *Pediatrics* **2016**, *137*, e20151226. [CrossRef]

47. Sawyer, S.M.; Afifi, R.A.; Bearinger, L.H.; Blakemore, S.J.; Dick, B.; Ezeh, A.C.; Patton, G.C. Adolescence: A foundation for future health. *Lancet* **2012**, *379*, 1630–1640. [CrossRef]

48. Deeg, D.J. Searching for roots in mid-life. Reflections on 'Mid-life. Notes from the halfway mark' by Elizabeth Kaye. *Patient Educ. Couns.* **1998**, *34*, 83–86. [CrossRef]

49. Zurakowski, T.L. Health promotion for mid- and later-life women. *J. Obstet. Gynecol. Neonatal. Nurs.* **2004**, *33*, 639–647. [CrossRef]

50. Fanelli, S.M.; Kelly, O.J.; Krok-Schoen, J.L.; Taylor, C.A. Low Protein Intakes and Poor Diet Quality Associate with Functional Limitations in US Adults with Diabetes: A 2005-2016 NHANES Analysis. *Nutrients* **2021**, *13*, 2582. [CrossRef]

51. Krok-Schoen, J.L.; Archdeacon Price, A.; Luo, M.; Kelly, O.J.; Taylor, C.A. Low Dietary Protein Intakes and Associated Dietary Patterns and Functional Limitations in an Aging Population: A NHANES analysis. *J. Nutr. Health Aging* **2019**, *23*, 338–347. [CrossRef]

52. Chen, J.; Pi, X.; Liu, W.; Ding, Q.; Wang, X.; Jia, W.; Zhu, L. Age-related changes of microbiota in midlife associated with reduced saccharolytic potential: An in vitro study. *BMC Microbiol.* **2021**, *21*, 47. [CrossRef] [PubMed]

53. Chen, L.; Ye, M.; Kahana, E. A Self-Reliant Umbrella: Defining Successful Aging Among the Old-Old (80+) in Shanghai. *J. Appl. Gerontol.* **2020**, *39*, 242–249. [CrossRef] [PubMed]

54. Vogt, D.; Berens, E.M.; Schaeffer, D. Health Literacy in Advanced Age. *Gesundheitswesen* **2020**, *82*, 407–412. [CrossRef] [PubMed]

55. Shinan-Altman, S.; Werner, P. Subjective Age and Its Correlates Among Middle-Aged and Older Adults. *Int. J. Aging Hum. Dev.* **2019**, *88*, 3–21. [CrossRef] [PubMed]

56. WH0. Ageing and Health. Available online: https://www.who.int/news-room/fact-sheets/detail/ageing-and-health (accessed on 12 May 2022).

57. Brivio, P.; Paladini, M.S.; Racagni, G.; Riva, M.A.; Calabrese, F.; Molteni, R. From Healthy Aging to Frailty: In Search of the Underlying Mechanisms. *Curr. Med. Chem.* **2019**, *26*, 3685–3701. [CrossRef] [PubMed]

58. Robinson, S.M. Improving nutrition to support healthy ageing: What are the opportunities for intervention? *Proc. Nutr. Soc.* **2018**, *77*, 257–264. [CrossRef] [PubMed]

59. Ekmekcioglu, C. Nutrition and longevity—From mechanisms to uncertainties. *Crit. Rev. Food Sci. Nutr.* **2020**, *60*, 3063–3082. [CrossRef] [PubMed]

60. Wang, L.; Si, L.; Cocker, F.; Palmer, A.J.; Sanderson, K. A Systematic Review of Cost-of-Illness Studies of Multimorbidity. *Appl. Health Econ. Health Policy* **2018**, *16*, 15–29. [CrossRef]

61. Pearson-Stuttard, J.; Ezzati, M.; Gregg, E.W. Multimorbidity-a defining challenge for health systems. *Lancet Public Health* **2019**, *4*, e599–e600. [CrossRef]

62. Kingston, A.; Robinson, L.; Booth, H.; Knapp, M.; Jagger, C.; MODEM Project. Projections of multi-morbidity in the older population in England to 2035: Estimates from the Population Ageing and Care Simulation (PACSim) model. *Age Ageing* **2018**, *47*, 374–380. [CrossRef] [PubMed]

63. Kitada, M.; Ogura, Y.; Monno, I.; Koya, D. The impact of dietary protein intake on longevity and metabolic health. *EBioMedicine* **2019**, *43*, 632–640. [CrossRef]

64. Barber, T.M.; Kabisch, S.; Pfeiffer, A.F.H.; Weickert, M.O. The Health Benefits of Dietary Fibre. *Nutrients* **2020**, *12*, 3209. [CrossRef]

65. Coelho, M.O.C.; Monteyne, A.J.; Dunlop, M.V.; Harris, H.C.; Morrison, D.J.; Stephens, F.B.; Wall, B.T. Mycoprotein as a possible alternative source of dietary protein to support muscle and metabolic health. *Nutr. Rev.* **2020**, *78*, 486–497. [CrossRef]

66. Cherta-Murillo, A.; Lett, A.M.; Frampton, J.; Chambers, E.S.; Finnigan, T.J.A.; Frost, G.S. Effects of mycoprotein on glycaemic control and energy intake in humans: A systematic review. *Br. J. Nutr.* **2020**, *123*, 1321–1332. [CrossRef]

67. Derbyshire, E.; Delange, J. Fungal Protein—What Is It and What Is the Health Evidence? A Systematic Review Focusing on Mycoprotein. *Front. Sustain. Food Syst.* **2021**, *5*, 581682. [CrossRef]

68. Colosimo, R.; Mulet-Cabero, A.-I.; Warren, F.J.; Edwards, C.H.; Finnigan, T.J.A.; Wilde, P.J. Mycoprotein ingredient structure reduces lipolysis and binds bile salts during simulated gastrointestinal digestion. *Food Funct.* **2020**, *11*, 10896–10906. [CrossRef] [PubMed]

69. Colosimo, R.; Warren, F.J.; Finnigan, T.J.A.; Wilde, P.J. Protein bioaccessibility from mycoprotein hyphal structure: In vitro investigation of underlying mechanisms. *Food Chem.* **2020**, *330*, 127252. [CrossRef] [PubMed]

70. Derbyshire, E. Fungal Protein with Fibre. Can Quorn's mycoprotein Contribute to Children's Fibre Intakes? *Complet. Nutr.* **2021**, *21*, 33–35.

71. Bull, C.; Belobrajdic, D.; Hamzelou, S.; Jones, D.; Leifert, W.; Ponce-Reyes, R.; Terefe, N.S.; Williams, G.; Colgrave, M. How Healthy Are Non-Traditional Dietary Proteins? The Effect of Diverse Protein Foods on Biomarkers of Human Health. *Foods* **2022**, *11*, 528. [CrossRef] [PubMed]

Journal of
Fungi

Review

Extracellularly Released Molecules by the Multidrug-Resistant Fungal Pathogens Belonging to the *Scedosporium* Genus: An Overview Focused on Their Ecological Significance and Pathogenic Relevance

Thaís P. Mello [1], Iuri C. Barcellos [1], Ana Carolina Aor [1], Marta H. Branquinha [1,2] and André L. S. Santos [1,2,*]

1 Laboratório de Estudos Avançados de Microrganismos Emergentes e Resistentes (LEAMER), Departamento de Microbiologia Geral, Instituto de Microbiologia Paulo de Góes (IMPG), Centro de Ciências da Saúde (CCS), Universidade Federal do Rio de Janeiro (UFRJ), Rio de Janeiro 21941-901, Brazil
2 Rede Micologia RJ—Fundação de Amparo à Pesquisa do Estado do Rio de Janeiro (FAPERJ), Rio de Janeiro 21941-901, Brazil
* Correspondence: andre@micro.ufrj.br

Abstract: The multidrug-resistant species belonging to the *Scedosporium* genus are well recognized as saprophytic filamentous fungi found mainly in human impacted areas and that emerged as human pathogens in both immunocompetent and immunocompromised individuals. It is well recognized that some fungi are ubiquitous organisms that produce an enormous amount of extracellular molecules, including enzymes and secondary metabolites, as part of their basic physiology in order to satisfy their several biological processes. In this context, the molecules secreted by *Scedosporium* species are key weapons for successful colonization, nutrition and maintenance in both host and environmental sites. These biologically active released molecules have central relevance on fungal survival when colonizing ecological places contaminated with hydrocarbons, as well as during human infection, particularly contributing to the invasion/evasion of host cells and tissues, besides escaping from the cellular and humoral host immune responses. Based on these relevant premises, the present review compiled the published data reporting the main secreted molecules by *Scedosporium* species, which operate important physiopathological events associated with pathogenesis, diagnosis, antimicrobial activity and bioremediation of polluted environments.

Keywords: *Scedosporium*; *Lomentospora*; emergent fungi; extracellular molecules; enzymes; secondary metabolites

Citation: Mello, T.P.; Barcellos, I.C.; Aor, A.C.; Branquinha, M.H.; Santos, A.L.S. Extracellularly Released Molecules by the Multidrug-Resistant Fungal Pathogens Belonging to the *Scedosporium* Genus: An Overview Focused on Their Ecological Significance and Pathogenic Relevance. *J. Fungi* **2022**, *8*, 1172. https://doi.org/10.3390/jof8111172

Academic Editor: Laurent Dufossé

Received: 30 August 2022
Accepted: 2 November 2022
Published: 7 November 2022

Publisher's Note: MDPI stays neutral with regard to jurisdictional claims in published maps and institutional affiliations.

1. Introduction: An Overview on the *Scedosporium* Genus

The *Scedosporium* genus is constituted of saprophytic filamentous fungi frequently isolated from human impacted environments, such as sewers, polluted waters, sediments, decaying vegetation, agricultural soils, hydrocarbon-contaminated soils, gardens, urban parks, playgrounds and hospital areas, compared to habitats with low human activity [1–6]. The *Scedosporium* genus is composed of the following species: *Scedosporium angustum*, *Scedosporium apiospermum*, *Scedosporium aurantiacum*, *Scedosporium boydii*, *Scedosporium cereisporum*, *Scedosporium dehoogii*, *Scedosporium desertorum*, *Scedosporium ellipsoideum*, *Scedosporium fusoideum* and *Scedosporium minutisporum* [7]. *Lomentospora prolificans*, formerly *Scedosporium prolificans*, was renamed due to its phylogenetic distance from the *Scedosporium* genus, as judged by both molecular and genetic parameters [8]. However, *L. prolificans* has been historically studied together with *Scedosporium* species; so, in this context, we decided to refer to scedosporiosis as the infection caused by both fungal genera in order to facilitate and to simplify the information.

Scedosporium species are emerging, opportunistic pathogens able to cause localized infections in immunocompetent individuals and disseminated infections in immunocompromised individuals [2]. Over the last few years, the number of cases of scedosporiosis has increased considerably, which may reflect, at least in part, an improvement in the diagnosis of its etiologic agents. For instance, the incidence of *Scedosporium* infection in a tertiary care cancer center in Texas (USA) per 100,000 patient–inpatient days increased from 0.82 cases between 1993 and 1998 to 1.33 cases from 1999 to 2005 [9]. Reviewing the literature, several publications have reported that the cases of disseminated scedosporiosis typically occurred in individuals undergoing hematopoietic stem cell transplantation (HSCT) and solid organ transplantation (SOT). In this context, the number of infections caused by these fungi accounted for approximately 25% of all non-*Aspergillus* mold infections in SOT recipients [10] and 29% of those in HSCT recipients, in which 75% of the infections in HSCT recipients and 61% of the infections in SOT recipients occurred within 6 months after transplantation [11]. The infection was disseminated in 69% and 46% of HSCT and SOT recipients with scedosporiosis, respectively [11]. Furthermore, the mortality rates in patients with disseminated *L. prolificans* infections are higher, up to 87.5% [12]. A study conducted by Heng and co-workers [13] revealed that scedosporiosis in hematology patients exerts a substantial impact on hospital resource consumption, length of stay and patient mortality, with the total costs (U$ 26,500.00 per patient) driven by ward stay and antifungal drug costs.

Scedosporium species also show a marked neurotropism and a high propensity to cause central nervous system (CNS) infections [2,14]. In human immunodeficiency virus (HIV)-positive patients colonized by *Scedosporium* spp., invasive scedosporiosis was proven in 54.5% of patients, with a mortality rate of 75%. In patients with CNS manifestations the mortality rate increases to 100% [15]. In an analysis of 99 cases of CNS infection caused by the *Scedosporium* genus, a similar percentage of mortality in immunocompetent and immunocompromised patients was reported (76% and 74%, respectively) [16]. Interestingly, CNS infection was preceded by near drowning or trauma in immunocompetent patients. Regarding the CNS infection in immunocompromised individuals, it was described as rapidly progressive disseminated lesions at various degrees of evolution [16]. Moreover, *Scedosporium* species rank second among the filamentous fungi most frequently isolated from cystic fibrosis patients, constituting a great risk factor for invasive infections for lung transplanted patients [17–19].

Infections caused by *Scedosporium* species are extremely difficult to treat because of the low susceptibility profile to all classes of antifungal drugs available for clinical use (e.g., azoles, echinocandins and polyenes). For *L. prolificans*, the scenario worsens since this species is pan-antifungal resistant [20]. At the moment, the treatment indicated for scedosporiosis is voriconazole together with surgical debridement when possible [21]. However, even when the recommendation is followed, the mortality rate is higher than 65% [20]. Thus, the relevance of *Scedosporium/Lomentospora* in the clinical scenario is obviously alarming due to both multiple antifungal-resistance and high morbimortality profiles.

Scedosporiosis usually starts with the inhalation or traumatic inoculation of conidial cells, which then germinate into hyphae that promote host cell/tissue invasion (Figure 1) [20,22]. The mycelial biomass formed by *Scedosporium/Lomentospora* species during the infection process resembles a typical biofilm structure, formed by a robust mass of hyphae surrounded by an extracellular polymeric matrix (Figure 2) [23–29]. The ability to form biofilms is essential for microbial cells to cope with environmental stress, host immunological responses and antimicrobial drugs [30]. The *Scedosporium* and *Lomentospora* biofilms are 2- to 1024-times more resistant to azoles (e.g., voriconazole), echinocandins (e.g., caspofungin) and polyenes (e.g., amphotericin B) than that observed in planktonic conidial cells. The increase in the resistance profile to antifungal drugs observed in biofilms is mainly due to the presence of the extracellular matrix, efflux pumps and the highly adapted response to oxidative stress [24,26,27,29]. The successful colonization by *Scedosporium* and *Lomentospora* species is partially due to the secretion of extracellular molecules that participate in nutrient acquisition, competition with other microorganisms, germination of conidia into hyphae and invasion of host cells and tissues, among other essential events [20,22].

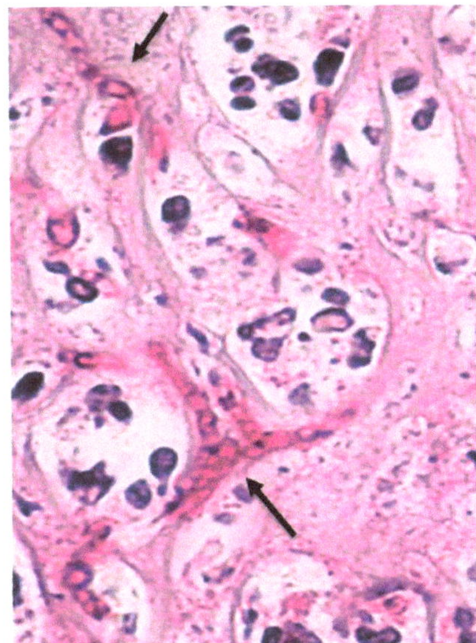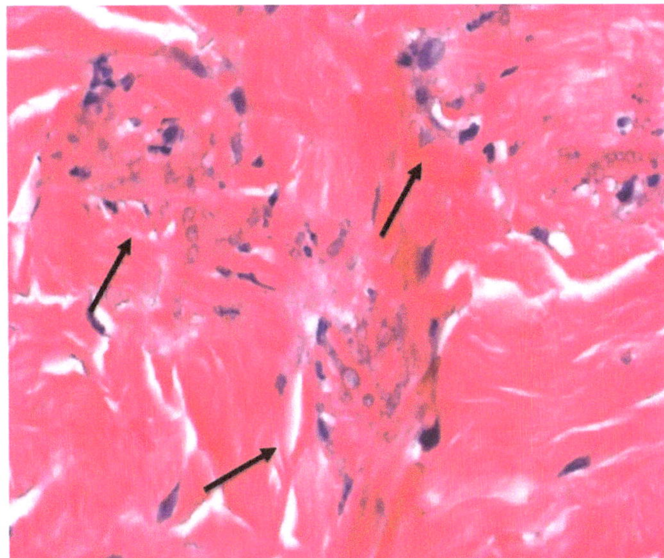

Figure 1. Histopathological sections evidencing *Scedosporium* fungal particles in human tissue. **Left image**: periodic acid-Schiff stain evidencing thin irregular, septate hyphae of *S. boydii* (black arrow) on a background of necrotic detritus and neutrophils in eviscerated ocular tissue (kindly donated by Dr. Virginia Vanzzini-Zago and Dr. Abelardo Rodrıguez-Reyes at Hospital Asociacion para Evitar la Ceguera en Mexico and Dr. Sonia Corredor-Casas at Instituto Mexicano de Oftalmologıa IAP Queretaro, Mexico). **Right image**: hematoxylin and eosin stain showing many hyphae of *Scedosporium* in the dermis of a skin biopsy (kindly provided by Dr. Stacy Beal, Assistant Professor, University of Florida, College of Medicine, Department of Pathology, Immunology and Laboratory Medicine, Gainesville, FL, USA). Original magnification of the images is 400×.

The previously published reviews about *Scedosporium* species have addressed the pathogenesis mechanisms, immunology, treatment options, epidemiology, taxonomy and/or use of those species for bioremediation [6,20,22]; however, these works have never focused only on the different roles that extracellularly secreted molecules can play. In this context, herein we examine the available information about the extracellularly released molecules by *Scedosporium* species, including polysaccharides, non-peptide small-molecule metabolites, non-ribosomal peptides and (glyco)protein-nature molecules (Figure 3), as well as their potential roles in environmental colonization, successful host infection, nutrition, diagnosis and stress response. In addition, we reported for the first time on the effects of *Scedosporium* secretions on *Tenebrio molitor* larvae used herein as an in vivo model of infection.

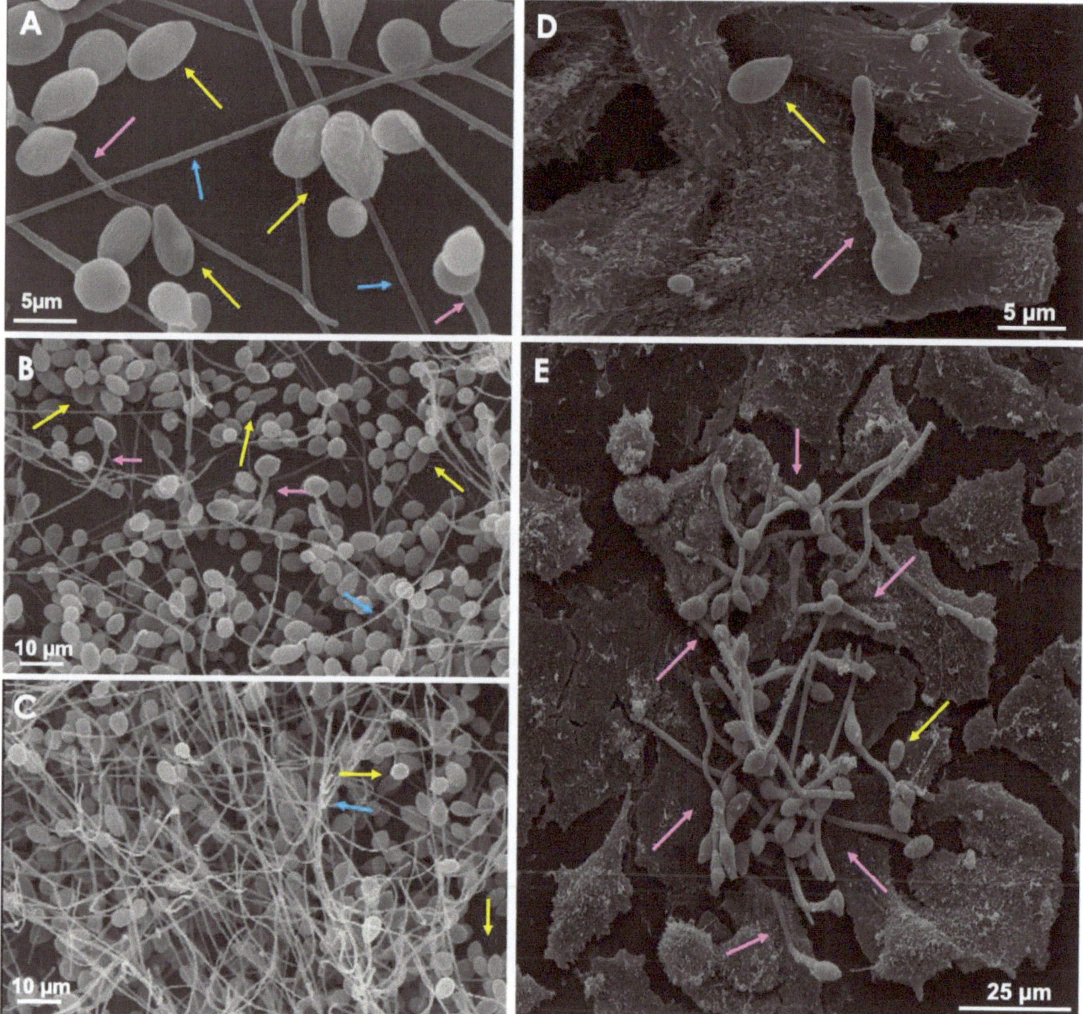

Figure 2. Distinct morphologies of *Scedosporium* evidenced by scanning electron microscopy. The images demonstrate the conidia (yellow arrow), germinated conidia (pink arrow) and hyphae (blue arrow) of *S. apiospermum* on glass substrate. Note in the micrographs (**A–C**) the mycelia formed by *S. apiospermum* on glass surface. (**D,E**) Interaction of *S. apiospermum* with A549 epithelial cells with conidia, germinated conidia and the filamentous network of hyphae cells on top of A549 cells.

DIFFERENT CLASSES OF EXTRACELLULAR MOLECULES OF *Scedosporium*

Non-Peptide Small-Molecule Metabolites

- 2-dimethylamino-1-(4-hydroxyphenyl)-8,10-dimethyl-6-dodecene-3-one (YM-193221)
- 6-6'-bis (2H-pyran-3-carbaldehyde) (pseudallin)
- polyketides (boydone A, boydone B, botryorhodine F, botryorhodine G)
- fusidilactone A
- (R)-(−)-mevalonolactone
- (R)-(−)-lactic acid
- Tyroscherin
- *N*-methyltyroscherin
- (R)-2-(2-hydroxypropan-2-yl)-2,3-dihydro-5-hydroxybenzofuran
- (R)-2-(2-hydroxypropan-2-yl)-2,3-dihydro-5-methoxybenzofuran
- 3,3'-dihydroxy-5,5'-dimethyldiphenyl ether
- 3-(3-methoxy-5-methylphenoxy)-5-methylphenol
- (−)-regiolone
- 2,4-dimethyl-2-hydroxy-5-(1,3,5,7-tetramethyl-nonyl)-3(2H)-furanone (AS-183)

Polysaccharides

- β-glucan

Non-Ribosomal Peptides

- boydine A, boydine B, boydine C, boydine D
- cyclic peptides (pseudacyclin A, pseudacyclin B, pseudacyclin C, pseudacyclin D, pseudacyclin E)
- gliotoxin
- siderophores (dimerumic acid, N$^\alpha$-methyl coprogen B)
- dehydroxybisdethiobis(methylthio)gliotoxin
- bisdethiobis(methylthio)gliotoxin
- isobenzofuranone derivatives (pseudaboydins A, pseudaboydins B)
- sesquiterpenes (boydene A, boydene B)

(Glyco)Protein-Nature Molecules

- peptidases (serine-, aspartic- and metallo-type peptidases)
- Phosphatase
- 120-kDa immunodominant antigen
- energy and/or carbohydrate metabolism (phosphomannomutase, triosephosphate isomerase, malate dehydrogenase, fructose-1,6-bisphosphate aldolase, phosphoglycerate mutase, mannitol-1-phosphate 5-dehydrogenase, aldose-1-epimerase)
- sterol metabolism-related protein
- Glucanase
- Ran-specific GTPase-activating protein 1
- nucleoside diphosphate kinase
- initiation factor 5a
- drug elimination/detoxification (major facilitator superfamily multidrug transporter, peroxiredoxin, manganese superoxide dismutase, translationally controlled tumor protein
- heat shock 70-kDa protein
- cytoskeleton/movement proteins (actin, cofilin, profilin, tropomyosin, tubulin)
- Asp f13-like protein
- haloacid dehalogenase-superfamily hydrolase

Figure 3. Overview of the different classes of molecules secreted/released by environmental and/or clinical strains of *Scedosporium* species.

2. Secretion: An Essential Biological Process in the Fungal Cell Cycle

The production of extracellular molecules is a universal process with fundamental importance in many aspects of the cellular physiology of all living cells, particularly fungi [31]. Throughout evolution, fungi have adapted their secretion machinery in order to perform a great number of specialized functions during different stages of the infectious process, allowing these microorganisms to cause illness [32,33]. The extracellularly released molecules play critical roles related to virulence and act at different stages of interaction, allowing fungal survival, multiplication and dissemination inside the infected host [33,34]. Moreover, it is known that secreted molecules can modulate the host immune response, helping the fungal cells to escape from the antimicrobial properties of antibodies, complementing proteins and antimicrobial peptides produced by the infected host [35]. For instance, the galactosaminogalactans secreted by mycelial cells of *Aspergillus fumigatus* are responsible for promoting fungal growth and survival inside the host due to their ability to trigger a Th2 immunosuppressive response [36]. Secreted/released molecules by fungi do not only interact with components and cells of the host immune system but can also induce different degrees of cytotoxicity on mammalian cells, being capable of activating distinct death pathways [37,38]. Some studies have shown that the exposure of host cells to fungal secretions is by itself sufficient to cause host cell death [39]. Schindler and Segal [40] showed that *Candida albicans* secreted metabolites directly affect the host cell cytoskeleton, inducing a rearrangement in actin filaments, which caused a decrease of 63% in the phagocytic activity of murine macrophages, resulting in the activation of apoptosis in these cells. Our research group showed that molecules secreted by mycelia of a clinical strain of *S. apiospermum* caused a significant loss of viability (CC_{50} = 0.24 µg/µL) in the confluent monolayer of pulmonary epithelial cells (A549 non-small-cell lung cancer cell line), inducing irreversible damage that begins with the rounding of the epithelial cells followed by their detachment from the plastic substrate (Figure 4) [35].

In a proteomic analysis performed before genomic sequencing of *S. apiospermum*, proteins involved in metabolic pathways (malate dehydrogenase, phosphomannomutase, triosephosphate isomerase, fructose-1,6-bisphosphate aldolase, phosphoglycerate mutase, mannitol-1-phosphate 5-dehydrogenase, aldose-1-epimerase, sterol metabolism-related protein), protein degradation/nutrition (aspartyl protease, haloacid dehalogenase-superfamily hydrolase), nucleotide metabolism (nucleoside diphosphate kinase), RNA processing (Ran-specific GTPase-activating protein 1), translation machinery (initiation factor 5a), morphogenesis (glucanase), transport (major facilitator superfamily multidrug transporter, Forkhead associated domain involved in signaling events and ABC-type transport system region), protection against stress (peroxiredoxin, heat shock protein, translationally controlled tumor protein, manganese superoxide dismutase), movement (cofilin, profilin and tropomyosin) and the allergen Asp f13-like protein were identified in *S. apiospermum* secretome [35]. In this regard, Figure 5 reveals the richness of polypeptides/proteins secreted by *Scedosporium*. Moreover, some of the secreted proteins were recognized by antibodies present in the serum obtained from a scedosporiosis patient, validating the role of secreted proteins during human infection [35]. The proteomic analysis of secreted molecules is currently being redone in our lab since the genome sequencing of *S. apiospermum* has become available [40,41]. It should be noted that the actual secretome analysis accounts for more than 120 distinct proteins (unpublished data), which drastically contrasts with the 25 previously identified proteins [35].

The larvae of the *Tenebrio molitor* beetle is currently being used as an in vivo model for studies on fungal infections, together with other invertebrate models such as *Drosophila melanogaster*, *Galleria mellonella* and *Caenorhabditis elegans*, due to the ease of manipulation and maintenance and low price, and as a valuable alternative to animal models [42]. In order to add more data about the effect of *Scedosporium* secreted molecules in the host, herein we present unpublished data about the cytotoxic effects of secreted proteins by three different strains of *S. apiospermum* in an in vivo model of *T. molitor* larvae; the experimental methodology is detailed in Figure 6. Proteins obtained from the strains after 7 days of mycelial culture were able to interfere with the viability of *T. molitor* in a typically dose-dependent manner (Figure 6B,C). These results demonstrate that *Scedosporium* secreted virulence factors are per se able to cause significant damage in the invertebrate host model.

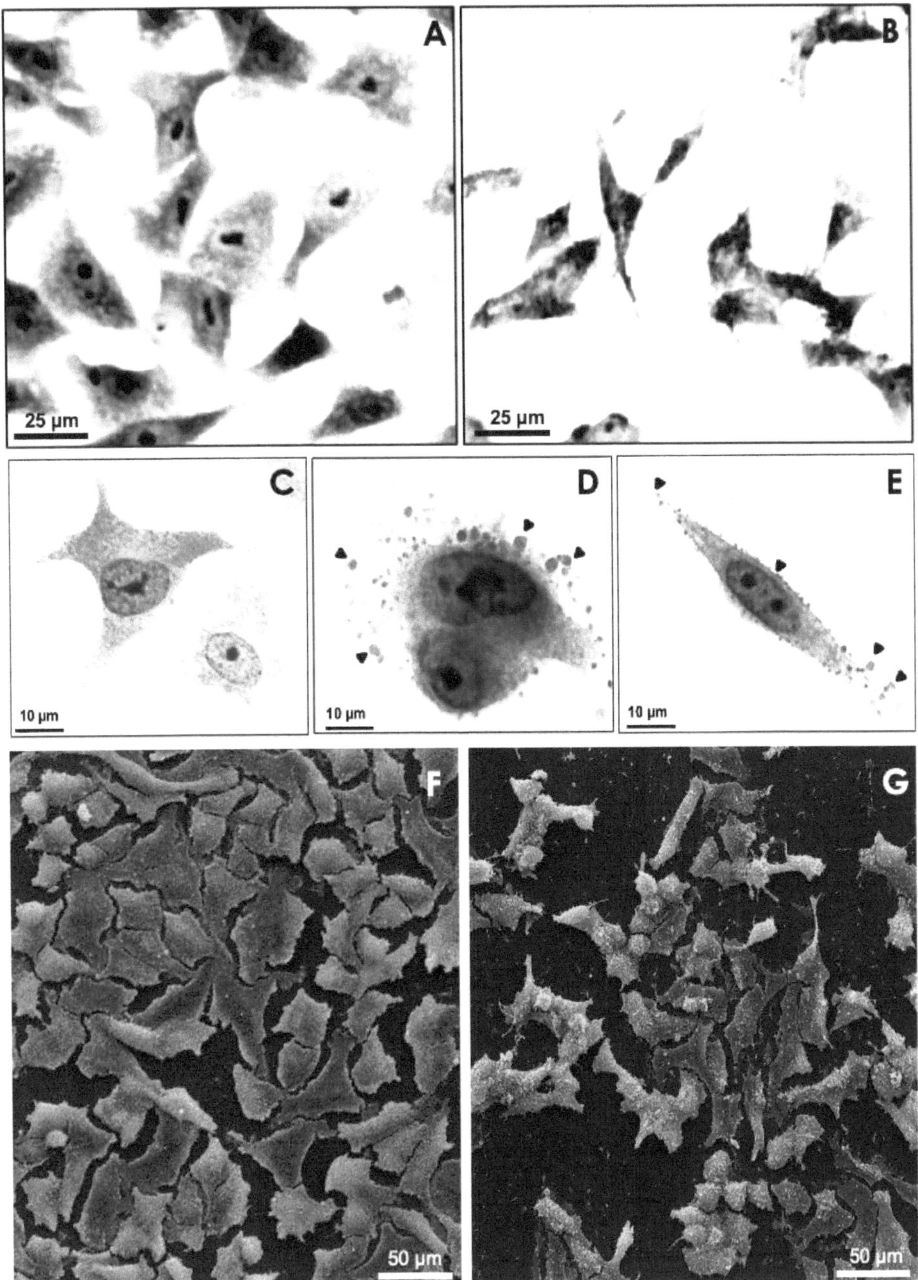

Figure 4. Light microscopy (**A–E**) and scanning electron microscopy (SEM) (**F,G**) images representing the monolayer of A549 epithelial cells before (**A,C,F**) and after (**B,D,E,G**) incubation with secreted molecules from *S. apiospermum* mycelial cells. Note the morphological changes in A549 cells, such as the release of the monolayer (**B,G**) and the presence of bubbles (arrowheads in **D,E**) on the epithelial cell surface. (Adapted from Silva et al., 2012 [35]).

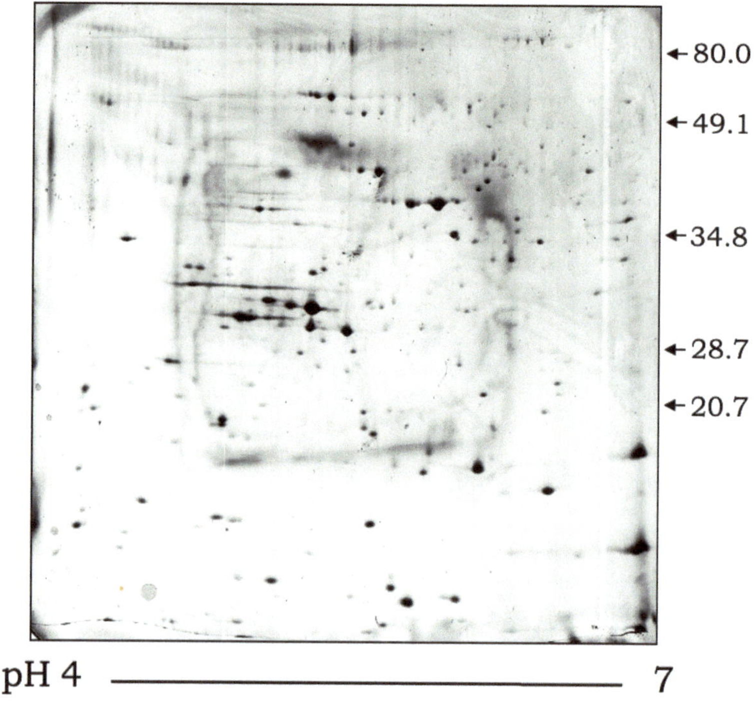

Figure 5. The 2-dimensional SDS-PAGE stained with Coomassie Brilliant Blue G-250 demonstrating the diversity of proteins extracellularly released by *S. apiospermum*. Molecular mass markers expressed in kilodaltons are expressed on the right.

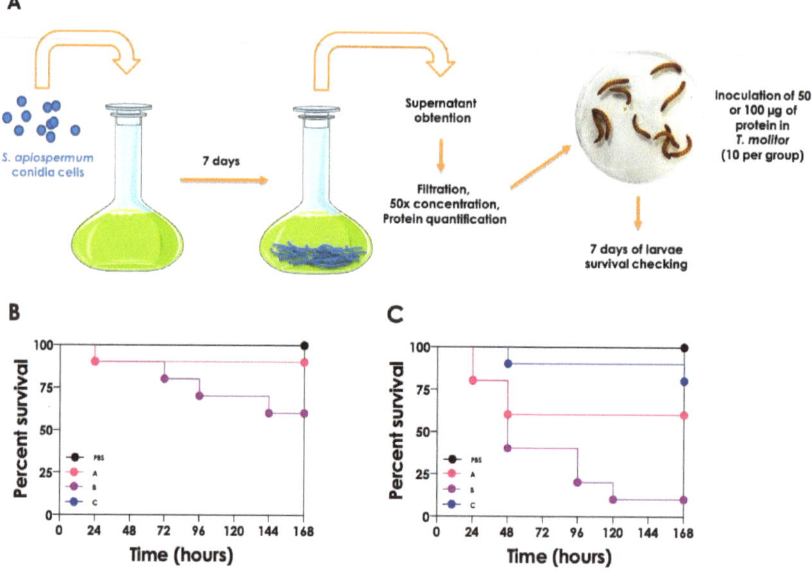

Figure 6. (**A**) Experimental scheme of *S. apiopermum* supernatant cytotoxicity test in *T. molitor*. Conidia

of *S. apiospermum* (strains (**A**–**C**)) were inoculated in Sabouraud media and incubated for 7 days at 37 °C in constant agitation (120 rpm). Subsequently, supernatants obtained from 7-day cultures of different strains (**A**–**C**) were filtered in order to withdraw the remaining cells, concentrated 50 times utilizing a 10-kDa Amicon membrane, and the protein concentration was determined through the methodology developed by Lowry and co-workers [43]. Finally, *T. molitor* larvae (10 per group) were inoculated with cell-free supernatant containing either (**B**) 50 μg or (**C**) 100 μg of proteins, and the larvae survival was checked daily for 7 days. The survival data were plotted using the Kaplan-Meier method using GraphPad Prism 8.

3. Secretion of Specific Molecules by *Scedosporium*: Ecological, Physiological and Pathological Perspectives

3.1. Extracellular Vesicles: A Biological Carrier of Active Molecules

Despite the large number of studies on the detection and characterization of extracellular molecules in fungi, the secretion pathways in these microorganisms are particularly complex and not yet well understood, being largely even unknown [44]. In general, the secretion of molecules to the extracellular milieu by eukaryotic organisms can occur via conventional or unconventional/alternative pathways [45]. In the conventional pathway, proteins that will be secreted have an *N*-terminus-linked signal peptide to be incorporated into transport vesicles within the endoplasmic reticulum (ER) lumen and then directed to the cell surface through Golgi apparatus [46]. Proteins secreted in alternative routes lack the *N*-terminus-linked signal peptide and can reach the cell surface by multiple mechanisms, most of them by vesicles [46]. In fungal cells, molecules to be secreted must go through the entire thickness of the fungal cell wall, which provides additional complexity to the secretion process [47]. In recent years, the number of studies on the mechanisms developed by fungal cells to promote molecular transport through the cell wall has greatly increased. Nowadays, there are three non-exclusive hypotheses about how vesicles cross the fungal cell wall. First, enzymes could remodel the cell wall creating passages for vesicles through this physical barrier. Second, the existence of small pores in the cell wall suggests that these holes can also be used in vesicular transport. Another hypothesis would be mechanical pressure through cell wall pores [44,48,49]. Rodrigues and coworkers [50] described for the first time that fungi produce extracellular vesicles in vitro and in vivo, which are secreted through the cell wall, revealing that vesicular secretion is a key mechanism of extracellular delivery. So far, the extracellular vesicles have been described and characterized in several fungal species, such as: (i) yeasts of *Cryptococcus neoformans*, *C. albicans*, *Candida parapsilosis*, *Histoplasma capsulatum*, *Malassezia sympodialis*, *Paracoccidioides brasiliensis*, *Saccharomyces cerevisiae* and *Sporothrix schenckii*, (ii) protoplasts of *A. fumigatus*, (iii) conidia of *Aspergillus flavus*, (iv) hyphae of *Alternaria infectoria*, *Trichophyton interdigitale*, *Trichoderma reesei*, *Rhizopus delemar* and *Fusarium oxysporum* f. sp. *vasinfectum* [49–60]. The importance of producing vesicles by pathogenic fungi has been constantly proven. It is known that extracellular vesicles act as virulence pockets that deliver a concentrated load of fungal products directly to host cells and tissues (Rodrigues et al. 2008). For instance, extracellular vesicles produced by *C. neoformans* carry capsular components such as glucuronoxylomannan (GXM) and glucuronoxylomannangalactan (GXMGal), which are antigenic polysaccharides, as well as other virulence-associated components such as enzymes (e.g., urease, laccase and acid phosphatase), heat shock and antioxidant proteins, lipids and nucleic acids (DNA and RNAs) [61,62]. It has been reported that the vesicle components are recognized by serum antibodies from patients with cryptococcosis, histoplasmosis, paracoccidioidomycosis and candidiasis [51,53,61,63]. Moreover, the vesicular content is capable of inducing the production of several cytokines, such as tumor necrosis factor alpha (TNF-α), transforming growth factor beta (TGF-β) and interleukin-10 (IL-10), as well as activating nitric oxide (NO) production in murine macrophages in a typical dose-dependent manner [64].

In *S. apiospermum*, our research group observed the presence of extracellular vesicles through transmission electron microscopy images [35]. That report added *S. apiospermum* to the list of human pathogenic fungi capable of secreting extracellular vesicles, being the first filamentous fungal pathogen to join this list [35]. The secreted vesicles were detected leaving the fungal cells and entering the extracellular space in different parts of the cell wall, including the areas closest to the plasma membrane and in direct contact with the extracellular medium, in both conidial and mycelial forms of *S. apiospermum* (Figure 7). Moreover, the secretion of vesicles by *S. apiospermum* was also observed during the interaction of fungal particles with A549 lung epithelial cells (Figure 8) [65].

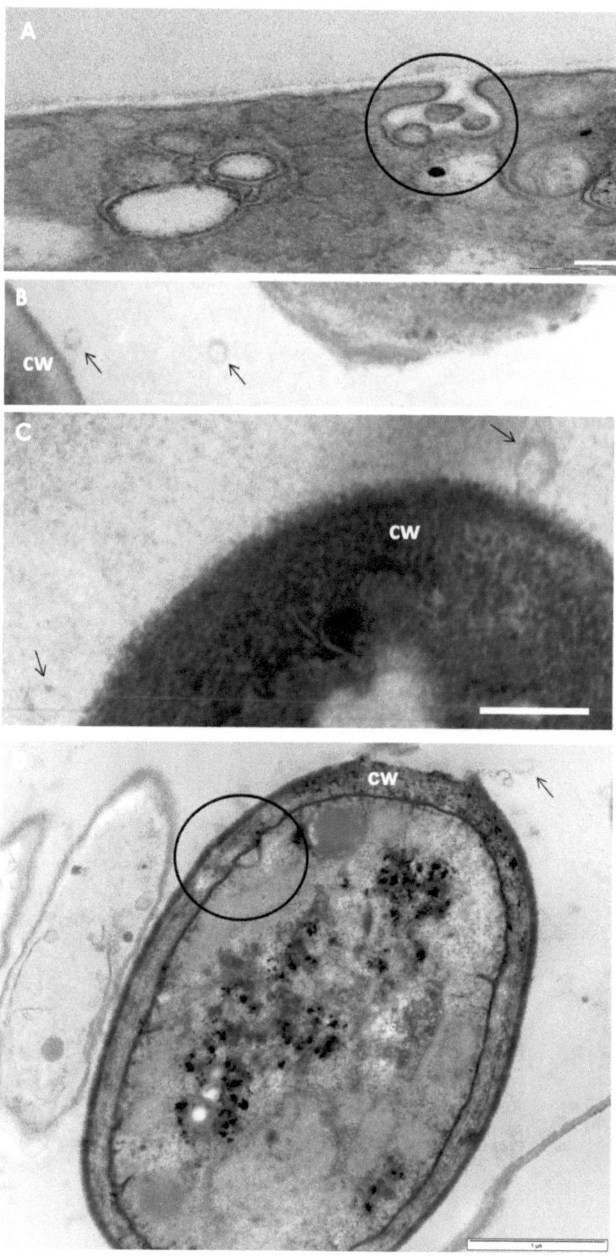

Figure 7. Transmission electron microscopy (TEM) showing vesicle-like structures in *Scedosporium* spp.

(**A**) TEM images demonstrated the presence of vesicles within a membrane invagination in *S. minutisporum* hyphae. The black circle evidences the invagination. Bar: 1 µm. (**B,C**) TEM micrographs showing vesicle bodies in the extracellular milieu of *S. apiospermum* hyphae. Arrows evidence the vesicular bodies. cw, cell wall. Bar: 250 nm. (**D**) TEM image of *S. apiospermum* conidia. The black circle evidences the membrane invagination, and the arrow evidences the vesicular bodies. Cw, cell wall. Bar: 1 µm.

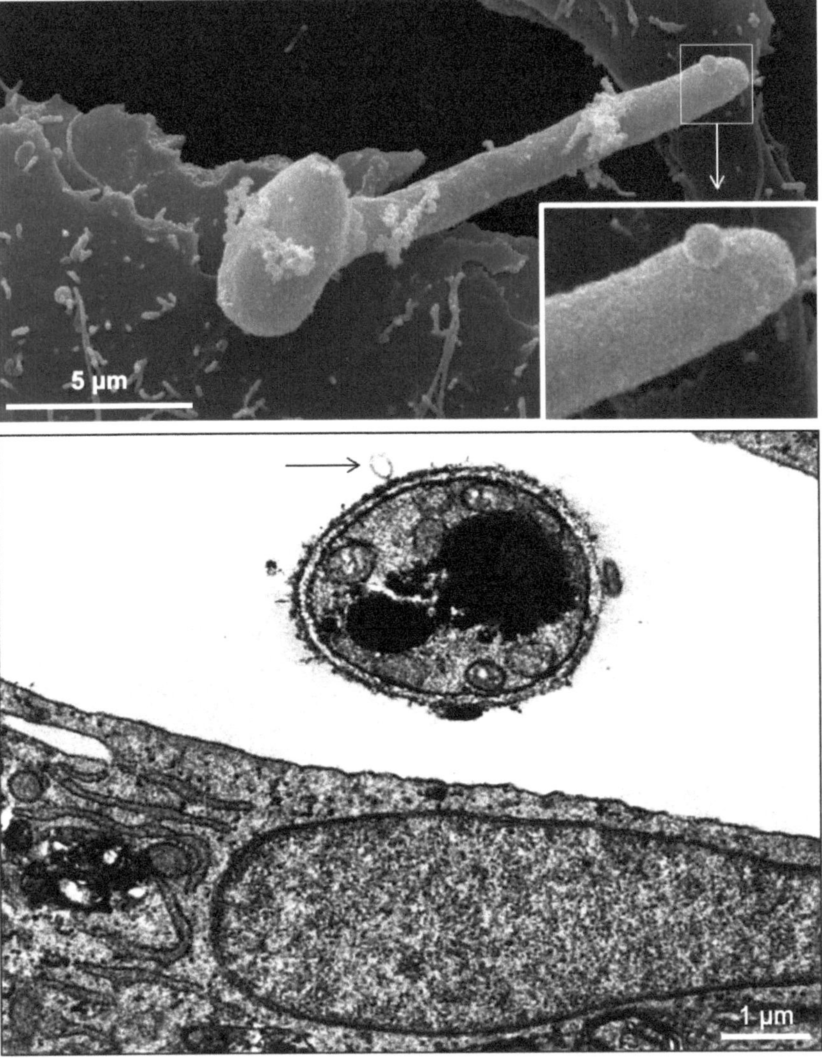

Figure 8. Scanning electron microscopy (SEM) demonstrating the presence of vesicles on the tip of the germinated conidia of *S. apiospermum* during interaction with A549 epithelial cells. Transmission electron microscopy (TEM) evidences (black arrow) a vesicle secreted by *S. apiospermum* in the presence of A549 epithelial cells.

3.2. Hydrolytic Enzymes

Hydrolytic enzymes are fundamental in multiple processes of fungal life cycle and pathogenesis, including cell morphogenesis, nutrition, stress response, adhesion, invasion of cells/tissues and escape from immune system attack [66,67]. *Scedosporium* species are able to secret a wide range of enzymes, including proteolytic enzymes (e.g., serine, metallo, cysteine and aspartyl peptidases), lipases (e.g., esterase), DNAse, phosphatases (e.g., phytase) and many others (Figure 9). In this way, in this section we will explore the known hydrolytic enzymes secreted by *Scedosporium* species.

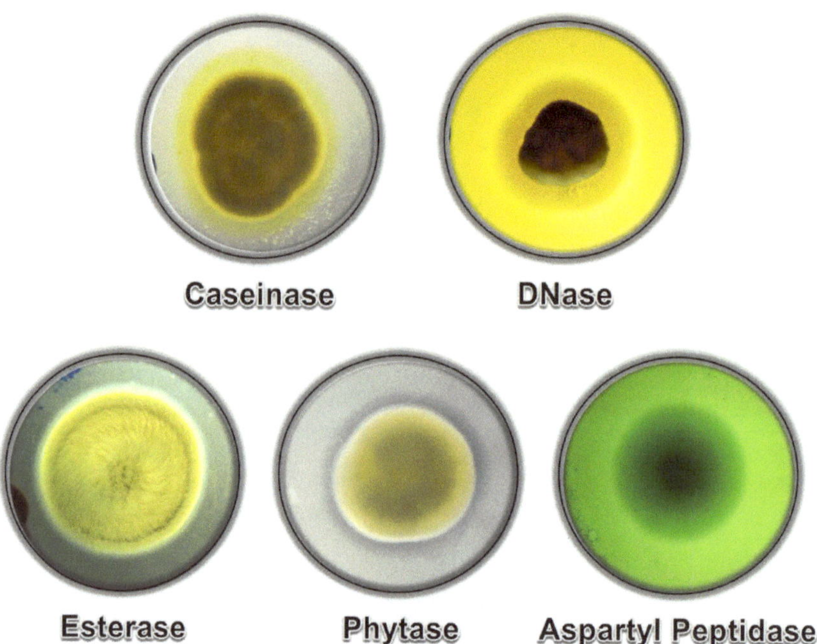

Figure 9. Examples of hydrolytic enzymes produced by *S. apiospermum*. The positive hydrolytic activity of caseinase, DNAse, esterase, phytase and aspartyl protease was determined through the detection of degradation halos around the fungal colony using distinct growth media containing specific substrates.

3.2.1. Peptidases

Peptidases (or proteolytic enzymes) are degradative enzymes that catalyze the cleavage of peptide bonds in macromolecular proteins and oligomeric peptides. Peptidases are the single class of enzymes that occupy a pivotal position with respect to their applications in both physiological and commercial fields [68]. They are responsible for the complex processes involved in the normal physiology (e.g., nutrition, growth, differentiation) of the cell as well as in abnormal pathophysiological conditions (e.g., degradation of key host molecules like surface receptors and proteinaceous humoral response molecules, including antibodies, antimicrobial peptides and complement proteins) [66]. For instance, the secreted aspartic peptidases (Saps) produced by *C. albicans* and several non-*albicans Candida* species have essential roles in the hyphal formation and invasion of tissues through the degradation of collagen, fibronectin, laminin and mucin [32,69,70], whereas the secreted allergens Asp f5 (a matrix metallopeptidase) and Asp f13 (serine peptidase) by *A. fumigatus* are important for the recruitment of inflammatory cells and remodeling of the airways' local immune response [71].

The first proteolytic enzyme described in *Scedosporium* spp. was a 33-kDa serine peptidase secreted by *S. apiospermum*, which presented optimum hydrolytic activity at pH 9.0 and temperatures between 37 to 50 °C, able to degrade human fibrinogen. The activity of this secreted serine peptidase was associated with the inflammation of the lungs of cystic fibrosis patients [72]. The production of distinct serine peptidases, belonging to the subtilisin-like, trypsin-like and elastase-like types, was also described in *S. aurantiacum* [73]. Interestingly, hypoxia conditions induced the secretion of these serine-type peptidases by *S. aurantiacum*; however, the main peptidases detected under these conditions were aspartic and cysteine peptidases, which presented optimum acidic pH for their full hydrolytic efficiency [74]. The aspartyl-type peptidase was also detected on the secretome of *S. apiospermum*, as corroborated with the use of either a specific substrate (cathepsin D) or peptidase inhibitor (pepstatin A) [35].

Metallopeptidases secreted by *S. apiospermum* presented differential profiles relying on the morphological type: mycelia were able to secrete six distinct peptidases ranging from 90 to 28 kDa, whereas conidia secreted a single peptidase of 28 kDa [75–77]. All metallopeptidases of *S. apiospermum* were active at acidic pH and completely inhibited by 1,10-phenanthroline, a classic inhibitor of metal-dependent enzymes [76,77]. These metallo-type peptidases are involved in the cleavage of key host proteins, such as immunoglobulin G, laminin, fibronectin and mucin [75–77], and they are also associated with the differentiation of conidial into hyphal forms [76,78]. Corroborating this last statement, 1,10-phenanthroline was able to completely block conidial germination, while ethylenediamine tetraacetic acid (EDTA) and ethylene glycol-bis(2-aminoethylether)-*N*,*N*,*N'*,*N'*-tetraacetic acid (EGTA), two well-known metal chelators, only partially inhibited the conidial germination [33,78].

Peptidases secreted by *Scedosporium* species were able to cause morphological changes in epithelial cells, detachment of the monolayer and reduce their viability [35,79]. In addition to planktonic cells, the production of hydrolytic enzymes has also been described in biofilm-forming cells of *Scedosporium* species. Biofilms formed for 72 h by *S. apiospermum*, *S. minutisporum*, *S. aurantiacum* and *L. prolificans* released peptidases able to hydrolyze albumin and casein at pH values ranging from 4.0 to 9.0 [80].

3.2.2. Lipases

Lipases are a class of enzymes that catalyze the hydrolysis of triglycerides to glycerol and free fatty acids, as well as the hydrolysis and transesterification of other esters [81]. These properties make microbial lipases relevant in several industries (e.g., food, chemical and pharmaceutical) and also for the infectious process through the hydrolysis of host components (e.g., plasma membrane) [81,82].

Studies about the secretion of lipases by *Scedosporium* species have focused on the use of this hydrolytic enzymes in industrial and bioremediation applications (which are better detailed in sub item 3.5). For instance, *S. boydii* secretes extracellular lipases able to biodegrade a triacylglycerol named tributyrin, as well as linseed, olive and soybean biodiesel [83,84]. Some studies have demonstrated that *Scedosporium* species are able to penetrate the membrane of mammalian cell lines [85–87], which indicates the secretion of phospholipase as observed in *L. prolificans* isolates from Mexican patients [88]. In addition, 72 h-biofilms of *S. apiospermum*, *S. minutisporum*, *S. aurantiacum* and *L. prolificans* produced lipases able to hydrolyze different lipid chain sizes, such as 4-methylumbelliferylbutyrate, 4-methylumbeliferyl heptanoate and 4-methylumbeliferyl oleate [80].

3.2.3. Phosphatases

Phosphorylation is an essential step for several processes in living cells, including cell cycle progression, central metabolism reactions, filamentation and gene transcription, among others [89]. Phosphorylation levels are coordinated by a fine balance between protein kinases and phosphatases in response to internal and external signals [89].

Acid and alkaline phosphatases were identified on the mycelial surface of *S. apiospermum* and also in the secretome of this species as demonstrated with specific inhibitors (levamisole, inorganic phosphate, sodium orthovanadate, ammonium molybdate, sodium fluoride, sodium β-glycerophosphate and sodium tartrate) and cytochemical localization [35,90]. The roles of these phosphatases in virulence are mostly unknown for *Scedosporium*, but for other fungal species these enzymes are involved in virulence traits, such as adhesion to epithelial cells and biofilm formation [35,90–92]. It is an open and promising area for future research.

3.2.4. Detoxifying Enzymes

To colonize the host, pathogens have to face toxic oxygen and nitrogen species produced by phagocytic cells and other threats [93]. One way to cope with such stressors is the production of antioxidant enzymes, such as catalase and superoxide dismutase (SOD). SODs are metalloenzymes that are on the front line of defense against oxidative stress, catalyzing the conversion of superoxide anion ($O2^{\bullet-}$) to H_2O_2 and oxygen molecules [93,94]. In fungal cells, SODs are found mainly intracellularly on both cytosol and mitochondria; however, few SODs are extracellularly detected [93].

Studies about SOD in *Scedosporium* species are scarce. Initially, a cytosolic Cu,Zn-SOD was characterized in *S. apiospermum*. The production of this enzyme was stimulated by iron starvation, and it was not detected on the fungal culture filtrates with the methodology used [94]. Subsequently, a glycosylphosphatidylinositol-anchored SOD detected on the surface of *S. apiospermum* conidial cells was described; however, the secretion of this enzyme was not evaluated [93]. Notably, a Mn-SOD was identified in the secretome of *S. apiospermum*, indicating that at least one SOD can be secreted and probably has an extracellular function [35].

Catalases act by decomposing H_2O_2 into molecular oxygen and water [27]. In *Scedosporium*, the catalases A1, A2 and A2′ were identified in mycelial extract, and the presence of catalase A1 in the extracellular milieu was also observed [95,96]. Due to its antigenic nature, the authors of that study proposed the use of catalase A1 as a diagnostic tool for *Scedosporium* species, which will be further detailed herein.

Interestingly, a peroxiredoxin was also detected on the secretome of *S. apiospermum* [35]. The gene encoding a peroxiredoxin (*SaPrx2*) is overexpressed when *S. apiospermum* is in a co-culture with phagocytic cells (THP1 and HL60), as well as upon exposure to menadione and H_2O_2 [97].

3.2.5. Mechanisms to Obtain Iron

Iron is an essential micronutrient for all organisms involved in several cellular processes; however, iron is not freely available because it is mainly under ferric state [98]. In this way, microorganisms have developed some mechanisms to obtain iron from the environment, including the secretion of hemolysins and siderophores.

Hemolysins are molecules that cause the lysis of red blood cells by disrupting the cell membrane, allowing iron acquisition from hemoglobin [99]. The hemolytic activity of *Scedosporium* species is most unknown, but herein we demonstrated the hemolysin activity of *S. apiospermum* through the presence of a halo around the fungal colony when grown on Sabouraud medium supplemented with 7% sheep blood (Figure 10). In addition, high hemolysis has been identified as a symptom of a patient with scedosporiosis [100].

Siderophores are small organic molecules able to scavenge iron in iron-restricted environments and/or environments with competition for this micronutrient [98]. *S. apiospermum* secreted two siderophores, dimerumic acid and Nα-methyl coprogen B, both from the hydroxamate type and coprogen family [98]. The production of the Nα-methyl coprogen B is higher for clinical isolates from respiratory samples compared to environmental strains, suggesting its use as a promising diagnostic tool, as will be detailed in Section 3.3 [98,101]. Moreover, the Nα-methyl coprogen B is essential for fungal growth and virulence, as demonstrated by its blockage synthesis due to the disruption of the *sidD* gene [102].

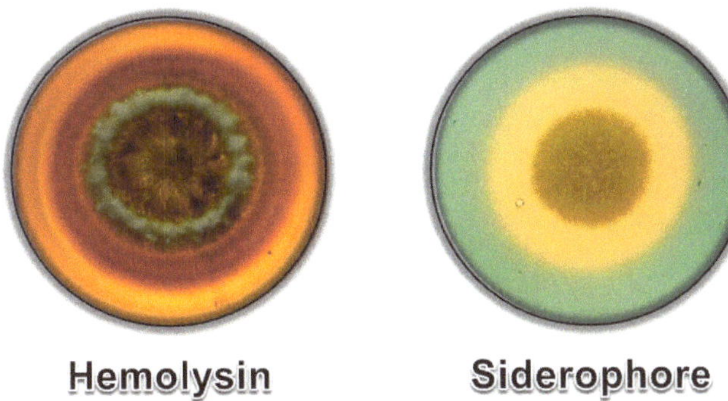

Figure 10. Production of hemolysin and siderophores by *S. apiospermum*. **Left**: Plate of Sabouraud medium supplemented with 7% sheep blood presenting a 7-day growth of *S. apiospermum* surrounded by a halo indicating hemolytic activity. **Right**: Plate of Sabouraud medium supplemented with chrome azurol S (CAS) and iron III solution presenting a 7-day growth of *S. apiospermum* surrounded by a halo indicating iron chelation.

3.3. Molecules Related to Fungal Diagnosis

Traditional diagnosis methods, such as mycological examination, can be ineffective for *Scedosporium* species in some cases, such as in polymicrobial clinical samples, due to their slower growth compared to other filamentous fungi (e.g., *A. fumigatus*). Another relevant issue in this regard is the morphological similarity of *Scedosporium* species to other hyaline filamentous fungal species (e.g., *Aspergillus* spp. and *Fusarium* spp.) in histopathological tissue sections [2,96]. For successful patient treatment, the correct *Scedosporium* identification is essential, since amphotericin B is the first line of treatment for several infections caused by filamentous fungi and *Scedosporium* species are intrinsically resistant to this antifungal agent [103]. In this context, immunological diagnosis emerged as a prominent option; however, immunological *Scedosporium* diagnosis has focused for long time on the use of polyclonal antibodies in counterimmunoelectrophoresis and immunohistological techniques, for which positive results can be due to a cross-reaction with antigens from other clinically relevant fungal species [2,104,105]. In this manner, the diagnosis of *Scedosporium* species through the identification of specific extracellular and/or secreted molecules arises as a promising research field. The peptide-polysaccharide called peptidorhamnomannan (PRM) was the first molecule from *S. apiospermum* (formerly *Pseudallescheria boydii*) suggested to be used in diagnosis [106]. PRM is composed of Rhap(1 → 3)Rhap on side chains, which may be (1 → 3) to (1 → 6)-linked mannose units, and is found on the surface of both conidial and mycelial cells of *S. apiospermum, S. boydii, S. minutisporum, S. aurantiacum* and *L. prolificans*; moreover, this molecule was also detected on the extracellular milieu of *S. apiospermum* growth, indicating its secretion [65,85,106–109]. PRM strongly reacts with an antiserum obtained against the whole *S. apiospermum* cell utilizing enzyme-linked immunosorbent assay (ELISA) and immunofluorescence techniques (Figure 11), but it reacts poorly against an antiserum obtained with *Sporothrix schenckii* (which also produces a PRM-like molecule), demonstrating that PRM can be used in the differential diagnosis of *Scedosporium* spp. [106]. In addition to its use in diagnosis, PRM is an essential molecule used by the fungal cells to interact with mammalian cells, such as larynx epithelial carcinoma (HEp2) and macrophages, as well as for the induction of pro-inflammatory cytokine production. Moreover, the use of monoclonal antibodies (mAbs) against PRM on the surface of conidial cells decreases phagocytosis and the chemical removal of *O*-linked oligosaccharides from PRM abolished pro-inflammatory cytokines [85,107,108,110]. Notably, the extracted PRM from the surface of *L. prolificans, S. apiospermum, S. boydii* and *S. aurantiacum* is able to inhibit the growth and biofilm formation of *Staphylococcus aureus, Burkholderia cepacia* and *Escherichia coli* [108], revealing its antibacterial activity.

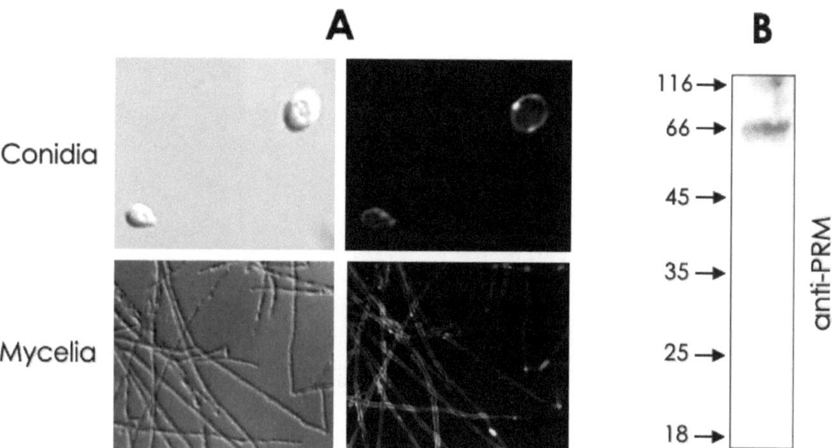

Figure 11. Detection of PRM produced by *S. apiospermum*. (**A**) Immunofluorescence microscopy evidencing PRM on the surface of both conidia and mycelia of *S. apiospermum* utilizing an anti-PRM antibody (kindly donated by Dr. Eliana Barreto Bergter from Instituto de Microbiologia Paulo de Góes and Universidade Federal do Rio de Janeiro). (**B**) Western blotting assay evidencing the presence of soluble PRM in the supernatant of *S. apiospermum* mycelial cells growth using anti-PRM antibody.

Subsequently, IgM and IgG1 K-light chain mAbs were developed. These mAbs recognize a carbohydrate epitope on an unknown extracellular 120-kDa antigen present on the *S. apiospermum* conidial and mycelial surface, and also present in fungal culture filtrates [103]. The mAbs specifically identified *S. apiospermum* and no other clinically relevant fungi, such as *A. fumigatus*, *C. albicans*, *C. neoformans*, *Fusarium solani* and *Rhizopus oryzae*, in immunofluorescence and double-antibody sandwich enzyme-linked-immunosorbent assays. However, these mAbs also react with *Graphium* and *Petriella* species [103].

In addition to diagnosis through the identification of cell surface components that can be secreted, another alternative approach to diagnosis is the detection of specific metabolites secreted by microorganisms during the infection course [101]. For instance, siderophores and pseudacyclins were suggested as good options for the identification of *Scedosporium* species on clinical samples [98,101,111]. As mentioned before, siderophores are secreted in environments with iron competition or scarce concentration, such as in cystic fibrosis (CF) sputum [98,101]. The siderophore Nα-methyl coprogen B was identified as specifically secreted by *Scedosporium* species among CF-related microorganisms and, consequently, a marker of *Scedosporium* colonization [101]. For this reason, its use has been suggested for *Scedosporium* diagnosis in CF patients. Utilizing high performance liquid chromatography, this siderophore was only identified in *S. apiospermum* supernatant, and not on *Aspergillus* spp. and *Exophiala dermatiditis*. In addition, the siderophore was only detected in sputum from CF patients colonized by *Scedosporium* spp. [101]. Likewise, five cyclic peptides with an unknown role in metabolism, named as pseudacyclins A-E, are produced exclusively by *Scedosporium* species and have been patented (International Publication Number: WO 2009/149675 A3) to be used as a diagnostic tool for the *Scedosporium* genus [111,112].

Enzymes are also molecules suggested for use in *Scedosporium* species diagnosis. As mentioned before, detoxification enzymes, such as catalase and SOD, are produced during all infection courses by microorganisms in order to face the reactive oxygen species produced by host phagocytic cells [96]. The catalase A1, a tetrameric protein of 460 kDa, is an enzyme present both in mycelium and in the culture supernatant, which is recognized by the serum of CF patients with scedosporiosis, but not with *A. fumigates*, through an ELISA assay [96,113].

The detection of 1,3-β-D-glucan is a traditional methodology in the diagnosis of invasive fungal infection [114]. This molecule is found in the fungal cell wall and is also secreted, which makes possible its detection in the blood of patients with invasive fungal infections [115]. The detection of 1,3-β-D-glucan has been reported in cases of brain abscess and invasive diseases caused by *Scedosporium/Lomentospora* species [115–118].

3.4. Secondary Metabolites

In fungi, secondary metabolites are derived from central metabolic pathways, being the acyl-CoAs the initial mole-cules for synthesis [119]. In contrast, non-ribosomal peptides (NRPs) are synthesized by NRP synthases initiating in amino acids [119,120]. An analysis of NPR synthases' gene clusters in *S. apiospermum* identified nine putative NRPS clusters involved in the synthesis of epidithiodioxopiperazines, siderophores, cyclopeptides or other still uncharacterized specialized metabolites [121]. Secondary metabolites play essential roles in cell development and interaction with other organisms [119]. For example, some of the most relevant secondary metabolites are the β-lactam antibiotics, such as penicillins and cephalosporins, produced by fungi belonging to the *Penicillium, Cephalosporium* and *Aspergillus* genera, as well as by bacteria of the genera *Streptomyces, Nocardia, Flavobacterium* and *Lysobacter* [120,121].

Secondary metabolites secreted by *Scedosporium* species have many biological activities, such as antitumor, antimicrobial, insecticidal and antidiabetics (Table 1). The tyroscherin produced by *Scedosporium* spp. Presented an in vitro selective antitumor activity against insulin-like growth factors-dependent cell lines, such as the human breast cancer cell lines MCF-7 and T47D, with half maximal inhibitory concentration (IC_{50}) of 9.7 ng/mL and 32 ng/mL, respectively [122]. The molecules ovacilin and boydone B produced by *S. boydii* present antitumor activity against the lung cancer cell line A549 with an IC_{50} of 4.1 and 41.3 μM, respectively [123]. The 3,3′-cyclohexylidenebis(1H-indole) is a secondary metabolite produced by *S. boydii*, which had its activity tested against a greater number of cancer cells: human lung cancer cell lines (A549 and GLC82), human nasopharyngeal carcinoma cell lines (CNE1, CNE2, HONE1 and SUNE1) and human hepatoma carcinoma cell lines (BEL7402 and SMMC7721), with IC_{50} values ranging from 18.69 to 27.53 μM [124]. Secondary metabolites produced by *Scedosporium* also have antidiabetic activity. The quinadoline A and the scequinadoline D, E and J promote triglyceride accumulation in 3T3-L1 preadipocytes cells, through induction of adipogenesis [125].

The worldwide emergence of fungal and bacterial resistant strains to antimicrobial drugs has highlighted the current low arsenal of antimicrobials to deal with some types of microbial infections [126,127]. For example, infections caused by resistant *Klebsiella pneumoniae* strains, as well as azole-resistant *Aspergillus* and *Candida* species, are associated with significant morbidity and mortality due to the lack of optimal treatment [126,128]. An estimate made in 204 countries showed that in 2019 the number of deaths associated with antimicrobial resistance was 4.95 million [129]. In this context, research into new antimicrobial compounds has intensified in recent decades, and secondary metabolites produced by fungal and bacterial cells are emerging as promising candidates, since microbial cells secrete bioactive compounds able to inhibit other microorganisms during the competition for nutrients [130]. Several secondary metabolites produced by *Scedosporium* species with antimicrobial activity have been identified; among them are compounds with antibacterial activity (Table 1; Figure 12). In this context, the *Scedosporium* metabolites gliotoxin, dehydroxybisdethiobis(methylthio)gliotoxin, *bis*dethio*bis*(methylthio)gliotoxin, fumitremorgin C, 12,13-dihydroxyfumitremorgin and boydone A had antibacterial activity against *S. aureus*, including methicillin-resistant strains [130–132]. The *Scedosporium* secondary metabolites also have antifungal and antiviral activities. The named "inhibitory compound" produced by *S. boydii* has a fungistatic activity against *Alternaria brassicicola*, reducing the disease incidence of black leaf spot of spoon cabbage [133]. Moreover, the secondary metabolites tyroscherin and *N*-methyltyroscherin were effective against *C. albicans, C. neoformans, A. fumigatus* and *Trichophyton rubrum* [134,135]. The secondary metabolites scequinadoline D and scedapin C displayed anti-hepatitis C virus activity with an effective concentration of 110.35 and 128.60 µM, respectively [136]. Moreover, some secondary metabolites (e.g., diketopiperazines pseudoboydone C, cyclo-(Phe-Phe), cyclopiamide E, 24,25-dehydro-10,11-dihydro-20-hydro-xyaflavinin and aflavinine) secreted by *S. boydii* had insecticidal activity against a major agricultural pest insect, *Spodoptera frugiperda* [137]. Other biological functionalities described for secondary metabolites secreted by *Scedosporium* spp are: inhibitor of acyl-CoA (e.g., AS-183), stimulator (e.g., pseurotin A) or inhibitor (e.g., (-)-ovalicin) of osteoclastogenesis; in addition to several other molecules that as yet have some undefined biological activity (Table 1; Figure 12) [123,124,137–151].

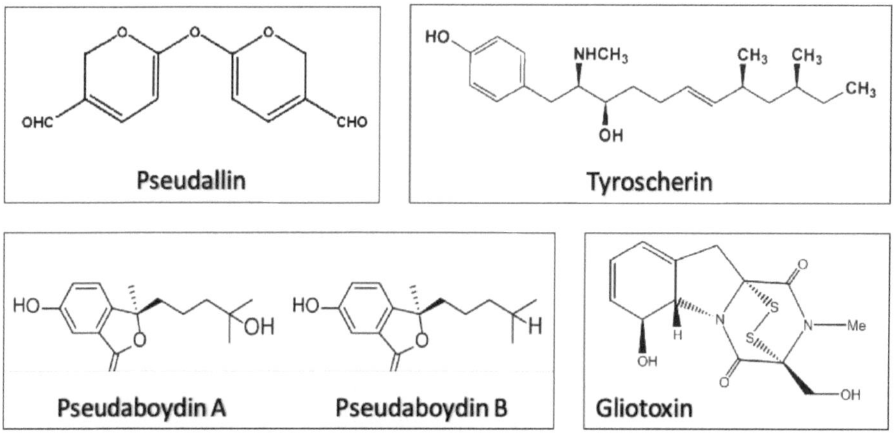

Figure 12. Some examples of non-peptide small-molecule metabolites extracellularly released by environmental and clinical strains of *Scedosporium* spp.

Table 1. List of secondary metabolites identified in *Scedosporium* species and their potential biological activity.

Species	Molecule	Activity	References
Scedosporium spp.	AS-183	Inhibitor of acyl-CoA	[138]
S. ellipsoidea	YM-193221	Antifungal	[134]
Scedosporium spp.	Tyroscherin	Antitumor; Antifungal	[122,135,139]
Scedosporium spp.	Dehydroxy*bis*dethio*bis*(methylthio)gliotoxin	Antibacterial	[131]
Scedosporium spp.	*Bis*dethio*bis*(methylthio)gliotoxin	Antibacterial	[131]
Scedosporium spp.	Gliotoxin	Antibacterial	[131]
Scedosporium spp.	12,13-hihydroxyfumitremorgin C	Antibacterial	[132]
Scedosporium spp.	Fumitremorgin C	Antibacterial	[132]
Scedosporium spp.	Brevianamide F	Antibacterial	[132]
Scedosporium spp.	(2*RS*,8*R*,10*R*)-YM-193221	ND	[139]
S. boydii	"Inhibitory substance"	Antifungal	[133]
S. boydii	Pseudallin	Antibacterial	[152]
S. boydii	Boydone A	Antibacterial	[123,130]
S. boydii	Boydone B	Antitumor	[123]
S. boydii	Botryorhodine F	ND	[123]
S. boydii	Botryorhodine G	ND	[123]
S. boydii	Fusidilactone A	ND	[123]
S. boydii	(*R*)-(-)-mevalonolactone	ND	[123]
S. boydii	(*R*)-(-)-lactic acid	ND	[123]
S. boydii	Ovalicin	Antitumor	[123]
S. boydii	Botryorhodine	ND	[123]
S. boydii	*N*-methyltyroscherin	Antifungal	[135]
S. boydii	Boydine A	ND	[140]

Table 1. *Cont.*

Species	Molecule	Activity	References
S. boydii	Boydine B	Antibacterial	[137,140]
S. boydii	Boydine C	ND	[140]
S. boydii	Boydine D	ND	[140]
S. boydii	Boydene A	ND	[140]
S. boydii	Boydene B	ND	[140]
S. boydii	Pseudaboydin A	ND	[137]
S. boydii	Pseudaboydin B	ND	[137]
S. boydii	(R)-2-(2-hydroxypropan-2-yl)-2,3-dihydro-5-hydroxybenzofuran	ND	[137]
S. boydii	(R)-2-(2-hydroxypropan-2-yl)-2,3-dihydro-5-methoxybenzofuran	ND	[137]
S. boydii	3,3′-dihydroxy-5,5′-dimethyldiphenyl ether	ND	[137]
S. boydii	3-(3-methoxy-5-methylphenoxy)-5-methylphenol	ND	[137]
S. boydii	(-)-Regiolone	ND	[137]
S. boydii	6-Chloro-2-(2-hydroxypropan-2-yl)-2,3-dihydro-5-hydroxybenzofuran	ND	[141]
S. boydii	7-Chloro-2-(2-hydroxypropan-2-yl)-2,3-dihydro-5-hydroxybenzofuran	ND	[141]
S. ellipsoidea	Pseudellone A	ND	[142]
S. ellipsoidea	Pseudellone B	ND	[142]
S. ellipsoidea	Pseudellone C	ND	[142,145,146]
S. ellipsoidea	Pseudellone D	ND	[145]
S. ellipsoidea	(5S,6S)-dihydroxylasiodiplodin	ND	[145]
S. ellipsoidea	Lasiodipline F	ND	[145]
S. ellipsoidea	(5S)-hydroxylasiodiplodin	ND	[145]
S. boydii	Pseuboydone A	ND	[137]
S. boydii	Pseuboydone B	ND	[137]
S. boydii	Diketopiperazines pseudoboydone C	Insecticidal	[137]
S. boydii	Diketopiperazines pseudoboydone D	ND	[137]
S. boydii	Haematocin	ND	[137]
S. boydii	Phomazine B	ND	[137]
S. boydii	Bisdethiobis(methylthio)gliotoxin	ND	[137]
S. boydii	Cyclo-(2,20-dimethylthio-Phe-Phe)	ND	[137]
S. boydii	Cyclo-(Phe-Phe)	Insecticidal	[137]
S. boydii	Ditryptophenaline	ND	[137]
S. boydii	Speradine B	ND	[137]
S. boydii	Speradine C	ND	[137]
S. boydii	Cyclopiamide E	Insecticidal	[137]
S. boydii	24,25-Dehydro-10,11-dihydro-20-hydro- xyaflavinin	Insecticidal	[137]
S. boydii	Aflavinine	Insecticidal	[137]
S. boydii	b-Aflatrem	ND	[137]
S. boydii	Pyripyropene A	ND	[137]
S. boydii	Pseudoscherine	ND	[137]
S. boydii	4-(1-Hydroxy-1-methylpropyl)-2-isobutyl-pyrazin-2(1H)-one	ND	[137]
S. boydii	4-(1-Hydroxy-1-methyl-propyl)-2-secbutylpyrazin-2(1H)-one	ND	[137]
S. boydii	O-methyl sterigmatocystin	ND	[137]
S. boydii	Asperfuran	ND	[137]
S. boydii	Pseudallicin A	ND	[147]
S. boydii	Pseudallicin B	ND	[147]
S. boydii	Pseudallicin C	ND	[147]
S. boydii	Pseudallicin D	ND	[147]
S. apiospermum	Scedapin A	ND	[148]
S. apiospermum	Scedapin B	ND	[148]
S. apiospermum	Scedapin C	Antiviral	[148]
S. apiospermum	Scedapin D	ND	[148]
S. apiospermum	Scedapin E	ND	[148]
S. apiospermum	Scequinadoline A	ND	[148]
S. apiospermum	Scequinadoline B	ND	[148]
S. apiospermum	Scequinadoline C	ND	[148]
S. apiospermum	Scequinadoline D	Antiviral	[148]

Table 1. *Cont.*

Species	Molecule	Activity	References
S. apiospermum	Scequinadoline E	ND	[148]
S. apiospermum	Scequinadoline F	ND	[148]
S. apiospermum	Scequinadoline G	ND	[148]
S. boydii	Pseurotin A	Stimulatory osteoclastogenesis	[143]
S. boydii	(-)-Ovalicin	Inhibitory osteoclastogenesis	[143]
S. boydii	Chlovalicin	ND	[143]
S. boydii	Dihydroxybergamotene	ND	[143]
S. boydii	AM6898B	Stimulatory osteoclastogenesis	[143]
S. boydii	Aspergiketone	ND	[143]
S. boydii	Pseudboindole A	ND	[124]
S. boydii	Pseudboindole B	ND	[124]
S. boydii	3,3′-Cyclohexylidenebis(1H-indole)	Antitumor	[124]
S. boydii	Indole alkaloids	ND	[124]
S. boydii	2-Hydroxy-2-(propan-2-yl) cyclobutane-1,3-dione	ND	[149]
S. apiospermum	Scetryptoquivaline A	ND	[125]
S. apiospermum	Quinadoline A	Antidiabetic	[125]
S. apiospermum	Scequinadoline D	Antidiabetic	[125]
S. apiospermum	Scequinadoline E	Antidiabetic	[125]
S. apiospermum	Scequinadoline J	Antidiabetic	[125]
S. apiospermum	Scequinadoline I	ND	[125]
S. apiospermum	Fumiquinazolines	ND	[125]

ND—not determined.

3.5. Molecules Involved in Biodegradation

Filamentous fungi can usually grow in xenobiotic-contaminated environments due to the production of extracellular enzymes (e.g., esterases) able to hydrolyze/oxidize the toxic compounds, transforming those molecules into intermediate metabolites, which can be further absorbed and metabolized by fungal cells [151]. In this context, *Scedosporium* species have gained attention from the scientific community due to their ability to thrive in decayed wood, manure, soils and heavily contaminated water, besides being able to adapt to high salt concentrations and low oxygen levels [153,154]. In 1968, the ability of *S. boydii* (formerly *Allescheria boydii*) to utilize *n*-alkanes (C-10; C-14; C-16 and C-18) and 1-alkenes (C-10:1; C-14:1 and C-16:1) as their only carbon source was described [155]. Subsequently, the ability of *S. boydii* strains isolated from oil-soaked soil in Canada to degrade linear aliphatic compounds was demonstrated [154]. Similarly, *S. apiospermum* is capable of utilizing aromatic compounds (diaryl ester phenylbenzoate, phenol, *p*-cresol, *p*-tolylbenzoate, 4-chlorophenylbenzoate and toluene) as the sole carbon source [156–159]. Other molecules/compounds degraded by *Scedosporium* species were reported: 2,3,7,8-tetrachlorodibenzeno-*p*-dioxin, polychlorinated biphenyl, acetaminophen, biodiesel, esters, diesel hydrocarbons, tetrahydrofuran and azo dyes (e.g., Reactive Yellow 145 and Remazol Yellow RR) (Table 2) [84,160–167].

A genome sequencing of the *S. apiospermum* environmental strain identified metabolic pathways that can potentially degrade ethylbenzene, xylene, dioxin, atrazine, styrene, naphthalene, fluorobenzoate, geraniol, chloroalkane, chloroalkene, benzoate, caprolactam, butanoate and aminobenzoate, among other hydrocarbons [168,169]. The proposed pathway of *p*-cresol degradation starts with its oxidation leading to 4-hydroxybenzylalcohol, which is converted into 4-hydroxybenzaldehyde and 4-hydroxybenzoic acid. The 4-hydroxybenzoic acid is converted into protocatechuate, which is metabolized in 3-carboxy-cis into cismuconate, which is converted into 3-carboxymuconolactone and 3-oxoadipate [6,156]. However, some enzymes proposed in this pathway, such as hydroquinone hydroxylase, 4-hydroxybenzoate 3-hydroxylase, hydroxyquinone 1,2 dioxygenase, protocatechuate

3,4 dioxygenase and maleylacetate reductase, were not found in a genomic study of *S. apiospermum* [168], indicating that these enzymes are classified among the hypothetical, with unknown function, or they are in the gap regions of the genome [168]. Rougeron and co-workers also suggested that the conversion of 4-hydroxybanzoate into gentisate, which is cleaved into maleylpyruvate, should be considered, as all genes necessary for gentisate catabolism are organized in cluster in the *Scedosporium* genome [6]. The phenol could be catabolized either by a catechol 1,2-dioxygenase or a phenol hydroxylase, leading both pathways to a 3-oxoadipate, which can enter the tricarboxylic acid cycle [6,156]. For phenylbenzoate, the catabolism pathway initiates with the hydrolysis of diaryl ester by an esterase leading to phenol and benzoic acid [6,157]. The 4-chlorophenylbenzoate is hydrolyzed into 4-chlorophenol and benzoate, whereas the *p*-tolylbenzoate is hydrolyzed into *p*-cresol and benzoic acid [6,157].

Table 2. List of xenobiotics degraded by *Scedosporium* species in the literature.

Species	Molecule/Compound	References
S. boydii	*n*-Alkanes	[155]
S. boydii	1-Alkenes	[155]
S. boydii	Hydrocarbons	[155,165]
S. apiospermum	Phenol	[156]
S. apiospermum	*p*-Cresol	[156]
S. apiospermum	4-Hydroxybenzoate	[156]
S. apiospermum	4-Hydroxybenzaldehyde	[156]
S. apiospermum	4-Hydroxybenzylalcohol	[156]
S. apiospermum	Protocatechuate	[156]
S. apiospermum	Phenylbenzoate	[157]
S. apiospermum	*p*-Tolylbenzoate	[157]
S. apiospermum	4-Chlorophenylbenzoate	[157]
S. apiospermum	Toluene	[158]
S. boydii	Rapeseed oil	[168]
S. boydii	Biodiesel	[83,84,168]
S. boydii	Diesel oil	[84,168]
S. boydii	2,3,7,8-Tetrachlorodibenzo-*p*-dioxin	[160]
S. apiospermum	Polychlorinated biphenyl	[161]
S. apiospermum	Polycyclic aromatic hydrocarbons	[159]
S. dehoogii	Acetaminophen	[162]
S. boydii	Diesel blend	[84]
S. boydii	Tetrahydrofuran	[163]
S. boydii	Dibutyl tin dilaurate	[164]
S. boydii	Di-*n*-butyl-oxo-stannane	[164]
Scedosporium spp.	Lignin	[169]
S. apiospermum	Olive mill wastewater	[166]
S. apiospermum	Reactive Yellow 145	[167]
S. apiospermum	Remazol Yellow RR	[167]

The fungi pathway of lignin degradation starts with its extracellular oxidative degradation, which produces a mixture of aromatic monomers that are then catabolized into the upper and lower pathways. In the upper pathways, the aromatic compounds are catabolized into hydroxyquinol, catechol, protocatechuate, gentisic acid, hydroxyquinone, gallic acid and pyrogallol; subsequently, these aromatic compounds suffer a ring-opening by dioxygenases, producing degradation products that can enter the tricarboxylic acid cycle [169]. In *Scedosporium*, the putative dioxygenases gentisate 1,2-dioxygenases, homogentisate 1,2-dioxygenases, hydroxyquinol 1,2-dioxygenases, catechol 1,2-dioxygenase and protocatechuate 3,4-dioxygenase were identified through sequence homology and bioinformatic analysis, suggesting the ability of *Scedosporium* to catabolize gentisic acid, hydroxyquinol, protocatechuate and catechol [169].

A study conducted by Janda-Ulfig and co-workers [170] suggested that oil-contaminated environments could favor the growth of *Scedosporium* species, since rapeseed oil and biodiesel oil stimulate *S. boydii* growth. This could explain the most frequent occurrence of *Scedosporium* in urban, industrial and agricultural areas rather than in environments with low human activity [5,168]. In this way, three patents have been deposited for the use of *Scedosporium* species: (i) as composting promoters, (ii) in the bioremediation of nutrient-rich effluents and (iii) in the bioremediation of livestock manure [6,171–174].

4. Conclusions

As summarized in the present study, the emergent and multidrug-resistant *Scedosporium* species are able to secrete a vast array of distinct molecules with a wide range of biological functions, including polysaccharides, non-peptide small-molecule metabolites, non-ribosomal peptides and (glyco)protein-nature molecules. The secreted molecules can act for the benefit of human kind, such as in the bioremediation of polluted environments and in the treatment of microbial infections. However, the molecules secreted by *Scedosporium* species are also essential for its pathogenesis during human infections, such as proteolytic enzymes. In this way, studies concerning these secreted molecules can help in the development of improved diagnosis techniques and treatment of scedosporiosis.

Author Contributions: Conceptualization, writing—draft preparation T.P.M., A.C.A., M.H.B. and A.L.S.S.; in vivo investigation, I.C.B. All authors have read and agreed to the published version of the manuscript.

Funding: This work was supported by grants from Fundação Carlos Chagas Filho de Amparo à Pesquisa do Estado do Rio de Janeiro (FAPERJ), Conselho Nacional de Desenvolvimento Científico e Tecnológico (CNPq) and Coordenação de Aperfeiçoamento de Pessoal de Nível Superior (CAPES—Financial code 001).

Institutional Review Board Statement: Not applicable.

Informed Consent Statement: Not applicable.

Data Availability Statement: Not applicable.

Conflicts of Interest: The authors declare no conflict of interest.

References

1. Guarro, J.; Kantarcioglu, A.S.; Horré, E.; Rodriguez-Tudela, J.L.; Estrella, M.C.; Berenguer, J.; Hoog, G.S. *Scedosporium apiospermum*: Changing clinical spectrum of a therapy-refractory opportunist. *Med. Mycol.* **2006**, *44*, 295–327. [CrossRef] [PubMed]
2. Cortez, K.J.; Roilides, E.; Quiroz-Telles, F.; Meletiadis, J.; Antachopoulos, C.; Knudsen, T.; Buchanan, W.; Milanovich, J.; Sutton, D.A.; Fothergill, A.; et al. Infections caused by *Scedosporium* spp. *Clin. Microbiol. Rev.* **2008**, *21*, 157–197. [CrossRef] [PubMed]
3. Gilgado, F.; Cano, J.; Gené, J.; Sutton, D.; Guarro, J. Molecular and phenotypic data supporting distinct species statuses for *Scedosporium apiospermum* and *Pseudallescheria boydii* and the proposed new species *Scedosporium dehoogii*. *J. Clin. Microbiol.* **2008**, *46*, 766–771. [CrossRef] [PubMed]
4. Harun, A.; Perdomo, H.; Gilgado, F.; Chen, S.C.; Cano, J.; Guarro, J.; Meyer, W. Genotyping of *Scedosporium* species: A review of molecular approaches. *Med. Mycol.* **2009**, *47*, 406–414. [CrossRef] [PubMed]
5. Kaltseis, J.; Rainer, J.; De Hoog, G.S. Ecology of *Pseudallescheria* and *Scedosporium* species in human-dominated and natural environments and their distribution in clinical samples. *Med. Mycol.* **2009**, *47*, 398–405. [CrossRef]
6. Rougeron, A.; Giraud, S.; Alastruey-Izquierdo, A.; Cano-Lira, J.; Rainer, J.; Mouhajir, A.; Le Gal, S.; Nevez, G.; Meyer, W.; Bouchara, J.P. Ecology of *Scedosporium* species: Present knowledge and future research. *Mycopathologia* **2018**, *183*, 185–200. [CrossRef]
7. Luplertlop, N. *Pseudallescheria/Scedosporium* complex species: From saprobic to pathogenic fungus. *J. Mycol. Med.* **2018**, *28*, 249–256. [CrossRef]
8. Lackner, M.; de Hoog, S.G.; Yang, L.; Ferreira, M.L.; Ahmed, S.A.; Andreas, F.; Kaltseis, J.; Nagl, M.; Lass-Flörl, C.; Risslegger, B.; et al. Proposed nomenclature for *Pseudallescheria*, *Scedosporium* and related genera. *Fungal Divers.* **2014**, *67*, 1–10. [CrossRef]
9. Lamaris, G.; Chamilos, G.; Lewis, R.E.; Safdar, A.; Raad, I.I.; Kontoyiannis, D.P. *Scedosporium* infection in a tertiary care cancer center: A review of 25 cases from 1989–2006. *Clin. Infect. Dis.* **2006**, *43*, 1580–1584. [CrossRef]
10. Husain, S.; Alexander, B.D.; Munoz, P.; Avery, R.K.; Houston, S.; Pruett, T.; Jacobs, R.; Dominguez, E.A.; Tollemar, J.G.; Baumgarten, K.; et al. Opportunistic mycelial fungal infections in organ transplant recipients: Emerging importance of non-*Aspergillus* mycelial fungi. *Clin. Infect. Dis.* **2003**, *37*, 221–229. [CrossRef]
11. Husain, S.; Muñoz, P.; Forrest, G.; Alexander, B.D.; Somani, J.; Brennan, K.; Wagener, M.M.; Singh, N. Infections due to *Scedosporium apiospermum* and *Scedosporium prolificans* in transplant recipients: Clinical characteristics and impact of antifungal agent therapy on outcome. *Clin. Infect. Dis.* **2005**, *40*, 89–99. [CrossRef] [PubMed]

12. Rodriguez-Tudela, J.L.; Berenguer, J.; Guarro, J.; Kantarcioglu, A.S.; Horre, R.; de Hoog, G.S.; Cuenca-Estrella, M. Epidemiology and outcome of *Scedosporium prolificans* infection, a review of 162 cases. *Med. Mycol.* **2009**, *47*, 359–370. [CrossRef] [PubMed]
13. Heng, S.C.; Slavin, M.A.; Chen, S.C.; Heath, C.H.; Nguyen, Q.; Billah, B.; Nation, R.L.; Kong, D.C. Hospital costs, length of stay and mortality attributable to invasive scedosporiosis in haematology patients. *J. Antimicrob. Chemother.* **2012**, *67*, 2274–2282. [CrossRef] [PubMed]
14. O'Bryan, T.A. Pseudallescheriasis in the 21st century. *Expert Rev. Anti. Infect. Ther.* **2005**, *3*, 765–773. [CrossRef]
15. Tammer, I.; Tintelnot, K.; Braun-Dullaeus, R.C.; Mawrin, C.; Scherlach, C.; Schlüter, D.; König, W. Infections due to *Pseudallescheria/Scedosporium* species in patients with advanced HIV disease—A diagnostic and therapeutic challenge. *Int. J. Infect. Dis.* **2011**, *15*, e422–e429. [CrossRef]
16. Kantarcioglu, A.S.; Guarro, J.; de Hoog, G.S. Central nervous system infections by members of the *Pseudallescheria boydii* species complex in healthy and immunocompromised hosts: Epidemiology, clinical characteristics and outcome. *Mycoses* **2008**, *51*, 275–290. [CrossRef]
17. Noni, M.; Katelari, A.; Kapi, A.; Stathi, A.; Dimopoulos, G.; Doudounakis, S.E. *Scedosporium apiospermum* complex in cystic fibrosis; should we treat? *Mycoses* **2017**, *60*, 594–599. [CrossRef]
18. Schwarz, C.; Brandt, C.; Whitaker, P.; Sutharsan, S.; Skopnik, H.; Gartner, S.; Smazny, C.; Röhmel, J.F. Invasive pulmonary fungal infections in cystic fibrosis. *Mycopathologia* **2018**, *183*, 33–43. [CrossRef]
19. Bouchara, J.P.; Le Govic, Y.; Kabbara, S.; Cimon, B.; Zouhair, R.; Hamze, M.; Papon, N.; Nevez, G. Advances in understanding and managing *Scedosporium* respiratory infections in patients with cystic fibrosis. *Expert Rev. Respir. Med.* **2020**, *14*, 259–273. [CrossRef]
20. Ramirez-Garcia, A.; Pellon, A.; Rementeria, A.; Buldain, I.; Barreto-Bergter, E.; Rollin-Pinheiro, R.; de Meirelles, J.V.; Xisto, M.I.D.S.; Ranque, S.; Havlicek, V.; et al. *Scedosporium* and *Lomentospora*: An updated overview of underrated opportunists. *Med. Mycol.* **2018**, *56* (Suppl. 1), 102–125. [CrossRef]
21. Tortorano, A.M.; Richardson, M.; Roilides, E.; van Diepeningen, A.; Caira, M.; Munoz, P.; Johnson, E.; Meletiadis, J.; Pana, Z.D.; Lackner, M.; et al. ESCMID and ECMM joint guidelines on diagnosis and management of hyalohyphomycosis: *Fusarium* spp., *Scedosporium* spp. and others. *Clin. Microbiol. Infect.* **2014**, *20* (Suppl. 3), 27–46. [CrossRef] [PubMed]
22. Mello, T.P.; Bittencourt, V.C.B.; Liporagi-Lopes, L.C.; Aor, A.C.; Branquinha, M.H.; Santos, A.L.S. Insights into the social life and obscure side of *Scedosporium/Lomentospora* species: Ubiquitous, emerging and multidrug-resistant opportunistic pathogens. *Fungal Biol. Rev.* **2018**, *33*, 16–46. [CrossRef]
23. Mowat, E.; Williams, C.; Jones, B.; McChlery, S.; Ramage, G. The characteristics of *Aspergillus fumigatus* mycetoma development: Is this a biofilm? *Med. Mycol.* **2009**, *47* (Suppl. 1), S120–S126. [CrossRef]
24. Mello, T.P.; Aor, A.C.; Gonçalves, D.S.; Seabra, S.H.; Branquinha, M.H.; Santos, A.L.S. Assessment of biofilm formation by *Scedosporium apiospermum, S. aurantiacum, S. minutisporum* and *Lomentospora prolificans*. *Biofouling* **2016**, *32*, 737–749. [CrossRef] [PubMed]
25. Mello, T.P.; Oliveira, S.S.C.; Frasés, S.; Branquinha, M.H.; Santos, A.L.S. Surface properties, adhesion and biofilm formation on different surfaces by *Scedosporium* spp. and *Lomentospora prolificans*. *Biofouling* **2018**, *34*, 800–814. [CrossRef]
26. Mello, T.P.; Branquinha, M.H.; Santos, A.L.S. Biofilms formed by *Scedosporium* and *Lomentospora* species: Focus on the extracellular matrix. *Biofouling* **2020**, *36*, 308–318. [CrossRef]
27. Mello, T.P.; Oliveira, S.S.C.; Branquinha, M.H.; Santos, A.L.S. Decoding the antifungal resistance mechanisms in biofilms of emerging, ubiquitous and multidrug-resistant species belonging to the *Scedosporium/Lomentospora* genera. *Med. Mycol.* **2022**, *60*, myac036. [CrossRef]
28. Mello, T.P.; Lackner, M.; Branquinha, M.H.; Santos, A.L.S. Impact of biofilm formation and azoles' susceptibility in *Scedosporium/Lomentospora* species using an in vitro model that mimics the cystic fibrosis patients' airway environment. *J. Cyst. Fibros.* **2021**, *20*, 303–309. [CrossRef]
29. Rollin-Pinheiro, R.; de Meirelles, J.V.; Vila, T.V.M.; Fonseca, B.B.; Alves, V.; Frases, S.; Rozental, S.; Barreto-Bergter, E. Biofilm formation by *Pseudallescheria/Scedosporium* species: A comparative study. *Front. Microbiol.* **2017**, *8*, 1568. [CrossRef]
30. Santos, A.L.S.; Mello, T.P.; Ramos, L.S.; Branquinha, M.H. Biofilm: A robust and efficient barrier to antifungal chemotherapy. *J. Antimicrob. Agents* **2015**, *1*, e101. [CrossRef]
31. Nosanchuk, J.D.; Nimrichter, L.; Casadevall, A.; Rodrigues, M.L. A role for vesicular transport of macromolecules across cell walls in fungal pathogenesis. *Commun. Integr. Biol.* **2008**, *1*, 37–39. [CrossRef] [PubMed]
32. Naglik, J.R.; Challacombe, S.J.; Hube, B. *Candida albicans* secreted aspartyl peptidases in virulence and pathogenesis. *Microbiol. Mol. Biol. Rev.* **2003**, *67*, 400–428. [CrossRef] [PubMed]
33. Silva, B.A.; Souza-Gonçalves, A.L.; Pinto, M.R.; Barreto-Bergter, E.; Santos, A.L.S. Metallopeptidase inhibitors arrest vital biological processes in the fungal pathogen *Scedosporium apiospermum*. *Mycoses* **2011**, *54*, 105–112. [CrossRef] [PubMed]
34. Hube, B. Extracellular peptidases of human pathogenic fungi. *Contrib. Microbiol.* **2000**, *5*, 126–137.
35. Silva, B.A.; Sodré, C.L.; Souza-Gonçalves, A.L.; Aor, A.C.; Kneipp, L.F.; Fonseca, B.B.; Rozental, S.; Romanos, M.T.; Sola-Penna, M.; Perales, J.; et al. Proteomic analysis of the secretions of *Pseudallescheria boydii*, a human fungal pathogen with unknown genome. *J. Proteome Res.* **2012**, *11*, 172–188. [CrossRef]
36. Fontaine, T.; Delangle, A.; Simenel, C.; Coddeville, B.; van Vliet, S.J.; van Kooyk, Y.; Bozza, S.; Moretti, S.; Schwarz, F.; Trichot, C.; et al. Galactosaminogalactan, a new immunosuppressive polysaccharide of *Aspergillus fumigatus*. *PLoS Pathog.* **2011**, *7*, e1002372. [CrossRef]

37. Miller, M.A.; Greenberger, P.A.; Amerian, R.; Toogood, J.H.; Noskin, G.A.; Roberts, M.; Patterson, R. Allergic bronchopulmonary mycosis caused by *Pseudallescheria boydii*. *Am. Rev. Respir. Dis.* **1993**, *148*, 810–812. [CrossRef]

38. Osherov, N. Interaction of the pathogenic mold *Aspergillus fumigatus* with lung epithelial cells. *Front. Microbiol.* **2012**, *3*, 346. [CrossRef]

39. Kogan, T.V.; Jadoun, J.; Mittelman, L.; Hirschberg, K.; Osherov, N. Involvement of secreted *Aspergillus fumigatus* proteases in disruption of the actin fiber cytoskeleton and loss of focal adhesion sites in infected A549 lung pneumocytes. *J. Infect. Dis.* **2004**, *189*, 1965–1973. [CrossRef]

40. Schindler, B.; Segal, E. *Candida albicans* metabolite affects the cytoskeleton and phagocytic activity of murine macrophages. *Med. Mycol.* **2008**, *46*, 251–258. [CrossRef]

41. Vandeputte, P.; Ghamrawi, S.; Rechenmann, M.; Iltis, A.; Giraud, S.; Fleury, M.; Thornton, C.; Delhaès, L.; Meyer, W.; Papon, N.; et al. Draft genome sequence of the pathogenic fungus *Scedosporium apiospermum*. *Genome Announc.* **2014**, *2*, e00988-14. [CrossRef] [PubMed]

42. El-Kamand, S.; Steiner, M.; Ramirez, C.; Halliday, C.; Chen, S.C.-A.; Papanicolaou, A.; Morton, C.O. Assessing differences between clinical isolates of *Aspergillus fumigatus* from cases of proven invasive aspergillosis and colonizing isolates with respect to phenotype (virulence in *Tenebrio molitor* larvae) and genotype. *Pathogens* **2022**, *11*, 428. [CrossRef] [PubMed]

43. Lowry, O.H.; Rosebrough, N.J.; Farr, A.L.; Randall, R.J. Protein measurement with the Folin phenol reagent. *J. Biol. Chem.* **1951**, *193*, 265–275. [CrossRef]

44. Wolf, J.M.; Casadevall, A. Challenges posed by extracellular vesicles from eukaryotic microbes. *Curr. Opin. Microbiol.* **2014**, *22*, 73–78. [CrossRef] [PubMed]

45. Zamith-Miranda, D.; Nimrichter, L.; Rodrigues, M.L.; Nosanchuk, J.D. Fungal extracellular vesicles: Modulating host-pathogen interactions by both the fungus and the host. *Microbes Infect.* **2018**, *20*, 501–504. [CrossRef]

46. Rodrigues, M.L.; Franzen, A.J.; Nimrichter, L.; Miranda, K. Vesicular mechanisms of traffic of fungal molecules to the extracellular space. *Curr. Opin. Microbiol.* **2013**, *16*, 414–420. [CrossRef]

47. Casadevall, A.; Nosanchuk, J.D.; Williamson, P.; Rodrigues, M.L. Vesicular transport across the fungal cell wall. *Trends Microbiol.* **2009**, *17*, 158–162. [CrossRef]

48. Brown, L.; Wolf, J.M.; Prados-Rosales, R.; Casadevall, A. Through the wall: Extracellular vesicles in Gram-positive bacteria, mycobacteria and fungi. *Nat. Rev. Microbiol.* **2015**, *13*, 620–630. [CrossRef]

49. Rizzo, J.; Rodrigues, M.L.; Janbon, G. Extracellular vesicles in fungi: Past, present, and future perspectives. *Front. Cell. Infect. Microbiol.* **2020**, *10*, 346. [CrossRef]

50. Rodrigues, M.L.; Nimrichter, L.; Oliveira, D.L.; Frases, S.; Miranda, K.; Zaragoza, O.; Alvarez, M.; Nakouzi, A.; Feldmesser, M.; Casadevall, A. Vesicular polysaccharide export in *Cryptococcus neoformans* is a eukaryotic solution to the problem of fungal trans-cell wall transport. *Eukaryot. Cell* **2007**, *6*, 48–59. [CrossRef]

51. Albuquerque, P.C.; Nakayasu, E.S.; Rodrigues, M.L.; Frases, S.; Casadevall, A.; Zancope-Oliveira, R.M.; Almeida, I.C.; Nosanchuk, J.D. Vesicular transport in *Histoplasma capsulatum*: An effective mechanism for trans-cell wall transfer of proteins and lipids in ascomycetes. *Cell. Microbiol.* **2008**, *10*, 1695–1710. [CrossRef] [PubMed]

52. Gehrmann, U.; Qazi, K.R.; Johansson, C.; Hultenby, K.; Karlsson, M.; Lundeberg, L.; Gabrielsson, S.; Scheynius, A. Nanovesicles from *Malassezia sympodialis* and host exosomes induce cytokine responses—Novel mechanisms for host-microbe interactions in atopic exzema. *PLoS ONE* **2011**, *6*, e21480. [CrossRef] [PubMed]

53. Vallejo, M.C.; Matsuo, A.L.; Ganiko, L.; Medeiros, L.C.S.; Miranda, K.; Silva, L.S.; Freymuller-Haapalainen, E.; Sinigaglia-Coimbra, R.; Almeida, I.C.; Puccia, R. The pathogenic fungus *Paracoccidioides brasiliensis* exports extracellular vesicles containing highly immunogenic. *Eukaryot. Cell* **2011**, *10*, 343–351. [CrossRef]

54. Silva, B.M.; Prados-Rosales, R.; Espadas-Moreno, J.; Wolf, J.M.; Luque-Garcia, J.L.; Gonçalves, T.; Casadevall, A. Characterization of *Alternaria infectoria* extracellular vesicles. *Med. Mycol.* **2014**, *52*, 202–210. [CrossRef] [PubMed]

55. Bitencourt, T.A.; Rezende, C.P.; Quaresemin, N.R.; Moreno, P.; Hatanaka, O.; Rossi, A.; Martinez-Rossi, N.M.; Almeida, F. Extracellular vesicles from the dermatophyte *Trichophyton interdigitale* modulate macrophage and keratinocyte functions. *Front. Immunol.* **2018**, *9*, 2343. [CrossRef]

56. Liu, M.; Bruni, G.O.; Taylor, C.M.; Zhang, Z.; Wang, P. Comparative genome-wide analysis of extracellular small RNAs from the mucormycosis pathogen *Rhizopus delemar*. *Sci. Rep.* **2018**, *8*, 5243. [CrossRef]

57. De Paula, R.G.; Antoniêto, A.C.C.; Nogueira, K.M.V.; Ribeiro, L.F.C.; Rocha, M.C.; Malavazi, I.; Almeida, F.; Silva, R.N. Extracellular vesicles carry cellulases in the industrial fungus *Trichoderma reesei*. *Biotechnol. Biofuels* **2019**, *12*, 146. [CrossRef]

58. Souza, J.A.M.; Baltazar, L.M.; Carregal, V.M.; Gouveia-Eufrasio, L.; Oliveira, A.G.; Dias, W.G.; Rocha, M.C.; Miranda, K.R.; Malavazi, I.; Santos, D.A.; et al. Characterization of *Aspergillus fumigatus* extracellular vesicles and their effects on macrophages and neutrophils functions. *Front. Microbiol.* **2019**, *10*, 2008. [CrossRef]

59. Brauer, V.S.; Pessoni, A.M.; Bitencourt, T.A.; De Paula, R.G.; De Oliveira, R.L.; Goldman, G.H.; Almeida, F. Extracellular vesicles from *Aspergillus flavus* induce M1 polarization in vitro. *mSphere* **2020**, *5*, e00190-20. [CrossRef]

60. Rizzo, J.; Chaze, T.; Miranda, K.; Roberson, R.W.; Gorgette, O.; Nimrichter, L.; Matondo, M.; Latgé, J.P.; Beauvais, A.; Rodrigues, M.L. Characterization of extracellular vesicles produced by *Aspergillus fumigatus* protoplasts. *mSphere* **2020**, *5*, e00476-20. [CrossRef]

61. Rodrigues, M.L.; Nakayasu, E.S.; Oliveira, D.L.; Nimrichter, L.; Nosanchuk, J.D.; Almeida, I.C.; Casadevall, A. Extracellular vesicles produced by *Cryptococcus neoformans* contain protein components associated with virulence. *Eukaryot. Cell* **2008**, *7*, 58–67. [CrossRef] [PubMed]

62. Rodrigues, M.L.; Nakayasuc, E.S.; Almeida, I.C.; Nimrichter, L. The impact of proteomics on the understanding of functions and biogenesis of fungal extracellular vesicles. *J. Proteom.* **2014**, *97*, 177–186. [CrossRef] [PubMed]

63. Vargas, G.; Rocha, J.D.; Oliveira, D.L.; Albuquerque, P.C.; Frases, S.; Santos, S.S.; Nosanchuk, J.D.; Gomes, A.M.; Medeiros, L.C.; Miranda, K.; et al. Compositional and immunological analyses of extracellular vesicles released by *Candida albicans*. *Cell. Microbiol.* **2014**, *17*, 389–407. [CrossRef] [PubMed]

64. Oliveira, D.L.; Freire-De-Lima, C.G.; Nosanchuk, J.D.; Casadevall, A.; Rodrigues, M.L.; Nimrichter, L. Extracellular vesicles from *Cryptococcus neoformans* modulate macrophage functions. *Infect. Immun.* **2011**, *78*, 1601–1609. [CrossRef] [PubMed]

65. Aor, A.C. Moléculas Secretadas por *Pseudallescheria boydii*: Detecção, Antigenicidade e Efeitos na Biologia de Células Hospedeiras. Master's Thesis, Universidade Federal do Rio de Janeiro, Rio de Janeiro, Brazil, 2013.

66. Santos, A.L.S. Aspartic proteases of human pathogenic fungi are prospective targets for the generation of novel and effective antifungal inhibitors. *Curr. Enzyme Inhib.* **2011**, *7*, 96–118. [CrossRef]

67. Pawar, P.R.; Pawar, V.A.; Aute, R.A. Role of extracellular hydrolytic enzymes in *Candida albicans* virulence. *Int. J. Pharm. Sci. Rev. Res.* **2014**, *2*, 1521–1532.

68. Silva, R.R. Bacterial and fungal proteolytic enzymes: Production, catalysis and potential applications. *Appl. Biochem. Biotechnol.* **2017**, *183*, 1–19. [CrossRef]

69. Pichová, I.; Pavlícková, L.; Dostál, J.; Dolejsí, E.; Hrusková-Heidingsfeldová, O.; Weber, J.; Ruml, T.; Soucek, M. Secreted aspartic proteases of *Candida albicans*, *Candida tropicalis*, *Candida parapsilosis* and *Candida lusitaniae*. Inhibition with peptidomimetic inhibitors. *Eur. J. Biochem.* **2001**, *268*, 2669–2677. [CrossRef]

70. Singh, D.K.; Németh, T.; Papp, A.; Tóth, R.; Lukácsi, S.; Heidingsfeld, O.; Dostal, J.; Vágvölgyi, C.; Bajtay, Z.; Józsi, M.; et al. Functional characterization of secreted aspartyl proteases in *Candida parapsilosis*. *mSphere* **2019**, *4*, e00484-19. [CrossRef]

71. Namvar, S.; Warn, P.; Farnell, E.; Bromley, M.; Fraczek, M.; Bowyer, P.; Herrick, S. *Aspergillus fumigatus* proteases, Asp f 5 and Asp f 13, are essential for airway inflammation and remodelling in a murine inhalation model. *Clin. Exp. Allergy* **2015**, *45*, 982–993. [CrossRef]

72. Larcher, G.; Cimon, B.; Symoens, F.; Tronchin, G.; Chabasse, D.; Bouchara, J.P. A 33 kDa serine proteinase from *Scedosporium apiospermum*. *Biochem. J.* **1996**, *126*, 119–126. [CrossRef] [PubMed]

73. Han, Z.; Kautto, L.; Nevalainen, H. Secretion of proteases by an opportunistic fungal pathogen *Scedosporium aurantiacum*. *PLoS ONE* **2017**, *12*, e0169403. [CrossRef] [PubMed]

74. Han, Z.; Kautto, L.; Meyer, W.; Chen, S.C.A.; Nevalainen, H. Growth and protease secretion of *Scedosporium aurantiacum* under conditions of hypoxia. *Microbiol. Res.* **2018**, *216*, 23–29. [CrossRef] [PubMed]

75. Silva, B.A.; Pinto, M.R.; Soares, R.M.; Barreto-Bergter, E.; Santos, A.L.S. *Pseudallescheria boydii* releases metallopeptidases capable of cleaving several proteinaceous compounds. *Res. Microbiol.* **2006**, *157*, 425–432. [CrossRef] [PubMed]

76. Silva, B.A.; Santos, A.L.S.; Barreto-Bergter, E.; Pinto, M.R. Extracellular peptidase in the fungal pathogen *Pseudallescheria boydii*. *Curr. Microbiol.* **2006**, *53*, 18–22. [CrossRef]

77. Pereira, M.M.; Silva, B.A.; Pinto, M.R.; Barreto-Bergter, E.; Santos, A.L.S. Proteins and peptidases from conidia and mycelia of *Scedosporium apiospermum* strain HLPB. *Mycopathologia* **2009**, *167*, 25–30. [CrossRef]

78. Palmeira, V.F.; Kneipp, L.F.; Alviano, C.S.; Santos, A.L.S. The major chromoblastomycosis fungal pathogen, *Fonsecaea pedrosoi*, extracellularly releases proteolytic enzymes whose expression is modulated by culture medium composition: Implications on the fungal development and cleavage of key's host structures. *FEMS Immunol. Med. Microbiol.* **2006**, *46*, 21–29. [CrossRef]

79. Han, Z.; Kautto, L.; Meyer, W.; Chen, S.C.; Nevalainen, H. Effect of peptidases secreted by the opportunistic pathogen *Scedosporium aurantiacum* on human epithelial cells. *Can. J. Microbiol.* **2019**, *65*, 814–822. [CrossRef]

80. Mello, T.P.; Barcellos, I.; Branquinha, M.H.; Santos, A.L.S. Hydrolytic enzymes (proteases and lipases) released by biofilm-forming cells of *Scedosporium/Lomentospora* species. *J. Microbiol. Exp.* **2019**, *7*, 62–65.

81. Singh, A.K.; Mukhopadhyay, M. Overview of fungal lipase: A review. *Appl. Biochem. Biotechnol.* **2012**, *166*, 486–520. [CrossRef]

82. Park, M.; Do, E.; Jung, W.H. Lipolytic enzymes involved in the virulence of human pathogenic fungi. *Mycobiology* **2013**, *41*, 67–72. [CrossRef] [PubMed]

83. Cazarolli, J.C.; Guzatto, R.; Samios, D.; Peralba, M.C.R.; Cavalcanti, E.H.S.; Bento, F.M. Susceptibility of linseed, soybean, and olive biodiesel to growth of the deteriogenic fungus *Pseudallescheria boydii*. *Inter. Biodet. Biod.* **2014**, *95*, 364–372. [CrossRef]

84. Boelter, G.; Cazarolli, J.C.; Beker, S.A.; de Quadros, P.D.; Correa, C.; Ferrão, M.F.; Galeazzi, C.F.; Pizzolato, T.M.; Bento, F.M. *Pseudallescheria boydii* and *Meyerozyma guilliermondii*: Behavior of deteriogenic fungi during simulated storage of diesel, biodiesel, and B10 blend in Brazil. *Environ. Sci. Pollut. Res. Int.* **2018**, *25*, 30410–30424. [CrossRef] [PubMed]

85. Pinto, M.R.; de Sá, A.C.M.; Limongi, C.L.; Rozental, S.; Santos, A.L.S.; Barreto-Bergter, E. Involvement of peptidorhamnomannan in the interaction of *Pseudallescheria boydii* and HEp2 cells. *Microbes Infect.* **2004**, *6*, 1259–1267. [CrossRef]

86. Aor, A.C.; Mello, T.P.; Sangenito, L.S.; Fonseca, B.B.; Rozental, S.; Lione, V.F.; Veiga, V.F.; Branquinha, M.H.; Santos, A.L.S. Ultrastructural viewpoints on the interaction events of *Scedosporium apiospermum* conidia with lung and macrophage cells. *Mem. Inst. Oswaldo Cruz.* **2018**, *113*, e180311. [CrossRef]

87. Mello, T.P.; Aor, A.C.; Branquinha, M.H.; Santos, A.L.S. Insights into the interaction of *Scedosporium apiospermum, Scedosporium aurantiacum, Scedosporium minutisporum,* and *Lomentospora prolificans* with lung epithelial cells. *Braz. J. Microbiol.* **2019**, *51*, 427–436. [CrossRef]

88. Elizondo-Zertuche, M.; Montoya, A.M.; Robledo-Leal, E.; Garza-Veloz, I.; Sánchez-Núñez, A.L.; Ballesteros-Elizondo, R.; González, G.M. Comparative pathogenicity of *Lomentospora prolificans* (*Scedosporium prolificans*) isolates from Mexican patients. *Mycopathologia* **2017**, *182*, 681–689. [CrossRef]

89. Albataineh, M.T.; Kadosh, D. Regulatory roles of phosphorylation in model and pathogenic fungi. *Med. Mycol.* **2016**, *54*, 333–352. [CrossRef]

90. Kiffer-Moreira, T.; Pinheiro, A.A.S.; Pinto, M.R.; Esteves, F.F.; Souto-Padrón, T.; Barreto-Bergter, E.; Meyer-Fernandes, J.R. Mycelial forms of *Pseudallescheria boydii* present ectophosphatase activities. *Arch. Microbiol.* **2007**, *188*, 159–166. [CrossRef]

91. Fernanado, P.H.; Panagoda, G.J.; Samaranayake, L.P. The relationship between the acid and alkaline phosphatase activity and the adherence of clinical isolates of *Candida parapsilosis* to human buccal epithelial cells. *APMIS* **1999**, *107*, 1034–1042. [CrossRef]

92. Bom, V.L.; de Castro, P.A.; Winkelströter, L.K.; Marine, M.; Hori, J.I.; Ramalho, L.N.; dos Reis, T.F.; Goldman, M.H.; Brown, N.A.; Rajendran, R.; et al. The *Aspergillus fumigatus* sitA phosphatase homologue is important for adhesion, cell wall integrity, biofilm formation, and virulence. *Eukaryot. Cell* **2015**, *14*, 728–744. [CrossRef]

93. Staerck, C.; Yaakoub, H.; Vandeputte, P.; Tabiasco, J.; Godon, C.; Gastebois, A.; Giraud, S.; Guillemette, T.; Calenda, A.; Delneste, Y.; et al. The Glycosylphosphatidylinositol-anchored superoxide dismutase of *Scedosporium apiospermum* protects the conidia from oxidative stress. *J. Fungi* **2021**, *7*, 575. [CrossRef] [PubMed]

94. Lima, O.C.; Larcher, G.; Vandeputte, P.; Lebouil, A.; Chabasse, D.; Simoneau, P.; Bouchara, J.P. Molecular cloning and biochemical characterization of a Cu,Zn-superoxide dismutase from *Scedosporium apiospermum*. *Microbes Infect.* **2007**, *9*, 558–565. [CrossRef] [PubMed]

95. Mina, S.; Staerck, C.; d'Almeida, S.M.; Marot, A.; Delneste, Y.; Calenda, A.; Tabiasco, J.; Bouchara, J.P.; Fleury, M.J.J. Identification of *Scedosporium boydii* catalase A1 gene, a reactive oxygen species detoxification factor highly expressed in response to oxidative stress and phagocytic cells. *Fungal Biol.* **2015**, *119*, 1322–1333. [CrossRef] [PubMed]

96. Mina, S.; Marot-Leblond, A.; Cimon, B.; Fleury, M.J.; Larcher, G.; Bouchara, J.P.; Robert, R. Purification and characterization of a mycelial catalase from *Scedosporium boydii*, a useful tool for specific antibody detection in patients with cystic fibrosis. *Clin. Vaccine Immunol.* **2015**, *22*, 37–45. [CrossRef]

97. Staerck, C.; Tabiasco, J.; Godon, C.; Delneste, Y.; Bouchara, J.P.; Fleury, M.J.J. Transcriptional profiling of *Scedosporium apiospermum* enzymatic antioxidant gene battery unravels the involvement of thioredoxin reductases against chemical and phagocytic cells oxidative stress. *Med. Mycol.* **2019**, *57*, 363–373. [CrossRef]

98. Bertrand, S.; Larcher, G.; Landreau, A.; Richomme, P.; Duval, O.; Bouchara, J.P. Hydroxamate siderophores of *Scedosporium apiospermum*. *Biometals* **2009**, *22*, 1019–1029. [CrossRef] [PubMed]

99. Ramos, L.S.; Figueiredo-Carvalho, M.H.G.; Silva, L.N.; Siqueira, N.L.M.; Lima, J.C.; Oliveira, S.S.; Almeida-Paes, R.; Zancopé-Oliveira, R.M.; Azevedo, F.S.; Ferreira, A.L.P.; et al. The threat called *Candida haemulonii* species complex in Rio de Janeiro State, Brazil: Focus on antifungal resistance and virulence attributes. *J. Fungi* **2022**, *8*, 574. [CrossRef] [PubMed]

100. Riddell, J., 4th; Chenoweth, C.E.; Kauffman, C.A. Disseminated *Scedosporium apiospermum* infection in a previously healthy woman with HELLP syndrome. *Mycoses* **2004**, *47*, 442–446. [CrossRef] [PubMed]

101. Bertrand, S.; Bouchara, J.P.; Venier, M.C.; Richomme, P.; Duval, O.; Larcher, G. N(α)-methyl coprogen B, a potential marker of the airway colonization by *Scedosporium apiospermum* in patients with cystic fibrosis. *Med. Mycol.* **2010**, *48* (Suppl. 1), S98–S107. [CrossRef]

102. Le Govic, Y.; Havlíček, V.; Capilla, J.; Luptáková, D.; Dumas, D.; Papon, N.; Le Gal, S.; Bouchara, J.P.; Vandeputte, P. Synthesis of the hydroxamate siderophore Nα-Methylcoprogen B in *Scedosporium apiospermum* is mediated by sidD ortholog and is required for virulence. *Front. Cell. Infect. Microbiol.* **2020**, *10*, 587909. [CrossRef]

103. Thornton, C.R. Tracking the emerging human pathogen *Pseudallescheria boydii* by using highly specific monoclonal antibodies. *Clin. Vaccine Immunol.* **2009**, *16*, 756–764. [CrossRef]

104. Kaufman, L.; Standard, P.G.; Jalbert, M.; Kraft, D.E. Immunohistologic identification of *Aspergillus* spp. and other hyaline fungi by using polyclonal fluorescent antibodies. *J. Clin. Microbiol.* **1997**, *35*, 2206–2209. [CrossRef]

105. Jabado, N.; Casanova, J.L.; Haddad, E.; Dulieu, F.; Fournet, J.C.; Dupont, B.; Fischer, A.; Hennequin, C.; Blanche, S. Invasive pulmonary infection due to *Scedosporium apiospermum* in two children with chronic granulomatous disease. *Clin. Infect. Dis.* **1998**, *27*, 1437–1441. [CrossRef]

106. Pinto, M.R.; Mulloy, B.; Haido, R.M.; Travassos, L.R.; Barreto-Bergter, E. A peptidorhamnomannan from the mycelium of *Pseudallescheria boydii* is a potential diagnostic antigen of this emerging human pathogen. *Microbiology* **2001**, *147*, 1499–1506. [CrossRef]

107. Xisto, M.I.; Bittencourt, V.C.; Liporagi-Lopes, L.C.; Haido, R.M.T.; Mendonça, M.S.A.; Sassaki, G.; Figueiredo, R.T.; Romanos, M.T.; Barreto-Bergter, E. O-Glycosylation in cell wall proteins in *Scedosporium prolificans* is critical for phagocytosis and inflammatory cytokines production by macrophages. *PLoS ONE* **2015**, *10*, e0123189. [CrossRef]

108. Oliveira, E.B.; Xisto, M.I.D.S.; Rollin-Pinheiro, R.; Rochetti, V.P.; Barreto-Bergter, E. Peptidorhamnomannans from *Scedosporium* and *Lomentospora* species display microbicidal activity against bacteria commonly present in cystic fibrosis patients. *Front. Cell. Infect. Microbiol.* **2020**, *10*, 598823. [CrossRef]

109. Meirelles, J.V.; Xisto, M.I.D.D.S.; Rollin-Pinheiro, R.; Serrato, R.V.; Haido, R.M.T.; Barreto-Bergter, E. Peptidorhamanomannan: A surface fungal glycoconjugate from *Scedosporium aurantiacum* and *Scedosporium minutisporum* and its recognition by macrophages. *Med. Mycol.* **2021**, *59*, 441–452. [CrossRef]

110. Lopes, L.C.; Rollin-Pinheiro, R.; Guimaraes, A.J.; Bittencourt, V.C.; Martinez, L.R.; Koba, W.; Farias, S.E.; Nosanchuk, J.D.; Barreto-Bergter, E. Monoclonal antibodies against peptidorhamnomannans of *Scedosporium apiospermum* enhance the pathogenicity of the fungus. *PLoS Negl. Trop. Dis.* **2010**, *4*, e853. [CrossRef]

111. Pavlaskova, K.; Nedved, J.; Kuzma, M.; Zabka, M.; Sulc, M.; Sklenar, J.; Novak, P.; Benada, O.; Kofronova, O.; Hajduch, M.; et al. Characterization of pseudacyclins A-E, a suite of cyclic peptides produced by *Pseudallescheria boydii*. *J. Nat. Prod.* **2010**, *73*, 1027–1032. [CrossRef]

112. Krasny, L.; Strohalm, M.; Bouchara, J.P.; Sulc, M.; Lemr, K.; Barreto-Bergter, E.; Havlicek, V. *Scedosporium* and *Pseudallescheria* low molecular weight metabolites revealed by database search. *Mycoses* **2011**, *54* (Suppl. 3), S37–S42. [CrossRef] [PubMed]

113. Mina, S.; Staerck, C.; Marot, A.; Godon, C.; Calenda, A.; Bouchara, J.P.; Fleury, M.J.J. Scedosporium boydii CatA1 and SODC recombinant proteins, new tools for serodiagnosis of *Scedosporium* infection of patients with cystic fibrosis. *Diagn. Microbiol. Infect. Dis.* **2017**, *89*, 282–287. [CrossRef] [PubMed]

114. Karageorgopoulos, D.E.; Vouloumanou, E.K.; Ntziora, F.; Michalopoulos, A.; Rafailidis, P.I.; Falagas, M.E. β-D-glucan assay for the diagnosis of invasive fungal infections: A meta-analysis. *Clin. Infect. Dis.* **2011**, *52*, 750–770. [CrossRef] [PubMed]

115. Chen, S.C.; Halliday, C.L.; Hoenigl, M.; Cornely, O.A.; Meyer, W. *Scedosporium* and *Lomentospora* Infections: Contemporary microbiological tools for the diagnosis of invasive disease. *J. Fungi* **2021**, *7*, 23. [CrossRef]

116. Cuerara, M.S.; Alhambra, A.; Moragues, M.; Gonzalez-Elorza, E.; Ponton, J.; de Palacio, A. Detection of (1,2)-β-D-glucan as an adjunct to diagnosis in a mixed population with uncommon proven invasive fungal diseases or with an unusual clinical presentation. *Clin. Vaccine. Immunol.* **2009**, *16*, 423–426.

117. Levesque, E.; Rizk, F.; Noorah, Z.; Ait-Ammar, N.; Cordonnier-Jourdin, C.; El-Anbassi, S.; Bonnal, C.; Azoulay, D.; Merle, J.-C.; Botterel, F. Detection of (1,3)-β-D-glucan for the diagnosis of invasive fungal infection in liver transplant recipients. *Int. J. Mol. Sci.* **2017**, *18*, 862. [CrossRef]

118. Nishimori, M.; Takahashi, T.; Suzuki, E.; Kodaka, T.; Hiramoto, N.; Itoh, K.; Tsunemine, H.; Yarita, K.; Kamei, K.; Takegawa, H.; et al. Fatal fungemia with *Scedosporium prolificans* in a patient with acute myeloid leukemia. *Med. Mycol. J.* **2014**, *55E*, E63–E70. [CrossRef]

119. Keller, N.P. Fungal secondary metabolism: Regulation, function and drug discovery. *Nat. Rev. Microbiol.* **2019**, *17*, 167–180. [CrossRef]

120. Felnagle, E.A.; Jackson, E.E.; Chan, Y.A.; Podevels, A.M.; Berti, A.D.; McMahon, M.D.; Thomas, M.G. Nonribosomal peptide synthetases involved in the production of medically relevant natural products. *Mol. Pharm.* **2008**, *5*, 191–211. [CrossRef]

121. Le Govic, Y.; Papon, N.; Le Gal, S.; Bouchara, J.P.; Vandeputte, P. Non-ribosomal peptide synthetase gene clusters in the human pathogenic fungus *Scedosporium apiospermum*. *Front. Microbiol.* **2019**, *10*, 1–14. [CrossRef]

122. Hayakawa, Y.; Yamashita, T.; Mori, T.; Nagai, K.; Shin-Ya, K.; Watanabe, H. Structure of tyroscherin, an antitumor antibiotic against IGF-1-dependent cells from *Pseudallescheria* sp. *J. Antibiot.* **2004**, *57*, 634–638. [CrossRef] [PubMed]

123. Chang, Y.C.; Deng, T.S.; Pang, K.L.; Hsiao, C.J.; Chen, Y.Y.; Tang, S.J.; Lee, T.H. Polyketides from the littoral plant associated fungus *Pseudallescheria boydii*. *J. Nat. Prod.* **2013**, *76*, 1796–1800. [CrossRef] [PubMed]

124. Yuan, M.X.; Qiu, Y.; Ran, Y.Q.; Feng, G.K.; Deng, R.; Zhu, X.F.; Lan, W.J.; Li, H.J. Exploration of indole alkaloids from marine fungus *Pseudallescheria boydii* F44-1 using an amino acid-directed strategy. *Mar. Drugs* **2019**, *17*, 77. [CrossRef] [PubMed]

125. Li, C.J.; Chen, P.N.; Li, H.J.; Mahmud, T.; Wu, D.L.; Xu, J.; Lan, W.J. Potential antidiabetic fumiquinazoline alkaloids from the marine-derived fungus *Scedosporium apiospermum* F41-1. *J. Nat. Prod.* **2020**, *83*, 1082–1091. [CrossRef]

126. Silva, L.N.; Mello, T.P.; Ramos, L.S.; Branquinha, M.H.; Santos, A.L.S. Current challenges and updates on the therapy of fungal infections. *Curr. Top. Med. Chem.* **2019**, *19*, 495–499. [CrossRef] [PubMed]

127. Larsson, D.G.J.; Flach, C.F. Antibiotic resistance in the environment. *Nat. Rev. Microbiol.* **2022**, *20*, 257–269. [CrossRef]

128. Arnold, R.S.; Thom, K.A.; Sharma, S.; Phillips, M.; Kristie, J.J.; Morgan, D.J. Emergence of *Klebsiella pneumoniae* carbapenemase-producing bacteria. *South. Med. J.* **2011**, *104*, 40–45. [CrossRef]

129. Antimicrobial Resistance Collaborators. Global burden of bacterial antimicrobial resistance in 2019: A systematic analysis. *Lancet* **2022**, *399*, 629–655. [CrossRef]

130. Staerck, C.; Landreau, A.; Herbette, G.; Roullier, C.; Bertrand, S.; Siegler, B.; Larcher, G.; Bouchara, J.P.; Fleury, M.J.J. The secreted polyketide boydone A is responsible for the anti-*Staphylococcus aureus* activity of *Scedosporium boydii*. *FEMS Microbiol. Lett.* **2017**, *364*, fnx223. [CrossRef]

131. Li, X.; Kim, S.K.; Nam, K.W.; Kang, J.S.; Choi, H.D.; Son, B.W. A new antibacterial dioxopiperazine alkaloid related to gliotoxin from a marine isolate of the fungus *Pseudallescheria*. *J. Antibiot.* **2006**, *59*, 248–250. [CrossRef]

132. Zhang, D.; Noviendri, D.; Nursid, M.; Yang, X.D.; Son, B.W. 12, 13-Dihydroxyfumitremorgin C, fumitremorgin C, and brevianamide F, antibacterial diketopiperazine alkaloids from the marine-derived fungus *Pseudallescheria* sp. *Nat. Prod. Sci.* **2007**, *13*, 251–254.

133. Ko, W.H.; Tsou, Y.J.; Ju, Y.M.; Hsieh, H.M.; Ann, P.J. Production of a fungistatic substance by *Pseudallescheria boydii* isolated from soil amended with vegetable tissues and its significance. *Mycopathologia* **2010**, *169*, 125–131. [CrossRef] [PubMed]

134. Kamigiri, K.; Tanaka, K.; Matsumoto, H.; Nagai, K.; Watanabe, M.; Suzuki, K. YM-193221, a novel antifungal antibiotic produced by *Pseudallescheria ellipsoidea*. *J. Antibiot.* **2004**, *57*, 569–572. [CrossRef] [PubMed]

135. Nirma, C.; Eparvier, V.; Stien, D. Antifungal agents from *Pseudallescheria boydii* SNB-CN73 isolated from a *Nasutitermes* sp. termite. *J. Nat. Prod.* **2013**, *76*, 988–991. [CrossRef]

136. Huang, L.H.; Xu, M.Y.; Li, H.J.; Li, J.Q.; Chen, Y.X.; Ma, W.Z.; Li, Y.P.; Xu, J.; Yang, D.P.; Lan, W.J. Amino acid-directed strategy for inducing the marine-derived fungus *Scedosporium apiospermum* F41-1 to maximize alkaloid diversity. *Org. Lett.* **2017**, *19*, 4888–4891. [CrossRef]

137. Lan, W.J.; Wang, K.T.; Xu, M.Y.; Zhang, J.J.; Lam, C.K.; Zhong, G.H.; Xu, J.; Yang, D.P.; Li, H.; Wang, L.Y. Secondary metabolites with chemical diversity from the marine-derived fungus *Pseudallescheria boydii* F19-1 and their cytotoxic activity. *RSC Adv.* **2016**, *6*, 76206–76213. [CrossRef]

138. Kuroda, K.; Yoshida, M.; Uosaki, Y.; Ando, K.; Kawamoto, I.; Oishi, E.; Onuma, H.; Yamada, K.; Matsuda, Y. AS-183, a novel inhibitor of acyl-CoA: Cholesterol acyltransferase produced by *Scedosporium* sp. SPC-15549. *J. Antibiot.* **1993**, *46*, 1196–1202. [CrossRef]

139. Katsuta, R.; Yajima, A.; Ishigami, K.; Nukada, T.; Watanabe, H. Synthesis of (2RS,8R,10R)-YM-193221 and an improved approach to tyroscherin, bioactive natural compounds from *Pseudallescheria* sp. *Biosci. Biotechnol. Biochem.* **2010**, *74*, 2056–2059. [CrossRef]

140. Wu, Q.; Jiang, N.; Han, W.B.; Mei, Y.N.; Ge, M.H.; Guo, Z.K.; Weng, N.S.; Tan, R.X. Antibacterial epipolythiodioxopiperazine and unprecedented sesquiterpene from *Pseudallescheria boydii*, a beetle (coleoptera)-associated fungus. *Org. Biomol. Chem.* **2014**, *12*, 9405–9412. [CrossRef]

141. Yan, D.F.; Lan, W.J.; Wang, K.T.; Huang, L.; Jiang, C.W.; Li, H.J. Two chlorinated benzofuran derivatives from the marine fungus *Pseudallescheria boydii*. *Nat. Prod. Commun.* **2015**, *10*, 621–622. [CrossRef]

142. Liu, W.; Li, H.J.; Xu, M.Y.; Ju, Y.C.; Wang, L.Y.; Xu, J.; Yang, D.P.; Lan, W.J. Pseudellones A-C, three alkaloids from the marine-derived fungus *Pseudallescheria ellipsoidea* F42-3. *Org. Lett.* **2015**, *17*, 5156–5159. [CrossRef] [PubMed]

143. Liu, D.H.; Sun, Y.Z.; Kurtán, T.; Mándi, A.; Tang, H.; Li, J.; Su, L.; Zhuang, C.L.; Liu, Z.Y.; Zhang, W. Osteoclastogenesis regulation metabolites from the coral-associated fungus *Pseudallescheria boydii* TW-1024-3. *J. Nat. Prod.* **2019**, *82*, 1274–1282. [CrossRef] [PubMed]

144. Wang, D.; Neupane, P.; Ragnarsson, L.; Capon, R.J.; Lewis, R.J. Synthesis of Pseudellone analogs and characterization as novel T-type calcium channel blockers. *Mar. Drugs.* **2018**, *16*, 475. [CrossRef] [PubMed]

145. Wang, K.T.; Xu, M.Y.; Liu, W.; Li, H.J.; Xu, J.; Yang, D.P.; Lan, W.J.; Wang, L.Y. Two additional new compounds from the marine-derived fungus *Pseudallescheria ellipsoidea* F42-3. *Molecules* **2016**, *21*, 442. [CrossRef] [PubMed]

146. Sathieshkumar, P.P.; Latha, P.; Nagarajan, R. Total synthesis of Pseudellone C. *Eur. J. Org. Chem.* **2017**, *22*, 3161–3164. [CrossRef]

147. Sorres, J.; Nirma, C.; Eparvier, V.; Stien, D. Pseudallicins A-D: Four complex Ovalicin derivatives from *Pseudallescheria boydii* SNB-CN85. *Org. Lett.* **2017**, *19*, 3978–3981. [CrossRef]

148. Macheleidt, J.; Mattern, D.J.; Fischer, J.; Netzker, T.; Weber, J.; Schroeckh, V.; Valiante, V.; Brakhage, A.A. Regulation and role of fungal secondary metabolites. *Annu. Rev. Genet.* **2016**, *23*, 371–392. [CrossRef]

149. Liao, J.L.; Pang, K.L.; Sun, G.H.; Pai, T.W.; Hsu, P.H.; Lin, J.S.; Sun, K.H.; Hsieh, C.C.; Tang, S.J. Chimeric 6-methylsalicylic acid synthase with domains of acyl carrier protein and methyltransferase from *Pseudallescheria boydii* shows novel biosynthetic activity. *Microb. Biotechnol.* **2019**, *12*, 920–931. [CrossRef]

150. Bills, G.F.; Gloer, J.B. Biologically active secondary metabolites from the fungi. *Microbiol. Spectr.* **2016**, *4*. [CrossRef]

151. Su, H.J.; Lin, M.J.; Tsou, Y.J.; Ko, W.H. Pseudallin, a new antibiotic produced by the human pathogenic fungus Pseudallescheria boydii, with ecological significance. *Bot. Stud.* **2012**, *53*, 239–242.

152. Deshmukh, R.; Khardenavis, A.A.; Purohit, H.J. Diverse metabolic capacities of fungi for bioremediation. *Indian J. Microbiol.* **2016**, *56*, 247–264. [CrossRef] [PubMed]

153. de Hoog, G.S.; Marvin-Sikkema, F.D.; Lahpoor, G.A.; Gottschall, J.C.; Prins, R.A.; Guého, E. Ecology and physiology of the emerging opportunistic fungi *Pseudallescheria boydii* and *Scedosporium prolificans*. *Mycoses* **1994**, *37*, 71–78. [CrossRef] [PubMed]

154. April, T.M.; Abbott, S.P.; Foght, J.M.; Currah, R.S. Degradation of hydrocarbons in crude oil by the ascomycete *Pseudallescheria boydii* (Microascaceae). *Can. J. Microbiol.* **1998**, *44*, 270–278. [CrossRef] [PubMed]

155. Markovetz, A.J.J.; Cazin, J.; Allen, J.E. Assimilation of alkanes and alkenes by fungi. *Appl. Microbiol.* **1968**, *16*, 487–489. [CrossRef] [PubMed]

156. Claußen, M.; Schmidt, S. Biodegradation of phenol and p-cresol by the hyphomycete *Scedosporium apiospermum*. *Res. Microbiol.* **1998**, *149*, 399–406.

157. Claußen, M.; Schmidt, S. Biodegradation of phenylbenzoate and some of its derivatives by *Scedosporium apiospermum*. *Res. Microbiol.* **1999**, *150*, 413–420. [CrossRef]

158. García-Peña, E.I.; Hernández, S.; Favela-Torres, E.; Auria, R.; Revah, S. Toluene biofiltration by the fungus *Scedosporium apiospermum* TB1. *Biotechnol. Bioeng.* **2001**, *76*, 61–69. [CrossRef]

159. Reyes-César, A.; Absalón, Á.E.; Fernández, F.J.; González, J.M.; Cortés-Espinosa, D.V. Biodegradation of a mixture of PAHs by non-ligninolytic fungal strains isolated from crude oil-contaminated soil. *World J. Microbiol. Biotechnol.* **2014**, *30*, 999–1009. [CrossRef]

160. Ishii, K.; Furuichi, T.; Tanikawa, N.; Kuboshima, M. Estimation of the biodegradation rate of 2,3,7,8-tetrachlorodibenzo-p-dioxin by using dioxin-degrading fungus, *Pseudallescheria boydii*. *J. Hazard. Mater.* **2009**, *162*, 328–332. [CrossRef]

161. Tigini, V.; Prigione, V.; Di Toro, S.; Fava, F.; Varese, G.C. Isolation and characterisation of polychlorinated biphenyl (PCB) degrading fungi from a historically contaminated soil. *Microb. Cell Fact.* **2009**, *8*, 5. [CrossRef]
162. Mbokou, S.F.; Pontié, M.; Razafimandimby, B.; Bouchara, J.P.; Njanja, E.; Tonle, K.I. Evaluation of the degradation of acetaminophen by the filamentous fungus *Scedosporium dehoogii* using carbon-based modified electrodes. *Anal. Bioanal. Chem.* **2016**, *408*, 5895–5903. [CrossRef] [PubMed]
163. Ren, H.; Li, H.; Wang, H.; Huang, H.; Lu, Z. Biodegradation of tetrahydrofuran by the newly isolated filamentous fungus *Pseudallescheria boydii* ZM01. *Microorganisms* **2020**, *8*, 1190. [CrossRef] [PubMed]
164. Ribas, R.; Cazarolli, J.C.; da Silva, E.C.; Meneghetti, M.R.; Meneghetti, S.M.P.; Bento, F.M. Characterization of antimicrobial effect of organotin-based catalysts on diesel-biodiesel deteriogenic microorganisms. *Environ. Monit. Assess.* **2020**, *192*, 802. [CrossRef]
165. Atakpa, E.O.; Zhou, H.; Jiang, L.; Ma, Y.; Liang, Y.; Li, Y.; Zhang, D.; Zhang, C. Improved degradation of petroleum hydrocarbons by co-culture of fungi and biosurfactant-producing bacteria. *Chemosphere* **2022**, *290*, 133337. [CrossRef] [PubMed]
166. Martínez-Gallardo, M.R.; Jurado, M.M.; López-González, J.A.; Toribio, A.; Suárez-Estrella, F.; Sáez, J.A.; Moral, R.; Andreu-Rodríguez, F.J.; López, M.J. Biorecovery of olive mill wastewater sludge from evaporation ponds. *J. Environ. Manag.* **2022**, *319*, 115647. [CrossRef]
167. Kumaravel, V.; Bankole, P.O.; Jooju, B.; Sadasivam, S.K. Degradation and detoxification of reactive yellow dyes by *Scedosporium apiospermum*: A mycoremedial approach. *Arch. Microbiol.* **2022**, *204*, 324. [CrossRef]
168. Morales, L.T.; González-García, L.N.; Orozco, M.C.; Restrepo, S.; Vives, M.J. The genomic study of an environmental isolate of *Scedosporium apiospermum* shows its metabolic potential to degrade hydrocarbons. *Stand. Genom. Sci.* **2017**, *12*, 71. [CrossRef]
169. Poirier, W.; Ravenel, K.; Bouchara, J.P.; Giraud, S. Lower funneling pathways in *Scedosporium* Species. *Front. Microbiol.* **2021**, *12*, 630753. [CrossRef]
170. Janda-Ulfig, K.; Ulfig, K.; Cano, J.; Guarro, J. A study of the growth of *Pseudallescheria boydii* isolates from sewage sludge and clinical sources on tributyrin, rapeseed oil, biodiesel oil and diesel oil. *Ann. Agric. Environ. Med.* **2008**, *15*, 45–49.
171. Berthon, J.Y.; Grizard, D. Fungal Inoculum, Method for Its Preparation and Methods for Its Utilization for the Treatment of Waste Water with High Content in Organic Material. European Patent EP 1352953, 15 October 2003.
172. Nomoto, T. Promoter for Composting. Japan Patent JP 3485345, 25 August 2003.
173. Laugero, C.; Tillier, D. Method for Biological Treatment of Animal Breeding Effluent and Device Therefor. European Patent EP 1242318, 9 October 2001.
174. Santos, A.L.S.; Bittencourt, V.C.; Pinto, M.R.; Silva, B.A.; Barreto-Bergter, E. Biochemical characterization of potential virulence markers in the human fungal pathogen *Pseudallescheria boydii*. *Med. Mycol.* **2009**, *47*, 375–386. [CrossRef]

Journal of *Fungi*

Article

Toxic Indoor Air Is a Potential Risk of Causing Immuno Suppression and Morbidity—A Pilot Study

Kirsi Vaali [1,*], Marja Tuomela [2,3], Marika Mannerström [4], Tuula Heinonen [4] and Tamara Tuuminen [5]

[1] SelexLab Oy, Kalevankatu 17 A, 00100 Helsinki, Finland
[2] Co-op Bionautit, Viikinkaari 9, 00790 Helsinki, Finland; marja.tuomela@helsinki.fi
[3] Department of Microbiology, University of Helsinki, 00014 Helsinki, Finland
[4] The Finnish Centre for Alternative Methods, Faculty of Medicine and Health Technology, Tampere University, Arvo Ylpön katu 34, 33014 Tampere, Finland; marika.mannerstrom@tuni.fi (M.M.); tuula.heinonen@tuni.fi (T.H.)
[5] Medical Center Kruunuhaka Oy, Kaisaniemenkatu 8B a, 00100 Helsinki, Finland; tuuminen@gmail.com
* Correspondence: kvaali43@gmail.com; Tel.: +358-50-550-1131

Abstract: We aimed to establish an etiology-based connection between the symptoms experienced by the occupants of a workplace and the presence in the building of toxic dampness microbiota. The occupants (5/6) underwent a medical examination and urine samples (2/6) were analyzed by LC-MS/MS for mycotoxins at two time-points. The magnitude of inhaled water was estimated. Building-derived bacteria and fungi were identified and assessed for toxicity. Separate cytotoxicity tests using human THP-1 macrophages were performed from the office's indoor air water condensates. Office-derived indoor water samples ($n = 4/4$) were toxic to human THP-1 macrophages. *Penicillium*, *Acremonium sensu lato*, *Aspergillus ochraceus* group and *Aspergillus* section *Aspergillus* grew from the building material samples. These colonies were toxic in boar sperm tests ($n = 11/32$); four were toxic to BHK-21 cells. Mycophenolic acid, which is a potential immunosuppressant, was detected in the initial and follow-up urine samples of (2/2) office workers who did not take immunosuppressive drugs. Their urinary mycotoxin profiles differed from household and unrelated controls. Our study suggests that the presence of mycotoxins in indoor air is linked to the morbidity of the occupants. The cytotoxicity test of the indoor air condensate is a promising tool for risk assessment in moisture-damaged buildings.

Keywords: indoor air water; dampness and mold hypersensitivity syndrome; mycotoxins; clinical toxicology; sick building syndrome; urine

Citation: Vaali, K.; Tuomela, M.; Mannerström, M.; Heinonen, T.; Tuuminen, T. Toxic Indoor Air Is a Potential Risk of Causing Immuno Suppression and Morbidity—A Pilot Study. *J. Fungi* **2022**, *8*, 104. https://doi.org/10.3390/jof8020104

Academic Editor: Laurent Dufossé

Received: 22 November 2021
Accepted: 4 January 2022
Published: 21 January 2022

Publisher's Note: MDPI stays neutral with regard to jurisdictional claims in published maps and institutional affiliations.

1. Introduction

There is convincing scientific data that the exposure to dampness microbiota and the decay products of construction materials may cause the development of polymorbidities and systemic inflammation [1,2], now collectively called dampness and mold hypersensitivity syndrome (DMHS) [3], although this designation has not yet received official recognition. The causality between the exposure to dampness microbiota (the source) and the clinical outcome (the effect) is still being questioned. Polymorbidity in the occupants of moisture-damaged buildings with indoor air toxicity has been extensively studied in Finnish schoolchildren and the occupants of a moisture-damaged hospital and a police station [1,4–8]. These studies have demonstrated that the prolonged exposure to dampness microbiota and the decay products of a building's construction materials may cause a plethora of non-respiratory symptoms such as profound fatigue, neurological, gastrointestinal, and muscular-skeletal problems in addition to the already acknowledged respiratory symptoms. Because no single biomarker to diagnose DMHS has yet been devised, the diagnosis remains clinical and is based on interview. The patients often recall some form of water leakage in their homes or workplaces and a gradual worsening of their symptoms.

Initially, the symptoms are reversible and related to the so-called sick building syndrome, (SBS) [9]. However, if there is continuous or cumulative exposure (e.g., starting already in the kindergarten, and then in school, high school and then workplaces) the symptoms may become aggravated and progress into irreversible multi-organ DMHS [3]. Importantly, DMHS is associated with an increased rate of respiratory infections [3] probably indicating decreased immunologic tolerance to environmental viruses and bacteria. Moreover, deprivation of the immune system has been proposed by us when reviewing the long-term morbidity among the occupants of one of the moisture-damaged school [10]. A higher prevalence of different oncological diseases, unexplained succeptibility to sepsis and a higher than normal rate of autoimmune diseases have been reported [10].

The investigations of the indoor air quality do not always recognize existing problems within buildings; for example, the impact of air toxicity has been overlooked. Most of the toxic metabolites produced by fungi or bacteria recovered from moisture-damaged buildings have a molecular weight of 300–2000 g/mol, i.e., they are non-volatile. However, mycotoxins may be present in indoor air either attached to fungal spores or to fragments of these microorganisms, e.g., broken or fractured conidia and hyphae. These particles may remain suspended in the air for a long time [11]. Recently, it was reported [12–14] that some species of dampness microbiota produce vesicles, or the so-called "guttation droplets", or exudates on a culture dish. These droplets may contain substances which are toxic to eukaryotic cell types at dilutions of 1/100–20,000 [12,15,16]. The most efficient solvents for mycotoxins are relatively polar solvents, such as methanol or ethanol, only a few mycotoxins are water soluble [17]. The relative humidity (RH %) of the air becomes elevated when the environment is crowded with occupants (e.g., schools during the working hours) which allows these droplets to move around aerodynamically and thus the toxins can spread throughout the environment [18]. For example, *Penicillium expansum* recovered from gypsum boards produced these kinds of toxic droplets, i.e., exudates that were released into the air and were >100-times more toxic in the cell culture assays than indoor air isolates of *Aspergillus, Chaetomium, Stachybotrys* and *Paecilomyces* [13]. These droplets may represent an even greater health hazard than the spores and nanoparticles of a microbial mass growing in the environment.

To address the issue of air toxicity, we have devised a novel method of collection of indoor air water vapour by adopting a newly patented technique with the so-called E-collector: (US Patent 10,502,722 B2; www.sisailmatutkimuspalvelut.fi, accessed on 21 November 2021) [18,19]. Briefly, the air condensate is collected on a cooled surface of a steel plate, the condensate then can be tested for its cytotoxicity against living cells [19]. Boar sperm cells [20,21] or several eukaryotic cell lines, such as porcine kidney cell line (PK-15) [13,22,23] or murine neuroblastoma cells (MNA) [13,23,24], have been used in toxicity studies instead of experiments in production animals. In this publication, we have successfully used boar spermatozoids and baby hamster kidney (BHK) cells for the same purpose.

Here, we describe a plausible pathway of how the mycotoxins present in the indoor air water samples gain access to the exposed individuals. We demonstrate the usefulness of urinalysis for the detection of mycotoxins as biomarkers of DMHS. A novel method [19] for the sampling and testing of indoor air toxicity is presented and shortly discussed.

2. Materials and Methods

2.1. The Patients

This working community comprised six members, and their work premises which were located partly below ground level in an office building (Figure 1). Within approximately three months after the move into this office, every member of this workplace community started to suffer from symptoms compatible with their exposure to indoor air dampness microbiota. Finally, the symptoms became unbearable. The patients wanted to remain anonymous and therefore their gender and age are not reported.

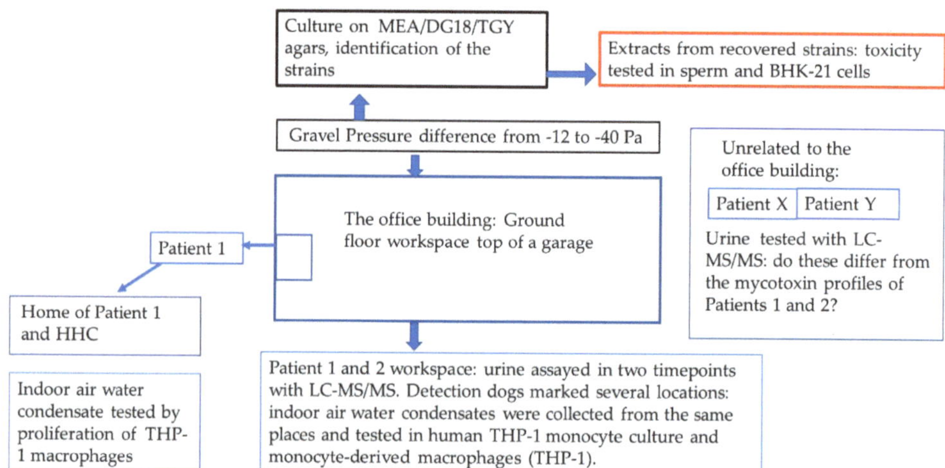

Figure 1. The study design.

Five (Patients 1–5) out of six occupants of the problematic building were examined by one of us (TT) and one (Patient 6) was interviewed over the telephone. All were healthy before the move into the problematic office. Their life histories and the symptoms were recorded during their visits and then retrieved with the patients' written consent.

For comparison of urinalysis profiles, we investigated two unrelated patients (Patient X and Y) who volunteered to have their urine samples tested because their symptoms were compatible with an exposure to dampness microbiota. They were unrelated to each other, they lived and worked in different cities than the working community in question. Neither microbiological nor toxicological investigations in their homes or workplaces had been performed at the time of our investigation. Patient X moved with all his/her belongings to a new home but continued to experience a general feeling of sickness, fatigue, and mucosal irritation.

2.2. Detection of Mycotoxins in the Patients' Urine Samples

The detection of mycotoxins from the patients' urine was performed in the Great Plains Laboratory Inc. (Lenexa, KS, USA) using a liquid chromatography tandem mass spectrometry (LC-MS/MS) technique. The urine specimens from Patients 1 and 2 of the working community and a sample from a household control (HHC) of Patient 1, and from Patients X and Y were analyzed. Follow-up samples from Patients 1 and 2 were taken 2.5 months later after the working community had moved to a clean building.

The detection of mycotoxins was performed according to the standard operational procedure of the laboratory. The presence of the mycotoxin mycophenolic acid was confirmed after extraction of urine and LC/MS/MS testing of the extract with a deuterated mycophenolic acid internal standard [25], indicating that a peak with all the characteristic ions of authentic mycophenolic acid [26,27] and the identical retention time of mycophenolic acid was present in many patients [28]. In addition, patients with high mycophenolic acid in urine has been tested as positive for mycophenolic acid by a commercial specific immunoassay (Cedia) from ThermoScientific for mycophenolic acid in urine. Mycophenolic acid can be produced by multiple species of Penicillium and other molds [29–31].

2.3. The Office Building

The working community had their office located on the ground floor of a building with a cellar. It was built from concrete elements and had a mechanical ventilation system. One of the walls in the office faced a non-heated hollow area which was below the soil surface and was lined on the bottom with a layer of gravel and on the top with a closing plate mostly without waterproofing. Thus, water from rain and melting snow had leaked inside the hollow area enabling microbial growth. There were several air leaks in the wall of the office adjacent to the hollow area. Because of the negative pressure difference (-12 Pa on average, momentarily even -40 Pa) in the office compared with the hollow area, the air from this area was able to penetrate the office. The hollow area remained moist all year round.

The office was also examined by sniffer dogs which are trained in mold detection. The dogs marked six locations, with five of these being beside the wall facing the hollow area. In some locations, the characteristic *odors* of *microbes* were noticeable also to the occupants.

2.4. Sampling of the Indoor Air Condensate Water

The device and the technique to collect indoor air water samples have been described elsewhere [19]. Briefly, the principle of the collection is based on the phenomenon of the condensation of water molecules on the top of cold surfaces of metal plates. After melting of the frozen water to room temperature, the condensate is collected from the tray of the device into Eppendorf tubes that are transferred to the laboratory where the condensate is analyzed in a cell culture assay. This collection technique enables the harvesting of airborne toxic substances in indoor air vapor that may contain mycotoxins [13] as well as various large molecules [32].

The condensing water samples from the six locations marked by the sniffer dogs were collected with the E-collector from the office building. Samples were also collected from the home of Patient 1.

2.5. Indoor Air Toxicity Studies

The toxicity of the indoor air water condensates collected from different locations of the office and from the home of Patient 1 was studied using the human acute monocytic leukemia cell line (THP-1) using water-soluble tetrazolium salts (WST-1) in a cytotoxicity assay [19,33]. The WST-1 assay is an indicator of mitochondrial activity and commonly used to assess cell viability; decreased mitochondrial activity reflects a loss of cell viability, i.e., cell death. Increased mitochondrial activity reflects an increased cell number, i.e., proliferation, or increased cellular respiration induced by mitochondrial uncoupling reactions [34,35]. The WST-1 Cell Proliferation Reagent was obtained from Roche (Basel, Switzerland). Human THP-1 monocytes (Cat. No. TIB-202) were from ATCC (LGC Promochem AB, Boras, Sweden), and were verified to be *Mycoplasma*-free (MycoAlert™ kit, Lonza Basel, Switzerland) prior to use. THP-1 monocytes were maintained in RPMI 1640 Medium supplemented with 10% fetal bovine serum (FBS), and when differentiated to macrophages, challenged with 25 nM phorbol 12-myristate 13-acetate (PMA) (Sigma Aldrich, Steinheim, Germany) for 48 h followed by a 24 h recovery period.

The cells were seeded into 96-well plates at a density of 10^5 cells/well and exposed for 24 h to the indoor air condensates at 10% and 25% concentrations (in RPMI supplemented with 5% FBS). Cells exposed to an equal volume of distilled water (10% or 25%) served as negative controls, and to nickel II sulphate hexahydrate (2.0 and 20.0 µg/mL) as a positive control. All samples and controls were tested in six replicates. In the assessment of cell viability, 10 µL/well WST-1 reagent was added to the cells for 2 hours, and subsequently absorbance was read at 450 nm. The absorbance is directly proportional to the mitochondrial activity (cell viability). The absorbances were normalized, i.e., the untreated control was set as 100%, and all other data were calculated relative to the control absorbance as either % decrease in cell viability (negative values) or % increase in proliferation (positive values). The statistical significance of the changes compared to the untreated control was tested with Student's *t*-test.

The indoor air condensates from the office were tested using both THP-1 monocytes and THP-1 macrophages at a 10% sample concentration. The condensates from the home of Patient 1 were tested at two concentrations, i.e., 10% and 25%, but using only THP-1 macrophages.

2.6. Detection of Mycotoxins from the Condensed Indoor Air Sample

A left-over aliquot of 0.5 mL of the condensed indoor air water sample in the Eppendorf tube was sent at ambient temperature for LC-MS/MS analysis to the Great Plains laboratory. This sample had not been stored protected from light.

2.7. Indoor Air Relative Humidity Measurements and Estimation of Respiratory Mycotoxin Exposure

The relative humidity was measured with Gann Hydromette BL RH-T (Gann Mess-u. Regeltechnik GmbH, Gerlingen, Germany). To show the probability of the inhalation exposure route instead of the more improbable oral exposure, we estimated the amount of water inhaled in the prevailing indoor air conditions (Table 1). The estimation is approximately 14–15 g (mL) of inhaled water/day, resulting in a challenge of 730–760 g of water/2.5 months exposure time.

Table 1. Estimation of the inhaled water during the stay in a problematic building.

	Inhaled Air Vol (L)	Number of Inhalations	Inhaled m^3 of Air/8 h	RH (%)	Temp (°C)	Humid Ratio	Inhaled Water/8 h/mL/g Water
Space 1 (Pat 1)	0.5	13	3.12	30.8	21.4	4.877	15.22
Space 2	0.5	13	3.12	30.2	21.4	4.781	14.92
Space 3	0.5	13	3.12	30.6	21.6	4.905	15.30
Space 4 (Pat 2)	0.5	13	3.12	29.1	21.6	4.663	14.55

The estimated volume of inhaled air by adults is 0.5 L/inhalation, the frequency of inhalations was estimated to be 13 times/min (the average 12–16 times/min, depending on the physical activity). The estimated frequency of inhalations is 13/h in office workers. The humid ratio was calculated from the measured relative humidity (RH) and room temperature (°C), according to the formula: available online: http://www.flycarpet.net/en/PsyOnline (accessed on 21 November 2021).

2.8. Collection of a Sample for Microbiological Analysis

A gravel sample was collected from the hollow area in the proximity of the floor of the office. There was no visible fungal growth in the sample. The sample was cultured on three agar media, namely malt extract (MEA) with the following formulations: malt extract, 20 g; agar, 15 g; and distilled water to 1 liter [36], dichloran-glycerol (DG18) with the following formulation: glucose, 10 g; bacteriological peptone, 5 g; KH_2PO_4, 1.0 g; $MgSO_4 \times 7H_2O$, 0.5 g; agar, 15 g; distilled water to 1 / L; 220 g of glycerol (analytical reagent grade) having a final concentration of 18% (wt/wt); and 1 mL of a 0.2% (in ethanol) solution of dichloran (2,6-dichloro-4-nitroaniline) to a final concentration of 2 mg / L [36] and tryptone glucose yeast extract (TGY) with the following formulation: tryptone, 5.0 g; yeast extract, 2.5 g; glucose, 1.0 g; agar 15 g; and distilled water to 1 L [37]. All the media were autoclaved for 15 min at 121°C, and pH values for ready medium were 5.5, 5.6 and 7.0, respectively.

Two inoculation methods with two replicates were used for each medium and dilution: (1) Direct inoculation on the agar plates (0.5 g/plate). (2) Serially diluted suspensions from the sample were pipetted onto agar plates (0.1 mL/plate) according to the instructions of National Supervisory Authority for Welfare and Health in Finland [38,39]. The composition of the diluent is as follows: 0.0425 g KH_2PO_4, 0.25 g $MgSO_4 \times 7 H_2O$, 0.008 g NaOH /1 L deionized water. The diluent was prepared as follows: adjusted to pH 7.0 ± 0.2; add 0.2 mL Tween 80 as the detergent, then autoclaved at 121 °C, 15 min. The plates were incubated at room temperature and the cultures were examined after one, two or three weeks with the fungal genera being identified by microscopy [40–43], and furthermore *Aspergillus* spp.

was identified at a species or section level [44]. The fungal identification was based on the inspection of the morphology as recommended by the Finnish authorized bodies, although the current nomenclature has been recently changed in accordance with the advances in molecular methods [38]. The bacterial colonies were counted and *Streptomycetes* were identified [38].

2.9. Detection of Toxicity from the Microbial Growth

The colonies of all the fungal and bacterial species (n = 32) that grew on the agar plates were tested in the boar sperm test according to Andersson et al. [20] and Castagnoli et al. [21], and in a mammalian cell test according to Rasimus et al. [45]. A colony or a part of it (20–30 mg) was extracted into 96 % ethanol (0.2 mL). In the toxicity tests, boar spermatozoa that are used in artificial insemination (Figen Oy, Seinäjoki, Finland) and baby hamster kidney cells (BHK-21 [C13] ATCC® CCL10™) were exposed to the ethanol extracts (n = 32). After incubation with the ethanol extracts, the inhibition of BHK-21 cell growth was assessed using the resazurin reduction assay [46], while toxicity to boar spermatozoa was assessed as a lack of motility observed with the CellSens computer program [21] and visually with a microscope [20]. Pure ethanol was a negative control in both tests. The BHK cells were maintained in Dulbecco's modified Eagle's medium (DMEM) supplemented with 10% fetal bovine serum (FBS), 2 mM L-glutamine, 100 U/mL penicillin and 100 µg/mL streptomycin.

2.10. Ethical Considerations

All participants of this study provided their written informed consent to use their records.

3. Results

3.1. Clinical Picture in Persons Exposed to Toxic Indoor Air

The symptoms reported by Patients 1–5 are summarized in Table 2. Patient 6 started to experience a very severe headache that did not become better during the weekends. This person was given a diagnosis of tension neck, and the symptoms were not interpreted as being associated with an exposure to poor indoor air quality. The patient was often on sick leave and changed the workplace before this investigation started. The work ability returned after changing the employer, and the headache did not reappear.

Table 2. The symptoms experienced by the working community. The symptoms often reported by patients with chronic mold illness are marked with *.

	Patient's History	Symptoms	Findings during the Visit
No 1	No previous exposure to dampness microbiota	* Irritation of eyes	* Symptoms became worse when entering the problematic building
	No change in the diet	* Very severe headache	* Symptoms did not relieve during weekends
		*Recurrent sinusitis	Status uneventful The lesions were indurated,
		* Multiple courses of antibiotics	watery and had been scratched except for dermatitis on hands, legs, stomach, knees and back
		* Thermoregulation problems	
		* Itching of the skin	
		* Exanthema	
		* Sore throat	
		* Gastrointestinal reflux for months	
		* "Brain fog"	
		* A flu-like feeling	

Table 2. *Cont.*

	Patient's History	**Symptoms**	**Findings during the Visit**
No 2	Lived in a town house with his/her family	* Dyspnoea	* Symptoms had started to ease in 2 weeks if the person was absence from work
	All the other family members were asymptomatic	* Migraine	* Large psoriatic lesions especially on the elbows, back, scalp and chest, lips were cracked open
		* Phlegm * Muscle pain * Fatigue * Concentration problems * Prolonged cough * Sleeping problems	* The blood pressure was elevated
No 3	Previously suffered from gastrointestinal problems and allergy	* Sore throat	*Redness of eye conjunctiva, unrelated to season
		* Blisters on the oral mucosa	Exanthema in the lower neck region
		* A flu-like feeling * Blurred vision * Severe fatigue * Dyspnea * Joint pain Irregular peristaltic action and menorrhagia	
No 4	Gastrointestinal dysbiosis, leaky gut	* Thermoregulation problems	Otherwise, the status uneventful
		* Two flu-like episodes	Axillary temperature 36.8 °C, exanthema in the lower neck region
		* Increased sputum production	
No 5	Previously healthy, the individual had not been exposed to dampness and mold	* Two courses of antibiotics for tonsillitis * Sore throat * Evenings: eyes dry and voice hoarse * Occasional instances of fatigue * Memory problems	Status uneventful
No 6	Did not attend the doctor's office	Was interviewed by telephone	

3.2. Detection of Mycotoxins from the Urine Samples

The mycotoxin profiles of Patients 1 and 2 and HHC, Patients X and Y, the controls, are shown in Table 3. It is noteworthy that the HHC for Patient 1 showed evidence of the excretion of ochratoxin A in the urine but the profile was different from that of Patient 1 and there was no mycophenolic acid detectable in his/her urine. The LC-MS/MS plots of mycophenolic acid are presented in Figure 2A,B. The mycotoxin profiles of Patients X and Y did not contain mycophenolic acid, and were different from those of Patients 1 and 2 who represented the work community exposure cohort.

Table 3. Detection of mycotoxins in urine by LC-MS/MS.

Mycotoxins Reference Values in ng/g Creatinine	Patient 1 First Test	Patient 1 Follow-Up Test	HHC to Patient 1 (Time-Point as for Pat 1 Follow-Up)	Patient 2 First Test	Patient 2 Follow-Up Test	Patient X	Patient Y
Aflatoxin M1 (3.5–20)	0	0	0	0	0	0	0
Ochratoxin A (4–20)	11.23	9.82	* 63.52	7.01	10.96	8.58	18.36
Gliotoxin (200–2000)	0	0	0	0	0	0	* 910.98
Sterigmatocystin (0.2–1.75)	1.14	0	0	0.74	0	0	0
Mycophenolic acid (5–50)	* 284.64	30.45	0	* 50	* 130.98	17.79	7.11
Roridin E (1–6)	0	0	0	0	0	0	0
Verrucarin A (1–10)	0	0	0	0	0	0	0
Enniantin B (0.07–1)	<0.07	0	0	0	0	0.43	0
Zearalenone (0.5–10)	0	0	0	4.36	* 15.54	5.55	0
Chaetoglobosin A (20–80)	0	0	0	0	0	0	0
Citrinin (10–50)	<10	<10	14.9	17.47	21.56	0	0

The * indicates excessively high values.

3.3. Evaluation of the Office before Our Investigations

In addition to the notable negative pressure difference and air leaks (from -12 to -40 Pa), the assessment of the office revealed insufficient ventilation and a water damaged area in the ceiling covered with gypsum board. The cultured ceiling material contained 10^7 colony forming units/gram (cfu/g) (on MEA plates) or 7×10^6 cfu/g of fungi (on DG18 plates), and 3×10^7 cfu/g of bacteria (on TGY plates). The identified fungi were *Aureobasidium* sp. (88–97%), and yeasts. It should be emphasized that these counts exceed the cut-off values, which are 10^4 for fungi and 10^5 for bacteria [38].

The sniffer dogs marked six locations, five of which were beside the wall facing the hollow area from where a sample was taken for microbiological analysis.

3.4. Detection of Mycotoxins from the Water Condensate

After the toxicity studies, a left-over sample from the condensed air was analyzed by the LC-MS/MS. No other peaks of mycotoxins were detected except for mycophenolic acid that was however below the limit of accurate detection. The absence of mycophenolic acid could not be ascertained beyond doubt from the indoor air sample because this sample had not been protected from light and this might have led to the degradation of mycophenolic acid.

3.5. Toxicity Studies from the Water Condensates

Toxicity studies from the office's indoor air are shown in Table 4, those from the home of Patient 1 and the HHC are presented in Table 5. All indoor air condensates (tested at a 10% concentration in the final cell culture test system) collected from the office caused adverse effects on cells: Samples from 2 locations out of 4 induced THP-1 monocyte proliferation, whereas samples from all 4 locations were toxic to the THP-1 macrophages. None of the indoor air water condensates collected from the home of Patient 1 and HHC were toxic to THP-1 macrophages tested at the same i.e., 10% concentration. However, after increasing the indoor water condensate amount from 10% to 25% in the test system, toxicity was observed in the home of Patient 1 and the HHC. The samples taken from the bedroom located on the 1st floor, and that from the living room located on a 2nd floor induced a slight proliferation of THP-1 macrophages. Samples collected from the home were not tested on THP-1 monocytes, Table 5.

(A)

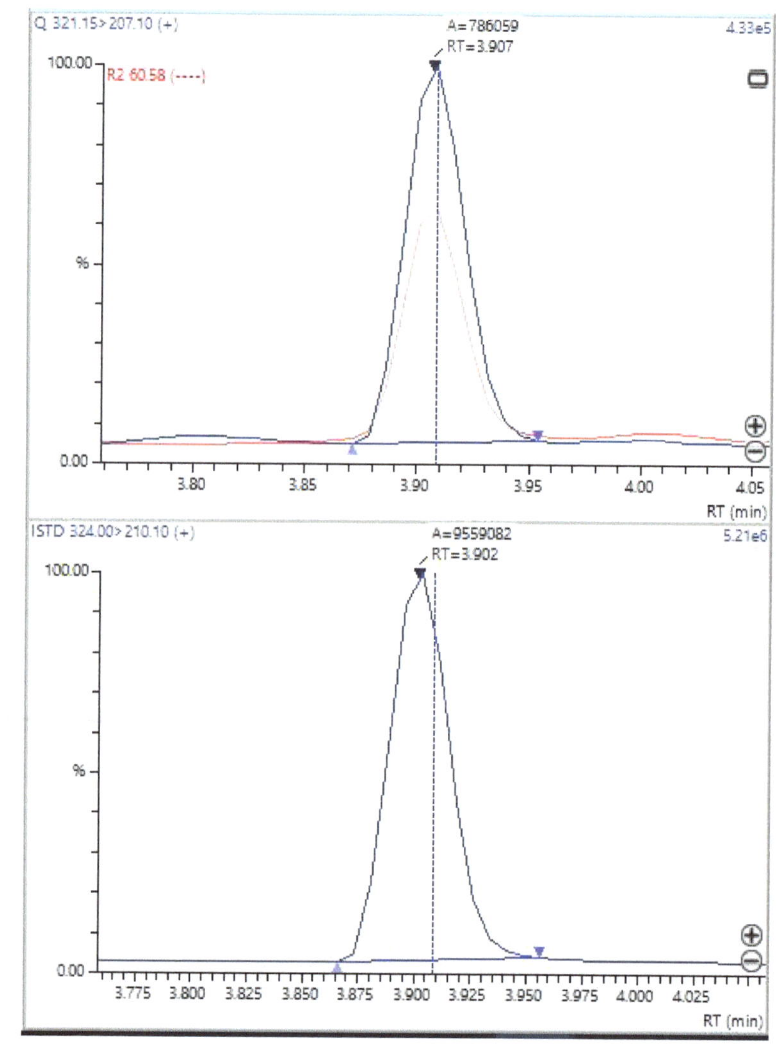

Figure 2. *Cont.*

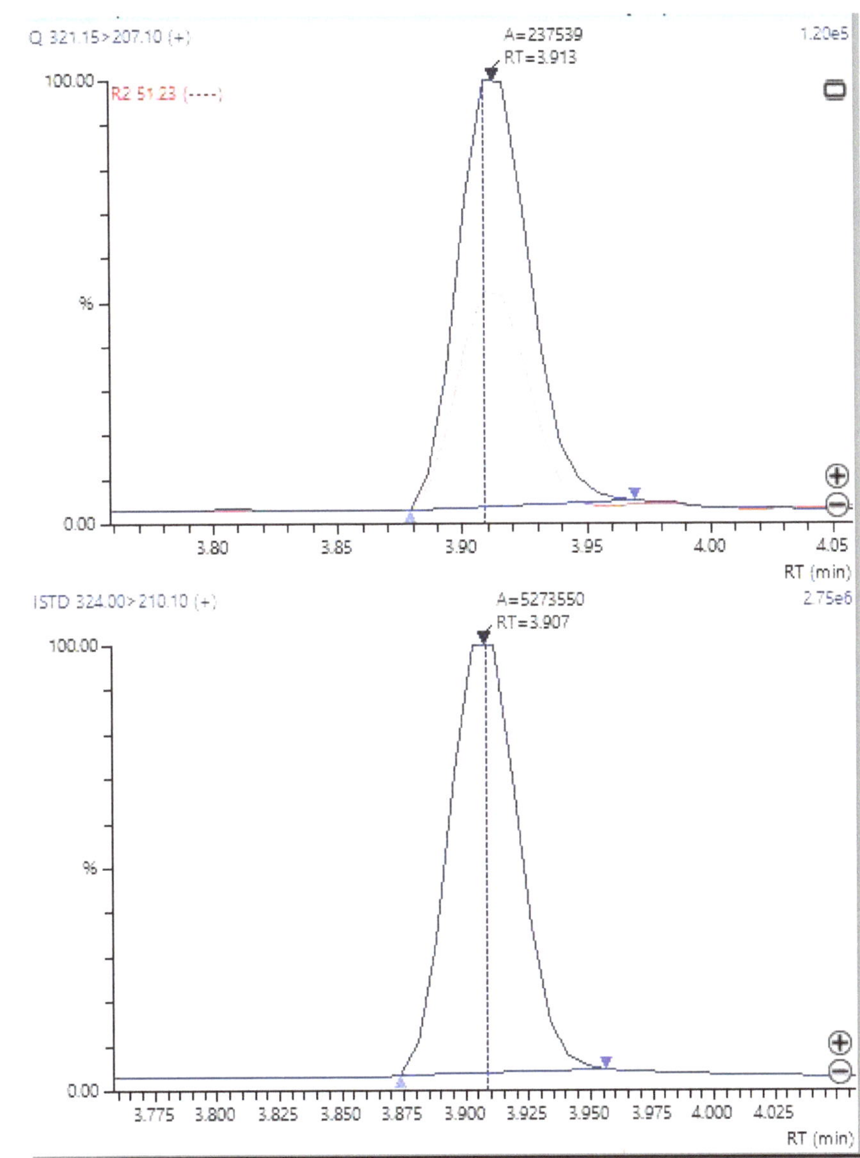

Figure 2. (A) LC-MS/MS plot of mycophenolic acid in the urine sample of Patient 1 above and the internal standard below. The graph shows a tracing in blue which is the quantifier ion (mass = 207.1) and a tracing in red which is the qualifier ion at mass = 159.1. The peak below with only a single blue tracing is the tracing of the deuterium-labeled mycophenolic internal standard (mass = 210.1). The mycophenolic acid internal standard has a virtually identical retention time to mycophenolic acid in the patient, but the quantifier ion is mass = 210.1 because of three deuterium atoms in the internal standard; **(B)** LC-MS/MS plot of mycophenolic acid in the urine sample of Patient 2. The upper graph of the patient shows a tracing in blue which is the quantifier ion (mass = 207.1) and a tracing in red which is the qualifier ion at mass = 159.1. The peak below with only a single blue tracing is the tracing of the deuterium-labeled mycophenolic internal standard (mass = 210.1).

Table 4. Testing of indoor air condensates collected from the office.

Sample	RH %	T °C	Change in Cell Viability %, Mean ± stdev, p	
			THP-1 Monocytes	**THP-1 Macrophages**
Working room 1	30.8	21.4	22.26 ± 20.81 *	−9.54 ± 1.25 ***
Sample 2	30.2	21.4	18.12 ± 16.6	−2.87 ± 2.06 *
Sample 3	30.6	21.6	7.66 ± 10.70	−6.26 ± 2.40 ***
Sample 4	29.1	21.7	28.40 ± 12.59 ***	−6.62 ± 1.26 ***

Patient 1 worked in room 1, sample 4 was taken close from working area of Patient 2. The toxicity of the condensates was tested at a 10% concentration on THP-1 monocytes and THP-1 macrophages. The results are normalized against control and expressed as % change in cell viability, mean ± stdev, as compared to the control. Negative values refer to a decrease in cellular viability, positive values refer to an increase in mitochondrial activity or proliferation. Both are considered as adverse effects. Each was tested in six parallel samples. The statistically significant changes in viability are bolded and indicated as * $p < 0.05$; and *** $p < 0.001$.

Table 5. Testing of indoor air condensates collected from the home of Patient 1 and the HHC. Toxicity of condensates on THP-1 macrophage when tested at 10% and 25% concentrations. The methods and explanations are as in Table 4.

Sample	RH %	T °C	Change in THP-1 Macrophage Viability %, Mean ± Stdev, p (Significance)	
			10% Condensate	**25% Condensate**
Dormitory 1st floor (Patient 1 and HHC)	35.6	22.4	−2.13 ± 4.55	−13.35 ± 5.84 ***
Bedroom (son) 2nd floor	37.8	22.2	1.78 ± 3.02	1.56 ± 3.72
Bedroom (daughter) 2nd floor	38.7	22.0	6.33 ± 10.30	1.07 ± 8.56
Living room 2nd floor (HHC worked in this space)	39.6	22.0	2.83 ± 6.17	4.50 ± 4.65 *

The statistically significant changes in viability are bolded and indicated as * $p < 0.05$; and *** $p < 0.001$.

3.6. Microbiological Analysis and the Toxicity of the Cultured Microbial Colonies

The microbial growth from the office gravel sample was abundant on all the plates: the total fungal colonies were 13,000 cfu/g on MEA and 3900 cfu/g on DG18, and the total bacterial colony count was 42,000 cfu/g on TGY. *Penicillium* and *Acremonium sensu lato* constituted 97% of the fungal colonies on MEA plates, and on DG18 plates 92% of colonies were only *Acremonium sensu lato*. The other fungi, namely *Aspergillus* section *Aspergillus* (formerly *Eurotium*), *Aspergillus ochraceus* group, *Cladosporium* and *Trichoderma* grew only as a few colonies. No *Streptomyces* were detected.

Altogether, 11 microbial colonies (10 fungi and one bacterium) out of 32 colonies tested were toxic to the cultured cells (Table 6). All 11 colonies were toxic in the boar sperm test, and four of them were also toxic to the BHK-21 cells. The species that were toxic in both tests belonged to the fungal genera *Acremonium sensu lato*, which grew on MEA plates, and *Aspergillus* section *Aspergillus*, which grew on DG18 plates. The fungi which were toxic in the boar sperm test were only from the *Aspergillus ochraceus* group, which grew on DG18 plates, as well as from several *Penicillium* species, which grew on DG18 or MEA plates.

Table 6. Toxic colonies: A total of 32 colonies were tested for toxicity to cultured cells.

Sample	Fungal Genus or Group	Culture Plate	Toxicity in Sperm Test	Toxicity in BHK-Test
Sample 2	*Penicillium*	MEA	Toxic	No
Sample 4	*Penicillium*	MEA	Toxic	No
Sample 9	*Acremonium sensu lato*	MEA	Toxic	Toxic
Sample 16	*Acremonium sensu lato*	MEA	Toxic	Toxic
Sample 18	*Penicillium*	DG18	Toxic	No
Sample 19	*Penicillium*	DG18	Toxic	No
Sample 21	*Aspergillus ochraceus-group*	DG18	Toxic	No
Sample 27	*Aspergillus ochraceus-group*	DG18	Toxic	No
Sample 28	*Aspergillus section Aspergillus (Eurotium)*	DG18, direct cultivation	Toxic	Toxic
Sample 29	*Aspergillus section Aspergillus (Eurotium)*	DG18, direct cultivation	Toxic	Toxic
Sample 32	Bacterium	TGY	Toxic	No

4. Discussion

This pilot study revealed the presence of mycotoxins in the urine of the occupants who had been exposed to toxic indoor air, and emphasizes that a urinalysis for mycotoxins, when combined with a careful patient history and medical check-up, is a valuable tool in the diagnosis of DMHS. Until today, exclusively gaseous and particle exposure of toxic indoor air has been investigated, but the condensed water component of the air has been largely overlooked [7,19]. Therefore, our condensed water approach followed by cytotoxicity assays provides a rationale of conducting a risk assessment producing a numerical outcome. Recently, a call for such health risk-based indoor air methods has been published [47].

We detected a high concentration of mycophenolic acid in the first and follow-up urine samples of the two occupants of the problematic office. Moreover, traces of mycophenolic acid in the condensed water sample were also evident. It is assumed that the content of mycophenolic acid would have been higher if the sample tubes been protected during their sampling, storage, transportation and analysis; unfortunately, this was not the case. The fungal species recovered from the office are a potential source of mycophenolic acid production. Although the amount of toxin might be small in terms of absolute values, the exposure is, however, significant because large volumes of indoor air are inhaled each day. This exposure is dependent on the relative humidity (RH %) of the air, i.e., when the RH % and temperature increases, the quantity of inhaled mycotoxins will increase. Toxins can also be inhaled along with fungal particles.

All occupants of this working community experienced symptoms that were compatible with the advanced SBS, or DMHS (Table 2). The mycotoxin profiles of the two occupants exposed to the same dampness microbiota were similar but were different from the profiles of two unrelated patients (Patients X and Y, Table 3). Patients X and Y also suspected that they had been exposed to dampness microbiota. They had also complained of symptoms, but the composition of dampness microbiota at the species level in their homes or workplaces must have been different. Therefore, these patients were taken as putative disease-controls for Patients 1 and 2.

Mycotoxins are secondary metabolites, and their production is highly species specific. Although mycotoxin production is common among indoor and outdoor fungi, some species are not thought to be mycotoxin producers, for example *Cladosporium* spp. [48]. Mycophenolic acid can be produced by the genera *Penicillium* [27] and *Aspergillus* section *Aspergillus* (formerly *Eurotium*) [49]. It was observed that several *Penicillium* species grew abundantly from the gravel sample taken from the hollow area in the proximity to the office; this was a location from which there was an air flow into the office. Four of the tested *Penicillium* species were found to be toxic in the boar sperm test as was the *Aspergillus* section *Aspergillus* which also grew from the gravel samples; this mold was not only toxic in the boar sperm test but also in the BHK-21 test. Half of the colonies of *Acremonium sensu lato*, which was a fungal group thriving in the gravel sample, was also shown to be toxic in both tests. There are only a few reports on the mycotoxins produced by *Acremonium* species. For example, it has been demonstrated that *Acremonium exuviarum* isolated from the building material produce a toxin called acrebol [24].

Previously, fungal identification was very difficult but now DNA-aided identification has proved to be beneficial. Here, we used conventional identification techniques established for the standard operational procedure of the accredited laboratory. It has been reported that a few fungal species may produce mycophenolic acid [50] and the species identified in this study belonged to the potential producers of mycophenolic acid. Importantly, we collected indoor air water condensate samples from other moisture-damaged offices, and six out of ten were also positive for mycophenolic acid (data not published). These samples were also toxic in the THP-1 tests. These findings suggest that mycophenolic acid is a common contaminant in Finnish moisture-damaged buildings.

The route of entry of mycotoxins into the body has been a topic of debate. For example, the urine of 3000 Swedish adolescents was recently analyzed with the presence of 35 different mycotoxins being identified [51]. It was speculated that children had been exposed by oral ingestion of contaminated foodstuffs. Instead, the exposure via indoor air was not considered although it is recognized that there are moisture-damaged buildings in Sweden and therefore the possibility of inhalation exposure should not have been excluded.

The oral intake of mycophenolate mofetil by Patients 1 and 2 was ruled out because they were not receiving any immunosuppressive medication. Taking into account the growth of potential mycophenolic acid producers in the office building, it does seem that the exposure occurred via inhalation of the office's air supply. Mycophenolate mofetil is a first-line immunosuppressive drug used in transplantation immunology [52]. It is used to inhibit the proliferation of B and T lymphocytes and to prevent graft rejection and is also administered in the therapy of lupus nephritis [53]. If our pilot results are corroborated in a larger trial, we may be able to provide mechanisms by which some individuals exposed to dampness microbiota often experience recurrent infections [3]. These infections point to a dysfunction of the individual's immune system. This can be either local through the inhibition of innate immunity of the cilia cells lining the mucosal epithelium, or systemic involving acquired immunity. Prolonged exposure to mycophenolic acid may further result in a dysregulation of immune checkpoints leading to uncontrolled cell proliferation e.g., cancer. We have already reported an epidemiological observation that a prolonged exposure to dampness microbiota is associated with higher oncological morbidity [10].

Another interesting finding emerging from this study was the excretion of citrinin by Patient 2. This toxin can be produced by *Acremonium* and *Penicillium* species [54]. Both genera grew abundantly from the gravel sample. Fungi of the *A. ochraceus* group are known to produce ochratoxin A (OTA) and this potent mycotoxin was detected in minor amounts in the urine of Patients 1 and 2. This fungus was present only as a few colonies on the agar plates.

OTA was found also in the urine of the HHC to Patient 1. This finding is consistent with the toxicological studies from the bedroom of both Patient 1 and the HHC (Table 5), however, no microbiological studies were undertaken in their home. The HHC worked from home in their apartment, whereas Patient 1 worked in the office building, in the tested room 1.

Ochratoxin and citrinin have been identified in raisins, coffee, cereals, wine and beer; enniatin and zearalenone have also been commonly detected in cereal product. The symptoms experienced by our occupants related to the workplace and their diets were unchanged. Therefore, the inhalation route of entry remains as a possible option. In our view, even minute amounts of mycotoxins inhaled from indoor air may be a potential health hazard, since the levels of inhaled mycotoxins are dependent on relative humidity (RH %) and temperature.

We performed the urinalysis twice in Patients 1 and 2 and detected changes in the levels of mycophenolic acid; in Patient 1, the concentration decreased, whereas in Patient 2, it increased after 2.5 months. Many mycotoxins are better extracted and dissolved in relatively non-polar solvents thus leading to the assumption that they are soluble in fatty tissues (reviewed in [2]). Therefore, their excretion kinetics would be expected to depend on an individual's mass of body fat as well as their genetic background [55]. Unfortunately, we did not examine the home of Patient 2. We emphasize that it is important to compare the mycotoxin profiles of the occupants who have had similar exposures, since it is very likely that in persons exposed to the same ecological system, their mycotoxin profile will display similarities. As an example, Patient Y (unrelated to the working community) excreted high levels of gliotoxin, whereas Patients 1 and 2 excreted mycophenolic acid, and the levels remained high in the follow-up.

The strength of this study is the comprehensive and holistic multidisciplinary approach to assessing not only the patients but also the environments where they lived and worked. The environment was studied not only by accepted conventional microbiological techniques but also by a novel indoor air condensation technique supported by the functional cytotoxicity tests.

The importance of this communication is the idea of adopting a holistic approach to help the occupants to identify health risks of indoor air e.g., due to the presence of molds. The cumulative or prolonged exposure to toxic indoor air will inevitably be detrimental when the reversible SBS will be transformed into an irreversible DMHS with the so-called loss of tolerance to many unrelated compounds. This may lead to the development of multiple chemical sensitivity (MCS), chronic fatigue syndrome (CFS), autoimmune diseases, debilitating neurological functions, and disturbances in peripheral nervous system, to mention only a few hazardous outcomes.

The limitations of this study are as follows: Firstly, we have investigated a small working community and therefore this study is a pilot. Secondly, we were unable to explicitly demonstrate that the condensate of the workplace indoor air contained the same mycotoxins as the urine from Patients 1 and 2, but the findings were suggestive that this was the case. In the future, it would be valuable to apply our holistic approach as presented in Figure 3 as a roadmap to explain the morbidity caused by toxic indoor air. This approach might prove beneficial to encourage collaboration and cooperation between the treating physicians, environmental scientists, public health care providers and insurance companies.

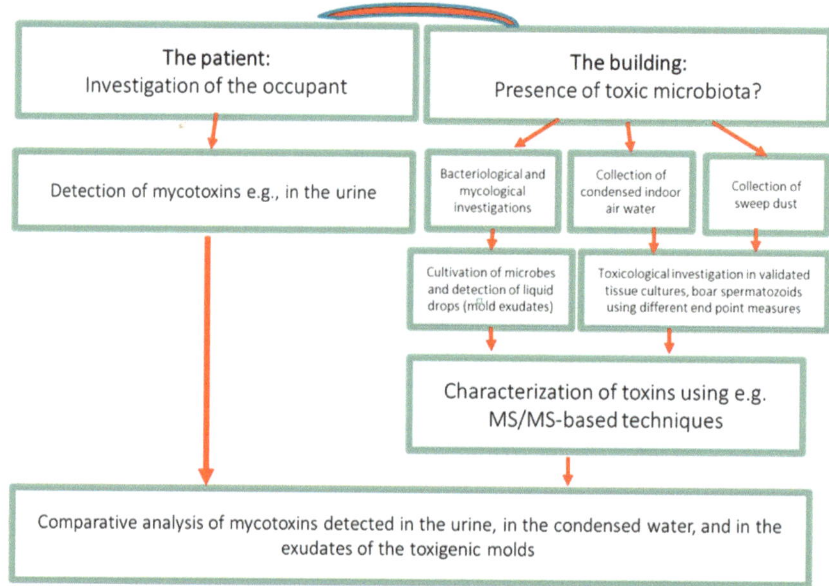

Figure 3. The roadmap to prove causality between the symptoms experienced by the occupant and the exposure to toxic indoor air.

5. Conclusions

We utilized a multidisciplinary, etiology-orientated approach for studying the morbidity associated with the toxic indoor air. We used (1) a technique for the collection of condensed indoor air water samples, (2) functional cytotoxicity tests to assess the toxicity of the indoor air condensate, (3) clinical evaluation of a working community exposed to poor indoor air, and (4) urinalysis for mycotoxin detection. We detected mycophenolic acid in the urine in both of the two studied occupants of the same moisture-damaged building. This potentially immunosuppressive mycotoxin can cause immune dysregulation that in the long run may be related to increased oncologic morbidity and susceptibility to infections.

Our recommendations are: 1. Examine the patients keeping in mind the possibility of a systemic and a multi-organ disease with high inter-individual variability; 2. Perform appropriate laboratory tests for toxicosis, depending on the clinical symptoms; 3. Undertake a urinalysis for mycotoxins; 4. Study the patient´s environment using microbiological and toxicological methods, especially techniques capable of detecting particulate matter, volatile organic compounds (VOC) of gases and indoor condensed water; 5. Recommend that patients should avoid the inhalation of toxic indoor air and provide prompt rehabilitation. If possible, one should attempt to identify the source of the health risk and in this respect, it is recommended that the toxicity of the indoor air should be investigated and a urinalysis also performed on their household contacts (HHC).

Larger studies are needed to test the validity of our holistic approach.

Author Contributions: Conceptualization, K.V. and T.T.; methodology, M.T. and M.M.; software, M.M. and M.T.; validation, M.M. and T.H.; investigation, M.T. and M.M.; resources, T.H.; writing—original draft preparation T.T.; writing—review and editing, K.V., M.T., and M.M; visualization, K.V. and T.T; supervision, T.T.; project administration, T.H.; funding acquisition, T.H. All authors have read and agreed to the published version of the manuscript.

Funding: Sisäilmatutkimuspalvelut Elisa Aattela OY funded the assessment of toxicity of the condensed indoor air samples. M.M. and T.H. were funded by FICAM Institute, University of Tampere, Finland.

Institutional Review Board Statement: These were waived in this study due to the fact that all of the cases provided written permission and that no invasive procedures were involved that could have harmed the participants.

Informed Consent Statement: Informed consent was obtained from all subjects involved in the study and no invasive methods were in use.

Data Availability Statement: Not applicable.

Acknowledgments: We thank Great Plains Laboratory for discussion and advice, all the patients who participated in this study and emeritus lecturer Ewen MacDonald for his help with the paper's style and grammar.

Conflicts of Interest: KV is owner of the SelexLab laboratory providing assays to the Great Plains Laboratory. The other authors' declarations of interest: none. The funders had no role in the design of the study; in the collection, analyses, or interpretation of data; in the writing of the manuscript, or in the decision to publish the results.

References

1. Salin, J.T.; Salkinoja-Salonen, M.; Salin, P.J.; Nelo, K.; Holma, T.; Ohtonen, P.; Syrjala, H. Building-related symptoms are linked to the in vitro toxicity of indoor dust and airborne microbial propagules in schools: A cross-sectional study. *Environ. Res.* **2017**, *154*, 234–239. [CrossRef] [PubMed]
2. Tuuminen, T.; Valtonen, V.; Vaali, K. *Dampness and Mold Hypersensitivity Syndrome as an Umbrella for Many Chronic Diseases—The Clinician's Point of View*, 2nd ed.; Science Direct: Reference Module in Earth Systems and Environmental Sciences; Elsevier: Amsterdam, Netherlands, 2019.
3. Valtonen, V. Clinical Diagnosis of the Dampness and Mold Hypersensitivity Syndrome: Review of the Literature and Suggested Diagnostic Criteria. *Front. Immunol.* **2017**, *8*, 951. [CrossRef] [PubMed]
4. Hyvonen, S.; Lohi, J.; Tuuminen, T. Moist and Mold Exposure is Associated with High Prevalence of Neurological Symptoms and MCS in a Finnish Hospital Workers Cohort. *Saf. Health Work* **2020**, *11*, 173–177. [CrossRef] [PubMed]
5. Hyvönen, S.; Poussa, T.; Lohi, J.; Tuuminen, T. High prevalence of neurological sequelae and multiple chemical sensitivity among occupants. *Arch. Environ. Occup. Health* **2020**, *76*, 145–151. [CrossRef]
6. Vornanen-Winqvist, C.; Jarvi, K.; Andersson, M.A.; Duchaine, C.; Letourneau, V.; Kedves, O.; Kredics, L.; Mikkola, R.; Kurnitski, J.; Salonen, H. Exposure to indoor air contaminants in school buildings with and without reported indoor air quality problems. *Environ. Int.* **2020**, *141*, 105781. [CrossRef]
7. Hyvönen, S.M.; Lohi, J.J.; Räsänen, L.A.; Heinonen, T.; Mannerström, M.; Vaali, K.K.; Tuuminen, T. Association of toxic indoor air with multi-organ symptoms in pupils attending a moisture-damaged school in Finland. *Am. J. Clin. Exp. Immunol.* **2020**; in press.
8. Salin, J.; Ohtonen, P.; Syrjala, H. Teachers' work-related non-literature-known building-related symptoms are also connected to indoor toxicity: A cross-sectional study. *Indoor Air* **2021**. Available online: https://onlinelibrary.wiley.com/doi/full/10.1111/ina.12822 (accessed on 21 November 2021). [CrossRef]
9. Tuuminen, T. The Roles of Autoimmunity and Biotoxicosis in Sick Building Syndrome as a "Starting Point" for Irreversible Dampness and Mold Hypersensitivity Syndrome. *Antibodies* **2020**, *9*, 26. [CrossRef]
10. Tuuminen, T.; Rinne, K.S. Severe Sequelae to Mold-Related Illness as Demonstrated in Two Finnish Cohorts. *Front Immunol.* **2017**, *8*, 382. [CrossRef]
11. Lu, R.; Pørneki, A.D.; Lindgreen, J.N.; Li, Y.; Madsen, A.M. Species of Fungi and Pollen in the PM1 and the Inhalable Fraction of Indoor Air in Homes. *Atmosphere* **2021**, *12*, 404. [CrossRef]
12. Castagnoli, E.; Marik, T.; Mikkola, R.; Kredics, L.; Andersson, M.A.; Salonen, H.; Kurnitski, J. Indoor Trichoderma strains emitting peptaibols in guttation droplets. *J. Appl. Microbiol.* **2018**, *125*, 1408–1422. [CrossRef] [PubMed]
13. Salo, M.J.; Marik, T.; Mikkola, R.; Andersson, M.A.; Kredics, L.; Salonen, H.; Kurnitski, J. Penicillium expansum strain isolated from indoor building material was able to grow on gypsum board and emitted guttation droplets containing chaetoglobosins and communesins A, B and D. *J. Appl. Microbiol.* **2019**, *127*, 1135–1147. [CrossRef] [PubMed]
14. Andersson, M.A.; Salo, J.; Kedves, O.; Kredics, L.; Druzhinina, I.; Kurnitski, J.; Salonen, H. Bioreactivity, Guttation and Agents Influencing Surface Tension of Water Emitted by Actively Growing Indoor Mould Isolates. *Microorganisms* **2020**, *8*, 1940. [CrossRef]
15. Gareis, M.; Gareis, E.M. Guttation droplets of Penicillium nordicum and Penicillium verrucosum contain high concentrations of the mycotoxins ochratoxin A and B. *Mycopathologia* **2007**, *163*, 207–214. [CrossRef] [PubMed]
16. Gareis, M.; Gottschalk, C. Stachybotrys spp. and the guttation phenomenon. *Mycotoxin Res.* **2014**, *30*, 151–159. [CrossRef] [PubMed]
17. Rahmani, A.; Jinap, S.; Soleimany, F. Qualitative and Quantitative Analysis of Mycotoxins. *Compr. Rev. Food Sci. Food Saf.* **2009**, *8*, 202–251. [CrossRef] [PubMed]
18. Salo, J. *Development of Analytical Methods for Assaying Metabolites of Molds in Buildings*; Aalto University: Espoo, Finland, 2014.
19. Mannerström, M.; Toimela, T.; Ahoniemi, J.; Makiou, A.S.; Heinonen, T. Cytotoxicity of Water Samples Condensed from Indoor Air: An Indicator of Poor Indoor Air Quality. *Appl. Vitr. Toxicol.* **2020**, *6*, 120–130. [CrossRef]

20. Andersson, M.A.; Mikkola, R.; Rasimus, S.; Hoornstra, D.; Salin, P.; Rahkila, R.; Heikkinen, M.; Mattila, S.; Peltola, J.; Kalso, S.; et al. Boar spermatozoa as a biosensor for detecting toxic substances in indoor dust and aerosols. *Toxicol. Vitr.* **2010**, *24*, 2041–2052. [CrossRef]

21. Castagnoli, E.; Salo, J.; Toivonen, M.S.; Marik, T.; Mikkola, R.; Kredics, L.; Vicente-Carrillo, A.; Nagy, S.; Andersson, M.T.; Andersson, M.A.; et al. An Evaluation of Boar Spermatozoa as a Biosensor for the Detection of Sublethal and Lethal Toxicity. *Toxins* **2018**, *10*, 463. [CrossRef]

22. Mikkola, R.; Andersson, M.A.; Hautaniemi, M.; Salkinoja-Salonen, M.S. Toxic indole alkaloids avrainvillamide and stephacidin B produced by a biocide tolerant indoor mold Aspergillus westerdijkiae. *Toxicon* **2015**, *99*, 58–67. [CrossRef]

23. Salo, J.M.; Kedves, O.; Mikkola, R.; Kredics, L.; Andersson, M.A.; Kurnitski, J.; Salonen, H. Detection of Chaetomium globosum, Ch. cochliodes and Ch. rectangulare during the Diversity Tracking of Mycotoxin-Producing Chaetomium-Like Isolates Obtained in Buildings in Finland. *Toxins* **2020**, *12*, 443. [CrossRef] [PubMed]

24. Andersson, M.A.; Mikkola, R.; Raulio, M.; Kredics, L.; Maijala, P.; Salkinoja-Salonen, M.S. Acrebol, a novel toxic peptaibol produced by an Acremonium exuviarum indoor isolate. *J. Appl. Microbiol.* **2009**, *106*, 909–923. [CrossRef] [PubMed]

25. Muth, W.L.; Nash, C.H., 3rd. Biosynthesis of mycophenolic acid: Purification and characterization of S-adenosyl-L-methionine: Demethylmycophenolic acid O-methyltransferase. *Antimicrob. Agents Chemother.* **1975**, *8*, 321–327. [CrossRef] [PubMed]

26. Puel, O.; Tadrist, S.; Galtier, P.; Oswald, I.P.; Delaforge, M. Byssochlamys nivea as a source of mycophenolic acid. *Appl. Environ. Microbiol.* **2005**, *71*, 550–553. [CrossRef]

27. Vinokurova, N.G.; Ivanushkina, N.E.; Kochkina, G.A.; Rinbasarov, M.U.; Ozerskaia, S.M. Production of mycophenolic acid by fungi of the genus Penicillium link. *Prikl. Biokhim. Mikrobiol.* **2005**, *41*, 95–98. [CrossRef]

28. Bentley, R. Mycophenolic Acid: A one hundred year odyssey from antibiotic to immunosuppressant. *Chem. Rev.* **2000**, *100*, 3801–3826. [CrossRef]

29. Regueira, T.B.; Kildegaard, K.R.; Hansen, B.G.; Mortensen, U.H.; Hertweck, C.; Nielsen, J. Molecular basis for mycophenolic acid biosynthesis in Penicillium brevicompactum. *Appl. Environ. Microbiol.* **2011**, *77*, 3035–3043. [CrossRef]

30. Ismaiel, A.A.; Ahmed, A.S.; El-Sayed el, S.R. Optimization of submerged fermentation conditions for immunosuppressant mycophenolic acid production by Penicillium roqueforti isolated from blue-molded cheeses: Enhanced production by ultraviolet and gamma irradiation. *World J. Microbiol. Biotechnol.* **2014**, *30*, 2625–2638. [CrossRef]

31. Lafont, P.; Debeaupuis, J.P.; Gaillardin, M.; Payen, J. Production of mycophenolic acid by Penicillium roqueforti strains. *Appl. Environ. Microbiol.* **1979**, *37*, 365–368. [CrossRef]

32. van Oss, C.J.; Giese, R.F.; Docoslis, A. Hyperhydrophobicity of the Water-Air Interface. *J. Dispers. Sci. Technol.* **2005**, *26*, 585–590. [CrossRef]

33. Salonen, H.; Heinonen, T.; Mannerström, M.; Jackson, M.; Andesson, M.; Mikkola, R.; Kurnitski, J.; Khurshid, S.; Novoselac, A.; Corsi, R. Assessing indoor air toxicity with condensate collected from air using the mitochondrial activity of human BJ fibroblasts and THP-1 monocytes. In Proceedings of the Conference of International Society of Indoor Air Quality and Climate, Philadelphia, PA, USA, 22–27 July 2018.

34. Mosmann, T. Rapid colorimetric assay for cellular growth and survival: Application to proliferation and cytotoxicity assays. *J. Immunol. Methods* **1983**, *65*, 55–63. [CrossRef]

35. Ost, M.; Keipert, S.; Klaus, S. Targeted mitochondrial uncoupling beyond UCP1—The fine line between death and metabolic health. *Biochimie* **2017**, *134*, 77–85. [CrossRef] [PubMed]

36. ISO Standard. *Indoor Air—Part 21: Detection and Enumeration of Moulds. Sampling from Materials*; ISO 16000-21; ISO: Geneva, Switzerland, 2013.

37. American Public Health Association; American Water Works Association. *Standard Methods for the Examination of Water and Wastewater*; APHA: Washington, DC, USA, 1980; Volume 15.

38. Valvira, Asumisterveysasetuksen soveltamisohje, Osa IV. In *Asumisterveysasetus §20*; Updated 2020; National Supervisory Authority for Welfare and Health, Valvira: Helsinki, Finland, 2016; Volume Ohje 8/2016, Available online: https://www.valvira.fi/documents/14444/261239/Asumisterveysasetuksen+soveltamisohje+osa+IV.pdf/cdfaaa39-d2e5-4bd6-b9e9-6d9c0f60bff6 (accessed on 21 November 2021).

39. Pessi, A.-M.; Jalkanen, K. *Laboratorio-Opas*, 1st ed.; Suomen Ympäristö-ja Terveysalan Kustannus Oy: Pori, Finland, 2018.

40. Gravesen, S.; Frisvad, J.C.; Samson, R.A. *Microfungi*; Special-Trykkeriet Viborg a/s: Copenhagen, Denmark, 1994.

41. Institut Scientifique de Sante Publique & BCCM. *Moulds in the Indoor Environment and Outdoor*; Institut Scientifique de Sante Publique & BCCM: Bruxelles, Belgium, 2017.

42. Bequin, H. *Moisiss ures de L'Environnement Intérieur et Extérieur*; Section of Mycology & Aerobiology, Ed.; Institut Scientifique de Sante Publique: Bruxelles, Belgium, 2004.

43. Larone, D.H. *Medically Important Fungi. A Guide to Identification*, 2nd ed.; American Society for Microbiology: Washington, DC, USA, 1993.

44. Klich, M.A. *Identification of Common Aspergillus Species*; CAB Direct: Wageningen, The Netherlands, 2002.

45. Rasimus, S.; Mikkola, R.; Andersson, M.A.; Teplova, V.V.; Venediktova, N.; Ek-Kommonen, C.; Salkinoja-Salonen, M. Psychrotolerant Paenibacillus tundrae isolates from barley grains produce new cereulide-like depsipeptides (paenilide and homopaenilide) that are highly toxic to mammalian cells. *Appl. Environ. Microbiol.* **2012**, *78*, 3732–3743. [CrossRef] [PubMed]

46. Bencsik, O.; Papp, T.; Berta, M.; Zana, A.; Forgo, P.; Dombi, G.; Andersson, M.A.; Salkinoja-Salonen, M.; Vagvolgyi, C.; Szekeres, A. Ophiobolin A from Bipolaris oryzae perturbs motility and membrane integrities of porcine sperm and induces cell death on mammalian somatic cell lines. *Toxins* **2014**, *6*, 2857–2871. [CrossRef]
47. Morawska, L.; Allen, J.; Bahnfleth, W.; Bluyssen, P.M.; Boerstra, A.; Buonanno, G.; Cao, J.; Dancer, S.J.; Floto, A.; Franchimon, F.; et al. A paradigm shift to combat indoor respiratory infection. *Science* **2021**, *372*, 689–691. [CrossRef]
48. Nielsen, K.F.; Frisvad, J.C. *Mycotoxins on Building Materials*; Wageningen Academic Publishers: Wageningen, The Netherlands, 2011.
49. Seguin, V.; Gente, S.; Heutte, N.; Verite, P.; Kientz-Bouchart, V.; Sage, L.; Goux, D.; Garon, D. First report of mycophenolic acid production by Eurotium repens isolated from agricultural and indoor environments. *World Mycotoxin J.* **2014**, *7*, 321–328. [CrossRef]
50. Samson, R.A.; Houbraken, J.; Thrane, U.; Frisvad, J.C.; Andersen, B. *Food and Indoor Fungi*; Westerdijk Fungal Biodiversity Institute: Utrecht, The Netherlands, 2019; Volume 2, p. 481.
51. Warensjo Lemming, E.; Montano Montes, A.; Schmidt, J.; Cramer, B.; Humpf, H.U.; Moraeus, L.; Olsen, M. Mycotoxins in blood and urine of Swedish adolescents-possible associations to food intake and other background characteristics. *Mycotoxin Res.* **2020**, *36*, 193–206. [CrossRef]
52. van Gelder, T.; Hesselink, D.A. Mycophenolate revisited. *Transpl. Int.* **2015**, *28*, 508–515. [CrossRef]
53. Mok, C.C. Mycophenolate mofetil for lupus nephritis: An update. *Expert Rev. Clin. Immunol.* **2015**, *11*, 1353–1364. [CrossRef]
54. Johannessen, L.N.; Nilsen, A.M.; Lovik, M. The mycotoxins citrinin and gliotoxin differentially affect production of the pro-inflammatory cytokines tumour necrosis factor-alpha and interleukin-6, and the anti-inflammatory cytokine interleukin-10. *Clin. Exp. Allergy* **2005**, *35*, 782–789. [CrossRef]
55. Genuis, S.J.; Kyrillos, E. The chemical disruption of human metabolism. *Toxicol. Mech. Methods* **2017**, *27*, 477–500. [CrossRef] [PubMed]

MDPI AG
Grosspeteranlage 5
4052 Basel
Switzerland
Tel.: +41 61 683 77 34

Journal of Fungi Editorial Office
E-mail: jof@mdpi.com
www.mdpi.com/journal/jof